The Bakers of Paris

and the Bread Question

1700–1775

fig. 5.

The Bakers of Paris and the Bread Question

1700–1775

Steven Laurence Kaplan

Duke University Press   Durham and London

1996

© *1996 Duke University Press*

All rights reserved

Printed in the United States of America on acid-free paper ∞

Designed by Cherie H. Westmoreland and typeset in Caslon by Keystone Typesetting, Inc.

Library of Congress Cataloging-in-Publication Data appear on the last printed page of this book.

This book has been published with the aid of a grant from the Hull Memorial
Publication Fund of Cornell University.

For Karen and David Blumenthal

Contents

Illustrations

The Baker's Outfit, a caricature portraying numerous bread-making operations, dominated by the task of bolting flour, a product that the baker might obtain in rough form from the water mill in the background   339

Popular retribution for a baker's betrayal: baker François is summarily executed for hoarding fancy loaves reserved for wealthy customers—bread aristocrats—in October 1789   534

The women of Paris returning triumphantly from their expedition to Versailles on 6 October 1789, with their royal trophy, the "baker," Louis XVI, whose link to subsistence is figured by the carts full of sacks of flour that precede and follow him   578

Figures

.

Tables

Acknowledgments

I am deeply grateful to scores of individuals and institutions for their intellectual, moral, and material assistance in this project, which has stretched out over many years. Two mentors from my Princeton days, Charles C. Gillispie and David Bien, pressed me at the outset to clarify my thinking and make choices, and not to worry overly about how to label what I was doing. Jeffry Kaplow, whose departure from the ranks of eighteenth-century historians impoverished us all, shared with me his impressive knowledge of and feel for Parisian life. On the French side, I got off the ground with help from one of France's most distinguished and tentacular old boys' networks, the normaliens of the rue d'Ulm. Francis Léaud, a specialist in English literature from Poitiers, and Henri Peyre, a Renaissance man from Yale, each wrote in my behalf to Jean Meuvret, their erstwhile school librarian, who had become one of Europe's most insightful economic historians and the leading specialist on France's subsistence question. Meuvret welcomed me into his seminar, allowed me to accompany him on the marathon weekly promenade from the Sorbonne to the Luxembourg station, and helped me to articulate my *problématique* and design my research strategy. Maurice Laugaa, a normalien of more recent vintage, put me in touch with schoolmate François Billacois, a superb historian of enormous generosity who did much to help me get started. Through Billacois I became connected with the Nanterre clan, two of whose members became dear friends. Pierre Goubert looked after me, fed me, criticized me, and enriched me in countless ways. The late Erica-Marie Benabou shared with me her insights and discoveries, her hospitality, and her affection: I miss her very much.

Virtually all the players on my implausible basketball team composed of graduates of the Ecole nationale des chartes, the institution that trained the curators for France's great libraries and archives, proved to be far more precious off the hardwood than on it, with the possible exception of Jean Dérens, an excellent rebounder and a wonderful friend. Dérens unlocked myriad doors, gave me new ideas for sources, and challenged me to rethink my premises. Fatally tentative under the hoop, François Avril was absolutely sure of himself in the manuscript room of the Bibliothèque nationale, where he helped me decode texts, unearth muddled references, and surpass my daily allotment of documents. A hustling player frustrated by bad hands, Michel Melot found striking images for me in the Cabinet des estampes. (Many years later, the wife of another player, Françoise Jestaz, helped me to complete my iconography in that same depository.) Alain Erlande-Brandenburg was unequivocally the least gifted of the lot on the basketball court, but now that he has taken over as head of the Archives de France, I intend to give him another chance—provided the American Church of Paris will still rent us their basement gymnasium for fifty francs an hour.

Two less athletic Chartistes facilitated my work in the Archives nationales and became dear friends and searching interlocutors. Presiding jointly over the reading room, governing with authority and sympathy, Odile Dresch Krakovitch and Yves-Marie Bercé enabled me to see as many as fifteen *liasses* of the Y series every day, a favor of incalculable benefit. Nicole Felkay initiated me into the rich holdings of the Archives de la Seine, as they were then known. (The staff of this establishment struck a blow for Gaullism and erudition by remaining open throughout the entire "uprising" of May 1968; my wife and I discreetly flew a banner from the window looking onto the river: "Salle occupée par les Kaplan.") I would never have made sense of the labyrinthine holdings of the Minutier central of the Archives nationales without the gentle initiation of Monsieur Collard. Those were the days before the famous computerized "listing" of the notarial acts of 1751, which meant endless hours of fruitless fishing. Certain of Professor Roland Mousnier's research aides, despite explicit instructions to eschew altruism at all costs, surreptitiously passed me priceless references to baker marriages and inventories. Bless them.

Over the course of the years, I discussed my work with numerous colleagues on both sides of the Atlantic, and presented it in seminars or lectures in France, Italy, Germany, and Holland. For their probing suggestions and their encouragement, and also for a panoply of references and research materials, I am very grateful to Daniel Baugh, Louis Bergeron, Luc Boltanski, Alain Boureau, Maarten Brands, Serge Chassagne, Jean-Michel Chevet, the late Michel de Corteau, Maurice Garden, George Grantham, Alberto Guenzi, Alain Guéry, Mary and Philip Hyman, Jean Jacquart, Christian Jouhand, Jacques Le Goff, Bernard Lepetit, Yves Lequin, Günther Lottes, Peter McClelland, Philippe Minard, Diego Moreno, Anne-Marie Piuz, Carlo Poni, Daniel Roche, the late Jacques Roger, Michael Stürmer, H.-U. Thamer, Laurent Thévenot, Jean Tulard, Lucette Valensi, John Weiss, and Pierre-Etienne Will. Douglas Van Houweling toiled patiently to teach me how to use the computer, and he vetted virtually all of my statistical runs. I have benefited over the years from the rigor and generosity of two reference librarians at Cornell, Susan Szasz Palmer and Bob Kibbee, who answered myriad queries. At short notice, Anne Varet-Vitu kindly and adeptly drew a crucial map. I am grateful to Jean-Michel Roy for locating one of the markets on it.

Jacques Revel engaged me repeatedly on issues that I had failed to resolve, and provided luminous counsel. Gilles Postel-Vinay pointed out numerous ways to deepen and sharpen my analysis. Claude Grignon opened provocative theoretical and comparative avenues. Darlene G. S. Levy and I argued for countless fruitful hours about the ways to construe the social, the cultural, and the political in the eighteenth century. Maurice Aymard showed me connections I had not seen and shades of meaning I had not appreciated. He and Clemens Heller welcomed me repeatedly to the Maison des Sciences de l'Homme, at which site I also benefited time and again from the support of the Ecole des hautes études en sciences sociales and its Centre de recherches historiques. I am one of the rare historians at home on the boulevard Raspail who is also familiar with the Quatrième Section of the Ecole pratique des hautes études, which was kind enough to invite me for a year's teaching. Emmanuel Le Roy Ladurie offered me the august forum of the Collège de France to present a series of lectures on the bakers and in addition supplied me with endless clues for improving my arguments.

Among other nurturing institutions, Cornell University has multiple claims upon my gratitude. The "job talk" I gave there twenty-five years ago already concerned the bakers. Various units of the university furnished real and funny money for all sorts of purposes, and the deans made it possible for me to spend a great deal of time in France. Parts of fellowships from the National Endowment for the Humanities and the American Philosophical Society sustained my fieldwork. The Grands Moulins de Paris enabled me to launch my unfinished career as a baker. I am indebted to many bakers for opening their bakerooms to me, teaching me, and drinking with me—in particular to two practitioners, now deceased, on the rue Saint-Jacques and the rue de la Roquette. The Deutsches Brotmuseum at Ulm/Donau warmly received me and provided me with invaluable guidance and beautiful illustrations. Editions Fayard, in the person of their president, Claude Durand, have enthusiastically supported this enterprise. Thanks to Rachel Toor, an exigent and thoughtful editor, I have deepened my bond with Duke University Press, whose staff have been wise and kind. Among managing editor Jean Brady's excellent decisions was the one to hire Joan E. Howard to copyedit the manuscript, a herculean task that Joan accomplished adroitly and congenially.

Three people have been with this project from the beginning to the end. My wife, Jane, got to know the archives of the police and of business failures rather intimately, working at my side after she finished her own thesis. She proved that she was more than a mere amanuensis by inserting politically incorrect comments on the notes she took that still make me laugh uproariously when I come across them. More recently, she has proposed stylistic modifications that have improved the text. Two of my earliest undergraduate students at Cornell and ablest research assistants, Mary Ann Quinn and Nan Karwan Cutting, worked on the bread question in the seventies and returned to it in the nineties. Mary Ann did the index with her usual brio, and Nan, now my colleague, helped me in countless ways, intellectual and material, to complete and enrich the manuscript. Following in their daunting footsteps, recent student aides Josh Friedman and (especially) Jennifer Gaffney have placed me in their debt with their imaginative problem solving and energetic "gophering." Three of my graduate students, Clare Crowston, Janine Lanza, and Sydney Watts expertly verified numerous endnotes. With infinite patience and cunning, Nancy Davenport, my current student assistant, helped me to wrap things up.

Far from academic life, a number of close friends contributed mightily and more or less unknowingly to the final stretch. Eliane and Frank Simon, Claire and Philippe Luxereau, and Isabelle and Henri Taieb: *je vous embrasse.* For similar ineffable reasons, I dedicate the book to two bread-loving Francophile Cornellians, Karen and David Blumenthal.

Introduction

Provençal dictum:
Whatever happens
Let there be bread
—Maurice Lelong, *Célébration du pain*

Cry of a crowd pillaging a Paris bakery, May 1775:
We must have bread!
—Archives Nationales, Y 10550, 3 May 1775

Mid-nineteenth century popular song:
One does not stifle the grumbling
Of the people, when they cry: I am hungry!
For it is the cry of nature:
　　We must have bread!
—Pierre Dupont in Benigno Cacérès, *Si le pain m'était conté*

Foodways:
I think it could be plausibly argued that changes of diet are more important than changes of dynasty or even of religion.
—George Orwell, *The Road to Wigan Pier*

Ferment:
In *Archives du Nord* I felt as though I was kneading a very thick dough. . . . I know about dough, mind you: remember that I make my own bread. You can never be sure that it will work. There are stages in bread-making quite similar to the stages of writing. . . . Finally, you sense that the yeast has begun to do its work: the dough is alive.
—Marguerite Yourcenar, *With Open Eyes*

I.

The mass of the population in eighteenth-century France derived the bulk of its calories from bread and sundry grain products. These people were primordially consumers: their lives turned on the urgent need to wrench an adequate bread ration for themselves and their family each day. They inhabited a world of agonizing insecurity as well as chronic poverty. Not only was their diet narrow and monotonous, but it remained hostage to a host of variables beyond their control, variables that permanently imperiled both production and distribution. Imagine a world in which the majority lived marginally, in fear and uncertainty. Yet this dependence and this vulnerability emphatically did not mean that the population at risk—say around half a million Parisians, to cite the case that directly concerns me here—ate virtually anything. On the contrary, precisely because bread constituted the measure of their survival, these consumers demanded that it be of good quality as well as of sufficient quantity, and the public authorities did everything in their power to fulfill this solemn expectation.[1]

To eighteenth-century Parisians the presence of bad bread in the markets or shops was an intolerable affront and menace. It signified either an act of social crime or a mark of social breakdown, two different levels of crisis. As sure as rising prices, inferior loaves announced disarray. During the Flour War subsistence riots in May 1775, various reports cited consumers brandishing ugly "black bread" at the central metropolitan market called the Halles—a toxic substance infecting the belly of Paris. So unusual was it to see such a repugnant and insulting loaf in the capital even in the midst of crisis that certain police officials suspected it must have been imported from the countryside rather than produced in the city. (The friends of Anne-Robert-Jacques Turgot, the minister at the time who practiced deregulation of the grain and flour trade combined with brutal repression of disorder, cultivated a conspiracy thesis that attributed the obscene bread samples to evil plotters determined to discredit his liberal policy, a policy dictated by the self-evident laws of nature.)[2]

When during the Flour War Parisians summoned the bakers to sell them bread at two sous per pound—the "ordinary" price, allegedly consecrated by Louis XVI himself, who was said to have ordered it to be delivered at this price to his "bon peuple"—the loaf in question was unequivocally white, not the top or elite quality, but not far from it. Nor is the story of Jean-Charles Leguillier, a journeyman cloth maker arrested on suspicion of pillaging a white six-pounder during the bread riots, implausible. Leguillier insisted that the loaf had been distributed to his aged mother as part of the parish charity system. "It was pointed out to him [by the interrogating commissaire] that one does not ordinarily give as charity the sort of white loaf that is represented here." The journeyman replied, according to police files, that "his mother has a white loaf because she gives extra money to the baker in order that he furnish her white bread." Mother Leguillier was not the only indigent Parisian, then or later, to cobble a mechanism for upgrading her eating status. The dearth and accompanying dislocation did not keep a twenty-eight-year-old laundryman called "Four Pockets," who supported a seventy-eight-year-old infirm father, his own children, and at least one sibling, from complaining that he was obliged to look elsewhere for an appropriately white loaf since "his baker had only dark loaves."[3]

It is no wonder that Jacques Necker, in certain ways an anti-Turgot, exercising the same responsibilities as his liberal predecessor on the eve of the French Revolution, worried so much about the bread that Parisians might be forced to eat if he did not unearth good quality domestic supplies somewhere in the realm. Though he could have made available Polish whole wheat flour mixed with some rye, capable of producing only a dark loaf that both looked and tasted strange, he feared that "the inhabitants of Paris would strongly feel the difference on both accounts between this sort of loaf and the bread to which they are accustomed." The political impact of that double alienation, physical and cognitive, could have turned out to be worse than shortage itself. Certainly a leitmotif of the pamphlets of 1789 was the vivid memory of bread that was both cheaper and better. By 1793 there was no more powerful mark of the limitations of the Revolution, or the subversion by the Counterrevolution, than the fact that "we have been eating for some time a gray loaf of bad quality that smells dusty and afflicts most citizens with bouts of colic and diarrhea."[4]

Today we are attached to bread, to adapt Pascal's terms, not by the "cords of necessity" but by the "cords of imagination."[5] Bread derived its empire in the Old Regime from its location at the intersection of material and symbolic life. Emancipated from economic and biological dependence on bread, we remain at the end of the twentieth century susceptible to bread's cultural claims. For some of us, bad bread is still a provocation if not an outright act of aggression—especially in a context (in this instance the French one) where it is supposed to be of the finest quality. Partly this reaction bespeaks the lingering aesthetic and spiritual vocation of bread, which, in some sense or other, we feel needs to be perfect, or close to it. Our expectations are predicated on diffuse yet compelling assumptions, historical and mystical, about the role of bread. In the broadest manner, bread seems to recapitulate the human experience. It rehearses the grand continuities, transformations, and tensions of the world: creation and fertility, reproduction and rebirth, tradition and transcendence, unity and multiplicity, individuality and community, nature and culture. It bears the indelible and ineffable mark of holiness, regardless of confessional orientation, though it acts upon most of us most dramatically through its Eucharistic guise. Assimilated commonly to the human as well as the divine body, bread mediates between sacred and profane, life and death, here and beyond.[6]

For me, as a participant in and observer of French life, bread carries a specifically French mark, which is of relatively recent origin, measured against bread's long trajectory through time. Particularly in its elongated, baguettelike form, it has become so intimately identified with France that it is practically a metonym for the nation and its civilization. It would be worthwhile studying the evolution of the bread-as-France image in the various media. Like many cultural stereotypes, though it reduces, it also "works." And surely for the uninitiated outside the hexagon, the image of a ruddy-cheeked man in a beret with a Gauloise cigarette dangling from his lips and a loaf of bread under his arm evokes France with more authority than either Astérix or Marianne. Finally, the most immediately compelling prod of bread is hedonistic: the singular pleasure of a delicious loaf.

One of the sad by-products of recent French modernization has been the drastic decline in the quality of bread. It has paralleled the *trente glorieuses,* the period of exuberant postwar economic growth that definitively replaced the *gagne-pain* with the *gagne-bifteck.* It is as if there were a corollary to Engel's law stating that even as families spend less on basics such as bread as their living standard improves, so they care less about the quality of those basics as their well-being increases and their cultural horizon expands. It almost seems that the French had to affirm their embrace of socioeconomic modernity by repudiating their attachment to bread—cutting the chords of necessity with symbolic violence. It seems only fair to observe that the French subsequently got to eat the bread that they deserved.

After years of adamant denial—it was almost an act of *lèse-nation* to talk openly about the matter—French observers of the bakery now generally agree, in the words of the doyen of academic experts on the bakery, Raymond Calvel, that "French bread has fallen very low." He attributed "the dramatic fall in the gustatory qualities of bread beginning in the years 1957–1960" to economic and technological factors (though I would be inclined to read these variables in terms of a broader cultural mutation in the profession and in the society at large). In order to reduce the time and the costs required to produce bread, bakers turned to rapid kneaders ("massacring the dough"), accelerators such as broad-bean flour and ascorbic acid whose oxidizing and enzymatic actions resulted in record fermentation times as well as more voluminous and whiter loaves, and later to techniques of cold leavening that spared the need to knead on demand ("the art of making bread almost without fermentation"). The bakers wrenched significant gains in productivity from these techniques at the prohibitive cost, in Calvel's calculus, of "sacrificing the essential: the taste."

Their "modern" loaf was "insipid," washed out, without texture, impossible to keep; in a word "bread was no more than a disguised misrepresentation."[7]

As if to expiate for the tardiness of the national reaction, the journalist-gastronome Jean-Pierre Coffe crusades tirelessly against the horror of eating a tasteless baguette ("without joy, without emotion, without appetite"), a manifestation of "the slow but inexorable degeneration of bread quality."[8] "Up till the industrialization of the bakery at the beginning of the sixties, French bread was considered a model for which foreigners envied us," recently wrote Michel Montignac, a diet reformer and foodie entrepreneur. Citing Coffe but moving beyond him—what a short distance from national denial to national masochism—Montignac concluded: "Since then it has not stopped losing merit, deteriorating to the point where today it is *one of the worst in the world*."[9] A number of prominent chefs queried by Montignac endorsed his view with more or less moderation. For Charles Barrier, the resilient restaurateur from Tours, "Bread today? 99 percent is sh . . . !" For Taillevent's Philippe Legendre, "in general terms I find the quality of bread quite lamentable."[10]

Many of today's bakers share in the harsh assessment of bread quality in the present era. A baker from a town in Dordogne put it brutally: "From the '50s, the profession has educated the public in bad taste by making bread of deplorable quality."[11] The refrain of the Parisian bakers with whom I have spoken was the complacency and laziness of many of their colleagues. Whether or not this understanding is valid and sufficient, it is interesting to note its widespread purchase. During the baker's *trente glorieuses,* "the trade had become too easy," observed Jean Jeudon of the rue de Ménilmontant. One had only then "to bend down in order to gather up the bucks. . . . why break your back?" A naive fetishization of machinery, a lack of rigorous competition, and the easy-street climate combined to "destroy the quality." Bread degenerated, according to Paul L'Hermine of the place Brancusi, because bakers "no longer had any desire to take the trouble [to do the job right]." They were able to attain a comfortable level of income without much effort: "everything is dandy." Echoing the same viewpoint, Michel Legallet of the rue Raymond Losserand traces the stupefying inertia of bakers to the postwar years of manna, practically unearned, that bred habits of indolence, myopia, and risk aversion. If 90 percent of bakers produced "shit," argued Jean-Luc Poujauron of the rue Jean-Nicot, the reason was moral rather than technical. They lacked not knowledge, which was readily available, but "the will to exercise their craft." Bernard Ganachaud, one of the most prosperous and thoughtful Parisian bakers, suggested that both bakers and their clients had suffered a joint loss of "values" concerning good bread. Unschooled and undemanding, they both settled for a poorly kneaded, lightly cooked, thin-crusted, overly white baguette: "What a shame!"[12]

These bakers are among the vanguard of reforming artisans and entrepreneurs who have toiled to rekindle the brilliant reputation that the Parisian bakery once boasted. The leitmotif that links all of their individual trajectories is what Ganachaud calls "getting back to basics." Excellence cannot be attained by applying the recipe of scientific theory and technological advance first articulated by the Enlightenment "doctors of flour," though the bakers I consulted by no means rejected "progress" in a spasm of artisanal Ludditism or moralizing postmodernism. Quality could only be *restored*. The road to a fulfilling future detoured through the past. In the heady postwar expansion, the past had not disappeared, but it had discreetly receded. With his usual blend of puckishness, insight, and sarcasm, the journalist Alain Schifres described in 1987 the unwonted onset of nostalgia fever among the bakers:

Two years ago, apart from their distressing taste for Gothic letters, most of them had as little regard for tradition as they had for their first *ficelle*. It was the era of bread shame. The only things in the shop window were biscuits, candy, and bottles of Cointreau. Now they are in a new phase, in Giono style, presenting to the public a flour-covered face in the midst of the stoneware, the cloths, the sheaves, and the tools. Little posters show us the hands of the baker busily kneading the dough. False. All the kneaders are mechanical. Elsewhere one reads: bread baked over a wood fire [cuit au feu de bois]. False. Contact with an open flame is forbidden. It is the oven that is heated by wood. In this craft the winners are sensible nostalgics.[13]

The bakers' relationship to the past was not devoid of ambiguity. Most of them knew very little about what I would call the historical past, whose claims required critical scrutiny and documentary corroboration. They were interested in a more or less legendary past, informed by certain partial realities, consecrated by certain moral verities, and shaped by pressing needs of various kinds. This past was both a source of absolving/guiding wisdom and a treasury of alibis. Like other groups, the bakers turned to the past when they were uneasy about the present. They unearthed or invented traditions that could prove useful. They drew on the past not merely to (re)orient themselves but also to (re)educate their public. In the past they would recover innocence and authenticity, and a capital of credibility that they had dissipated. The past would provide grounds for a new covenant with their customers. In the past the more candid bakers would refurbish their self-esteem as well as their public image.

"Faire à l'ancienne"—do it the old-time way—was L'Hermine's advice to any prospective baker. For him and many of his colleagues, this was a talismanic doctrine. Among others, L'Hermine advertised "his whole line of special breads made the old way [à l'ancienne] with sourdough starter as in the days of yore [comme *autrefois*]." The Fleuriane mills' baguette was made "à l'ancienne." Numerous bakeries have been baptized "Au bon pain d'*autrefois*," an appellation that pointedly evokes "the good bread *of yore*." The dough of the didactically dubbed baguette Rétrodor "is simply composed as it was *autrefois*." The publicity strategy of Rétrodor's developers, the Viron millers of Chartres, was to bludgeon: "for a few centimes more, give yourselves the guarantee of *autrefois*" and "to *re*discover all the authenticity of the true French baguette. Rétrodor is made as *autrefois*." The pointy, *bâtard* like loaf marketed under its name by the Banette mill consortium issued from "a traditional process characterized by artisanal kneading" and "hand fashioning." It aimed at consumers "who were lovers of true bread and who wished to rediscover the natural aroma of the bread of *autrefois*." Whereas Ganachaud specified his *autrefois* (the years around 1900) in reference to his *poolish* method of fermentation and Jean-Pierre Coffe located the juridical articulation of his *autrefois* in the fourteenth century, "the good bread of *autrefois*" embodied in the Banette country loaf transported us all the way to "a lost paradise regained."[14]

Autrefois—Erewhon and right here—was thus an all-purpose, all-encompassing recourse and metaphor, at once rooted in time and timeless, fixed in space and siteless, a concrete solution and a romantic fantasy. Autrefois was a state of mind as well as business connected with various positive images: robustness, steadfastness, honesty, virtue, performance. The allure of the unvarnished, uncorrupted professional tradition betrayed a deeper longing, hardly confined to bakers, or to the French, for the more solid values that were believed to have held sway at one time. (Right and left, from Philippe de Villiers to Bill Clinton, values became a hot mainstream item.) One of the striking features of this version of tradition is its lability, and its capacity to accommodate.

The bakers groped to forge a usable past, not one that would confine or cripple them. Far from rejecting the intricacies of modern technology, Professor Calvel hinted that the culture of the *ancienne* offered the best and safest way of mastering and domesticating them. Neither L'Hermine nor Ganachaud worried that the use of cold fermentation, so starkly at odds with traditional habits, could be construed as treachery or could undermine one's commitment to the vital force of the *ancienne*. Seeking to strike the right balance, a Banette publicity brochure portrayed the baker as a "guardian of traditions" but insisted that he "is nonetheless an artisan of today." The symbolic power of (good)will bridged the chasm between theory and practice. So Lionel Poilâne had the decorative ironwork of his father's artisanal oven reproduced on each of the twenty-three ovens at his Bièvres factory, ovens constructed in the consoling spirit of "*rétro*innovation." And the frankly industrial Fournil de Pierre, its central factory at Saint-Denis awash in wistfulness for "the rediscovered bakery [la boulangerie retrouvée]," placed its ambitions in the same tradition of sensible marriage, evoking "a happy balance between the utilization of perfected matériel and a perpetuation of certain gestures of *autrefois*."[15]

Meg Bortin, an American journalist based in Paris, was doubtless right to warn almost twenty years ago that the evocation of *l'ancienne* was more often a "coup du décor"—a decorative ploy—than the harbinger of better bread. And Alain Schifres was equally on the mark a decade later in debunking the "enchantment with autrefois. The original bread does not exist. . . . The fashion for the authen*tic* is governed by the authen*toc*"—gaudy imitation.[16] Yet our confidence in the past as a guide to primal virtues seems boundless.

2.

This book is about *autrefois,* but not inescapably the autrefois of our wistful bakers. Transported there, they would (receptively) discover a time when bread was at the very center of social, economic, and political life and when most people ate very large amounts of it every day. As long as bread was the ration of survival for the majority, however, there was nothing quaint or folkloric about it, even if matters of quality never relinquished their salience.[17] The autrefois of the Old Regime was no Cockaigne, no lusty Jan Steen tableau in which ruddy-faced consumer-citizens partook of a permanent secular communion, in which the era's famous *douceur de vivre* trickled down sloppily from the top. Yet it would be equally misleading to paint an unremittingly miserabilist landscape that portrays eighteenth-century France as an anachronistic sub-Saharan space of starvation, its capital a dense enclave of paradigmatic suffering.[18] By and large people got by in this world, though most of them had to struggle at times and adjust to a terrifying uncertainty. For it was as much the unsettling precariousness of things as the actual privation and the chronic penury that set the tone.

The subsistence imperative held this society hostage. The result was an agonizing subjection to caprice masquerading as contingency, to an environment that could be neither controlled nor propitiated. The structuring role of fear is striking in its triangular Parisian incarnation: at one end of the base stood the people dreading their dependence on bakers who seemed prepared to see them "mourir de faim"—die of hunger—to satisfy their greed; at the opposite pole cluster the bakers, acutely conscious of their vulnerability, unnerved by "cette haine si long-tems enracinée"—this long-rooted hatred—in the hearts of the consumer-people; and at the apex, attentively monitoring, negotiating, repairing, loomed the police, the *grands trembleurs* of the Old Regime.[19]

I do not want to suggest, in light of the ostensibly pitiless dominion of the material, that women and men could live by bread alone in a world of sinister reduction. The problem is not just to work out the material determinants but to remember that these determinants are rarely mechanical in origin or application, and are always culturally mediated and socially transposed. Inclement weather, insufficient manure, and primitive plows on their own explain little. The *mercuriale,* registering grain prices in the markets, is neither a self-sufficient nor an autonomous causal agent. Desperately hungry Parisians did not eat everything that omnivarousness theoretically permitted and/or required. Like today's Rwandans in Goma who bartered their strange protein- and vitamin-enriched biscuits for paltry rations of familiar beans or sorghum flour, eighteenth-century Parisians rejected rice and other ersatz in similar ways. Nor was it unusual or illogical for them voluntarily to pay for bread they seized from bakers in the tumult of rioting—at prices they deemed just, to be sure. Taste and custom still matter, notions of right and wrong still make a difference, even in drastically unfavorable material circumstances. Nevertheless, allowing for the incalculable autonomy of culture and the profound complexity of social life, the unyielding dependence on bread still shaped every aspect of daily existence. (And, in any event, the tyranny exercised by bread was as much symbolic as material; even as the chords of necessity yoked consumers, the chords of imagination tightened as well.)

Though Parisian society was highly stratified, no one evaded the bread nexus, directly or indirectly, in one way or another—if not as consumer, then as master, manager, official, neighbor, rentier, supplier, creditor, and so on. Broadly construed, no one lived outside the symbolic and material orbit of bread exchange. No matter how lofty one's station, no one could remain indifferent to the fact that everyday activities for most people focused on the acquisition of daily bread. Given the intimacy of everyday relations in the apartment building, the neighborhood, the work site, and sundry other places of commercial and sociable exchange, even those wholly insulated from personal or familial penury could not distance themselves from this predominant fact of majority existence. No matter how elaborate the masking or the displacement or the sublimation, or how segmented the society, or how inured the actors, at bottom the economy was about the allocation of bread (as was much of the politics).

Not a "people" problem that could be socially quarantined, neither was bread a sporadic preoccupation that could be temporally confined. Most historians seem to have operated on the premise that if it works, it is not worth examining. Thus they acknowledge the subsistence issue only in the breach, when a full-blown crisis flowers abruptly and temporarily disrupts the ordinary course of things. Languishing in the rut of routine, that ordinary course rarely attracts attention. For multiple reasons those crises, especially the ones that seem built into the system despite their exceptional nature, are crucial—not merely because they sometimes issue in grand, transformative crises, but because they loom large in the imagination (and the imaginary) of every community and every generation, a constant absent presence that influences thinking, behavior, and decision making, and because they wreak mighty enough havoc when they actually erupt in their evanescent and recurrent rhythm.

Still more important, however, if one is to practice the historical craft seriously, is the ordinary course of things itself, for its own sake, the daily rut for its weary ruttedness. With what sort of confidence can we evoke our scholarly autrefois when we concentrate exclusively on the poetry at the expense of the prose? The everyday cannot merely be extrapolated (or interpolated) as a thick residual, a backdrop meant to sharpen the fancy chiaroscuro. Though the banalities are paradoxically more difficult to get at, we must recover them if we are to make sense of the (privileged) rest.

The relentless everyday demands of bread construct the pattern of daily life in Paris as tellingly, and as subtly, as the elaborate protocol of Versailles governs social life in the gilded cage. In certain ways, this subsistence tropism, this collective habitus (resembling in its acute phase what the sociologist Pitirim Sorokin might have called bread *taxis,* the compulsive riveting of attention upon the task of food procurement) participated in the civilizing process.[20] Far from articulating a steep hierarchy of rigorously discrete and unbridgeable strata, it forged curious bonds of mutual dependence even as it reinforced cleavages within the structure. The shared subsistence anxieties of the haves and the have-nots, the governors and the governed, attached them to each other and acted as warrants of mutual responsibility. Bread dependence helped to define the relation between nature and culture. It accentuated traits of common humanity that no one could escape. It placed into relief the limits of social knowledge and power. It reminded everyone of the primitiveness that still inhabited advanced civilization. It underlined the precariousness of social arrangements, their fragility. Bread need organized, choreographed, and ritualized social practices that focused on the goal of assuring stability (or reproduction). It led to the elaboration of norms and rules that were neither brutally reducible to a calculus of economic rationality nor mechanically applicable without regard to circumstances. In the end the system of regulation seems to reside less in the avowed principles embodied in statute than in the logic of practice that accrued over time.

On the local level, a large part of the population, drawn from above as well as below, subscribed to what one might call the *everyday order*.[21] The everyday order was a homeostatic exchange system that integrated Parisians of diverse activities, ranging from sheer idleness at the two extremes, otiose aristocrat and down-and-outer, to various forms of wealth creation through labor or investment in production and distribution of goods or services in the overlapping public and private sectors. The everyday order manifested a collective disposition to diminish the costs of disruption and dysfunction. Its diffuse yet mobilizing objective was to sustain the life and relative equilibrium of the sprawling and motley metropolis. Partly a matter of daily reinvention or improvisation, the everyday order was at the same time the more durable product of an unarticulated consensus that crystallized in certain spatial and institutional nodal points throughout the city: guilds, marketplaces, churches and religious houses, taverns, neighborhoods, the very accessible *hôtels* of the police commissaires, two or three of them per quarter, and the far more remote albeit more numerous *hôtels* of the rich and powerful.

This order was not a procrustean apparatus of social control, though it contained important incentives and sanctions geared to favor collective tranquility. Nor was it an idyllic charter of social collaboration, although it thrived on all sorts of reciprocities and cascades of clientage. It was not codified or formalized in a legal grammar but allowed to evolve constantly in the vernacular of custom. Its norms were quite elastic, especially at the margins. The everyday order groped not toward universal rigor but rather toward a viable and visible middle road. It precluded neither conflict nor disorder in many spheres, but it consecrated provisioning as a nonpartisan issue subject to debate perhaps on means but never on ends. This order was founded on the instinctively felt conviction that the city could not survive without a degree of collective self-restraint, a shared sense of limits, and a will to reduce its vulnerability.

It was widely understood that everyone suffered when the everyday order broke down, when the city's capacity for self-government collapsed, when its regulatory mechanisms—the foundation of its autonomy—ground to a halt. The everyday order was above all predicated on the city's capacity to assure the minimal physical and psychological well-being of its inhabitants. The subsistence question tested the city's moral and material resources ceaselessly. In varying degrees of magnitude and urgency, all Parisians were consumers of bread. Allowing for similar variations, all these consumers were producers of social calm or social turbulence.

The everyday order—blandly commonplace, self-consciously discreet, in the best of all possible worlds widely internalized, locally enacted—reposed ultimately upon what I have called the social contract.[22] In this solemn exchange, which defines and undergirds the relation between state and society in the Old Regime, the consumer-people submit to fiscality, conscription, and other forms of extraction, and pledge a more general fealty and quiescence in return for the assurance that they will not be left to starve. The king, as nourishing prince, embodying the state, becomes not the victualer-of-everyday, though he oversees the modalities instituted to do this job, but the baker of last resort. The women who marched to Versailles in October 1789 to enforce the contract by bringing back the baker, the baker's wife, and the baker boy to the capital understood the significance of the king's role. So did the Parisians, and the peasants of the hinterland, during the Flour War who shouted "long live the King, and may the price of bread go down!"—in epitome, a whole political doctrine turning on a mingling of deference and reciprocity. The more chilling side of the same expectation was the poster threat as the riots petered out and the newly ascendant monarch prepared to receive his ceremonial legitimation at Reims that "if the price of bread does not go down, we will exterminate the king and all the blood of the Bourbons."

It is not surprising that throughout the riot-torn regions, between the heart of Burgundy and the Norman frontiers, the rumor spread that Louis XVI had set the price of wheat at twelve livres or twelve livres ten sous per Parisian setier (the countryside version) or bread at two sous per pound (the urban discourse). A water carrier of the rue Mazarine accused of entering a bakery and refusing to pay more than eight sous four deniers for a four-pounder insisted that "such was by the will and order of the king." And it is true that, under intense pressure from irate and nervous demonstrators, the young governor of Versailles had ceded and fixed the price "in the king's name" at two sous per pound, the mythohistoric just price inscribed in the memory of the century. In this context it is hard to condemn the conduct of a twenty-three-year-old Parisian paver, Jacques Lemarchand, who proffered twelve sous for a six-pounder for which the baker demanded sixteen sous. He had learned matter-of-factly from his butcher that the day before in solemn circumstances "bread had been taxed at two sous per pound, and this is what determined him to pay it according to this standard." Such was the alibi of Marie-Catherine David, who argued with a *forain,* or country baker, in a Neuilly shop that the king's word was final.[23]

When the police failed in its ordinary work of evening out fluctuations in the price of bread and parrying short-term dearths, the king had no choice but to step in. That was the very definition of kingship, to take the word of the Assembly of Police convened by the Parlement of Paris during the jolting crisis of 1768, a crisis seriously aggravated if not detonated by the government's revolutionary decision to liberate grain and flour from police regulation (or, to put it in more familiar terms, by the king's decision to "emancipate" his children from their sempiternal dependent status). Be a king, exhorted the magistrates, who were normally inclined to reduce rather than enhance the monarch's range of intervention, and usually quite ready to provide surrogate fathership in the king's place to the orphaned nation. So delicate was this issue, so telling the test, that the ordinarily opportunistic magistrates did not dare play politics with it. Be a king, enjoined the judges, even as the posters in the streets denounced him for being a speculating, hoarding, liberty-abusing grain merchant instead of a royal personage.

Being a king meant assuming the traditional responsibilities of kingship, notably the obligation to honor the social contract of subsistence. When the consumer-people faced disaster, the king had to intervene in the first instance as regulator and avenger and then if necessary as supplier. He could not coldly tell them, in the stark idiom of scientific verity, that it was beyond the monarch's province to flex the old muscles, that the lean king had superseded the bloated prince of old, that the enlightened leader could not arbitrarily violate the absolute rights of property and liberty and wantonly trespass the rational boundaries imposed by nature's law. The naive avowal of helplessness was no more convincing than the slicker disqualification by nature. The ardent effort by the comte de Saint-Priest to persuade a band of Parisian women in the early days of the Revolution that the dearth was the product of accidents beyond Louis XVI's control ("that the King could no more make grain grow than he could make it rain") was simply not credible, given the widely held view of potent kingship and the hallowed social contract. Voltaire felt the bind in the immediate aftermath of the Flour War when he wrote in the royal voice: "The Good Lord made me king of France, he did not make me head baker." (Added the modern royal reasoner, at one with Turgot's liberalism: "I want to be the protector of my nation and not its regulatory oppressor.")[24]

The pathos was sharpened in an anonymous memorandum written in the fall of 1789 in the midst of a subsistence crisis and a nasty polemic between Jean-Jacques Rutledge, a former noble warrior who had exchanged his sword for a vitriolic pen, representing the Paris bakers, and the Leleu brothers, whom Turgot had appointed in the aftermath of the Flour War to maintain an emergency grain and flour reserve for the capital. Incensed that these entrepreneurs would brandish the king's sacred name, representing inviolable commitments to his people, to mask their odious speculative maneuvers and scapegoat innocent bakers, the nameless author wrote: "Is it because these men (I am mistaken, for they are Hydras), these men, as I was saying, use the sacred name of the king, a name that they use lavishly and far too frequently, to shout everywhere, in all times and places: this grain is for the king's storehouse? But do they really not know or could they be unaware of the fact that the King is the father of the people, and how could a father put himself in a position where he could cause his children to die of hunger? That is not believable, it is an obviously false contention."[25]

3.

I have developed these themes and others in four earlier books dealing with various aspects of the subsistence question.[26] In them I have made a series of arguments about the nature and evolution of the state and the society, and their relations; the character and practice of politics, high and low, including the role of the monarch, the parlements, and the mass of ordinary subjects called the people; the desacralization of kingship; the multiple vocations and deep cleavages of the Enlightenment; the incapacity of the Enlightenment to deal with the "people problem"; the ambivalent attitudes toward capitalism; the perils of Manichean conceptions of modernization; the rise and fall of reformist ideologies and lobbies, and the galvanizing impact of grain liberalism; the crystallization and the missed opportunities of an Enlightenment-era conservative or traditionalist ideology, prior to and beyond raw antiphysiocracy; the socializing power of what I call the subsistence mentality, with particular attention to the play of one of its ramifying threads, the famine plot persuasion, a symptom of the broader propensity for imagining the world in conspiratorial terms; collective action, violent and nonviolent, ephemeral and structural, spontaneous and orchestrated; the organization and operation of administration, royal and parlementary, at the center and the periphery; (the) police as administration, worldview, and strategy of governance; shifting patterns and competing notions of social control; relations between Paris and the provinces; conceptions of exchange in general and of the grain and flour trade in particular; markets and how they work, with specific reference to the Parisian supply system and its imperialist ambitions; the relations and tensions between what I call the marketplace, as site and regulatory *idée force,* and the market principle, as the idea of laissez-faire, a policy so powerfully corrosive of the traditional social and economic systems (in Tocquevillean terms, this can be seen as the dialectic

ie reach of the extant data. I use the notion of breadways, borrowed
now fully established domain of foodways, itself an analogue of the
f folkways, first used in the early twentieth century to describe ways of
feeling, or behavior common to a given social group. The study of
s explores the tyrannical hold of bread on consumers' diet and dis-
id their rejection of alternatives. It looks specifically at the tastes of
virtually all of whom, despite the hierarchical range of distinctions,
wheaten and white loaves. It shows the ways in which moralists and
joined administrators in various more or less futile efforts to modify

section I examine closely the vectors of supply and the conditions in
ey engage the demand: the different sorts of bakers who furnish bread
ensions that often mark relations among them, their technical and
ial practices, and the sites and protocols of exchange with customers.
shops and marketplaces, assess the bakers' variable contributions to
l measure the enormous importance of the sale of bread on credit.
simonious sources, I try to elicit as much as possible about the buying
d expectations of consumers, and their relations with bakers both in
.borhoods and the slightly more impersonal marketplaces.

:cond part of the book is a history and sociology of the bakers. It is one
e modern studies of a guild, and of the profession or trade that extends
and beneath, the corporate demarcations. I explain how the guild
s relations with tutelary authorities, how it articulates interests and
resources, and how it responds to criticism of both internal and
provenance. The story of the baker's trajectory begins but does not end
ie guild. I take stock of the experience of apprenticeship and then
e licensed apprentice through his more or less protracted itinerary as
an. The struggle for control of the labor market rages throughout the
etween the often aggressive journeymen and the increasingly anxious
I join the journeymen at work in order to discover what they do, how
le and execute tasks, and how they get along with each other and with
ses. I also pay heed to the journeyman's (highly circumscribed) life
ie shop: sites and practices of sociability, bonds and enmities, relative
distress, and the prospects for founding a family.

areers of bakers who do not permanently remain journeymen culmi-
hat I call establishment, which could mean mastership but must mean
a business in some form or another. Marriage, inheritance, and various
patronage can assist in establishment, which can also take place il-
show that mastership was far more accessible than commonly be-

between equality and liberty); the conditions fo

this world, illustrated by the experience in the

credit in social as well as economic relations; a

books constitute a viewpoint about how to appr

Regime in many of its most significant aspects.

That viewpoint frames and informs this final

tion of the subsistence complex. I have quite consc

arguments that I have formulated elsewhere, save

ence did not suffice. I respectfully invite the reade

the impact of the subsistence question on the rela

his subjects to look at *Bread, Politics, and Political E*

the moral economy and paranoid ideation to scrut

suasion, or who is frustrated that I say so little herein

flour trade to have recourse to *Provisioning Paris.* 1

bakers is a wholly new study with an agenda of its o

could not complete it (having essayed a start some tv

researched, conceptualized, and written the others. 1

picture more or less intelligible to me. That permitte

basic of the basics. No single, overriding thesis govern

claim that the bread question relentlessly hedges an

activity.

I am tempted to say of the history of bread what c

of the eighteenth-century subsistence scientists obs

bread making: "by a singular fatality, the arts of first n

in all times, have been the most neglected."[27] It remai

historians know so little about a subject/problem of

cance, one that insinuates itself willy-nilly, at one time

degrees of penetration or inexorability, into our refle

whether our chief interest lies in social or political or i

genre of history. This book is about the elementary str

about what we could loosely call the normal or natural,

ary interest in the exceptional or preternatural. It is th

reconstruction of an intricate puzzle undertaken with

and rigor. It attempts to account for both very simple

habits, and relations. It is divided into three overlapp

which privileges one of the three major subjects of the bo

police.

Part I treats the demand for and supply of bread, 1

cultural, and economic perspective rather than a statistic

beyond

from th

concept

thinkin

breadw

course,

Parisia

focus c

scienti

habits

In

which

and t

comn

We v

each,

From

habit

the r

T

of th

beyo

wor

mar

exte

wit

foll

jou

cer

ma

th

th

ou

w

n

c

f

I

lieved, even before the liberalizing reforms of the 1770s, though the majority of bakers still did not attain it.

The next stage of the baker trajectory addressed in part II begins with a study of marriage strategies and patterns of alliance. Along with more subtle forms of advantage, wealth at marriage helps to launch careers, but it is not a reliable predictor of fortune later in life, a subject disaggregated and analyzed in another chapter. Most bakers gravitate around a middling level of well-being, with the fat and the lean anchoring the extremes of the bell curve. The lean are sometimes structurally feeble, sometimes victims of errors in judgment, accidents of the business cycle, and/or apparent character flaws. Even as credit keeps the healthy bakers alive, debt envelops the vulnerable bakers in a spiral of pathology that is sometimes fatal. In two chapters I consider the ways in which bakers handle debt and the recourse to formal declaration of failure. Both a means of achieving success and an emblem of its attainment, a good reputation was a form of capital that bakers managed with as much care as they handled business transactions or financial investments. The final chapter in part II surveys the ways in which reputation could be preserved or enhanced, tainted or reclaimed.

The last section of the book inspects the police of bread and bakers. I assess first the police understanding of supply and demand, attempting to extrapolate a usable set of estimates for yield from wheat through flour into bread and for both aggregate and per capita consumption of wheat and of bread. I argue that the perceptions and the methods of the police underwent a subtle process of modernization, but I begin with the premise that the police, even before they were formally struck by Enlightenment, had rarely behaved in a benighted and tropismatic fashion. However much underlying assumptions evolved, the aims of the police remained largely the same: to assure enough bread of good quality and correct weight at accessible prices in conditions of exchange considered legitimate and susceptible to surveillance. I show how price policy remained the most complicated and controversial part of the police of bread and bakers. I analyze the debate over the principle and strategies of price-fixing (called *taxation*).

In the final chapter I survey the policing of the price through much of the eighteenth century, following the grassroots practices of the commissaires, who varied sometimes quite dramatically in approach, as well as the dialogue that the lieutenants of police sought to sustain with the bakers. I take note of the dramatic impact of radical changes in national subsistence policy on the capital in the sixties and seventies that called into question the regulatory system in the grain and flour trade while tolerating, in precarious counterpoint, a rigorous police of bakers. It was not in the interest of the police, however, to bludgeon the bakers into submission or ruin, though they pressured them, and compensated them, in a variety of ways. Across the century, I suggest, the police toiled to lower the social and economic transaction costs of sustaining subsistence homeostasis.

For reasons that were at once pragmatic and strategic, I chose to halt my investigation in the 1770s. One does not go to the archives and look under "B" for bread and bakers to bring a study such as this to fruition. There is no coherent corpus of materials. One works largely in an oblique manner, using highly dispersed, often submerged sources that shed light indirectly on the subject. The research is extremely arduous and long. Given the canonical scission of documents into Old Regime and Revolutionary deposits, and given the explosive density of the Revolutionary materials, I felt at the outset that it would require more lifetimes than I could marshal to attend to both tableaux.

No less significant, albeit evidently more controversial, was my intellectual rationale. I did not want to subject the entire project to the overwhelming freight of the Revolutionary telos. Too often the Old Regime is implicitly reduced to a mere rehearsal for the monumental eruption of 1789–94. Viewing the world through a prism that disfigures, one passes one's time cataloguing harbingers of the events and mutations that one knows subsequently took place. I might be tempted, for instance, to interpret parlementary contestation of royal authority as an indicator not of political disintegration and/or sclerosis but rather of political vitality and creativity. Similarly, the fact that the price of bread reached its secular zenith on 14 July 1789, the day of the taking of the Bastille, is interesting, but not intrinsically worthy of dominating (and deforming) my long-run narrative. In any event, in terms of Labroussean cycles or according to political or cultural reckoning, I believed that the period after Turgot opened a new conjuncture and that it had to be examined on its own, though any such study, I felt, would logically build on my work. Today, I still maintain that the Revolution, despite its enormous importance, weighs far too heavily on the history of the Old Regime, though I think that with extreme circumspection one can do justice to both without becoming ensnared in teleological traps. (Let me add, however, that my own recent work on writing about the Revolution provides me with absolutely no reason to think this guarded optimism is well founded.)[28]

Concerning the subsistence question, what should be inscribed on the research agenda for the future? For Paris, I believe that we now have a reasonably thorough and reliable picture of both the morphology and the evolution of the provisioning system, and its critical political, social, economic, cultural, and psychological connections and implications. The structure of the provisioning trade, from top to bottom, from hinterland to *halle* and hearth, has become familiar to us. We have charted the startling shift from grain to flour, the rise of commercial milling, the role of the brokerage, and the pervasive influence of credit. We know who the bakers were, how they worked, and on what terms they conducted business with consumers. We have taken the measure of the socializing impact and mobilizing power of bread in daily city life. We have a good idea about how markets functioned, particularly within the metropolitan collection net, though we still need local studies to illustrate the conventions that facilitated exchange and the frictions that inhibited it.

In the near future, I expect important contributions in two domains: the study of foodways and the articulation of the nexus between production and distribution. After the perspicacious suggestions of Pierre Legrand d'Aussy in the eighteenth century and Armand Husson in the nineteenth, surprisingly little has been written about the food habits of Parisians and the concrete ways in which patterns of consumption changed, and resisted change.[29] Historians need to venture the sort of work about taste, distinction, knowledge, social classification and conflict, and popular culture that sociologists Pierre Bourdieu and Claude Grignon, each in his own way, have undertaken for the recent period.[30]

The debate on production and productivity, with its demographic corollaries, has been reopened, signaling a salutary renewal of rural history in France. Future inquiry will help determine the veritable carrying capacity of Old Regime France as well as the utility of the long-lived Malthusian model.[31] As the agricultural revolution makes a comeback and the demographic profile becomes even more subtle in its configurations, we will confront the problem of connecting the work of agriculture with the work of commerce. Jean-Michel Chevet's colorful exasperation with historians who underestimate the real increase in agricultural yields and overestimate the institutional aspects of economic development is symptomatic of the tonic debate that is brewing: "The market [is made to seem like] Jesus multiplying the breads or changing water into wine. . . . it is the philosopher's stone."[32]

Finally, we need a recasting of the periodization of the subsistence question and the larger foodways complex. Reading the grand *enquêtes* of the late 1850s on the provisioning trade and on the bakery, one has the impression that "plus ça change . . ." One encounters the same anxieties, a familiar preoccupation with famine plots, the state in a terrible double bind that makes action and inaction equally perilous and problematic, a working-class fixation on white loaves, grain and flour dealers of dubious or odious reputation, bakers more deeply than ever in the grip of millers, markets blocked by ideology rather than bad roads or impassable rivers, and so on. Once again the Revolutionary refraction jeopardizes the clarity of our sight, this time by exaggerating discontinuity and obscuring inertia. It would be enlightening to look closely at the first half of the nineteenth century in order to discover to what extent the old subsistence mentality survives, the notion of a social contract persists, an integrated national market system has really emerged, the administrative regulation of the bakery has fundamentally changed, the recruitment and itinerary of bakers have taken new directions, the labor market remains a site of sharp controversy, the techniques of baking have substantially evolved, the bread buying and selling protocols have taken on new gestures, words, and meanings, and the symbolic hold of bread has withered or tightened. In the long view, from Colbert through Zola, I hope that historians will strive for a richer, deeper melding of the social and the economic with the political and the cultural. I still endorse the hubristic conviction I held as a graduate student that a problem as total as the subsistence question merits something resembling a totalizing approach.

I

Bread:

Demand and

Supply

Chapter 1

Breadways

Psychologically, culturally, politically, and economically, bread was one of the most powerful "structuring structures" that governed private and public life in Old Regime France. It was at the core of both the material and symbolic organization of everyday existence. France was not merely "panivore," following the picturesque contemporary idiom; it was obsessed with bread.[1] Bread was its primary means of survival, its paramount vector of sacrality, and its most comforting trope. It was impossible for the French to conceive of their well-being, here and now or hereafter, outside the confines imposed by the bread paradigm, at once tutor and tyrant.

Even if there were other foods available, the *Encyclopédie méthodique* reported, "the bulk of the people believe that they are dying of hunger if they do not have bread." For them the loaf contained something more than calories and nutrients. Whatever its particular form, texture, and composition, it conveyed an assurance of continuity and fidelity and it served as a measure of diverse sorts of legitimacy. Crystallizing both collective identity and individual destiny, bread forged the complicated links between sacred and profane, hope and anguish, whole and part, mother and child, prince and subject, producer and consumer, seller and buyer, justice and injustice.[2]

If the French were "the biggest eaters of bread in the whole world," it was less for mystical than for rational reasons. While bread was evidently the worldly substance closest to divinity, it was believed to be also "the food most analogous to the human species." It was "the healthiest of all foods" and "the most essential to life." In the words of a distinguished physician, "it is appropriate to all sexes, ages, ranks, and temperaments, to the rich as to the poor, to the King as to the last of his subjects."[3]

The "panivorous" rationality, however, fed on itself in a perilously circular way, foreclosing certain strategies of survival by preventing the optimal utilization of all available foods. Passionately interested in finding ways to attenuate subsistence crises, the food scientist Antoine-Augustin Parmentier despaired of making his case for surrogates and alternatives in France. "It will be hard," he wrote, "to convince the French accustomed to eating one kind of bread or another that three-quarters of the world gets by without it." Reviewing Parmentier's *Examen chymique des pommes de terre*, the *Journal de l'agriculture* warned that "the current furor to make everything into bread can only prove dangerous." It referred to Parmentier's efforts to circumvent the extraordinarily conservative disposition of Parisians in particular and of the French in general ("impassioned ignorance and ridiculous prejudice") by presenting the potato as a friendly farinaceous vegetable to be used in (procrustean) bread. It would be virtually impossible to induce the French to change their basic foodways, observed a commentator in the *Journal économique:* "there are persons so stubbornly installed in their old habits in this respect that it would take an infinity of examples, and require almost miracles before their very eyes, to persuade them of their error."[4]

Grafted on a widespread pagan culture that paid special deference to grain, the Christian tradition, no less deeply rooted in the agrarian world, invested bread with a powerful and miraculous spirituality. Christian texts and liturgy abound in bread imagery. Among many other things, Jesus, the "Living Bread," was "a nourishing god who addressed a population suffering from chronic shortage." Instituting the Eucharist, Christ said of the bread: "This is my body." There is no more moving moment in Christian practice than the communion, mediated by bread and wine, that joins the believer and his or her God. So meaningful was the very act of breaking bread that until the second century the words *fractio panis* referred exclusively to the Eucharist. It seems very likely that for a great many French women and men the mood of the Eucharist echoed in the communion of everyday bread breaking in the family or at work. Surely the Eucharistic mold reinforced the conviction that only bread could sustain life in its deepest sense; that food acquired providential power and status only when it took the form of bread.[5]

Quasi-liturgical practices spilled over into daily bread-breaking routine. The best-known example is the custom of tracing the sign of the cross on the loaf with the tip of a knife. Bread's holy nature required a special protocol of respect; fears of retribution impelled believers to rigor. To avoid misfortune, they avoided placing the loaf bottom up, as if this were an act of desecration. It was a sacrilege to waste bread, to fail to utilize all the leftovers in soup or some other form, or to offer them as alms. (The Christian horror of waste served the mundane needs of the civil order, where economy in subsistence became the watchword of administrators and scientists, and Voltaire and Rousseau, despite their antagonistic views on the social utility of luxury, joined hands to assail the prodigal waste of precious flour in wig preparation and cosmetics.) One had to repair the insult to bread allowed to fall to the ground by kissing it. Stepping on bread was a damnable act. The prescriptive glissando from liturgy to folklore was not always coherent. Thus in some parts of France it was considered a sin to give bread to animals, while in other places mixing chicken feed with bread crumbs was a fecund recipe for increasing the number of eggs.[6]

The familiar act of the offering of holy bread dramatized the connection between material and spiritual concerns, casting into relief both the sanctified status of bread and its social obligations. It was a sacrificial ritual with a redistributive function, though critics claimed that the bulk of the holy bread never reached the needy to whom it was destined. Theoretically each Parisian was bound to make the offering when his turn came within the parish ("we are all equal in the eyes of the Lord"). While certain observers saw moral advantage in inducing Parisians to acknowledge publicly the pivotal role of bread in their individual and collective lives, Louis-Sébastien Mercier deplored the ostentatious aspect of the ritual ("a spectacle of vanity") and its tendency to issue in unsavory disputes regarding rank and magnitude of contribution ("these pious trivialities"). The parishes frequently summoned the courts to condemn recalcitrant citizens. (Lieutenant General of Police Hérault personally reprimanded a woman baker who had less excuse than anyone for failing to make the offering.)[7]

To describe a well-merited disgrace in the eighteenth century, one called it "holy bread." Apart from its sacred metaphorical vocation, as quasi-ubiquitous everyday trope bread impinged on everyone's consciousness. Bread conveyed notions of health, fecundity, fortune, cleverness, wisdom, home, family, love, work, pleasure, joy, comparative worth, and so on. For Dr. Paul-Jacques Malouin, the eighteenth century's first great baking expert, drawing on the Greek and the Hebrew as well as on colloquial usage, life and bread were "synonymous terms." According to Voltaire, a Jesuit adversary "paid me the honor of publishing two volumes against me in Lyon in order to earn some bread (I don't think he got white bread for it)." Asked why he carried a hammer, a master locksmith arrested by the police in May 1775 explained "it's my bread." A fugitive journeyman denounced in May 1739 "is not where he is supposed to be, earning his bread." With the Enlightenment, noted *Le Babillard*, a late-eighteenth-century periodical, "the art of thinking and writing has become a breadwinner." The shift from the image of *gagne-pain* (breadwinner), common in the eighteenth century, to the formula *gagne-bifteck* (steak-winner), a post–World War II locution, marks the enormous socioeconomic distance traversed in two centuries.

Direly ill, a person has lost "the taste for bread." An individual who has already enjoyed many years of life "has already baked more than half of his bread." A person devoid of malice is "good like bread." A wonderful man or woman is "better than good bread." A taciturn or doleful person "has lost his bread in the oven." A young woman who became pregnant before marriage "has borrowed a bread from the next ovenful." In nineteenth-century Alsace couples in a hurry "take loaves from the coming ovenful," but certain men learn at their expense that "it is not always he who heated up the oven who is the first to bite into the warm bread." Drawing on a classical binary metaphor at the end of 1994, in a front-page article *Le Monde* noted that Prime Minister and soon-to-be presidential candidate Edouard Balladur "must have suspected that after having eaten his white bread, he would sooner or later be obliged to swallow his black bread,"[8] that is to say, tougher times would surely follow his early success. In early 1995 President François Mitterand remarked ironically on candidate Jacques Chirac's leftward drift: "a new social contract for employment has become [his] daily bread."

These tropes turned imperceptibly into myriad proverbs that characterized a wide range of comportment. "To eat one's white bread first" referred to an individual who enjoyed tranquility at the outset and then endured troubles. It is difficult to remedy an affair that one has begun badly: "If one puts them in the oven incorrectly, one gets bedeviled loaves." It was lamentably selfish to "to eat one's bread in one's bag." To express one's boredom with a given experience, one remarks that "that is as long as a day without bread." A person who arouses vain hopes "promises more butter than bread." To signify one's right to partake of one's neighbor's bread, one averred that "a bread that is sliced open has no master." Long before the Revolution people dreamed of "Liberty & baked bread," an ideal/idyll that tended to be individual rather than collective and suggested that happiness was having some property and not being subject to anyone. In the mid-eighteenth-century edition of his dictionary, Pierre Richelet suggested that it was possible to have a rich conversation without ever abandoning one's loaf, linguistically speaking.[9]

Not by Bread Alone

One of the rare eighteenth-century thinkers who did not associate bread with either liberty or felicity was Simon Linguet, a lawyer, muckraker, and pungent social commentator. His case is interesting because he contested the constrictive bread archetype vehemently, without any regard for its sacral immunities, in terms of the rational requirements for social production and reproduction. Beyond the play of paradox in which he delected, he raised many crucial questions that others repressed or ignored concerning the logic of the "bread-centered" (*panivore*) subsistence system, its legitimizing and mystifying discourses, and its perverse effects. These reflections informed Linguet's fierce campaign against the Enlightenment's self-proclaimed economists, the physiocrats, for whom grain was at the very core of the socioeconomic system and the liberty to dispose of it freely was the first principle of politics.

This Montesquieu (or perhaps this Persian) of French foodways began by exposing the overbearing ethnocentrism of hegemonic assumptions about bread—"prejudices" equally shared by the elite and popular cultures. "First, we have the idea that it is the only food suitable to our nature and that humankind would perish without it," wrote Linguet. "It is, however, a fact that the majority of mankind is unfamiliar with its usage," he added, echoing Voltaire who had made the same observation without pondering its implications, "and that among those who adopted it it produces only pernicious effects." The French needed to look beyond their navel: for Americans, who ate cassava, plantains, bananas, and corn, bread was "a delicacy and not a regular food"; sustained by rice, which was infinitely easier to cultivate, conserve and prepare than grain-wrought farinaceous foods, Asians viewed the latter as luxuries; in Africa, where people survived through hunting, fishing, and fruit consumption, bread and its panoply were unknown.

A Copernican revolution away from the truth, the French had to strip away the sophistic rhetoric and see bread for what it really was: "a tedious and costly compound, a nuisance in every sense" that was consumed exclusively and exceptionally in a "little corner of the planet, . . . our little Europe." Even there, "where it seems so necessary," a vast number of people were excluded from its so-called benefits: the Spaniards who survived on chestnuts or acorns; the French who lived on a buckwheat or millet porridge or a sort of corncake; or the Germans who managed with boiled potatoes. The numbers spoke eloquently in Linguet's judgment: barely 5 percent of the world's population ate any form of bread. "There's the universal food! There's the important object of subsistence to which governments must sacrifice everything, on which all political speculations must be founded, and whose form we must attempt to give, by every violent manipulation, to all the foods with which nature bountifully endowed us [precisely] to turn us away from this one."[10]

Far from being a harbinger of democracy, bread, especially in its wheaten form, was a marker of despotism, distinction, and disdain. Far from being an authentic commodity of first necessity, bread, especially wheat bread, was called into being by "luxury." To account for the "slavery" to which bread reduced (European) mankind, Linguet elaborated a sort of dependency theory:

"There is no type of food that holds men more tightly in a state of dependence" than bread. The need to provide a regular supply of bread held the people—the poor who were the overwhelming majority—"in chains," subjected them to arbitrary power, reduced them to contemptible status.

The "bread-as-the-only-possible-food" ideology justified this tyranny, and the exploitation "of the little people" by "the big." Given the enormous intrinsic difficulties of cultivating bread-making grain, conserving it, transporting it, transforming it into flour, and then converting it into bread, and given the temptations to which the greedy and manipulative easily succumbed, it was inevitable for the system to fall prey to disorders, in particular "monopoly," whose victims were the little ones. The bitter irony of this subsistence system was that it engendered "the secret of perpetuating famines" that it was supposed to prevent. "Bread," implored the beggar. It was bread, however, that begot the beggar, and the misery in which the majority languished. "When bread returns to assassinate you," admonished Linguet, "the image of peace and quiet, of liberty and of abundance" associated with bread will rapidly recede.

Anterior to its political ramifications, "this cruel dependence" derived from the natural and social requirements of bread production. Agriculture subjected the cultivator and then the consumer to abominable "constraints." No work was physically harder and morally more exigent than growing bread grains; no activity was less certain of issuing in success, given the risks (weather, disease, predators). In the cycle of application and vigilance, there was no repose, for once a harvest came in, another had to be prepared. And the harvest was not the end of the tale but the beginning of a new chapter. For the grain had to be threshed, winnowed, stored, conditioned, transported; then cleaned, ground, bolted, graded, stored, transported, and (re)conditioned. Only then was it ready for the killing task of nocturnal bread making that fell to the bakers. And a few days after it was baked the bread was no longer fit for eating.

For Linguet, there was no rationality in this exhausting, debasing, and aleatory process. It was neither in the interest of the state, which needed to develop wealth and stimulate population growth ("the plough opens the tomb of our species"), nor in the interest of consumers, for whom bread offered security no purchase, to persist in such an exorbitant and ultimately self-destructive course. It made infinitely more sense to turn to the inexhaustible supply of fish, harvestable without planting and incessant labor ("the seas are a vast countryside," which could not be turned into "a prison"), or to the production of rice, which was cheaper and easier to grow, more abundant and more certain, largely impervious to the elements, extremely durable, infinitely simpler to prepare, at least as tasty and more nourishing, in a word "the most beneficent of all foods."

The truly "economic" food, rice did not have the nefarious sociopolitical consequences of bread. Linguet located the proof in the life experience of Asia, which he idealized in familiar Enlightenment terms as simple, just, and sweet, devoid of the "degrading dependence" and the other criminal abuses "disguised under the name of the progress of the arts."[11]

In Linguet's idiom, bread was not merely a metaphor for a diseased body social. It was a veritable "poison" that menaced the health of many consumers. It was a "killing drug" whose primary ingredient was "corruption." Extremely difficult to digest, no food was "more dangerous" to the stomach nor a graver menace to the circulatory system, for bread thickened the blood and freighted its movement. The rich enjoyed a permanent antidote to the poison: they could afford to diversify their diets. Because they ate a monotonous and narrow diet built around bread, the poor stood to suffer the most. They wore their pathology on their faces, faces that betrayed systemic atrophy and premature aging in a leaden complexion.[12]

Bread induced moral as well as physical degeneration, according to Linguet's diagnosis. It led the rich on the road to decadence through self-indulgence. The *mollesse,* or flaccidity, of their loaves communicated itself to their spirits and cankered them. The abasement to which they reduced the poor further drained their moral capital. Flaccidity and cruelty were hardly the virtues on which to nourish an elite. Excessive refinement, reflected in the culinary arts as well as in bread making, undermined the strength of the nation. No wonder that "our *Frenchmen,* [who] attached about as much importance to an above-average chef as to a *Turenne* or a *Bossuet,* are also the greatest bread eaters in the universe." For different yet ultimately complementary reasons, virtually no one in French society (save those strategically assigned to the margins) can do without bread. "Thus," concluded the lawyer-writer, "the bakery is doubly criminal toward humankind: it increases the slavery of the poor and serves only the sensuality of the rich."[13]

On the surface Linguet's frontal assault on the material and spiritual nucleus of French civilization seemed so outrageous that it did not have to be taken seriously. Like Rousseau, who seemed to work at not fitting in, Linguet could be dismissed as a freak of nature (even as he fought against the rule of nature imposed by the physiocrats). All the more so because Linguet himself admitted to suffering from a lifelong physical incapacity to enjoy bread, whose consumption caused him unfailingly to experience disagreeable "heartburn" and "indigestions." The argument "that renders me invincible," he boasted playfully, "is my stomach." On one level, then, his whole discourse could be interpreted as mere vengeful projection, his vindictive manner of dealing with an unyielding, idiosyncratic revulsion that at bottom may have mortified him. It seemed perfectly appropriate that the leading physicians concurred that "the loss of taste for things is a sure sign of bad health, and the distaste for bread is the most unfortunate of all distastes."[14]

Not even the intellectuals engaged Linguet: his position was too farfetched. Parmentier merely rehearsed the claim for the merits of bread as a universal food of incomparable value, fully worth the efforts required to produce it. He considered the case to be self-evident, and he concluded in a reassuring, therapeutic tone that "wheat and bread are pernicious neither from the perspective of politics nor from those of morality and physics, as a famous philosopher has tried to show." Worried much less about Linguet's fundamental objections to the entire bread system, which he felt would exert no influence whatsoever, than about his specific critique of the perils of an unpoliced grain trade, which could reinforce the still robust regulatory camp, the minister-philosophe Turgot encouraged the économistes to riposte. Yet even the most substantial of these liberal refutations, Abbé Morellet's, was a superficial enterprise, largely predicated on raillery for Linguet's apparent extravagance (he would have us "live off fish like the Greenlanders"), his allegedly flawed anthropology (the Turks were Asians and they used bread), and his inability to move beyond facile paradox and uninformed reverie (he would console us with the precious knowledge that the famine produced by his system would thankfully not be "the work of monopoly").[15]

One of the rare contemporary social commentators to endorse Linguet's view was another abbé, Jean-Baptiste Briatte, partisan of the moral economy rather than the market economy, friend of the poorest consumers rather than the most affluent producers, oblivious to the sacral moorings of bread despite his tonsure, but not to its presumptive social mission. His stomach tolerated bread very well indeed, unlike his heart or his head. Bread was treachery itself, promising more than it could deliver, manipulating emotions and only partially nourishing bodies, undependable and inconstant yet jealous and demanding undivided commitment. "Nothing better proves the imprudence of having adopted it as the most precious of foods than its inadequacy in fulfilling the function to which it was assigned," contended Briatte. It was to prove very difficult, however, to convince eighteenth-century Parisians that they could not do just fine by bread alone.[16]

Wheat

Historians associate the extension of the wheat frontier with the dramatic mutations of the nineteenth century: modernization, industrialization, democratization, the diffusion of higher living standards among the poorer sectors of the population. But eighteenth-century Paris was a traditional preindustrial royal city filled with a teeming mass of poor people who, like their more fortunate compatriots in private townhouses and their less fortunate brethren in jails and hospitals, ate wheaten bread. In this habit they did not differ from Londoners (and nearly three-quarters of all Englishmen in the second half of the century) or citizens of other European metropolises such as Geneva. As for the rest of France, according to agronomist Henri-Louis Duhamel du Monceau, "the inhabitants of the cities use a wheaten bread almost exclusively." "The people here," an official in Toulouse asserted, "know no other food than wheat; it would not be possible to make them adopt another." Bordeaux reported similar breadways, though it is not clear for either city how far the wheat line descended toward the base of the social structure.[17]

Psychologically and materially, there were few more decisive moments in the apprenticeship to city life than the repudiation of inferior grain in favor of the grain of the Host (nonwheaten bread could not be validly consecrated, in the eyes of the church), the grain of cleanliness as well as holiness, of opulence as well as nourishment. In Paris, one learned quickly, wheat was taken for granted. The rich hinterland around the capital—the plain of France, the Vexin, the Valois, the Brie, the Beauce, and so on—focused on producing wheat for the Paris market. When necessary, the capital drew wheat from the more distant provinces and from abroad. The city consumed very little rye, maslin, or barley. "Paris wants only pure wheats," according to a memorandum prepared for the controller general in the early years of the century.

Does one need to appeal to taste and mentality to account for the phenomenon of wheat consumption in Paris and other big cities? It could perhaps be ventured that the choice of wheat is an income effect of higher urban living standards, even among the laboring poor. The income elasticity of the Parisian demand for wheat, as a substitute for inferior grain, would be considered quite high. Another possible explanation calls attention to the equal bulkiness of wheat and (say) rye. The markups per unit value for transport and transaction costs would be about the same. Since rye is lower-priced to begin with, this markup equality makes wheat relatively cheaper and could be thought to encourage its substitution for the inferior cereal, apart from taste considerations. We lack the data to test either hypothesis.

Neither seems readily applicable to the tens of thousands of Parisians who were "objectively" and comparatively too poor for wheat. Their claim on wheat cannot be explained in purely market terms. Nor can either argument account for the tenacity of resistance to any move away from wheat even in calamitous times, when victualers such as the Swiss banker Isaac Thellusson and philosophes such as Voltaire deplored the "dainty refinement" of urban consumers. In practice, it is not clear what constraints this choice implied and what other consumer goods ordinary Parisians had to renounce in order to accede to and retain the wheat preference. Wheat had an institutional anchoring in the capital that only partly reflected economic considerations. The city invested all its denizens regardless of rank with a sort of cereal franchise. Parisians reckoned the right of wheat eating and the status it conferred as indefeasible. The full story and chronology of this process of acculturation, which probably began in the seventeenth century, remain to be constructed.

Modern authorities ascribe the supremacy of wheat not to its nutritional properties, which are similar to those of the other cereals (wheat holding a certain advantage in quantity and quality of proteins), but to its unique ability to be made into a light, voluminous, finely textured, readily digestible, and savory loaf (owing to the fact that its proteins take the form of gluten, the elastic substance that enables the dough to retain the carbon dioxide generated by yeast fermentation). Rye is the only other grain that contains significant amounts of gluten-forming proteins, but they are incomparably weaker, and thus produce a dough that wants for spring and elasticity. Unlike rye's slightly sour, bitter taste, wheat conveyed a sweetness that seemed to be the natural cognate of its whiteness, a whiteness largely resulting from the fact that the wheat berry embraces a much higher proportion of starchy endosperm and a smaller proportion of skin and bran. Eighteenth-century cereals experts emphasized that wheat was not a frill or a delicacy but the most suitable element for the nourishment of man, the one that most fully and harmoniously corresponded to the makeup of the human body.[18]

The significance of this somewhat banal assertion, which could hardly have surprised the simple consumers of the time who knew from instinct and experience the marvels of wheat, is that it was made with scientific authority. As a result of recent discoveries in botany and food chemistry and the interminable debates on the nature of digestion, Polycarpe Poncelet, Parmentier, and the other subsistence scientists believed that for the first time they could explain why wheat provided the body with all of its necessities in excellent quality and quantity, and why its flour was weightier, better rising, and more nutritious than that of any other cereal. Once they fully fathomed its secrets, they believed they could apply this knowledge to the improvement of agricultural, conservation, milling, and baking techniques and thus contribute, even more directly than moral philosophers and social reformers, to the happiness of mankind.[19]

White

Parisians vaunted a second conceit that significantly complicated the city's requirements for (relative) serenity. They demanded not only wheaten bread but a reasonably fine and decidedly *white* loaf. For the Parisian consumer of rural origin, attaining the promised land of wheat might very well have seemed tantamount to the end of hierarchy (the same sort of optimism/illusion afflicted a certain number of artisans who presumed that ascension to mastership signaled the end of invidious distinctions). In fact, however, wheat did not necessarily entail white. It lent itself readily to the confection of a highly differentiated range of breads constituting a potentially rigid pecking order. Indeed, a frugal and politic use of wheat implied a pattern of stratification that in many ways would mimic the rank differences that separated and classed breads composed of wheat, of rye, of barley, of buckwheat, or of other, still "lesser" cereals. But that brand of social rationality had little purchase in the capital. White did not become the appanage and the escutcheon of an elite. If white bread was not yet "whiter in the popular quarters than in the rich ones," as it would become in the nineteenth century, it was widely consumed, in one form or other, by a very large part of the population. And the intermediate loaf in the capital was whiter than the top white elsewhere in the realm.[20]

There was no particular originality in the Parisian fixation on white. White was the universal (and apparently timeless) symbol of goodness, purity, certain kinds of power, and a host of (other) status claims. Parisians in this regard were no different from the Roman aristocrats described by Pliny (who notes that chalk was sometimes used to whiten the flour consumed by the patriciate) or from modern Americans whose consumption of bread increased when flour was bleached and decreased when an absence of bleach slightly darkened the end product. (One must keep in mind, however, that grain seed type and milling technology rendered the white wheaten loaves of Bourbon Paris very different from the breads that this label identifies today, less white and coarser, more like the country-style loaves that we know than the familiar urban baguettes.)[21]

It would be interesting to trace precisely when *pain blanc* became a trope in popular French to signify a happy or positive notion or the end of hard times, and how the metaphor fared over time (its brief eclipse/discredit during the Revolution serving as a fruitful counterpoint to a largely heroic career). From Commissaire Nicolas Delamare at the end of the seventeenth century through Lieutenant General Jean Lenoir eighty years later, the Parisian police remarked repeatedly on the obsessive preoccupation with whiteness that marked consumer behavior, in times of penury as well as abundance. Consumers presumed (then as they still do now) that a good-looking bread was bound to be good-tasting. Exchanges in the shops and markets demonstrate that quality mattered enormously, but there is some evidence to suggest that numerous Parisians were even more sensitive to color than to taste. In any event, the self-image of many Parisians seemed directly linked to their breadways.[22]

The obstinacy with which ordinary Parisians insisted on white bread struck contemporary commentators. A zealous grain liberal and bread reformer, Abbé Nicolas Baudeau deplored the fact that in the capital "the people have the bad habit of judging bread [only] by their eyes, they look only at its color, they want only white bread." César Bucquet, an innovative miller, complained that "the people of Paris have become accustomed to eating a white loaf, a bread that is too white." As for the dark or *bis* loaf, "only a dearth can induce the artisan to eat it," noted the miller-reformer. Equally preoccupied with "the habit of eating white bread that the workers and servants of this city have contracted," a ministerial adviser on provisioning matters writing in the midst of the grave subsistence crisis of 1767–70 suggested that only legislative fiat—the infamous two-species decree limiting bread production to *bis-blanc* and *bis*—would enjoy any short-term sway. It hardly seems surprising that a group of Parisian workers, lured to Lyon by substantially higher wages several years later, returned disgusted, "explaining that they absolutely could not adjust to the habit of eating the excessively dark bread that is usually made there." When wheat was reasonably priced, observed Bucquet's collaborator, the writer Edme Béguillet, "the Bakers have trouble making bread white enough to satisfy the false delicacy of our lowest classes of citizens."[23]

The lowest classes included prisoners. Demanding "white bread"—the only sure sign that it would be "good bread"—in early December 1751, inmates at Fort Levêque, led by a woman, rejected the dark and heavy loaves distributed to them ("bad bread," "unfit to enter the human body," "not edible," "the dogs would not want to eat it"). They hurled rocks and bottles at the guards and refused to return to their cells. The authorities called for reinforcements, who fired on the mutineers, killing at least three and wounding two in their initial volley, apparently not preceded by warning shots. Though the investigating officials preferred to believe that a riot could not occur on account of bread alone ("that the bread was not good was not a sufficient reason to launch a revolt"), it is clear that discontent with the bread had been building up for a week or more. To show that he was not indifferent to repeated complaints, the day before the uprising the head jailer had summoned the baker to account in person for the quality of his bread. Doubt clearly sapped the baker's confidence, for he accorded six liards to each prisoner who accepted his bread at Fort Levêque as well as at the Châtelet prison. Their umbrage allegedly overheated by wine and the hope of extorting another cash indemnity, the prisoners had agreed among themselves not to accept the next day's bread if it was "as bad as the previous days."[24]

Although we cannot date the beginning of the empire of white, looking forward it is striking to note the continuity in the expectations and practices of Parisian consumers. During the interlude of the Revolution, for political, economic, and ethical reasons it rapidly became both difficult and unfashionable to think and eat white ("Citizens, is it merely to have a more or less white bread we have taken up arms?" asked the journalist Prudhomme, trembling with indignation, as early as August 1789). In terms of its breadways, however, mid-nineteenth-century Paris seems in many ways indistinguishable from the city a hundred years earlier. "The population of Paris has a very marked repugnance for bread of inferior quality," the lawyer Bethmont attested before a commission investigating bakery matters in 1859. "The people, in all classes, want to eat the same bread." If one did not know that the following testimony belonged to baker Berger standing before this same panel, one might easily have imputed it to one of the articulate bakers of the 1760s, a Pierre-Simon Malisset or a Toupiolle:

Why is the production of second-quality bread so restricted? Because the public is accustomed to eating white bread. If you send for white bread from the [working-class] faubourg Saint-Marcel, I am sure that it will be better than you will find in the [fashionable] Chaussée-d'Antin. The worker wants good white bread; he tells himself this: I eat only bread, I want good bread; I do not want dark bread.

The indigent day laborer preferred three pounds of white to four pounds of *bis*, and the itinerant mason, who ate *bis* "in his home region," demanded white in Paris. Like his eighteenth-century counterpart, the nineteenth-century worker was willing to pay the same price as a bourgeois for bread of quality: "Am I not worth another?" he asked, in justification of his prodigality and his amour propre.[25]

During the eighteenth century the Parisian fixation on white seemed perfectly consonant with the image of arrogance, self-indulgence, and Babylonian corruption that the capital projected in the eyes of many contemporaries. Those who spoke in the voice of the provinces deeply resented the tribute that the periphery was obliged to pay to the unchaste consumers of the center. Local officials questioned the right according to which Parisians ate more bread and finer bread than the inhabitants of the provinces. Nourished on black bread cooked once a week in Corsica, the young Napoléon bespoke this same sense of shock and outrage when he first discovered that the humblest Parisians ate fresh white loaves. Beyond the question of fairness, another commentator writing toward the beginning of the eighteenth century decried the anomalous and demoralizing situation in which "the lowliest artisans [in Paris] eat a finer loaf than the best bourgeois in the provinces." While Revolutionary leaders in Year II denounced as calumny ("it's once again a perfidy of the agents of the tyrants") the bruit that Paris ate "very white" at the expense of their brothers in the countryside ("one leaves them the bran"), provincial citizens with any memory reaching into the Old Regime knew very well that the charge was plausible.[26]

Moralists and Molletists

Subsistence experts joined moralists and public officials in condemning the extravagant tastes of the consumers of the capital. Bread became the innocent vector through which "sensual pleasure" conquered (and cankered) the lower reaches of the body social. It exercised a worrisome carnivalizing effect, blurring distinctions that structured the social order and undermining the sturdy values that had protected the "little people" from the ravages of "refinement." (The Paris case changed the terms of the social debate: more familiarly, moralists such as Mercier, anticipating Revolutionary egalitarians, of '89 as well as '93, denounced the consumption of white bread—"too delicious" and produced "at the expense of the bread of the poor"—by the *dominant social groups,* theoretically and symbolically the natural beneficiaries of this "odious" distinction.)

In most product domains, luxury was simply beyond the ken and grasp of the vast majority of the population. But bread was much less remote and inaccessible. Infiltrating the tastes of the popular milieux, the doughy "sensual man" led the people astray. "One of the misfortunes of the realm is this delicacy that the lowest people affect in the choice of bread," the banker-victualer Isaac Thellusson admonished the procurator general of the Paris Parlement during the grave dearth of 1740. During a similarly ominous crisis a generation later, Séguier, the parlement's advocate general, deplored "the false delicacy of eating white bread." In his (macro)critique of the misallocation of social resources, Rousseau was perhaps right in blaming wigmakers, among other purveyors of criminally wasteful luxury, for depriving the poor of the precious flour needed to make their bread. Yet it was clear that flour destined exclusively for bread making was not safe from misappropriation entailed by "our vanity." Everyone understood that the whiter the flour, the smaller the number of consumers who could be fed by a given amount of grain.[27]

In the eyes of the comte d'Essuile, the "monstrous dissoluteness" of great cities such as Paris, which gradually extinguished the "race of healthy and vigorous citizens," fed on the unwonted lust for bread purity. Baudeau lamented the way in which urban civility dulled the people's natural (animal?) sense of what was intrinsically good, for their obsession with "color" had blinded them to true measures of "quality." "It is the well-founded reproach to the people of Paris that all authors rightly make," concluded Béguillet, "of sacrificing their well-being in order to live from white bread, whereas a good household loaf would be more healthful, more substantial, more savory, and more suitable to people given to heavy work and would have the double advantage of being cheaper and more economical, one eating infinitely less of it because it remains longer in one's stomach."[28]

Nothing influenced more decisively the way in which eighteenth-century Parisian bread was to be made than the late-seventeenth-century quarrel over *pain mollet*, a soft and light "fantasy" bread prized by a substantial segment of the wealthy population (by "most persons of rank," according to a notable who sat on the board of the city's biggest hospital). Apparently introduced into France shortly after the wars of religion, the mollet loaf became the favorite of Marie de Médicis, the queen of Henri IV, whence one of its common labels, *pain à la reine*. Often made with milk, which rendered the dough heavy and inhibited its capacity to rise, the mollet loaf overcame these burdens and acquired its legendary lightness through the substitution of brewer's yeast, or barm (*levure de bière*), for the traditional leaven generated by the dough itself (the *levain franc* that we associate with sourdough fermentation). This resurrection of a practice widely used in antiquity, complained Delamare, introduced "sensual pleasure" into the theretofore healthy breadways of the capital, generating an escalation of refinement that he feared would misuse resources and confound priorities (in particular, "thereby to draw the common people into error," and tempt the bakers to reduce the range of common breads). Aside from the larger profit margins afforded by the mollet, many bakers appreciated the barm because it rendered the dough much easier to work and accelerated the proofing process.[29]

As a result of a conflict between bakers of luxury breads and forain bakers over the provisioning of taverns, coupled with the desire of the first lieutenant general of police, La Reynie, to reassess the healthfulness of soft-dough baking practices, in the late 1660s *pain mollet* went on trial. In response to La Reynie's request for counsel, the Paris Faculty of Medicine, after listening to reports by four physicians, voted by a substantial majority (either 45 to 30 or 47 to 33) to ban brewer's yeast as deleterious to human health. Hearing the baker–tavern keeper case on appeal, the Paris Parlement shortly afterward solicited the opinions of six physicians and six notables, who divided on the issue, with a majority favoring a prudent usage of the barm.

The adversaries of brewer's yeast, led by the celebrated medical doctor and writer Gui Patin, maintained that, from the time of Hippocrates, healthful and agreeable bread contained only two ingredients: flour of good quality and clean water. Prompted by a synergy of (producer) laziness and (consumer) dissoluteness, the resort to barm produced a bread that was lighter, to be sure, but less nourishing and capable of transmitting disease. For the brewer's yeast was the "écume" ("froth" but also "skum") of a quintessentially "doleful drink" capable of damaging the nervous system, provoking urological difficulties, even causing leprosy. Hot, bitter, and corrosive, this yeast originated in the disgusting "rot" of barley and water, and the bread it begot communicated these detrimental qualities to other foods it touched as well as directly to the body that incorporated it.

Defenders of the beer by-product, who included a number of enthusiastic consumers of mollet loaves, emphasized precisely its widespread use in both contemporary Europe and antiquity, seeing this as proof of its efficacy and its innocuousness. They depicted the barm as structurally analogous to the leaven but more intelligent and vigorous. "Coarse," "ponderous," and the fomenter of "sourness," traditional leaven fermented slowly, weakly, and imperfectly, generating a flawed loaf that lacked "viscosity," staled rapidly, and taxed the digestive system. Brewer's yeast, the calumnied product of a purified beer rather than the epitome of an "unhealthy beverage," was inherently "more subtle and more penetrating," deeply and evenly impregnating the dough and issuing in a light and pleasant bread that both satisfied the consumer's taste and enhanced his well-being. In a spirit of lucidity and compromise, several of the yeast partisans conceded that certain regions diffused suspect barm, often quite old and in decay, which should be avoided, and that it was judicious for the baker to use only a small portion of the powerful catalyst, to be mixed with the more sluggish leaven, which would be energized by the marriage.[30]

The debate aroused considerable passion and set against one another authorities who were supposed to agree on what was right for the social body (the first president of the parlement, a convinced and practicing *pain molletiste*, who may have engaged Molière to write in favor of his cause, versus the more austere and wary lieutenant general of police) and experts who were supposed to provide incontrovertible scientific tests. In its narrowly construed decision on the lawsuit, the parlement allowed the use of certifiably pure brewer's yeast, specifically banning its importation from Picardy and Flanders, sources of dubious reputation. "One hundred years ago the medical faculty banned mollet bread," observed Mercier in the last quarter of the eighteenth century. "Today there is no physician who does not dine with a little mollet bread."[31]

A zealous proponent of inoculation in the 1760s, the scientist and writer La Condamine said the same thing in a celebrated verse in which he used the mollet controversy as the allegorical ammunition for an attack on putatively scientific resistance to vaccination. While Patin "had concluded that death flew on the wings of mollet bread," since those benighted days millions of écus' worth of the decried substance have entered the capital, where year round,

> tous les matins nos magistrats,
> les procureurs, les avocats,
> les ducs & les marquis à l'ambre,
> le bourgeois, le petit collet,
> jusques à la femme de chambre,
> en prenant leur caffé-au-lait,
> rendent hommage au pain mollet.[32]

Beyond the anecdote lies the fact: barm had quickly become the object of a brisk and lucrative commerce. Very few bakers failed to use it, on its own for luxury breads and in alliance with leaven for ordinary breads. Parisian loaves were rarely of *pâte ferme*. The standard white and intermediate breads were made of a soft or bastard dough that further enhanced their "delicacy" and their elitist cachet.[33]

A deputation of baker guild wardens, called *jurés*, correctly inferred that the loaves they seized from a *porteuse* could not be for her personal consumption as she claimed, "being mollet loaves and [therefore] not being suited to the usage of the said individual"—that is to say, servant girls did not commonly eat luxury breads. By and large, however, Parisian consumption habits defied the formal/traditional social structure of breadways. The normative ascription of bread type to socioprofessional status did not apply in practice. Even within the elite wheaten spectrum, it was not true that hard-toiling workers characteristically ate an undifferentiated loaf composed of flour extracted at the highest possible rate while the cream of the crop ("the flower"), consisting of some of the finest meal, was reserved for those "who did not work much with their bodies." Obviously, factors such as disposable income, health, age, sex, employment, and even "temperament" (which usually turned out to be a muffled surrogate for social rank—though the recourse to euphemism is in itself instructive) helped to determine the bread one ordinarily ate. But the range was far narrower in the capital than elsewhere, and heavily weighted in the direction of "delicacy" rather than "robustness."[34]

Parisian Bread Types

Within the wheaten register, Paris became familiar with "an astonishing number" of breads of different vocations, known under different names, which sometimes varied from quarter to quarter. Long the loaf of first choice, initially produced by the itinerant country bakers known as forains and then domesticated by city bakers, the *pain de Chailly* (from the town later called Chilly-Mazarin) had practically disappeared by the eighteenth century. "He ate his *choine* first," noted the proverb in an enthusiastic testimonial to the *pain de Chapitre*, first prepared by the baker of the chapter of Notre-Dame for its canons. "Good and tasty," heavily kneaded, long resistant to molletization, it was known as "the best of things." As late as 1658 the *blanc de Chapitre* was listed as the first of current breads in the statutes of the guild of master bakers of the faubourg Saint-Germain, and it remained the masterpiece loaf required of bakers entering the city guild until the eighteenth century. The traditional names for middle and dark loaves, respectively *coquillé* and *brode*, did not survive.[35]

Commissaire Delamare's supple tripartite taxonomy remains the most useful guide to the eighteenth-century Parisian bakery. Among the *gros pains*— that is to say, the ordinary loaves generally weighing three pounds or more, in contradistinction to the *petits pains*, luxury breads usually weighing a pound or less—there were three broad categories: white (*blanc*); intermediate, mid-white for optimists, mid-dark for pessimists (*bis-blanc*, also known as *demi-blanc*); and dark (*bis*). Within these categories, however, there was considerable variation in the quality and provenance of the wheats, in the kinds of flour used and their doses, in the kneading and fermenting techniques, and in the forms in which they were fashioned. Concomitantly, the nomenclature fluctuated, and it was not always easy to determine what precisely a label signified.

Thus the *pain blanc de Gonesse* could either be a veritable Gonesse white baked by the forain corps, a considerable number of whom came from the village of Gonesse located northeast of the capital, or a loaf *à la façon de Gonesse* that did not necessarily replicate the original recipe. A *bis de Paris* was unequivocally lighter and whiter than a generic dark loaf, but manifestly it was less attractive than a *pain bis clair*, which almost certainly was as white as the bourgeois breads of most towns and perhaps little different from certain top loaves. Presumably a *pain moyen* was an intermediate bread, euphemistically baptized to exorcise the painful ambivalence of the *bis-blanc* marker. Theoretically, a *pain de cabaret* was a white three-pounder of a bastard or soft dough, but the tavern clientele had been leaning increasingly toward mollet, and it is possible that the bakers followed, despite the regulations prohibiting luxury doughs in this venue.

Even the *pain à soupe* (frequently called *pain de souper* in baker registers), a large, flat bread with little crumb, destined to be consumed in a soup that often constituted the entire meal, increasingly was made with a mollet dough. (The cheaper *croûtes à potage* were made of stale mollet loaves.) In the fantasy or luxury realm (today's code word is "specialty"), Guillaume César Rousseau dubbed "à la reine" a mollet that contained butter as well as milk, salt, and egg yellow for coloring, whereas historically the reine eschewed butter, which was the distinguishing trait of a *pain de festin*. (Meanwhile the pastry-makers guild continued its multisecular campaign against baker encroachment on its terrain: breads containing butter and/or eggs were in fact cakes, the pastry makers contended, and thus bakers had no right to make them.) And the *pain de sol*, fixed in price but variable in weight, which used to signify necessarily a fantasy loaf, could denote merely a refined white, a sort of elegant roll.

Innovations in milling technology in the second half of the eighteenth century further muddled the scheme of grading and labeling. The flower of the flour lost some of its prestige even as middling flour, once largely confined to dark loaves, became ennobled. By the end of the century, certain commentators distinguished between *gros pain*, by glissando now considered *grossier*, or coarse, and thus outside the mainstream of the capital's breadways, and *pain de Paris*, presumably the finest whites, probably molletized. At the same time, certain bakers became more or less narrowly specialized in luxury loaves, including new varieties laced with muscatel and other sweet wines that fell under the generic rubric of petits pains.[36]

If it seems certain that most of Paris ate "mostly very white"—remember that the pallete of reference is eighteenth-century—it is virtually impossible to convert this characterization into precise sociodemographic terms. Shop bakers increasingly did not bother to make a frank *bis*, save in periods of subsistence stress and then usually as a result of official pressure. Shop bakers such as Gilbert sold more *pain commun* than the loaf he labeled *blanc*, but the commun was a very good intermediate bread, surely more than "half-white" in both technical and symbolic terms. Market bakers posted slightly cheaper prices and were supposed to serve in the first instance a popular clientele (i.e., "the most indigent class"). In ordinary times, however, they proferred two sorts of white, "the better loaf [le beau]," which they dispatched to the houses of their most favored customers, and "the other [l'autre]," a slightly less handsome loaf that only curmudgeonly critics would have dubbed medium, sold at merely 5 to 10 percent less than the home-delivered breads. (Dr. Malouin censured the forains precisely for baking less and less intermediate bread and no *bis* whatsoever, and for Parisianizing their practice by putting mollets in their stalls.)[37]

The evidence suggests that most Parisians shared Restif de la Bretonne's horror for "the odious distinction" that separated the *blanc* from the *bis*. "It is cruel to condemn the most laborious and socially most useful class to take as the base of its diet the loaf that the well-off and unoverworked class disdains," echoed the *Courrier de France* in 1790, which already in the early Revolution imagined a single loaf, morally and materially much closer to the *bis* than to the *blanc*. Far from satisfying La Reynie's late-seventeenth-century wish that the *bis* serve as "the food of all the people," it was known as the bread "of which the people of Paris do not wish to partake," as Bucquet put it. Like Madame de Maintenon and Voltaire, they were willing to eat a dark loaf during a serious dearth, but only under duress, and without the enthusiastic spin that the celebrated philosophe retrospectively put on this heroic experience of 1709–10.

Denouncing the bakers for refusing to offer enough *bis* since it brought them lower returns, the authorities more or less implausibly blamed them for causing the people to reject this stigmatized loaf. It seems more likely, however, that the supply withered in response to the demand. If the Parisian worker had been as upset as one observer suggested because he could no longer find the whole wheat loaf prepared with a firm dough "so necessary for his strength and so salutary to his health," it is reasonable to suppose that he would have clamored for it. It is possible that the bakers made a point of making it carelessly, as Malouin suggested, "not wishing for the consumer to take to it." But it is dubious that they would have dared to impart to it a dusty or rotten flavor if consumers avidly coveted a good quality dark loaf. The desperately indigent consumers, about whom we know very little as long as they remain outdoors, who needed and wanted a so-called *pain de ménage,* or everyday loaf, were the ones who paid the price for the anti*bis* complicity of the bakers and the laboring poor.[38]

An argument frequently made in favor of a dark, whole wheat loaf was that it was "more nourishing than the Paris kind," and for this and other reasons better suited in particular for people who labored hard. A number of assumptions informed this judgment. First, the partly scientific, partly poeticomystical view of whole meal bread as the expression of a natural concord, a perfection marked by Providence—"this lovely harmony" invoked by Poncelet, the "union" of complementary principles celebrated by Abbé Antoine Pluche. Wheat was *intended* as man's food as it grows naturally; to remove any constituent was an act of spoliation, morally and hygienically deleterious. Second, as the eighteenth-century English chemist H. Jackson contended, "brown bread simply made merits Preference, as it is easier to digest" in addition to providing better nutrition. Third, it was more filling, sustaining a sensation of "plenitude" that was psychologically as well as socially useful. Finally, a *bis* loaf did not need to be fresh in order to be delicious; on the contrary, it improved in quality in the first few days after baking. And, if necessary, the dark loaf could be conserved far longer than any white/light bread (the Swiss and German "Bompernikel" was touted to last nineteen years, but it was not a primarily wheaten loaf).[39]

The debate over the relative merits of white and dark, refined and whole meal, was to rage for the next two centuries. Indeed, it is not yet fully resolved, emotionally, politically, or scientifically. During the last third of the eighteenth century, Parmentier implicitly questioned the dark meal wisdom by severely criticizing the idea that bran, much more densely present in a *bis* than in a *blanc,* introduced positive properties into the loaf. While Dr. Malouin suggested that some bran (but how much?) facilitated digestion, purified the system, and fortified one's strength, Parmentier attempted to demonstrate, on the basis of recent discoveries concerning the inner structure of the wheat berry, that bran "damaged the alimentary effect" of the wheaten flour. In a celebrated apothegm that food scientists still like to cite, the pharmacist–subsistence savant charged that bran "makes weight and not bread." A pound of bread without any bran nourished better than a pound and a quarter with it. Nor did a *bis* "last longer in the stomach" as was often affirmed. Less "massive" and more generous in volume, a less coarse bread both "sated" and nourished more efficiently.

Parmentier urged enterprises that had to feed large numbers of people to furnish a smaller amount of bread of a higher quality—less dark and thus less contaminated with bran. Fellow subsistence commentators César Bucquet and Edme Béguillet agreed that bran by itself had nothing to recommend it: it had no nutritive properties, it was acidic and indigestible, and it hampered the conservation of flour. Nevertheless, they observed pragmatically, even when present in fairly large doses, as in military rations and in many country breads, bran seemed not to do any damage. So long as "the Troops and the people are in the habit of eating bread in which all the parts of the grain entered without any visible harm resulting," it was pointless to use scientific arguments to embarrass the government and create needless difficulties and anxieties. For various reasons, Parmentier came to moderate his assault on bran, admitting that small amounts could provide a certain premium of satisfaction, especially for manual workers. But he continued to contest the persistent claim that *bis* could in any fundamental way, according to any utilitarian criterion, supersede *blanc*.[40]

Given the general allergy to wheaten *bis* in the capital, it goes without saying that breads of lesser (and mixed) grains were hardly welcome. Social commentator Pierre Legrand d'Aussy optimistically claimed that rye could produce a loaf "as white as the bread of wheat," but Philippe Macquer, a physician and chemist, more realistically described rye as "the food of poor folk," capable of making an agreeable but hardly a sumptuous bread if combined with wheat. While Dr. Le Camus concurred that a rye loaf could never aspire to be more than bread "for the poor people," he invited the larger public not to overlook its salubrious effects when used occasionally (e.g., emptying the intestinal canal of impurities). Among standard dearth tactics was the recourse to rye, a step bakers would never take, remarked Delamare, unless "obliged."

In the early 1690s the reassuringly named Sieur Vérité, a supplier of luxury loaves to magistrates, also baked a rye bread. It is not clear whether this was a curiosity, a serious effort at innovation, or a response to dearth. During the crisis of 1725, Pontoise mealmen refused to buy cheap rye because "this sort of flour is not proper for Paris" and the bakers would consequently not use it. At that very moment Commissaire Duplessis considered creating a special rye day at the market in order to encourage bakers to buy this "excellent" grain for incorporation into their *bis* loaves. "We know the difficulty of getting Parisians to taste rye," acknowledged the procurator general to the lieutenant general of police during the acute shortage of 1740, "but there are certain extreme moments when everything is worth trying." Banker Isaac Thellusson, the capital's major supplier during this crisis, argued that rye was "an excellent food" preferred by "entire peoples" to wheaten bread. Sensitive to the weight of "habits" and "prejudices," however, he proposed a gradual and quasi-surreptitious introduction of rye into the wheaten loaves baked for Parisians (whom he did not want to see behave like the soldiers dispatched by Louis XIV against the bishop of Münster in 1665, who refused to march unless a wheaten bread was substituted for the rye loaves distributed to them).

At the beginning of the 1760s several seigneurs commissioned a bread composed of one-third wheat and two-thirds rye. Described as "a bit white, . . . very refreshing and substantial," the loaf appears to have remained a philanthropic fancy. Faubourg Saint-Antoine baker Piochart was one of several bakers who produced a *bis* loaf consisting of equal parts of wheat, rye, and barley during the grave subsistence crisis of the late sixties. Appreciated by the "country folk," it was overwhelmingly rejected by the Parisian public. Troubled by the suffering around him, a baker from the rue des Martyrs developed a mixed-rye loaf that he offered to teach other bakers to make "for the relief of the poor people and the day workers." But there do not appear to have been many takers—many martyrs?—despite the acuity of the dearth.

A whole grain loaf with a dose of barley aroused "the mistrust" of the "little people" during the subsistence difficulties of 1775. Although Turgot experimented with a mixed-grain "controller general's loaf," he entertained no illusion about the prospects of employing rye, "it not being a consumer good for the city of Paris." The great success of rye bread in Germany ("it is made to ultimate perfection," noted the Swiss subsistence commentator Jean-Elie Bertrand), and the utilization of both barley and oats in Sweden, did not influence Parisians. The authorities in the capital tested a dearth bread composed of one-quarter oats in 1709, but oats never again appeared on the agenda of hypotheses even in crisis periods, so deeply ensconced was the conviction that it produced a black and bitter bread that was "distasteful and unhealthy."[41]

Ménage à Many: Toward a People's Loaf

During the second half of the eighteenth century the call for the development of a viable people's bread quickened. The argument fed on a blend of moralistic, paternalistic, and scientific criticism and proposition. Aware that an appeal to health and hygiene alone would carry insufficient weight to generate reform, Dr. Malouin struck a political chord: the governors could not afford to be indifferent to what the people ate. In terms redolent of the campaign against the ravages of luxury, the physician called on authorities to "put a halt to the creeping softness in all the ranks of society, for that is what makes for the feebleness of a State." The specific *mollesse* Malouin had in mind was of course both the source and the product of the connivance between bakers and consumers who refused to give up the soft-dough habit in favor of a robust "bread of the people."

Cankered by the manners of the rich, "the poor person" had to be induced to do what was truly best for him or her economically and morally, and in order to achieve this objective the bakers had to be "forced" to bake old-fashioned, hard-dough *gros pain*—that bread being "more suitable to sustain the strength and the propagation of mankind," a goal common to both Colbertists and physiocrats. The requirements of social and psychological control, suggested Malouin, should inhibit bakers from openly displaying the luxury breads they baked—whose discreet sale would be tolerated provided the baker supplied a sufficient quantity of people's loaves.[42]

Echoing Pliny's criticism of Roman bread stratification, Abbé Pluche wondered aloud whether "the taste for distinctions does not equally damage the rich and the poor?" *Mollesse* denied the poor the elegant parts of the flour that would render the loaf lighter and more subtle, but it also deceived the rich by providing them with a bread whose "grand merit is for the eye," a loaf bloated with water and "almost without body." The physiocratic Abbé Baudeau never tired of repeating that "that which makes the bread beautiful or white is not that which makes it good to the taste and healthful." For Pluche, nature imperiously demanded an alimentary solidarity that implied a certain social solidarity, which in turn brought with it accessory political advantages.

Most commentators, however, like Baudeau, did not question the hierarchical model of consumption, and they were far less concerned about what the rich ate per se than with what they left for the poor. The former could afford to favor a loaf "without good taste" despite its whiteness, capable of lasting only one day, incapable of providing optimal nourishment, and produced in a highly wasteful manner. The rich were expected to compensate society for their self-indulgence by subsidizing the people's bread—paying more for luxury loaves in order that the mass of consumers could pay less for a high quality "common bread." This was the only practicable form of solidarity that linked rich and poor. The poor had no prerogative for profligacy. Their moral right to existence was hedged by an obligation to eat intelligently and economically—that is, according to the standards elaborated (and eventually imposed?) by savants and public administrators.[43]

Thus Malouin affirmed, without ironic intent, that the people should never be "forced to eat white bread," a situation as unacceptable as a social contract that forced them to be free. Bread conveyed social identity, and the people should never be tempted to forget who they were and what their real needs were. They had a right to good bread, "but good bread as would suit the poor folk." Because they were poor, the people had to defer to a rational strategy of resource allocation. The new people's loaf would not bear the stigma of *bis* in its title: in homier and more neutral language, it would generally be called a *pain de ménage*. But this ostensibly rural idiom should not mislead: this was to be primordially a city people's bread, and the ultimate test of its legitimacy and efficacy would be its success in winning over Parisians. The agronomist-farmer Cretté de Palluel deemed it a "great misfortune" that Parisians did not benefit from the revivifying connection with nature that a household bread provided. Tapping into the bucolic romanticism of the Rousseauphiles, he contrasted the robust, red-cheeked peasant child nourished on a *ménage* loaf containing rye and perhaps some barley with the feeble white-bread-eating urban child whose face was "almost always scrawny, pale, and rachitic."[44]

From the perspective of public rationality and civic-mindedness ("As a citizen and for the good of the country," Bucquet wrote solemnly, "I would like to see the people of Paris accustomed to eating a household loaf"), the first advantage of a *pain de ménage* was that "one can extract about twenty-five pounds more bread per Paris setier" than with the standard Parisian practice. Though admittedly this bread would not be "as white to the eyes," all the experts agreed that the composite structure of this hybrid loaf, marrying the best of the dark and the white meal, would make it "good and healthful." Imbued with the right values, the consumers of the Old Regime did not need to await the lessons of the Revolution in order to learn to scorn this "external makeup of color."

The absence of flour drawn from the richly laden middlings crippled white loaves. These *gruaux* made the bread more savory and nutritious because they contained the part of the wheat berry that was "sweetest, creamiest, and most substantial." The development of the (quasi-new) technology of gradual reduction, baptized economic milling, now made it possible to recover virtually all the middling richness, which could then be invested in the household loaf. In a transport of enthusiasm, Bucquet affirmed not only that the popular bread of blended meals would be good but that, well made, it would be "better than the bread of the rich made of the fine flower of the wheat." Nor must we lose sight of the essential, remarked a Paris baker advising the lieutenant of police in the late sixties: "in buying the *ménage,* or household, loaf, it's the *ménage* that the poor consumers seek," that is, the savings in cost.[45]

Differences persisted among the specialists on the exact composition of the household loaf and the best way to make it. In the early 1690s several bakers in the cour des Quinze-Vingt offered "a household loaf composed of all the types of flour [de toutes farines]." For want of elucidation, however, it is impossible to tell just how "whole" this loaf was likely to have been, for the concept of "toutes farines" remained highly relative, covering extraction levels that probably varied by as much as 15 percent. Pluche called for an ideal balance that resulted less in a popular loaf than in a universal bread of the *juste milieu* epitomizing the structure of the wheat berry. Aiming strictly at the poorest group of consumers, a handful of equally quixotic commentators still envisioned putting barley, oats, and various beans in the meal. Mercier wondered whether the hardy potato could not become the hale basis for a new bread.

A practicing baker argued that only a firm dough could produce a household bread possessing both savor and body, capable of pleasing the consumer and imparting a durable sensation of repletion. "Bread has no built-in support of its own," he explained, "and it is not with water that one can put in that solidity," despite the pretensions of the molletists. "Flour is what is needed"; it alone can guarantee a hardy dough. But Parmentier challenged the conventional wisdom that a popular loaf had to be "heavy, tight, firm, and dark" in order to fill and nourish. "The more volume the bread has, the better it sustains," the food chemist contended. Kneaded more gently and with more water, the same quantity of meal would produce 25 percent more bread occupying twice the volume.[46]

For exponents of the household loaf, the experience of dearth was both an opportunity and a source of abiding frustration. Promoters did everything they could to avoid the taint of crisis. First of all, they did not conceive of the *ménage* as a dearth ersatz. On the contrary, they imagined it as a function of the glaring inadequacies of everyday consumption habits; it was meant to address and resolve in the first instance structural problems rather than conjunctural ones. Second, its proponents knew that if it acquired a reputation as a mere surrogate, a loaf of desperation, there would be no chance of integrating it into the daily diet. Yet it was clear that in the subsistence sector, innovation was crisis-driven. Until the 1760s, the *ménage* project was associated almost exclusively with the disastrous memories of 1709, when parts of the realm faced faminelike conditions and myriad kinds of loaves "where everything in the grain enters" abounded. If scarcity opened the way to a certain conceptual audacity, it was conversely the time least hospitable to the practical application of new ideas. For in their anxiety, consumers were inclined to view innovation as a form of subversion, oppression, and/or manipulation. Mistrust of government and trade peaked with the soaring *mercuriale*, the grain price register. Thus, for instance, when a ministry official named Trudaine de Montigny, in the name of the economics ministry (contrôle général), proposed the adoption of a "very good household loaf" begotten by a form of economic milling, in the light of the volatile character of public opinion he warned that "this operation must be undertaken with great care and precaution."[47]

So it was hard to introduce a household bread in the midst of crisis, and it was perhaps even harder to insinuate it after the end of crisis. During the protracted crises of the sixties and seventies, the government and private individuals pressed the case for a household loaf with long-term implications as well as the case for dearth palliatives. If Paris area intendant Bertier de Sauvigny's "beggars' loaf" clearly fell into the latter category, the efforts husbanded by police chief Antoine de Sartine belonged primarily to the former. The lieutenant general ordered a series of five tests (*essais*) conducted in the capital, in the meal market of Pontoise, and at the milling center of Corbeil. A final experiment in this series took place under the direction of Bricoteau, the highly regarded head of the Hôpital général's Scipion public-assistance bakery, in the oven room of the baker of the first president of the Paris Parlement, in the presence of several jurés from the guild. The bakers criticized various aspects of the procedure and recommended further inquiry, predicated on the assumption that a plausible household bread in Paris would have to be very close to what they called "a common bread"—a *bis-blanc* loaf. At about the same time, in collaboration with Bucquet, probably through the Hôpital général, Bricoteau actually marketed a loaf proclaimed to be *pain de ménage* composed of all the meal with only the coarse bran removed.[48]

During the mid-seventies, when he operated a huge bakery at Vaugirard in the nearby suburbs, Bucquet baked a household loaf described as between a *bis* and a *bis-blanc*. He made more of this bread, composed of "economic" meal, than of standard white. In 1774, a "Société des philosophes économistes," under the direction of Abbé Baudeau, produced at the abbey of the Prémontrés a bread hailed by the bookseller Hardy as "the true household loaf." It was a whole meal loaf, containing elite, white, medium, and dark flour. (Perhaps in order to discredit it, it was also rumored that barley joined the wheat to make the loaf.) To those who purchased it, Baudeau distributed a brochure explaining the merits of the bread and praising the buyers for their good sense. For a brief period, the physiocratic abbé seems to have found a number of bakeshop outlets, identified with large signs over the door heralding "economic bread." The guild bakers were said to be vexed at this unlikely intrusion. Reports suggest that consumers were intrigued but suspicious. This real-life experiment seems to have lasted no more than several months.[49]

So hospitable to innovation was Sartine that he did not recoil from meticulously scrutinizing propositions that he himself regarded as outlandish. La Plaine de Moulinot, a former infantry captain, and his wife proposed a "secret potion" that they promised would increase production from meal to bread by a quarter, accelerate proofing, quicken the cooking process, and generate a bread that was more nourishing, savory, and healthy, and that would last at least four times longer than the ordinary loaf. Applied to institutional provisioning (armies, hospitals, schools, etc.), this "economic secret" would save the king a fortune. The technique could also be used in Paris to beget an economic household bread and to increase the quality and reduce the costs of other loaves, saving the capital at least one hundred thousand pounds of bread a day. Moulinot's wife transmitted the secret; she was the daughter of Pierre Brodin de la Jutais, a physician apparently attached to the court, and an inventor and improver, who had proposed in the early sixties a "febrifuge" powder guaranteed to multiply the yield of a wide variety of grain, fruit, vegetables, and flowers.

Sartine ordered a series of tests involving his inner circle of experts and practitioners: master baker Toupiolle, hospital baker Bricoteau, baker-miller-entrepreneur Malisset, and Commissaire Machurin. They supervised controlled experiments in which the bakers converted the same quantities of the same type of flour. Following the Brodin de la Jutais "méthode oeconomique," large quantities of an extremely viscous "liquor" replaced and/or complemented water in the leavening/kneading process. The liquor treatment produced between 20 and 35 percent more bread in several different tests. We know nothing of the quality of the bread produced, but since it elicited no further mention, it is safe to presume that Moulinot did not obtain the very substantial "gratification" that he was seeking.[50]

None of the "economic" breads of the sixties and seventies entered into the daily repertory of Parisian bakers and consumers. The bulk of the laboring poor seemed to have been able to afford mid-white and white loaves—at least during times of subsistence ease. Economic milling improved the taste and quality of the white and mid-white and probably contributed to the maintenance of prices at more or less their midcentury level. Composed of the "flour of the wheat" (*la farine de bled*), as the first-yield flour was called, and the first through the fifth grades of the middlings ground (and reground) "economically" and blended together, a household loaf would have been a very attractive alternative in terms of taste, nutritiousness, sense of satisfying satiation, conservability, and cost. Its "look," however, continued to repel Parisians: no matter how carefully made, it was never white, but a yellow flecked with gray. For the eighteenth-century Parisian worker, no less than for his counterpart a century later, it was "a matter of self-esteem. . . . it would be a humiliation for the worker to buy a bread that he could not show in his workshop." Malouin was probably right in hinting that the only way to introduce a *pain de ménage* into the bakeshops and stalls of the capital was through regulatory fiat. Another method, equally implausible, was to free frugal consumers from the putative tyranny and the monopoly of the baker by enabling them to bake at home, or at least to prepare their dough at home: "the good women of the people would buy their own flour, knead and shape it into loaves, and take it to be baked in the oven of their choice," including several royal ovens established just for this purpose.[51]

The Flawed Art of Baking; or, Science to the Rescue

The art of the baker, observed the philosophe-mathematician Condorcet in his funeral eulogy for fellow academician Malouin, "is the most necessary to the people." Yet, as virtually all subsistence commentators concurred, it had been utterly neglected until quite recently. Most pointed to the dwarfing impact of the empire of habit and the lack of *true* knowledge. Baudeau blamed the police regime, with its corporate vanguard, which deprived bakers of an "interest" in amelioration by severely confining their opportunities to profit from innovation. Without focusing his indictment on the regulatory apparatus, Parmentier agreed with the core diagnosis: "as long as men are not convinced that in changing methods they will increase their profits, they remain deaf to the improving advice that one dispenses." Yet he underlined that improvement would not merely gratify the avid appetite of the bakers, "because in doing things better, they would be able to sell more cheaply and earn more at the same time." By cultivating their self-interest, in the best liberal-utilitarian manner, the bakers would contribute directly to the well-being of society.[52]

Praised for over a century by a stream of travelers, literary observers, officials, and certain subsistence experts for the quality of their breads of various types, the bakers of Paris had reason to think that they were the best. Surely they were the best in France, and John Evelyn was but one of many discerning Englishmen who considered that it was in France "that, by universal consent, the best bread in the world is eaten." Le Camus of the Paris Faculty of Medicine concurred: "One eats in Paris the best bread that can be found in all of Europe." His voyages abroad, particularly in Italy, rapidly converted the reformer-writer Abbé Gabriel-François Coyer to French bread chauvinism. Asked to evaluate comparatively, Lacombe d'Avignon, who had serious quarrels with Parisian baking practices, nevertheless conceded that "there are few countries where one makes bread as good as in France." Although he was the first to insist on the urgent need to seek improvement, Sartine's successor, Lenoir, did not hesitate to affirm that "France is the country of Europe where the best bread is prepared, [and] in Paris it is made better still than in the provinces." Who could blame the corps of itinerant bakers (*forains*) that supplied the capital for boasting in a petition written in 1738 that Paris "has always eaten the finest bread in Europe"?

Those Paris bakers who became aware of the strictures of Malouin, Parmentier, and others affirming that "their art is still in the cradle" must have been fairly jolted. These experts assured them that the fault was not entirely their own: millers bore a large part of the responsibility for the primitive state of things, for they, too, operated in ignorance and prejudice. "A good milling industry and a good bakery would constitute an unheard-of source of wealth for the kingdom," proclaimed Parmentier, "because it would then be possible to save a third on the grain used in production." While the Baudeaus (and the Turgots) insisted on the need to introduce correct principles of political economy, founded on deference to nature, in order to transform the subsistence theatre, Malouin, Parmentier, and their partisans argued for the primacy of Science, founded on the conquest of nature, as the only sure guide to modernization.

Whereas a self-taught albeit imaginative and risk-taking practitioner like César Bucquet believed instinctively that "in truth a baker would have to be signally maladroit not to be able to make a good loaf provided he had good raw materials," the laboratory-trained Parmentier contended that sorely lacking "enlightened labor power accounts for infinitely more than the quality of the raw materials that are used." It was up to the scientists to teach the bakers how to bake. As Louis-Sébastien Mercier put it in his pithy and brutal fashion: "Making wheat into bread is a chemical operation that must be enlightened by chemists; blind routine denatures the process." Yet he himself realized that such stark affirmations from outside and above were absolutely incapable of persuading "all the baker boys, all the servants, or even all their mistresses, who league together to assert that nothing can be added to the perfection of bread as they make it, which is just as their grandfathers ate it."[53]

In Abbé Pluche's ideal-typical scenario, Physics descends "from the top of the celestial spheres where it is pleased to reside" to spend some time in the bakery, not to wrench its practitioners from Gothic ignorance but to learn from and with them ("the common arts are the true helpmate of philosophy"). The subsistence scientists, however, did not share Pluche's premise of humility. As the miller-baker César Bucquet described it, Physics arrived on horseback to vanquish—in the guise of Parmentier "mounted on his Pegasus" unable to conceal his contempt "for ignoramuses like me." Before the arrival of "the doctors in milling and baking," there was darkness; once they began to profess, there was (allegedly) light. Milling and baking had been practiced for centuries, but it was not until the savants subjected them to scrutiny that these arts began to flower. The "prejudices" deplored by Condorcet ("the most numerous, the most absurd, the most harmful"), the "blind custom" decried by Lieutenant General of Police Lenoir, and the "blind routine" excoriated by Parmentier's collaborator, the chemist Cadet de Vaux, were nothing other than the accumulated experience of generations of more or less skilled bakers. The "popular errors" that Parmentier wanted to extirpate were the "secrets of the trade" passed from father to son, the trade's capital of wisdom.

The scientific discourse was one of the strategic instruments deployed by the savants to displace and dominate the practitioners of the crafts. Despite certain mollifying gestures, the Encyclopedic culture bent on imposing its brand of Reason on the world of work had little use for the deeply rooted popular culture of artisanry, viewing the latter as idiosyncratically oral, narrowly practical, stubbornly unreflective. Science connoted first emancipation and then progress; it required a sustained effort of theorization as a prelude to perfection. The savants set about studying and classifying the crafts in order to translate them into the scientific language as part of a grand project of rationalization and control (which ultimately issued in Taylorism in a rather different technological and political context).

Parmentier contended that "this language is not as distant as is supposed from the men of the craft . . . for whom this language will one day become very familiar." There was a reasonable likelihood, however, that the bakers would react in the manner of Bucquet, for whom the language of "systematic chemistry" practiced by Parmentier and Cadet de Vaux was a "foreign language," a language of mystification meant to subjugate the workers, a discourse that reeked of vanity even as it "smelled of the pharmacy." Bucquet resented less the science of the savants than their arrogance, their refusal to learn *his* language, to inform their theory with *his* experience, and to honor *his* kind of knowledge, method, and creativity, in a word, *his* culture. Rather than suffer the humiliation implicit in the "ennoblement" of their professions promised by Cadet, many bakers and millers preferred to remain commoners.[54]

A School for Baking

With the creation of the Ecole gratuite de la boulangerie in 1782, Science occupied the bakery institutionally. This establishment showed "how much Chemistry can advantageously influence the most venerable of the Arts," boasted the *Journal de Paris*. Put more concretely, it presaged the end "of bad bread"— whose alleged ubiquity the bakers of the guild and Gonesse bitterly contested. The evocation of bad bread resulted largely from Cadet's double effort to justify (and command credit for) the existence of the school and to settle his scores with Bucquet. Cadet hinted that official anxiety over the poor quality bread distributed in the hospitals and particularly in the prisons prompted the authorities to found the school as a vehicle for the immediate improvement of institutional baking. These "stealthy and shadowy insinuations," denounced as "vile calumnies" by Bucquet, were aimed at him and his family, who operated, or had operated, the provisioning concessions for the Hôpital général and several prisons. Later Cadet implicitly held Bucquet, and unschooled artisans like him, responsible for the "frequent revolts" that had wracked the prisons of Paris throughout the eighteenth century (*"There's the bread* was the rallying cry"). Thanks to the momentum generated by the school, and above all to his "zeal," Cadet claimed to have been able swiftly to introduce "economy" as a method and ethic in public milling and baking operations.[55]

In fact, the school issued from several years of deliberations among the highest authorities, two of whom, Jacques Necker, the banker-philosophe and economics minister, and Police Chief Lenoir, were especially preoccupied with subsistence questions. Impelled by the latter's passion to marry Science and Administration, a bakery committee planned the school, geared to address all the stages of bread production, beginning with the cultivating, harvesting, and storage of grain, placing particular emphasis on milling technology and flour grading and preparation, and ending with the process of baking and the channels of distribution. In order to assure rigor but also accessibility, the lieutenant general of police emphasized his determination to unite "practice with theory." The administration of the school reflected this objective: Jean-Baptiste Brocq, a master baker and chief of the bakeries of the Hôtel Royal des Invalides and the Ecole Royale Militaire, where he had met Parmentier and tutored him in the mysteries of baking, was named director of the school, assisted by another baker, Bricoteau, a veteran of hospital baking and an expert for the police on baking issues; Parmentier and Cadet de Vaux were to serve as the two professors. They would be monitored by a surveillance committee chaired by Lenoir and including both eminent subsistence specialists such as the scientists and academicians Duhamel du Monceau and Mathieu Tillet, and senior master bakers.

Despite the familiar "free school [école gratuite]" appellation, this was not to be a trade school focused on training bakers for the capital. The school's charge was to conduct research on all aspects of the bread question and to disseminate the results throughout the entire kingdom. Students would come from all over France ("seminarians," in Cadet's quaintly clerical idiom), and the professors, seconded by their finest pupils, who would earn masterships as a reward for their diligence, would take to the road to spread the word. Intendants could pose specific problems for resolution, such as grain diseases or defects of mill architecture or uncertainties about the quality of the water used in kneading.[56]

The school's Parisian course consisted of a dozen lectures, usually presented on the bread-market days of Wednesday and Saturday, in the bakeries of the Scipion hospital annex in order to permit the students both to observe and to participate in the work of bread making.[57] The lectures were organized around major themes (e.g., leavening and kneading, flour conservation, the oven) that elicited both theoretical and practical dimensions of the task. For a discussion and demonstration of economic milling, Cadet took the class to the La Courtille mill at Saint-Denis.

With biting irony, Bucquet described one of the public lectures by Cadet, whose "male eloquence" attracted fashionable women not afraid to soil their dresses with flour in the granary. "This clever Professor is not unaware how important it is to dispose one's audience toward the sublime, to heat its imagination with grand and marvelous images." Thus Cadet began his talk on the bakery by showing how a new microscope revealed the genitalia of a male flea to be relatively almost as large as those of a man. Whence the speaker moved on to a discussion of the air, humidity, germination, conservation, and so forth. "Those who were able to follow the orator enjoyed the benefit in less than half an hour of a complete course of Physics, Chemistry, Agriculture, Milling, and Baking." According to his critic, Cadet plagiarized much of his material, claimed credit for inventions with which he had nothing to do, and incorrectly explained numerous important processes.[58]

In addition to their cycle of courses in Paris, at the invitation of both royal and municipal officials, Parmentier, Cadet, and occasionally Brocq gave instruction in provincial towns such as Amiens, Beauvais, Chartres, Orléans, and Montdidier. Predictably, Cadet found their services desperately needed. At Amiens "the art was [still] in its infancy," and in Beauvais "milling and bread making were in the state in which they found themselves two centuries ago." The scientist-promoter claimed huge audiences ("the assembly was brilliant [at Amiens]. . . . three or four hundred spectators came regularly to listen to the economist-professors") and a "festive" atmosphere. Other observers reported that the numbers were far below the expectations of the professors and the hopes of the organizers, bespeaking "a muted resistance" on the part of local practitioners. According to Mercier, the professors made a conscious effort to reach out to their audience: "those who teach use the popular language, and the lectures they give are within the reach of the baker boys."

A correspondent from Amiens (writing in a tone that could have been

Cadet's) evoked the skepticism the organizers felt about the prospect of a positive reception on the part of "these coarse and stupid artisans," individuals wedded to their immemorial and erroneous routine. To the general surprise of the academic elite of this provincial town, however, "the luminous instructions of Messrs. Parmentier & Cadet convinced them. . . . prejudices gave way to interest." The result of this initial success was an institutional legacy. A student of Cadet's, a pharmacist-chemist named Lapostolle, under the sponsorship of the intendant, offered a course regularly at Amiens, where he assumed the title of "Professor and Demonstrator of the Course of Milling and Baking."[59]

The school served as laboratory for a welter of research studies and trials. Parmentier continued his work on flour chemistry, principles of nutrition, fermentation, mixed-grain breads (including potato meal alloys), and cooking methods. Extracts of his earlier work, in the form of analytical recommendations to practitioners, sometimes revised, appeared under the school's imprint. At Necker's request, in collaboration with his colleagues, the pharmacist-chemist experimented with a multigrain household bread targeting provincial consumers. The team also worked on problems of preventing and rehabilitating sprouted wheat. Cadet de Vaux induced Benjamin Franklin to share his secret for baking bread on a coal-burning stove. The school supplied the American inventor with flour and he transmitted sample loaves and an explanation of his techniques to Cadet, who also expressed interest in his work with "Indian corn." Beyond publications financed directly by the government, Cadet diffused the school's accomplishments through the *Journal de Paris* and the *Journal de physique* as well as several provincial weeklies.[60]

It is virtually impossible to measure the impact of the school in the fields, the mills, and the bakeries. Its promoters spoke in lyrical and nebulous terms of their successes, successes measured in part through the intense administrative demand for their services. We know that municipalities, intendants, societies of agriculture, academies, provincial estates, parlements, and the ministries of war and the navy expressed interest in the instructional activities and pastoral mission of the school. Lenoir and Cadet proudly called attention to the rapid international resonance of their institution: "the royal prince of Prussia, several electorates, the grand duke of Tuscany, the king of Sweden, the king of Naples, [and] the grand master and the order of Malta had recourse to the enlightened views of the nascent School of Baking."

But we know extraordinarily little about the artisanal reception and application of "scientific" principles and procedures. The practical results are particularly hard to assay in the bakery as opposed to the milling industry, because the reformers made far more modest capital and structural demands on the bakers than on the millers. Save for oven design and construction, which involved dramatically less money than, say, the conversion to economic milling, the changes demanded of bakers were relatively subtle and had less to do with equipment and overall organization than with issues that were at once conceptual and commercial (e.g., renunciation of grain buying in favor of flour purchasing, the kinds of flour to use, the sorts of bread to make) or with the refinement of more or less familiar techniques (kneading, fashioning, proofing, choice of combustible, oven insertion and removal). One could identify an economic miller by visiting his mill or by examining his graded meal; one could not readily discern a "scientific" baker through an examination of his bakeroom; and it would have been singularly pretentious and unscientific to suppose that every good loaf of bread owed its genesis to a Cadet or a Parmentier.

We do not know who attended the courses in Paris where the school was based. No serious effort to negotiate with the guild seems to have been made, though the reforms of 1776, which changed the rules of recruitment and raised questions about the obligations of apprenticeship, might have opened a window of opportunity. Evocations of the course refer vaguely to an audience of "garçons" rather than established bakers. There are hints of some active resistance on the part of bakers who regarded the blandishments of Science as needless intrusion, and more palpable signs of widespread indifference. It was very hard to convince bakers, especially Paris area bakers, that they were not already quite good at what they did.[61]

Despite some dissension among the members of the school's steering committee, who resented the total control exercised by Parmentier and Cadet over the scientific agenda, and some friction between Parmentier and the Société d'agriculture, some of whose members also served on the committee, the school survived into the early years of the Revolution. In 1789 it was administratively amalgamated with the Société d'agriculture, but this does not appear to have affected its program or its autonomy. It was called upon to address issues of political economy and politics rather than more ostensibly neutral matters of technology, in particular to offer advice on price-fixing policy and on dearth-combating strategies, including substitute household breads. Parmentier found himself the object of several groundless attacks launched by critics of economic milling, who associated his doctrine either with aristocratic nostalgia because he allegedly sought to institutionalize a tripartite hierarchy of bread types or with a callous strategy of adulteration because he allegedly sought to drown good meal in a sea of bran. The school also had supporters among the

Revolution's expanded corps of subsistence commentators, including one who wanted to make it a veritable policy-making institution, driven by Science and insulated from Politics, and another who called on the school to sponsor national competitions for new forms of economic bread.

By 1793–94, the school appears to have lost its headquarters on the rue de la Grande-Truanderie, where its ovens in any case were in a state of dilapidation. An order of the government's finance committee promulgated in October 1795 transferred the school to the former Convent of the Assumption, where it was to rebuild its teaching and laboratory facilities in quarters shared with the Magasin des Subsistances, one of the city's major storehouses for flour and grain. For reasons that are not fully clear, the head of the granary vigorously opposed the installation of what he called the "Ecole publique de Boulangerie et de Meunerie," as did a number of influential administrators, who felt that all available resources had to be invested in feeding the people immediately rather than working out reforms for the future. Parmentier and Cadet defended themselves with equal vehemence, calling attention to the very concrete progress the school had already made toward resolving subsistence imbroglios and injecting "economy" into all domains, and to ongoing experiments on a bread made with one-third corn meal, redolent of Franklin's experiments, that could constitute the long-awaited breakthrough in the quest for a workable household loaf.

In terms familiar to governments of old and new regimes throughout history, the following year the Ministry of the Interior regretfully—and "temporarily"—suppressed the school. "In an ordinary time, the government can and should diffuse with a sort of profusion the means for the encouragement of and instruction in all the branches of industry," it declared; "but when its financial situation barely allows it to meet its most pressing expenses," it could not sustain a valuable but not indispensable institution. Parmentier and Brocq vainly drafted a petition in which they protested the injustice and shortsightedness of the decision in the light of the school's achievements and the courage of its personnel, "who during the Revolution never ceased to court the dangers inseparable from the duties that they had to fulfull."

The government's response was not graceless. It awarded Brocq the concession to supply bread to several public institutions, and it named Cadet as "inspector of the Bakeries, hospices, and prisons of the commune of Paris" at one thousand francs a quarter. It is true that Brocq had immense difficulties collecting monies that were still due him in 1800. In Year XII, convinced that a resurrection of the old school was not imminent, Cadet appealed for the creation of a new school dedicated to economic milling.[62]

Emerging from the Revolution, France remained economically and mentally entrapped in the same prison from which Simon Linguet quixotically had yearned to free its people. Bread was one of the rare pivotal institutions of the Old Regime that revolutionaries openly respected. Not even in the social equivalent of its most dechristianizing phase had the Revolution dared to challenge the absolute necessity and desirability of the Eucharistic model of daily nourishment. Certain leaders envisaged not new sources of food but merely new forms of (putatively egalitarian) bread. Even these rather modest adjustments, driven as much by chronic penury and disorganization as by ideology, aroused considerable anxiety. "Bread," Parisians cried at numerous junctures, and not just metaphorically. "Bread," but not just any loaf. Few consumers seemed prepared to accept the idea that the price of the Declaration of the Rights of Man and the other democratic gains or promises of the Revolution had to be a less satisfying four-pounder than they had known in the bad old days.

Bread nostalgia was of course a permanent feature of monarchical propaganda. The loaf became the oriflamme of a conservative populism even as it remained at the center of radical symbolics. The bakers often felt themselves squeezed from both directions. The Parisian bakery suffered a traumatic shakeout during the Revolutionary years. In the end, it affected personnel more than practice. The bakers continued to work much of the night, drink too much, fight among themselves, and hoard on occasion. They also clung to their way of doing things: the Revolution had not succeeded in impelling them to abjure their practice and culture of work. But the anti-intellectual currents that briefly drove certain Jacobins to contest the tyranny of Science quickly receded. The specter of regulation and reeducation rapidly returned to haunt (or more precisely, to hector) the bread makers. Ever mindful of genuinely indomitable traditions, Napoléon overcame his horror for Parisian wheat-and-white exclusivism. In the grand manner of the *princes nourriciers* of yore, he made sure that ordinary Parisians had little to complain about in the shops and markets. Meanwhile taste, finally purged of its aristocratic cachet, could now become an openly avowed subject of daily politics. One of the leading arbiters of post-Revolution taste, Grimod de la Reynière, himself the product of a certain (anti-Revolutionary) Enlightenment, announced the bad news in 1806: "the Bread of Paris, formerly so renowned, and which quite rightly passed as the leading bread of Europe, has taken a steep decline."[63]

Heretofore, historians have tended to discuss bread in extremely nebulous and remote terms. They have evoked it more often in the breach, as an absent presence: as long as there was enough of it available at an accessible price, there was no reason to excavate further. They had little knowledge of the provisioning nexus and little feel for the practices and the attitudes of bakers and consumers. Now we have a much richer and more concrete picture, although it is far from complete. We know a great deal about the kinds of breads that were baked and eaten, even if we cannot historicize and sociologize demand and taste as rigorously as we would like. We can demonstrate the triumph of wheat and white even if we cannot date precisely the process of conquest.

The *relatively* confined range of Parisian breadways stands in stark counterpoint to the highly stratified character of urban society. Important bread distinctions existed, to be sure, but within much narrower boundaries than was generally supposed. Breadways helped to construct identity, but in a more complex way than a crude (but plausible) elite-people dichotomy might suggest. In this way bread unified perhaps even more than it divided (even allowing for macabre images of the hoarding mollet baker François dangling from the *lanterne* in the summer of 1789).

Breadways were not immutable, but they evolved slowly and more or less organically. Molletization, dramatized by the famous trial as if it were a genuine novelty, in fact proofed in gestation for over a century. The impatient savants of the Enlightenment learned firsthand just how resistant breadways could be to the paternalistic-reformist impulse. Dazzled by their own command of the science of it, the reformers then, like economists today, paid too little attention to cultural forces and too much attention to technical manipulation and moralization. The politics of regeneration after 1789 (or 1792) proved no more corrosive of food habits than public-spirited science had been. Ultimately it took long-term economic mutation and sociocultural transformation to dislodge bread progressively from its timeless *rente de situation*.

Chapter 2

Bread Making

For the bakers, by and large, bread making was onerous but not hard. Technically it was merely a matter of certain rules of thumb (as the millers had inadvertently baptized them) and of imitating what one had learned from one's father or one's masters. For the eighteenth-century bread reformers, bread making was onerous and, correctly accomplished, it was not simple. Beyond just a traditional recipe and a certain dexterity, it required a veritable dose of chemistry and physics, a proper setting, the right equipment, favorable atmospheric conditions, and exquisite timing, not to mention raw materials of fine quality. Even today the bakers and the experts have not agreed on what the right mix is.

Site and Tools

"Rarely convenient, always badly situated, . . . most often dark and poorly ventilated"—Parmentier's description of the typical bakeroom seems entirely plausible. Characteristically it was located either behind the shop or in the cellar. Usually, but not always, the baking itself took place there. In some instances, because of extreme exiguity or fire risk, the oven was located nearby, and the bakeroom was reserved exclusively for the confection of the dough. Parmentier complained in particular about basement bakerooms so confined that the baker could barely manipulate the peel, so hot that the dough "melted" during proofing, so dark that one could hardly see, and so stifling that one could barely find air for a deep breath. This description corresponds to the bakeroom operated by master Maroteau beneath the rue de Grenelle, or to the establishment of his colleague Foubert on the rue Saint-Denis, located deep in the bowels of the house, without ventilation and requiring "[artificial] light day and night." On the other side of the spectrum was the capacious area in back of the shop on the ground floor of the house owned by one master baker and occupied by another on the rue Dauphine, or the raised bakeroom, ample in

Bakeroom with tools, nineteenth century. *Source: courtesy, Grands Moulins de Paris.*

size albeit dilapidated in condition, located above and behind a shop on the rue Mouffetard.[1]

Most bakers maintained supplies of flour (and, much less commonly as the century progressed, of grain). The size of the stock depended in the first instance on the magnitude of the operation and the baker's capital standing. But the provision also varied as a function of the availability and size of storage facilities. César Delmarre baked ten ovenfuls a day, maintaining an athletic cadence, but kept a surprisingly small stock, because he had nowhere to put it, his landlady refusing to rent him a room on the second floor next to where she slept for fear that his activities would disturb her peace. Fellow master Pierre-Augustin Gillet was similarly circumscribed: he kept his flour in the hallway outside his bedroom on the first floor over his shop. Other bakers had large, efficient systems that seem conceptually if not technologically ahead of their time. Guillaume Rousseau installed a pipe to conduct flour directly from his granary into his kneading trough. A commissaire described the installation of master baker Barbaro of the rue Saint-Jacques as "two machines for drawing flour" connecting the spacious storage area to the bakeroom.[2]

Flour storage facilities were important not only to buffer against shortage but also in order to enhance prospects for high quality production. Flour that "rested" for a time after grinding, for example, was easier to work, absorbed more water, and rendered a livelier dough than "hot" meal immediately put to use. It was also significantly easier to bolt and mix flour that had reposed. Before the development and diffusion of economic milling, many bakers preferred to do their own bolting and mixing at the bakery. The stronger bakers boasted separate bolting rooms or fitted their flour rooms with bolters; the weaker ones set up and dismantled bolting devices in the bakeroom itself or in the hallway outside the shop. The most primitive devices consisted of wool and/or silk cloths used as filters and sifters. The most sophisticated were veritable machines, armed with gears, levers, and planes, some of them requiring a space larger than an average bakeroom, built by specialists such as Claude Roucin, "bolter maker [faiseur de bluteaux]." A handful of other bakers, with adequate space and resources, utilized other "machines" to clean and ventilate grain or to convey sacks of grain or flour directly from wagon to granary.[3]

Bakerooms varied considerably in the range, elegance, and sheer bulk of equipment with which they were endowed. It is not difficult, however, to identify the core items without which not even a resourceful baker could function. A kneading trough—two or more in the largest establishments—usually made of a hard, nonporous wood was the single most important instrument of production next to the oven. Covered with a wooden top, the trough became a worktable upon which the dough could be weighed and fashioned into loaves. Scrapers, spatulas, dough cutters, and knives accompanied the trough. The dough proofed in baskets of various designs and materials or in cloth wrappings on a special set of shelves. The baker required several cauldrons and basins to contain (and sometimes heat) water, scales to weigh the loaves, brushes for gilding and cleaning the loaves, a panoply of instruments for stoking the wood and maintaining and cleaning the oven, and a receptacle for the charcoal. Unlike most of the tools, esteemed durable for ten or twenty years, every year the baker had to replace several of his peels, used to insert the dough into and remove the bread from the oven.[4]

Water

Bakers employed only four ingredients to make ordinary white and dark wheaten breads: flour, water, leaven or yeast, and salt. Everyone agreed that good bread could only be made from first-rate grain expertly ground into a sprightly, absorbent, and glutenous flour. Serious disagreement arose, however, on the importance of water quality. "Good water," affirmed the mitron de Vaugirard, the eponymous baker boy in Lacombe d'Avignon's late-eighteenth-century mock autobiography, "is the soul of bread." "The quality [of bread]," wrote a leading member of the Paris Faculty of Medicine, "depends most often on the quality of the water that one introduces in the process of kneading." Another physician and bread specialist, Malouin, also insisted on the critical importance of "pure" and "light" water. As proof of this contention several commentators pointed to the excellence of the bread of Gonesse. Highly competent Paris bakers imitated the Gonesse style, using the same flour and the same methods. Yet the resulting loaves "never succeeded either in color or goodness." "This is wholly imputed to the excellency of the [Gonesse] water," remarked an English observer in 1683. Dr. Le Camus attributed the "more exquisite taste" of the Gonesse bread exclusively to "the difference of the water."[5]

The quality of the bread "does not at all depend on that of the water with which it is made," asserted Parmentier, who put the counterargument most forcefully. The water thesis was nothing more than an alibi invoked by bakers eager to divert attention from "vices in the bread-making process." Fusion with flour and fermentation radically modified the nature of the water used and thus made its original quality "indifferent" to the success of the bread-making operation. The amount of water added, its temperature, and the way it was used were far more significant than its quality, Parmentier contended. He derided the claim that rainwater was infinitely superior to fountain or well water because it was lighter and combined with the flour and leaven more gently and completely. He pointed to the fact that three-quarters of Parisian loaves were made with a heavy, saline well water, yet there was no evidence of variation in quality from "well" neighborhoods to river or rainwater quarters, and Paris bread was rightly considered to be one of the best made in Europe. To clinch his case, Parmentier himself baked loaves made with rain-, well-, river, fountain, and distilled water and found no difference in taste, whiteness, or lightness.[6]

Few contemporaries appear to have subscribed to Parmentier's revisionism. Subsistence specialists, seconded by public authorities, urged bakers to turn away from the wells in favor of rainwater or water drawn from the Seine (whose "goodness" Parmentier himself had defended against "calumny"). "Heavy, cold, and raw," well water was also potentially dangerous, for the wells were frequently contaminated by latrine seepage and other waste materials that could engender "popular and often endemic diseases, unfamiliar to physicians." The oven temperature was generally not high enough to kill all the microorganisms introduced into the loaf by well water. A campaign to ban the use of well water by bakers was launched in the sixties. It was considered of sufficient importance to command a place in the discussion of *doléances* for the Parisian grievance lists (*cahiers*) in 1789. Bakers such as François Sivry (who installed conveyer pipes leading directly from his well to his kneading troughs) and Louis Toiret continued to favor well water because it was directly at hand and spared them the cost of carting water from the Seine. A number of commentators suggested that baker objections could be overcome by making Seine water available to them free of charge as a measure of public security.[7]

Beyond the question of quality, there were less strenuous disagreements over the quantity and temperature of the water to be used in bread making. The amount depended on many variables: the nature of the wheat, the conditions of its harvesting, the type of milling and conditioning, the kind of bread sought. Some flours drank as much as three-quarters of their weight in water; middlings absorbed even more, while so-called wheaten flour took much less. All flour drank more in the winter than in the summer, and soft dough was thirstier than firm dough. It was less unfortunate to err on the side of too much water than to err on the side of too little. Too much gave the crumb aberrantly irregular and swollen eyes, a flat taste, and a crust that separated from the crumb, whereas too little produced a loaf that was at best hard to digest and at worst a source of colics.

Water tended to diminish whiteness, a fact that may have induced certain bakers to use very conservative doses. It was generally recommended that the water be warm, especially in winter and particularly in the early stages of kneading. Bakers worried, however, about overheating, for excessively warm water "cooked" the flour and irreversibly checked fermentation. Water that was too cold inhibited fermentation momentarily, but this damage could be repaired. Hot water also caused the dough to turn bitter and retain a slightly unpleasant taste, and it thickened the crust at the expense of the crumb.[8]

Fermentation

Parmentier located "the soul of the bread-making process" in the leaven. The preparation and insinuation of the leaven into the dough required "the most care and skill" of all the stages. Parmentier esteemed that the whiteness, lightness, and volume of the loaf "depended absolutely" on the leaven. Nor was this a purely theoretical or savant perspective. A master baker castigated his head journeyman "for preparing leavens so bad that they would ruin the mass of flour that he was going to put in the kneading trough, . . . *the leaven being the key to bread making.*" The starter leaven (*levain de chef*) that the baker put aside from the previous kneading was "the foundation for all the work." It was usually stored in a special vessel designed to protect it from the air, the sun, odors, and the vicissitudes of the temperature.

A good leaven was one that was not perfectly sour but always in the process of gradually souring. Excessive acidity remained a chronic peril to bakers. If the starter was too strong, it would communicate a sharp, disagreeable taste to the bread. There was no absolute standard, however; the leaven had to be matched with the quality of the flour and had to be composed (one is tempted to say improvised) as a function of other conditions, such as the weather. Dry and glutenous flour, for example, required a "young" leaven in large amounts, while humid meal that was freshly milled or drawn from soft wheat needed strong leaven to protect the dough from softening and resisting fermentation.[9]

Leaven bakers in Paris ordinarily "refreshed" their starter three or four times during the kneading phase. Through the process of fermentation that it triggered, the leaven enabled the baker to draw out the finest aspects of a good flour and to impart a new and superior quality to a mediocre one. Leaven preparation tested the baker's intelligence and experience. He had to know how to nurture the leaven development and how to read the *apprêt,* or state of readiness and transition. Periodic "refreshing" renewed and promoted fermentation while at the same time taming the dough and rendering it softer, less sour, and more analogous in consistency and taste to the bread it would soon become.

Bakers used three or four leavens. The starter took twelve or fifteen hours to ripen, though it could be allowed to fortify for several days. The so-called "first leaven," which followed the starter, theoretically sat for six or seven hours before being joined to the second leaven, which matured for four or five hours, though in practice it is doubtful that most bakers found enough time to meet these rigorous norms. The final, or *toutpoint,* leaven rarely "asked" for more than an hour. Each phase took longer in the colder than in the warmer months. It was better for the early leaven to be too strong rather than too weak, and for the toutpoint to be too "green" rather than too "old."

After each refreshing, the leaven batch had to be covered, for it was extremely sensitive to drafts and changes in air temperature. The leaven became increasingly light yet bulkier in volume as it grew. A satisfactory toutpoint had to pass the swimming test: it floated to the top of the water when immersed. A first leaven that sank was either beyond its proper readiness or not yet there. Bakers repaired an overripe toutpoint by drowning it in water in order to weaken its force ("to tire it out"). Malouin proposed that the same effect could be achieved by using salt, which dampened fermentation. Bakers tended to favor a large batch of leaven. The toutpoint generally contained one-third of all the flour, and sometimes even half. The better the wheat and the more absorbent the flour, the less leaven was necessary.[10]

The chief liability of kneading on leaven was that it took an exorbitant amount of time. Parmentier, who strongly supported the leaven method, nevertheless deplored "the overbearing slavery" it imposed on the bakers, allowing them never more than three consecutive hours of quiescence. He envisioned a process of fermentation based on a single leaven batch, but he never worked it out convincingly. The only clear alternative to the multistage leaven process was the use of barm, or brewer's yeast. Widely practiced by Parisian bakers, the barm method appealed to many because it hastened fermentation and rendered the dough much easier to work. Barm spared the baker the need to "mount" the dough in the trough, a gesture incidentally symbolic of the life-giving character of bread, and one that exhausted him. Under the influence of brewer's yeast—"like a spark set to combustible material"—with much less work the dough rose more rapidly than in the sourdough process. By substituting barm for starter leaven, bakers estimated that they could increase their productivity by one-third. Those who were raised in the barm method did not hesitate to affirm that a starter batch simply could not produce a soft dough.[11]

Though Dr. Malouin was not an advocate of barm, he conceded that when properly used it yielded "a light and better-tasting bread" than the leaven. He warned, however, that it had to be employed with dexterity, for an error in barm fermentation had a far more deleterious impact on bread quality than a mistake in leaven preparation. A flawed bread kneaded on brewer's yeast was simultaneously sour and bitter as well as sticky. Moreover, Malouin cautioned, barm should never be used for making dark bread because it made the loaf taste too good and feel too light, disorienting consumers and causing them to complain that it did not last as long as it should have (in their stomachs and in their cupboards). Moreover, since dark dough fermented relatively quickly on its own, it needed no artificial stimulation. In order to profit from the advantages of each technique, Malouin endorsed a leaven-barm synthesis, which he said was already commonly applied in the Paris bakerooms. A pound of brewer's yeast mixed in with the first or second leaven made a vigorous toutpoint sufficient to cover dough for four hundred pounds of bread. The resultant loaf allegedly tasted better and staled much less rapidly than a simple-leaven bread.[12]

Uninhibited by Malouin's pragmatic reserve, Parmentier launched a veritable crusade against the use of brewer's yeast. He had a fund of mistrust on which to draw, dating back to the great debate over *pain mollet* in the 1660s and

1670s, which prompted a contingent of physicians from the Paris Faculty of Medicine to argue that brewer's yeast was not fit for human consumption. A police report in 1725 echoed this idea: barm tempted the bakers and flattered their customers because it gave the loaf a pleasant appearance ("des yeux et bonne mine") and a delicate aspect, but it was at bottom "a piece of garbage, very disgusting and capable of harming the human body." Certain eighteenth-century commentators continued to regard brewer's yeast as a sort of adulterant. In his dictionary on the crafts, Macquer wondered about its impact on the health of consumers. Numerous medical doctors, joined by moralists, believed that bread made from brewer's yeast provoked the same ill effects on the nerves, the urinary tract, and the skin as beer.

Parmentier suggested that it would have been definitively banned by the parlement in 1670 had the magistrates known as much about its chemistry as was understood a century later. He favored a total prohibition on its use, yet he was not hopeful on the chances of changing minds on the subject, "given the interests at stake." As the *Encyclopédie* recalled, barm owed its introduction and wide diffusion to "bakers' greed," to an indifference to the reservations of the scientific and police communities. Alert to the perils of brewer's yeast, the Delafosse brothers developed a method of extracting it that they contended removed "its natural disposition toward putrefaction." Successful tests under the scrutiny of Lieutenant of Police Lenoir resulted in the award of a fifteen-year privilege to the brothers to market their "incorruptible leaven."[13]

The leaven process was burdensome, Parmentier allowed, but "the [barm] remedy was worse than the disease." To make good bread, dough had to be fermented evenly and gradually. Whereas a starter was gentle, leaven yeast jolted the dough violently and forced fermentation before the component parts of the dough were ready to form "a perfect and homogeneous whole." According to Parmentier, a baker could never be certain of success with a bread kneaded on brewer's yeast. At best it yielded a bread that was passable on the first day and then became uninviting in taste as the crumb rapidly decomposed. At worst, the dough softened suddenly because the fermentation proved too brusque, rendering an insipid bread, or it collapsed in the oven and gave a loaf with a bitter taste. The barm required hot water in all seasons, Parmentier pointed out, thereby attacking the glutenous quality of the flour that was so critical for the development of the dough, and the taste and texture of the bread. Since it demanded less work, the barm loaf was less white as well as less savory and smooth than the sourdough bread. The "onerous yoke" of barm, warned Parmentier, would inexorably cost the baker his customers, revolted by the inferior bread he produced.[14]

Subjected to the empirical "use" test that Parmentier himself employed on other occasions, his objections were not very compelling. Hundreds of bakers kneaded on barm, without provoking consumer outrage. There was a brisk commerce in barm that bakers, brewers, and independent merchants struggled to control. The bakers preferred to buy from dealers who were linked to them in some way, and whom they could thus trust. The surest connection: a baker's relative, such as Jeanne Sourdille, the wife of a master baker and the associate of a Flemish barm merchant. Bakers often bought an annual subscription for barm from a merchant, in the hope of procuring a guaranteed supply of a more or less constant price and quality. Parmentier contended that the system worked so long as demand for barm was slack; whenever it intensified, the merchants passed on spoiled or adulterated yeast to their fixed-rate clients in order to profit from the higher prices on the open market. Yet Parmentier probably exaggerated the bakers' inability to discern good from dubious barm, as well as the means of redress available to them in case of problems concerning quality or quantity.

When barm became difficult to obtain, or exorbitantly expensive, as a result of maneuvers by dealers, the baker guild called on the police to intervene in order to "regularize" the market. In 1725, for example, the jurés joined a commissaire in rounds of visits to the "laboratories" and lodgings of merchants and regraters suspected of hoarding barm. Hundreds of pounds of dry and liquid yeast were seized. The brewers' guild tried more or less abortively to reinforce its competitive position by exercising a right of inspection over independent barm dealers and by excluding cheaper yeast from Picardy and Flanders on the grounds that it was of equivocal quality. On the whole, the bakers seem to have found "foreign" yeast dealers to be easier to deal with and more reliable.[15]

Salt was the final ingredient in the bread-making recipe. However excellent the quality of the flour, water, and leaven yeast, Malouin proclaimed, the bread could not attain perfection without the action of salt. Salt elicited the best qualities of the dough, inducing it to absorb more water and air, and thus to yield more bread, and imparting a lightness, a better taste, and a greater capacity to keep. Parmentier disputed Malouin's claims. In his view, salt retarded fermentation, hardened the dough prematurely, retained humidity, and destroyed that wonderfully fruity and nutty taste that a good loaf communicated.

Most informed opinion seemed to fall between the positions of Malouin and Parmentier, acknowledging both advantages and disadvantages in the use of salt. Parmentier counted on the very high price of salt to keep bakers away from it, but he surely underestimated the extent to which it was commonly used. Bakers found numerous commercial and quasi-commercial sources for the purchase of salt, as they did for barm. A number of them became implicated in illicit salt trading.[16]

Kneading

The baker combined all the ingredients for the bread in the stage called kneading. First, he doused the toutpoint in water, to which he added more flour. He mixed, pressed, and divided vigorously so that all the lumps capitulated. Second, he introduced more flour and thickened the dough with rapid gestures but in a gradual manner. He had to act quickly enough to prevent the dough from "languishing," yet gently enough to prevent it from "burning." The resultant *frase* formed an elastic, flexible body that the baker worked lightly and continually. Third came the *contre-frase*, in which the baker worked the dough smartly, combining it in one great mass and then breaking it into pieces and hurling it left and right. He pulled and massaged with his fingers so that every part of the dough felt the impulsion. The fourth step, *bassinage*, prepared the dough for the most critical working by adding water in order to mix in the particles of flour that somehow managed to avoid integration until this point.

Refreshed, the dough was now subjected to "beating." The baker plunged into the mass, punched, pulled, turned, stretched, pounded, cut, and tossed chunks of ten to twenty pounds of dough around the trough. His cries and groans signaled the enormity and intensity of physical effort. He had barely three-quarters of an hour to knead two hundred pounds: if he took any longer, the dough would weaken as the fermentation lost its peak. The beating made the dough "long" and "even," and resistant to separation. Near the end of the operation, the baker had to moderate his aggressiveness. If he continued to punch instead of turning and folding the dough toward the center, he could render it "bucky" and hard to handle. The baker knew it was ready when it no longer stuck to his hands. Some bakers then removed it piece by piece to a clean trough or table.[17]

"The stronger, livelier, and more adroit the kneader-baker," wrote Baudeau, "the better the bread." The work was daunting, however, and bakers and workers, according to Malouin, were increasingly reticent about making the necessary investment. Bakers compensated for less work by relying on stronger yeast, a trade-off that saved energy at the cost of denying the bread its maximal whiteness and consistency and its most exquisite taste. Whereas it had once been common for baker boys to knead with their feet, Malouin noted that in the eighteenth century they often insisted as a condition of accepting a post

Kneading by hand.
*Source: © cliché Bibliothèque
nationale de France, Paris.*

that they be spared from "mounting the dough." Nor have I encountered in
after-death inventories or sealings a single bakeroom equipped with the long-
armed lever kneader that once was used to bash the dough for *pain brié*. The
first machines for kneading appeared in the 1760s, to be followed by scores of
others across the next hundred years. Despite the enervating labor of kneading,
however, bakers staunchly resisted mechanization, perhaps in part because of
the repugnance that consumers evinced for what they derisively styled "mechanical bread." In 1840 only 130 mechanical kneaders existed in all of France.[18]

The burgeoning popularity of soft-dough breads, along with the growing
allure of barm, reinforced the trend toward a less violent kneading. In the
seventeenth century, *mollet* usually designated a bread that contained milk as
well as the standard ingredients, and that was more often than not the size of a
roll. In the eighteenth century, any bread of soft dough, regardless of size, form,
and composition, was often called mollet. Containing more water and air than
a firm dough, the mollets were light, viscous, and easy to digest. Firm doughs,
which required much more vigorous work, became increasingly uncommon,
especially in white loaves. For the bulk of their ordinary production, most
bakers turned to a "bastard dough" midway between the firm and the soft.[19]

Certain bakers, now as in the eighteenth century, claim, not entirely facetiously, that "physical" kneading enhances the taste of the bread by incorporating the baker's richly nitrogenous perspiration. Physicians, however, have usually been appalled by the unhygienic conditions in which kneading took place
and the consequent danger to public health. Thus, wrote Dr. Léon Petit in 1900:

There is no species more repugnant than that of the *geindre,* naked to the waist, pouring
out sweat, gasping in the last throes, spilling and mixing into the dough that you will eat
several hours later all the secretions of his overheated body and all the excretions of his
lungs, congested by the impure air of the asphyxiating bakeroom.[20]

Kneading in an eighteenth-century bakeroom. *Source: from Paul-Jacques Malouin, Description des arts du meunier, du vermicellier, et du boulanger, ed. Jean-Elie Bertrand, 1771. Courtesy Kroch Library, Cornell University.*

The mitron of Vaugirard warned that the sweat and vile breath of kneaders imparted a bad taste to the bread. A faubourg Saint-Jacques baker learned just how sensitive the dough was to atmospheric conditions. His batch became infected beyond repair by the odors emanating from nearby septic trenches that were being emptied just as he was kneading. If the bread of the faubourg Saint-Marcel was "bad," noted Mercier, it was because of "bad air," itself a product of the quarter's "misery."[21]

Fashioning the Loaf

A ripe, winelike odor and a certain degree of swelling alerted the baker that the dough was ready to be weighed and fashioned into loaves. As always, timing was crucial. If a dough was underproofed, the gluten would be insufficiently stretched and the resulting loaf would be cramped in volume and rather heavy. If the dough reposed too long, excessive fermentation would sap the potency of the gluten and deprive it of its elasticity. An overproofed loaf suffered from inferior volume, irregular shape, and uneven texture.

Shape and weight were intimately connected, for the contours of the loaf helped to determine the margin of extra dough ("bonus") that had to be allowed to compensate for loss of weight in baking. By the middle of the eighteenth century, long loaves had supplanted round loaves in popularity for all weights of six pounds and under, especially for white bread. In the twenties all of master Joseph Herbert's four-pound breads were round. By the late forties, master Utroppe Constant gave more than half his production the long form. The round shape had been associated with firm dough that boasted a compact crumb and as little crust as possible. The soft dough required a flatter shape and more crust, and thus became elongated. The change in fashion did not serve the cause of economy, for the round loaves proofed more rapidly and took up about one third less oven space for the same weight, kept better, and, most important, lost significantly less weight in the oven because they offered less surface for evaporation and contained less water. A baker had to allow about as much extra dough (13 or 14 ounces) for a four-pound long loaf as for a six-pound round one. Moreover, the longer soft-dough loaves tended to become lighter in weight, and suffered further comparative disadvantage vis-à-vis the big round loaves, since heavier breads lost proportionately much less than lighter ones. (A one-pound loaf sometimes lost as much as five or six ounces, while a twelve-pound bread lost only one and a half pounds.)[22]

Bakers were tempted to undercook (though the threat of repelling their clients served as a deterrent) or to cheat on the weight. Bakers were also known to wet down the loaf just before placing it in the oven, a tactic that I witnessed in several present-day Paris bakeries. Nor was good faith enough to spare the baker the risk of being denounced for short-weight bread. He had to anticipate the impact of other variables that defied precise calculation. They included the type of oven, the temperature of the oven, the season, the quality of the flour, the skill and state of fatigue of the worker responsible for weighing the loaves, and the state of repair and accuracy of the scale.[23]

It was recommended that one worker divide and weigh the dough and another shape it until its surface was smooth and continuous. If the job was not completed rapidly, part of the dough would overripen and produce badly flawed loaves. Two experienced workers could weigh and shape two to three hundred pounds of bread in a half hour. Breads under a pound and round loaves of all sizes were placed *sur couche* on wooden racks or shelves that were covered with moistened sheets of flour sack in the summer and a woolen cloth in the winter. Most bakers seem to have preferred to proof long loaves in wicker or lined wicker containers shaped like the loaves themselves. These *pannetons* (or *bannetons*) helped to hold the form and prevent "bruising" by containing the fermentation. The lining tended to retain moisture and dirt, however; if the pannetons were not dried and cleaned regularly, they could ruin

Weighing and forming loaves, mid-twentieth century. *Source: courtesy, Grands Moulins de Paris.*

the product. Generally, the formed loaves rested *sur couche* for about three-quarters of an hour.[24]

The Oven and the Combustible

Even before the dough loaves were put down for final proofing, the baker had to begin to prepare the oven, for once they were ready they could not wait. There was no more room for temporizing or adjusting or repairing. If the dough did not begin to cook "at the summit of its exaltation," the bread could not attain optimal quality. This was the baker's most anxious moment. As he rushed to ready the oven, it was not rare for him to "curse in the midst of the smoke that surrounded him, . . . to hurl his tools, to accuse his coworkers, as a sudden change in atmospheric conditions pushed the dough to the point of no return before he was ready."[25]

Bakers considered the oven to be their most important tool. A well-built and maintained oven heightened their chances of producing excellent bread and helped to limit their energy costs. Ovens were located in basement or rear bakerooms, though a few bakers operated them in open courtyards, presumably only in the warmer weather. Not infrequently, bakers had to have ovens built when they entered lodgings not previously inhabited by a baker or when the baker-predecessor dismantled his oven upon leaving in order to meet the

Preparing the oven,
mid-twentieth century.
*Source: courtesy, Grands
Moulins de Paris.*

conditions of his lease. A new oven cost between 100 and 150 livres in the first half of the century. Prevented from setting up ovens beneath their rented shops by landlords and fellow tenants who coveted the basement space or trembled at the risk of fire, a number of bakers had to install their bakerooms in other buildings, sometimes hundreds of meters away.[26]

Paris ovens were generally made of refined clay, though bricks were occasionally used. A slightly flattened hemispheric shape was preferred because it distributed heat more or less evenly. The most delicate part of the oven, the floor, was built last to protect it from damage, after the vault and the dome (or chapel). From top to bottom the oven rarely measured more than a foot and a half; a "taller" oven would have wasted heat, inhibited the swelling of the loaves, and dried out the crusts. The bakers at Troyes estimated the life of an oven to be twelve years, while one of their critics claimed that it lasted for several baking generations. Malouin fixed oven life at nine years. Parmentier granted twenty-four years to the most robust, brick-vaulted ovens. Virtually all the practitioners and commentators agreed, however, that the floor of the oven had to be replaced every year.[27]

In order not to disappoint, and thus risk losing, his customers, a vigilant baker foresaw the need for reconstructing, arranged well in advance for the oven builder to come, and baked heavily over the preceding days in order to cover demand. A floor could be installed in a morning and be ready for baking by evening. In September 1771 master baker Chantard of the rue des Cordeliers paid twelve livres plus two bottles of wine for a new floor. The oven makers had a notorious and virtually unquenchable thirst, ostensibly because they had to work in intensely hot environments. Provident bakers furnished refreshment on the spot in order not to lose precious time. Baker Regnault committed the imprudence of leaving the ovenman alone for a quarter hour. When he returned, the bakeroom was deserted. An hour later Regnault found the artisan in a tavern, the fourth that the oven maker had visited.

Complaints about the fragility of ovens were frequently voiced, but most attempts to improve oven design and operation focused on ways to lower energy costs. Bakers disagreed on the tactics most apt to prolong floor life. Some avoided washing out the embers after heating, while others maintained that such a cleaning fortified the clay. Nor did circumspect bakers dry their wood in the oven, a practice that wore out the floor and at the same time dissipated heat.[28]

According to the Paris municipality, "wood and wheat are inseparably linked" and must be equally monitored. Items designated "of first necessity," they were both needed in order to prepare bread for "almost a million inhabitants." Wood, like grain, required a vigilant police in order to sustain a regular flow and prevent and/or repress "monopolies" geared to create artificial shortages. Entrusted with responsibility for monitoring waterborne wood traffic, the city fathers deployed the carrot and the stick to maintain an orderly commerce. On the one hand, they protected merchants against seizure for debts (thus acknowledging their quasi-public trust) and accorded them free storage areas in the ports. On the other, they imposed a price (or "tax") schedule to which the merchants were expected to adhere strictly save in periods of glut or subsistence ease.

Inclement weather, producing ice or drought or flood, constituted the major threat to the provisioning of wood (as it did to the supply of flour). From the police perspective, the worst moments arose from the convergence of wood and grain/flour dearths, as occurred in 1739. The municipality dealt with crises by establishing a special baker reserve, to which the merchants were obliged to subscribe. The authorities had to police the bakers as well as the merchants, for the former were tempted to use their privileged access to wood (sometimes managed via ration tickets) to conduct an illicit regrate trade.[29]

The Paris bakers preferred a "white" wood to the "new" wood and the "hard" wood that was coveted for private use. The white burned more slowly and evenly, providing the energy for almost twice as many bakings. Most bakers, however, could not afford the preferred birch woods and were obliged to settle for other varieties. Inspired by the Prussian example and the experiments of Benjamin Franklin, Parmentier looked into the possibilities of using coal as an alternate combustible, but his research did not yield any significant results. The experts did not agree on how heavily the price of wood weighed on the price of bread (though we know in retrospect that few goods rose as steeply in price as wood across the century). The author of an "Avis économique sur la boulangerie" dismissed the wood variable as "insignificant," while the bakers of Troyes considered it "the most important of their daily expenses." Commenting on a project to establish a royal bakery during the crisis of 1725, a government adviser estimated that it cost two livres in energy to convert a setier of wheat into bread. Rarely did authorities isolate the cost of combustible in calculating their price schedules.[30]

Debts to wood merchants, who habitually extended credit, helped drive a number of Paris bakers into commercial failure. The bakers complained intermittently that the cost of wood was undermining their capacity to sell bread at a reasonable price that still afforded them a reasonable profit. In the mid-1780s a handful of bakers, dissatisfied with the representation of their syndics, launched a vigorous lobbying campaign to obtain a government-imposed reduction in the price of wood. Their petition was signed by scores of confreres, each of whom was asked to contribute a sum to defray the costs of the campaign. Enraged by this usurpation of their authority, the syndics attacked the band of mutineers for "extorting" funds and purveying untruths. Impressed by the case against the insurgents, initially the lieutenant general of police favored their indictment. Upon examining their presentation of evidence, which called into question the efficacy of guild leadership, he called for further investigation. Apparently the parties reached some sort of compromise out of court. In any event, royal letters patent responded to the alarm of the bakers over the rising price of wood, which the government committed itself to control more tightly.[31]

It was uneconomical to bake only one or two batches (*fournées*) in a day since the additional energy needed to bake five or six more was almost derisory, yet only a minority of bakers could sustain the optimal rhythm. The oven had to be heated for between forty-five minutes and an hour before removing the remnants of combustible and putting in the dough loaves. The baker knew the oven was hot enough when a pinch of flour held over it reddened immediately. If it turned black, the oven was too hot; if it remained white, it was not hot enough. Flour of excellent quality generally required a less hot oven than less outstanding merchandise, and thus favored fuel economy.

The degree of heat necessary also depended on the number of pounds to be loaded and on the shapes and kinds of dough. The task of "conciliating" the baking dispositions of different types of loaves was extremely difficult. The big loaves were put in first, placed in back on the left side in order to draw the heat and begin cooking. After each row was completed, moving left to right and back to front, the oven mouth was closed for a minute in order to allow the loaves "to puff out." An oven seven to nine feet in diameter took about three hundred pounds of *gros* and no more than two hundred pounds of *petit* bread. It required about a half hour to load. Breads that did not fit were relegated to the leaven for the next oven. Master bakers assigned the loading of the oven (*enfournement*) to the head journeyman (known as the *geindre*) or an experienced hand, for it was easy "to break" or "to fatigue" the formed loaves if they were not handled gingerly on the peel, or shovel. Sometimes, just before loading, the baker "gilded" the dough forms with honey or egg yolk or milk "in order to give them color," a technique still used today.

Baking

Parmentier observed that of all the cooking processes that foods had to undergo, bread baking was the most treacherous. The baker could not remove the bread to check on its progress. He had to *know* when to remove it, for once out it could not go back in without suffering serious damage to color, crumb, and crust. Baking time varied with flour and dough type, shape, weather, and so on. As a rule, a four-pound loaf took an hour, an eight-pounder two hours, and a twelve-pound bread three hours. "The well-cooked bread," ran the bakers' proverb, "is always good." A baker knew he had cooked well when the color satisfied his eye, when the ring of the crust was sonorous, when the crumb sprung back after being depressed, and when the loaf had lost its "extra" weight. The breads had to be removed from the oven just as carefully as they were loaded to prevent cracking and splitting. If the loaf cooled slowly, it would stay fresher longer: thus the baker stacked the loaves against one another to keep them warm and covered them if they were a bit overcooked in order to catch the escaping vapor and moisten the bread. Once cooled, the breads were brushed with gloss (*fleurage*), covered, and stored in a dry place.[32]

A good loaf projected a sort of proud fullness. Golden yellow in color (called "bloom"), neither too pale nor too striking, the crust had a uniform thickness and a smooth surface, devoid of fissures, a leathery touch, or an overglazed cast. The crumb could not be either pasty or gritty or sticky or dry or crumbly. It should exhibit the springy elasticity that was the proof of "a good liaison, itself the product of skillful kneading and well-controlled fermentation." A well-fermented crumb emitted a slight winelike odor and possessed hundreds of small eyes that testified to a vigorous kneading and aeration. This regular texture or vesiculation should be complemented by a creamy color and a crumb of a fine pile, more or less smooth and silky, which readily stripped off into thin flakes.

Just as wheat needed a year and flour a month to be ready, Malouin contended, so bread needed a day's respite after baking in order to reach perfection. Like Sylvester Graham in the 1830s, a number of physicians warned against bread that was too fresh: it caused gas and colics, and it swelled the stomach. The hard protective crust that formed after a day appealed to certain consumers. Soft-dough bread, by contrast, clearly tasted best the first day, and loaves that were stale increased "the melancholic humor." As the eighteenth century progressed, more and more Parisians, of all ranks, looked forward to fresh bread every day. Forains, who used to bake two or three days prior to market, found themselves obliged to bake at least a few batches on the eve of market days.[33]

It is easy for well-fed people today to take (a somewhat precious) delight in the beauty and pleasure of the well-turned loaf. The cognitive and material field of reference of the bulk of eighteenth-century consumers, who lived in a world of chronic subsistence penury and uncertainty, was fundamentally of a different order. Yet we know that they were habitually exigent, and sometimes even fastidious about their bread. Professional pride aside, bakers had compelling commercial incentives to please their customers with high quality products. Although the fermentation stage offered some play, there were few shortcuts in baking. The basic steps were the same for all practitioners. (Yet today's egregious unevenness in quality from shop to shop makes us wonder whether disparities in dexterity [tours de main], assiduity, and rigor did not issue in similar discrepancies in the eighteenth century.)

Beyond psychological impalpables and esoteric knowledge, what advantaged certain bakers vis-à-vis others were both superior conditions of production and high quality raw materials. In the first category one could note the size and comfort of the bakeroom (with attendant implications for ventilation, temperature control, etc.), excellence of equipment, and the talent, motivation, and number of the journeymen (called garçons). In the second rubric, one places the quality of the flour, water, and leavening agent. Time management mattered enormously. It depended on the briskness of demand, the multiplicity of the baker's commitments, and the manpower capacity of the shop, but also on the organizing skills of the master and his wife.

In discussing the grueling, endless toil in the Old Regime bakery, one must guard against the usual perils of anachronism, ethnocentrism, and romanticization. One is tempted to believe that even blasé bakers experienced a sense of abiding mystery in the processes of fermentation and proofing, a mystery that still does not seem reducible to mere scientific account. The awe-inspiring aspect seems enhanced in retrospect by the absence of machinery and (most) additives. One is also struck by the fragility of the bread-making enterprise, the manifold conditions that had to obtain for success, and the ease with which everything could fall and fail. Nor can one overlook the sharp contrast between the image of a somewhat brutish baker capable of lifting hundreds of pounds and kneading mounds of dough at a terrific pace and the need for delicacy, agility, and grace in so many of the other critical gestures. Finally, it is hard *not* to imagine the master or the geindre taking a certain satisfaction when the bread coming out of the oven (for the nth time) seemed just right. And could wearying routine have rendered them utterly indifferent to that warm, slightly humid, nut-fruit-and-wine aroma that gently engulfed the bakeroom as the just-baked batch ripened?

Chapter 3

Baker Shops and Bread Markets

In the domestic domain as in so many others, the Enlightenment (if one can dare speak of it momentarily in monolithic and practically anthropomorphic terms) broadcast contradictory cues to the denizens of the everyday world. Some elementary life gestures it urged them to reclaim and perform on their own. Breast-feeding is probably the best-known example. Others it urged them to abandon, to farm out in the name of a more rational division of social labor. Bread making fits into this category. The message of progress was, let the professional baker do it. It is surely true that many more people baked their own bread than dispatched their babies to wet nurses during the eighteenth century. This autarchical practice prevailed not only in the countryside but also in many towns. Where bakers did serve, often they were no more than *fourniers,* that is, artisans who received a proofed dough and provided merely the office of the oven, not the integral process of bread making. Reformers called on consumers to renounce this toilsome task; benefits would accrue to the household and also to the bakery, following this line, for the more clients they served the better bakers they would be obliged to become. Paris was one of the cities in which the vast majority of the population already depended on full-service baking professionals for their daily bread.

Three formal strains of bakers served the capital: masters, who belonged to the guild; *faubouriens,* who operated in the sprawling faubourg Saint-Antoine, whose "privilege," or special royal franchise, spared them from the need for corporate affiliation; and *forains,* also outside the boundaries of the guild, who brought bread to the city from their hinterland bases twice a week. How many bakers regularly participated in the provisioning of the capital? The testimony is not unequivocal. In his *Traité de la police,* Delamare claimed that the capital was served by a total of 1,810 bakers, distributed as follows: 250 bakers of *gros* and *petit* bread within the traditional city limits (*enceinte*), 660 bakers of *gros* in the faubourgs, and 900 forain bakers from the suburbs and the hinterland. Thirty years later the author of the article "Baker" in the *Encyclopédie,* which is

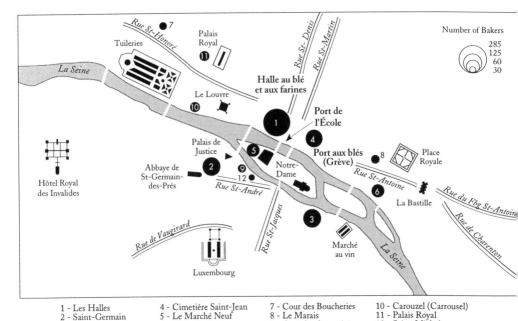

Number of Bakers

285
125
60
30

1 - Les Halles
2 - Saint-Germain
3 - Place Maubert
4 - Cimetière Saint-Jean
5 - Le Marché Neuf
6 - Saint-Paul
7 - Cour des Boucheries
8 - Le Marais
9 - Quai des Augustins
10 - Carouzel (Carrousel)
11 - Palais Royal
12 - Saint-Michel

supposed to have been particularly sensitive to artisanal life, used these same figures, as if they were fresh, without any effort of verification.[1] Delamare's calculations are credible, but his categories are somewhat confusing because they ignore socioprofessional status ("quality") in favor of geographic origin. Bakers of *petit* (who also produced *gros* loaves) were masters, a substantial number of whom lived in the faubourgs, sometimes on the same streets as the faubourg bakers of *gros*.

In the *Traité* Delamare presented another figure that also exercised unwonted influence. He reported that 1,534 bakers furnished the fifteen Paris bread markets; five to six hundred of them were from the city and the faubourgs, while the rest were from suburbs and the region. Once again it is impossible to determine "quality" because we do not know what portion of the city and faubourg bakers in the markets were guild members. On the basis of his own market-by-market enumeration, Delamare should have counted 1,524 bakers rather than 1,534 bakers. The error was rectified in one of the early-eighteenth-century editions of Sauval's *Histoire et recherches des antiquités de la ville de Paris*, which merely reprinted Delamare's list. Other so-called authorities did not even have the merit of checking Delamare's arithmetic. Despite the fact that there were no longer fifteen official markets and that their composition had changed substantially over the years, the *Journal économique* slothfully copied Delamare's list (with 1,534 as the total) as if it were the latest survey in 1754, and Des Essarts, author of the *Dictionnaire universel de police*, which was meant in part to be an aggiornamento of Delamare, followed suit in the 1780s.

In a manuscript note dating from about 1700 in the Delamare papers, which the commissaire himself may have written, it is estimated that 2,393 bakers served Paris: 1,534 at the markets (the *Traité* figure) plus 859 others, presumably masters and simple bakers from the city and the faubourgs located exclusively in shops.[2] This number is more than 32 percent higher than the published reckoning, though it concerns the same epoch. From the very beginning of the eighteenth century it is perhaps prudent to set the total number at around 2,000.

The first evidence we have that distinguishes professional status dates from about 1720. A well-informed correspondent of the procurator general of the Parlement of Paris—probably a police official—fixed the baker universe at 1,607: 489 masters in the city and faubourgs (195 in the city proper, 244 in nine faubourgs, including four in the faubourg Saint-Antoine, and 50 widows whose residence is not indicated); 252 bakers in "privileged places," most of whom were in the faubourg Saint-Antoine; 16 "privileged bakers following the royal court"; and 850 forains.[3] At about the same time, another study submitted to the procurator general evaluated the global number of bakers at 1,509. In this survey there were two hundred fewer forains (650 v. 850) and one hundred more city and faubourg bakers (859 v. 757).[4] An "accounting of the bakers" prepared in 1721 confirms the estimates for master bakers (explicit in the first of the surveys of 1720 and implicit in the second) with a slight rectification. In addition to the 195 guildsmen in the city, the 244 in the faubourgs, and the 50 widows, there were also 33 *anciens* (former jurés or guild elders) and 6 current jurés, for a grand total of 528 masters. An enumeration drawn up in 1724 lists 512 masters distributed in the city and twelve faubourgs (and including 48 widows).[5] According to an "accounting of the guilds" established by the police in 1743, there were then 580 masters.[6] During the next thirty years, their number did not vary significantly.[7] There is no doubt that the guild expanded as a direct consequence of the reforms provoked by the efforts of Controller General Turgot, chief minister and ultraliberal reformer, to abolish the corporate system in 1776, but I have found no authoritative enumeration of guild size for the last few decades of the Old Regime.[8]

The Delalande registers, a report of grain, flour, and bread entries into the capital recorded under the aegis of a police official of this name, can help us ascertain the number of nonmasters who ministered to Paris. The first of the

two 1720 evaluations claimed 252 bakers in the privileged enclave of the faubourg Saint-Antoine; without specifying the number, the second indicated
that there were more. In the 1727 police register, 275 Saint-Antoine bakers sold
their bread in the markets (representing 30 percent of all market bakers). In 1733
their number dwindled to 247 (but they now constituted fully a third of the
market baking force).[9] Now, it is very likely that the total number of Saint-
Antoine bakers was larger than the market enumeration suggests (say, by 50 to
100), because there were bakers there who were unable to procure a space in the
market or who preferred to concentrate their energies in their shops.

The most startling testimony provided by the registers concerns the forain
corps. All the estimates of the forains (900, 850, 650, etc.) appear far too high.
Doubtless there were some forains who distributed their bread illegally, outside
the marketplaces. But the vast majority supplied the markets; indeed, contemporaries define forains not only as "foreigners" but as outsiders who furnished
bread in the designated places on Wednesdays and Saturdays. In 1727 there
were at most 385 forains (almost 42 percent of all market bakers), of whom 280
were class-two bakers (who purchased their grain or flour in the hinterland)
and about 105 were class-one bakers (forain in their geographical moorings and
in their relationship to the clients but purchasers of supplies on the Paris grain
and flour markets). Six years later the forain universe had shrunk to 291 (still 40
percent of the global market force), composed of 211 class-two bakers and 80
class-one bakers. One reason there were fewer forains around 1730 than there
had been in 1700 was that the number of marketplaces had declined during the
period from fifteen to twelve. (Even allowing for this contraction, Delamare's
1,524 market bakers seems unduly high, fully 37 percent more than the total
number of market bakers in 1725, the top year in the register series.)

On the basis of this appraisal, I believe that there were at most 1,400 bakers
serving Paris around 1730: 550 masters, 350 nonmasters in the privileged places
(the enclaves free of guild jurisdiction, especially the faubourg Saint-Antoine),
400 forains, and 50 to 100 illicit bakers operating in the cities, faubourgs, and
outskirts. Their number increased in the course of the century, but not dramatically, perhaps not even after the post-Turgot reorganization. For if the guild
expanded, it is possible that it drew some of its new masters from the Saint-
Antoine reservoir. It also seems likely that the forain contingent continued to
erode slowly but irreversibly. If one imagines, conservatively, a population of
550,000 (underrepresenting floaters, rural inhabitants who crossed into the
capital to buy bread, etc.), then there was approximately one baker for 393
inhabitants. Mid-nineteenth-century estimates put the ratio at 1700 or 1800
per baker, but we know nothing about the composition of these figures.[10]

Bread Supply by Type of Baker

In their polemic against the police and the corporate system, the physiocrats—the cadre of liberal *économistes*—and their friend Turgot depicted the guildsmen as enjoying a stranglehold—"a monopoly"—over the Parisian bakery. The portrait was very much overdrawn. It reflected not the data, which were available at least in terms of orders of magnitude, but the ideological and programmatic needs of the group, who lobbied for a sustained higher level of grain prices and an unencumbered trade in provisioning merchandise. In addition to suffering the stigma of all holders of exclusive economic privilege, the guild bakers unfairly found themselves scapegoated as the singular nemesis of the consumer-people.

Though they exercised considerable leverage collectively and individually, the master bakers had a decidedly limited kind of monopoly: the exclusive right to bake and sell in certain places within the city and faubourgs and the exclusive possession of certain privileges, as well as certain servitudes. But the masters had nothing like a corner on the bread market, and they faced relatively brisk competition from nonguildsmen in the twelve marketplaces (and in some cases off the market as well, for many forains delivered to the homes of their clients).[11] Master bakers furnished much less of the capital's bread than we have heretofore been led to believe, even as the shops in general accounted for less of the total supply than had commonly been supposed.

On the basis of the raw figures in the Delalande registers, the masters contributed 34.7 percent of *market* bread in 1727 and 30.1 percent in 1733. If one assumes, with the police statisticians, that all shop bread was guild bread—an erroneous assumption as we have argued—then one calculates that masters were responsible for 51.2 percent of the total uncorrected bread supply in 1727 and 50.6 percent in 1733. If one relies on the corrected figures proposed above (minus the margin-of-error adjustment), one finds that the master portion fades to less than half the aggregate supply: 45.2 percent in 1727, 45.3 percent in 1733.

Nonguild bakers, then, played a relatively greater role in furnishing Paris with bread than did the masters. In 1733 bakers from the faubourg Saint-Antoine composed one-third of the entire market baking corps (as opposed to 29.7 percent in 1727).[12] They accounted for 31.4 percent of the *market* bread, more than the forains of the second class. These 247 bakers produced 22.2 percent of the total supply according to the 1733 police register and 27.4 percent of the total corrected supply (which includes an estimated shop component in addition to the market portion). The forain bakers, a contingent of about 295

Fifteenth-century bakery.
Source: Photo Collection Roger–
Viollet.

Woman buying part of a loaf in
a time of scarcity, eighteenth
century. *Source: courtesy,*
Grands Moulins de Paris.

(drawn from the first as well as the second class of bakers in the police register) contributed 38.5 percent of the *market* bread and 27 percent of the unadjusted total bread supply. The forains furnished 24.5 percent of the corrected aggregate supply. Thus the nonmasters—that is, Saint-Antoine bakers, the forains, and the illicit bakers (2.8 percent)—contributed 54.7 percent of the total Parisian bread supply.

Shop versus Market

The neighborhood shop was not the major outlet for bread distribution. Following the computations in the police register, one finds that in 1727 over one-quarter of all the bread consumed in Paris was meted out in the shops. Thus the markets outsold the shops two to one. Six years later the share of shop bread climbed slightly to a bit over 29 percent. The shops fare better in terms of the corrected figures, shown below, but the markets clearly remain the preponderant force in distribution.[13]

Year	Market	Shop
1727	67.1%	32.9%
1733	63.7%	36.3%

In 1730 there were about eight hundred shops in Paris and the faubourgs (including the privileged places and the illicit establishments). In addition, there were probably another fifty shops on the outskirts of the capital more or less regularly frequented by Parisians. Every street in Paris could not boast a shop, but many could vaunt a half dozen or more, and there was not a single quarter in which consumers did not have a number of nearby shops from which to choose. Bread outlets framed neighborhood organization; one situated one's address in reference to a nearby shop, which served as a humble yet notorious monument in the social landscape of everyday life.[14] The neighborhood bakers were familiar figures to all Parisians, and virtually all Parisians had recourse to the shops for one reason or another even if they were habitual market buyers. The shop was more convenient than the markets, for it offered bread every day of the week—fresh bread (though it ought to be said that not everyone appreciated fresh bread). The shop proposed a much wider selection of breads than the markets. At the market the buyer found only *gros pain* (theoretically six-, eight-, and twelve-pound loaves, but the four-pounder commonly appeared at the markets in the eighteenth century) made of a hard or firm wheaten dough in two or sometimes three varieties (white, medium, dark).[15]

Alongside the standard *gros* the shop bakers offered an array of small breads, from one-half pound up to four pounds and often beyond, some of which were made from soft dough or with milk, butter, and/or eggs ("fantasy" bread). Of course the market buyer had no interest in the elite breads because s/he could not afford them and because s/he found them less filling, for both material and psychological reasons. And the reason the buyer went to the market for the *gros* loaf was because s/he would generally get it more cheaply there than at the shop. (The bakers of Gonesse, a hamlet famous for its bread making, would protest, rightly, that some of their buyers, even the poor ones, may have preferred the forain loaves because they tasted better.)

Haggling was the rule of the market. The police insisted on it, the public expected it, and the bakers complied within certain limits. All the bakers knew more or less what the "current," or "common," price would be, in light of the preceding market, the *mercuriale*, the weather and the season, the political climate, orders or recommendations from the authorities, or the lack of them, and so on. The current, when it was not imposed by the police, a relatively rare occurrence that we shall consider below, had a quasi-magical quality to it. It seemed to be the fruit of a collective epiphany, for virtually all the bakers, widely dispersed throughout the metropolis, seemed to know it at more or less the same time. The news passed on something resembling what the French later called the "Arab telephone" and Americans the "ghetto wire." When a baker claimed he got the current wrong, invariably he got it too high; given the practical efficacy of the system of communication, despite its tinkered character, it is likely that the baker in question had failed to listen carefully.

In bargaining with potential clients, bakers started out by trying to get a little more, but ended up accepting a little less. A servant at the Marais market "haggled all he could to get his loaves at twelve sous each," but he finally settled for twelve sous six denier. Louis Passavant, a Saint-Antoine baker at Saint-Paul market, asked twenty sous for his eight-pound loaf. A woman named Brodchy offered seventeen sous and walked off to bargain elsewhere when Passavant refused to come down below nineteen sous.[16]

This sort of haggling generally did not occur in the shop. Some consumers tried to bargain and play one shop off against the other. A game warden named Antoine Delmarre, for example, went into the shop of baker Houdouard on the rue Saint-Antoine in quest of a one-pound loaf. The baker's wife asked four sous six deniers. Delmarre, an alert consumer, offered four sous on the grounds that this was the price at which the bread had sold at the last market. When Madame Houdouard proved obdurate, Delmarre went down the street to Charon's shop. Charon, too, asked for four sous six deniers but finally yielded to Delmarre's ardor and agreed to sell (he insisted) "at a loss, exceptionally." No wonder Charon's wife refused him the same bread at four sous the next day when Delmarre returned, convinced that he had found a good thing.[17]

Haggling sometimes led to misunderstandings and violent bickering. A construction worker named Louis Lefreyon stopped at the Palais-Royal market on his way to work to pick up a six-pound loaf. He offered the wife of forain Jean Denomberet eight sous. She asked for either a liard (three deniers) or a sous more, depending on whose testimony one believes. She claimed he grabbed the bread and tried to run off. Lefreyon protested that Dame Denomberet had agreed to give him the bread for eight sous and then changed her mind, keeping his money and denying him the loaf as well. They shouted at each other, exchanged blows and drew a crowd clearly sympathetic with the buyer. This was just the sort of confrontation the police feared most, for it could easily swell into a riot if feelings became intense enough. The guard arrived in time to prevent further disorder.[18]

As a rule—we shall note later one intriguing exception—the shopkeepers did not budge from their "current" price, and their current price was usually six deniers above that of the market. This was the tolerance that police allowed to help them defray their higher overhead costs and to encourage the laboring poor to haggle at the market. There was some price variation from shop to shop, but it appears to have been idiosyncratic: there were no shops that were notoriously cheaper than others.[19] At the market, the poor consumer could count on a lower price at the end of the day, if there was any bread left, for the police compelled the market bakers, in particular the forains, to sell their remaining loaves at a discount.

Nevertheless, the shop had some allure even to the poor consumer. It was usually easier to obtain credit from a shop baker than a market baker. The serenity of the shop ambience, the everyday accessibility of the shop, and its fixedness in the neighborhood encouraged the establishment of relations of confidence between seller and buyer. The shop baker had the impression that he *knew* his clients and that they in some sense belonged to him because they lived in the area he claimed as his own. Very rarely did he refuse credit. In the beginning it was a way of winning a new client and later it was a lever for assuring the client's fidelity.

The market baker had regular clients and he often granted credit as well. "That makes two market days [of bread] that I owe you," a woman told a Chaillot forain, as she took two six-pound loaves and one fourteen-pound bread. Widow Herbert from the faubourg Saint-Antoine, Vaugeois from Noisy, and masters Huron and Leguay from the city were among the others who gave credit at the market. Yet, given the nature of the market—the physical chaos, the swarms of buyers, many strange faces, the frenetic pace of transactions, and the haggling—it was difficult for the market baker to build the same sort of relations as his shop-rooted colleague. Without money, then, there was a good chance a consumer would go to a shop (or he might use what little cash he had to buy a four-pound loaf at the market and still go to the shop between market days).[20]

In some shops the poor consumer received preferential treatment, though it is not always clear what the criteria of discrimination were. In April 1755 master baker Rousseau charged a seamstress eight sous six deniers for the same loaf for which he demanded nine sous six deniers from the marquis de Chanzeron. In November he made the marquis pay nine sous, but a number of other clients of humbler conditions paid only eight sous. During one week in December 1761, Chenaut charged Gombault eight sous and "Mademoiselle" seven sous six deniers, while "Monsieur qui demeure ché Monsieur Grangant" paid nine sous several days later, apparently for the same four-pounder. Bakers Tayret and Modinet also practiced a stratified price policy, but the principle of the differentiation, as in the case of Chenaut, does not always seem to have been "social" in the sense of a transfer from well-off to poor.[21] Cases such as these are the only instances in which shopkeepers abandoned the current price. In this sense preferential treatment can be viewed as an analogue to market haggling. In some circumstances, however, it is possible that the preferential price was indeed the current price and that the higher price was simply abusive.

As a rule shop bakers seem to have had a very broad range of customers. In the course of several months Tayret served, among many others, six wine merchants, two unspecified guild masters, thirteen other artisans (a glazier, coach makers, carpenters, a knife maker, a mason, a cooper, a blacksmith, a butcher, a plasterer) who were either masters or journeymen, three transport workers, three cooks, one servant, four office workers, two lodgers, one schoolmistress, a gardener, a soldier, a sergeant in the guard, an officer of the watch, and two minor officials. Selling on credit, bakers registered buyers both nominally (with or without socioprofessional status) and simply by status and occupation, despite the risk of duplication and confusion (thus Rousseau sold loaves to "la ravadeuse," "la blanchisseuse," and "la brodeuse" and Tayret to "la blanchisseuse" and "le soubrigadié"). Dehais furnished a client "who touches to cure scrofula"—almost as exalting as supplying a king. Certain customers purchased the sort of loaves that one might expect given their socioeconomic status. Thus, for instance, Tayret's driver bought only twelve-pound dark breads. In many cases, however, this sort of reductive grid does not work. Baker Andriot sold mollet breads to unskilled day workers and to lawyers.[22]

Consumers were free to frequent whatever shop or market they desired. They were bound only by the *pratik,* or most favored, relations they had established with suppliers and by the daily work obligations that may have shaped their spatial and temporal itineraries. Yet from the perspective of the police there existed something resembling a geography of expected/legitimate consumption habits. Bread resonated with work and family, and bespoke values of rootedness, visibility, and predictability that the authorities fostered. One took one's bread normally more or less near where one lived and/or labored. Bread-buying patterns constituted a marker of community, and mapped out a set of social and perhaps moral boundaries. It often seemed to the police that looking for bread elsewhere was tantamount to looking for trouble. "Why has she come from so far away to bargain for her loaf, . . . and what lured her to this quarter?" asked a commissaire of a sixty-year-old nurse's aide (and widow of a shoemaker) who wandered into the Saint-André-des-Arts section, where she was not known. Accused of stealing a ten-pound loaf at the Cimetière Saint-Jean marketplace, Victoire Brodecher, the wife of a Swiss Guard, was a priori suspect: she had no reason to be here since she lives "at the other end of Paris, on the rue neuve Saint-Sauveur." Similarly, the police questioned the intentions of a twenty-one-year-old water carrier who was found behind the counter of a bakery in the rue des Cordeliers. The first thing the commissaire wanted to know was why he came all the way from the faubourg Saint-Antoine, where he resided and toiled, to buy bread here. "I was hungry," he replied, not very convincingly.[23]

A Closer Look at the Shop

An ancillary advantage of the shop was that there one could buy a variety of bakery-related goods in addition to bread. Shopkeepers sold small portions of flour, despite the fact that retail flour sales were forbidden in the bakeries. The municipality complained that shopkeepers sold the flour "at about double the legitimate price." Some consumers probably used the flour for home baking, a practice that was marginal but not wholly extinct. Equally illegal but a bit less common was the sale (in fact the regrating) of wood, eggs, butter, and, less frequently in the faubourgs, courtyard poultry. Shopkeepers also sold bran and other flour by-products either to individuals for feeding animals or to starch

A late-eighteenth or early-nineteenth century bakery with the baker's wife weighing bread (a practice that suggests perhaps a later dating), while journeymen tend to the oven and the fashioning of loaves in the adjacent bakeroom. *Source: © cliché Bibliothèque nationale de France, Paris.*

makers. Mealmen purchased bran and middlings for regrinding. Consumers could obtain stale crusts for dipping in their soup and bread crumbs for use in cooking from the shop.[24]

Virtually all shopkeepers conducted a large and profitable commerce in charcoal—the "first residue" of the wood used to heat their ovens, which they removed before putting in their dough. They collected substantial amounts— master Jean-Baptiste Côte had two hundred bushels in his cellar—in heavy metal containers that were sealed off "hermetically" to extinguish the embers. The retail price varied enormously with the quality of the wood, the season, and the demand: in the fifties it sold at half a sol, 3 sous, and five sous a bushel; in the sixties at three, four, and five sous; in the seventies at four and five sous; and in the eighties at five sous.[25]

Though the shops differed significantly in magnitude of production and clientele, physically they were very much alike. In arrangement and decoration the bakeshop was rather austere, even somber. It did not yet boast the heavy iron door—the "fortress"—installed in the early nineteenth century to defend against "the sudden invasions of the people." Nor did it have anything in common with the overwrought taste for ostentation that came with the innovations that Viennese-trained artisans introduced into the Parisian bakery after

1840.[26] The front of the shop consisted of a large opening covered by a convex grill, and frequently by a screen as well. Breads were displayed behind the grill—a constant temptation to thieves, vandals, and the hungry poor.[27]

The interior of the shop was usually crowded with equipment and goods. Against one wall was the counter, generally made of wood and presided over by the baker's wife (or, more rarely, his mother). On the counter rested a scale and weights for weighing bread and a knife for cutting portions of the large loaves—exactly the same instruments that one should find on the modern bakery counter today.[28] In the drawer on the side of the counter the baker's wife kept the registers in which she noted the sale of bread on credit and the arrival of flour or grain. Behind the counter on the wall hung the *tailles* on which were entered the breads delivered to clients on credit at home, and sometimes also bread sold on credit in the shop. On the walls were also shelves on which the bread was stacked. Along the sidewall were containers for crusts, bread crumbs, flour, bran, other flour by-products, and charcoal. In the larger shops a table for work and storage stood against a sidewall perpendicular to the counter wall. Various implements hung on the wall in the space free of shelves.

Next to the counter a door led to the back shop, which served as a storage space in the bigger shops and as a kitchen in the more modest ones. Sometimes a trapdoor in the shop itself opened to the basement, where the bakeroom was located, though more commonly one reached it via a staircase in the court-yard.[29] In a few cases, the bakeroom was in a back room at street level—a rare luxury for the journeymen. Usually the baker's family lived in the building in which the shop was situated. A considerable number of bakers seemed to have kept large dogs, as pets and perhaps as protection, if one judged by the irate complaints of customers or neighbors who fell victim to excesses of canine zeal.[30] Shopkeepers were extremely jealous of the open area in front of their shops, which they wanted to keep clear at all times in order to exhibit their bread and facilitate the entry of clients. They quarreled bitterly and sometimes violently with other artisans who violated their sacrosanct commercial space.[31] Apparently the bakers cared less about keeping the street clean than about keeping it unobstructed, for they were fined frequently by the police for failing to sweep or remove garbage or snow.[32]

The Bread Market

In the middle of the sixteenth century there were only four markets in the capital that offered bread for sale. By the end of the seventeenth century, however, Delamare listed fifteen bread markets manned by over fifteen hundred bakers.[33] The number of bakers appears extraordinarily large, for within a decade of the publication of the *Traité*, there were under a thousand bakers supplying the market. Since the population did not suddenly decrease, baker productivity did not increase brusquely, and the non-market-baker universe did not change drastically in size, it is difficult to account for the disproportion.

Not long after the *Traité* went to press, Delamare himself began to revise his estimates, in response to queries from the procurator general. In 1721 he wrote that bread was no longer sold in three of the markets on his list: in the halle du faubourg Saint-Antoine; opposite the Temple; in front of the hostellerie des Bastons royaux.[34] The Delalande registers indicate that there were only twelve markets in the capital (1725–33). Nine of these markets appear on Delamare's revised list. The Quinze-Vingt market appears to have shifted to the Carouzel, but with a drastically reduced level of activity. The huge market in front of the Jesuit church on the rue Saint-Antoine seems to have been replaced by a far more modest one near the Church of Saint-Paul. The police registers amalgamate the grandes halles and the halles de la Tonnellerie into one Halles market and indicate the existence of a new market at the cour des Boucheries.

In 1723 a group led by the influential Daguesseau of the family of distinguished magistrates was granted the privilege to found a market in the faubourg Saint-Honoré for the sale of many articles, including bread. This market does not appear on the police registers, nor is it mentioned on the procurator general's correspondence in the twenties. According to Narbonne, a police commissaire from Versailles writing in 1735, "this market did not succeed; merchants did not supply it, and it no longer exists." This obituary appears to be premature, however, for in 1752 a police report refers to bread sales here. Eight years later two forain bakers petitioned the lieutenant general of police to investigate the fiscal "abuses" of the concessionaire of the Daguesseau market, who, they charged, exacted "excessive fees" for the use of the commercial space. Many merchants have been driven away already and others will soon leave, the bakers hinted, but it is impossible to infer from their petition how many bakers (if any) still furnished the market regularly.[35]

It is reasonable to hypothesize that the police ignored the Daguesseau market because it functioned badly or intermittently and thus was counted on by neither the public nor the authorities. The same might have been true of the market of Saint-Nicolas-des-Champs on the rue Saint-Martin, a space listed by Narbonne as a bread market but absent from the police record. It is possible that handfuls of bakers "squatted" at Saint-Nicolas or in other markets not officially recognized as bread markets, without authorization but in response to

Bread market at the Quai des Augustins, circa 1670. *Source: Musée Carnavalet, photo by Lauros-Giraudon.*

public demand. Where this demand was intense and persistent, markets were more or less spontaneously generated. A bread market, for example, was improvised for many years on the rue Montmartre. One suspects that other de facto markets sprung up sporadically, sometimes for relatively brief periods of existence, in other quarters, for the twelve official markets were concentrated at the center of old Paris, leaving the bulk of the periphery—now a relatively densely settled, blooming area—without local market service. Another market without de jure status but with a long tradition took place on Sunday mornings in front of Notre-Dame. It was frequented by the very poor, for bakers brought for sale at a reduced price their poorly cooked, burnt, or damaged breads.[36]

In response to new patterns of demand, Inspector Poussot, in the late fifties and early sixties, proposed an extension and a rationalization of the bread-market system. Certain of the older markets, such as the one at Place Maubert, were sufficiently capacious and well organized, though some of the bakers were still obliged to sell from their wagons for lack of available stalls. One of the newest, the "little" market of the faubourg Saint-Germain with a covered halle and individual vending areas for bakers, was a model of efficient spatial arrangement. But other markets, in Poussot's view, needed to be restructured. He called for the fusion of the Carouzel, which was seriously underutilized, with the Palais-Royal, a market bristling with traffic. He proposed the resuscitation of the Quinze-Vingt, albeit one cannot tell from his suggestion whether in fact bread was being sold there in 1760.[37]

Overcrowded and in disarray, the Halles commanded top priority. Poussot imagined relieving some of the pressure by adjoining to it part of the space once occupied by the Hôtel du Soissons and the Cemetery of the Innocents. At the same time he proposed shifting some of the bakers to a new market, which in fact amounted to the resurrection of the old Saint-Nicolas market. This decentralization would provide a buying place for the inhabitants of La Villette, the faubourg Saint-Martin, La Chapelle, and the faubourg Saint-Denis, who had been obliged to come to the Halles for want of a local market of their own. Similarly, the inspector urged the creation of a new market to serve the faubourg Montmartre, la nouvelle France, and Les Porcherons, and for the revival of the halle in the faubourg Saint-Antoine as a bread market.[38]

Officially nothing ever came of Poussot's decongestion and "regionalization" plans. Wildcat markets may have formed here and there, but among the half-dozen markets created or revitalized after 1760, none was even partly consecrated to bread. Within the existing bread markets, however, commercial and administrative inertia could not undermine Poussot's ambitions. In collaboration with commissaires such as Dubuisson at the Cimetière Saint-Jean and Machurin at the Halles, the inspector saw to it that each market baker had a license to a space, that each space had a stall, and that each stall had a number posted on it so that the public could easily denounce infractions to the police and so the market bakers would be constantly reminded of their accountability. By banning vehicular traffic during the markets, aggrandizing space for the bakers at the expense of the other sellers, and encouraging bakers to increase their weekly supply, Poussot facilitated transactions and probably increased the level of business. There is no doubt that most of the official bread markets were better managed after the early sixties than before. From the early sixties until the Revolution got under way, the organization of the bread markets does not seem to have changed. ("It is very important for us to know, Monsieur, how many bread markets there are in this city, what days of the week they are held, and the number of bakers who habitually supply them," wrote the department of police to the commissaire of the Palais-Royal section on 4 June 1791.)[39]

Allocating and Regulating the Use of Market Space

The police conception of the provisioning trade as a public service rarely appeared more clearly than in the distribution of stalls to bakers who wanted to sell bread at the market. "The baker who takes a stall in a market for his business," intoned Delamare, "contracts a kind of obligation toward the public to supply the stall with a sufficient amount of bread each market day."[40] There was no fixed quota, though each baker indicated at the outset how much he could supply per week, and he was expected to fulfill that engagement within reasonable limits.[41] The regulations warned that a baker who failed to supply his stall "adequately" for three consecutive market days automatically forfeited his right to do business at the market and faced other penalties as well. The parameters of "adequately" changed from moment to moment, depending on specific supply conditions, so that bakers enjoyed some leeway and occasional breathing room. Yet bakers who did not take their "contract" seriously came to regret it. The police attached utmost importance to the idea that the baker had a public duty, and they punished baker apostates, not only at times of subsistence crisis but in periods of tranquility as well.

The regulatory point of reference for the eighteenth-century police was the ordinance issued by the lieutenant general on 22 September 1725, on the recommendation of the Assembly of Police, a more or less weekly council of luminaries including the first president and the procurator general of the Paris Parlement, the lieutenant general of police, and the *prévôt des marchands* who headed the municipality. The capital was suffering from a serious dearth that had already provoked a violent uprising against the bakers in the faubourg Saint-Antoine and, perhaps partly in a reaction to popular disorders, a signal lack of ardor on the part of bakers to supply their shops and stalls "adequately." The police viewed this baker lethargy as a "criminal affectation," and the lieutenant general ordered all bakers to supply regularly and "sufficiently" or face three thousand livres in fines and possible loss of market space and masterships as well as corporal punishment. The penalty was extremely severe if one considers that first-offense fines for selling above the current price in 1725 were rarely over thirty livres. In the course of the next few months more than a dozen bakers had their business spaces confiscated and suffered three-thousand-livre fines for interrupting market supply.[42] Some bakers received heavy fines but were allowed to continue market service. The Assembly of Police recommended forgiving the fines of the bakers who had good reputations and who pledged to redeem themselves (for signs of contrition were obligatory) by furnishing their stalls zealously. There was an impulse to treat offenders particularly sternly in the midst of a crisis because the social stakes were so high.[43]

It was also true that bakers often had more justification for failing to meet their obligations during a dearth than during normal times. Victims of the crisis, the bakers—some of the bakers—had trouble finding grain, converting grain into flour, obtaining credit, and so on. The police recognized that there were bakers "genuinely unable to continue supplying Paris." At the Saint-Paul market, for example, during the crisis of 1725–26, a large number of bakers simply quit out of helplessness. With proof of goodwill, the police usually allowed these involuntary delinquents to return when they recovered.[44]

Though the decade following the dearth of 1725–26 was a relatively quiet provisioning time, the police struck harshly against baker irresponsibility, especially when it was of a flagrant sort. For failing to supply his space at the Cimetière Saint-Jean for six consecutive market days in October–November 1727, Michel Ferry lost not only his market stall but also his mastership, on top of a three-thousand-livre fine. Commissaire Divot detected the makings of a plot when two Gonesse area bakers brusquely stopped supplying in July 1729. Their "manifest disobedience," Divot contented, was "the upshot of the threats made by and in the name of the bakers of Gonesse to quit supplying if they suffered constraints [e.g., police controls] in the sale of their bread, and this with the aim of increasing the price [of the bread] without any regard to its relation to the price of grain." The lieutenant of police reprimanded the bakers and fined them each three thousand livres. In the fall of 1731, after a two-month absence, a baker from Gonesse escaped, "by grace and without setting precedent," with a one-thousand-livre fine, perhaps because he had been in good odor with his commissaire or because the police did not wish to discourage the Gonesse area baking community. In February 1734, and September 1735, four other bakers, apparently all from the faubourg Saint-Antoine, each endured a fine of three thousand livres and confiscation of their spaces. During the subsistence crisis of 1740, two forains from Goussainville and the faubourg Saint-Antoine were fined three thousand livres each and banned forever from the market.[45]

Bakers were not allowed to stop and start market supply at their convenience. But, unlike the bakers of classical Rome, they had the right to renounce their profession, or at least their market service, definitively. The police considered such a decision irrevocable. In order to escape penalties the bakers had to give notice to the commissaire of the market at least two weeks in advance of his expected departure in order to allow time to find a replacement and maintain "a perfect continuity of supply." The retiring baker was supposed to explain his motives to the commissaire. If the latter was not satisfied with the baker's story, or if he feared he could not find a replacement quickly enough to maintain the service, he could pressure the baker to stay on. Save in crisis periods, the police usually had no difficulty recruiting replacements. When Germain Prevost of Tillet informed Commissaire Divot that he could no longer fulfill his market obligations, Divot relieved him and assigned his space "immediately" to the son of a master baker.[46]

In principle bakers coveting a stall at the market petitioned the commissaire of the quarter in which the desired market was located. The commissaire made inquiries into the baker's reputation, his business practices, and his solvency. If he was satisfied that the baker would and could serve faithfully, and if he had a vacant space, he granted it by written ordinance and made the award contingent on his continued satisfaction with the baker's performance. On 17 May 1730, Courcy, the senior commissaire of the Halles district, named Nicolas Grinon to the stall numbered forty-three, situated between bakers Félix and Provots. On 8 January 1734, he assigned a space on the *carreau* to a master baker from the faubourg Saint-Lazare. Another commissaire allotted a spot in the Saint-Paul market to a faubourg Saint-Antoine baker.[47] Each certificate of allocation reminded the baker of his obligation to furnish the market twice a week and to sell directly to the public without any intermediary, fixed the size of the space (in terms of *panniers,* or basketsful, of bread), and forbade the baker to dispose of the spot without the commissaire's consent.

Despite an explicit prohibition, there was a commerce in market stalls. Pierre Châtelain included his space in the Augustins market in the sale of his business concern (*fonds*) in 1738. Many years later, for a space in the same marketplace, Jean-Pierre Longprez paid six hundred livres, a very high price. Widow Paillard of the faubourg Saint-Antoine sold her spot in the Halles for four hundred livres. She had the temerity to promise, in the notarial contract, "to guarantee [the buyer] from any trouble or difficulty whatsoever."[48]

Once they acquired market spaces, bakers considered them as property that they could transmit or dispose of as they saw fit. More common than outright sales were transfers through marriage. In honor of a double marriage uniting the Delamarre and Félix baker dynasties, each father vowed to transmit to his son a space in the Halles "that he could dispose of in the future as his own property." The daughter of a faubourg Saint-Antoine baker brought her husband a spot in the Saint-Paul market evaluated at two hundred livres, while the daughter of another Saint-Antoine baker received as part of her dowry a stall in the Marché-Neuf appraised at three hundred livres.[49]

The police seem to have tolerated certain transfers, especially when they took place within families or when they involved bakers high in their esteem. Prudent bakers took the trouble to ask permission for a transfer—requests that were rarely denied. Widow Jombert, a mistress baker, obtained Courcy's approval to give a space in the Halles (evaluated at only one hundred livres) to her daughter, who was about to marry a master baker. Seller Claude Jacquet and buyer Petition, the first a forain and the second a master, each solicited permission from the Saint-Germain market commissaire before having their transfer notarized. Doubtless to protect himself—he must have gotten cold feet—Bonnet Chantry, a Saint-Antoine baker, sought Courcy's permission "to keep" a spot at the Halles that he had acquired by purchase seven months before.[50]

When they felt that bakers were taking advantage of their indulgence, the police could respond severely. Without informing Commissaire Divot, baker Jean Legrand conveyed the two spaces he had at the Cimetière Saint-Jean to his daughter, a woman of dubious reputation. She sold one of the spots to a master baker for 120 livres. The lieutenant general of police annulled the transfer (calling it "abusive and fraudulent"), confiscated the spaces, and condemned the father and daughter together to a three-hundred-livre fine.[51]

Periodically the police required all bakers in the bread market to justify their presence by submitting for verification the ordinances by which they were awarded their stalls. These threatened purges sent a tremor of fear through the marketplaces. During the seven-day grace period by the end of which the bakers were required to produce their titles, scores of bakers rushed in expiation to confess their irregular status to the commissaires and to implore them for legitimation.[52]

A number of bakers, like Legrand, appeared to have more than one market stall. Ironically, the practice seems to have been more common in the late seventeenth century, when there were more bakers (but, it is true, also more markets), than in the eighteenth century. Over a dozen bakers at the Marché-Neuf, for example, had two spaces, and several had three. Widow baker Lapareillé had several spots at the Halles in the 1740s. Beset with difficulties, a woman forain named Lacouture asked the police in 1772 for permission to abandon one of two stalls that she managed to maintain in two different markets.[53]

Extremely crowded conditions impelled the bakers to seek more space. An ordinary space at the Halles (usually styled to accommodate two *panniers*) was about ten feet across and not as deep. At Saint-Germain an uncovered spot was six feet by four feet. For want of room bakers could not bring all the bread they were capable of producing, or they were obliged to sell from their wagons, which obstructed the road. Bakers fought acrimoniously to protect every foot of space, sometimes among themselves but more frequently against the outside enemy: any other merchant or artisan selling anything else at the same market.[54]

Space was enormously valuable at the marketplace. Theoretically, it was all royal space, unless the king specifically conceded it to one of his subjects. Yet at some markets the "bourgeois" who inhabited the houses in front of which the stalls were situated tried to extort fees from bakers (and others) on the pretext of allowing them to encroach on *their* space. It is possible that these same bourgeois, in addition to claiming suzerainty over official spots to which the police had assigned bakers, also created new spaces that they rented out illicitly to bakers who thus joined the market clandestinely. In some of the bread markets the bakers had to pay fees authorized by the police. In 1726 at Saint-Germain bakers in covered stalls paid twenty sous a week and those in un-covered stalls twelve sous for each *pannier*. The farmers in the Daguesseau market, where all the spots were apparently the same size, had royal permission to collect six deniers from each baker for each market day. According to several irate bakers, however, the farmers abused their privileges by exacting far higher fees.[55]

To Market! To Market!

To reach the market in time to set up, Gonesse bakers started out before midnight. So congested were the arteries leading to the belly of Paris that a faubourg Saint-Antoine baker regularly left for the Halles at 2:00 A.M., though it was only a few miles distant. Many of the bakers reached their stalls as early as half past four in the morning. Few arrived later than 6:00 A.M. The streets were crowded as one approached the center, not only with bakers but with thousands of other itinerant merchants, fatigued and impatient to reach their destination. Tempers flared easily over rights of way, road priorities, and parking privileges. Claude Debis, a Saint-Antoine baker, was whipped and knocked down from his wagon by three butcher boys who wanted to pass him. (By and large, however, the road home, when cash replaced loaves, seemed more perilous than the road to market: Nicolas Félix was one of numerous forains during the century who heard ambushing malefactors utter the chilling injunction "la bourse ou la vie," your money or your life).[56]

A surprisingly large number of market bakers, including the forains, did not own their own horses and wagons. They borrowed vehicles and animals or hired carters or drivers for the twice-weekly trips. Of 111 bakers in our inventory pool, many of whom had stalls in the market, 81 had no horses, six had one, and four others had two or more. When Françoise Josset's single horse was run over and killed by a three-horse wagon driven by a forain at the Halles, Josset feared that her commerce was ruined, for she could not afford to replace the animal.[57]

There was no less tension in the bread markets than on the roads. Each baker had an acute sense of the territorial imperative, which was a question not only of physical space but of commercial space as well. Commercial space was bounded by moral considerations rather than by cold economic calculations: notions of fidelity, precedence, prescriptive right, honesty, honor. Each baker had his habitual claims—for instance, claims on certain customers. If these customers were deterred or diverted from their usual supplier, the baker blamed a colleague. The empire of habit weighed heavily. There was no place for the unexpected. If Saturday's market did not go like Wednesday's, in the baker's mind it was because another baker had trespassed somehow on his rights. Each baker suspected the others of seeking advantage at his expense (in terms of reputation or of money). Surprisingly formalistic on one level, market relations remained intensely personal, one could even say passionate. In this moral climate, where physical and emotional enervation rendered the actors even more excitable than they ordinarily were, there was a great temptation to mete out justice on the spot as one perceived it, that is, to settle scores directly and sometimes violently.

Thus it is not astonishing to hear of two women forains scratching each other's eyes out at the Palais-Royal or two baker wives fighting at the Halles over clients, one of the wives stabbing a master baker who imprudently tried to make peace. Both affairs concerned "professional jealousy," which probably meant a conflict over customer loyalties. Accused by a fellow woman baker of stealing some of her bread, a Gonesse forain slapped her colleague and threatened to kill her. Dupille, from Gonesse, brutally beat up a faubourg Saint-Antoine baker at the Halles, in a quarrel over sales practices. Another forain assaulted a sixty-three-year-old bread deliverywoman, ostensibly because she knocked over some loaves in his stall. In fact, he was avenging himself against her employer, a rival baker with a stand nearby. A long-simmering quarrel between two Saint-Antoine bakers erupted in violence in the Marais market when the wife of one attacked the wife of another with a stick.[58]

Sometimes a prelude to violence, verbal rage served on other occasions to discharge aggression without physical confrontation. The bakers Fricourt, father and son forains, quarreled biliously over the son's desire to assume part of his father's business. They impugned each other's honor in the Halles in front of their clients. The father called the son a "rogue" and a "scoundrel" and his daughter-in-law a "bitch" and a "whore." They replied that he was "un vieux Jean Foutre, un voleur, un coquin, reste de Bicetre." The wife of master baker Cousin was distraught to see the space next to her stall at the Halles taken over by the Tournays, a newly married master-baker couple, perhaps because she had hoped to annex it to her own. As soon as Dame Tournay arrived, Dame Cousin began to insult and harass her. She threw her bread on the ground, called her "thief," "showed her the [cuckold's] horns," announced aloud that "Tournay got her spot only because she was laid and licked by the official in charge of giving out spaces." Dame Tournay complained not only that these calumnies damaged her business but that they also "caused discord in her household." Perhaps because the police judged these verbal flare-ups to be inevitable and even salutary, they rarely intervened. Only when there was a real risk of subsequent disorder in the market did Inspector Poussot make the offending baker "promise to be tranquil."[59]

Daily Baker Production

Concretely, what did it mean to say that a baker converted thirty-six setiers or one setier each day of market? What did it mean in terms of bread and in terms of effort? Given our uncertainty about the yield of wheat for flour into bread, we cannot translate setiers into loaves with real confidence. Basing calculations on the average weekly provision of each baker in 1733 and on a yield of 180 pounds per setier, we can imagine a stall displaying 1,530 pounds of bread distributed as follows: 26 twelve-pound loaves, 40 eight-pound loaves, 101 six-pound loaves (mostly round), and 73 four-pound loaves (mostly long).

As a result of baker turpitude, guild vigilance, and police diligence, we can obtain a less hypothetical and more vivid idea of the baker provision. In 1735 the jurés denounced Simon Guitton, a faubourg Saint-Antoine baker selling in the Maubert market, on the grounds that a large part of his bread was "mollet," of soft rather than firm dough, and thus not permitted in the markets. The police confiscated all his bread. Since the seizure occurred at 5:00 A.M., it is safe to imagine that he had not yet had time to sell many loaves. Guitton brought 639 loaves (only thirty-nine of which were over four pounds and none of which was over eight pounds) weighing 2,026 pounds.[60] On the basis of an estimated yield of 160 pounds per setier (a figure lower than it would be for most market bakers because so much of the bread was mollet and thus finer), Guitton converted twelve and one-half setiers. In fact, this baker appears in the 1733 police register, where he is listed as supplying two and one-half muids a week or about fifteen setiers per market, a figure close enough to our calculations to allow us to conclude that Guitton was living up to provisioning expectations quantitatively even as he was cheating qualitatively.

Two years later, Nicolas Barron, a forain from Ruelle selling at the Carouzel, also suffered an early-morning confiscation. He brought 350 breads weighing 2,342 pounds, the majority in loaves of six, eight, and twelve pounds. Basing our figures on a yield coefficient of 180 to 185 pounds, we can calculate that Barron used twelve and one-half to thirteen setiers of wheat to make his bread.[61] He, too, appears in the 1733 register. Measured against the one setier he was listed as bringing in on an average market day, Barron's performance in December 1737 was nothing short of spectacular. Another day that same year a Versailles baker named Michel Daru placed 2,474 pounds of bread on sale. His stock consisted of 52 twelve-pound loaves, 92 eight-pound loaves (marked as being ten-pound loaves), 54 six-pound round loaves and 444 long breads of the same weight, 70 five-pound round loaves (which weighed on the average less than four pounds each), and 44 long loaves of four pounds.[62]

Bakers usually reckoned their provision not in terms of pounds or setiers/muids but in *fournées,* or ovenfuls (or batches). This convention was not inspired by a concern either for the standardization of weight and measures or for market data. Bakers had different types and sizes of ovens, they used different combustibles, and they baked different kinds and shapes of bread. (And certain bakers, such as Moncouteau on the rue Mouffetard, operated two ovens simultaneously.) An average fournée, then, is a highly parlous concept. Nevertheless, contemporaries believed that the range of variation was sufficiently limited to allow for useful comparisons. Pierre Dupont de Nemours claimed that bakers—he must have been thinking primarily of shop bakers—usually baked 110 four-pound loaves in a fournée and that these 440 pounds of bread were the product of two setiers of wheat.[63]

Dupont surely exaggerated on both scores. At most a handful of bakers obtained this enormous yield in 1764 when Dupont wrote. As for the fournée, Malouin's estimate of between 150 and 300 pounds is far more convincing. A fournée of *gros* bread was as much as one-third larger than one of *petit,* though the shape also influenced output. Widow Herbert's experience was probably typical of many bakers: she rarely obtained more than 200 pounds from a fournée. Her colleague Simon Coispeau reported a fournée of four-pounders averaging 170 pounds.[64]

How many fournées did a baker normally cook a day? On average Malouin proposed four or five a day (presumably for Paris shop bakers cooking every day with no allowance for market-day supplements). If one assumes a production of 210 pounds per fournée and a relatively high bread yield of 205 pounds, Malouin's suggestion does not appear unreasonable for the 1760s. The closest one comes to a survey concerns 238 masters who solicited loans or grants from the state at the end of 1789 in the midst of a crisis that had threatened their ability to operate. Thus we are dealing with bakers who may be maimed in varying degrees in their productive capacity, or who may have exaggerated their cadence since grants were supposed to correspond to the number of ovenfuls habitually baked. Outliers symmetrically clustered at the extremes: two bakers who produced one and two fournées a day respectively were matched by two others accounting for twelve and fourteen fournées. The majority fell into the broad range of plausibility: 15 bakers at three fournées, 47 at four, 59 at five, 54 at six, 20 at seven, 15 at eight, and 5 each at nine and ten.[65]

François, the baker who lost his head to infuriated revolutionaries in 1789, cooked ten fournées a day, but his was a very large enterprise specializing in *petit* bread. Another baker during the Revolution, Buérin of the rue Mouffetard, baked ten to twelve a day "in order to satisfy the brisk sales done at his shop." More representative is Frambois, a faubourg du Temple master who baked six fournées a day earlier in the century, or the bakers in the section of the Arsenal in 1793, who averaged about five a day. In light of a steadily declining marginal cost of production, Parmentier considered it inefficient and uneconomical for a baker to do fewer than five or six fournées a day. Yet it is well to keep in mind the contention of Gibert, a substantial Saint-Antoine baker, who said that materially he could barely handle sixty setiers a week—anywhere from six to nine fournées a day—in a shop in which he was assisted by three journeymen, a servant, and his wife.[66]

Market bakers of course produced more bread on the eve of market days than bakers usually cooked. Malouin related that forains normally did nine or ten fournées before Wednesdays and Saturdays. Simon Guitton appears to have baked eight to nine fournées for market and Michel Daru and Nicolas Barron nine to ten. Fiacre Lescuyer needed seven fournées for his market provision. Malouin claims that a few forains did up to "fifty-three fournées in two days and two nights before each market day." This would have required a weekly provision of around ten muids and a rate of more than a fournée an hour, a pace inconceivable in a single-oven bakeroom. If Malouin's giants are hard to find in the markets, the runts pullulate. In the forties widow Thevenet from Saint-Antoine brought four fournées, and in the fifties Charles Plé of Bonneuil also marketed four.[67]

Porteuses: *Bread Deliverywomen*

Immortalized in fiction, celebrated in a famous statue that stood in the center of the capital, the *porteuse*, or bread carrier/deliverer, was a familiar sight in the streets of Paris from Saint-Louis until the ascension of de Gaulle. Her primary function was to distribute bread to her baker's customers. Given the fact that many bakers also operated in the marketplace and thus had a widely dispersed clientele, the porteuse often had to walk long distances. (Disposing of a cart and horse to accomplish her rounds, porteuse Marie-Magdelaine Guenard was a rare case.) The typical porteuse needed strength as well as endurance, for the basket she carried on her back was freighted with up to one hundred pounds of bread, with which she often had to mount four or five floors. (One porteuse, barely able to control her heavy load, knocked over several bottles of costly syrup in a grocery store, whose owner confiscated her basket and loaves as security for restitution.)[68] In part for this reason, many bakers assigned their journeymen to this task. Others calculated, however, that it was more economical to send an unskilled woman who commanded a lower wage and whose presence was never missed in the baking room.[69]

Frequently porteuses were full-time servant girls who labored in the household as well as the business of their baker-masters. Numerous bakers dispatched their daughters to deliver, secure in the knowledge that they had thus sharply reduced the chances of being cheated. (If "portage" was not yet the "organized robbery" that it became in the nineteenth century, it was nevertheless common enough for carriers to siphon off loaves through various accounting and rhetorical tricks.) Finally, there were professional porteuses who lived outside the baker's household and were difficult to distinguish socially and economically from the mass of low-wage dayworkers (*journalières* and so-called *gagne-denières*) who worked at odd manual jobs. Some, such as Françoise Le Sueur, the wife of a journeyman ribbon maker, or Marie-Jeanne Dulars, the widow of a journeyman baker, were freelancers who worked for several different bakers at once. Others, like sixty-year-old Dame Regnault, the wife of a gagne-denier, and Marie-Magdelaine Guenard, the wife of a journeyman cabinetmaker, labored full-time for one master. Catherine Lescuyer was too old to do all the work, so her boss, baker Vassou, employed several porteuses, as did baker Vaugeois. Marie-Louise Picard came to the job more or less accidentally, through necessity and family connection. When her husband, a journeyman baker, was drafted into the army, she replaced him as delivery person for his master.[70]

In addition to delivering bread, certain porteuses had other functions. Those who enjoyed the confidence of their bakers collected cash payments. Baker Leprince entrusted his porteuse with various coins in order to enable her to make change easily. Marie-Magdelaine Guenard collected at the same time she delivered, while the porteuse of a Saint-Antoine baker went back to solicit

Statue of the bread *porteuse,* destroyed by the Nazis (in the square of the Saint-Jacques Tower). *Source: Photo Collection Roger-Viollet.*

payment at the end of the day. Vaugeois's porteuses were frequently unsuccessful in obtaining payment, and thus became responsible for keeping fairly complicated accounts for credit sales. Baker Garmont's porteuse reckoned with him periodically rather than daily, accumulated a large debt to him, and then "delivered to him a bankruptcy of seventy-three livres." Several porteuses helped to take the bread to market, set up the stalls, and sell it in behalf of the baker or alongside his wife. Besides delivering bread, Anne Charlotte Robert served as grain- and flour-buying agent for her master—a task requiring considerable technical and commercial skill.[71]

Given the keen competition for customers, it was logical that bakers counted on their porteuses to develop friendly relations with them. There was of course a risk that an unscrupulous client might try to take advantage of the congenial disposition of the porteuse. Geneviève was the porteuse (and one of several *domestiques*) of Saint-Antoine baker Jean-Baptiste Hornet. On market days, from Place Maubert, she delivered several loaves to the house of Maître Dobet, a lawyer attached to the Paris Parlement living in the Ile de la Cité. The problem began when Dobet noticed her one day and began subsequently to take delivery of the bread himself in his study in order to be in a position "to jest with her." One day he asked her "if the cross that she wore around her neck was of real gold and used the pretext of examining it in order to caress her breasts." Another time he offered her money "to buy trinkets" and then attempted to kiss her, an advance she found difficult to resist because she was weighed down by her bread basket. She related the story to her master, who did not seem surprised or alarmed. Nor were her fellow servants, who chided her over her diffidence: "she should take his money [for a kiss], and in case he wants other things beyond a kiss, she should scream and call for help."

On yet another occasion Dobet promised the porteuse a louis d'or if she would "se laisser baiser." She pocketed the money and let him kiss her. Then he "tried to put his hand under her skirt and raise it up." She resisted, "telling him that he was asking too much; he wanted a baiser and she accorded it to him." But "that was not what he meant by a baiser," retorted Dobet; he could not be expected "to pay a louis d'or for so little in return." Then he "tried to take her by force." She screamed, as she had been coached. He panicked and begged her to quiet down, before his servants and family heard. Geneviève returned home and again confided in her master. This time Hornet drew the line. The "honor" of his servant had been threatened. No customer was worth that. The baker called on Dobet in order to rebuke him. The lawyer insulted Hornet, denounced his porteuse as "a mere whore," and tried to throw him out of his house.[72]

The porteuse was vulnerable in her honor (and a fortiori in her sexuality) merely because she was a domestic servant, or viewed as such. Thus she was presumed to lack education, upbringing, and moral sense. If she was not the customer's whore, she was her master's whore. "Go knock up your porteuse" seems to have been a standard insult/provocation directed against bakers. ("I'd rather do it to your wife" appears to be the conventional riposte.) Numerous bakers (and their wives) rewarded faithful porteuses by taking the trouble (and the business risks) of defending their honor during their lifetime and bequeathing them gifts and money in their wills "in recognition of good service." Other bakers, such as Patrimoux, a "mauvais sujet" who fathered the offspring of three of his servants, or François Houy, "a debauched subject" accused of sexually abusing his own daughter, treated their porteuses with brutality.[73]

Raw physical violence, however, was not the only form of tyranny exercised by the bakers. Marie Lalouette, the underage daughter of a water carrier, went to work for the baker Jean-Baptiste Dubuisson in the faubourg Saint-Antoine, serving both in the oven room as a journeywoman and in town as a porteuse. Though she did not lodge strictly speaking with the baker, many long nights tending the dough and preparing to bake kept her in his house. As she labored under Dubuisson's gaze, he often flattered her on her professional virtuosity, complained of the uselessness of his old and infirm wife, and expressed a desire to marry her if his wife should die. All the neighbors described Lalouette as a quiet and industrious worker who was both "très sage et très simple." It is unclear what effect these mellifluous words of seduction had on her.

The witnesses related that Dubuisson pursued her tirelessly in his ardent "badinage," caressing her and trying to lift her skirts on the staircase. She claims that she protested vigorously against his efforts to "get her to grant him her ultimate favors," warning him that she would leave his employ and extracting from him a pledge that he would "leave her alone." Dubuisson did not keep his word, nor did Lalouette fulfill her threat, for on two different occasions, surprising her at night, the master "overwhelmed the said girl [and] ... enjoyed her carnal company, as a result of which she became pregnant." (On one of these occasions the baker took her while she was "asleep in a spot in the house called la Couche," the place where one gave life to the dough that became bread and the name of the state of a woman giving birth to a child.) While she ceased working in a journeyman's role, she continued to carry Dubuisson's bread, apparently as a result of an understanding that he would "pay the delivery costs and support the baby." Well into her ninth month, however, Dubuisson reneged, denying that he had had sexual relations with her and refusing to make financial amends. At that juncture, in desperation, she filed charges with the police.[74]

Illicit Bakers

Among the persons who supplied Parisians with bread there were a certain number who did not have the right to do so. A significant proportion of these illicit bakers operated under the protection and sometimes the direct guidance of established bakers. Following the classic script, a master, sometimes in or near retirement or sadly lacking in business skill or perhaps quite successful and entrepreneurial in disposition, "lent" his name, usually in return for a rental price or a variable share of the profit. This practice found favor in many occupations, and it was repeatedly prohibited in both guild statutes and police regulations and sentences. A woman named Thérèse Billot "conducted the commerce of the bakery under the name of the said [master Vincent] Barbacois," who apparently lacked the capital to get his own operation off the ground. Master Guillaume Grand and his son, also a baker, rented out a house to a couple; attached to their notarial lease was Grand's certificate of mastership, the quasi-official gesture by which he granted them his "coverage" to exercise the baking trade by lending his name. From the renters' point of view, the price of their legitimacy was included in the price of the lease. While he conducted his business in one shop, master Jean Theveneau rented out another to various bakers who counted on his protection; in between leases, he used the second shop as an outlet to sell his own bread. Numerous bakers lacking official status ("without quality") justified their operations by affirming "that they held their right" from a member of the guild.[75]

"No master can associate himself with another person who is not a master," enjoined the guild statutes. Yet the notion of a "société," or business association, appealed to certain bakers in part because it seemed a more dignified or less reprehensible solution than vulgar name lending. It made great good sense to Marie-Anne Fleury, the widow successively of two master bakers, who dowered her daughter with a 50 percent share of her bakery business. The arrangement amounted to an annuity for the mother, who was in any event ready to retire. In this instance the association was all the more interesting and illicit because the son-in-law was a flour merchant, a trade juridically incompatible with the profession of baking, which precluded speculative dealing. The "société" between master Jacques Pasquier and an untitled baker named Galland flourished until the former died, gravely jeopardizing the latter's chances of continuing the business, perhaps even of reclaiming his share of their common possessions vis-à-vis a united front of Pasquier's three nephews, all bakers in their own right. Forains adopted the "société" device to their needs by using it as a vehicle to organize the shared furnishing of market spots.[76]

Other masters found ways to turn a profit while incidentally rendering service to fellow masters. In this scenario, despite the formal corporate injunction requiring each master to maintain his own oven, a baker "lent" not his name but his baking facilities. Albeit master Gaspard Thevenot had not kept a shop for four years, he regularly supplied his stall at the Halles market by baking one or two batches a week in Rousseau's shop in the faubourg Saint-Denis. Highly erratic in his commerce because of habits of dissipation, master Blanvillain amortized the costs of his bakery by sharing it with a paying colleague.[77]

A considerable part of the contingent of illicit bakers consisted of persons who went at it pretty much on their own, without any borrowed or rented cover or formal protection. These included the proliferating "people without title or quality," quite frequently women, pursued in the streets by the jurés for selling a variety of breads, allegedly "badly made." The women often worked for forains without licenses. The jurés seized their merchandise and when they could, their tools and equipment as well. The police court usually confirmed the confiscation of the bread but restituted capital goods such as wagons and horses. The jurés made an early-morning descent on the rooms of Antoine Pigal, a journeyman joiner exercising the profession of "baker," according to their intelligence. Pigal's wife insisted that her husband really was a joiner who was off on a job. The fact that her hands were encrusted with dough suggested, however, that something illicit was going on, a suspicion reinforced by the wafting odor of a near-ready fournée. The jurés discovered a back room with an oven, and a smaller one with a kneading trough and other utensils. They seized the bread and distributed it to charity, and placed the tools and flour stock under seals.[78]

In their rounds, the jurés also found a carter named Leduc who baked an ovenful now and again and delivered the loaves himself. Although he lived in the same building as a widowed mistress baker whose husband also conducted an illicit bakery business, in an ironic inversion of familiar city-country relations, Leduc announced that he functioned "under the protection of Michel Leduc of Gonesse," almost certainly a relative. Marie-Angélique Allingre was a classical baker without credentials, a kind of internal forain, who operated without protection, dependent on her own ingenuity and perhaps indirectly assisted by her husband, a wine merchant who sent customers her way. The police were much more concerned about the many starch makers uncovered with bakeries on their premises who, it was feared, might traffic in cereals, divert grain from the bread supply, and/or produce poor quality loaves. Unlike the vast majority of illicit bakers, the starch makers had their ovens destroyed on judicial order.[79]

Madame Martinique, the widow of a plasterer and now married to a carpenter, conducted first a modest street traffic in small loaves and then advanced to the supply of a stall at the Halles market. She arranged to have her bread baked here and there as a favor by indulgent bakers. Smitten by ambition, she transformed her status from a transient and opportunistic part-timer into a fixed and protected competitor by renting a shop in the rue du faubourg Montmartre under the name, and patronage, of a master baker. Had she behaved with discretion and not transgressed the unwritten rules of the trade (after having violated the written ones), she might have made the transition uneventfully. But she irritated Françoise Lapareillé, the widow of a master, who ran a shop directly across the street and had in the past baked Martinique's bread for her, by openly soliciting business from merchants and others, among them her established clients. "Bakers are not like used-clothes dealers; they must not hail and summon each other's customers," Lapareillé remonstrated. Martinique responded violently, denouncing her rival as a drunkard and the latter's daughter as a thief, and celebrating her inalienable right to "call on everyone to come get Good Bread, and not the stuff of the complainant, which is no more than a drug." Lapareillé refused to stand by while a person "who has no title destroys my business."[80]

Widespread regrating operations generated a sort of secondary market system of improvised but quasi-permanent stalls as well as itinerant sales. Unlicensed sellers regularly set up stands near the abbey of the faubourg Saint-Antoine, a symbol of trade freedom. Widow Grisot, a former nurse's aide, sold bread there that she acquired from a baker on the rue de Reuilly, in competition with other "resellers," mostly women, who either bought their bread outright from bakers (Cretté of the rue de Montreuil supplied two such dealers) or worked on a commission basis for bakers who thus acquired another outlet not directly associated with their shops or with their porteuses (affording them a certain deniability when the police or the guild struck). Operating on her own, Marie Catherine Begonnan specialized in the illegal distribution of little loaves. Simon Demartin did the same thing, but as an agent for a baker who spared him the capital expense and the risks.[81]

Baker Pluralism: Parallel Professions

Baking was an onerous and demanding profession. As a rule, it was virtually impossible to combine it with another job—unless one followed the pattern of such part-time forain bakers as Nicolas Lamarre, who was a *laboureur* at Dugny, northeast of Paris. A handful of ostensibly full-time bakers, however, engaged in complementary or parallel professional activities at the same time that they supplied bread. They managed this feat because they were extremely well-off and thus able to delegate responsibilities, or on the contrary because they were rather marginal in the baking business and therefore free to undertake other tasks. In both cases, they had to be reasonably good at cutting corners.

Despite the formal interdiction to conjoin the two professions, baker Pierre Félix seems to have been a more or less overt grain trader. Roubaix operated as a small-time grain trader (*marchand bladier*), but he was also a forain baker, which made him much more likely to escape detection than a city colleague who bought grain to resell it. Living in the grain-trading enclave of the rue de la Mortellerie next to the port, master baker Louis Chevallier brandished the title not of grain merchant but of *marchand grenier*, which meant that he specialized in the small cereals that rarely entered the bread supply.[82]

Another profession that a well-capitalized or merely entrepreneurial baker could undertake simultaneously was that of carter or deliveryman. A faubourien named Guay amortized the cost of his investment in cart and horses by renting out his services and equipment to pick up and deliver the grain and flour of confreres. Another faubourien, Freneau, a baker and carter, ran a similar operation. His neighbor, Antoine Secret, sported the same double title but apparently did not own his own driving stock. A master baker living in the city center, André Leroux, may not have been able to turn stones into loaves, but he did practice the occupations of stone merchant and deliveryman even as he baked bread.[83]

In the Eucharistic spirit, Marie-Angélique Allingre, who styled herself a "boulangère" though she operated in corporate territory in the faubourg Montmartre, shared a wine business with her husband. A forain at Monceaux was also a wine merchant, while another at Montreuil ("marchand boulanger") directed a large catering business ("marchand traiteur"). A forain at Sceaux ran a tavern. Selling food and/or beverage along with bread seemed logical enough. Less obvious was the connection between the trades of baker and *mercier* (seller of all sorts of wares), simultaneously exercised by forains Jean-Baptiste Lacour and Toussaint Bontemps (the bakery as general store? the general store as bakery?).[84]

Certain bakers held venal offices, the fruit of either inheritance or investment in search of prestige and assured annuity income. Most of them did not require actual work or else permitted the owner to farm out the work, especially when it was menial, to paid hands. A few bakers seem to have served as juré measurers or porters in the seventeenth century, offices theoretically incompatible with their status as players in the provisioning game and that gave them an inside track on grain and flour acquisition as well as an influence on price making. Toward the middle of the next century master bakers Nicolas Terai and Georges Baudouin each served as *juré porteur de sel,* a venal office in the salt administration (the latter having paid 9,700 livres for his post). Similarly engendered by royal fiscal needs, the post of *officier controlleur vendeur de volailles,* a task of poultry inspection, seduced Marin Laurence, a former guild juré. Also an erstwhile juré, master François Brivot, the son of a laboureur from Verdun, purchased the post of *officier planchéeur*, dealing with building materials. Bread and wine proved an irresistible combination to two other masters, respectively *juré jaugeur de vins* (1717) and *officier inspecteur sur les vins et* autres liqueurs" (1754), posts pertaining to the inspection of wine stocks. The master baker and bourgeois de Paris Claude Lapareillé capped his upward trajectory by becoming in the mid-thirties one of the bumptiously named *conseillers du Roy, inspecteurs, controlleurs et visiteurs sur les vins, eaux de vie, liqueurs et autres boissons*— another post dealing with alcoholic beverages—for the bagatelle of 35,000 livres. Three fellow guildsmen held the titles *officier porteur de charbon, messager juré de l'Université*, and *officier de la panneterie du Roy*, dealing respectively with coal, higher education administration, and the inspection of the bakery. Masters Simeon Delamare and Nicolas Jacob sought to enhance their status by becoming *dixainiers*, more or less honorary city officials. But their ambitions were doubtless undercut by the price of the post: between four hundred and five hundred livres. Paramilitary municipal functions allured a number of other masters, who served as archers in the *guet à cheval*, or mounted watch (in 1751 this charge commanded 1,700 livres), and as *gardes-archers de l'Hôtel de Ville*, members of an elite brigade of municipal archers.[85]

Heretofore, historians have had no clear idea where, how, and in what conditions Parisians acquired their daily (and biweekly) bread. Nor have they possessed a graphic sense of the organization of the bakery and the behavior of the bakers. This chapter has cast into relief the central institutional status of the bread markets and the bakeshops. In terms of frequentation and fidelity, no other urban institutions could compete with them. One cannot infer from the published ordinances and statutes the way these institutions functioned. Though hedged by a honeycomb of rules and monitored by multiple authorities, they afforded the bakers (and the consumers) considerable latitude for solving problems.

Interacting with extraordinary frequency, bakers and buyers constituted the consumer culture of every day. In many instances, they traded more than bread against money (or the promise of its future remittance). They exchanged expectations and commitments, working out a relation of sociability (implying certain kinds of accountability) as well as of commerce. Each vigorously pursuing his or her own interest, the baker and the customer worked to transform a moment's flirtation into a protracted romance. Once they had achieved a certain stability, exchange relations took on a quasi-theatrical aspect, marked by coded, ritualized gestures on both sides. Commercial rupture implied a social breakdown that went beyond a dispute over commodities, even those of first necessity. A core of habitual consumers came to feel that they had claims on their bakers just as many bakers came to feel that they had claims on consumer loyalty. In the most articulated relationships, that meant there were certain things that each party could not readily refuse the other. Well before it was eaten, bread served as an agent of communion (and symmetrically of disunion). Alongside the church and the tavern, the bakeshop or stall exercised community-making (and breaking) functions.

The guild was not the utterly dominant force in provisioning depicted by the ideologically driven analyses of the physiocrats. Far from exercising a monopoly, masters did not even account for the bulk of the capital's daily supply. Yet the masters were strategically situated, politically and socially as well as commercially, and they exerted considerable influence in what was nevertheless a decidedly pluralistic landscape. Apart from the allocation of social and political capital that differentiated them, bakers of all sorts forged a professional identity in common. As makers and purveyors of bread, they competed against one another, sometimes quite aggressively, but they learned to recognize their shared interests beyond their different categorical prerogatives and individual strengths and weaknesses. Masters rubbed shoulders with faubouriens and forains in the markets. Many faubouriens operated shops just like masters. All of them devised home-delivery schemes. In the interstices, various genres of authorized and unlicensed bakers did business in the streets and in improvised marketplaces. Consumers thus could encounter bakers in a wide array of sites and circumstances. Different protocols governed each type of encounter, resulting in buying and selling practices and baker-customer relations peculiar to each arena. Like the institutions that framed them, those practices and relations remained remarkably durable across the century. Save for significant personnel changes and a certain rationalization of operations, especially in the forain contingent, the markets and shops of 1789 resembled strikingly those of 1720.

Chapter 4

The Forain
World

Forain, according to Delamare, referred to persons who brought their goods "from the outside" to be placed on sale in the capital. In the Parisian bakery, forain commonly designated country bakers who purchased their grain or flour locally and transformed it into bread (*usually* of three pounds and over of a firm dough) for sale at the Wednesday and Saturday markets. This definition corresponds to the category of bakers of the second class in the Delalande registers, the purest sort of forains. Yet there were also forain bakers in the first class, if we take as criterion geographical location rather than grain and flour buying. These bakers lived in the suburbs or the hinterland, but they purchased the bulk of their supplies in the Paris market, so that technically speaking they did not bring "from the outside" a new product but instead transformed merchandise that had already reached the capital in its raw form.[1]

If instead of rural location and supply pattern one uses as a litmus exclusion from the guild, then the bakers of the faubourg Saint-Antoine can also claim the title of forain.[2] Though the Saint-Antoine bakers crossed paths with the masters every day in the markets of Paris and the *plat pays,* and though many of them did business from shops as well as from the bread markets, the guild assimilated them to the forains for most purposes.[3] The police perceived the Saint-Antoine bakers as androgynous creatures: in their role as market suppliers the police regarded them as forains, but as shop bakers the police viewed them in very much the same way they considered the masters. Partly in deference to the special juridical (and thus socioprofessional) status of the faubourg, the police often were inclined to deal with them as a class unto themselves.

The faubourg bakers saw themselves as city types rather than as itinerant rural artisans. They frequently called themselves "merchant bakers," a juridically empty but socially gratifying title that they appropriated as a sort of riposte to the bakers who brandished their masterships. Some Saint-Antoine bakers, like their guild counterparts, called themselves bourgeois de Paris, an elusive and protean title that sometimes connoted social pretension and/

or economic well-being. Second- and third-generation Saint-Antoine bakers often sought to become masters (even as certain country forains moved into the faubourg Saint-Antoine)—less out of strategic necessity than out of gourmandise. Still, while they sometimes treated the country bakers with the same disdain with which the masters treated them, they repeatedly joined hands with forains to combat what they labeled guild oppression. The faubourien capacity to mobilize large numbers of bakers in order to examine matters of common interest, to join in various actions against the guild, and to negotiate with authorities suggests that they enjoyed a quasi-corporate existence.[4]

As we have seen, the three categories of forains supplied well over half of the capital's total bread needs. The police esteemed that their contribution was invaluable. Delamare at the beginning of the eighteenth century and Des Essarts at the end insisted on the importance of treating them "favorably" in order to "attract them" to furnish the city. In fact, the police were so confident that the forains would continue supplying that they did relatively little, in a positive way, to encourage their fidelity. The police often punished them severely when they failed to fulfill their responsibilities, and the authorities afforded them little immunity from corporate harassment.[5]

The forain's best friends appear to have been their seigneurs, who had a fiscal interest in their well-being. Nicolai, president of the Chambre des Comptes and lord of Goussainville, one of the main forain centers, vehemently protested against police measures that limited the freedom of his client-subjects to bake what they wished and sell it as they saw fit. Among the forains' other influential protectors were the cardinal d'Estrées and Jean-Baptiste Machault d'Arnouville, who served first as controller general and later as keeper of the seals, both seigneurs of the fief of Gonesse. The seigneurs extracted dues from the bakers as landowners and renters and as market users, but they very rarely collected oven fees (*banalités*). Only a handful of forains did not own or utilize independent baking facilities. A political incentive also prompted the seigneurs to shelter the forains. They did not want to concede to the royal government the dangerously elastic right to intervene in traditional relations of contractual subordination or lord-vassal subjection in the name of the so-called public interest.[6]

The only time the Paris police rushed to the aid of the forains was when they judged that forain supply was in imminent jeopardy. In 1709 and again in 1775, during outbreaks of popular violence, they sent troops to escort the bakers from their homes to Paris. When nine bakers from the Versailles area told Commissaire Delavergée in October 1725 that they could no longer supply the capital because the Versailles police threatened them with huge fines if they failed to furnish their home markets on Wednesday and Saturday, the Paris authorities intervened at the highest levels to neutralize that order and to keep the bakers in the Parisian sphere of influence. When individual bakers abandoned their service peremptorily, the police punished them sternly. But when a large and sudden attrition occurred, suggesting demoralization or economic difficulty, the authorities tried "to excite" the bakers to return with promises of short-term financial aid or other compensations.[7]

As in the Paris bakery, there was a tendency among the forains toward the constitution of family dynasties and family alliances. These families placed many of their sons in the business, married their daughters to well-established sons of comparable baker families, and sometimes extended their influence through investments in arable land and in mills, thereby achieving a primitive sort of vertical integration. We will have more to say about these familial phenomena when we discuss baker marriages and wealth. For the moment let us note that the Destors family was implanted in Gonesse (two units), in Bonneuil (one), and in the faubourg Saint-Antoine (two); the Félix family in Gonesse (two), Bonneuil (two), Tillet (two), and Saint-Antoine (two); the Destuvigny family in Gonesse (two), Bonneuil (one), Tillet (one), and Dugny (one). Three generations of the Bethmont family were active in baking at Gonesse at midcentury. The Beton family dominated the baking industry at Chaillot, as did the Vavasseurs at Issy. Members of baker families sometimes held important public positions. A member of a large Goussainville baker family, Nicolas Lapeslier, was judge in the local prevotal court and also had the title of royal counselor. Françoise Delamare, the fiscal procurator at Arnouville, was the scion of a prominent laboureur-baker family.[8]

Forains in the faubourg Saint-Antoine and in the immediate suburbs were usually full-time bakers (though they may have kept some animals and raised a garden). More often than not country forains had more than one occupation. Many of them were active cultivators. Baker Joly was the renter-manager (*fermier*) of a large farm near Villars-le-Bel. Many of the bakers at Gonesse were simultaneously laboureurs, and in some of these cases baking was only the auxiliary profession. Nicolas Raflon, a baker at Saint-Germain-en-Laye, served as a carter for much of the week. Baker Charles Houlet was also a wood and coal merchant, and a grain, oats, and hay dealer. Other forains owned or worked vineyards, served as millers, operated taverns, and manufactured beer and starch.[9]

In 1740 sixteen forains at Bonneuil, many of whom owned arable and other property, paid an average of 76.4 livres in taille, the direct royal tax that fell most heavily on peasants (median = 66.5). Twelve forains, not all of them carryovers, paid an average of 113.75 eight years later, almost double the mean amount paid by all inhabitants (n = 138). In 1770 only six bakers remained on the Bonneuil list, paying an average of 92 livres within an extensive range from 9 to 171 livres. A decade later they numbered three, taxed respectively at 12, 34, and 71 livres. The Riberettes, who remained on the rolls until 1770, were enumerated as bakers, fermiers, and laboureurs. One of their neighbors operated as miller and baker and held some land. Among the four forains on the Dugny list in 1748, two Lamares, characterized as bakers and laboureurs, paid 358 and 365 livres. The bakers at nearby Goussainville were generally much less affluent: in 1765 three paid less than 12 livres in taille while the fourth, a laboureur, paid 137 (and 68 livres in capitation). The sole baker on the roll for 1765 at Stains, also a wine merchant, was assessed at 75 livres for the taille and 36 for the capitation.[10]

The country bakers did not constitute a stable population. Of the 208 forains who supplied the official bread markets in 1727, only 114 were still furnishing bread in 1733 as bakers of the second class. Some of them suffered business failures, while many others left the bakery in favor of other occupations. A small number, however, merely shifted from the second to the first class. Throughout the century there seems to have been a slow movement of forain bakers from the countryside to the faubourg Saint-Antoine or to corporate Paris. Jean Ponol moved from Créteil to the faubourg Saint-Antoine, as did Gabriel Mouchy and Pierre Bethmont of Gonesse. Nicolas Fresnot of Gonesse, Jean Houdouard of Montreuil, Jean-Baptiste Bequet of Le Bourget, the Reverard brothers of Saint-Germain-en-Laye, and Nicolas Veron of Paliseau started as forains and later became Paris masters.[11]

In most cases the shift to Paris represented an abandonment of other occupations or a total commitment to the baking profession. Sometimes the move resulted from the division of a family estate, which constrained a son to choose between baking and farming. In other instances the forain married into a Paris family or responded to the invitation of friends or relatives to establish himself. It is likely that most forains regarded the movement toward Paris as a form of upward mobility.

Shrinkage of the Market Corps

There were many fewer market bakers in 1725 than in 1700, if Delamare's figures were correct, and still fewer in the 1760s than in 1725, though our data for the second half of the century are extraordinarily fragmentary. Table 4.1

Table 4.1 Market Bakers in Paris, 1700–59

Market	Delamare Ca. 1700	Police Registers 1727	1733	Poussot Ca. 1760
Boucheries	—	23	16	—
Palais-Royal	40	33	22	42
Place Maubert	159	125	88	53
Halles	446	285	247	220
Marché Neuf	89	60	43	53
Carouzel	—	34	23	6
Saint-Germain	147	122	100	114
Saint-Michel	36	14	14	—
Quai des Augustins	92	26	20	—
Cimetière Saint-Jean	158	117	96	95
Saint-Paul	—	59	51	—
Marais	46	29	27	—
TOTAL	1,213	927	747	583

shows how things changed over time in the twelve markets that the police considered the official bread outlets for the bulk of the century.

The column totals are misleading (save for 1727 and 1733) because there were in fact 1,524 bakers on Delamare's list (five of his fifteen markets are not represented here), and there were surely more than 583 in 1760, but we lack data for the others. To measure the decline in market bakers, it is more instructive to compare groups of markets for which our information is complete. The nine markets that in the early 1700s welcomed 1,213 bakers in 1727 received only 811 and in 1733 only 657.[12] Between 1700 and 1733, then, there is an erosion of almost 50 percent. Between 1733 and 1760, however, there seems to have been a relative stabilization. In the seven markets for which we have data, there are only thirty-six fewer bakers in 1760 than in 1733.

Without a continuous run of information on baker presence and reliable data on production, it is very difficult to make sense of this radical retrenchment. It is easy enough to say that there were too many bakers at the beginning of the century (given the acute congestion described by Poussot in 1760, one marvels at the capacity of Delamare's markets to contain twice as many bakers in the same space). It may be that the police, obsessed with the fear of shortage and with recent memories of the disasters that took place between 1660 and

Table 4.2 Bakers at the Bread Markets, 1725–33 *(Supply in Muids)*

Year	1st-Class Bakers	1st-Class Annual Supply	2d-Class Bakers	2d-Class Annual Supply	Total Number of Bakers	Total Official Annual Market Supply
1725	687	42,480	271	17,697	958	60,177
1726	669	39,840	271	17,657	940	57,407
1727	647	39,958	280	18,382	927	58,340
1728	602	39,355	248	17,069	850	56,424
1729	571	39,030	239	17,442	810	56,472
1730	582	39,793	233	16,285	815	56,078
1731	584	41,522	231	16,527	815	58,049
1732	577	38,506	221	16,150	798	54,656
1733	536	38,229	221	17,065	747	55,294

1700, authorized many more bakers than necessary to come to market. It is likely that a higher percentage of shop bakers furnished the market at the end of the seventeenth century than in the 1730s and that there were more forains, including a large number of part-time bakers, especially laboureurs who sought to supplement their incomes in hard times by exploiting the captive Parisian market. The suppression of the jurisdiction of the Grande Panneterie, a quasi-seigneurial jurisdiction that exercised a costly right of surveillance over the masters, and the absorption by the city guild of the masters of the faubourgs in 1711–16 probably encouraged many bakers to concentrate their energies and capital in the shops and to abandon the market.[13] Market supply was far more atomized in 1700 than it was thirty years later. With more space and less competition, the smaller market-baker corps compensated for its contraction by bringing larger amounts of bread to each market. And to a limited extent compensation was effectuated, too, by the spasmodic appearance of de facto bread markets and squatter bakers, analogues to the illicit shop bakers.

Thanks to the Delalande registers, we can study in depth the striking shrinkage that occurred between 1727 and 1733 (see the broader supply context in table 4.2). Between these two years the baker corps diminished by 19.4 percent, or 180 bakers. There were 37.7 percent fewer city masters and 26.5 percent fewer faubourg masters in 1733 than in 1727. All told the body of masters declined by eighty-one bakers, or 28.6 percent. The forain contingent suffered a loss of 19.2 percent, or seventy-one bakers, a loss much heavier among the second-class bakers of the *plat pays* (24.6 percent) than among the first-class forains from the immediate environs of the capital (2 percent). The faubourg Saint-Antoine bakers were far more resistant than any of the other major groups; their numbers decreased by only twenty-eight, or 10.2 percent.

Not only did the baker force diminish between 1727 and 1733, but it also underwent something like an internal purge. Fewer than half the bakers who furnished the market in 1727 were still there six years later. During this period 205 newcomers blended in with the market veterans. Some of the markets were veritably decimated. Only 38 percent of the Halles' first class persisted from 1727 to 1733, 49 percent of the Halles' second class, 46 percent of Saint-Germain's first class, 43 percent of Saint-Germain's second class, 27 percent of the Augustins' first class, 39 percent of the Marais's second class, and so on. Saint-Michel, the smallest market, was the only one with a persistence rate of over 70 percent for both classes. It is hard to imagine that such turnover occurred regularly across the century; the pool of potential baker substitutes simply was not large enough. Such a high rate of turnover must have seriously disrupted baker relationships with clients and generated an element of uncertainty in the weekly provisioning picture.[14]

Market Provisioning, 1727 and 1733

To get a sharper idea of the nature of market-bread provisioning, we shall focus successively on the years 1727 and 1733 (see tables 4.3 and 4.4). Of the 927 bakers who supplied the markets in 1727, 283 were masters (30.5 percent), 275 were Saint-Antoine bakers (29.7 percent), and 369 were forains (39.8 percent). Over 80 percent of the masters were recruited from the faubourgs. Core city masters apparently were too busy in their shops to enter the market in large numbers. The average weekly supply of the city masters, 1.16 muids, was not as high as that of their faubourg confreres, 1.34. The combined master average weekly provision was 1.3 muids, and the masters as a group accounted for 34.5 percent of the total official market supply. The Saint-Antoine bakers contributed a little more than a quarter of the market provision with an average weekly supply of 1.14 muids per baker. Accounting for the largest chunk of the supply, 40 percent, the forains averaged a weekly contribution of 1.19. That average was depressed by the poor showing of the first-class forains, who brought under a muid apiece (versus 1.26 muids for the class-two forains). Drawn from the immediate environs, the class-one forains had a more desultory commitment to the marketplace than their country counterparts. I suspect that many of

Table 4.3 Bread-Market Supply, 1727 *(in Muids)*

Market	Number of Bakers	Average Weekly Provision per Baker	Total Annual Supply per Market
Halles	285	1.44	21,281
Saint-Germain	122	1.23	7,817
Place Maubert	125	1.17	7,588
Cimetière Saint-Jean	117	1.15	7,025
Marché-Neuf	60	1.31	4,095
Saint-Paul	59	1.05	3,220
Boucheries	23	1.27	1,517
Marais	29	0.91	1,378
Quai des Augustins	26	0.96	1,304
Carouzel	34	0.66	1,174
Palais-Royal	33	0.64	1,092
Saint-Michel	14	1.17	849
TOTAL	927	1.21[a]	58,340

[a]This figure, obtained by dividing the total annual supply of bread per market by the total number of bakers and then breaking that quotient down to a weekly average, seems more accurate to me than the straight mean of the market averages in this column, 1.08, because it weights all bakers' output equally, regardless of the size of the market they serve.

them kept shops that may have attracted "frontier" Parisians. Nor were the nearby forains as aggressive businessmen as the others. One indication is that they bought their grain and flour almost exclusively in Paris.

Among the 927 bakers were a surprisingly large number of women: 131, or 14.1 percent of the total market corps. One hundred one of these women were widows who operated businesses they had inherited from their husbands. More intriguing are the thirty women who were not baker widows. Despite a strong traditional prejudice against investing women with responsibility and despite the enormous physical burden of managing not only a bakeroom but an itinerant commerce as well, many women bakers succeeded in acquiring title to market stalls in their own names. Equally surprising is that more than two-thirds of all the women were from Paris (city and faubourgs), and thus involved in baking operations shoulder to shoulder with the male majority that dominated the trade. On average, the women supplied about 2.5 setiers' worth less bread per week than the entire baker universe (1 muid versus 1.21 muids). The

Table 4.4 Bread-Market Supply, 1733 *(in Muids)*

Market	Number of Bakers	Average Weekly Provision per Baker	Total Annual Supply per Market
Halles	247	1.60	20,497
Cimetière Saint-Jean	96	1.63	8,151
Saint-Germain	100	1.34	6,968
Place Maubert	88	1.41	6,448
Saint-Paul	51	1.20	3,184
Marché-Neuf	43	1.34	2,990
Marais	27	1.07	1,499
Boucheries	16	1.77	1,469
Palais-Royal	22	1.14	1,309
Quai des Augustins	20	1.18	1,222
Carouzel	23	0.75	893
Saint-Michel	14	0.91	663
TOTAL	747	1.42[a]	55,293

[a]This figure, obtained by dividing the total annual supply of bread per market by the total number of bakers and then breaking that quotient down to a weekly average, seems more accurate to me than the straight mean of the market averages in this column, 1.28, because it weights all bakers' output equally, regardless of the size of the market they serve.

widows produced slightly more than the global women's average (1.08) and the nonwidows substantially less (0.7).

The Halles was by far the strongest market; it had the greatest number of bakers, average provision per baker, and total weekly supply. As a rule, the markets with the largest weekly provision were also the ones whose bakers had the highest average supply per person. For example, the Marché-Neuf, Saint-Germain, and Place Maubert are in the upper half of both accounts. The bakers of Saint-Michel, by contrast, are fifth in average provisioning while the market is last in total supply. Predictably, the second-class bakers, who concentrate their commerce in the market, contribute a larger average supply than the first class does (1.26 muids versus 1.19 muids). The Saint-Paul forains hold the record for the highest weekly average, 1.6 muids, followed by those of the Halles (1.54) and the Marché-Neuf (1.47). Singled out by Poussot for their relative feebleness, in 1727 Palais-Royal and the Carouzel were already among the weakest markets. The first-class bakers in each furnished less than three

Table 4.5 Second-Class Forain Supply, 1725–33 *(Wheat in Muids)*

	1725	1726	1727	1728	1729	1730	1731	1732	1733
Number of bakers	271	271	280	248	239	233	231	221	211
% of total, Paris supply	21.7	21.3	23.5	21.9	21.3	20.0	20.3	20.3	21.8
Total amount per year	17,524	17,447	18,382	17,069	17,282	16,250	16,347	16,150	17,065
Average per baker per year	65.3	64.8	65.7	68.8	73.0	69.9	71.5	73.1	80.9
Average per baker per week	1.26	1.25	1.26	1.32	1.40	1.34	1.38	1.41	1.56

setiers per market day. Only one market in the second class—the Augustins—failed to average at least a muid per baker per week. The largest single supplier was a first-class baker at the Halles, who averaged an enormous six muids per week. No forain furnished more than three. The least robust bakers—all in the first class—brought only one setier a week, which hardly made the twice-weekly trip to market worthwhile.

Of the 747 bakers who filled the twelve markets in 1733, 202 were masters (27 percent), 247 were Saint-Antoine bakers (33.1 percent), and 298 were forains (39.9 percent). The total market force declined by almost 19.5 percent between 1727 and 1733; in relative terms, the forains held about the same place, while the masters (down 3.5 percent) lost ground to the bakers of the faubourg Saint-Antoine (up 3.4 percent). City masters continued to withdraw from the markets, in relative as well as absolute terms. Whereas in 1727 there were four faubourg masters for every master rooted in the city, by 1733 the ratio had surpassed five to one. While the number of second-class, or country, forains decreased by almost one-quarter between 1727 and 1733, the contingent of first-class, or suburban, forains grew slightly.

The amount of bread furnished each week by bakers in all categories increased, compensating for the contraction of the supply corps. The average size of the baker's weekly provision in 1733 was 1.42 muids, more than two and one-half setiers above the 1727 figure. The masters of 1733 accounted for 29 percent of the total official market supply, well below their performance level of 1727. Nevertheless, they were the mightiest individual suppliers, averaging 1.5 muids a week per baker, with the faubourg masters outprovisioning the city confreres by three and one-quarter setiers a week. The masters located in the faubourg Saint-Lazare converted 1.79 muids of wheat into bread a week, which was more than any other category of bakers. Though diminished in number, the Saint-Antoine bakers took on a significantly larger burden by supplying 32.1 percent of the market, more than 7 percent above their 1727 achievement. On average each baker expanded his weekly share by over two and one-half setiers to 1.35 muids. The forains continued to provide almost as large a weekly dose as they had in 1727 (see table 4.5).[15] The average weekly provision per baker, which rose by three setiers a person to 1.44 muids, reflected their assiduity. As in 1727, in 1733 the country forains converted substantially more wheat than their suburban counterparts (1.56 versus 1.1). Because the average annual supply per baker increased almost a quarter between 1725 and 1733, the second-class forains were able to continue furnishing the same proportion of the total Parisian supply. The bakers from Gonesse were the most productive forain subpopulation with a weekly supply average of 1.73 muids.

Between our two reference years, the female ranks were decimated. There were only fifty-six women in 1733, representing under 7.5 percent of all bakers, or about half their relative number in 1727. Only four of them were now widows, presumably operating as bakers for their own account, and none of the four was a forain. A considerable number of the independent women in 1727 had been practicing the craft illicitly. It is possible that the police, seconded or prodded by guild officials, cracked down on these unauthorized bakers and drove them virtually out of existence. The erosion in the number of widows is still more baffling. A number doubtless retired, others died, still others remarried in these seven years, but why was the widow pool so much larger (relatively as well as absolutely) in 1727 than in 1733? Like the whole baker universe, the women of 1733 were more productive than they had been in 1727, averaging over 1.2 muids a week per woman, which is just about the global market average in 1727.

Market stratification did not significantly change between 1727 and 1733. The six markets that had the most bakers and the largest supply in 1727 were also the leaders in 1733, and five of the six markets that had boasted the highest average provision per baker retained their preeminence. Saint-Michel was the only market that did not suffer a baker loss, but it was also the only one whose production per baker diminished instead of increasing. Again, the second class outstripped the first in average weekly supply, 1.56 versus 1.37 muids. The Boucheries and the Cimetière Saint-Jean bakers led the way with provisions of 2.24 and 2 muids respectively. Two other second-class markets, Saint-Michel and the Augustins, by contrast, fell beneath a muid in weekly production per baker. With a 1.58-muid average supply, the first-class bakers at Saint-Jean were also the strongest in their category. No baker in the second class brought less than four setiers' worth of bread each week, while in the first class there were several bakers who brought only one setier. Several bakers of both classes at the Boucheries and the Halles averaged six muids a week, or thirty-six setiers of baking for each day of the market, a remarkable accomplishment. The global diminution in the number of bakers amounted to a rationalization of organization. The bakers had no trouble in expanding their service. The prospect of a larger market share induced them to make more efficient use of their labor and their capital equipment.

Gonesse

By far the most renowned forain center was Gonesse, a small town located between three and four leagues (twenty kilometers) northeast of Paris. Once a cloth-milling center, by the seventeenth century its biggest mills had been converted to grain grinding.[16] The extremely fertile arable land of the area produced more than enough wheat to keep the four major water mills and the two largest windmills busy year-round. A thriving grain market drew dealers from well beyond the periphery of the town and bakers from Paris despite the interdiction on purchases within an eight-league radius of the capital. "The large majority of its inhabitants are bakers," wrote Delamare, an opinion that was widely shared by eighteenth-century writers. It is hardly likely that at any given moment a majority of Gonessiens were bakers, but there is no doubt that a substantial number of them baked bread for the Paris markets. At the end of the sixteenth century at least sixty bakers operated in the town. In the registers of one Gonesse parish that listed professions, twelve of the thirty-seven persons who died in 1658 were bakers, seven of twenty-three in 1659 and twenty-two of fifty-four in 1652. Two baker confraternities functioned in Saint-Nicolas parish. There were more than enough bakers to sustain an active practice of mutual aid redolent of the corporate world.[17]

Many of the Gonesse bakers were at the same time laboureurs.[18] It is a curious occupational combination, for the tasks of baker and farmer have very little in common. The intense metropolitan bread demand drew these peasant families into the baking business. Ambitious peasants always tried to diversify their sources of income and, for the most part, the farming and baking calendars were compatible. Bakers who did not have to rely exclusively on the market for their wheat obviously had an advantage over their rivals. Laboureur-bakers were not recruited among the farm families with the largest holdings, for whom the economic incentives of baking were simply not powerful. Small or middling laboureurs, the Gonessien baker-peasants nevertheless claimed a higher socioeconomic status than the plain bakers.

Between the handful of big bakers and the swarm of little ones, there was a great range in economic well-being. Taille assessments in 1740 among forty-six bakers listed ranged from three to 230 livres (mean = 53, median = 44). A baker and large landowner, Louis Gouffe, was the outlier, paying over a thousand livres. The roll at mid-century showed a general decrease in affluence with a similar pattern of dispersion: forty-seven bakers averaged thirty-four livres (median = 17). In 1765 the state levied a mean taille of thirty livres on twenty-nine bakers (the median of twenty-nine suggesting a marked attenuation of dispersion). There were almost surely fermiers/laboureurs on this list who also engaged in baking, though they cannot readily be identified. The tax rolls identify only twelve bakers on the 1781 list, seven of whom also bear the designation of laboureur, with a range from six to 432 livres (mean = 58, median = 15).[19] While the after-death inventories confirm the heterogeneity of fortunes in the baker corps, the major differences appear to reside neither in real property holdings nor in professional capital investment (the bakerooms look similar to middling Parisian shops, though virtually every Gonessien had his own heavy wagon and horse) but in the magnitude of grain and flour stock held.[20]

These families tended to marry among themselves and to constitute remarkable dynasties across several generations. At one point in the eighteenth century almost 70 percent of the sons of bakers married daughters of bakers. About half the daughters married bakers or sons of bakers, while over 15 percent married merchants, in particular flour dealers, and about 14 percent married laboureurs.[21]

The capital's notorious need for the bread of Gonesse seemed to render the city extremely vulnerable. In 1590 the future Henri IV set up camp in Gonesse, "convinced that if two weeks passed without bread from Gonesse or elsewhere reaching the Halles in the habitual quantities as it always had, . . . then the city would not be able to hold out any longer against famine." According to Condé, who also endeavored to cut off the supply routes to Paris during the Fronde, the way to conquer the capital was to deprive Parisians of their Gonesse bread for a week. "As long as the bread of Gonesse is not in short supply (as was said 140 years ago during the Fronde)," wrote Mercier in the 1780s, "the commotion will not be general; but if Gonesse fails to appear for two consecutive markets, the revolt will be universal." Nostalgically and hopefully, a conservative newspaper,

L'Ami du Roi, in 1791 evoked Condé's conviction that to take Paris "it sufficed to deny the city Gonesse bread for a full week." If these remarks exaggerate the real dependence of Parisians on Gonesse bread (especially in the eighteenth century), they nevertheless suggest the extraordinary reputation and allure of bread made in Gonesse and sold at the Wednesday and Saturday markets.[22]

What made the Gonesse bread, in the words of the Englishman Evelyn, "the best bread of France"? What was at the source of the Old Regime aphorism "as good as the bread of Gonesse"? At the end of the seventeenth century Dr. Lister described it "as purely white and firm and light" in contrast to Paris bread, which he found "coarser and much worse." The standard explanation invoked by contemporary observers to account for the excellence of Gonesse bread was that the water with which it was made—drawn from the Crould river—was of exceptional quality. An Italian subsistence specialist submitted negative evidence in favor of this thesis. He reported that Gonesse bread could be duplicated neither by Gonessiens who were brought to Paris to bake it in the capital nor by Paris bakers who used Gonesse wheat, also reputed for its outstanding properties. Contemporary skeptics intimated that the water explanation amounted to nothing more than a marketing device geared to discourage outsiders from trying to duplicate the process and thus compete with the (unreplicable, peerless) authentic product.[23]

Another idea, put forth before the Revolution, was that the Gonessiens made their bread with middlings, the bits of rich flour attached to the bran that most bakers, before the introduction of economic milling, disposed of as animal food. The Gonessiens allegedly separated the middlings from the bran by hand, using various grades of bolters and sifters. They enriched and enlivened a flour that fed only on a natural starter yeast without the addition of barm. This is an intriguing thesis but one that requires much more serious proof than is now available. If this indeed was the Gonesse "secret"—given the peculiarly nutty taste and appearance of middling bread, it would have been an impossible professional secret to keep—it seems incredible that it was not adopted on a wide scale by Paris bakers, especially in light of the savings in flour and the potential market it promised.[24]

Unwilling to concede a substantial part of their market to the Gonessiens, Paris bakers themselves made a bread "à la façon de Gonesse." The statutes of the bakers of Saint-Germain-des-Prés in 1659 (before the fusion of all Paris masters into a single guild) authorized them to bake the Gonesse-style bread. Façon-de-Gonesse bread, in some cases baked by ex-Gonessiens now established in the faubourg Saint-Antoine, passed for the real thing in a number of Parisian taverns. According to Dr. Malouin, however, the quality of the real thing declined drastically by the mid-eighteenth century, even as imitations proliferated. Locating the halcyon days of the real Gonesse bread in the sixteenth century, Malouin maintained that "today it is no longer made." The original bread was produced from a hard dough kneaded with the feet and prepared "with the greatest care," wrote Malouin. This Gonesse bread was widely adopted by Parisians and rebaptized *pain de Chapitre*. The skill test ("masterpiece") required by the guild of candidates for membership supposedly required the fabrication of *Chapitre* bread until sometime in the seventeenth century—an ironic revenge, if Malouin is right, of the forain-bumpkin on the city slicker.[25]

Malouin suggested that either the Gonesse cult had virtually died out by his time or that Gonesse forain and Paris bread so resembled each other that the old distinction served only to nourish Parisian nostalgia. Indeed, Parmentier hinted that the Gonessiens, who he claimed were "no more numerous in the markets than the other forains who supply this city," had to conceal the fact that theirs was Gonesse bread in order to compete for customers in the markets. Similarly Legrand d'Aussy, an acute observer of material culture, noted in the last decade of the Old Regime that "very little bread came from Gonesse to Paris these days" and that "most of what is sold under this name is made in the faubourgs Saint-Denis and Saint-Martin." Already around midcentury Abbé Lebeuf found little Gonesse bread in Paris, in part because many of its bakers had infiltrated into the Paris faubourgs.[26]

Only a systematic study of the Gonesse economy in both its agricultural and its artisanal inflections across the eighteenth century will enable us to understand the phenomenon of the disappearing Gonessiens. All we can say with confidence is that the Gonesse contingent, like the entire forain corps, dwindled markedly in the first third of the eighteenth century. Between 1727 and 1733, the number fell from 105 to 67. In 1727 over half of the second-class forain bakers at the Halles, more than three-quarters of those at the quai des Augustins market, and almost one-third of those at the Saint-Germain market were Gonessiens, whereas in 1733 these proportions slipped to 42.8 percent, 61.5 percent, and 10.8 percent respectively. They dominated other markets, however, such as the Cimetière Saint-Jean (from 48 percent to 61.5 percent) and the Marché-Neuf (from 50 percent to 67 percent), more completely than they had before, and they remained 9 percent of the total of all bakers in the market and almost one-third of the forains of the second class. Moreover, next to the twenty-one masters from the faubourg Saint-Lazare (1.79 muids), Gonessiens boasted the highest weekly bread supply in the markets (1.73 muids) in 1733.

It is safe to assume that only the fittest Gonessiens survived this commercial purge of the marketplace. Given their reduced number and their enhanced productivity, from the vantage point of the mid-1730s there is no reason to expect these Gonessiens to suffer further decimation of their ranks. Yet only forty-nine bakers appear on the town's taille rolls in 1742, and fewer than half that number remain in 1780. By 1790 only ten forains from the legendary supply town appear to frequent the markets of the capital.[27]

The Forain Revolt against Corporate Tyranny

In the eyes of the forains the guildsmen were engaged in a relentless crusade, masked and legitimized by their wide-ranging police powers, to reduce them to subjection and marginality. The forains resented their second-class status and the swaggering interference of the masters in their professional lives. They felt the same rancor toward Parisian colonialism as did many other provincials in other socioeconomic categories. Sharp tensions between forains and city bakers dated back as far as the time of Saint Louis. The latter lobbied hard to restrict the former to offering their bread on market days only (then just once a week) and to engaging exclusively in retail sales of a narrow range of *gros* loaves. In the fifteenth century the guild complained bitterly that the forains were usurping city clients. They obtained measures again confining the forains to market days (now twice a week), allowing bourgeois home delivery only for bread ordered in advance (a condition hard to verify and invariably subject to contention), and forbidding opportunistic street sales. Commenting on another corporate offensive the following century, one of Delamare's collaborators noted dryly that "the bakers of Paris . . . sought only to ruin the business of the forains." (In the 1850s jealous Paris bakers still sought to stifle forain competition.)[28]

Against the guild the forains had few weapons. Given their geographical dispersion and their lack of organization, it was very difficult for them to resist corporate aggression. Periodically, groups of outspoken forains sought redress in the courts. Long, costly, and often demoralizing, litigation nevertheless enabled the forains to mobilize themselves momentarily in an association of common interest and to air their grievances at the highest instances. Paradoxically, the authorities from whom the guild solicited and generally obtained regulations directed against the forains also proved to be the forains' surest allies. For the government had no appetite to enhance the guild's monopoly merely to gratify the masters' self-regard. On the contrary, while enforcing measures that served the public interest as it construed it, the government desired to lure as many bread suppliers to the capital as possible: that was the supreme law of provisioning. So it tried to draw the line when it felt that the city bakers "wore down the forain bakers" gratuitously, "which caused them to desert the markets."[29]

The forains pressed what amounted to a class action suit against new guild statutes first proposed in 1746, statutes contested within the guild as well as outside, debated by police authorities, and not registered by the parlement until 1757, and even then in provisional terms that did not satisfy masterly demands or decisively seal the fate of the forains. Indeed, one intriguing piece of evidence suggests that the case took on popular political ramifications. Paris would have been "in combustion," argued a contemporary observer, had the court not opted for a prudent tergiversating tactic: "there was talk of nothing less than burning out the Paris bakers after having looted them, and perhaps [the avengers] would not have stopped there." According to this witness, the initial news was that the guild bakers had succeeded in obtaining a measure that prohibited the unincorporated bakers from remaining in the marketplaces after noon and from commissioning their porteuses to deliver bread.

The "forains ranted and raved" but apparently went a step further, either by withholding bread or by conforming precisely to the new regulation, for on the next market day "the markets were short of bread for an hour and [the people] were on the brink of revolt." The parlement saved the day by issuing a decision that temporarily returned things to the status quo ante until the fundamental issues could be judged. A large crowd awaited the announcement of the court's finding, which it greeted with enthusiastic applause. The judgment was printed within the hour, and it was posted during the night throughout the city. The next day "the forains brought the necessary quantity of bread" and the porteuses decorated their baskets with colorful ribbons. Yet the perspicacious commentator wondered aloud whether the story was as simple as it appeared on the surface. The urgent yet banal bread question seems to have gotten mixed up in complicated Eucharistic politics. As part of the ongoing, acrimonious quarrel over the denial of sacraments, marked by the refusal of the episcopacy to allow dying persons who had not been confessed by anti-Jansenist priests to receive last rites, "for the last six months, they [the unnamed] have tried different ways to push the people to revolt. . . . it is to be feared that those who have a vested interest in confusing all the issues will not stop at this last attempt." Nor did Damien's attempt to assassinate Louis XV in that same year allay this inquietude.[30]

Encouraged by the suspensive decree, the forains redoubled their juridical "opposition" to the registration of the new statutes on the grounds that the regulations perpetuated basic elements of corporate tyranny. Initially ninety-six Saint-Antoine bakers were joined by fifty-six from the Gonesse area and thirty others, but the number reached close to two hundred by the time the parlement issued its decision, after several more years of maneuvers and appeals, in July 1760.[31] The forains pled for a measure of liberty and dignity. Like virtually every other petitioner for redress in the tradition-driven Old Regime, the forains cast their argument in conservative terms. The right we are demanding, the forains insisted, we had enjoyed "from time immemorial" until "recent times"—apparently located in the seventeenth century—when the guild began to attack them. Thus the corporate tyrants were the innovators and the forains, faithful to the old ways, were the victims of this "revolution." Beyond this wistfulness loomed a more trenchant and pragmatic line of reasoning. These bakers exploited the only real advantage they had: the conviction of the police that the capital depended heavily on the forain contribution for its survival. The forains contended that corporate imperialism seriously hampered the provisioning of Paris, and they hinted darkly that nonguild bakers would be forced to abandon the service if corporate power were left unchecked. The forains vehemently contested the jurés' claim to exercise a grand police over all aspects of the bakery. Let the jurés mind corporate business and we will mind our own, proposed the forains. They asked that the jurés be forbidden from "harassing forain bakers with heavy-handed searches and visits," and more generally they sought guarantees "against being persecuted in the trade by the jurés."

To achieve a certain independence and security, the forains had a number of concrete demands. They wanted the unqualified right to deliver bread to their clients, a right that they claimed to have exercised "for centuries" before the onset of the corporate reaction. They denounced the obligation to reduce their bread prices, which they claimed the jurés tried to force early in the afternoon. They first suggested that they be allowed to sell as late as 9 or 10:00 P.M. before being required to lower the price. In a later brief they offered to discount the bread at 6 or 7:00 P.M., without, however, being obliged to dispose of it all. They would lower prices until 8 or 9:00 P.M.—it was unclear by how much—and if they failed to attract buyers, they would be free to store their leftover bread in depots nearby or to take it home. To mollify the police, they agreed to fix a limit, upon consultation with the commissaire, on how much bread each forain could withdraw at the end of the market.

The forains made two other points, but with less intensity, perhaps because they were not issues about which all of them felt strongly. They criticized the restrictions on the kinds of bread forains could make. The country bakers were willing to settle for the right to make loaves of soft dough that weighed over three pounds, but the faubourg Saint-Antoine bakers, perhaps because they thought they could successfully compete with the city masters in the semiluxury market, wanted authorization to bake mollet loaves under three pounds.[32] Finally, the forains denounced regulations governing sales to *cabarets*, without, however, frankly demanding permission to sell bread to tavern and innkeepers. This affair was of much less concern to the Saint-Antoine bakers than to the Gonessiens.

The jurés denied "ever having presumed to exercise any right of superiority in the public places of the market." Yet they insisted on their right and duty, which they traced back at least to the thirteenth century, to exercise a general police "on *all* bread"—that is to say, on all bakers—entering the capital. Naturally, they presented their police as a public service geared to protect consumers. The jurés nevertheless insisted on their prerogative to "seize [forain bread] in case of encroachment on the profession of the [master] bakers of Paris."

The jurés began with a haughty, hard-line position, but they fully understood that they were on the defensive. After having awarded the masters very favorable statutes, the Parisian authorities would now have to appease the forains in some ways, the jurés correctly reckoned. Having opened the debate with an immoderate stand, the jurés would subsequently be able to offer certain concessions without in fact giving very much up, they hoped. As an opening bid, they denounced all forms of home delivery as a cover for widespread off-market commerce and as a source of consumer fraud. They later agreed, however, to accept home delivery if bakers first offered the bread for sale on the market until 8 or 9:00 A.M., if they delivered it themselves without allowing outside middlemen to intervene, if they did not "cry out their bread" on the streets and in the courtyards, and if they had lists of clients to prove that delivery was not a pretext for speculative selling.[33]

The jurés wanted to retain the monopoly on the manufacture of soft-dough bread, they solemnly averred, not because it brought the masters considerable profits but because this type of bread was less filling and somehow less "nourishing" than bread made of firm dough. The markets are "for the people," they argued, and it is thus "good police" to prevent the people from being tempted by a bread that would not serve their needs. This was a view that reflected the canonical police notion that since the masters preferred a bourgeois clientele and catered to its tastes, the forains had to assume responsibility for baking "a good household bread for the artisans and the little people."[34] Yet it was evident that the jurés really were concerned about protecting their exclusive right to make small "fantasy" loaves (mollet) and that they could tolerate forain competition only in simple soft-dough bread above three pounds. The jurés continued to insist on discount sales beginning at 3 or 4:00 P.M., but they intimated that they would agree to lowering prices later in the day provided the forains were still forbidden to remove bread from the market. As for the tavern police, the jurés refused to cede ground to the "special pretensions" of the Gonessiens.

The parlementary decision of 1760 was presented as a compromise measure. The forains won a few concessions, but at bottom the jurés lost little authority. The decree permitted the forains to deliver to customers of all social origins without any restriction on how delivery was to be made. The bread had first to be placed on the market, but the ruling did not indicate for how many hours. Discount selling was not to be imposed until 6:00 P.M. in the winter and 7:00 P.M. in the summer, and bakers could remain in market until 8 or 9:00 P.M. They could not, however, store bread or take it home under any circumstances. The forains were authorized to bake soft-dough bread weighing over three pounds provided it contained no salt, milk, or butter. They were not allowed to supply the inns and taverns. Finally, while the jurés were forbidden "to interfere" with the rights of the forains, the decree nevertheless confirmed the jurés in their exercise of a general surveillance in all the shops and markets to make sure that both police regulations and guild statutes were respected.

Psychologically, in the short run, the parlementary ruling may have buoyed forain morale, for it seemed to prove that the unincorporated bakers were capable of collective action and that the authorities were not unresponsive to their needs. The masters confirmed their right to inspect all bread destined for Paris and to compel forains to comply with the new statutes. Yet the jurés could only conduct "visits" in the presence of a commissaire who would arbitrate on the spot in his mixed judicial and administrative capacity. More generally, the judgment "forbids the bakers of Paris to trouble the forain bakers and those of the faubourg Saint-Antoine in everything ordered by the present decree."

There is no evidence that the decision genuinely reduced tensions between the guild and the forains. The marketplace and its outposts were to remain a battleground between rival races of bakers until the Napoleonic reorganization of the bakery. Yet the forains continued to make small gains, nibbling away at the fetters that constrained them. In the spirit of the Turgotian liberalism that briefly issued in the disappearance of the guilds, Lieutenant of Police Albert told the forains in 1776 that they could sell small loaves in the markets without restriction and deliver to customers as they saw fit. In response to forain opposition to the project for new statutes for the resurrected guild, in 1785 the parlement, on the advice of the lieutenant of police, quashed an article forbidding tavern keepers to buy bread from bakers other than the masters. The court denounced this clause as "contrary to freedom [of trade]" and likely to occasion a multitude of useless and costly seizures and lawsuits.[35] Yet by the time the playing field became more or less (juridically) even, the capacity of the forains to exploit the new opportunities had diminished considerably. The forain contribution remained important through the first part of the nineteenth century, but these baker-outsiders never recaptured the prominence, pragmatic and folkloric, that they had enjoyed during all of the seventeenth and much of the eighteenth centuries.

Chapter 5

Bread on Credit

"Europe buys bread on credit," observed Abbé Galiani, one of the eighteenth century's leading authorities on the subsistence question.

The rich [buy on credit] out of ostentatiousness, the poor out of indigence. Now, calculate the time lost keeping track of the *tailles* [tally sticks], the delays on recovery of money expended, the total losses, and the partial losses. Calculate the disappearance of impecunious customers and the interminable waiting list of creditors with claims against the estate of a great lord, and *Pity the Bakers*.[1]

Advantages and Constraints of Credit Relationships

There is considerable truth in Galiani's remark. In a highly competitive arena, bakers extended credit initially in order to win the custom of buyers. Credit was an expression of trust and a gesture of congeniality. It implied a bond of reciprocal obligation: the buyer would remain (more or less) exclusively faithful to the baker who cultivated him. The relationship could be extremely delicate, for once he obtained a custom a baker could not demand payment of sums due without risking friction or alienation or loss. Press too early and the client might embrace a competitor; wait too long and the client might owe so much that in a sense the baker became dependent on his goodwill. When a Saint-Antoine baker named Jacques Betemont insisted that Dame Regis, an artisan's wife, settle her account, she walked across the street and obtained immediate credit from baker Lamare.[2] Once master Jean Rousseau allowed the bill of future royal finance minister Le Pelletier des Forts to reach several thousand livres, he realized that he had no choice but to continue to extend credit, lest his chances of collecting anything be substantially compromised.[3]

One customer, Planquier, most likely a dayworker, bespoke an attitude that was probably widely held by consumers, poor as well as affluent. He believed that he had a right to credit, and to continued credit once the line had been opened. If the baker cut him off without reason, he violated a sort of contract and relieved his client of the obligation to repay. Planquier had not yet accumulated seven livres in bread debt when his baker, François Petit, peremptorily decided "to discontinue extending credit to him" (probably because he had learned something worrisome about Planquier). Petit sent his wife—such was the baker's courage and the prevailing division of labor in the bakery—to announce the news to Planquier and to demand payment closing out his account. The startled customer told her that he would not pay in retaliation for this unwarranted decision to cut him off. She replied "that she was not in the slightest obliged to furnish him his bread without cash in hand, all the less so because he had no fixed address." "Infuriated," Planquier insulted her, kicked and sword-whipped her, and sent her home bleeding. Petit's fear that Planquier might run off without paying was not unfounded. His colleague Denisot asserted that he had lost four thousand livres' worth of bread supplies to individuals who moved away without notice.[4]

Furnishing clients of power, wealth, and/or birth seemed to flatter the baker (and may, in certain circumstances, have been shrewd business). With these customers there was no question of asking for cash. Bread was delivered by the baker's staff or picked up by the client's servant. Thus master baker Gilles Pasquier extended credit to the duc d'Orléans, Prince Beauvau, the duc de Penthièvre, and the duchesse de Bourbon (whose bill, 610 livres, was the highest of the lot). Saint-Antoine baker Lepage had no mastership, but what prestige he could claim for serving Prince Charles (694 livres), the prince de Rohan (920 livres), the duc de Sully (313 livres), and Advocate General Daguesseau (535 livres). Such credit, wrote Mercier, "is the privilege of the nobility." Nobles were never harried to pay: "One waits when it is a question of a titled man." If a baker cut off his credit, a noble would denounce his impertinence, withdraw his "protection," and find another source.[5]

In Mercier's model of the social structure, "the opulent [family] inhabits the ground floor, the rich are on the floors immediately above, the [laboring] poor are on the fourth floor, and the indigent are in an attic just under the roof." Baker's credit "never goes above the fourth floor." In fact, there was no security for the baker on the fourth floor, in part because many of its inhabitants teetered on the brink of "falling" into the attic. Nevertheless, there is some evidence that bakers granted credit to the attic as well. Nor is there any doubt that this credit was absolutely critical to the survival of many Parisians—and to the public tranquility so dearly prized by the authorities.[6] As a mid-nineteenth-century miller put it, "the baker is something like the treasurer of the indigent."

Bakers then were expected to give credit to the poor—at least to the "honest" and "deserving" poor. Bakers who refused "would provoke an outcry against themselves."[7]

In the world of goods, there was no generally felt moral imperative in the eighteenth century. Bakers could be incredibly mean in all the senses of the word. (There is no shortage of tragic vignettes such as the one related by the bookseller-diarist named Hardy about a thirty-year-old pregnant woman who slit her own throat outside a baker's shop after the proprietor refused her a loaf of bread on credit.) There are indications, however, that some bakers did feel an obligation to carry impoverished neighborhood clients whom they knew to be "notoriously insolvent" and whom they more or less expected to write off. But the bulk of the credit that they extended to "workers" and "working people" they counted on collecting. If the habit of extending credit to "the populace" drove many bakers to "ruin," as Restif de la Bretonne suggested, it was owing less to any charitable disposition than to a combination of bad judgment, bad management, and bad luck.[8]

It is not surprising that the shop was the main source of bread transactions on credit. Rooted in a physical and moral community, the shop baker knew the people in his neighborhood. He could follow the vagaries of their daily lives by direct observation, and through gossip with other clients or with friends at the tavern down the street. He could continually reassess his strategy and his tactics vis-à-vis specific clients. Much less expected, however, is the considerable extent of credit sales in the bread markets. From the market stall, the baker had neither the vantage point nor the leverage of the shop baker. The clientele was shifting, the competition brisk and boisterous, the bargaining often intense, the atmosphere sometimes rough and impersonal and almost always chaotic. Yet many market bakers developed a faithful following. Widow baker Herbert of the faubourg Saint-Antoine accorded credit to "her regular customers." Master Guy Leguay appears to have offered credit more liberally in his stall than at his shop. A forain from Noisy offered credit on home deliveries as well as market sales. Another from Chaillot gave credit on bread she hawked (illegally) from her wagon in the street. Baker Huron appears to have charged a slight premium for credit sales at the market, making his customers pay the shop price, which was a little higher. The police demanded that bakers adhere to the current (or, during crises, the officially set maximum) price for cash sales at the market but tolerated a sort of penalty increment on credit sales, a penalty that hurt the poor far more than the rich.[9]

Occasional "treasurers of the indigent," the bakers were sometimes also neighborhood bankers. Master Tayret made small cash advances ranging from eight to forty sous to a dozen customers, most of whom were working poor. His colleague Marin Picot operated on a loftier plane: he lent much larger sums to a clientele of nobles. Another baker paid the butcher, washerwoman, and shoemaker bills of several customers. In many accounts it is impossible to disentangle money due for loans from accounts receivable for bread, combustible, or flour. It is likely that these loans were made as favors and rewards, to cement the bond of clientage, rather than as moneymaking ventures.[10]

Tools for Keeping Track

How did the bakers keep track of transactions on credit? They used the *taille* or a written register or both together. (Claude Pampelune, a Saint-Antoine baker, was an anomaly: his credit transactions were purely verbal, based on mutual "good faith" and thus impossible to manage rigorously.)[11] The taille was a wooden tally stick that came in pairs, one for the baker and the other for the customer. Each time the customer received a loaf of bread, the baker simultaneously notched both sticks with a sawlike knife. A baker who felt that his client was abusing his credit said that the taille was "too long." The tailles varied in actual size. Usually they accounted for between forty and a hundred loaves. When the baker pronounced a taille to be "plaisnne," or full, he called it in and asked for payment, or some form of more formal acknowledgment of indebtedness, such as a notarial note.[12]

Master Lazare Joquain explained the major reason why bakers depended on the taille, declaring that, "not knowing how to read, he kept an account book of supplies, but to take its place [on sales] he kept a scrupulous reckoning of tailles, following the custom of his fellow bakers." In another statement Joquain avowed that he also relied on his memory, which suggests one reason why quarrels with customers did not issue in easy resolution. Illiteracy seems to have been the major justification for not keeping books ("barely knowing how to sign my name . . ."). But bakers also used the taille because it was convenient and because it offered the chance for a double system of accounting. Widow baker Armand preferred to relieve her delivery boy of the burden of remembering the distribution and oblige customers to acknowledge receipt by using the taille. Bakers Berthelot, Panier Chappet, Chalons, and Canois all used the taille conjointly with account registers.[13]

Theoretically, there could be no dispute between customer and baker over the number of loaves of bread furnished, since each mark was notched in the presence of the buyer. (In fact, consumers occasionally complained that there

Eighteenth-century bakeshop; credit offered at shop on register or at home on *taille*. *Source: from Malouin, Description, ed. Bertrand, 1771. Courtesy Kroch Library, Cornell University.*

were too many notches—without explaining how this came about.) Discord erupted, however, concerning the weight of the loaf notched (though certain bakers had a different set of *tailles* for the different weights, or a differential notching code) and especially regarding the price, which was never indicated. How was the baker to calculate the price of bread notched over a period of a few months? Either he chose a conventional price reflecting a central tendency over the period, or he turned to his register, into which he (or more likely his wife) transferred *taille* entries daily (or weekly), marking the date and price next to them.[14]

Good accounting served the bakers well in the collection and litigation processes, but their bookkeeping was highly idiosyncratic and uneven in quality.[15] Certain bakers pretended to keep books but in fact did not use them on a daily basis to record transactions and manage affairs. Instead, they used them as levers for collection. Master Modinet seems to have entered only periodic totals of monies owed (a congeries of *billets*, or promissory notes) rather than ongoing dealings. His colleague Huron inscribed the number of loaves sold but noted no prices. Baker Pointeau deployed a combination of the Modinet and Huron approaches: occasional entries of sums due with a single price coefficient artificially applied to all transactions. Thus his client Madame Laurent was charged the same amount for all 887 four-pound loaves of different varieties purchased between August 1767 and June 1768, a time of acute price fluctuation. Master Fleury kept a running history of how much each customer owed rather than a sales and credit register, but he omitted dates and rarely stipulated real prices.[16]

Huron listed his clients as "le bossu [the hunchback]," "le fourbisseur [the sword maker]," "le procureur [the attorney]," and "le maitre macon [the master mason]," which would make identification difficult if not impossible should he or his wife not be available to decipher the register. Widow baker Huin wrote down street names only ("an individual living on the rue des Augustins") but later could not recall names or professions. Sébastien Lapareillé's *livret des fournitures de pain* (register of bread sales) was useless because "it is not possible to make out anything, . . . neither names of debtors nor amounts remaining due." Vavasseur and his wife discredited their bookkeeping because "they sometimes neglected to enter [items] received." A failure to keep effective books left master baker Voitreu wholly unaware of the fact that he was robbed of 3,500 livres over the course of several months.[17]

Some bakers did better. Rousseau was rather meticulous: he entered a price for each loaf sold, recorded dates in most cases, and frequently described the type of bread as well as the weight (*pain de potage, pain de table,* and so on). Philibert Motot boasted similar rigor. He listed the merchandise sold, names of clients, and sums due, and he noted when customers made payments. Still, even the most complete and faithful registers were as a rule in egregious disorder: accounts were begun here and continued there; dates were neglected; numbers were slurred; receipt entries were obscured. The only thing that bakers rarely forgot to do in their books was to advertise a reward in case of loss.[18]

Reckoning

The baker's ideal was to reckon frequently and regularly with customers. Widow Constant waited till the taille was used up. Huron asked for some payment every week. Pigeot tried to keep debts below fifty livres; indeed, most of his accounts totaled less than twenty livres at any given moment. Installment payments afforded an opportunity to reconcile the needs and interests of creditor and debtor. In return for a steady flow of payment, the baker accepted modest sums and a relatively long period of restitution. Accustomed to dealing with customers from the little people, Saint-Antoine baker Jacques Betemont readily agreed to an *ouvrière's* proposal to make weekly payments of thirty sous.

One customer agreed to pay master baker Michel Fleury six livres every two weeks, another "sis livre par moy jus qua de finisson de paiemon" (for obligations respectively of 80 and 334 livres).[19]

Short of obtaining remittance, the prudent baker sought to convert the bread debt into a surer form of obligation that would prove more compelling in court or that might itself be negotiable. A customer-debtor who signed a note (*billet*) openly owned up to his or her debt (and provided bakers who did not keep sound books with a powerful substitute). Baker Modinet converted his tailles into notes payable on call ("a la vollontee"). Baker Fleury accepted promises that were open-ended:

Conte de conte aveque Monsieur Fleury maistry boulanje pour pan fournis et livre la somme de trois sen quarantte quate livre don je prommay lui paier.[20]
[Reckoning of account with Monsieur Fleury master baker for bread furnished and delivered amounting to the sum of 344 livres, which I promise to pay him.]

Another baker insisted on a specific date for reimbursement. His customer Rival drafted the following statement on 19 October 1766: "Je reconet devoir a madame Oudar la somme de 24+ 15s done je lui promet lui payer au qatre doctobre prochain sans delais" (I acknowledge owing Madame Oudar the sum of 24 livres 15 sous, which I promise to remit on 4 October next without delay). Clients such as the comte de Jaucourt always tried to exact favorable terms—delays for a year or two—in return for some immediate mollifying payment. The baker obtained the largest dose of legal protection by having a *billet* converted into a notarized obligation, but this step cost him money and was generally viewed as extravagant.[21]

Another sort of guarantee that some bakers sought or accepted was the deposit of personal goods as pledges of future payment. In many cases this security was purely symbolic; in others it was of considerable value and could easily be transformed into cash in case the debtor failed to redeem it on schedule. One client left widow baker Constant a green ring with two small diamonds, another ring mounted in gold, and a pair of earrings. Master baker Denis Larcher received a dress from a clockmaker and a skirt and jacket from a shoemaker. A customer consigned a clock and a watch movement to baker Chantard. Master Patineau had enough clothes and linen from diverse customers to do business as a secondhand clothing dealer. In return for a four-pound loaf, a "mulatto from Bengal working as a mason" gave a baker a back-basket that he could reclaim upon payment.[22]

In numerous cases, bakers actually accepted goods as full payment rather than as security. (One wonders how much of the artisanal/small merchant world operated as a rule on barter.) Baker Leduc received 168 livres in wine from a tavern keeper. A bourgeois paid baker Chantard partially in wheat. In return for an unspecified number of four-pounders and luxury "milk breads," tailor Lasperle made baker Pigeot two pairs of *culottes* and provided cloth for a jacket and vest for his son and a *culotte* for his journeyman. Baker Michel Beburre claimed to have had "carnal commerce" with a woman who owed him money for bread, but it is not clear whether he considered this transaction as payment in kind for his merchandise.[23]

Much of what we know about the accounts receivable of individual bakers comes from business failure (*faillite*) records. This information tends to be biased in the direction of bakers who did not pursue collection vigorously enough (or who had particularly renitent or debilitated customers). Various clients owed baker Lemerle the following amounts, in livres: 1,292, 883, 798, 630, 623, 560, 494, 478, 436, 395, 373, 335, 273, 270, 253, and 230. Few other bakers I encountered allowed customers to amass such prodigal levels of indebtedness. Baker Mussant was in deep trouble, but he had only one account over three hundred livres (an innkeeper) and three over two hundred livres. Debais had claims, in livres, of 474, 239, 171, 116, 94, 72, and 44. His clients included a duchess and a count, a curate and a vicar, a lawyer and a bourgeois de Paris, and an artisan.[24]

Of course bakers who kept a relatively tight rein on individual accounts also experienced difficulties. Bernier insisted on frequent payments and allowed only one customer to go over the one-hundred-livre mark. On average, his twenty-nine customers each owed him twenty-five and a half livres. To accumulate accounts receivable as imposing as Lemerle's, one either had to have quasi-institutional clients or to extend credit across a very long period of time. Master baker Gilles Pasquier allowed a pensioner named Bertault to build up almost nine thousand livres in debts over a seven-year span. Apparently Pasquier cut him off several times but reinstated the line of credit each time Bertault came up with a good-faith payment.[25]

It was the goal of bakers to sustain long-term relationships with customers. The trick was to navigate between the Scylla of ruin and the Charybdis of rejection. The ideal was to have clients such as Modinet's Sieur Pichot, who bought a three- or four-pounder on credit every day (or occasionally a smaller mollet loaf) for years and years but paid every month or two, or Dugland's Lafrancet, who remained faithful for many years and rarely allowed his account to surpass the ten-livre level at any given moment. It is impossible to determine at what point baker Chantard suspended credit to each of his debtor-customers, but in the early seventies he was still lamenting:

1743:   Un apele Le quatres peruquier 24+ [a wigmaker owes 24 livres]

1743:   Un autre peruquier apelle aquaire 18+ [another wigmaker owes 18 livres]

1744:   Un apelle Resgrier maitre tailleur me doit 17+ pour pain fournis [a master tailor owes me 17 livres for bread supplied]

1745:   un apele dosire postillon me doit 7+ 15s [a carriage driver owes 7 livres 15 sous]

1751:   un soldat suisse me doit pour pain fourni 33+ 4s [a Swiss soldier owes me for bread 33 livres 4 sous]

1751:   Le perre Anguin me doit 22+ [Father Anguin is 22 livres in debt]

1752:   Une veuve chez le md episier me doit 50+ [a widow living at the merchant grocer's owes me 50 livres]

1759:   Un suice de maizon me doit 40+ pour pain fournis [a Swiss house servant owes me 40 livres for bread]

1761:   Monsieur le core marchand de vin me doit 40+ [a wine merchant owes me 40 livres]

1761:   Monsieur Ravin un peti marchand de vin en gros me doit 12+ 12s [a small wholesale wine merchant owes me 12 livres 12 sous]

1762:   Plusieur garcons tailleurs ensemble me doive 150+ [several journeymen tailors together owe me 150 livres]

1764:   Monsieur Richar marchand de canne me doit 22+ 14s [a cane merchant is obliged to me for 22 livres 14 sous]

Several of the people to whom Chantard had granted credit "in their dire need" died in the Hôtel-Dieu, sure proof that they were poor, or had suffered bankruptcy.[26]

Our studies of baker after-death inventories and commercial failures underscore the major role played by credit in the baker business. Almost 90 percent of inventory bakers had accounts receivable for bread, the highest amounting to 33,245 livres. Most of these debts were modest and spread among a large number of persons. One baker had 153 debtor-clients. Masters had more customers in debt (mean = 40, median = 31) than faubourg bakers (30 and 20) and forains (23 and 17). Of the 20 percent of total baker assets that consisted of paper claims, the vast majority were accounts receivable for bread (roughly three thousand livres per baker around midcentury, more for the merchant and master bakers, less for the forains). The inventory bakers optimistically expected to collect most of them.

Accounts receivable for bread made up the largest component of total assets in the failures, representing on average half of declared fortunes. Bread debts claimed by bakers ranged from 170 to 19,311 livres, averaging 4,511 livres (median

= 3,301). Bread makers parading the title "merchant baker" had the largest stake in bread accounts receivable, almost nine thousand livres, twice the level of the guildsmen. Even the forains extended substantial amounts of credit, averaging accounts receivable of over four thousand livres. In many cases bread debts more than made up the difference between hard assets and liabilities. There is no doubt that many bakers (but emphatically not all and perhaps not even a majority) could have avoided catastrophe had they been able to recover sums due.

Collecting

The first stage in the collection process was usually a visit by the baker or his wife to the debtor-customer. The latter had somehow intimated that he or she would or could not meet his (periodic) obligations: by failing to show up for a "count," by rebuffing a porteuse or a journeyman, by switching bakers, by losing a job, or by being expelled from a lodging. The baker meant to accentuate the gravity he attached to the client's default by this personal démarche. Sometimes he was gratuitously aggressive, as when baker Mignet erupted, drunk, into the vicarage of the curate of Saint-Hilaire, "yelling, cursing, and swearing," or when baker Prudhomme attacked a wine merchant's virility (kicking him savagely in the genitals) and his bourgeoisness (mutilating his wig). Yet the baker-collector was rarely welcome, even when his manner was solemn ("Redeem your honor," he exhorted) and conciliatory ("Let us transact a settlement," he proposed). Master Joseph Colombel called on Dame Beauregard to request that she pay for six months' worth of bread, representing 383 livres, that he had furnished eighteen months earlier. While her lover, dressed in a nightshirt, beat the baker with a cane, she called him "thief" and "murderer" and screamed for the watch. Her aim apparently was to cause a "public scandal" that she could blame on the baker and use to discredit him.[27]

The Saint-Antoine baker Jean Mallet had even less luck. He presented himself at Sieur Paulin's apartment to collect the payment for fifty-two four-pounders that he had supplied more than a year before. Paulin's wife hotly declared that they owed nothing, punctuating her affirmation with a kick to Mallet's groin that, according to a surgeon's report, produced "a partial hernia of the left side requiring bleeding and medication."[28]

Women emissaries did not fare better than men: gender offered restraint no purchase on either side. Widow baker Chaudron, seeking almost two hundred livres for bread furnished, was brutally beaten ("My days are threatened," she sighed melodramatically) by a less than kind Sieur Gentil, who also "sullied her honor" with insults ("bitch," "sodomite"). Similarly, the wife of master baker Pierre Félix, soliciting payment from a dressmaker, was insulted and beaten, jeopardizing her five-month pregnancy.[29]

Frustrated bakers and their wives sometimes took their revenge when they spotted delinquent customers passing in the street. Master Jean Baurin rushed out and jumped a journeyman roofer and his wife. The crowd that gathered evinced no sympathy for the creditor-baker. Eyeing an innkeeper who owed him money, master Bontemps shoved her around and let passersby know that she was "a thief, and whore, and that she cheated all the bakers in the neighborhood." On rare occasions, a nasty quarrel served as a sort of mutual catharsis, preparing the way for settlement. A brawl between forain Charles Loree and his wife on the one side and innkeeper Marie Mareau on the other had resulted first in a police investigation, a surgeon's report, and a lawsuit and then in a quiet meeting before a notary. In exchange for payment, Loree agreed to drop his legal actions. The parties "acknowledged each other as persons of honor and probity, not durably smirched by the insults that were uttered, and discharged each other from all claims and litigation."[30]

Rebuffed by the customer-debtor, did the baker have any alternative to litigation in the pursuit of recovery? A long shot was an appeal to the lieutenant general of police or to the local commissaire to intervene in the name of fairness and the public interest. The chance of inducing the police to get involved in a struggle over money was remote, but it was worth asking, for administrative mediation would spare the baker legal and court expenses. Jean Guerin, a merchant baker from La Croix-Rouge, was an almost perfect candidate for police solicitude. He was poor, operated a small and vulnerable business, supported eight children, and helped out several indigent relatives. A wine merchant owed him 809 livres for bread supplied to his tavern. Despite Guerin's reiterated pleas, the wine dealer refused to pay and now was "hatching secret measures," the baker believed, to foil his attempts at recovery. Guerin was sure that a little pressure from Lieutenant General Hérault would induce the wine merchant to pay. Without this money, Guerin claimed that he would be "utterly ruined," by which he meant he would be forced to shut down his bakery. The police had to decide, first, whether the baker was telling the truth and, second, whether his contribution to the provisioning of Paris warranted extrajudicial action.[31]

Another low-cost step depended on an equally uncertain contingency: the death of the debtor—provided there was some property in his estate. Normally seals would be placed by a commissaire on the deceased's lodging and effects. A baker alert to the news of the death (which families often tried to conceal for a time) could formally "oppose" the lifting of the seals on the grounds of a creditor's claim. Thus the family of the defunct debtor would be obliged to make an arrangement with the creditor(s) before it could dispose of the estate, or to engage in a suit. In this manner the widow of merchant baker Estienne Delarue obtained 223 livres from the widow of a royal official who served as guarantor of someone else's bread debt. A Saint-Antoine baker commissioned a mounted process server (*huissier à cheval*) to present his opposition to lifting the seals of his late customer, perhaps with the idea that he could thus intimidate the deceased client's family.[32]

The baker's professional "quality" enhanced his position far more than the heroic flourish of an emissary on horseback. For in the competition among creditors and in all attendant litigation, the law favored the baker by according him a "preference" vis-à-vis other claimants on the personal effects (including cash, furniture, clothes, and other goods) of his customer-debtor, dead or alive. Theoretically, this "privilege" covered only an amount equal to the previous six months' supply of bread and had to be executed within six months, but both of these limitations were customarily ignored. A baker could even use this privilege to attack a will.[33]

Litigation

The ultimate recourse for debt collection was litigation. Usually bakers sued in either the Châtelet or the consular jurisdiction, depending on the nature of the case (the socioprofessional "quality" of the debtors, the sums involved) and their inclination (previous success, guild recommendations, advice of colleagues or legal professionals). The consular court was an alluring arena because it was a businessman's venue run by judges elected from the leading guilds, who tended to sympathize, from personal experience, with the business of debt recovery and who often solicited expert reports from the baker-jurés. In this court the plaintiffs could save a great deal of money by pleading for themselves: the language and procedure were simplified and accessible. Even if the bakers opted for summonses delivered by huissiers and representation by counsel, they still saved money on court costs, which were quite modest. Certain bakers felt, however, that the consular jurisdiction depended too much on mediation and consensus and that for debt recovery its sentences *par corps* carried less authority than those of the prevotal-presidial court.

In the Châtelet the baker had to have a lawyer and faced higher procedural expenses. In one such suit Saint-Antoine baker Pierre Lepage obtained a sentence ordering a Sieur Daquin to pay 444 livres—at the cost of 101 livres in legal fees. The court ordered the defendant to pay all the court costs (*frais de procédure*), but it remains questionable whether Lepage could collect costs over and above the principal of the debt. In any event, he was liable to his lawyer. Huissier fees also mounted up. One huissier named Chambolin went after baker Claude Gillet sword in hand because the latter contested a bill of fifty-eight livres five sous for various collection steps. Whatever the legal venue, collection costs were often onerous. In eight years of debt recovery efforts, master Denis-Claude Lecocq Senior claimed he spent nine thousand livres in "expenses and actions undertaken by his huissier and his lawyer."[34]

All of the suits that I have seen in the civil chamber of the Châtelet involved relatively small sums. Master baker Veron won the largest award, 111 livres 6 sous, against a bookdealer for bread debts several years old. The smallest was twenty-seven livres plus interest that master Lemerle obtained against "a worker." In most cases the defendants were not even present to hear the condemnation.[35]

Some of the judgments in the consular court were also quite modest. Master baker Delanoue won thirty livres against a furniture maker; master Morisson gained thirty-six livres eight sous against a boardinghouse keeper; widow baker Lecoeur secured a judgment of fifty-seven livres against a carriage maker; baker Bire had a painter sentenced to remit sixty livres.[36] Yet there were many sentences for quite substantial sums: master baker Petit was awarded 214 livres against an upholsterer, while his colleague Cousin procured a judgment of 292 livres against a locksmith.[37] The biggest suits by far were directed against quasi-institutional clients, the innkeepers. Many of these cases were clearly worth litigating: 1,222 livres, 1,041 livres, 1,061 livres, 668 livres, 664 livres, 600 livres, and 588 livres. The average claim for nineteen suits against innkeepers was 455 livres.[38] The biggest single suit for collection I have uncovered amounted to almost ten thousand livres, but I do not know in which court it was filed.[39]

Winning a favorable ruling in court was only half the battle. Armed with a sentence of "condemnation," the baker-plaintiff still had to collect the sum awarded him by the law. Debtors could choose to appeal the suit. Even if they did not really intend to appeal, an announced intention to do so might buy time, or open the path to a negotiation resulting in a more favorable outcome than the court prescribed. In the face of judgments that were executable immediately, regardless of eventual appeal, a debtor could hide, flee, or seek to protect his property through various ruses.

A certain number of customers found to be in default must have deferred to the sentences and made payment—because they played by the rules, because they wanted to avoid further trouble, or because they were worried about their reputations. Those settlements are too discreet to find their way into the archives (though I am surprised not to have found any trace of them in the notarial minutes: prudence should have impelled the debtor-reimbursers to insist on unassailable proof of remittance). The real denouement of the suit was very frequently not the verdict but the act of property seizure performed by the baker-victor. Imagine his state of mind. After so many rounds in the recovery process, beginning with friendly exhortation and ending up in court, he finally triumphed. Yet it was in many ways a hollow or incomplete victory that frustrated and angered him, for collection was not automatic. Nor did the justice system assume responsibility for collection. It was up to the baker to mobilize his forces and descend on the debtor (if, as in most cases, the court authorized a *saisie,* or distraint, in a judgment *par corps*). By now the baker was seething. He was indignant, vindictive, and merciless in the pursuit of his prey—like Barnabé Grezel who tormented his debtors and had them followed and watched and harassed by a motley host of domestics, journeymen, process servers, repossessors, and lawyers.[40]

Unable to gain entry to the apartment of Isidore Chatelain, whom the court had ordered to reimburse him, master baker Nicolas Herbert called on a commissaire to supervise a legal break-in (*ouverture des portes*). Herbert seized all the furnishings and effects despite Chatelain's insistence that none of it (save six paintings of religious subjects and a bed) belonged to him. Baker Sauvegrain had all the property and annuities of the Religieuses bernardines du sang precieux seized to cover bread debts, jeopardizing their very survival according to the plaintive sisters. Unwittingly true to his name, master baker Jacques Lesimple did not move dexterously enough. When he arrived to carry out a seizure of the effects of wine merchant Gigier, who owed him over seven hundred livres, he learned that Gigier had slipped away the night before with all his belongings. Widow baker Reverard arrived in time to seize the possessions of a master saddler, but the debtor's other creditors challenged her right to dispose of his property. Having won and executed the judgment of one suit, she now faced another.[41]

No single social fact is more significant for an understanding of the provisioning system than the dense credit nexus. It enveloped virtually everyone, at one time or another, from the producers to the myriad intermediaries and then to the bakers and to all manner of consumers. When it worked, it wrought wonders, keeping the cumbersome and rickety machine running on an astonishingly modest amount of fuel. When it faltered, even at the margins, it provoked enough friction to disrupt the flow of goods and generate a paroxysm of anxiety in the tributary branches of the system. And when it failed close to the core, it threatened to paralyze the entire supply operation.

Historians have learned a great deal about credit at the level of the enterprise or the bank. They have uncovered much less about how it engulfed daily life. The dominoes bearing the greatest freight in this long and delicate chain concern the grain and flour trade, whose characteristics are mapped out elsewhere.[42] Yet it is striking to see how deeply the consuming public was touched, and how intricately bakers and customers became bound to each other. Buyer and seller each sought to use the (mutual) obligation in an advantageous manner. The city lived as much on credit as on calories. In order for it to survive, both had to be oxidized at a regular, more or less smooth pace. Credit was a coinage of sociability, a cement of clientage, and a ray of hope. It was also a source of acrimony and an incubator of disorder. At bottom both creditor and debtor would have liked to find a way to protract their relationship indefinitely. The consumer hoped to reckon as infrequently as necessary (or possible). The baker accorded as much leeway as necessary (or possible). In few other social domains was mutual accommodation as much a matter of public as of private interest.

II

Bakers:

Social Structure

and

Life Cycle

Chapter 6

The Guild

With roots reaching back as far as the Gallo-Roman period, the baker guild was one of the oldest Parisian corporations. The statutes it received in 1270 continued to govern the guild for the next four and one-half centuries.[1] Yet this statutory stability did not spare the guild serious troubles, especially during the reign of Louis XIV. Those disorders diminished considerably in 1719 with the promulgation of new statutes. Though the new regulations resembled the old in many ways, they mark a genuine beginning for the guild, for it emerged from its difficulties with far greater strength and coherence than it had enjoyed before.

The aim of the guild was to protect the commercial, moral, and political interests of its master-members. Its defense of those interests led it to clash violently in the course of the seventeenth century with the grand pannetier, the seigneurial supervisor of the trade, with the bakers of the faubourgs who were organized in rival guilds, with the faubouriens without organizations of their own who either sought incorporation with the Paris masters or affirmed the right to exercise prerogatives claimed exclusively by the city guildsmen, and with the forain bakers, who had always resisted and would continue to chafe against the restrictions that the guild tried to impose on their commercial freedom. By 1719 the first two conflicts were definitively resolved, for the jurisdiction of the grand pannetier was suppressed and the faubourg corporations were absorbed by the city guild. Relations between the guild and the faubouriens who remained unincorporated (located virtually exclusively in the "privileged" space of Saint-Antoine) and with forains from the environs remained strained until the Revolution dismantled the corporate regime.

The Grand Pannetier

In the early days of the monarchy, the grand pannetier was one of the first officers of the crown. Among his powers was the right to exercise, usually through a lieutenant, *basse justice* over the bakers of the city and the faubourgs.

In practical terms, this meant that the grand pannetier regulated the baking industry by controlling recruitment, conducting inspections to check for quality and honesty in production and sale, and settling disputes within the profession. The jurisdiction was a lucrative one, for the grand pannetier collected fees for "receiving" new masters and for "visiting" each master once a year, and he also profited from the confiscations of merchandise, tools, draft animals, and so on, that he pronounced against bakers who violated the rules. From the beginning, the grand pannetier faced a constant jurisdictional challenge from the Prévôté de Paris (better known later as the Châtelet), which claimed a general police over the different crafts and trades. Over the years the sphere of influence of the grand pannetier alternately expanded and contracted, depending on the aggressiveness of the prevotal officials, the talent of the grand pannetier's own lieutenant, the mood of the parlement, and the concourse of circumstances.[2]

At first, the bakers appear to have favored the ambition of the grand pannetier, for he seemed to afford them protection against the sometimes brutal sentences of the prévôté as well as a special status and prestige that few other corporations could boast. Yet it became increasingly clear to the guildsmen that the grand pannetier was much less interested in preserving their prerogatives than in increasing his own. Successive grand pannetiers tried either to win control of the guild (by manipulation of its statutes or its members) or to destroy it. By the seventeenth century the guild was tacitly allied with the Châtelet against the grand pannetier in order to salvage its corporate integrity and autonomy.[3]

The conflict between the baker guild and the grand pannetier was complicated and aggravated by two other developments in the seventeenth century: periodic offenses by faubourg and forain bakers against guild exclusiveness and multiple efforts by the royal government to rationalize corporate organization and the police of work for political, economic, and ideological reasons, or for fiscal motives, or both. The guild strenuously resisted the efforts of the faubouriens and forains to win the right to make and sell *petit pain,* to sell in the streets, to improvise squatter markets, and to sell on nonmarket days, and the Châtelet systematically supported the Paris masters on these issues by pronouncing stiff measures against the baker-intruders. Unable to beat the city guild, a substantial number of faubourg bakers demanded the next best thing: the right to join it. All faubourg artisans save those from Saint-Antoine (whose franchises were not definitively accorded until 1657) were required to join or form a guild as a result of an edict of 1581.[4]

Some of the city guilds accepted artisans who had set up in the new quarters called into being by the growth of Paris. Others, like the bakers, saw no interest in diluting the monopoly by extending its benefits, and they obliged the faubouriens in their trade to form their own corporations. But the new corporations (e.g., the master bakers of the guild of the faubourgs Saint-Michel and Saint-Jacques, the master bakers of the guild of the faubourg Saint-Germain, etc.) aroused little enthusiasm, for they had very little to offer their members—at least by comparison with the plums that the city corporation ostensibly reserved for its members.[5]

In 1634 parts of the faubourgs Saint-Denis, Montmartre, and Saint-Honoré were absorbed into the city as a result of the new city boundary (*enceinte*) that was drawn. Citing a royal declaration of February 1635 ordering the city corporations to admit artisans from these areas, many bakers clamored for admission to the guild, and some, without waiting for an answer to their demand, began baking petit pain and exercising other "Paris" privileges. The grand pannetier encouraged these fusionist bakers, ostensibly to irritate the guildsmen, to win the gratitude of the future masters, and to benefit from the unexpected increase in the pool of admission fees. The guildsmen rebuffed these applicants, shrewdly basing their case not on their right to exercise the monopoly but on the technical point that these faubouriens, unlike artisans from other crafts that had established their own faubourg guilds, had never obtained any sort of mastership and thus were not entitled to assimilation consequent upon the 1635 declaration. After almost a decade of litigation, an arrêt du conseil found in favor of the city guild.[6]

The next major episode in this hundred years war erupted in March 1673 when the king issued an edict ordering all faubourg artisans to join city guilds. A second edict, published a year later, aimed directly at "exceptional" jurisdictions such as that of the grand pannetier, for it assigned the exercise of all seigneurial justice to the prévôté in the Châtelet. Both measures bespoke Colbert's determination not to impose homogeneity but to clarify lines of authority and demarcation in the light of a structuring rationality that retained a real suppleness. Guilds, like other corporations that constituted the social structure, derived not from the unction of nature but from the action of the state; thus all economic enterprise, grand or paltry, did not necessarily have to occur within a corporate framework. The government would incorporate artisans and merchants in certain circumstances (e.g., within the city of Paris and most of its faubourgs) and not in others (e.g., the faubourg Saint-Antoine) in order to test the underlying hypotheses concerning economic outcome and, no less significantly, in order to eradicate any doubt about the political authority for doing so—about the uncontestable political basis of social organization, the legitimacy of society depending on political will and not the inverse. In this same spirit of rationalizing reconfiguration, there was absolutely no place for privileged (that is, more or less private) jurisdictions such as the grand pannetier's, which rendered no benefits to either state or society, deriving neither from nature nor from unencumbered civic roots but from a sort of tolerated usurpation whose time had run out.

Thus Colbert announced the extension of the corporate model and at the same time, through other steps, implicitly acknowledged his willingness to consider other solutions on a case-by-case, or territory-by-territory, basis. It should be added that while the government spoke in a single voice once Colbert made his view known, there reigned no unalloyed consensus on how to utilize the guild system (or, more grandly, on what political economy to follow). The multiple nuances that inhibit us from reducing Colbert to a straightforward dirigiste encouraged a handful of would-be liberals to argue for the gradual undoing of monopoly practice. Their mercantilist colleagues saw the guilds as vehicles of a regulatory strategy without which French success in the international trade arena could not be assured. A traditionalist faction advanced the absolute necessity of corporate structures as constitutive of the social and political system on which everything rested, not for the (mere) realization of economic objectives. And there were fiscalists for whom the corporations may have had incidental or residual economic, social, and/or political advantages, but those advantages paled in significance next to the contribution of the guilds as instruments of financial extraction. In this context, it was plausible to perceive the measures of the 1670s as experimental.

The baker guild bitterly opposed the application of the merger order. The position of the grand pannetier was ambivalent, though in the end he supported the efforts of the faubouriens to infiltrate the city guild. If the grand pannetier hesitated for a moment, it was because he understood the close link between the edict on fusion and the edict on seigneurial justice, and he feared that if the former measure were successfully applied, it would be very much easier for the government to execute the latter, which would result in the suppression of his post. Nevertheless, the lieutenant of the grand pannetier ceremonially received as masters several faubouriens who posted their candidacies on the basis of the edict of March 1673.

The guild attacked those inductions in court and applied for an injunction to block further admissions. The royal procurator at the Châtelet intervened on the guild's side on two grounds. The first, juridical in character, was a direct assault on the Grande Panneterie. The procurator claimed that only he was empowered to receive the oath of new masters and to grant masterships. The second point, whose rationale hardly flattered the master bakers' sense of professionalization, concerned the police of provisioning and was surely more difficult to contest. The procurator argued that it was in the public interest to allow bakers to set up freely in several of the faubourgs since Paris urgently needed a large corps of bakers and "experience" proved that by subjecting bakers to "formalities of apprenticeship and mastership" one discouraged their establishment.[7]

The creation of the Lieutenance générale de police in 1667 strengthened the position of the Châtelet vis-à-vis the Grande Panneterie. Operating out of the Châtelet, the lieutenance was a modernized, pugnacious and omnicompetent version of the prévôté, which had been grappling with the Grande Panneterie from the time of Saint Louis. Given its newly defined prerogatives, the lieutenance was sharply disinclined to share the police of Paris with anyone, especially in the provisioning domain where it was most delicate.

Litigation over the case dragged on for many years in a climate heavy with acrimony. The guildsmen tried to execute a Châtelet order for destroying the oven of a new master from the faubourgs. The lieutenant of the grand pannetier ordered a guildsman jailed for "irreverence" and attempted to remove two jurés from their leadership posts. Though an edict published in 1678 confirmed the general dispositions of the edict of 1673 for all the crafts, two decrees, or arrêts du conseil, in August 1682 and April 1684, quashed the new masterships, ordered the new masters to withdraw from the guild, and effectively scuttled the fusion idea by banning the induction of new masters by the grand pannetier. In the meantime—another provocation—the grand pannetier had encouraged the pretensions of the faubourg Saint-Antoine bakers, not to join the guild but to exercise the craft on the basis of strict equality with the city masters in their privileged territory. This meant that they could bake petit pain, sell in the streets, and supply inns and wine shops. An arrêt du conseil of February 1677 had explicitly denied the Saint-Antoine bakers these rights.[8]

The fusion idea survived, however, and ultimately triumphed. It remained attractive to the fiscalist circle in the royal government, because the baker's guild, given the special jurisdiction to which it was assigned and as a consequence of its internal disorders, had escaped the series of extortionate measures (creation of offices of controllers, visitors, auditors, etc.) imposed on the other corporations. In a revitalized and enlarged form, the guild would be in a position to contribute more generously to the royal treasury.[9] The Châtelet had objected to fusion in the name of the public interest and as a means of chastening the arrogance of the grand pannetier. At bottom, however, the notion of a single guild, bound by a modernized set of rules, appealed to the rationalizing, codifying police of the prévôté. Multiple and rival corporate systems in the same craft created confusion and made the profession difficult to regulate. Moreover, by the 1700s, it was clear that the faubourg Saint-Antoine offered sufficient opportunity for new or outside bakers to set up without corporate interference. Finally, fusion this time explicitly meant the suppression of the anachronistic and troublemaking exceptional jurisdiction of the Grande Panneterie and on this ground alone was a measure of good police.

By now even the grand pannetier perceived the futility of resisting the combined forces of the guild and the Châtelet, not to mention the ministry. The Cosse-Brissac family, "owners" of the panneterie, were prepared in 1711 to accept what they had energetically rebuffed in the 1670s, because the king now offered a substantial financial indemnity. The edict of August 1711 ordered all faubouriens, save those from Saint-Antoine and other "privileged places," to join the city guild, which was to become the unique baker community for greater Paris and which was to be placed under the exclusive jurisdiction of the lieutenant general of police. Current faubourg masters had to pay 220 livres for (re)incorporation. Faubourg journeymen with apprenticeship and journeymanship time completed could earn admission by remitting 330 livres. An ambiguous clause invited other masters "without quality"—presumably self-appointed masters, often stigmatized as "false workers," who had practiced the craft for some time but who had never bothered to legitimize their professional status—to purge their taint and join the guild for 440 livres.

In theory, virtually no one was to be excluded, though in fact the guild leaders worked hard to keep out elements judged "unhealthy" (and the police did not object to this sort of obstructionism despite the remonstrances of the grand pannetier, who stood to lose fees as a result of such discrimination). All new masters were to exercise the profession as the strict equals of the original city guildsmen.[10] In return for abandoning the Grande Panneterie, the Brissac family obtained the right to collect all sums paid for reception to mastership by city and faubourg bakers for a period of seven years.

The merger/assimilation proved to be a long and slow process, in part because the city guildsmen dragged their feet. They saw no reason to hurry things along, especially since the mastership fees would not accrue to the community for seven years, and they resented the continued interference in their internal affairs. They quarreled not only with the grand pannetier and the faubourg applicants but also with royal officials, who became directly involved in corporate management in the period after the suppression of the Grande Panneterie and prior to the granting of the new statutes of self-government anxiously awaited by the guild. In May 1712 the government named the first jurés who were to administrate the reorganized guild. The old jurés vehemently opposed this measure, which deprived them of their posts and flouted the will of the city masters, who habitually elected their own leaders. The Brissac family denounced their move as part of a ploy meant to prevent many faubouriens from joining the guild and thus to deny the grand pannetier the fruits of the indemnity.

In 1716 the government apparently authorized the guild to hold free elections for new jurés in the hope that the city masters had by now learned a lesson in submissiveness. The masters, however, were still in a defiant mood, for they overwhelmingly voted for two dissident candidates, Vatre and Pelet, the leaders of the resistance movement since 1712. Denouncing this "spirit of cabal to elude the execution of the edict of August 1711," the government annulled this election and named the jurés, including Michel Decq, one of the new arrivals from the faubourgs. Henceforth, until the award of statutes, the guild was required to choose one of its jurés from among the faubourg masters.[11]

Tensions persisted between the city masters, who still dominated the guild, on the one side and the government, the newcomers, and the Brissac family on the other. The Brissacs wisely withdrew from the affair, selling their "rights" to a group led by tax farmer and financier Dufour for 108,000 livres. Dufour must have calculated either that a fresh consortium unencumbered by past hostilities could take advantage of the desire to put this conflict to rest or that the government would now back his pristine and reasonable claims for a profitable denouement. In 1718, when the seven-year concession was supposed to expire, the guildsmen petitioned the government to accord its long-awaited statutes. But Dufour complained angrily that the city masters had conspired to prevent some faubouriens from becoming masters and to induce others not to make the payments, and he demanded either an indefinite extension of his concession that would permit his group to recoup its investment or a cash settlement. In January 1719 the royal council ordered the guild to pay Dufour seventy-five thousand livres in three installments over the next fifteen months, after which it would be free of all claims emanating from the ex–grand pannetier.

Faced with this huge debt, the guild was now highly motivated to admit new members and to press for the payment of fees due for masterships. In the next few months corporate officials uncovered eighty-three faubouriens who exercised the profession illegally ("sans qualité") because they had failed to apply for admission and pay the dues. Under guild pressure, the authorities now threatened delinquent bakers with loss of mastership, closure of their shops, and demolition of their ovens. To assist the community in meeting its obligations, the king authorized it to collect four hundred livres for all new masterships and twenty livres for each apprenticeship license. In order to enable the guild to meet its first two installments to Dufour, the council's ruling authorized it to borrow fifty thousand livres.[12] The jurés were supposed to keep all these fees separate from other revenue in a special amortization fund and to present an accounting to the lieutenant general of police every three months. In practice the jurés failed to keep a careful record of collection and, as we shall see, corporate finances remained muddled through much of the eighteenth century.

The Jurés

In May 1719 the bakers won royal approbation for statutes that restored their corporate franchises and established the regulations that would govern the profession, with relatively few changes, until the end of the Old Regime.[13] The guild was to be administered by six jurés, three of whom would be elected each year by a plurality of votes for two-year terms by an assembly composed of the current jurés and all the former jurés, or elders (*anciens*), and by a revolving sampling of corporate commoners: twenty "moderns" (masters with more than ten years' seniority) and twenty juniors, or *jeunes* (masters with less than ten years' membership), drawn in alphabetical order from the list, or *tableau*, of masters. Electors absent without legitimate excuses were subject to fines of ten livres. Jurés were supposed to be chosen from "persons known for their experience and probity . . . among the most notable of the masters [moderns only] of the guild."

It is unclear whether candidacies were discreetly arranged in advance by the jurés and the elders or whether nominations were commonly received from the floor. The official voting report and the apparent dispersion of votes (e.g., in 1745: 91 for Tribouillet, 76 Bourguin, 55 Feret) suggest that there were normally more than three men standing for the three openings, though it is possible that there were many blank ballots. In only two cases between 1745 and 1775 did the official report mention an unsuccessful candidate: Rougelin received only twenty votes in 1756 and disappeared from view and Delahogue attracted only eighteen votes in 1760 but was elected the following year with sixty-eight votes, the highest total.[14]

Elders constituted the largest and most influential voting bloc and they tended to choose candidates from among their own group, which resulted in a double tendency, oligarchic and dynastic. In the course of one corporate generation, from 1745 to 1775, eleven families had at least two of their members elected juré (Lapareillé, Decq, Fleury, Cousin, Chappe, Félix, Genard, Fourcy, Leroux, Legrand, and Feret). The son of a juré who was himself elected juré automatically enjoyed "precedence" over his cojurés, even if the latter were older or had obtained more votes than he. The dynastic tendency was reinforced by intermarriages among juré families—the Leroux and the Lapareillés, for example.[15] By midcentury, it is clear that the process of faubourg assimilation had been fully completed. There were as many jurés from the faubourgs as there were from the city (and some of the city jurés had in fact moved in from the faubourgs).

The jurés did not all have the highest intellectual and moral credentials. They had a multitude of more or less complete accounting and record-keeping tasks, yet not all of them could write or read.[16] Nor were they all able to boast of reputations for honesty and irreproachable behavior. Jacques Mouchy was one of several masters condemned for selling short-weight bread. Nicolas Driancourt was sought by the police for sneaking off without paying his rent and for leaving his house and shop in a dilapidated condition.[17] We shall have occasion to meet other jurés whose conduct was not exemplary. The revised baker statutes of 1785 called for the immediate dismissal of any guild officer convicted of an "infidelity" and disqualified masters with such convictions from standing for office. In retrospect, under the burgeoning assault of guild critics, this eleventh-hour reform seems strained, a concession extracted under pressure.

The first juré elected was to become the accountant, responsible for all guild finances. He could disburse no funds without a deliberation of the corporate bureau, a sort of executive council composed of jurés and elders that was to meet every Monday and Thursday afternoon. He was to record all transactions in a special register kept in the guild's safe. The juré-accountant was to report on the state of guild finances before 1 November every year to an assembly composed of jurés, elders, ten moderns, and ten juniors. If he owed the guild any money according to the assembly's decision, he was expected to pay it back immediately. Similarly, if he had advanced money, the guild would reimburse him as soon as it could. Any dispute that could not be settled in the assembly was to be resolved by the lieutenant general of police.

Corporate administration was not merely a question of financial management. In executing the police of the baking industry, the jurés exercised solemn public functions that were meant to serve the general interest. The welter of privileges they cultivated and protected for their colleagues' advantage could be rationalized as the counterpart to this public service. From the members' point of view, the mission of the jurés was to defend those privileges, as enumerated in the statutes, and more broadly their interests, as they unfolded in daily practice. Prior even to this daunting task, however, was a more diffuse yet urgent responsibility. Membership in the guild invested the master, in both concrete and palpable ways, with a substantial allotment of social capital. Mastership conferred prestige and honor; it was a powerful marker of distinction. It made these bakers the interlocutors of the king by joining them to the Great Chain of Being that led directly to him at the earthly summit. It integrated (or incorporated) them into a social system that remained highly selective. Mastership was both a potent source of self-esteem and an instrument of leverage in the everyday world. The masters counted on the jurés to harbor and fructify this capital.

The most important statutory privileges guaranteed the masters against excessive or unauthorized competition of any sort (e.g., from forains, faubouriens, journeymen, starch makers, remarried baker widows, small-grains dealers, or others). The statutes accorded exclusive prerogatives to bake and sell certain breads in certain times and places. They also invested the masters with the right to be policed by their own colleagues rather than by outsiders; to employ apprentices, to determine hiring conditions for journeymen, and thus to control the labor market; to favor the professional ascension of their sons; to practice their craft anywhere in the kingdom without soliciting special permission; to organize a *confrérie* and various forms of mutual assistance. In addition to safeguarding these titles and immunities, more generally the jurés argued the cause of the masters, individually and collectively, in remonstrances to the lieutenant of police, or to higher instances, on such issues as price policy, supply responsibilities, access to raw materials, debt recovery, credit terms, and so on. Formally or informally, the corporate leaders defended bakers who fell into trouble. They aided those who were ill or temporarily victimized by accident or family disaster. Beyond material succor, they organized symbolic expressions of recognition calling public attention to the masters' status. They defended professional honor even as they fended off rival interests. The more energetic officers, taking notions of community very seriously, attempted to reconcile bickering bakers, though they frequently found themselves at the center of quarrels. In a word, successful and highly regarded jurés paid close attention to their constituents and assisted them in many of the problems, ordinary and extraordinary, that they encountered day to day.

The jurés unabashedly used their broad police powers to entrench and enhance the privileged position of the masters. Their rivals and victims— unincorporated bakers and other artisans and merchants over whom the jurés exercised a "right of inspection"—regarded the guild privileges as unfair advantages or abuses. In the view of the lieutenant general of police, however, those privileges, in most cases, were not merely legitimate but necessary and functional aspects of the organization of the provisioning trade. In theory, there was no discordance between guild and public interests: each was defined in terms of the other. In a sense the jurés earned the guild's right to enjoy a (limited) monopoly and certain prerogatives in return for satisfying the imperious demands of the lieutenant general that the shops and markets be supplied at all times with good quality bread in adequate quantities at a reasonable price.

The jurés' police powers, to be exercised over the whole bread industry, not just over the guild, made them the arbiters of the profession. They acted as the direct auxiliaries of the lieutenant general: they launched investigations that included spying on suspects; they conducted "visits," usually in the company of police officials, that involved searches and seizures; they participated in the issuing of summonses, in arrests, and in trials; they were named as mediators by the commercial courts in cases involving provisioning issues. It was rare that the authorities failed to heed a juré complaint or preferred the word of an ordinary baker to that of a guild officer. The leaders fully understood that the more successfully the guild policed itself, the less internal interference it faced from royal authorities.[18]

One of the most demanding and controversial of the police functions of the jurés was the surveillance of the forain bakers. It was the responsibility of the jurés, in conjunction with the commissaires, to make sure that the forains supplied the market adequately, that they did not leave the market prematurely, or remove bread from the market for the purpose of storing it or taking it home, that they sold bread only on market days in the appropriate places, that they baked no luxury breads, and that they abstained from furnishing taverns and inns. The jurés participated actively in the police of work: they regulated apprenticeship arrangements, tried to impose conditions governing job placement, imposed a code of discipline on journeymen, tracked down fugitives and illegal workers in the inns as well as in the bakerooms, and sought to prevent or expose all forms of labor insurgency or work associations. The jurés labored to purge from the industry bakers without "quality" or "title"—not only forains or journeymen who set up illicitly in the corporate territory on their own or behind a borrowed name but also artisans from allied professions, most notably the starch makers, who took up unauthorized baking as a sideline.[19]

Of crucial importance in the eyes of the lieutenant of police was the mission of the jurés to ascertain the quality of the bread produced by all sorts of bakers, as well as to verify its weight. Below-standard bread was seized and usually distributed to the poor, while the bakers in question received summonses that could result in serious punishment, especially in times of scarcity and stress. The jurés were charged with the task of keeping unauthorized persons from dealing in flour seconds and by-products and of tracking down bakers who used their bakeries as covers for wholesale trade in grain and flour (that is, reselling merchandise they purchased ostensibly for conversion into bread by them).[20]

Article 36 of the statutes of 1719 empowered the jurés to visit all the mills serving Paris in order to search for "abuses" (e.g., defects in quality or cheating on transactions) in the milling process or in the mixing of flour, and to check on the reliability of the weights and measures used by the millers. Throughout the eighteenth century, as before, the bakers remained convinced that the millers as a group were dishonest ("thieves," in the words of Porcherons baker Jean Dumas) and that the mill was a trap. In one of many indignant petitions they addressed to the Châtelet, the jurés in 1726 asked for a more sweeping authority to police mills in light of a proliferation of "violations and frauds." The lieutenant general encouraged the jurés to investigate the mills thoroughly, and "even to seize merchandise."[21]

The jurés toured the milling areas at least once a week. On one day in September 1739 the chambre de police judged thirty-nine millers who were denounced by the jurés for failing to have proper scales and weights in their mills. That was the most common miller offense and it was considered a grave affair by the bakers, for it prevented them from verifying on the spot the amount of "loss" claimed by the miller in the grinding process. Less frequently the jurés uncovered flour or wheat "unfit to make bread" in the mills. Though the statutes did not mention mealmen—this subspecialty was relatively new in the early eighteenth century—the jurés inferred from article 36 the authority to "surprise" them in the storerooms and examine their merchandise.[22]

As a rule the jurés gave the statutes a loose construction, thereby assigning to themselves a broad authority to take initiative in areas not explicitly charted out on paper. For example, to control the recruitment and placement of journeymen more effectively—a problem evoked summarily in the statutes—the jurés named clerks and established special procedures (which were, incidentally, contested not only by many journeymen but also by some masters as well). Similarly, though the statutes remained silent on the question of business (*fonds*) transfers, the jurés took it upon themselves to oversee transactions in order to make sure that corporate interests were not prejudiced.

The statutes defined the responsibilities of the masters as well as those of the jurés. They composed a list of dos and don'ts that placed limits on their ambitions, created mechanisms to avoid and/or reduce friction among them, informed them how to protect their interests, and how to set up general standards of conduct as well as specific procedures for concrete problems such as hiring and firing personnel, training sons, and forming partnerships. A number of articles treated commercial practices in the shops and markets, while other clauses considered a master's obligations to the guild. The jurés were, for instance, supposed to be "honored and respected" by their colleagues, and they did not hesitate to have irreverent masters arraigned in the procurator's court in the Châtelet.[23]

At least one juré could be found every day of the week at guild headquarters, a modest two-room office that was located for many years on the quai des Grands-Augustins. The larger of the two rooms served as an assembly hall. Decorated with paintings and engravings of Saint Honoré, one of the profession's religious patrons, Christ, Louis XIV, and Louis XV, it contained seventy-two chairs plus one armchair, a large oak table, several writing desks, and two cupboards or wardrobes. Framed lists of the elders and of the masters also adorned the walls, along with a tapestry and several mirrors. The jurés worked and received in the smaller room, which also boasted a tapestry and several paintings as well as a framed memorial list of deceased masters. The room was furnished with a table, several chairs, forty stools, and two cupboards. The jurés stored guild records and valuables in the four cupboards. In one cupboard were kept the guild's silver plate, a reliquary of Saint Honoré, a Christ on an ebony cross, several representations of Saint Honoré and the guild's other patron saint, Saint Lazare, a holy water basin, and an offering plate. Another contained corporate registers, papers dealing with litigation, royal legislation, and police sentences, regulations and guild statutes, and a safe that held some papers (receipts for purchases of offices imposed by the king and for annuities) and cash (3,395 livres in 1752, 710 livres in 1760). Guild seals, other registers, scales and weights, and a clock were found in a small cupboard.[24]

Confréries

Another obligation that the statutes imposed on the masters concerned the baker *confréries*, or brotherhoods. Article 14 required each new master, as part of his admission ceremony, to donate twelve livres for "confrairie" plus a three-pound candle in honor of Saint Lazare. Article 15 stipulated that during their first three years of mastership, each guild member would contribute a one-pound candle for Saint Lazare's feast day. Article 16 taxed each member forty-five sous a year for "confrairie," to be used for the celebration of the divine service on the feast days of Saint Lazare and Saint Honoré and in services for departed masters and their wives.

Saint Honoré, patron of the bakers, nineteenth-century
depiction. *Source: courtesy Deutsches Brotmuseum,
Ulm/Donau.*

Little is known about the precise origins or the evolution of the baker
confrérie(s) or the functioning of confréries in the eighteenth century. Delamare suggests that the baker guild probably began as a religious or pious
association. Over the years the bakers apparently had considerable difficulty in
finding funds to support the brotherhood. The first patron of the bakers was
Saint Pierre aux Liens, perhaps because his feast day falls on what was traditionally esteemed to be the first day of the wheat harvest. He was displaced
later by Saint Lazare, who the bakers believed could help protect them from
leprosy, to which their craft exposed them, so they feared, as a result of constant contact with heat and fire. Though the bakers never renounced a commitment to Saint Lazare, at some point they moved the headquarters of their

confrérie to the Church of Saint-Honoré, making its namesake their patron, it is claimed, in honor of the fact that he had been a baker before he became a prelate.[25]

Though Saint Honoré may well have eclipsed Saint Lazare, both brotherhoods appear to have subsisted in the eighteenth century because the dues notices sent to the masters indicated that half of the forty-five sous of annual fees was for the confrérie of Saint Honoré and the other half for that of Saint Lazare. In addition, each master was expected to pay five sous five deniers a year to Messieurs les Prieur et Prêtres de la Congrégation de la Mission du Prieuré et Couvent de Saint-Lazare, the money commutation of the petit pain the bakers had contributed to the religious each week since the Middle Ages in return for the congregation's pledge to care for stricken bakers in its leper house. In the eighteenth century the guild maintained a chapel in the Church of Saint-Lazare, where a perpetual weekly low mass had been founded for the baker dead. Most religious activities, however, took place in the church that housed the Saint Honoré brotherhood.[26]

Since the sixteenth century the royal government had become increasingly critical of the conduct and management of many confréries in the various métiers. Some "degenerated" into circles of libertinage and irreligion, others served as a pretext for a kind of political subversion, while still others drained the guilds of their resources. In the early eighteenth century, the government may have suspected the baker confraternities of falling into this third category of abuse. There was some evidence that the juré-accountants, who were ordinarily heads of the confréries, failed to keep separate the finances of the brotherhoods and those of the guild, and covered the former's deficits with the latter's revenue.

The royal council pressured the bakers, and other corporations as well, to abandon the practice of treating the brotherhood as a branch of the guild (as the statutes implied it was) and to set up an entirely distinct and indeed independent confraternal administration. When royal commissioners began to review and revise the books of the jurés, they systematically disallowed any claims for disbursement made for confraternal purposes. Thus, for example, in 1763 the jurés were told that the 260 livres that they spent for an engraving of Saint Honoré would have to be paid for by them or by the brotherhood, but not by the guild.[27]

As long as the corporate statutes required masters to pay confraternal dues, a genuine administrative separation between guild and confréries appeared inconceivable. The royal government did not force the guild to eliminate the statutory connection until after the major corporate reorganization in 1776.

Until then, one finds that the jurés spent considerable time battling refractory masters who had no appetite to pay the confraternal dues. In some instances, they squabbled vehemently and the jurés sought not merely back dues but "reparations" for "insults." In some cases the litigation lasted for years. The royal procurator habitually condemned the masters to pay the dues and expenses but refused to gratify the jurés' vanity and taste for vengeance. The jurés in turn were pressed and frequently sued by the mission of Saint-Lazare for the recovery of arrears. In one case in 1759 the religious referred to uncollected masters' dues from as far back as 1721.[28]

The confrérie was probably the vehicle through which the guild managed its mutual-aid activities. It helped to organize funerals for members and their wives for which it provided ceremonial silver plates and the pall. This rite of passage signaled to the world that the baker family, through its participation in the guild, had amassed a certain social capital. As a marker of honor, and perhaps also of prestige, and as the announcement of important familial and commercial transitions, the guild funeral often marshaled sizable numbers of masters, and sometimes issued in informal conclaves dealing with business and with corporate politics. The guild also arranged for commemorative masses to be celebrated through the confréries.[29]

It is not clear what social services the guild and the brotherhoods rendered to the living. One encounters myriad vague allusions to assistance for old and/or indigent masters, but there is no mention of how this system may have operated. It was funded, at least in part, by fines levied against masters for transgressing the corporate regulations.[30] We do not know whether the baker brotherhoods, like other corporate auxiliaries, provided sick care, forms of unemployment or professional disaster insurance, or facilities for credit. Nor can we say to what extent the confréries served as poles of sociability, providing opportunities for interaction among masters and master families apart from funerals or official ceremonies.

New Statutes

For reasons that are not clear the guild sought new statutes in the mid-forties. They needed this revision, according to the royal government, because the statutes of 1719 could not "foresee" all the measures that later proved to be "necessary for the good order of the guild." But in fact astonishingly little was modified or innovated in the new statutes approved by the king in May 1746. In order to obtain standardized and detailed records, one article enjoined the jurés to keep at least seven separate registers (e.g., for deliberations, apprentice licenses, uncompleted apprenticeships, masterpieces, admissions, litigation, and police sentences)—a measure without any policy implications. Another article, equally limited in significance, proposed rewarding jurés and elders with twenty sous in tokens "in order to excite them to attend [assemblies] regularly and to deliberate more sagely on the said business of the day."

Of more importance was article 54, which ordered the jurés to draft a tax roll for the purpose of imposing the capitation on the members when notified each year by the lieutenant general of police of the global sum due from the bakers.[31] This was an unpleasant and onerous task that infringed upon the jurés' time and created tensions between them and their master-colleagues. Often masters quarreled with the jurés over the tax bracket in which they were placed, many were several years late in making payments, and a handful refused to pay despite threats of having to lodge royal troops and pay fines and interest. The jurés frequently had to take them to the courts, and in numerous cases they were still trying to collect from the estate after the death of the master involved.[32] On the other hand, this tax assessment power increased the leverage of the jurés over the masters. In order to induce masters to obey or support them, the jurés did not hesitate to threaten them with an increase in their capitation ranking.[33]

The capitation responsibility turned the attention of the jurés outward as well as inward, broadening their political contact with Parisian authorities. In the hope of obtaining a favorable capitation quota, the jurés courted the counselors in the Châtelet who had responsibility for corporate impositions. The jurés also had to collect the *vingtième de l'industrie,* though it appears that they did not have the liberty to draft this assessment list themselves.[34]

Only one new article was of genuine portent, and it provoked a torrent of protest that delayed the legal promulgation of the statutes for eleven years. The dispute turned on article 36, which initially forbade all masters to wheel carts or drive horse-drawn wagons full of bread through the streets for distribution to customers (and potentially for spontaneous spot sales on the street) and subsequently exempted from its strictures all masters save those from the faubourgs, who were "maintained" in the "traditional" right to serve their clients in this fashion. Now the (preamalgamation) guild membership had always been against this practice on the grounds that it constituted a form of surreptitious and illegal competition, since bakers from the faubourgs would be able to encroach on the terrain of the city shop bakers by bringing their wares to the very door of consumers. The guild leadership had also been hostile to street commerce on the grounds that it was extremely difficult to police the quality and weight of bread distributed outside the normal channels. The issue had been hotly debated when the faubouriens began to enter the city guild in large numbers after 1711, and it remained a source of tension—one of the few persistent notes of discord—between the city and the faubourg masters throughout the first half of the century.

Periodically, the jurés cracked down on the faubourg masters by seizing the bread sent around in carts or wagons. The faubouriens invariably appealed and in most cases they seem to have avoided judicial condemnation. In the hope of putting an end to further doubt and further wrangling, Lieutenant General of Police Marville issued a sentence in 1744 confirming the right of the faubouriens to deliver by cart or horse and wagon and forbidding the jurés "to trouble them in any way." (Remember that we are dealing here with *masters* located in all the faubourgs, not with the nonmasters clustered in the faubourg Saint-Antoine and habitually dubbed faubouriens.) The jurés appealed, but in vain, for an arrêt of the parlement ratified and reinforced the police sentence favoring the faubourg bakers. As a result of this decision Marville compelled the guild to modify the original article 36 from the proposed new statutes that prohibited all bakers from delivering with carts or horse-drawn wagons.[35]

This modification, imposed without consulting them ("a surprise to us") and contrary to their will (in a formal deliberation, the guild assembly voted against incorporating the parlementary arrêt into the statutes) infuriated a large number of masters, who felt that it was not only "an injustice" but a source of "serious damage" to their business. One hundred ninety-one insurgents joined together to sue to block the implementation of the new statutes. Their argument was familiar, featuring the usual combination of jealous self-regard and enlightened devotion to the general good. The exception clause, the plaintiffs contended, authorized the masters of the faubourgs to operate "mobile bakeries" with which they would "steal" the customers of the city bakers "by the facility they would offer the citizen of getting his bread without having to go to fetch it." This system would also "favor frauds" by placing transactions beyond "the vigilance of police officers."[36]

Rejecting their case out of hand, Marville affirmed that the modified article 36 did not endanger the public interest in the slightest. On the contrary, the lieutenant general applauded the faubourg exception precisely because it made it easier for consumers to obtain bread and he hinted that he would not be averse to seeing the delivery prerogative extended to all masters. The insurgents tenaciously pursued their opposition for years, until they gradually became convinced either that it was futile to hope for a change or that the clause was not as damaging as they had feared. Prompted by two elders, a substantial number of the "opposants" abandoned their suit in 1752. Another group, led by a master baker who had since dropped out of the profession, negotiated its withdrawal in exchange for an indemnity of 450 livres to cover legal expenses. It is likely that some of the insurgents never formally accepted the new statutes, which were finally registered by the parlement in the fall of 1757.[37]

The real rupture in continuity, however, occurred in the corporate reforms of August 1776, which were in general more drastic than historians have supposed. "Re-created" in 1776, after a brief period of outright suppression, the guild underwent a democratization of recruitment, a far-reaching reform that destroyed much of the old corporate ethos. In place of the discriminatory Manichean system of admissions, the royal legislation prescribed that all candidates would be treated in the same way. Not surprisingly, annual admissions sharply increased from 1776 until the Revolution.

The statutes of 1783 nuanced and codified the changes presaged by the reform law of 1776. Formal apprenticeship was no longer a prerequisite for upward mobility, though journeymen with completed apprenticeships could become masters at age twenty, while those without apprenticeship had to wait until age twenty-five. The only statutory advantage accorded masters' sons was the authorization to stand for mastership at the age eighteen if they had worked in their father's shop for two years. Instead of preparing a masterpiece, candidates had to submit to an examination in all matters pertaining to the profession lasting at least two hours. To stand for examination the applicant had to be recommended by two master bakers and two "notable bourgeois," who were familiar with his reputation and his character. There was still room for abuses on the part of the leadership, but much less than before. The new system was far more open and straightforward. It provoked very few complaints of unfairness or corruption.

There were no seismic changes in the way in which the guild was to be administered. Syndics and their assistants, called *adjoints,* replaced jurés in title, but their functions were very much the same, especially in the "police" domain. Collegial responsibility was stressed. Unlike the juré-accountant, the syndic-receiver, or treasurer, did not precede his colleagues in authority. The statutes of 1783 did not give the same primacy to finances that the earlier ones had, perhaps in part because the royal government had succeeded to a considerable extent in clarifying corporate finances by forcing a systematic reckoning of accounts and had assumed guild debts as part of the grand reform. In general internal corporate management seemed to be less complicated than before. There seemed to be less to do now; as a result the statutes made many fewer specific and detailed prescriptions on how to conduct business. The old guarantees against abuses in expenditures remained in force, but there was less preoccupation with books and records. Theoretically the elders would be less dominant than they had been before, though they were still bound to form an aristocratic pressure group. In place of the assemblies that they once controlled, the guild was to be governed by an assembly of "deputies," drawn from among the moderns—there was still a Genevan reticence toward real democratization—and elected by a general assembly of all masters. Officers or deputies convicted of "infidelities" were to be removed from their positions, and masters with sullied backgrounds were to be disqualified from standing for office.

The traditional professional dos and don'ts persisted. City and faubourg masters seemed fully reconciled in interest and status. Together they still lorded it over the forains and the Saint-Antoine faubouriens, though the lieutenant of police, backed by the parlement, excised an article prohibiting tavern keepers from buying certain categories of noncorporate bread. These unincorporated bakers suffered the same restrictions on their freedom to bake and to sell on and off the market. Masters were now to pay syndics twelve sous for each of two mandatory visits they would make to their shops each year, though the syndics retained the right to make unremunerated surprise inspections whenever they wished. Masters were henceforth required to post their names in large letters over their shops. Now that mastership had become more accessible, illicit bakers ("without quality") were to be pursued even more energetically than before.[38]

Under Royal Surveillance: Guild Financial Management

Along with the police functions of the guild, the corporate issue that seemed of greatest concern to the royal government was financial management. Financial disorder, it was believed, would discredit the guilds and indirectly stigmatize the government, since the guilds were considered semipublic organizations. Mismanagement would also provoke resentment and perhaps rebellion within the membership, further crippling the guild's ability to perform its duties. Heavily indebted communities tended to sink into lethargy or to resort to expedients that were dangerous and dishonest. Moreover, financial disarray sometimes led to, or was caused by, corruption.

The government wanted the corporations to remain in honest and able hands for the sake of good order and right principle. But its attitude was not wholly disinterested. A healthy corporate system was a precious fiscal resource for the government as well as a credit to society, as the ministers of Louis XIV had

vividly demonstrated. If the government failed to supervise corporate administration, the guilds might not be in a position to contribute when called upon.

The government's policy, then, was profoundly ambivalent. On the one
hand it reproached the guilds their indebtedness, an emblem of incompetence
or malfeasance, and on the other it periodically aggravated that indebtedness
by imposing extraneous and often trying burdens on them. (Arguably, the
corporate posture was also ambivalent, at least concerning royal fiscal exactions, for as oppressive as the resultant indebtedness was, the funds that the
guilds advanced to the monarch imperceptibly constrained his freedom of
action in their regard, and made him subtly dependent on them.) Louis XV's
ministers were not comfortable with this contradiction. On the whole they
emphasized housecleaning, retrenchment, and balanced budgets for the sake of
corporate integrity, efficacy, and longevity. They hectored the guilds for over
half a century until many of them settled their old accounts, more or less
faithfully, and placed their new ones on a sound footing. But even reform-
minded ministers needed money; not all of them were able to resist the blan-
dishment of fiscality and easy prey.

The bakers had been extraordinarily fortunate to escape unscathed from the
epidemic of barely disguised fiscal extortions that struck the guilds between
1691 and 1710, a period of great strain for both Louis XIV and his twenty
million subjects. Apparently baker immunity resulted not from the need to
shelter these agents of subsistence fulfillment but from the exceptional seigneurial jurisdiction to which the guild was still subject and from the uncertainties over the future composition and administration of the city corporation.
In the legislation ordering reorganization of the whole profession, the king
frankly expressed the hope that the new, enlarged community would be able to
purchase the many offices created between 1690 and 1710 and meet the obligations that other corporations had discharged.[39]

For some reason, however, this retroactive adjustment never seems to have
come about, perhaps because the bakers took so long to settle with the grand
pannetier and obtain statutory autonomy. The government imposed two
charges on the bakers during the Regency, and neither seems to have been a
legacy of Louis XIV's regime. The first was a sort of forced loan—and the
obligatory purchase of a 2 percent annuity at a principle value of 5,267 livres 10
sous raised between 1720 and 1723. The second was a so-called *droit de confirmation* in 1721 for an undetermined amount. This *droit* seems to have subsisted for
a number of years, for in 1729, master baker Mongueret received notice from
the lieutenant general that he had three days in which to pay his share of seven
livres eighteen sous.[40]

The major fiscal blow against the guild during the reign of Louis XV was struck in 1745. Justified as an extraordinary war-linked measure, an edict of February created offices of "inspector-controller of the jurés" in each of the guilds and authorized each of the guilds to buy up these posts in order to avoid suffering a humiliating and idiosyncratic internal supervision. The speed with which the jurés moved bespeaks the horror with which they contemplated the prospect of falling hostage to unknown authority. To pay off this debt the masters had to tax themselves. In 1753 the annual *droit de rachat* per member was three livres; it may have been higher in the years immediately following the edict of February 1745. We do not know how long this special tax persisted, though there is evidence that a number of masters tired of paying it rather quickly.[41]

In 1758, in the midst of another war, the king attached a heavy supplement to the sale price of 1745, wresting another sizable sum from Paris guildsmen. The bakers' share was thirty-two thousand livres. Those guilds that paid quickly would benefit from certain tax remissions; those that failed to pay faced (the threat of) the loss of all their privileges and immunities. The bakers freed themselves from the new mortgage on their corporate future in less than a year, again by means of loans. At least some of the masters had full confidence in the financial solvency of the guild, for the lenders were five bakers including the current juré-accountant and three elders.[42]

Another piece of legislation pertaining to the guilds appears on the surface to have been motivated by fiscal considerations, but in fact it may have had more substantive aims. In an edict of March 1767 Louis XV announced that twelve persons would be named by him to masterships in each of the guilds. On the basis of traditional practice, it was probably not unfair to presume that the government wanted to raise money by selling these would-be masterships back to the guilds, which would readily buy them in order to protect their control over recruitment. The parlement denounced this law as a purely fiscal device; historians have followed its lead.[43]

Very little is known, however, about the application of the measure. The king claimed his aim was to make mastership available to deserving journeymen who had been denied upward mobility more or less arbitrarily as a result of exclusivist guild policy. A close reading of the parlementary remonstrances suggests that the court was not sure of itself, that it did not know precisely what the government intended. For most of the argument focuses not on the burden of fiscality but on the dangers of short-circuiting the usual recruitment pattern that assures craft competence and of degrading mastership by admitting untrained and unsocialized outsiders. The parlement also warned that the competition from twelve new masters would ruin many members of weaker guilds.

These were years during which the controller general prepared many measures meant to liberalize and invigorate economic activity. The government proved that it was not indifferent to the stringent criticisms of the friends of liberal economists Gournay and Quesnay, among whom were a number of philosophers and public officials who campaigned for guild reform. A priori there is no reason not to take Louis XV at his word in his preamble justifying the creation of the twelve extraordinary masterships. While it is not certain that the twelve "royal" masters were actually named to the baker guild, there is no evidence to suggest that Louis XV gave the guild an opportunity to buy up this reform.

The financial history of the guild in the eighteenth century turns much less on royal fiscality than on the direct and indirect consequences of the suppression of the grand pannetier's jurisdiction. At issue was not the question of the guild's obligation to the concessionaires of the Brissac family, though in fact the latter had to sue the guild in 1720 in order to obtain payment of the last third of their indemnity.[44] Rather it was the way in which the jurés handled the financial business arising from the need to pay back large debts contracted to satisfy the grand pannetier's successors.

The edict of 20 January 1719 had authorized the guilds to levy a twenty-livre fee on new apprentices and a four-hundred-livre fee on new masters (in addition to the 220-livre fee paid by faubouriens for admission to the guild). The money raised from these sources had to be placed apart from all other revenue in a debt amortization fund. These fees were construed as extraordinary and provisional levies that the government would reduce or abolish once the bakers' corporate debt had been reimbursed. In order to stay closely informed, the ministry enjoined the jurés to provide detailed reports on collection at regular intervals.[45]

In arrêts du conseil of 5 April and 3 May 1729, the government complained that the jurés had not filed periodic financial statements. Moreover, it suspected that the seventy-five-thousand-livre debt had long been paid off and that the jurés had nevertheless continued to collect the special fees and to use them for unauthorized purposes. The arrêts ordered the jurés to present records of all receipts and expenditures in the last decade within two weeks. The jurés refused to cooperate despite threats of arrest and seizure of their papers. Reiterating its suspicion that the jurés had failed "to keep an honest and precise record" of collections and that in addition they had diverted funds "for their personal profit," the government again enjoined them to submit their books.

The government also devised a method for verifying juré claims. It ordered all masters, journeymen, and apprentices to submit their letters of mastership and licenses of apprenticeship, along with receipts of fees paid, so that auditors could compare the *real* sums collected with the fees acknowledged by the jurés. Bakers who did not comply within a month faced the threat of loss of mastership (or disqualification from rising to mastership in the case of delinquent apprentices or journeymen), the closing of their shops, and the demolition of their ovens. What were the masters to make of this menace? The government was hardly going to cripple the ranks of the provisioning corps. Yet it certainly could afford to make examples. Moreover, since the concerned masters had all presumably paid, they might reasonably be supposed to have been curious about the utilization of the funds they contributed.[46]

In an arrêt of 28 January 1731 the royal council concluded that the jurés' failure to submit their records to scrutiny was proof that they had embezzled some funds and used other monies for purposes other than the mandated reimbursement of loans. The government charged the ex-jurés (elders) with "a criminal conspiracy" to cover up their corruption by electing jurés "devoted to them." The current jurés acceded to the elders' demands for banquets and cash bonuses and then doctored their accounts to conceal these expenses. Determined to punish the guilty, the council named the lieutenant general of police to try (both present and former) jurés for administrative abuses as well as for disobedience of royal command.[47]

The rank and file had proved to be no less recalcitrant than the leadership. The government conceded in April 1731 that "the great majority of masters and apprentices" had failed to present their titles and receipts for inspection. It gave them a "final" two weeks' grace period and renewed the dire threats uttered in the previous legislation.[48] The royal council presumed that the bakers acted in solidarity with the elders—further proof of the latter's perfidy. More likely what moved them was a distaste for outside police tutelage of any sort. In any case, it is unclear why the government did not execute the initial order more vigorously and why it imagined the second arrêt would inspire more obedience (or more fear) than the first.

No bakers appear to have been punished for failing to show their papers, despite all the blustering. But one could not infer from this that all the bakers complied. On the contrary, it is quite likely that large numbers continued to ignore the injunction and that the government simply gave up more or less discreetly when it realized that its bluff was called. The government continued, however, to press the ex-juré-accountants, though it arrested no one. A handful appear to have submitted their books in 1731. Faced with the "obstinate refusal" of the others, the government threatened them in May 1732 with a one-thousand-livre fine and loss of their rank of elder.[49] More ex-jurés seem to have complied, perhaps because they took the threat of a fine more seriously than the menace of prison.

As late as 1737 there were still refractory ex-jurés, for the government raised the amount of the fine to three thousand livres.[50] We do not know ultimately how many years of records the bakers submitted or how much evidence of malversation the government uncovered. The fact that the royal council did not formally obtain the indictment of a single elder suggests that it could not make a compelling case or that restitution was quietly made.

In any event in an arrêt du conseil of 12 June 1740, the ministry officially declared that the Grande Panneterie debt had been wholly acquitted and that as a result—"in order to make it easier for those who aspire to mastership to attain it"—the fees for entry were reduced from four hundred livres to two hundred livres for outsiders climbing up as apprentices with masterpieces and from one hundred livres to fifty livres for sons of masters. The special apprenticeship license charge was lowered from twenty to five livres.[51] It is not certain, however, that this arrêt was ever applied. If it was executed, it did not survive a long time, for the guild statutes proposed in 1746 set the mastership fee at four hundred and two hundred livres. Apparently the guild leadership succeeded in convincing the government that it needed the larger dues in order to meet its obligations, including the amortization of new debts contracted to purchase the offices created in 1745 for the purpose of raising money to pay for war.

The inquiry into baker finances should not be viewed as an isolated case or an investigation limited specifically to problems arising from the suppression of the grand pannetier's jurisdiction. The examination of the baker guild was part of a vast enterprise launched by the government in 1716 "to revise" and "audit" the accounts of *all* the guilds, even as the grand pannetier debt affair was only one episode, albeit an important one, in the story of baker management. Citing the chronic indebtedness of the guilds and their inability to meet their obligations, an arrêt du conseil of 3 March 1716 appointed a special royal commission to verify both the claims of creditors and the books of the corporations. The arrêt suggested that one of the chief reasons for the financial predicament of the guild was the failure of the jurés to present accounts of their administration. The jurés were given one month to present all their records since 1689 to the commission, which would then evaluate them and rule on their validity.[52]

As we have seen, the jurés delayed and resisted as long as they could. But the tenacity and determination of the commission never faltered. Ultimately a majority of the jurés in most communities capitulated. The commission sat for three-quarters of a century. Understaffed and faced with an enormous labor of research and verification, it was able to review the accounts of only about half of the guilds. If one judges by the baker experience, the commission's "revision" or audit was thorough and conscientious.

It examined the jurés' books item by item. When a juré-accountant described an expense he could not document or justify, or when the commission regarded his justification as spurious, it peremptorily disallowed the claim. Thus, for example, the commission repeatedly refused to legitimize expenses for taxi-carriages on the grounds that such transportation was a luxury for which the guild could not be asked to pay. It disallowed countless claims for the payment of premiums to jurés or elders for various so-called services. It refused to honor claims for "visits" in the faubourg Saint-Antoine and other privileged places because such inspections were an integral part of everyday juré duty. It rejected demands for reimbursement for nighttime inspections when the jurés could not produce the Châtelet ordinance authorizing this police. One juré's request for a disbursement of twelve livres to pay for a basket in which to place masterpieces was rebuffed because expenses such as this were supposed to come from the general office fund. Considered outside the ambit of the guild's administration, confraternal expenses were stricken from the books.

Other claims were reduced rather than denied out of hand. "Exorbitant" lawyers' fees on two occasions were trimmed from 1,164 to 600 livres (in a case regarding the forains) and from 437 to 200 livres (bakers v. measurers). Reproaching the jurés for "superfluous expenditures" in their inspection rounds, the commission approved only 69 of 417 livres entered into the accounts. The greatest single extravagance of the baker guild appears to have been its penchant for investing enormous sums in testimony of its devotion to the royal family, a form of conspicuous consumption rationalized as an investment meant to enhance its sociopolitical capital. Twice on the occasion of births in the royal family and once when the dauphin was ill, the jurés sponsored Te Deums for which they asked for 1,780, 2,395, and 3,196 livres respectively. Unmoved by the bakers' lavish royalism, the commission reduced each of these items to 450 livres, the standard fee for Te Deums.[53]

It was much more difficult for the commission to revise the revenue entries than the expense column. On numerous occasions it strongly suspected that the jurés had understated income, but it was not always able to prove its case. Several times the commission charged that the jurés had not recorded all payments for admissions to masterships (1712–18). For the forties the commission increased the revenue side of the registers on the grounds that the jurés had understated the revenue (*gages*) received for the purchase of offices of inspectors and controllers. Twice in the next decade the commission found that the books did not acknowledge all the fees collected for apprenticeship.

The audit turned out to be a protracted, two-stage procedure. On the basis of its review of the records the commission issued a judgment fixing the officially recognized totals for income and expenditures. With one exception, every juré regime from 1744 (when our series begins) to 1780 (when it ends) was found to have received more than it spent and thus to be in debt to the guild. In most cases the jurés had owed much more than they had been willing to acknowledge when they initially settled accounts upon leaving office. The ex-jurés apparently contested the fines and refused payment. A number of years later—usually at least a decade—the commission rectified its audit either in the light of new evidence presented by ex-jurés or on the basis of a more indulgent reading of the books motivated by a desire to reach a compromise. In every case this "revised revision" lowered the claim against the ex-jurés. On the basis of the documents, however, we cannot determine how many of the former guild leaders met all or part of their obligation.

In any event the juré "deficits" were rarely inordinately large in either the first or the second revision. The guild administration of 1773–74 was charged with the biggest first-audit debt—10,568 livres, or more than a quarter of total revenue for the year. The commission later reduced the sum owed to 3,068 livres. The largest second-audit debt amounted to 3,749 livres, assigned to the jurés of 1755–56. The average first-audit sum due by ex-jurés was 4,061 livres (median = 2,714), while the mean second-audit debt was 1,280 livres (median = 953).

If we could assume that the delinquent ex-jurés reimbursed all the sums due, then we could affirm that the guild enjoyed a budgetary surplus during every single year between 1744 and 1780.[54] Ranging from a paltry 171 livres in 1767 to a hefty 13,285 livres in 1774, this paper surplus averaged a fairly substantial 3,465 livres (median = 3,244). The old stereotype of corporate financial disarray and desperation does not hold true for the bakers. To be sure, the guild was indebted (like many healthy public and private enterprises, then and now). But debt service did not sink the bakers by absorbing the totality of their resources. Though we cannot reconstruct the guild budget, it seems clear that the jurés rarely lacked cash to meet their current obligations and could probably have begun to extinguish the principal of some of their "constituted" annuities (especially if they had been able to collect more expeditiously the money owed by their predecessors).

Both the revenue and the expenditures of the guild remained quite modest throughout the century. Average annual income amounted to 23,274 livres (median = 22,056), while mean expenses were 16,677 (median = 15,890). Both average revenue and average expenditures were higher in the second half of the period than in the first.

Years	Average Revenue	Average Expenditures
1744–62	21,588 livres	15,726 livres
1763–80	24,867 livres	17,575 livres

If one removes from the calculation the years following the reorganization of the guild, as a result of which both income and expenses plummeted (to 9,077 livres and 6,927 livres respectively for the years 1775–80), then the disparity between the first and second periods is even more striking, for mean income for 1762–74 amounts to 30,940 livres and mean expenditure to 21,671 livres. In part this rise bespeaks the heavy inflation of the sixties and seventies, but it probably also reflects more aggressive leadership and expansion of corporate activities. Ironically, the baker guild was never more vigorous than on the eve of its (ephemeral) suppression and restructuring. Not only was the budget larger in the later period than in the earlier one, but ex-juré delinquency was less acute (by almost 19 percent), even though jurés left office owing on average about 20 percent more. Thus in the second period the guild boasted significantly larger immediate and paper surpluses.[55]

To change the managerial habits of the baker leadership, the commission did not count exclusively on their goodwill. In 1749 it imposed new rules and procedures far more specific and constrictive than ever before (though once again it is not clear what precise penalties the jurés would risk by defying the regulations). Henceforth there was to be order, uniformity, and regularity in bookkeeping. Nothing was to be left to baker caprice. The commission specified the size of the book, the quality of the paper, and the kind of columns to be drawn and mandated the juré-accountant "to enter in it consecutively and without any blank spaces or skipped lines all income and expenses." Within four months after the completion of his term, he was to submit his accounts for approval to the guild and for auditing by the commission. Even if he had advanced his own money for corporate operations, the juré-accountant would not be reimbursed until after the royal audit.

In order to try to prevent complicity between successive head jurés, regulations now held the new accountant responsible for collecting any sums owed by his predecessor. A more careful record was to be kept of monies collected from fines and confiscations. Fees and dues exacted by the jurés from individual bakers had nothing to do with guild finances and could not be drawn against guild accounts. Exact lists were to be kept for amounts collected for inspection visits in bakeshops. Without written permission from the lieutenant general of police, as well as corporate approbation, the jurés could not borrow money in the guild's name. Nor could the jurés demand reimbursement for expenses incurred in seizing illicit merchandise unless the police authorized the operation.

To reduce onerous litigation costs, the jurés were forbidden to appeal adverse decisions of the Châtelet without a vote of approval by the guild. Detailed billings were required to justify payments to lawyers. The commission fixed a ceiling of fifteen hundred livres for all baker office expenditures, including rent, wood and wax for heat and light, paper, pens, ink, printing, and clerical wages. The jurés were allowed to take taxi-carriages only if they could demonstrate the urgent nature of their business, and in any case they could not exceed three hundred livres a year for this purpose. For miscellaneous needs, the jurés had a contingency fund of only two hundred livres.

By imposing statutory spending limits, the commission freed itself from the obligation of haggling with the jurés over many rubrics and spared the guild from any unhappy surprises. Never before had the jurés been so tightly monitored. Finally, if a head juré felt unable to fathom and apply all these instructions on his own, the commission granted him sixty livres to spend on professional bookkeeping help.[56]

It is impossible to extrapolate from the audits and regulations a full sense of the nature of the revenues and expenditures of the baker guild. In addition to fines, interest and damage penalties, *gages* from the offices they owned, and the host of fees for the various steps leading to masterships, for apprenticeship licenses, for the registration of journeymen, and for inspections, they probably had other sources of income. Beyond office and field expenses, the largest single expense must have been debt amortization. But there are extremely heavy extraordinary expenses: public celebrations, lawsuits (4,816 livres paid to one lawyer in the bitter campaign against the grain and flour porteurs who toiled to constrict their freedom to buy raw materials), the "purchase" of substitutes for militia conscription (3,096 livres for the price of nine militiamen in 1745–46), and damage caused by collective popular violence (155 livres as a result of the *tumulte* at the grain markets at the Halles and the ports in 1747–48).[57]

Discord within the Guild

Financial management was an object of internal debate as well as external scrutiny. No issue caused greater discord within the guild. Nor was it characteristically a feud that pitted more or less impotent rank-and-file masters against a tightly knit aristocracy of former and present jurés. On the contrary, this friction was significant precisely because it divided the jurés and led to bilious squabbles among them. Despite the intimations of the royal government, reaffirmed and widely broadcast by the anticorporate publicists, that the juré elite was a clan of connivance and solidarity-in-corruption, it is clear that the jurés did not systematically cover for one another, that they often distrusted or despised one another, and that at least some of them genuinely cared about the state of the guild. Former juré Charles Berton, for example, sued his comanagers to recover the confraternal dues they allegedly embezzled. Their quarrel persisted for over a decade. Juré-accountant Denis Larcher, himself later the target of suspicion for malversation, decried his predecessor's desultory bookkeeping and irregular practices.[58]

At an executive council meeting in January 1735, elder Edme Hugot questioned "the probity" of his colleague Pierre Nezot and refused to approve his account of his tenure as head juré. The men came to blows after Hugot charged Nezot with extortion and called him a "bougre" and "Jean Foutre." Seven years later Nezot's record was still hotly discussed. Elder Gabriel Legrand accused Nezot, whom he characterized as "a thief," "a dog," and a "scoundrel," of having stolen money from the guild safe during his administration. Perhaps to divert attention from his trouble, Nezot joined a baker friend in spreading the word that Jean-Baptiste Grimbert, another elder, was "cartouchien," an evocative trope for bandit. A decade later Nezot wrangled with a head juré whom he hinted was "faithless and corrupt."[59]

Gabriel Legrand also led the attack against Nicolas Driancourt, who had succeeded Nezot as head juré. In 1735 the royal procurator's court, on the request of the guild leadership, ordered Driancourt to submit his books immediately for verification or to pay a three-thousand-livre fine. Driancourt appealed and managed to delay and effectively prevent a reckoning through continued litigation. Given his sullied reputation, the elders, marshaled by Legrand, officially ostracized Driancourt by having his name struck from the corporate tableau of anciens. Driancourt responded by obtaining a parlementary decision on 27 March 1741 ordering the guild to restore him to his status and seniority. This injunction did not prevent some elders from denouncing him as "a thief in front of the executive council" or Legrand from threatening him with his cane and trying to throw him out of the window of the community headquarters.[60]

Violence erupted in the council room in March 1745 when one elder brutally assaulted another apparently over a long-festering dispute concerning corporate finances. It is likely that the boisterous altercation between the elders Gilles Félix and Chappe Senior on the one side and François Genard on the other also turned on matters of honesty in guild dealings.[61]

On several occasions in the forties, the guild doyen Claude Larticle, a former juré, mobilized the elders to sue the outgoing leadership either for a more detailed accounting of finances than they had given or for the reimbursement of sums due. As far as one can tell the embattled jurés found refuge in dilatory counterlitigation.[62] Internal pressure was never enough to force these administrators to reveal or disgorge. No wonder the royal audit commission found a warm welcome on the part of many guild members, despite its violation of the quasi-sacrosanct principle of corporate independence.

Nor were the outgoing jurés always on the defensive. Unable to obtain reimbursement of sums they claimed to have advanced on behalf of the community during their tenures, these jurés sought redress in the courts. Guillaume Neveu, for instance, sued Larticle and the other elders in 1739 for the return of 848 livres. For years ex-juré Pierre Fremin abortively fought the guild in an effort to recover nine hundred livres he claimed to have advanced during his administration in 1710–11. Unable to meet payment for his wheat supply in 1726, Fremin tried to extricate himself by offering his creditors the claim he had against the guild as payment.[63]

It is quite likely that there were numerous episodes of rank-and-file protests against financial mismanagement, but we have encountered only one case. Summoned to pay his share of the *droit de confirmation* imposed by Louis XV toward the beginning of his reign, a master baker named Nicolas Houdouart, who was the son, grandson, nephew, and brother of master bakers, spat defiance at Denis Larcher and his fellow jurés: they were "buggers of thieves" and he would "never pay them a cent." Nor was this outburst simply displaced fury against royal fiscality. For Houdouart, it seemed to be an issue less of unjust taxation than of misappropriation of taxes, for he specifically complained that Larcher and his colleagues were "devourers of guild wealth" who "[had] siphoned off this tax for their own profit." A year later a brawl erupted in the guild offices apparently over the capitation assessment fixed by Larcher.

There may indeed have been grounds for questioning Larcher's integrity, for an innocent indiscretion in an after-death inventory informs us that he used 156 livres in capitation money plus another two hundred livres in guild funds for his baking business—with the intention, it was underlined, of repaying "these loans" at the expiration of his mandate as juré.[64] One suspects that the practice of borrowing was fairly widespread among the jurés, and it was probably at the root of many of the disputes between outgoing jurés and the guild. Several lieutenants of police raised questions about the signs of a pervasive climate of (more or less petty) corruption involving exactions committed upon masters, usually the younger ones, and upon relatively helpless apprentices and mastership candidates whose destiny depended on the goodwill of the guild administration.[65]

Even less is known about the strife of a more frankly political nature that jarred the guild periodically throughout the eighteenth century. At several junctures groups of masters, composed of elements of all the grades, tried to force overbearing and/or venal leaders out of office, by both statutory and judicial means. They also challenged the jurés on the collection of diverse fees and on the practice of seizures of merchandise and tools along with the disposition of funds issuing from them. Led by moderns and an occasional elder, segments of the more or less disfranchised block of masters (together making up the vast majority of the guild) sought vainly to open the electoral process. Twice in the fifties a modern tried to organize a boycott of juré elections. Another repeatedly bid the jurés to convoke special assemblies in which to explain their actions and allow ordinary masters to interrogate them. A jeune asked about the possibility of passing around a grievance list to be formally submitted to the leadership. On one occasion in the early sixties, a juré denounced the circulation of a document criticizing guild administration and calling for collective remonstrance to the royal council. The traditional power brokers did not remain locked in intransigent posture. They actively recruited some of the most vociferous members of the opposition to join the oligarchical complex, which broadened its boundaries in the second half of the eighteenth century. Finally, masters contested the corporate chiefs on concrete policy issues, including how to deal with royal fiscal manipulation and monitoring, what attitude to take vis-à-vis the faubourg Saint-Antoine and the forain corps, and how to respond to government price control of bread as well as of raw materials.

Frustrated with the failure of the guild administrators in the mid-eighties to obtain satisfaction from the government concerning the price of wood, one of their most important raw materials, a coterie of bakers usurped their representatives' authority, constituting themselves in a sort of vigilante delegation to deal with authorities through front- and back-channel approaches. Accomplices went from shop to shop soliciting voluntary contributions to finance their campaign "to get the price of wood lowered." They cast themselves as syndics but readily owned up to the imposture when challenged. They had no choice, they argued, but to assume the role of the syndics since the real syndics refused to undertake the task of obtaining a price adjustment. A half-dozen bakers, who had remitted between three and six livres each, filed charges against the self-appointed guild governors for "swindling." The latter insisted on the sincerity and the gravity of their mission, and explained that they had wisely used the monies collected for lobbying excursions to Versailles, the cost of several memoranda and briefs, and the purchase of a gift to thank the valet de chambre of the minister who was expected to order the reduction of the price of wood.[66]

As members of the same family, master bakers of all ranks were supposed to feel a bond of unity and solidarity. Yet the brotherly ethos that prevailed at the patronal festivities and funerals sometimes ceded to acerbic disputes over issues of governance. Masters bristled under the yoke of chiefs who treated them as inferiors and/or clients. Some of them demanded a real voice in corporate affairs, which, they advanced, were preeminently their affairs.

Intriguing fragments of evidence within the baker community, echoing behavior in other guilds, hint that the eighteenth-century corporate world may have experienced something resembling a democratic movement occurring in response to a sort of aristocratic resurgence (or "feudal reaction") within the guilds. Upstarts, even rebels in the eyes of the governing elite, the insurgent masters regarded their leaders as faithless traducers of the equality of rights on which corporate fraternity was theoretically predicated. In ways redolent of the parlementary revolt against royal absolutism, the reformist-masters wanted to cleanse guild corruption, force the rulers to render their accounts public, compel them to conform to the fundamental laws of the guild, and renew the moral basis of community by restoring an ambience of collective participation and responsibility.

Cumulatively, it seems reasonable to speak of the development of a kind of political consciousness on the part of some masters, a complement to what they had already learned as journeymen confronting a different brand of authoritarianism and exclusion. Resistance in the guild was tantamount to a lively civic apprenticeship. The insubordinate masters, like the idocile journeymen, deepened their understanding of the rhetoric and the meaning of rights and their command of the tactics of constitutional argumentation, electioneering, and lobbying and of other sorts of pressure and even subversion. In a word, they were introduced to politics. The experience prepared them for future leadership roles within the guild and, by the end of the century, beyond it. Although it does not appear to have hampered the guild's stewardship of corporate interests in the short run, in the middle term this insurgency may have weakened the coherence of the guild and undermined the collective identity that formed its most precious capital.

The Critique and the Defense of the Corporation

From the outside, the corporate system as a whole came under sharp attack in the eighteenth century, especially after 1750.[67] Given the central place of bread in French life, the baker guild was often singled out for reproach. According to the critics, the community was inherently evil because it was permeated by "the spirit of monopoly." Monopoly was barbarous and unjust because it violated man's natural right to work. This "exclusive privilege" permitted the guild to despise "ability" and "integrity" and to recruit its members in return for money and favor, thereby exacerbating social tensions as well as jeopardizing the consuming public's well-being.

Its adversaries concurred that the baker monopoly's gravest fault was that it prevented effective competition. "Free competition," they contended, not the spurious competition of fellow monopolists or of shackled forains, was the sole means of assuring the public honest, economic, and dependable service. By peer-group pressure, corporate regulation, and intimidation, the bakers enforced internal conformity. The monopoly subsidized inefficiency and slovenliness while discouraging and/or impeding innovation. The critics disparaged forain competition as not genuine, for the forains were hampered in myriad ways from selling their bread.

Without pressure from competition, the masters enjoyed excessive profits, these commentators contended. They passed on to consumers the costs of inefficiency, of corporate mismanagement, litigiousness, and prodigality, of the acquisition of mastership, of the so-called police of the monopoly and of guild-related royal fiscality right where it hurt them most: in the price of their staple food. The commerce of bread, "so precious to the people," was the very last that a wise government should allow to fall into the avid clutches of monopoly. "There is no doubt," Turgot wrote to an intendant during his ministry, "that if everyone had the liberty to bake and sell bread the People would be more abundantly and cheaply supplied."

By making "all of Paris a free, open, and continuous marketplace" and by allowing the forains to sell to whomever they pleased at all times, wrote the mitron de Vaugirard, "the Paris bakers would be forced to sell more cheaply." Nor was there any reason to worry about good order, assured Abbé Baudeau: "laissez faire and laissez passer. . . . that is all the police any commerce needs." A growing list of writers and public officials favored opening the baker's trade to all comers. The prize essay competition announced by the Agricultural Society of Lyon on the eve of Turgot's reforms bespoke the mood of the day: "Would it be advantageous to suppress the baker guilds in the major cities?"[68]

Save insofar as it concerns internal mismanagement, the critique of the guilds did not make a great impression on the Parisian police. Well before the liberal onslaught, the authorities had turned their attention to abuses in financial administration and recruitment. In the early thirties when he was lieutenant general of police, Hérault investigated reports that the jurés and elders were extorting money, drink, and gifts from masters and apprentice candidates. His successor, Marville, continued to watch the bakers attentively, sending his secretary to observe assemblies and executive meetings and to review deliberations. Later, as a member of the royal auditing commission, Marville pressed the former baker leaders to settle their accounts and collaborate with the new lieutenant general in establishing regular surveillance of current bookkeeping procedures.[69] By midcentury the baker guild was clearly more efficiently and honestly run than it had been, and though it was by no means free of defects, it was surely less corrupt, profligate, and tyrannical than the liberal writers suggested.

Nonetheless, the police remained firmly convinced of the utility of the baker guild as a means of assuring and regulating the provisioning of the city. The authorities felt they had tighter control of the bakers through the guild than they could have otherwise, especially in matters concerning the grain trade, bread marketing, pricing policies, and the comportment of bakery workers. The police understood that the bakery monopoly was far more limited in practice than critics intimated. Moreover, they did not share the liberal faith in the inexorably tonic effect of "free competition" and commercial self-regulation. From their perspective the baker guild was first and foremost an extension and instrument of the police in the service of a public subsistence interest.[70]

The Paris guild had "lay" defenders too, though they were less numerous and less vitriolic than its critics.[71] In 1774 an anonymous essayist with remarkable knowledge of provisioning questions disputed one of the main arguments of Turgot and the economists: "it is not because the bakers of Paris are master bakers that they sell their bread at a higher price than the forains." Rather it was because Parisians had a much higher overhead and produced "a better-made, more delicate, less heavy bread." Writing at about the same time, the subsistence scientist Parmentier made the same case: Paris bakers had to pay more for wood, labor, rent, and food; they were required to produce a far wider range of breads and to bake every day; they worked under close police scrutiny to assure that they followed all the rules meant to serve the public well. If the forains sell their bread more cheaply, Parmentier concluded, "it is because it is of an inferior quality."[72]

The anonymous essayist contended that opening the profession through unlimited competition would create many more problems than it would solve. A multiplication of bakers would mean reduced profits while costs remained the same. In order to indemnify themselves, bakers would be forced to diminish quality and raise prices, and the public would thus suffer doubly. They would be tempted to flee to the suburbs or countryside whence they would supply the city less regularly and less copiously and be subject to less effective control. With fewer resources, baker would not be able to maintain stocks that cushion the capital against sudden shortages.

This author insisted that it was too risky to disperse and atomize the bakery in this way and to put it in the hands of "unknown" men. Total liberty in the baking profession would produce chaos and would enable the bakers "to ransom the people." Competition was indeed necessary to check excessive prices, he allowed. But he felt that such competition already existed between master and forain and that in any case too much competition was always more dangerous than too little.

Like the police, this writer believed that the guilds were flawed institutions only in terms of corporate self-government. He favored a thoroughgoing reform featuring a severe constriction of the power of juré-syndics, public control of the corporate budget and a structure of accountability and verification of the books, a ban on litigation, suppression of the masterpiece trial, and a general easing of admission requirements, including a moderation of costs for apprenticeship and mastership. Certain of these ideas were implemented in the reorganization of the guilds following Turgot's efforts to abolish them.[73]

The liberal certitudes about the crippling defects and pullulating abuses of the guilds seem increasingly untenable. The more we learn about both corporation and trade, about the different sorts of capital invested in the system, and about the economy and the social structure of eighteenth-century Paris, the more we are inclined to nuance the picture. Of course the guilds were flawed in myriad ways, but they were also more supple, more culturally and economically diverse, and more complicated in their internal operations and in their relations with production and marketing outside the corporate world than has generally been admitted.

Given the vital merchandise in which they dealt, the bakers were in some ways an exceptional case. The police counted heavily on the guild and subjected it to relatively close surveillance. Yet the guild evolved on its own in ways similar to the (uncanonical) itineraries of many of its sister corporations. Despite its manifest oligarchical character, it accommodated a vigorous political life marked by tonic eruptions of dissension that accelerated the socialization of a substantial number of bakers into the (more or less) democratic practices of electoral campaigning and manipulation, parliamentary maneuvering, lobbying, and resistance. The guild prosecuted an aggressive foreign policy that brought its members into contact with other bakers and other tradespeople and merchants against whom they battled, to be sure, but with whom they also forged new and unconventional alliances.

Using its leverage to bargain and compromise, the corporation protected the masters from excessive royal intrusion. It recruited new members in a relatively flexible and open manner. It accorded in practice considerable freedom to its masters to conduct their businesses as they saw fit, thus enabling the most entrepreneurial bakers to collaborate with the most innovative millers to transform the elementary structures of the provisioning trade. The same sense of realism distinguished its performance in the struggle to control the labor market, where it simultaneously strained to impose uniform standards on both masters and workers and tolerated idiosyncratic and discreetly wrought transactions among the parties. The guild was rotten in many places, but it showed more resourcefulness and resilience than anyone could have plausibly expected. For both good and bad reasons, the guild continued to attract successive generations of young baker aspirants.

Chapter 7

From Apprentice to
Journeyman

The notion of apprenticeship was at the very heart of the corporate conception of work and hierarchy.[1] One had to *prepare* for eventual mastership. Apprenticeship was the first stage of initiation—and of selection and subordination. Ideally, it was to begin in early adolescence in order to shape the youth's values decisively as well as to train him professionally. Biologically and culturally, for the child-apprentice it was a liminal moment as he encountered the challenges of puberty at the same time that he abandoned (and was abandoned by) his natural family in order to enter an adoptive one. Substantially older apprentices may have chafed against the notion of reverting to a familial context. The apprentice was supposed to tremble in awe and fear before the authority and the dexterity of the master (and indeed the licensed journeymen as well). Apprenticeship was moral and political socialization as well as craft education. In some shops the apprentice was frankly not expected to acquire the fundamental skills his first few years, though an observant apprentice learned a great deal anyway.

Despite contractual promises to teach the apprentice all the secrets of the trade, the neophyte was often left with only the menial and marginal tasks that neither required nor developed craft skills: sweeping and cleaning, delivering goods, serving as a quasi-domestic. He was extraordinarily cheap labor and easy to exploit, a bargain for the master. Though the master may have neglected or postponed the craft lesson, he did not fail to teach the apprentice from the beginning about the primacy of subjection in the shop and the larger corporate world. The "good apprentice" was disciplined, reliable, loyal, tranquil, and above all obedient.

Apprenticeship was not merely preparation for practicing the profession within the corporate world. It was an absolute prerequisite for everyone who aspired to mastership (or who merely wished to work regularly as a journeyman), except for the sons of masters who were born after their fathers became masters. All guild rules were geared to favor heredity or dynastic recruitment. But if the master's son was exempted from apprenticeship, it was not merely to pay deference to the guild's aristocratic posture. It was quite rightly presumed that a son raised in a baker's household (where there was usually no clear demarcation between domestic and professional life) experienced a "natural" and inexorable apprenticeship, one that generally began very early in life. For that reason the 1719 guild statutes authorized the "reception of sons as masters" beginning at age twelve. Upon subsequent reflection, the guild decided that a twelve-year-old was too immature to assume fully the master's responsibilities, and so it raised the age to eighteen in a revision of its statutes at midcentury. Despite this dispensation from corporate apprenticeship for masters' sons, it is interesting to note that a number of masters formally apprenticed their sons, by contract, to colleagues, to relatives, or to themselves. In the first part of the seventeenth century all sons of masters were theoretically required to serve a three-year apprenticeship regardless of the date of their fathers' mastership.[2] Their sons may have been rebellious or maladjusted children. But it is possible, too, that the fathers believed in the efficacy of the training and discipline of the apprenticeship cycle.

In the eighteenth century a candidate for mastership had to be a Roman Catholic "of good reputation and manners," without any trace of contagious disease, who had attained twenty-two years of age (twenty years after the 1783 revision of the statutes). In addition, he had to have completed a three-year apprenticeship (vouched for by a *brevet*, or license, signed by his master and approved by the guild leadership) and a training/probationary period of at least three years as a journeyman. Finally, he had to submit a masterpiece, or *chef-d'oeuvre* (unless he was the son of a master), to the judgment of a guild committee, which could approve it, require that it be done all over again, or send the young baker back for "perfecting" as a journeyman. If the masterpiece was accepted, the candidate would have to acquit a host of monetary obligations to the guild, to the masters who served on his panel, and to the confrérie. (A master's son benefited from at least a 50 percent reduction on most of these charges and from a total exemption regarding some of them.) Until 1783 a boy could not begin a baker apprenticeship until he turned fourteen years of age. After that date, the age was lowered to twelve. In order to prevent the masters from transforming the apprenticeship into a simple captive labor market and to placate the objections of the journeymen, who bitterly resented the idea of competing with cheap apprentices for jobs, the guild forbade the masters from engaging more than one apprentice at a time.

The Contract

The guild required that a contract, signed under the auspices of a notary, govern the relationship between master and apprentice. There was a remarkable consistency across time in both the form and the substance of the 203 such contracts signed between 1708 and 1775 that I have examined. Characteristically the master promised to "show him [the apprentice] the baker's craft" or, more elaborately, "to teach him the craft of baking and everything concerned with it without hiding anything from him." The master pledged, furthermore, to give the apprentice board and room (including heat and light), to have his laundry washed, and "to treat him gently and humanely." In only eight of 203 cases were there departures from these conventional arrangements (e.g., the lodging engagement was omitted twice, the board pledge once, the laundry stipulation once), and it is not certain that these mutations were anything more than notarial oversights or idiosyncracies.[3]

In return for these services, the apprentice vowed to "learn as well as he possibly can everything that the Sieur his master will teach him, to obey him in everything proper and honest, to work to enhance his profits and attenuate his losses . . . without being allowed during the said period [of apprenticeship] to leave his master's house to go work elsewhere." The "runaway" clause was especially important to the masters. One contract makes clear the vulnerable fugitive status of the faithless apprentice: "in which case of absence he [the apprentice] consents to be sought out wherever it will be necessary in order, if he is found, to be returned to the shop of his said master to finish the time remaining until expiration [of the agreement]." In addition to submitting to the master's will and laboring to improve his business, in forty-eight cases the apprentice promised to supply his own clothes. It appears that this is a genuine variation in contractual arrangement rather than a mere notarial anomaly, for in several cases the masters agree to provide clothes in the second and third years after an initial year of probation.[4]

Clothes were not the only form of wage offered by certain masters. In twenty-seven of 109 contracts drafted before 1754, the masters promised to provide "in place of wine" sums varying from twelve sous to 110 sous a week. The mean wage amounts to slightly more than forty-eight sous a week, a sum ample enough to provide wine at virtually every meal or to afford pocket money for sundries. None of the youngest apprentices—fourteen- and fifteen-year-olds—received a wine allowance, but there was no rigorous correlation between age and wage. The highest- and lowest-paid apprentices were both eighteen years old.

It is striking to note that after 1754, not a single contract in our sample contains a wage clause. It might be that this result bespeaks a tainted sample. But it is also possible that the guild addressed the issue and established a new policy in the early fifties. More likely, the guild leadership decided to try to enforce the old doctrine, which almost certainly had disapproved of the use of wages, especially variable wages. In every aspect of professional life, the guild sought to mitigate competition among masters by prescribing uniform codes of conduct. For example, all masters were supposed to offer prospective journeymen the same wage, the "going wage" sanctioned by the guild bureau. Moreover, no master had the right to try to induce a colleague's workers to leave their master in order to go to work for him. The aim of these measures was to prevent the chaos that would result if masters could freely bid for the services of each other's workers. By the same token the guild must have regarded the practice of sweetening the contract of certain candidate-apprentices (especially the older ones who were physically capable of the most strenuous labor) as a dangerous kind of competition and business individualism likely to cause tension among the apprentices as well as friction among masters.

Not only was it relatively rare for a master to pay an apprentice, especially after 1754, but in a few instances an apprentice had to pay the master with whom he wished to work to take him in. Jean-Firmin Leblanc, priest-vicar of Saint-Roch, had to pledge to pay master baker François Louiset two hundred livres for agreeing to take on his ward, Pierre-François Rolland, all of whose property was mortgaged as collateral for the guarantee of this payment. In another case earlier in the century, a master baker demanded 120 livres, which he promised to restitute upon completion of the apprenticeship. Another master extracted a one-hundred-livre deposit to serve as a fugitive bond in case his apprentice fled.[5]

On occasion masters found themselves hotly courted, even harassed by perspective apprentices, or rather by their agents. Charles Eyduc, a domestic in the household of the comte de Clarmont, pursued Michel Driancourt, a master from the rue Saint-Antoine, in the hope of persuading him to "teach the craft of baking" to his brother. Eyduc took Driancourt from one tavern to another at the Basse Courtille and in the Marais in order to win his favor. It is impossible to say whether Driancourt had special qualities that appealed to Eyduc or whether it was increasingly difficult to find an opening in the second half of the eighteenth century.[6]

In the vast majority of cases, however, the masters sought no remuneration by way of either bond or *pot de vin;* many contracts specifically mention that no price was placed on the transaction. There were, to be sure, fees ("droits") to pay to the guild for the license and to the jurés who were parties to the contract and guarantors of its fulfillment. The *brevet* itself cost forty livres (as compared, say, with three hundred livres for the drapers, eighty-eight livres for grocers, and thirty livres for shoemakers), while the miscellaneous charges represented another twenty to thirty livres. This was *not* a negligible expenditure, especially for a young orphan or newly arrived provincial. In some cases the master advanced the necessary sum to the candidate-apprentice, who promised to pay it after expiration of the apprenticeship (but before the formal awarding of the license).[7]

One of "the most revolting" abuses of the corporate system, according to Bigot de Ste.-Croix, *parlementaire* and philo-physiocrat, was the extremely long term of the apprenticeships. Often six or seven years in duration, followed by a similarly protracted journeyman's probation, the apprenticeship, Bigot maintained, was "another kind of servitude." In fact, in the baker's guild, servitude was much less onerous if one measures it merely by the criterion of time. Of 203 contracts, only three failed to conform to the three-year period fixed by the corporate statutes in 1719 and reiterated in 1746/57 and 1783. All three exceptions occurred *before* 1719: two worked for three and one-half years and one for four and one-half years. In the seventeenth century one often encounters contracts for four years and more rarely for five (usually in the case of very young apprentices). Even during this epoch, however, the three-year apprenticeship was still the most common. In the baker guild of Rouen, apprentices had to serve for four years, while in Lille the term was two years. Other Parisian guild apprenticeships rarely exceeded four or five years in the eighteenth century.[8]

Portrait of the Apprentice

The contracts yield very little information about the backgrounds or the personal characteristics of the apprentices. We know the ages of only thirty-three of the 203 apprentices. The mean age was 20.27, the median 19, the mode 18: considerably older than most contemporary testimonies would have us believe.[9] The youngest was fourteen, while the senior apprentice was an astonishingly ripe thirty-two! The guild statutes themselves suggest that apprentices were often very much younger than our average, for they specify that time spent in apprenticeship before the age of twelve shall not count toward fulfilling the requirements leading to mastership. A very small sample for the 1640s suggests candidates for baker apprenticeship presented themselves at an earlier age in the seventeenth century. For seventeen cases, the average age was 16.35, the median 16, and the mode 15. It is not clear why apprenticeship began so much later in the eighteenth century. In theory, the guild wanted the masters to take the apprentices while they were very young and most impressionable. But the masters may have found it in their economic interest to take the apprentices when they were physically, emotionally, and intellectually more mature and thus able to work harder and more productively. A more intense competition for places may have given the masters this opportunity in the eighteenth century.

One sign of growing maturity of a kind is that many more apprentices manifested the rudiments of literacy in the eighteenth century than in the seventeenth. Of the 156 cases for which this information appears, almost 80 percent were able to sign their names. The ability to write was an enormous liberating force for the apprentice. It strengthened the apprentice vis-à-vis his master and the guild hierarchy, for it gave him a certain status and capacity to affirm his rights. In addition the ability to write certainly assisted him in his commerce, especially if he attained the mastership and had to deal in the grain and flour markets.

In 159 cases we know nothing about the geographical origin of the apprentices. Despite the massive dependence of the capital on the provinces to furnish it with citizens, judging from data in dozens of other guilds, it is probable that a very substantial number were either native Parisian or had been in Paris quite a while before contracting with a master. The contracts reveal that twenty-three apprentices were of provincial origin, seven from areas within the Parisian sphere of influence.[10]

In the seventeenth century an apprentice baker was usually presented (and represented before the notary) by a relative or sponsor. A century later—unless we are the dupes of notarial style—in the overwhelming majority of cases (180), the apprentice contracted in his own behalf without any patronage. This is the case even for self-proclaimed minors. Fathers spoke for sons in only nineteen instances, uncles twice, and a cousin and a parish priest once each. The trend toward older apprentices accounts to some extent for the extraordinarily high proportion of autonomous or unaccompanied candidates. Yet it seems surprising that the guild did not insist on a patronage system. A patron—relative or sponsor—served as both a moral and financial guarantee of the good conduct of the apprentices. Nor was this guarantee of purely heuristic or didactic value; it had very practical applications. In one contract, for example, a merchant—bourgeois de Paris promised to pay one hundred livres to master baker Pierre Nezot if his apprenticed nephew ran away. In another, a baker named Charles

Leger swore to search for and return his apprenticed son if the latter fled the shop of master baker Conty of the rue Montmartre.[11]

Thirty-five contracts permit a glimpse of the social origins of the apprentices. Ten were sons following in their fathers' footsteps or, rather, trying to move a step beyond their fathers. Four of this group were the sons of master bakers, three of whom were definitely established in Paris, while six were the sons of nonguild bakers. In terms of familiarity with the trade, the sons of bakers, even simple journeymen bakers, had an advantage over the sons of strangers to the craft. Of the latter, three were nevertheless the sons of master artisans (a nail maker and two shoemakers) who knew the corporate system well. The fathers of six apprentices were mere journeymen: two masons, two coopers, one shoemaker, and one weaver. Two were sons of men who boasted the slippery title bourgeois de Paris—a cachet that may conceal their real profession even as it ostensibly enhanced their prestige.

Two other fathers came from the opposite end of the social scale: they were unskilled dayworkers, or *gagne-deniers*. Two apprentices were sons of provincial tavern keepers. It is not implausible to imagine that these *cabaretiers* became acquainted with their sons' new masters while those bakers were on grain or flour-buying missions. It is even possible that the cabaretiers illicitly stored grain or flour for the bakers. Ten fathers of apprentices worked in agriculture, seven as *laboureurs* (did any of them sell grain—on or off the market—to the master bakers who took in their sons?), two as winegrowers, and one as a simple dayworker. Between five and ten of the other fathers' professions could have been exercised in a rural milieu as well.

On balance, more apprentices appeared to come from the towns and bourgs than from the villages. At least 25 percent of all these apprentices came from socioeconomic milieux that we can confidently describe as very poor and perhaps even marginal. Of the rest, a handful seemed to emerge from relatively comfortable (albeit not well-to-do surroundings), while the bulk appeared to belong to quite modest ranks.

Relations between Masters and Apprentices

The apprenticeship agreement was not a private labor contract between a master and a candidate. The contract was a corporate document even as apprenticeship was a specifically corporate experience. The guild was party to every contract. Without the signature of the jurés, the contract was neither binding nor legitimate. The guild issued the *brevet*, or license, for apprenticeship. It made the rules for apprenticeship and it mediated relations between masters and apprentices. The guild sought to impose a common discipline and code of conduct on masters as well as apprentices in order to ensure good order. It guaranteed each master that none of his colleagues would enjoy a competitive advantage in the apprenticeship market and that master and apprentice would honor their obligations to one another.

For reasons that remain inaccessible to us, a number of masters apparently accorded special treatment to their apprentices. Because they were irresponsible or indifferent or susceptible to bribery or personally linked to the families of their apprentices, these masters were willing to endorse licenses without providing the prestations they promised to furnish and without requiring the services that the apprentice pledged to perform. Indeed, some of these contracts appear to have been elaborate fictions. The guild refused to tolerate these frauds and these master-apprentice complicities. The jurés investigated the dubious relationships—usually denounced discreetly by jealous colleagues of the masters in question—and summoned suspect masters and apprentices to defend themselves in the chamber of the royal procurator, the court devoted to corporate matters in the Châtelet.

François Sequin, the apprentice of master baker Jean-Baptiste Limoges, was found to have been "absent from his master for a very long time." His apprenticeship license was confiscated, he was banned from ever becoming a master, and Limoges suffered a fine of twenty livres in addition to court expenses. The court found that apprentice Pierre Hennequin "has not been living with his master for a long while, though the time stipulated in his contract has not yet expired." He was barred from mastership and his master was ordered to pay the guild fifty livres in fines, twenty livres in damages, and legal expenses. Apprentices François Garouard and Angélique Rodet elicited the same harsh sanctions when their masters did not even appear to defend themselves. Perhaps because his case was less egregious, Pierre Ladit merely had his license declared null and void, and his master escaped penalty.[12]

The guild, it should be noted, did not always win its argument. In many instances the court decided that the alleged absences of the apprentices were not decisive and they could reasonably be expected to finish their time. Indeed, there are cases in which apprentices abandon their shop and master "for cause" and are later allowed to return. Pierre Lalande served only four months of an apprenticeship begun just before Christmas in 1744. He was permitted to "complete the time remaining on his license" seven and one-half years later.[13]

While in some instances masters apparently protected their apprentices, in others they abused them. When there was no doubt about culpability, the jurés supported the plaintiff apprentices without hesitation. The guild referred the case of master baker Germain Fleury to the procurator's chamber. After Fleury refused to obey the jurés' orders to honor his obligation to permit his apprentice François Fournier to complete his three years and earn his license, the court ordered the baker to pay damages and costs and the jurés to find the young man another master willing to allow him to complete his time. Apprentice François Buchillot won a similar judgment against master Pierre Houllier.[14]

For motives of vengeance or rancor, sometimes masters refused to grant their apprentices their brevets, despite the fact that the trainees had fulfilled their contractual obligations. Deprived of the title *garçon* or *compagnon* (journeyman) or *apprenti consommé* (graduate apprentice) that the license conferred, these baker boys would be obliged to accept work at much less favorable conditions and at a lower wage than they merited. The jurés usually referred these apprentices to the Châtelet, where the royal procurator, upon presentation of proof, ordered the masters to sign the license within a week or less. If the master failed to comply, the procurator authorized the guild to issue the license in his name. On occasion the guild utilized extralegal means to pressure recalcitrant masters to live up to their obligations. If one is to believe Charles Millot, a master baker from the rue Galande, the guild's *comptable en charge* (chief financial officer) threatened "to have him punished with a fine and to have his capitation tax increased" if he did not deliver his apprentice's brevet to the corporate office.[15]

A number of masters maintained apprentices without having signed contracts. Louis Marois, for example, a master located on the rue Saint-Denis, took in ("par charité," he claimed) a fifteen-year-old "beggar" from Picardy named Nesseler. Marois promised "to teach him the craft of baking if he was an honest man." It turned out that Nesseler was not honest; he stole twelve écus in cash and the clock that hung over the oven. (But it also turned out that Marois made little effort to instruct the boy; he had him wash dishes and run errands.) The guild regarded the practice of taking in clandestine apprentices as an illicit and unfair form of labor exploitation (unfair as much vis-à-vis the other masters as vis-à-vis the so-called apprentice). When the jurés suspected that a master employed an illegal apprentice, they summoned him to present the contract and brevet to the guild office; if he failed to exonerate himself, they had him arraigned in the Châtelet.[16]

In eighteenth-century London, masters are said to have systematically brutalized their apprentices, and—the two phenomena are surely not unrelated—apprentices are said to have been the single most turbulent socioprofessional group, judging from their participation in riots and disturbances. In Paris—at least among the bakers—the picture is much less clear. Relations between journeymen and masters, as we shall presently see, were frequently bitterly strained. Doubtless, in many instances, apprentices joined in actions against their masters, in which they were easily mistaken for and characterized as *garçons boulangers* in the police reports that have survived. There is no indication, however, that the apprentices, as a group, were in the vanguard of alienation or insubordination. The masters often treated their garçons harshly, but they do not seem to have singled out the apprentices. The example of Jean-Baptiste Fournier, a twenty-year-old apprentice from Meaux who was repeatedly bludgeoned with a broomstick by his alcoholic master, seems exceptional. Nor should one ignore the other side of the coin, perhaps equally as aberrant in the extreme. Take, for instance, the case of a certain Carrillon, who was apprenticed to master baker Louis Dugland. During the master's frequent absence, Carrillon attacked his wife in bed, threatened her, and called her son (the head *garçon* in the oven room?) names such as "rogue" and "rascal."[17]

Journeymanship

Upon completion of the training period specified in his contract, the apprentice automatically became a journeyman—a *compagnon* or *garçon*. The journeyman was located at the summit of the hierarchy of the laboring class in the Old Regime precisely because he had acquired an apprentice's license and become a part of the corporate structure. This license meant that he was a skilled worker (relatively speaking; even the journeymen in the least demanding and least differentiated trades had greater skills than water carriers, stevedores, or deliverymen) and thus that he was entitled to a higher wage (and status) than workers who lacked either his dexterity or his title. His credentials also signified that he had a greater chance to find a steady job.

Beyond the (overt) professional consideration, the journeyman was also distinguished by the milieu he frequented. He tended to associate with journeymen of the same trade. They wore the same kind of clothing, spoke a common argot, drank in the same taverns, often lived in the same inn, and shared a common perspective on many matters. Sometimes these intimate, everyday contacts acquired an institutional or quasi-institutional character: the journeymen formed illegal organizations, sometimes wholly sub-rosa (such as the *compagnonnage*), sometimes semiclandestine (such as the confrérie, which

Baker boy making home deliveries with *tailles* in hand to record sales on credit, circa 1737. *Source: courtesy Deutsches Brotmuseum, Ulm/Donau.*

projected a patina of legitimacy). It was as much this special form of sociability and solidarity—this networking—as it was his professional skill that made the journeyman an elitist in the world of work.

In the course of the eighteenth century, the former took greater and greater precedence over the latter in distinguishing the compagnon. For the strictly professional advantage enjoyed by journeymen became less and less compelling. Historically, the journeyman had been defined by the fact that "he was awaiting mastership." That is to say, he was generally considered upwardly mobile. It became increasingly difficult, however, for the journeyman to transcend all the barriers that obstructed the path to mastership—barriers that were partly material and partly institutional and political. In light of this ostensibly new socioeconomic reality, the term compagnon acquired a second meaning, one symbolic of the evolution of the corporate structure since the imagined golden age of fluidity and openness when journeymen were said to have participated fully in guild life and culture. Henceforth compagnon signified those who failed as well as those who succeeded, workers who accomplished their apprenticeships but who lacked "the means of rising to mastership or setting up shop." The journeymen in many trades were increasingly assimilated to the *alloués*, workers who could not even envision corporate integration or mobility because they lacked licenses and were destined "to work all their lives by the day." It is this journeyman that was never allowed to outgrow the sociobiological status of boy (*garçon*), as the journeyman was commonly called in the bakery and other professions.[18]

Portrait of the Journeyman

On the surface it is difficult to distinguish the journeyman bakers from the apprentices. Naturally the garçons were a little older than the apprentices, though not by very much if one is to judge by a survey of seventy-one examples drawn between 1723 and 1779. The median age was twenty-six, the mean was 26.34. More interesting, of course, is the range. Three journeymen were only fifteen, and one, André Lambert, was, at age sixty-four, easily old enough to be the grandfather of the three striplings. Obviously it would be rash to pretend that a deeply felt community of needs and interests could unite the very young and the old. The bulk of the journeymen were, however, between the early twenties and early thirties: garçons only in the wishful thinking of their masters.

Whatever their ages, from a distance garçon bakers looked alike, for they dressed alike. On numerous occasions they were identified by Parisians who recognized their distinctive costumes (so-and-so had the "air of a garçon baker" was the refrain of these testimonies). Characteristically they wore a sort of grayish-white tunic or overall, sometimes cut out of a grain or flour sack. The notorious part of the outfit was the white cotton bonnet, occasionally decorated with a red or white rosette ("coquarde").[19]

Unless we meet them as apprentices—even there, the documents are not loquacious—it is hard to learn anything about the background of the journeyman bakers. Of a sample of forty-one drawn between the years of 1728 and 1772, almost 37 percent—a much higher proportion than expected—were unable to sign their names. Journeymen sometimes boasted nicknames that give us clues about their geographical origins: Jean Farge dit [called] le Maconnois; Charles Wagon dit le Flamand; Barthelemy Mazet dit Dauphiné; Louis Peret also dit Dauphiné; Denis Morlet dit Montargis; Noel Savary dit le Contois; Lepere dit Langevin. (It would be impossible, however, to infer from his nickname that François Barcoud dit la Jeunesse [the Youthful] came, as he did, from the Berry.) Journeymen from the same "pays" tended to frequent one another. Though they worked in different shops, four German journeymen regularly drank together and chatted in their mother tongue. Hubert Chatelain, Jean Belanger, and Antoine Mathieu were journeymen who worked in different parts of Paris, but they all lived together in a fruit seller's apartment, linked by their common Burgundian origins. Several Auvergnats played cards, shared a girlfriend, and made trips home together.[20]

Job Placement

Theoretically, to obtain a position with a master baker, a journeyman had to apply to the guild office.[21] In order to prevent the labor market from becoming the scene of bitter rivalries among the masters—fissures in the corporate structure that could invite insubordination among the journeymen as well as treason among the masters—the guild claimed a monopoly on placement. All eligible journeymen were required to register with the guild upon arrival in Paris or upon leaving a master. When a master needed a journeyman, he asked the clerk of the guild to assign him one from the pool of registrants. Presumably, he could request a specific journeyman of his acquaintance, if the latter were free of all obligations to other masters and had filed for work with the clerk. Masters who hired journeymen on their own initiative were subject to censure, fine, and in the most extreme case, expulsion from the guild.

Many masters resented the guild's overmighty posture. They encouraged the journeymen to join them in black-market or surreptitious placement. This procedure spared the journeymen time and occasional registration fees and enabled them sometimes to exact a slightly higher wage than the going price, the uniform wage sanctioned by the guild. Such joint subversion tempted many journeymen. But others still seethed at the idea that placement was a sovereign prerogative belonging exclusively, *d'office*, to the masters. At bottom there was little difference between this monopoly and the one claimed by the guild. Since their hands were very often their only property, journeymen in all the guilds, not just garçon bakers, stubbornly defended their right to dispose of them as they saw fit. They wanted to decide for themselves where they would work and under what conditions.[22]

The result was incessant and impassioned conflict between journeymen and masters, between journeymen and guild, and, occasionally, between fellow masters. Because the police considered work to be the most fundamental form of social control, they were deeply concerned about the placement question. Throughout the eighteenth century the police vigorously supported the corporate effort to exclude the journeymen from any real role in the placement procedure. The authorities encouraged the guilds to include in their statutes explicit clauses prohibiting workers from "caballing among themselves in order to place one another with the masters." Sentences from the Châtelet, parlementary arrêts, and royal legislation reiterated and reinforced these restrictions.[23]

During the eighteenth century no corporation was more violently jolted by placement conflict than that of the bakers. The clerks in charge of placement quite often had an authoritarian swagger. The journeymen regarded them as tyrants and detested them. On Monday afternoon, 28 May 1742, Estienne Berton, a master baker and first clerk of the guild, entered a wineshop on the grande rue du faubourg Saint-Martin to have a drink with a colleague. Berton spotted a number of journeymen bakers in the garden of the shop. Since he needed a "master garçon," or *geindre,* for another colleague, he approached the table of journeymen. According to several witnesses, the journeymen suddenly turned on Berton, knocking him down and beating him violently with canes and broom handles until "his head was broken." Berton died the next morning.[24]

Berton's confrere Fontaine was also beaten in the head and groin because, he contended, the enraged journeymen mistook him for Bazile, the second clerk of the guild, who had acquired a reputation for callousness and arrogance in placement affairs. Fourteen years later a clerk named Bassille—perhaps the same one—filed several complaints with the police against two journeymen named Berry and Villenone. He had assigned them shops, but they had not stayed where they were placed because, charged Bassille, they were "very bad subjects and libertines." Since then the garçons had been "tormenting" him to give them new posts. Though Bassille assured them none was vacant, they refused to believe him. They threatened him, harassed him in the guild office, and followed him to his shop several times. One night Berry hit him in the head with a rock. The next evening Bassille encountered him again, along with other journeymen. Again, they demanded to be placed, insulted him, and threw him to the ground. Perhaps remembering Berton's experience, Bassille avowed that he feared for his life.[25]

Later that month another journeyman threatened to kill Bassille and the other guild clerk. Infuriated because the clerks refused—apparently without explanation—to place him at a shop near the pont Saint-Michel that he wanted badly and knew to have an opening, Jacques Fagand, a thirty-five-year-old native of the Beauvaisis, protested vehemently before the guild bureau. He insulted the anciens and the jurés, and he followed Bassille home on several occasions in a fervid but vain effort to make his case. The jurés warned him that he was courting imprisonment. Utterly indifferent, Fagand took consolation in the thought that at least "they would feed me" in jail. At the request of the jurés, he was incarcerated in the Grand Châtelet.[26]

Milsain, another guild clerk responsible for placement, took no chances. He brandished a gun to deter a band of twenty or more journeymen bakers who followed him home and threatened to kill him. One of these garçons, Renoult dit le Flamand, had already hectored him several times for a post and menaced him with a beating. For a number of years it had been common practice for "thirty or forty garçons" to congregate around the guild office several times a week. They yelled for jobs, and at the same time, according to a *regrattière* who sold fruit nearby, they were boisterous and undisciplined "to the point even of putting their hands up the skirts of women and girls."[27]

The jurés suspected that the incidents of journeyman rebelliousness were not isolated and spontaneous epiphenomena. Behind them the guild officials perceived what they called a "cabal," that is, a conspiracy hatched by a number of journeymen who bound themselves together for the purpose of defying their masters in pursuit of various moral and material ends. The fear of cabals was nothing less than an obsession, and it was shared by both the police and the masters. They tended to see "the spirit of cabal" everywhere. Nor was this preoccupation with conspiracies and revolts merely another manifestation of ruling-class paranoia, for these cabals proliferated in the course of the eighteenth century. Though they did not detonate any mass risings, they nevertheless seriously disrupted public order and jolted the working world almost every day.[28]

In 1756 the baker-jurés denounced to the police a suspected cabal that sought, inter alia, to challenge the guild's and the masters' control of placement. Even as large bands of journeymen gathered to bait and harry the jurés and some masters at various locations in the city, "several others of these journeymen who do not work"—according to a guild complaint filed with the police—"engage in the business of placing their comrades in jobs." The guild regarded this organized undertaking as far more ominous than the periodic assaults of individuals on the poor guild clerks because it was a collective and planned act of usurpation.[29]

The police promised the jurés to try to keep a tighter rein on the journeymen, especially in regard to placement, but expressions of discontent and episodes of mutiny continued to erupt. Swarms of journeymen went on infesting the Halles, particularly on market days, when their presence was more keenly felt than at any other time, in order to demonstrate against corporate placement policy and other guild constraints. The jurés told Poussot, the activist police inspector assigned to the Halles in the late fifties, that on some days more than three hundred unemployed journeymen assembled outside the guild office. Poussot urged the police commissaires to arrest them during their patrols: "this police would make the other journeymen behave better and would keep those who were serving masters within their duties." And frustrated individuals, like thirty-year-old Denis Fournier, an unemployed journeyman from Burgundy, continued first to supplicate, then to browbeat, and finally to assault the guild clerk upon the latter's refusal (or inability?) to locate an opening.[30]

In the last half of the eighteenth century, the guild system came under caustic attack. One of the grounds of criticism concerned the tyranny that the corporations exercised over their workers in terms of placement. Perhaps partly in response to this line of reproach and partly in reaction to chronic journeyman agitation (and grumbling by many masters as well), the lieutenant general of police promulgated an ordinance in 1781 that seriously diminished the baker guild's direct part in placement operations.[31] Henceforth masters could venture to engage garçons on their own initiative and garçons could seek work from masters without applying through intermediaries. The guild placement service was still available for the asking, but it was transformed from a requirement to a convenience. Still, the guild continued to play a major part in defining and mediating labor relations and in disciplining journeymen. All garçons were still obliged to register with the corporate office every time they changed positions, and in certain circumstances the jurés could constrain the journeymen to enter the shops that the guild designated.

By the time of the Revolution, the baker boys had attained a certain autonomy. They participated in organizations that appeared to have a durable existence. Their "rouleurs" negotiated job contacts and contracts. Yet, no less than the masters, even those who lived in a conflictual relationship with the guild, in the end the journeymen had trouble giving up the guild, which provided after all useful services of communication and coordination. (As common adversary to the journeymen, the guild also furnished a permanent incentive to union.) In an ironic postscriptum to their age-old battle against the corporate clerks, in July 1791, several months after the formal and definitive suppression of the guild, a group of garçon bakers submitted a petition to the municipality "asking that the clerk of the former corporation of bakers be preserved in his function in order that he might continue to dispense the service that he was charged with performing for [the journeymen]."[32]

Quitting: Work as Social Control

The act of quitting work in the bakeshop generated as much friction as the gesture of trying to find work. As a rule neither the master nor the guild nor the police wanted to see a journeyman leave work. The master invested a considerable effort in training the garçon after his fashion. Once he was integrated into the process of production and distribution, the master depended on him heavily, for the master's business frequently took him away from the shop and oven. The master was extremely reluctant to release a worker who was both honest and competent. The guild perceived turnover as a sign of disorder. Movement and change taxed the guild's administrative capacity even as it caused the jurés anxiety. The corporation clearly preferred to see workers remain for long periods with the same master. The statutes contained a potentially vicious ambiguity that seemed to give the masters license to "retain" their journeymen if they so pleased at the same time that it authorized the journeymen to request leave.[33] In the eyes of the journeymen, the controls placed on the right to quit infringed even more on their liberty than the restrictions on their freedom to place themselves.

As Voltaire put it, work itself was "one of the first elements of police."[34] Work was a form of disarmament and integration. Normally work implied submission to a master. Work—legal work, that is—identified a worker and fixed him in place. It was a sort of tracer dye that enabled the police to locate the worker in the body politic and social even as a radiologist tracks the forces that menace the human body. Socially reassuring, work was also economically useful, for the worker directed his energies to a productive goal. Moreover, work served the individual as well as the society. From his childhood the journeyman had heard the admonitory refrain "no salvation outside of work." Frankly associated with the idea of suffering, work had always been viewed as a punishment for original sin. Work represented "a duty" from which theoretically no one, since the end of the state of innocence, could win exemption. Necessary and efficacious, work-penitence had the power of redemption. This conception was doubly attractive, for while facilitating personal salvation, the obligation to work also guaranteed the social order. To constrain the people to work was to force them to save themselves and to tighten the social fabric.[35]

From this perspective, the vocation of the police was a beneficent despotism serving moral and religious as well as sociopolitical ends. It followed that, in the police view, to refuse to work was at once an antireligious and an antisocial gesture, an act of revolt that threatened both the society and the individual. To leave work even for a moment was to escape surveillance and subordination. Unemployed—*hors de condition*—journeymen risked falling into the infernal world of those who refused regeneration by work in favor of degradation through idleness and eventually rebellion through crime. Confronted with an already immense and still growing floating population, the police, from Dela-mare to Lenoir, worried deeply about the implications of nonwork. They viewed idleness as a disease and a sedition: contagious, dangerous, intoler-able.[36] The moment the worker was no longer usefully occupied in some approved form of labor, the police became alarmed. The police had little use for holidays, Sundays, nights, interludes between jobs—viewing them warily as intervals replete with uncertainty.

Virtually all employers in the eighteenth century, corporate masters or cap-tains of industry, required their departing workers to give notice. This manda-tory *avis* varied from profession to profession and sometimes from season to season. The shoemakers ordinarily demanded a week but asked for three weeks before the four great annual festivals; the nail makers two weeks; the potters a month; the ironworkers three months; and the royal glass factories up to two years! The bakers insisted on two weeks' notice until 1781, when the term was reduced by one-half.[37]

Giving notice was necessary but rarely sufficient for leaving one's master. There were other conditions. The master cobblers (*savetiers*) posed one of the harshest and most arbitrary: a worker could not leave "without legiti-mate cause"—to be determined unilaterally by the master (with no reciprocal guarantee against illegitimate dismissal). The same cobblers, the iron dealers, and many other communities retained the workers until they returned all the money that had been advanced to them.[38] The advance constituted a sort of insurance premium against a precipitate and inopportune departure. Many masters revealed themselves to be remarkably generous when a worker entered their shops for the first time. Though the bakers did not write this limiting clause into their statutes, many of them engaged in the practice of granting advances and using the indebtedness of their journeymen for leverage in deal-ing with them.[39] Other master bakers either borrowed from journeymen or else held the wages of their journeymen in so-called safekeeping.[40] In either case the journeymen would be unlikely to leave until they had obtained their due.

Most corporations required that the worker "finish the work begun" before leaving.[41] In the case of the bakers, this meant completing the time of service agreed upon. This condition provoked many venomous disputes for two rea-sons. First, regardless of the desires of the masters, the journeymen usually refused to bind themselves for a fixed period of work. Second, agreements were almost always oral, and the master's word in the absence of contradictory testimony from a third party was taken as proof of contract.

"Congé," according to Savary des Bruslons's *Dictionnaire universel de com-merce,* meant the permission that "a Superior" grants to "an inferior" to do something. When the journeymen fulfilled all the conditions, the master was theoretically obliged to give him congé, that is, a certificate authorizing his departure. Without a written discharge, the departing worker could not legally find work, because before entering a new shop he had to show that his papers were in order. At this juncture the police of placement fused with the police of departures: entering and leaving became part of the same process and objects of the same control. For many guilds one finds evidence of the practice of written congés fully developed in the seventeenth century. For some corporations there is the trace of an analogous protocol of inspection as far back as the thirteenth century. The congé system was generalized by an arrêt du conseil of 2 January 1749 and was reaffirmed after the successive reforms of the corporate system in 1776 and 1777.[42]

A complaint formulated by the bakers' guild in 1671 suggests that although the journeymen were required to give notice and take leave in writing, large numbers ignored these prescriptions. Even worse, certain masters were delighted to take in journeymen who had illicitly fled the shops of their confreres. This defiance of the rules brought disorder and "a trouble and divorce among the said master bakers." In response to guild protest, the lieutenant general of the Grande Panneterie—whose jurisdiction was later absorbed by the lieutenant general of police—enjoined all journeymen to request leave in writing (a stipulation that prejudiced the interests of the illiterate garçon) and prohibited masters from employing journeymen who did not present a written justification of departure from the previous master.[43]

Article 40 of the statutes published by the guild in 1719 specified that journeymen had to give two weeks' notice and request a certificate of leave and that masters could not receive journeymen without proper papers. The evidence suggests that jurés were not entirely successful in enforcing this article, for they felt obliged several times to turn to the police for help (a sign of either the fecklessness of the jurés or the magnitude of disobedience or both). In 1728 the jurés reported widespread transgression of the rules. Journeymen left not only without notice but often on the eve of market days, when there was the most work to be done. Masters continued to receive fugitive journeymen with open arms, while innkeepers also gave them shelter in contravention of the law. Emboldened, the garçons tried "to dictate the law to their masters." Lieutenant General of Police Hérault pronounced a sentence ordering the precise execution of article 40 and threatening fines and more extreme punishment for masters, journeymen, or innkeepers who failed to comply.[44]

Ten years later the guild denounced the journeymen in similar terms, adding that this time, after leaving their masters, they gathered in several inns and taverns to "cabal" against them for the purpose of winning higher wages and greater autonomy in placement. Aside from a clause that prohibited "assembly" and "caballing" in taverns or elsewhere, the sentence issued by the lieutenant general differed very little from the one trumpeted and posted in 1728. The police published similar prohibitions in 1749 and 1756, and it is very likely that there were other sentences that have escaped our attention and many more guild complaints that did not result in judicial action.[45]

As a result of a police sentence issued in 1769 (and immediately reiterated and thus powerfully reinforced in a parlementary arrêt), the hand of the guild (and the unincorporated "master" bakers) was significantly strengthened in the battle over article 40 (which in the new statutes of 1746/57 had reemerged virtually without change as article 46). There were certain "privileged" areas in Paris, however, over which the guild had no jurisdiction. One of these was the faubourg Saint-Antoine. In the eyes of the jurés, Saint-Antoine was a rebellious quarter well before 1789, for it was the place where journeymen who defied the guild statutes (whose crucial clauses, remember, were written into public law) could take refuge with relative impunity. Considering the congé system to be "a violation of their liberty," the journeymen quit their masters peremptorily and fled to the sanctuary of inns in the faubourg Saint-Antoine or to the shops of faubourg bakers who needed help and were not bound to inquire about its provenance. Some faubourg bakers profited from this easy and open labor market, but others fell victim to the alleged arrogance and capriciousness of the journeymen in the same way as the guild bakers. Indeed, because the faubourg masters had no corporate edifice to protect them, they may have experienced greater insubordination than their guild counterparts.

A large group of faubouriens complained to the police that journeymen came and left without notice, disrupting the bread supply, insulting the masters, and even physically abusing them before withdrawing to the inns. These masters asked the lieutenant general to empower two of their eldest and most accomplished colleagues to conduct a "police" of the shops and inns analogous to that undertaken regularly by the guild jurés. The guild was very anxious to deny the journeymen their faubourg refuge, but the leadership was not warm to the idea of assigning this mission to their unincorporated faubourg colleagues. In deference less to corporate prerogative than to administrative efficiency, the lieutenant general decided to authorize the guild itself to undertake the same police in the faubourg that it exercised in the rest of Paris. Now journeymen would no longer have a haven across the border. Everywhere they would be required to give notice, apply for a certificate, and show their papers before receiving work in the shops or shelter in public houses.[46]

If it is impossible to say what proportion of baker boys could document their entry and exit, it is clear that the jurés sought vigorously to expose the wrong-doers. Patrolling on the night of 3 July 1772, the jurés learned that master Lemerle had certificates for four of his five workers; the delinquent, he explained lamely, "had been in the shop for barely two months." His colleague, Sauvin, avowed that he had not yet pressed the issue of papers, for his two journeymen had entered service only four days earlier. Recent arrival is the most common excuse for lack of certificate and registration, intoned jointly by bourgeois and worker. But that hardly fit the case of La Harpe's two garçons in August 1770, both of whom had entered this baker's service more than two years before. Four of Garnier's five journeymen were in order; the fifth lacked a certificate because he was "in dispute" with his former master. Master Fillon told a cunning and implausible tale that put his wife, conveniently out of town, in charge of garçon documentation, keeping on her person the certificates of their two employees in a sack pinned to her skirt. A neighboring master played a variation on this theme: his journeyman's certificate was in a desk drawer, the unique key to which never left his absent wife's pocket. One of the garçons of master Lesueur naively presented a blank certificate that he had forgotten to fill out (or have filled out), almost certainly one of the many counterfeits that circulated on the substantial market in false papers that thrived not only in Paris but also in the provinces (where paper workers were said to be bearers of at least two certificates each, and where the clandestine compagnonnage associations were implicated in a traffic of illicit documentation). A thirty-three-year-old Auvergnat ended up in a Paris jail when he took the liberty of fabricating a certificate instead of complaining to the police when his previous master refused to provide an exit visa.[47]

Journeymen smarted at the lack of reciprocity concerning job notice prior to separation from the employer. In their individual and collective expressions, they repeatedly demanded some measure of balance on this matter and others that had a contractual inflection. Their notion of contract implied not necessarily a precise symmetry but a recognition of mutual obligation. Whether from principle or personal disarray and pique, when arbitrarily or capriciously dismissed, garçon bakers (like other journeymen) sometimes evinced violent anger. In uneasy expectation that he would be fired imminently for debauchery, Antoine dit le Flamand refused to sleep off his inebriation sagely. He remained unsteadily albeit vigilantly on the shop floor in a ritual effort to claim and protect his position and thus ward off a possible replacement (who, ironically, had not yet been summoned by his master). Fired for alleged drunkenness and absence, a garçon called Honoré resisted with words and blows, making a vague threat to call on his successor, who had already been engaged, to manifest solidarity by refusing the job. A proud and testy Auvergnat named Julien, fired for harassing a fellow garçon whom he disliked for his Picard origins, tried to impose single-handedly a sort of damnation of his master's shop to prevent his removal. He plied his replacement with wine, then passed to intimidation and insult, and finally appealed to his successor's sense of solidarity. To solidify his retreat to higher ground, Julien even apologized to the Picard: we can all resist the masters equally even if we can't all be Auvergnats.[48]

Liberalization and Insubordination

In February 1776, at the very moment that lead minister Turgot promulgated his bold edict abolishing the guilds, the baker-jurés were conducting what might have become their last rounds. They were visiting shops to determine whether the journeymen in service "entered work armed with certificates issued by their last masters." At widow Gresel's, at least one of the three garçons had dubious papers. According to police files, the bakeress, on the rue Saint-Jacques for more than forty years, "has been unhappy with her garçons for quite some time, she cannot even remotely control their comings and goings as she deems necessary; on the contrary, it is they who impose the law on her." (Nor did they contest her merely because she was a woman: scores of her male colleagues uttered the same lament.) Regardless of what transpired on the shop floor, however, it is likely that the guild, as the repository of official baker memory and the steward of the sanctified corporate way, would have continued to report an ever-growing menace of worker disobedience. Still, beyond ideological reflex and rhetorical imperative and beyond individual cases, no one was prepared for the seemingly real-life apocalypse of 1776. For a moment it seemed that any prospect of journeyman discipline would disappear with the controller general's violent stroke of reform. At the announcement that the guilds were suppressed, a convulsive fever of "joy" suddenly possessed "the little people" of Paris, according to bookseller Hardy. Journeyman bakers, among others, set off fireworks in several quarters. Hardy detected "a sort of political fanaticism that overheated . . . so large a number of heads." Pitched battles broke out between "journeymen of different professions drunk with their future liberty and the masters."

The guild reacted to these events, legislative and popular, in a panic that was at once authentically visceral and carefully calculated. Could the government really tolerate a situation that was objectively as well as subjectively cataclysmic? Claiming to speak not merely for the masters but for the faubouriens and the forains as well, the baker-leaders described a world stood on its head: the journeymen "see themselves already as the equals of their Masters." To be sure, things had not been perfectly in order in the good old days before the edict—a concession that implicitly raised questions about the capacity of the corporative instrument to perform the functions that the bakers insisted now more passionately than ever before that only it could accomplish. "For some time, the members of the guild, [and] even the forains, have no longer been able to govern them [the journeymen]," avowed the bakers. This was no ordinary labor problem, for at stake was the provisioning of the capital and thus nothing less than the tranquility of society and the stability of government: "it is obvious that a master cannot make on his own all the bread he requires to meet the demand."

Depriving the masters of any moral, political, or juridical leverage, Turgot's measure thrust them into utter despair even as it cast society into a parlous state of nature. Garçons refused to work, leagued together openly in the taverns where they used to meet more or less clandestinely, plotted to "compromise the public service," dreamed quixotically of "becoming Master[s] without ability, fortune, means." What passed yesterday for "seditious and tumultuous assemblies" appeared today to enjoy the legitimacy and protection conferred by "liberty." The sometimes anxious insolence of old-style insubordination had suddenly flowered into an impatient but confident "arrogance," for these workers "argue on the basis of this [Turgot's] edict" and on that of "the intention" expressed by the king.[49]

Turgot himself was horrified by these disorders; he believed in economic liberty, not working-class insubordination. Had his ministry survived, he would surely have disabused the intoxicated journeymen of their misconceptions. In the wake of the corporate reestablishment that followed Turgot's fall just a few months after the promulgation of the abolitionist edict, and in direct response to the master bakers' complaints regarding the boundless "insubordination of the garçons," the lieutenant general of police published an ordinance that reaffirmed the old rules, especially those regarding notice and leave. With few modifications these rules stayed in effect until the Revolution. Many of them reappeared in a slightly different guise in the nineteenth century, during which time the bakery workers remained "a population very difficult to steer," if not "an untamable race . . . dominating their bosses."[50]

The Livret

The police put the finishing touches on the structure of bakery labor control in August 1781 with the introduction of the *livret,* a tool of artisanal documentation. A number of guilds had already been experimenting with this device since the early seventies; several had utilized a sort of livret as far back as the early seventeenth century. Corporate leaders and police authorities believed that it could significantly enhance their ability to oversee and "contain" the working population. So promising did this system appear that one month after the publication of the ordinance enjoining the bakers' guild to adopt it, the royal government issued letters patent applying it to many others classes of workers throughout the kingdom.[51]

The written congé, or certificate of leave, was the direct ancestor of the livret. The imperial livret, often portrayed as a Napoleonic innovation, was nothing more than its second (or third) reincarnation. Beginning in 1781, every journeyman was required to obtain a livret, or small booklet, by registering with the corporate bureau. The clerk noted both in the guild record and on the livret the name, age, birthplace, current address, and last master of the journeyman. To enable the guild to verify his identity and to prevent instances of impersonation or the use of aliases, each journeyman was obliged to present his baptismal certificate. Current Paris journeymen had two weeks from the time of the publication of the ordinance to enroll. All "new" garçons had to register within three days of their arrival in Paris.

To encourage rapid registration, the guild offered to suspend the fees of the first six hundred garçons who applied. Normally the livret and registration together would cost eight sous (a lower amount than the ten sous demanded by the shoemakers). Every time a garçon changed shops, he had to report within twenty-four hours to the bureau, where the clerk inscribed the change on the livret and in the corporate record. Each entry cost the journeyman four sous. When he gave notice of intention to leave, the master was supposed to note the date on the livret. The garçon was to place each successive certificate of leave in the livret, which became a cumulative history of his working life as well as a passport for legal survival. Masters could not employ journeymen without livrets, tavern keepers could not serve them, and lodgers could not offer them rooms. The master who hired a journeyman kept his livret during his entire stay—another hedge against the inopportune or undesired departure of the worker. Some of the master bakers may also have followed the example of the master cloth makers (*fabricants d'étoffes*), who etched on the livret the sums of money owed to them by their journeymen.

The livret was the fullest expression of the policy of social control in the world of work—a control more ambitious, more rational, more total than before. It was born of the imperious need felt by the police to know better in order to contain better. The livret was inspired by the same preoccupation that prompted Guillaute, an officer in the constabulary, to envisage a fabulous and frightening system of individual and global surveillance supported by a data bank and a data-retrieval mechanism. The livret developed from the same sort of concern that spurred the police codifer Des Essarts to propose a project of bureaucratic surveillance for "classifying all the workers [into discrete hierarchical strata]," a project based on the fusion of the old corporatist idea with the new social science. "Strangers" and unknowns frightened the police; all the inhabitants of the world at work had to be identified. Like grain in the police of provisioning, the inhabitants of the world of work had to be constantly visible. In the midst of the dizzying confusion of Parisian everyday life, the livret gave the police the reassuring illusion of being able to immobilize workers in both space and time.[52]

Given the state of the documentation, it is practically impossible for us to measure how well the control system worked. After completing his apprenticeship in Paris, Jacques Deschamps, a native of Provins in the Brie, spent a number of years abroad as a baker. Despite persistent efforts, upon his return to Paris he was unable to find a shop. According to Deschamps, he "did not succeed [in getting work] for lack of a certificate." Deschamps was a man without civic existence, a socioprofessional pariah who could not reinsert himself into the world of work. Jean Guillet, a twenty-six-year-old journeyman baker, had more prudence and foresight than his colleague Deschamps. Before leaving for an indefinite sojourn in the provinces, he requested a certificate from the police that would facilitate not only his outbound journey but also his eventual professional reintegration.[53]

A Dangerous Complicity

It seems very likely, however, that Deschamps's experience was aberrant. It suggests much less wiliness and tenacity on the part of the journeyman and much more restraint and solidarity on the part of the masters than we know to have often been the case. Journeymen and masters both had incentives to disregard the rules; though the authorities reserved the stigma of "rebelles et mutins [rebels]" exclusively for the former, on many occasions it was equally merited by the latter. Masters hired on the black market for several reasons:

because it was faster and simpler, especially when they needed help urgently; because they had their eyes on a specially competent journeyman whom they were glad to "entice away" from their rivals (in French it is "débaucher" with the connotation of corruption); because they were hostile to guild policy or wished to express contempt for current guild leadership. Masters Lelarge and Jamin each employed two journeymen without certificates. Masters Filet, Nerodot, Gillet, and Monin each kept three journeymen without papers. A forain baker named Louin had four "black" journeymen in his service. The list of masters with one illicit garçon would be very long.[54]

These violations of police and corporate regulations deeply troubled the guild leaders. The defection of a single master could cause grave disorders. By a sort of chain reaction—by vengeance or emulation or self-defense—other masters might place their interests before their moral and legal obligations. Masters might be tempted to raid each other's shops for labor, to offer higher wages than the going rate, or to commit other uncollegial acts against both the spirit of the guild and the letter of its statutes.[55] Aware that they no longer faced a solid corporate and professional wall but one with gaping breaches, the journeymen would be tempted to take advantage of the disarray in demanding raises or other advantages. Many worker "cabals" in the eighteenth century were provoked by competition and jealousy among masters. What chaos and what a threat to the system when rebel masters became the protectors of rebel workers.

The guild and the police worked hard to expose and repress this dangerous complicity. The jurés, accompanied by a commissaire, made frequent surprise nighttime visits to the bakeshops of guild, faubourg, privileged, and forain bakers. The masters of garçons without papers were issued summonses to appear in the Châtelet; their "black" journeymen were either sent directly to jail (like twenty-eight-year-old Marc Batteur), returned to their "rightful" masters from whom they had failed to take leave or from whom they had been "seduced," or allowed to remain provisionally in the shop where they were illicitly employed. Sometimes the master who fell victim to an illegal departure filed a complaint either directly with the police or through the guild bureau. All five of master Pierre Mouchot's journeymen left him at once, "putting him in the greatest difficulty." The police arrested two of them within a day of his complaint and began searching for the other three.

Predictably, the masters got off much easier than their journeyman-accomplices. Convoked by the procurator's court, they suffered fines that were hardly dissuasive, save for the poorest among them: ten to twenty livres plus court expenses. In many of these cases, the masters themselves did not even appear to explain themselves, make moral amends, or suffer castigation. They were convicted in absentia or by procuration.[56]

Some of the masters were cavalier about the fate of their journeymen; they made no effort to keep them out of jail. Others tried to protect them, both before and after arrests. Lamarre, a baker from the faubourg Saint-Antoine, refused to open the door to allow the syndic to inspect. To cover his undocumented worker, Desmange, master Fleuret asked his friend and colleague Blondeau to furnish a certificate claiming to have been Desmange's previous boss. Master baker Jarry of the rue Saint-André-des-Arts hid his "black" journeyman so that the guard could not arrest him. Master Pierre Morin fought in court to keep his three workers despite their lack of valid papers. Master Jules Poton and widow mistress Dubreuil also contested the guild's charges against them and their journeymen.[57]

Masters Refuse to Issue Certificates

Without a certificate the journeyman's situation was extremely precarious, but it was not always easy even for the most diligent and subdued worker to obtain this precious document. The masters claimed that all they wanted was for the journeymen to give them sufficient notice so that their business affairs were not compromised. Yet there are a considerable number of cases in which the masters rebuffed their journeymen's legitimate request to depart. For one reason or other they were determined to retain them. The masters cynically wagered that these journeymen did not have temperaments to risk an illegal exodus with fugitive status and that they lacked the lucidity to seek legal redress. There probably were more than a few instances when intimidated journeymen fell into a kind of awkward shop serfdom.

Journeymen did have certain rights, however, and not all of them by any means were either too cowed or too ignorant to insist on them. They had the right to appeal to the jurés for justice, and if they failed to receive satisfaction, they could turn to the police. The Châtelet ordered master Louis-Victor Artin to deliver immediately a certificate to Jean Heussard who had served both as his apprentice and subsequently as his journeyman. If Artin refused, the royal procurator himself would issue a license authorizing Heussard to work where he pleased. Jacques Gartin successfully sued his master several years later, not only for a certificate but also for back wages.[58]

A number of masters argued that it was within their prerogative to withhold, at their discretion, certificates from journeymen who, in their estimation, behaved badly. Master baker Pasquier allowed that he had no complaint about the work performed by his journeyman Denne, but he resented his "bad faith." According to Pasquier, Denne demanded more money per week in salary settlement than they had agreed upon in the beginning. Because Denne refused to abandon his claims, Pasquier denied him a certificate. (For a price, then, Pasquier was willing to settle.) Master Gilbert Dubreuille fired his journeyman Berri for stealing salt. While he paid him his wages, Dubreuille refused "a certificate that he served faithfully, . . . [since he had] every reason to complain of him." Infuriated, Berri threw a stool at Dubreuille and called his wife "bitch, whore, and other [invectives]," which did not enhance his chances of getting his papers.[59]

Like many other corporations, the bakers' guild required each master not merely to authorize a journeyman's departure but also to grade the journeyman by declaring "succinctly . . . if he had been satisfied or not with the assiduity at work and the conduct of the journeyman." The idea was to ensure a continuity—lifelong—of control by generating a record on each worker that no one could escape. It was meant perhaps also to incite the worker to submission by threatening his professional future. Masters could use the certificate to blacklist journeymen in the same way that journeymen concerted to "damn" certain masters. It is easy to understand the anguish of journeyman Lejeune, who portrayed himself as the victim of a bullying master who beat him as an unredeemable scamp. Lejeune had reason to worry "that this will prevent him from being able to work for any Paris master."[60] It is impossible to say how many masters took vengeance on their journeymen by stigmatizing their competence and their character on the certificate.

Taverns and Inns

If journeymen without papers did not seek protection with new masters who were willing to employ them illicitly, it was quite likely that they took refuge in the labyrinthine underground of taverns (some of which gave lodging), inns (all of which served drink of sorts), and rooming houses (many of which were associated with taverns) that checkered Paris in the eighteenth century. Now, these establishments were by no means exclusively redoubts of outlawry. They served the needs of all journeymen bakers, those at work (who at any given point represented the majority of the journeymen universe) as well as those in flight, those who were jobless involuntarily, and those who were for one reason or another down-and-out. We shall discuss the social functions of these jour-

Parade of journeyman bakers, 1905. *Source: postal card, private collection.*

neymen institutions later. For the moment let us note that if fugitive baker boys frequented the inns and taverns, it was because they knew that they would find help there, at least for the short run: lodging, credit, fraternity, perhaps complicity in hiding, or assistance in finding work. At the inns and taverns, the journeyman without papers hoped to obtain a breathing spell and an opportunity to recharge himself with energy and rechart his course of action.

These establishments enjoyed little respite from surveillance. Police commissaires or officers of the guard, accompanied by the jurés, made frequent surprise visits—usually late at night—to inns and taverns reputed to harbor bakers. (The other guilds participated in similar inspections while, independently of the corporate structure, the police made their own "nocturnal descents" as part of their regular round.) In a raid upon Bottin's inn on the rue du Bout-du-Monde in September 1728, Commissaire Courcy surprised eleven garçon bakers without certificates. In six rooming houses on the night of 13 January 1739, attended by several jurés, Commissaire Regnard arrested eight journeymen who lacked certificates of leave from their former masters. One of them, a native of Languedoc, Bourjade, had been out of work for over seven months. Seven other journeymen benefited from the commissaire's indulgence: they suffered a fine of twenty livres and were ordered to obtain papers. Aptly named for the task, Commissaire Boullanger led a *huissier à verge* and an *exempt* on an expedition in January 1762 to "different lodging houses to which journeymen bakers withdraw." The majority of journeymen whom he aroused were able to show certificates. But he was obliged to jail five as "mutinous and rebellious" for want of papers. One of them, a twenty-eight-year-old Norman named Robert Leger, claimed that he had been "working in Paris for ten years without a certificate." Boullanger also issued summonses to three lodgers for having given beds to journeyman bakers without asking for their leave papers. The following year, in a single room of an inn operated by the wife of a former garçon baker, the jurés found five journeymen in violation. During the same period Pierre Hetard, one of the lodgers whom the baker boys preferred, himself probably a former bakery worker, persisted in welcoming journeymen without papers despite several prior convictions for this infraction.[61]

In 1771, a lodger received a fine of thirty-five livres and a stiff warning from the lieutenant general of police for having given shelter to eleven journeyman bakers without papers. Two Burgundian journeymen were arrested by an inspector in a wineshop at three in the morning when they failed to "justify their condition" with proper documentation. Two others were arrested at Les Porcherons, also at 3:00 A.M., for want of papers. One of them, a twenty-two-year-old Lyonnais named Benoist Gaillard, had been unemployed for five months. Baker boys were not even safe in the street if they encountered a passing patrol. The watch arrested Charles Baron, for "loitering and lacking papers." The guard caught another journeyman without a certificate wandering near his rooming house at 11:00 P.M.[62]

Because inns and taverns were places of passage and shadowy movement, eruptions of discontinuity in the theoretically unbroken line of surveillance and control, they were inherently suspect in the eyes of the authority. The police of the inns and taverns was considered a branch or an extension of the police of work. Police regulations forbade lodgers to provide bed and tavern keepers and other wine and beer sellers to offer drink "to any workers, lackeys, servants, and other domestics"—this sequence and association of abject conditions is in itself significant—without first demanding to see their certificates of leave. On occasion the punishment meted out for violation of the regulations was draconian. For example, in August 1727, January 1728, and June 1732 the police ordered the closing and walling up for either six months or a year of the shops of three tavern keepers who received *gens sans aveu* (apparent vagrants, without fixed residence, employment, or documentation). As a rule, transgressors—especially first-time offenders—faced nothing more serious than a fine.

Nor was a certificate of leave a durable guarantee of immunity.[63] In principle, after two weeks it expired and the worker fell into the netherworld of the floating population, of gens sans aveu and hors de condition, a world without legal existence. Lieutenant General of Police Sartine further narrowed the margin that separated the good worker from the vagrant. He instructed his commissaires to arrest all workers who have been out of work for more than one week.[64] The same instructions dealt with vagabonds, servants, and gens sans aveu; Sartine amalgamated them all into the same category of social control.

A number of other rules aimed at beaming light into the underground of inns and taverns. Lodgers of all sorts were required to keep a register, which an inspector was supposed to examine every month, containing the name, permanent address or place of origin, and "quality" of the client, along with the reasons for coming to Paris. Innkeepers and lodgers complained bitterly of police harassment and tyranny but, from the other side of the street, Lieutenant General of Police Lenoir sharply criticized his men for their *lack* of rigor in verifying the movement of people in the registration books. The police also fixed the hours of the taverns in order to limit opportunities for association and debauchery. (In Bordeaux, baker boys were prohibited from drinking in taverns after 7:00 P.M. in spring and summer and after 6:00 P.M. in fall and winter.) In addition, parlementary arrêts and police orders forbade cabaretiers to receive more than four journeymen (presumably in one group) at a time and "to favor the practices of the so-called Devoir [a compagnonnage rite] of the said journeymen" or other "associations of workers."[65]

Idleness and Criminality

With few exceptions, to be hors de condition for more than a very short time was as harrowing for the journeyman as it was disagreeable for the stewards of the social order. Few journeymen had the wherewithal to survive an extended period without wages. The inn provided real consolation, but it was not the *pays de Cocagne*—at least not for more than a few days. Nor was it sufficient to procure a certificate in order to find work; there seems to have been a substantial amount of structural unemployment in the baking ranks, especially in the second half of the eighteenth century. Some of the unemployed journeymen managed to find temporary (and illicit) work with master bakers who needed help on the eve of a market day or to replace a journeyman fallen ill. Others turned elsewhere for work. Jacques Bonnefoy labored as a dayworker (*journalier*) in the quarries. François Germain had prepared in advance: he was a tanner and leather finisher as well as a baker boy. André Gauthier became a part-time water carrier and gagne-denier.[66] For want of luck, tenacity, or enterprise, many journeymen found no work of any sort. These destitute fugitives became increasingly demoralized and desperate.

Those who prized order above all else in the eighteenth century regarded unemployed journeymen as a menace to their well-being. Idleness was the second fall of man. To be out of work was not only to escape surveillance but also to expose oneself ineluctably to all manner of moral and physical corruption. No longer linked to society, the jobless worker became willy-nilly its enemy. According to the ideology of order, the (unjustifiably) idle classes were the dangerous classes.

The idle journeyman was, by definition, a criminal. It was expected that he would turn to crime—or that he would *have to* turn to crime—and he was a priori suspect whenever a crime was committed in his milieu. The burden of proof was placed upon the journeymen to clear themselves. Charles Louis Wagon, a thirty-year-old journeyman from Arras, was denounced to the police when he brought several garments to a seamstress for refitting. Since it was known that he had been without shop for almost two months, the presumption was that he had to have stolen the clothing. In the end, however, the police acknowledged his innocence.[67] Joseph Desbain was a seventeen-year-old baker boy who left his home in Alsace because, in his own words, "he wanted to see France." Unable to find work, he lived in rooming houses until he found himself charged with "complicity in theft." In fact he was arrested because he was "sans condition," because he was a wanderer and a stranger. He, too, was eventually exonerated.[68]

Unemployed journeymen were not always unjustly accused. There is no doubt that many committed the crimes with which they were charged. One must, however, be extremely prudent in dealing with this evidence, for it comes from archives that tell us nothing about the host of unemployed journeymen who did not feel obliged or authorized to steal or otherwise break the law. (Nor must we forget that employed journeymen also committed crimes, drank too much, contracted venereal disease, etc.; work itself proved to be a much less effective shield against depravity than the preachers of social tranquility claimed.) It would be silly to presume that as a rule idle journeymen confirmed the darkest fears and predictions of the forces of order. At the same time, it would be equally wrong to avert one's eyes from a theater of serious criminality that was directly related to idleness or unemployment.

In some cases, journeymen seem to have been driven to crime—habitually theft—by more or less pressing need. Out of work for almost two months and penniless, baker boy Claude Picard stole two four-pound loaves of bread in broad daylight, with no cunning whatsoever, from the display behind the grill of a bakeshop. Arrested for stealing an empty flour sack that he pawned for a livre, Claude Henry explained that he had been without a shop for over eight months. Nor was this his first bout of ill fortune. This thirty-seven-year-old journeyman had already been in jail for debt. Out of work for only two weeks, Jean Perruche, a twenty-year-old Burgundian, confessed to stealing two handkerchieves. He told the court apologetically that "he planned to sell them in order to eat, that it was not at all by libertinage that he was led to commit these thefts." Claude Swire stole some clothing because he needed money to go look for work outside of Paris. Marin Guichard, a seventeen-year-old from Caen, had not worked for weeks when he was found sleeping in a private garden whose fruit he had "pillaged" for his dinner.[69]

Theft was the most common crime committed by idle baker boys. From a sample of thirty-eight journeyman crimes, twenty-seven fall into this category, six deal with various kinds of drunken behavior, two concern counterfeiting, and three remain unidentified (they were probably robberies). Most of the theft was petty. Five journeymen stole handkerchieves, thirteen purloined items of clothing or linen, three filched some pieces of silverware or other household goods, and four took food. Virtually all the misdeeds appear to have been opportunistic crimes, crimes of circumstance rather than of planning, clumsy offenses rather than imaginative transgressions. The only thefts involving substantial sums of money were cash robberies of 308 and 312 livres. The two cases of fabrication and distribution of "false monies" and forged notes were the only sophisticated felonies. One of the journeymen implicated had the hubris to style himself "garçon baker by profession, at the moment go-between and solicitor of deals."[70]

The only flagrantly ugly crimes were those triggered by drink. A twenty-three-year-old journeyman named Farcy, out of work for five weeks, got into an argument with a blacksmith after some heavy drinking in a tavern. The blacksmith called him a "useless bugger" [Jean Foutre] and slapped him. Farcy replied with the cane that he and many other journeymen carried despite the fact that they were prohibited by law from bearing any arms. This gesture was meant to salvage his "honor," a notion probably as dear to a garçon as to a seigneur, and he killed his opponent. Inebriated, Jean-Baptiste Colinet and another journeyman broke up a barroom, assaulted customers, and sexually molested an eight-year-old girl. The other crapulous crimes were far less grave: fighting, brawling over a card game, urinating in a tavern, passing out in the streets.[71]

The idle journeymen who committed offenses were neither among the youngest nor among the most recently settled in Paris. The average age for twenty-eight known cases is almost twenty-seven years old with a range from seventeen (only two were under twenty) to forty-eight. Only two were freshly arrived from the provinces; all the others had been in Paris for at least several years. (Still, it ought to be underlined that it was possible to be utterly rootless even after a decade or more in a city so boundless and so motley.) In the minds of most of these workers, however, Paris was not their home. They still instinctively identified with the provinces from which they came. Only one journeyman of the twenty-two for whom we have information was born in Paris. Among the others, five came from Burgundy, three each from Flanders and Auvergne, two from the southeast, two from the center near the Loire, three from the west, and one from the east. On average, the journeymen had been out of work for three months, a very long time considering their limited resources. Several had been idle for as long as seven or eight months, while a few had been out of work for only a few weeks. The latter seemed to have panicked once their savings and/or lines of credit were exhausted or to have tried to extricate themselves rapidly before becoming mired in pauperization. The former—the veterans—obviously had found a way to get by. A few of them seem to have become self-conscious delinquents; the three recidivists in our sample came from this group of hard-core idle. One of them was sentenced to be strangled and hanged. As a rule, the sentences were more moderate, banning the journeymen from Paris for three to nine years (in one instance in perpetuity) or relegating them to several years in the galleys.

Nothing seems longer or more uncertain than the trajectory from apprenticeship to mastership. We do not know what percentage of apprentices completed their training; we do know that only a minority of journeymen became masters. Despite formal guarantees, the apprentice was never sure of receiving the instruction he was due. The older and the more autonomous the apprentice, the greater the chance he had of acquiring professional skills. For the older apprentices, entry into journeymanship was not a jolting break. Nevertheless, the journeyman had both freedoms and responsibilities unknown to the apprentice.

The journeyman had to make choices. He had to decide how to play the game. Journeymen did not all share the same habitus: those who wagered confidently on integration into the guild and/or on professional establishment probably calculated quite differently in their relations with bosses and fellow workers than did those destined to meander forever in the ranks of subordination. Allowing for the marked bias of our chief sources, it is still clear that relations between masters (of all provenance) and journeymen were frequently strained. More than wages or even working conditions per se, the paramount object of contention between employers and employees appears to have been the right claimed by the latter to dispose freely of their labor. This was one way in which they insisted on being treated as men rather than as sempiternal garçons.

Work was an overarching matter of public concern: the social question in the bakery was not a straightforward issue of labor relations in the modern sense of the term. The authorities reinforced the corporate predisposition to control entry and exit tightly in this institutional labor market. Many journey-men—skilled workers with a certain consciousness of their position vis-à-vis other workers in French society—vigorously resisted these efforts to confine and discipline them. Unfortunately, we cannot measure the impact of this individual and collective experience on their behavior as workers and as citizens later in their own lives or later in the century. It may be that the intensity and incidence of conflict mounted steadily throughout the century; or this impression may be the artifact of the potent Revolutionary teleology.

Chapter 8

At Work

One reason why journeymen fled the bakerooms was that their work was overwhelmingly hard and often debilitating. It required an enormous and relentless investment of physical energy ("one had to be strong . . . and dumb to be a baker," went a nineteenth-century dictum). The baker-laborer had to handle sacks of flour weighing 325 pounds and knead 200 kilograms or more of dough with hands and sometimes with feet. The garçon in charge of bakeroom labor was called the *geindre,* according to one view because he uttered deep groans, or *gémissements,* while kneading. George Sand called it "a painful and savage cry, . . . one would think that one was witnessing the last scene of murder."[1]

The baker boys also prepared wood for the oven, lit the fire, drew water, weighed and fashioned the loaves, and baked them. Nor were they finished once the breads were ready for distribution. If they were not involved directly in sale or delivery, then they were expected to winnow grain, bolt and/or condition flour, gather charcoal for sale, clean the oven, cut and stack the wood, tend to the leaven for the next baking, and so on. The pace in the bakeroom was often frenetic, which compounded their fatigue. They had to rush to knead, rush to proof, rush to bake, rush to sell in order to meet the demands of customers and parry the ambitions of rival shops.[2]

The journeymen bakers complained that they were always tired. The law obliged them to remain in working garb six days a week. In fact, many of them worked virtually every day of the year, generally at least sixteen hours a day. "He burns the midnight oil for me," wrote Mercier of the baker, "let us salute him." The constant night work was the worst part of it; it remained an incandescent issue in bakery labor politics through the nineteenth century. "The night, a time of rest, is for us a time of torture," grieved the journeymen bakers in a long poem anonymously published in 1715 to call attention to their "misery." In the eighteenth and nineteenth centuries, bakery workers protested vehemently that it was "against the laws of nature" to live only at night, to live like "bats," to live a "nocturnal slavery."[3]

Night work was morally as well as physically devastating. According to the spokesman for the bakery-worker cause, it made family life virtually impossible or at least unstable, and it led to a drastic deterioration of behavior and values. Night work segregated bakery workers from the rest of society. They formed a class apart, a class deemed "vulgar and brutal." At work, they were absorbed and thus contained. Once they escaped, however, they exploded into "crapulous frenzy." Journeyman bakers would drink, brawl, and whore less if night work was suppressed, it was argued. For the journeymen of mistress baker Lapareillé, the working day began at half past eleven in the evening. For the baker boys of masters Mareux and Barre, it was midnight. Journeyman François Reims began at 2:00 A.M., while his colleague Jean-Baptiste Langneon started a half hour later. One master worked with his garçons without pause from 8:00 P.M. to 7:00 A.M., while another established in the same quarter slept the night through secure in the knowledge that his workers labored in his interest.[4]

Constantly on call, obliged to stop and start several times in the breadmaking procedure, and desperately in need of repose that they could take only sporadically, journeyman bakers commonly slept in the bakeroom itself. Journeyman Martin Macadrez went to bed at 7:00 P.M. "atop the oven," as the bakers liked to put it. Mistress baker Goizet kept two mattresses "on the oven" for her journeymen, as did master baker Jean-Baptiste Côte. Master baker Savalle kept two "little beds filled with feathers" on the oven, while master Boucher fashioned "a little niche" above the oven in which he placed two sleeping bags made of flour sacks for his workers. Jean Blain, a wealthy master, had all three of his journeymen sleep on a single mattress, each, however, with his own pillow.[5]

The bakery worker lived most of his life at work, asleep, and in the stuporous moments in between in the cramped, oppressive space of the bakeroom. This meant that, in many cases, the journeyman was a "veritable troglodyte," as an early-nineteenth-century police report put it. Probably a large majority of Parisian bakerooms were located in basements or in subbasements—"caves" as they were called. Conditions had improved by the second half of the nineteenth century, but George Sand was still warranted in styling them "somber dungeons." Characteristically, the bakeroom was crowded with utensils, tables, and supplies. There was barely enough room to maneuver to accomplish the simplest operations. In some rooms the journeymen could not stand up straight. Take, for example, the *fournil* belonging to master baker Briquelot on the rue Saint-Martin, in a house he rented for the substantial sum of 850 livres a year. It was a tiny, windowless room in dilapidated condition, made even more confining by the need to improvise repairs. The ceiling had to be rein-

A sixteenth-century bakeroom with a servant collecting fresh loaves in the background next to a journeyman who is kneading dough, while a powerfully built baker, half-naked and sweating profusely, removes breads from the oven. *Source: © cliché Bibliothèque nationale de France, Paris.*

forced with a jerry-built trellis of boards and poles. "The ceiling being low," wrote a police official, "it is very difficult to knead without banging one's head."[6]

The lack of space encouraged sloppiness and carelessness. Journeymen washed and even urinated (as I have seen them do today) in the pails they used to carry water to the flour. The water system itself was sometimes contaminated from waste. Outside seepage drained or dripped into the flour; encrusted dirt from the walls chipped into the flour. Insects fell into the dough even as rats nested in the flour-storage areas. The utensils were rarely ever washed. The crusty, golden bread made in these caves was appetizing only because one did not have to contemplate the often squalid process of its production.[7]

The cellars rarely had adequate ventilation. The air was thick sometimes with flour dirt, sometimes with a hot and choking humidity. When the oven burned, the heat was searing. The worker labored in shorts, sweating profusely. In between bakings, especially in winter, the bakeroom was generally damp and very chilly. The ambience was as unhealthy as the work was enervating. Mercier was struck by the contrast between the hardy, ruddy-cheeked butcher boys and the haggard, sallow baker boys who stood forlorn in the doorway like scarecrows covered with flour.[8]

Broken Health

A journeyman baker had to be strong to do the work and dumb to embrace a profession that more often than not broke his health, left him both infirm and impecunious, or killed him prematurely. As a result of the grueling physical burden and the noxious conditions, bakery workers frequently fell ill. "In my experience bakers are more often ailing than other workers," wrote a leading student of comparative occupational disease. They were highly susceptible to a startling array of serious maladies. The most common were respiratory illnesses, ranging from a chronic bronchial cold to pleurisy, pneumonia, asthma, emphysema, and tuberculosis. Bakers inhaled floating particles of flour that fermented in contact with saliva and stuffed up the throat, lungs, and digestive tract with a sort of paste that caused hoarseness, coughing, and shortness of breath. Overheated and sweating, journeymen regularly had to leave the bakery to deliver bread or run errands outdoors. The brusque change in temperature favored the gestation of colds and other infections.[9]

Strenuous toil produced hernias, ulcers, and varicose veins and, according to some commentators, predisposed the journeymen to cardiovascular problems. Hard labor in a damp milieu made it likely that bakers would suffer at one time or another both rheumatism and arthritis, swelling of the hands (from kneading), and deformation of the knees and legs (from working ceaselessly on one's feet). Baker's eczema, apparently caused by the absorption of flour dust and yeast spores, afflicted a large number of workers. They may have misinterpreted its appearance as a harbinger of leprosy, a disease whose specter obsessed them and one that they tried to exorcise with various religious practices. The intense heat and light of the fire—one of the alleged causes of leprosy—along with the irritating presence of flour dust was supposed to have provoked a much higher incidence of eye difficulties among bakers than in other professional groups.[10]

The stress and strain of arduous exertion performed through every night and part of every day was said to have generated grave "nervous" disorders—psychological problems of personality adjustment and equilibrium. Their work made bakery workers characteristically "misanthropic," "morose," and "very unstable." These problems, in turn, encouraged various forms of brutal "home" therapies—especially drinking and whoring—which resulted in further physical and moral deterioration. Though we have no statistics to verify these or any other claims, it was widely believed that alcoholism and venereal disease reached near epidemic levels among the bakers.[11]

Mercier recommended that the bakers be "compensated" in some fashion for what they lost in their health while working to satisfy pressing public needs. Citing examples from Greece and Rome, Macquer in his *Dictionnaire des arts et métiers,* argued that "the best-policed nations always accorded some privileges [in compensation] to the bakers." A century later, George Sand maintained that in return for providing us with the most nourishing and holiest of goods, the baker "should march just behind the priest in solemn civic celebrations."[12]

In fact, few concessions were made to the bakery workers in the Old Regime. The authorities refused to permit them to eat meat during Lent in order to enable them to recover the strength they needed to work. When a journeyman fell ill, there was often a powerful disincentive to his caring for himself properly. While convalescing, baker boy André Pochon had to pay out of his pocket for a replacement: twenty sous a day in wages plus twenty sous a day for food. No wonder that the leitmotifs in the baker boys' poem of misery are "purgatory," "hell," "captivity," "torture," "prison," "forced labor."[13]

Bakers repeatedly bemoaned the fact that few of their colleagues reached old age. The prudent and fortunate ones, after wasting their strength and their youth, quit the bakery for less taxing occupations ("one does not die a baker," went the adage). Those who remained in the bakery were said to die typically between the ages of forty and fifty as a result of exhaustion, disease, or dissipation or from a combination of these elements. Master baker Philibert Rouget, for example, died in his bed in the middle of the night at age forty, "worn out." In the late nineteenth century it was maintained that the suicide rate of bakery workers in some European countries was far above the mean rates for the entire population.[14]

Division of Labor

In the bakerooms of the relatively large bakers there was a clear-cut hierarchy and division of labor. The *geindre,* or *brigadier,* was the "master journeyman," or foreman.[15] Delamare claimed that the name geindre derived from the Latin word for apprentice or aide. Another explanation was that geindre was a corruption of *gendre,* in recognition of the fact that the first journeymen often became sons-in-law of the master (certainly that was rarely true in the eighteenth century). The interpretation with the strongest purchase, as we have noted, linked the geindre with the verb *gémir* and the sound uttered by the journeyman as he kneaded. According to this view, masters found it reassuring (from their snug beds?) to hear the sound of kneading—both the slapping of

(Above) Shaping, weighing, and placing loaves in oven, eighteenth century. *Source: from Malouin, Description, ed. Bertrand, 1771. Courtesy Kroch Library, Cornell University.*

(Below) Removing and cooling loaves, eighteenth century. *Source: from Malouin, Description, ed. Bertrand, 1771. Courtesy Kroch Library, Cornell University.*

the dough and the groaning of the worker. When they did not hear the latter sound especially, the masters became uneasy, for a good journeyman "geint bien."[16]

By the eighteenth century, the geindre frequently no longer kneaded. He had broader responsibilities, largely because the master spent increasingly less time in the bakeroom. (Jean-Claude Desmeures may have been an exception, at least among bakers with three garçons, for he remained present in the bakeroom through the weighing of the dough, presumably to certify the absence of error, leaving "the rest of the work leading to the perfection of the bread to his said journeymen.") The first one up, the geindre prepared the day's (or rather the night's) work. He checked the weather outside and the barometer in order to know what temperature to make the water. The first journeyman determined how much bread was to be baked, measured out the flour and water, checked the development of the starter leaven. He supervised the kneading process—still the most delicate and critical task—which was now entrusted to the second journeyman, also known as "the first aide." It was the geindre, however, who judged when the dough was ready and who divided it and weighed it. He was also supposed to light the oven, place the molded dough inside it, and remove the cooked bread. The geindre had to have a talent for organization: the timing of the process was crucial. In addition to kneading, the second journeyman took care of the flour stock, brought in the water, cleaned the kettles and pails, and cut and placed the wood. He assisted the geindre in loading and unloading the oven, the most prestigious task in the bakeroom. The third journeyman collaborated in many of these tasks, did much of the heavy and dirty work, and delivered the bread to the market or to the houses of customers.[17]

In the small shops, the hierarchy was compressed and the division of labor less sharp. For example, in master Garden's shop the second journeyman (there were only two) delivered the bread. When he was severely beaten in a barroom brawl, master Vincent Lefevre had to hire a journeyman to take his place at the oven, for he had only one garçon. Widow Villemain had an even more modest establishment. While she took care of the commerce, her journeyman, Joseph Thevenin (who became her son-in-law, a rare stroke of good fortune for a journeyman in eighteenth-century Paris!), ran the bakeroom by himself.[18]

In seventeenth-century London the average bakery proprietor employed three or four journeymen and one apprentice. Malouin, the first scientist to specialize in bakery matters, wrote that masters needed at least two journeymen, that they "usually" had three, and it was common in Paris to have four.

For some thirty-five cases that I have counted in after-death inventories, the average and median were both 1.9. Only five masters had three journeymen, while nine had only one. The average number of baker boys per master increased substantially in the second half of the century, reflecting perhaps a general expansion in the Paris baking industry, or a shakeout resulting in stronger, more productive shops. Before 1745, the average was 1.53 (n = 18); for the period after, it jumped to 2.4 (n = 17). The average number of garçons per master in the Popincourt section during the Revolution was 1.7, whereas in the nearby Quinze-Vingt thirty-three bakers apparently worked alone, a situation that seems implausible unless the bakers employed undeclared family members or other illicit workers. Employing six journeymen in 1790, the widow Brille operated an exceptionally large enterprise. There is a fairly strong correlation between the tools in the bakeroom and the number of journeymen (r = .64 at .001), but a substantially less compelling association between the total number of rooms in the master's dwelling and the number of journeymen (r = .39 at .023).[19]

The masters with the largest number of journeymen were also the masters most inclined to employ domestics. Thus domestics do not seem to have been used as surrogate journeymen by masters employing few journeymen. Rather, they were characteristically assigned a wide range of tasks outside the bakeroom by masters presiding over relatively large staffs and establishments. I found thirty-two bakers each with one domestic. Only one master, Pointeau, kept two at a time. He paid his servants between twenty-five and thirty ecus a year, which he encouraged them to save by leaving them in his possession until they were ready to leave his employ. This master also furnished them with clothes and paid for writing lessons for one. Domestics tended to remain in service much longer than journeymen. Sometimes they were rewarded in wills. Charlotte Mulat, for example, received four hundred livres in cash plus bedroom furniture and accessories and Barbe Railly received 150 livres cash and the bed on which she slept.

Another Charlotte, this one named Villecour, was sentenced to be tortured and then strangled and burned at the stake, apparently for collaborating with her master, baker Jean-Pierre Jourdan, in the poisoning of his wife. As a rule the servants worked closely with the bakers' wives—on nontoxic affairs. Taking responsibility for running the household, they liberated the wives to manage the shop and tend the books. Only very rarely did servant girls work in the bakeroom. It was not unusual, however, to see a domestic delivering bread in place of a journeyman or helping to sell it at the market.[20]

Turnover and Wages

There is no way to measure with any confidence the average length of time that a baker boy remained with a master. The scattered fragments of information we have suggest a relatively rapid turnover, but these data reflect a bias in the documentation in favor of journeymen in motion. They tell us about Jean-Baptiste Colinet, age twenty-two, who worked for "several different masters" in a period of a few months; about Jean-Baptiste Beuzome, who asked for leave after only two months; about a journeyman named Clugny who worked for master Largy for a while, left him for four years, and then returned to work for him for another year before leaving a second time; about Alexis Cresson, who moved down the street to a new master after quarreling with Sieur Cousin, for whom he had worked less than a year; about baker boy Moreau, who left Paris on occasion for long sojourns in his *pays*. We know that Gilbert Mesag, a twenty-seven-year-old journeyman, left a shop on the rue de Charenton for the rue du faubourg Saint-Martin after a very short stay. He may have left not because he was temperamentally unstable but because he discovered an opportunity for professional advancement. Such was the case of seventeen-year-old Jean Regnier, who quit his position as third journeyman ("garçon en troisième") after only six months in order to move to another shop as "geindre." We know virtually nothing about the journeymen who may have found a home in the shop and amidst the family of a congenial or fraternal master or about the journeymen who remained in one place for many years by dint of habit or about those who stayed because they expected to succeed the master (sons, other relatives, ordinary journeymen who stayed for many years in order to reach mastership and acquire a business).[21]

One reason why certain journeymen stayed on and others moved frequently from shop to shop was the wage they received (and perhaps also the manner in which they received it). In principle, wages were not supposed to vary from shop to shop. That was the corporate ideal and the corporate rule. It was meant to preempt price wars, journeyman raiding, and internecine rivalries in general. In fact, the so-called current or going wage served only as a very rough guide to the masters. They paid less if they could get away with it, and more if they had to. Some of them offered blandishments in kind, such as wine, food, and clothing. In order to attract the best workers, other masters promised a higher wage after a probationary period of several weeks or months or a cash bonus after a year. A number of masters, such as Dufour, Martreix, and Berger, were willing to make advances to their baker boys when they first entered the shop, when they were in straitened circumstances, or when they fell ill.[22]

We lack the information necessary to construct a reliable baker wage curve for the eighteenth century. For thirty-nine instances, the average wage was 4.27 livres a week, but this sample does not permit us to distinguish with certainty the wages of the different ranks of journeymen. The highest recorded wages— eight and nine livres a week on the eve of the Revolution—surely belong to first garçons, though it is worth noting that a few geindres were receiving from five and one-half to six livres in the sixties and seventies, the same salary that some of their colleagues had commanded in 1723.

Our average wage in the second half of the century, 4.71 livres, was clearly higher than the average for the first half, 3.69. But wages were extremely sticky; they rose much less rapidly than prices. And one still finds geindres in the 1760s who received only four livres a week and baker boys in the 1770s who earned as little as ten sous and twenty-eight sous a week. In the bakery of the Hôtel-Dieu the geindres and the other garçons earned in 1761 exactly what they had earned in 1726. The journeyman bakers were probably among the lowest-paid skilled workers in eighteenth-century Paris. They had reason to be unhappy about their wages as well as about their working conditions.[23]

Given the low wage level, it is probable that journeymen also received a food allowance or were fed. In one case it was specified that a baker boy was to receive one livre extra a week for food. The fact that the royal procurator at the Châtelet estimated the daily wage of a journeyman baker, for purposes of litigation, at thirty sous in 1768—a nominal wage almost 50 percent higher than any of the salaries in our sample—suggests that he was allowing for a food (and perhaps also a lodging) component. Most of the journeymen could, as we have seen, live in the bakeshop more or less permanently. Most of them also received a wine supplement that was not reflected in the nominal salary.[24]

Wages were generally paid by the week, though it is not rare to find bi-monthly and even yearly salary agreements. In any event, journeymen had to ask for advances repeatedly since their wages were not due to be paid until after the "contract" period expired. Generally masters granted small "loans" willingly in order to win their employees' sympathy. Master Berger, who hired a journeyman Runnet at sixty livres a year, advanced him in the first few weeks two livres ten sous for clothing, one livre ten sous for shoes, six sous for a bedpan, four sous for a bottle and goblet, and sixteen sous six deniers for the lottery. Buffin, geindre to master Tayret, also played the lottery, which may help to explain why he was frequently in debt to his master. Tayret seems to have run a kind of company store, forcing his employees to buy from him by insisting that they take their advances in kind. He provided Buffin with a cotton cap, a bowl, stockings, and a shirt. The master also charged him six sous for breaking a tile—a gesture hardly likely to endear him to the new employee. Master Dufour advanced money to his journeyman when the latter was sick and could not work for a wage. Though the baker boy of Martin Martreix was relatively well paid, he was endlessly in debt to his employer.[25]

As attractive and useful as they were, advances were only a part of the wage story. Very frequently masters fell behind in their payments to their journeymen. So unusual was it for a master not to owe money to his baker boys that the after-death inventory of master Etienne Huin celebrated in a special entry the fact that "nothing is due the journeyman bakers." Master Tayret readily accorded advances and loans, especially in the beginning. But in short time he cashed in on his fund of goodwill by asking his journeymen to be satisfied with a partial rendering of wages on payday. Jacques Barbier's master owed him 164 livres, almost a year's salary. Widow Le Guay died before paying her journeyman 150 livres in wages due. Master Pointeau noted in his shop register on 27 January 1770: "I owe to Jean *compagnon* five weeks and the sixth begun on this day."

When another journeyman nicknamed Le Lionnois gave notice, Pointeau calculated that he owed him 450 livres 10 sous, over a year and a half's wages. It is possible that Pointeau was serving as banker upon Le Lionnois's request even as journeyman Louis Chator asked master Edme Polard to hold his earnings. Journeyman Joseph Thevenin lent his mistress six hundred livres "from his savings," which probably meant that she had been withholding a large part of his wages over at least several years. These journeymen may not have trusted themselves to save their wages, or they may have feared for the security of the money in their own hands.[26]

These primitive forms of payroll savings, wage withholding, and lending practices led inexorably to rancorous quarrels. The masters more often than not kept very summary accounts or no records at all. The journeymen rarely, if ever, asked for notes, private or notarized, in recognition of money owed by the masters. Sometimes a master promised a journeyman one wage to lure him into work at a critical time and then paid him at a lower rate when he left. Journeyman Jean-Baptiste Beuzome claimed that master Personne promised him three livres a week when he began but offered less than two livres a week when he left. Journeymen Jacques Brou (50 livres), Antoine Bailly (6 livres 12 sous), Jean-Simon Bricard (9 livres), Nicolas Guejay (10 livres 16 sous), and François Brunet (55 livres) sued their masters in the consular courts, and they all won judgments against them.[27]

Before filing suit for a very large sum of back wages, journeyman Clugny submitted his case for mediation to the consular courts. The testimony taken by the referee reveals a great deal not only about master-journeyman relations but also about baker business practices, the familial division of labor in the shop, and the precarious margin on which some bakers operated. Clugny claimed that he entered the shop on 21 April 1781 at three livres a week but that from 2 June until he resigned on 22 February 1782, his wage was doubled. During these ten months, Clugny claimed that he could get his master to pay him a total of only forty-six livres, "Sieur Largny disposing of no credit at all and being obliged to save his money for the purchase of flour." When the journeyman insisted on a final reckoning, Largny, according to Clugny, asked his wife to handle it, in light of her "good sense."

Apparently her accounting met Clugny's expectations, but her recommendations were rejected out of hand by her husband, who claimed that Clugny had profited from the fact that his wife had been drunk when she calculated his wages. Largny insisted that he had paid Clugny regularly every week and that he owed him wages for only five days. The master told the mediator that he kept no record of payment to his journeyman. The mediator examined a register kept by his wife, which did show payments to journeymen and which implicitly proved that Clugny had not been paid regularly as Largny contended. Moreover, Clugny asserted that he had worked for Largny once before, four years earlier, and that he had not received a regular wage payment then but, rather, a lump-sum settlement of 130 livres on his departure. Unable to persuade the master to acknowledge his journeyman's claims, the mediator proposed that the consular court order him to pay Clugny his back wages plus expenses.[28]

Journeyman bakers expressed their dissatisfaction with low wages in several ways. A number threatened to leave more or less immediately if they did not receive a raise. Some mentioned specific opportunities elsewhere that would allow them to earn a better living.[29] Other journeymen, as we have seen, took revenge on the masters who denied them a raise by leaving without notice at a moment when they were badly needed in the shop. The police and the corporate fathers considered such individual and more or less isolated cases of wage dissatisfaction as relatively innocuous and marginal events. What alarmed them were signs of a collective consciousness and of a collective strategy on the part of the journeymen.

Such signs flashed periodically throughout the century. From time to time they were blurred and uncertain—reports of "plotting" and "agitation" in the clandestine journeymen's organization called the confrérie and in various "clubs" located in the inns, especially in the faubourg Saint-Antoine. On other occasions, these markers flared vividly when subterranean rumbling gave way to overt collective action. In 1704 the journeymen in the faubourgs formed what their employers called "a league and a cabal to increase their daily wages." There is no doubt that the journeymen were organized: they had leaders, they had precise demands, they had well-rehearsed tactics, and they had fixed meeting places. They struck the shops of bakers who refused to meet their salary demands. Where necessary they used intimidation and force to prevent a journeyman from going to work in the boycotted shops. The leaders amassed a strike fund to help sustain those journeymen who left work in the name of their common cause. In the mid-fifties the jurés repeatedly denounced journeymen for holding illicit meetings, for gathering in large numbers in the streets near the guild bureau, for quitting without notice, for defying the placement rules, and especially for engaging in "cabals among themselves in order to impose their will [faire la loy] upon the masters and widows of said guild concerning the wages they wished to demand."[30]

Relations between Masters and Journeymen

Where masters made a special effort to treat their journeymen well, relations between them were excellent. Many journeymen inspired confidence and many masters relied heavily upon them to conduct everyday business. Master baker Bricard, for example, spent long weekends with his family at Corbeil, leaving Jacques Fortin, his fifty-three-year-old journeyman, to do all the baking. Nicolas the Bourguignon, a baker boy arrested for delivering bread of light weight, stubbornly languished in jail for over three weeks rather than reveal his master's name and address. The masters protected employees who served them well and with whom they got along. In 1750 Gervais Protais's employer helped save the adolescent baker boy from being "kidnapped" by bands of armed men who were certainly archers of the Hôpital général looking for youthful gens sans aveu. Some masters fraternized with their journeymen as if they were social equals. Master Robert Richard of the rue Buffroy drank with his journeyman André Basière. Master Paul Jacob of the rue Saint-Victor regularly took a *chopine* of wine with his journeymen in a tavern across the street from the shop. Master Debure's card game was open to journeymen as well as masters. Journeyman J.-J. Mirra appeared as a "friend" and "witness" at the wedding of master Etienne Ligou.[31]

Yet very often there were serious tensions between masters and journeymen. The major source of these tensions was the desire of the master to impose his will and the refusal of the journeyman to submit. These were the masters whose complaints against "the insubordination" of their journeymen became a refrain in the eighteenth century. Like the police and the corporate elders, they were alarmed by growing signs of the "arrogance," "independent spirit," and "indocility" of their workers. These were the journeymen who openly defied their masters, their guild, and the police by joining "cabals," assembling secretly, dropping out of the world of work without notice. These were the journeymen who were widely "intoxicated" by the vista of liberty that Turgot (momentarily) opened in 1776 with the abolition of the corporations, who quit their masters joyously and set fireworks and illuminations to celebrate their emancipation.[32]

The police from Commissaire Delamare through Commissaire Lemaire (and beyond) insisted on the need for the "subordination" of workers because they considered them naturally turbulent, mutinous, and dangerous. Their model of subordination was that of servant to master. The police assimilated the world of work to the underlife of domestic service. The state of being a worker was servile and degrading even as work itself was abject. The worker owed the same submission to his employer as a domestic to his master. Many of the features of the police of work were drawn from the police of domestics. The worker was bound to the shop in the same way the servant was to the household. Both owed "respect and obedience to their masters" and when they failed to "honor" them they had to make "submissions," a didactic and humiliating rite reaffirming their subjection. It was perfectly natural and proper for master bakers like Jean-Francois Aubillion of the rue Saint-Martin to refer to his journeymen as "my domestics." It was just as natural for many journeymen to resent this domestication acerbically.[33]

Masters such as Aubillion kept their workers at a distance. Propinquity did not breed even a semblance of equality. For many masters it was important to draw a clear-cut boundary, both professional and social, that separated them from their workers. Though it was a minor affair, the conduct of the masters of the faubourg Saint-Antoine in 1744 was symbolic of the voluntary estrangement that the relation of subordination first encouraged and then demanded. To pray for the recovery of Louis XV from the illness that nearly killed him, the masters and journeymen convened in separate places at different times.[34]

It would be wrong to imagine the bakerooms of Paris as battlegrounds in an open war between masters and journeymen. Where there was war, it was generally cold war. For the most part, the tensions between employers and employees remained latent, or they were channeled away through mechanisms of sublimation. Occasionally they bolted to the surface and found expression in several ways. The masters lorded over their journeymen and abused them, as if to punish them for their disobedience, and the journeymen avenged themselves by various acts of insubordination, subversion, and crime.

The journeymen repeatedly complained that their masters beat them brutally as a form of discipline. The jurés regarded these charges as fantasies or simply preferred to overlook them. The accusations doubtless were framed in terms too sweeping to be credible. Yet there were masters who had reputations for bullying their journeymen and the journeymen of their colleagues, apparently as a form of therapy both for themselves and for the garçons.[35]

From time to time the journeymen replied to the masters in kind. But when journeymen beat masters the offense was infinitely graver, given the legal and moral code that governed their relationship. Sometimes these incidents appear to have been wanton, senseless assaults, criminal in the narrowest sense, as when three baker boys hit and kicked a master baker's wife and tried to steal her necklace, or when two other journeymen jumped master baker Alexis Roger, bludgeoned him in the head and chest, and ran off with his jacket and hat. Other acts of physical aggression were the outgrowth of specific quarrels or grievances, gestures of retribution perceived as the only effective way to settle a score with superiors against whom the workers had absolutely no economic, social, or political leverage. Embittered over their dismissal, recently fired journeymen returned to salvage a bit of self-esteem and put the master in his place. Alexis Cresson ambushed his former master in the latter's habitual tavern. François Collin attacked his ex-master in his shop after having insulted his wife "as much as one can insult a woman of honor." Claude Berri might have been appeased if his master, Dubreuille, had been willing to give him "a certificate of faithful service" upon his dismissal (for theft). When Dubreuille refused, Berri assaulted him.[36]

Brandishing cockades in their bonnets—a badge of conspiracy, defiance, and mutiny in the eyes of authorities—a band of journeymen, perhaps as many as fourteen, laid siege to master François Moriceau's house. Several of them had worked for Moriceau earlier, one of whom, called Nantais, had apparently been fired. For a while he nurtured the ambition of breaking Moriceau's arm or leg because, he said, "I'm furious with him and he will pay me for it sooner or later." It was Nantais who struck Moriceau with a rock. Because it seemed to be symptomatic of a burgeoning current of intractability, the corporate officers made a special issue of this incident. The jurés convoked the elders of the guild and together they joined with Moriceau in an angry complaint not only against Nantais and the cockaded gang but against the large number of baker boys who "did not stop insulting and abusing all the Masters day after day," threatening to leave them ("which is totally contrary to the public good") and "even to burn down the houses of the Masters."[37]

Journeymen also avenged themselves in another manner, which, although much less violent, often produced more damaging effects. They attempted to discredit the master baker and his family in the community with the aim of striking at them where it hurt most: their business. Master Denis Le Coc and his wife of the rue Mouffetard denounced their former journeyman, recently discharged, for mounting a "cabal" to ruin them by spreading false rumors about their integrity and competence. A rival master from the rue Mouffetard, Muloteau, allegedly seconded the devilish baker boy. An unstable journeyman named Le Roux abandoned master Augustin Le Grand and spread the word in the neighborhood that Le Grand's wife was "a bitch and a whore" who wanted Le Roux to have sexual relations with her in the storeroom at the bakery. These calumnies threatened to bring "discord into the household" as well as "dishonor in the community," and that was why the master's wife denounced Le Roux to the police.[38] Honor was a marketable commodity as well as a crucial element in a master baker's conception of himself. We shall have other occasions to gauge his attachment to it.

Alleged Journeyman Dishonesty

If the master was brutal, in the stereotype of him fashioned by the disillusioned journeyman, the journeyman, in the stereotype of him forged by the cynical master, was dishonest. Some masters seemed to assume that it was in the nature of the baker boy to cheat or rob the master whenever he could. The advantage of the bleak expectation was that it enabled the masters to find a scapegoat whenever they needed to account for a failure, an error, or an accident. In his third failure proceeding (*faillite*) in eleven years, master Pierre Galmont conveniently blamed part of his heavy losses—in vague terms, without names, dates, or places—on "journeymen who robbed him." In his failure petition, master Nicolas Eloy had the courage to mention his journeyman-culprit by name. But Eloy, too, attributed to him a series of crises in an outrageously nebulous and self-serving manner. To enhance his position vis-à-vis his creditors, master Bristch did not hesitate to impute a large part of his substantial losses to the ill will and ineptitude of his journeymen.[39]

There were of course faithless baker boys even as there were bullying masters. The most common reason adduced for cashiering a journeyman was his "infidelities," and some of these dismissals were richly merited.[40] Many of these infidelities were petty matters that caused the master to lose confidence in his employee but did not impel him to call on the police. Others were flagrant acts of criminality ranging from minor larceny to murder. Juxtaposed to the everyday behavior of the bulk of journeymen, these crimes were clearly aberrant. Yet news of them traveled rapidly among the masters and helped to reinforce their suspicions and their reductionist image of the baker boy. In some cases these crimes were banal *faits divers;* the fact that they were committed against the masters was more or less a matter of convenience and accident. In other cases, however, it appears reasonable to think that the crime was not merely a blind deed of passion or material plunder but also a gesture of vengeance and reprisal.

The simplest and probably least risky bakeshop crime was the theft of bread. Master Legendre accused his journeyman of stealing four four-pound loaves. A journeyman named Nicolas Cuny tried rather clumsily to steal bread for two baker comrades who were out of work. After he was fired, he bribed his master's ten-year-old son to steal money from his father. A sixteen-year-old baker boy used bread he stole from his master to pay in kind for his "debaucheries" with an obviously pragmatic prostitute. This journeyman hit the jackpot: he was fired and jailed and he contracted venereal disease. Three baker boys joined together to rob the shop of their own master. They marketed the stolen bread for profit on the neighborhood streets.[41]

François Laurent, a third garçon responsible for delivering bread to clients, was a bit more cunning. He stole bread during the stressful fall of 1770 and sold it on the streets (for a bargain price, a concession that he did not need to make in the time of serious dearth). Laurent covered himself, however, by charging the cost of breads he diverted to his own pocket to his master's regular customers. He did so by marking extra notches on the *tailles* that he used to keep track of bread sales on credit. After working for his "bourgeois" for only six months, a twenty-seven-year-old journeyman baker also used the "false tally stick" to embezzle significant sums that he declared "to have eaten up with other baker boys."[42]

Because he was the "journeyman of confidence" of his master, Jean-Baptiste Debourges collected money from customers who purchased on credit. He siphoned off a substantial sum, which he apparently used in part to make loans to other journeymen. Debourges claimed that he had to account to his master only once each quarter—few bakers could afford such infrequent reckonings—and that he intended to make up the missing sums in due time. In similar fashion, François Marquet collected over one hundred livres in sums due from customers during one month, which he "dissipated in debauchery."[43] Laurent and Debourges were convicted of theft and sentenced to be banished from the capital for three years. Each also had to go on display in the pillory in order to dramatize publicly the lesson of their crime. Laurent wore a sign that proclaimed, "Journeyman baker unfaithful to the clients of his masters." Debourges denounced himself as a "Journeyman baker unfaithful toward his master."

Flour theft, another in-shop crime, was considered a more serious offense because it involved a more complex and rewarding operation. A journeyman baker named Champagne stole flour from his master on a regular schedule and disposed of it clandestinely with the help of friends. Champagne's achievement pales in significance next to the massive and well-organized pilferage of the "Desmeures ring," however. Desmeures was a master baker who invested in his three journeymen "his total confidence, with the principal keys to his house," including those to the flour-storage room. Desmeures could not have been in very close touch with his business, for the three garçons apparently stole dozens of 325-pound sacks of flour without his becoming aware. The journeymen removed the sacks in the middle of the night and usually carried them on their heads across the city fiscal barriers into the countryside, where they sold them to other bakers and perhaps to pastry makers. So busy were the three that they tried to recruit help among some unemployed baker boys for moving the stolen flour.[44]

Nicolas Gamain, a twenty-year-old journeyman from Picardy, also had his master's keys (after two years of faithful service). Impatient to establish himself independently, Gamain used the keys to rob his employer of over thirty-five hundred livres, which he planned to use in order to acquire a mastership. He averred that he would have returned the money later with generous interest. The Parlement of Paris considered Gamain's betrayal of his master so grave a crime that it increased the prevotal court's sentence from life in the galleys to death by public hanging and strangulation. Another journeyman was caught by his master in the act of removing three thousand livres from a cashbox, but we do not know what fate was reserved for him. One journeyman and his wife, who worked as a bakery countergirl, stole money from their master, the journeyman's own brother.[45]

Journeyman La Roche stole two chickens from his master, DuBois, only two weeks after he had been hired. DuBois fired him on the spot and deducted two livres from his wages (surely most of the money that was due him). "You are withholding forty sous from my wages, but you will pay them dearly," threatened La Roche. Several weeks later La Roche made good on his threat, robbing his master of five hundred livres and other items.[46]

Cases of other types of pilferage abound: furniture, silverware, watches, jewelry, clothing.[47] Baker Louis Cornaillon, the victim of the theft of eight shirts, a silver buckle, and a watch, believed that his ex-journeyman committed the crime in order to avenge himself for having been fired. In at least one instance the robbery was accompanied by a horrifying eruption of violence. Michel Houlier, a twenty-one-year-old journeyman, clubbed his napping master to death in the bakeroom and then went upstairs, where he slit the throat of his master's pregnant wife. From the police description, it appears that Houlier, who was a distant relative and countryman of the master, could have gotten away with the items he coveted without harming anyone had he not nurtured a virulent grudge. This journeyman was sentenced to be broken alive on the wheel after asking public forgiveness and suffering torture.[48]

Bakeroom Sabotage

Journeyman bakers served as scapegoats not only for the generality of crimes committed in the bakeshop but for crimes of craftsmanship perpetrated in the bakeroom as well. A master baker's commercial success ultimately depended on his bread. Bread of bad quality repelled clients and damaged the shop's reputation. When problems occurred, masters tended to blame them on their journeymen. There was no one else to accuse since many masters had virtually abandoned the bakeroom. The only question was whether the journeymen ruined the work out of carelessness or ill will. These masters did not stop to ponder the significance of their own abdication; they expected the journeymen to do the job. Given their daytime obligations (primarily buying grain and flour, seeking credit, and recovering debts), many masters simply could not stay up all night. One master supervised his journeymen till midnight and then left "the rest of the work until the perfect completion of his bread" to them.[49]

Masters claim that journeymen performed badly in part because of their inveterate debauchery. Bakery worker Lebesse stayed out all night and came to the shop in the morning unable to work.[50] Garçon François Germain was drunk and absent so often and was so feckless when he worked that master Pierre Aubin dismissed him. Martin Gillet charged that his journeyman was so drunk one day that he loaded the oven with wood incorrectly and almost burned the house down. Drunk on brandy, journeyman Hebert spoiled part of a batch of bread by placing the dough askew in the oven. Master Roze discovered that his bread was being undercooked by his journeyman Gilles Belon. Roze blamed his failure on Belon's fascination for the cleaning lady, Jacqueline Morel, with whom he conducted a "commerce of debauchery."[51]

Master Augustine Charon's garçons were not debauched. According to him, they were part rascally and part slovenly. They did not care as passionately as he did about making good bread. Endowed with a fiery temper, Charon exploded every time he detected a defect in workmanship (defects, one must add, that he might have prevented had he spent more time in the bakeroom). He beat his apprentice Pierre for overheating the oven, and he reprimanded Lejeune, a journeyman, for overloading the oven with dough. Another time he noticed that the dough for the *petit pain* (luxury or "fantasy" bread) "did not have its ordinary quality." An analysis of the cooked bread convinced Charon that his workers had "altered" the quality by withholding from the dough the necessary amount of butter and eggs. Moreover, Charon found some incriminating evidence: Lejeune had baked three pastries for himself and his coworkers using eggs and butter. (Lejeune later insisted that the ingredients for these cakes had not come from Charon's bakeroom supplies, but he did not venture to explain why the petit pain was flawed in composition.) Lejeune became "more insolent and less attentive to his work" after this incident, if one is to believe his master.

Charon reported another affair in which Lejeune, to spite his master, prepared the yeast and starter dough badly. The employer chastised the journeyman: such shoddy work "will ruin the whole mass of flour that he was going to put in the kneading trough, the leaven being the principle of all bread." Unable to contain himself, Charon accused Lejeune of seeking to sabotage his business. He knocked him down to the ground, inflicted a vicious wound to his head, and fired him on the spot. Lejeune ran to the open window, where he called for the watch to save his life ("I'm being murdered") and for the people outside to bear witness: "you see how my master fixed me," he shouted, pointing to his bloody face. The shopgirl tried to calm him down and induce him to forget the altercation. If you go to the police, she warned, "we will say that you hurt yourself by falling down and banging your head." Lejeune had the courage to go to the police and file charges against his former master. He insisted that he was a faithful worker and "an honest man and man of honor," and he expressed the fear that Charon would try to ruin his career by blacklisting him with other employers.[52]

The wife of master baker Jean Mathias pushed Charon's mistrust of the baker boys a step further. She hinted that they were instinctively antagonistic to the interest of the master. She charged that her journeymen were systematically "careless in the making of bread." As a result, they had "discontented our customers and driven them away from our shop." Louis Hiest leveled similar charges—instinctively—against his garçons. They had baked five hundred eight-pound loaves that proved unfit to eat. He blamed them for the debacle, though he would have come closer to the mark had he questioned the integrity of his flour supplier. As far as one can tell, a geindre named André was guilty of burning part of a fournée. Fearing the inevitable rebuke and perhaps worse, in a panic he threw the burned loaves into the sewer. Early in the century André might have gotten away with his little lie. But it was Year II, the bakeshop was the center of the universe for the police of each section, and poor André found himself indicted for "a counterrevolutionary crime."[53]

The wife of master baker Pierre Casson was completely at the mercy of the couple's journeyman, Jean Cabit, because her husband was gravely ill. Instead of stepping into the breach—one wonders whether Dame Casson gave him any incentives?—Cabit neglected his work, ruined many fournées of bread, and insulted and abused the master's wife. Commissaire Regnaudet, during a routine inspection, found all the bread in one bakery to be of light weight. The master placed the blame on his journeymen. Another journeyman committed a far more dangerous act of sabotage. "In order to avenge himself against his master," who had registered him for the draft lottery for the militia, he put soapy water into the dough, ruining the entire fournée and "endangering the public good" as well as the master's reputation.[54]

What to Do about Journeyman Indiscipline

The highly critical portrayal of the journeyman bakers that emerges from these cases is in many ways an overdrawn, tendentious, self-serving view. Some journeymen on some occasions were drunk, rushed, fatigued, negligent, inept, mischievous, vindictive. There is no reason, however, to believe that the bulk of this tribe of nocturnal troglodytes did not perform their job satisfactorily most of the time. It is interesting to note all the same that the disparaging and pessimistic view of the baker boys was not confined to a circle of dyspeptic masters. Many of the complaints of these masters were confirmed and codified by Parmentier, a scientist who boasted above all of his objectivity and disinterestedness. The masters would have been gratified to read in Parmentier's *Parfait Boulanger* that journeymen often ruin the preparation of the dough, the most delicate phase of bread making; that short-weight bread was more often than not the result of the "maladroitness and misconduct of the journeymen" ("How many times has it not happened that the weigher, to injure his master, testily reduced the weight in dough of several loaves?"); that when the master fell ill, the journeymen did their work "even more badly" than ordinarily.[55]

Indeed, Parmentier's stereotype was even more deprecatory than that of the masters whom we have met: the savant had a certain disdain for, along with a deep suspicion of, the baker boys. In his perception, they were inherently "undisciplined," "insubordinate," and "naturally fickle." He regarded them in the same way the police viewed the mass of little people: they had to be contained. "The journeyman bakers," wrote the subsistence specialist, "are so disposed to do ill, either by laxity of conduct or by a disposition that inclines them to do the opposite of what they are supposed to do, that the Master must never forget to examine attentively their doings, and to treat them precisely like clocks that must be rewound from time to time by adjusting the movement." Given the crucial role of the journeymen in the process of bread making, Parmentier regarded their indiscipline as one of the greatest obstacles to perfecting the art of baking. The master needed to be "seconded," but above all he needed to be "obeyed." The baker boy needed to have "strength, courage, and intelligence," but most of all he needed to submit.[56]

Despite his very harsh assessment, even Parmentier conceded "that the fault is not always on the side of the journeymen." On this point, Parmentier drew support from Brocq, the head of the bakery at the Invalides, where Parmentier was chief pharmacist. This executive baker pointed out that the masters "sometimes abuse their authority" and fail to show the proper consideration to their employees. Masters also were known to make excessive demands on their garçons. Masters who were "ignorant" in the art of baking virtually forced their journeymen to resist their will. Masters often attributed "to the negligence and incapacity" of their workers "vices that in fact resided in the master's own methods or in the nature of the raw material used." Finally, masters frequently failed to offer sufficient instruction and supervision, and therefore stood themselves in large measure responsible for the errors of their journeymen.[57]

Brocq helped to restore a semblance of balance to Parmentier's evaluation, but at bottom both men agreed that the most pressing issue remained the discipline of workers. "All the honest bakers of Paris are of one voice on this subject," reported Parmentier; "give us at least a new Regulation to contain and tame the indocility of our journeymen." Nothing could do more to improve the security of Parisian provisioning, the most critical of all police functions, and the quality of daily bread. As part of this new regulation, Brocq proposed that all Parisian journeymen be placed by an expert, according to skill and function, in one of four professional classes. The expert, or inspector, would keep a register noting both the capacity and conduct of each journeyman, and he would visit the bakerooms to verify both. The journeymen would not be able to quit without notice and the inspector, rather than the master, would "grade" the journeyman on his certificate of *congé*, thus preventing the frequent quarrels that opposed master and journeymen when the latter left. In this improved climate, Brocq felt that the incidence of "cabals to stop work" would decrease substantially.[58]

In response to pressure from the bakers, the Paris police finally accorded them new regulations on journeyman discipline in August 1781, less than three years after the publication of Parmentier's book. In fact, most of the twenty-three articles merely reaffirmed familiar strictures and reiterated rules that had existed for many years. Brocq's new classification and inspection system was not implemented, perhaps because the corporate fathers opposed it (it threatened to shift the locus of inspection from the guild to the police), perhaps because it was considered an unnecessary refinement on the *livret* system of control. Still, the fact that the old regulations were restated so energetically and reinforced by certain fresh provisions (such as institutionalization of the livret) seemed to bespeak a new determination on the part of the police to join the masters in a vigorous campaign to restore subordination and good order.

Surely the journeyman baker's sensitivity to overbearing masters was heightened by the particularly onerous nature of the work he had to perform. More demanding still than the harshness of the physical labor itself in a cramped and sometimes stifling site was the rhythm of the work: it was more or less endless, requiring the garçon to remain dressed in bakeroom garb and to sleep more often than not in the workplace. Living in a sort of servitude (that it was as much in fetter to the public as to the employer proved to be of no consolation), the journeyman adamantly refused to be assimilated in any manner to the servant.

Genuinely pathetic to the outsider, the bakery worker's condition seldom appears to have aroused the master's compassion. After all, the master himself had been there (though his psychological distance from subordination may have dulled the vivacity of his memory of the experience). The master depended heavily on his garçons and was inclined to scapegoat them when something went wrong. The journeymen keenly felt the incongruity between the responsibility imparted to them and the rewards they received. Heavily taxed in physical and moral terms, poorly remunerated, and sometimes abused, the journeymen moved readily from shop to shop, sure of their ability to integrate themselves rapidly, given the uniformity of skills and the familiarity of the division of labor. If George Sand and Gaston Bachelard are right that the mystical ferment of the bread-making process did not leave even the most jaded practitioners indifferent, it is nevertheless reasonable to think that journeymen at work fantasized a great deal about the more livable world outside the bakeroom.

Chapter 9

The Journeyman's World outside the Shop

Life "after" work was an elusive notion to a journeyman who remained more or less steadily in someone's employ, during which time he was more or less unremittingly at toil. He had to live his free moments rapidly, intensely, and in a staccato fashion. The tavern was the place to which he had chronic recourse, often several times a day. The inn became a focal point more often when he was *hors de condition* than when he was on a job. Male sociability was easier to organize and to sustain than sexually mixed conviviality, given the constraints on his time. Characteristically, male interactions had some connection with work. The secret associations to which a number of garçons belonged provided mutual assistance and camaraderie, but they stood primordially as vehicles for collective worker defense. Though occupational imperatives made family life extremely difficult to launch and sustain, a considerable number of journeyman bakers were married.

Marriage

The data available about journeyman baker families are extremely limited. Surprisingly, few marriage contracts turn up in the samplings in the Parisian notarial archives. This may be because journeymen married in relatively small numbers, because they married outside Paris, or because they had insufficient means to draft a contract.[1] From sixteen contracts and a dozen other documents we are able to glean a little information about the women the journeymen married. Three were daughters of bakers, two of whom appear to have been Paris masters. One was the daughter of a Parisian master gardener. Five others came from the provinces. The fathers of the first three were respectively a hotel keeper at Ponthierry, a stocking maker in Picardy, and an agricultural worker from Burgundy. The last two were daughters of a locksmith and a roofer. In four of the five cases the father was deceased at the time of the marriage and there is no indication of the professions exercised by the brides, if any.[2]

Five journeyman brides were widows. Two were widows of forain or itinerant bakers established in Parisian suburbs or slightly beyond (one of them, Elise Chauvin, was in fact twice the widow of bakers), and a third was the wife of a Paris master. It is virtually certain that all three women continued their husbands' commerce. Another widow, whose husband had practiced surgery near Choisy-le-Roi, had no profession of her own, while the fifth, whose husband's occupation is unknown to us, served as a domestic. In thirteen other marriages we know the professions of the brides: three dealers in secondhand goods, mostly clothing; three cooks in private houses; two bread delivery-women; one washerwoman; one dyer; one fruit seller; one dayworker; one perfume worker.[3]

In terms of the social composition of the partners, few of these marriages are startling. Journeyman baker was not an exalted métier, as we have seen, unless the journeyman in question had an unusually good chance of ascending to mastership. Thus it is not surprising to see journeymen marrying women of the "little people" (*menu peuple*), that is to say, women like themselves. Apart from the baker widows, to whom we shall return shortly, only three marriages betrayed on odor of *mésalliance*. On the surface it is not at all clear how Louis Leshopier, the son of a petty wine grower in Picardy, succeeded in marrying Geneviève Le Guay, the daughter of a deceased master baker and the sister of two bakers, one of whom was probably a master in their late father's place. Yet Geneviève's father must not have left a prosperous enterprise, for her dowry was a mere one thousand livres, all in furniture and clothing. Leshopier compensated for his humble rural background by bringing to the marriage a contribution of 1,050 livres, consisting of eight hundred livres in cash and 250 livres in rentes (capitalized value).

The marriage of another winegrower's son, Claude Grignon, is no easier to account for. His bride was a surgeon's widow. Now, surgery in a provincial locus was not a resplendent profession, but hometown surgeons had social pretensions, and in addition this surgeon's widow had a dowry of two thousand livres, which was very likely to bulge when the inventory of her late husband's property was completed. It is true that Grignon was able to offer one thousand livres, that the widow had no profession of her own, and that a widow's hand, all things being equal, was inherently less valuable than that of a maiden. Another curious marriage joined Jean-François Thevenin to Thereze Le Guay, a seventeen-year-old orphan whose father and stepfather had both been master bakers.[4] Thereze's dowry, three thousand livres, reflected a relatively comfortable position. With only a thousand livres contribution, Thevenin could hardly match Thereze's investment. But Thevenin was the son of a successful merchant baker at Saint-Dizier, and it is possible that Thereze's mother was seeking someone to take over her late (second) husband's shop.

For the journeyman bakers of humble origins, matrimonial congruence in status or rank may not have mattered very much. They had little to offer and could expect little in return. They did not view marriage as a means of establishing themselves. Indeed, affective ties may very well have been of greater importance in determining the marriages of these journeymen with little hope of upward mobility than they were in the case of journeymen who had the imagination and the wherewithal to calculate and wager on their destinies.

The story of Barbe Chantereau reveals attitudes toward marriage, money, status, and fate in this milieu. Chantereau was a twenty-year-old cook for a bourgeois de Paris when she first met Pierre Fremin, a baker boy in the shop where she bought bread for her master's household. Taken with her, Fremin visited her frequently, flattered her, and led her to understand that "his intention was to marry her." Barbe resisted his advances, in large measure because she did not believe, at least in the beginning, that their marriage was a real possibility. She protested to Fremin that she was "not well-to-do enough to hope to marry him." But Fremin allayed her doubts tenderly, suggesting—in her words as the court recorded them—"that there was not so much disproportion between her and him."

Fremin pressed his courtship vigorously in letters and visits. One day when her master was away they found themselves alone in the kitchen. "After several speeches and oaths [of love and fidelity]," Barbe testified, Fremin threw her down on the ground and took her by force. Nevertheless, she continued to see him, sometimes in taverns, sometimes in her mother's apartment. But, after more than a year of romance, when she announced that she was pregnant, Fremin recoiled. He would support the child and the mother, but he could not marry her, not at least while his parents were alive. For Pierre Fremin was not an ordinary journeyman baker. He was the son of a master baker. He would soon obtain mastership. His father would want to profit optimally from his own situation and his son's brilliant prospects by marrying him well, that is, obtaining a substantial dowry to help him launch his career and perhaps also cementing new relations and constructing new networks to facilitate his commerce. In the end, Barbe's instinctive doubts were vindicated. She was too lowly and too poor for him.[5]

Did marriage give the journeyman baker a substantial upward boost in economic and professional terms? Remember that we are dependent on a very thin base of sixteen marriage contracts from which we cannot generalize but only speculate. The first point is that journeyman bridegrooms do not, for the most part, come to the marriage destitute. They must have resources in order to attract resources. The equilibrium between the male and female dowries of this group is remarkable: the average contribution of the groom was 928 livres (median = 856) and the average dowry of the bride was 1,000 livres (median = 800). Each contributed approximately the same portion of his/her dowry in cash: 200 and 217 livres. They diverged sharply on the average amount of the other dowry components: the bride had almost twice as much value in furniture and clothing and nothing at all in annuities (*rentes*), land and property, or tools and goods for their commerce, whereas the groom had a relatively substantial capital in rentes, a sizable stock of baking tools and goods, and a small amount of real estate.

On average, the fortune with which they launched their married life amounted to 1,635 livres (median = 1,281), broken down as follows: cash, 23 percent; commercial goods, 6 percent; furnishing and clothing, 41 percent; real estate, 3 percent; and rentes, 26 percent. Even if this total fortune could have been converted into cash at its face value and invested as a whole in the journeyman's future, it would hardly have been enough to win for him a mastership (masters' sons excluded), let alone enable him to purchase a shop and clientele or launch a new business from scratch.

Which were the most promising marriages? As a group, the four journeyman bakers who were sons of bakers (three Parisian bakers of whom two were journeymen and one a faubourg Saint-Antoine independent baker, and one merchant baker from Champagne) had the highest average contribution: 1,402 livres, followed by the two sons of laboureurs, at 716 livres and then by the three sons of men exercising other rural professions at 675 livres. Far and away the most attractive brides were the two daughters of master bakers who brought an average of 2,000 livres each to the marriage, followed by the three daughters of men exercising rural professions with 1,265 livres. The couple who amassed the greatest marriage fortune was composed of a son of a baker and the daughter of a master baker (4,000 livres). Predictably, they had the largest portion—600 livres—in bakery goods and tools. In the next highest class the wife's father was a master baker also while the husband's was a winegrower (2,050 livres). This couple boasted the highest cash component, 800 livres. Then came a group of three bereft of any maternal or paternal link to the bakery with a fortune of 1,742 livres. The grooms and brides who resided in the same parish had a higher average fortune—2,150 livres—than those who lived elsewhere in different parts of Paris (1,739 livres).

Marriage with a baker widow usually constituted a dramatic upward leap for a journeyman baker. Widows often found it hard to manage the bakeroom and shop on their own. They were so dependent on their geindres that they sometimes married them (in some instances perhaps in part to avoid damaging their moral reputations, for they had to live in close touch day and night with their workers). Or they married a journeyman or a baker from outside their shop, if they had the time and leisure to pick and choose. Estienne Girard, the wine-grower's son, must have had a great deal of courage to marry Elise Chauvin, for she brought him five children from the beds of two successive baker husbands. But she must also have contributed a lusty dowry, featuring a working business, faithful clients, a bakeroom, a space in a Paris market, and so on. Anne Martin, the widow of another forain, must have provided her husband with a similar opportunity, but he enhanced her status and perhaps helped through his family relations to increase their business, for he was the son of a Paris master. Of course, not all baker widows were desperate to get married, despite the onus of going it alone, as a twenty-seven-year-old journeyman from the Saintonge named Mainguet found out. He thought he was courting (and winning) the widow Ouffroy, a thirty-seven-year-old woman who ran a forain bakery and an inn. Meanwhile she believed that she had merely hired him as her geindre. She settled the issue brutally by firing him.[6]

The data from the limited number of contracts we have suggest that journeymen tended to marry women they met near their shops rather than women from their *pays*, or home country. In four of the marriages both of the parties are from the same parish. In four other cases both the bridegroom and the bride were living in Paris intra-muros. In only two cases were the brides living outside the capital at the time of betrothal. At the moment of marriage we know precisely the age of only one bride: Thereze Le Guay, the daughter of a deceased master baker, was seventeen. The five widows in this sample would surely have swelled the average bridal age. It was relatively common for widows—especially widows with means—to marry younger men. With five children from two previous marriages, it was very likely that Elise Chauvin was considerably older than Estienne Girard. Two of the other journeymen who married widows were thirty-one years of age.

Thirteen contracts provide hints about levels of literacy. In seven cases both bridegroom and bride signed. In two cases neither signed. In three instances he signed while she did not, and on one contract the reverse occurred. The three journeymen who could not sign were all sons of Burgundian laboureurs. Of the five brides who could not sign, one was a servant, another the daughter of a rural dayworker, the third the daughter of a baker, the fourth the daughter of a locksmith, and the last the widow of a forain baker. The inability of these women to sign should have somewhat diminished their allure to the journeyman bakers who aspired to mastership or independence, for bakers relied heavily on their wives to run the shops, oversee relations with clients and sometimes with suppliers, and frequently to keep the books as well.

Family Life

The journeyman baker's bride was never his first wife. Prior to his marriage to her he had taken more or less binding vows on the altar of the oven. It was virtually impossible for him to be faithful to both his wife and his job at the same time. He was too busy, too fatigued, too brutalized. His profession was more exigent than the most amorous young bride, especially at night. Marriage with a troglodyte was not like most marriages.

The journeyman and his wife not only did not sleep together at night, but in many instances they did not even live together in a formal or nominal or legal sense. Each had his/her residence, however dismal or marginal. They visited each other, some times not even on a regular basis. The erosion of the marriage tie, moreover, seems to have begun right after the wedding. It was not the product of gradual atrophy and festering that is familiar to us in modern bourgeois society. One wonders in these instances why the couple bothered to marry in the first place. Familial pressure? Conformism? Religious scruple: to consecrate the physical union? Youthful optimism and enthusiasm? Loneliness and the need for some sort of mutual attachment and symbiosis, even at a persistent remove?

Perhaps neither bride nor groom realized or suspected how little marriage would change their lives. Once they were married there appears to have been no quixotic effort to reconcile marriage and professional life. Rather, the couple shared a more or less tough-minded resolution. Without any tinge of acrimony, Louise Le Peteur, the forty-three-year-old wife of journeyman baker Adam Hermand, explained why she and her husband had never set up a common household during their seven and a half years of married life: "It is not the usage among journeyman bakers for their wives to live with them." In fact, the custom was founded on or sanctioned by a number of police sentences, based on the presumption that journeymen had to reside where they worked, which forbade masters to allow journeymen to keep their wives with them in the shop. Catherine Miny had married journeyman baker Blaise Jacob when she was eighteen years old. Now at age thirty-two, she had never lived with him and indeed had not even seen him in the previous half year. She did not even know if he still lived and worked in the faubourg Saint-Antoine where she had last crossed his path. She lived in a furnished room (*garni*), sharing a bed with a young female servant. She had been wandering from rooming house to rooming house for the past few years.[7]

Both Louise Le Peteur (wife Hermand) and Catherine Miny (wife Jacob) were utterly destitute. Need drove each to commit a folly, a petty theft—the first time either one had ever done such a thing. In their interrogation, each woman had the opportunity—indeed, the invitation—to cast blame on her husband for ignoring his responsibility to support her. Yet neither one indulged in a vengeful outburst against her husband. Neither believed that her husband was at fault. In their relationship, Louise explained, Adam Hermand did not have to and was not expected to provide her with money because she herself had "a métier to earn a living." Louise once spun wool and later sold bread in the street. Most recently she hawked articles of hardware. Catherine Miny called herself a "*gagne-denière*," but in fact she had been eking out an existence from the time she was eight years old by selling "lavender and other things" on the street. Catherine and Louise accepted marriage to journeyman bakers for what it was in many cases: an occasional encounter, no implicit promise of happiness and no material consolation, no recrimination, and perhaps, after all, still a powerful affective and/or sexual bond.

Unencumbered by illusions and unfulfilled dreams, despite their misery, wife Hermand and wife Jacob were perhaps less badly off than other wives of journeyman bakers. Françoise Balin was only a cook, but she expected her husband to succeed professionally and build a family, for he was a mature fellow, a widower, and she brought him a dowry of family furnishings and clothing, plus a thousand livres in hard-earned savings. But he proved to be "a libertine" and a parasite. While she continued to work, he refused to find a job, devoured her entire dowry, and beat her. It took her four years to screw up enough courage to file a complaint with the police. Another cook, Nicolle Bideault, had a similar experience with her baker husband. Refusing to work, he tried to exploit her. He insulted and beat her and forced her into hiding.[8]

If some women were keenly disappointed by the marriages they made to journeyman bakers, other women were even more let down by their failure to contract such marriages. These women were the victims of broken promises or vague reassurances of the sort with which Pierre Fremin broke down his beloved's resistance.[9] Given the enormous difficulties of marriage and the large number of young, robust, celibate journeymen, it is not surprising to find instances of premarital sexual relations leading to illegitimate births. I have found not a single case in which the journeyman stepped forward as a matter of honor to marry the poor girl. More common was the reaction of bakery worker Jean-François Auger to the accusations of Marie-Joseph Testu, the twenty-one-year-old daughter of a master shoemaker. For about four years Auger and Testu had been seen together courting in public, kissing in stairways, walking arm in arm late at night. Then Marie-Joseph discovered that she was pregnant. Auger readily admitted that he had slept with her, but he claimed that everyone else had too. Auger's brother-in-law, who was also his master, stood ready to testify (for his family's honor and his shop's reputation were also at stake) that he had seen all of his journeymen engage in sexual relations with Testu on top of his oven. Auger's gallantry stopped with his offer to pay the cost of Marie-Joseph's delivery.[10]

Marie-Joseph demanded an immediate reckoning, for she esteemed that her situation was untenable. Other women were more patient with their suitors. Marie-Louise Gobelet, a fruit seller, was the wife of a grain merchant in Orléans.[11] Immediately upon the death of her husband, a journeyman baker named Pierre Breton, whom she had known for some time, came to importune her for her hand. But, according to the widow, "instead of keeping his word, he sought only the means to seduce me under the cover of reiterated promises of marriage." What a protracted and fruitful seduction it was! Breton must have been remarkably convincing and the widow as fecund as she was vulnerable, for out of their "carnal copulation" there were born not one but three children over a period of three years. Even as Breton continued "to elude her," he managed "to eat up her fortune" without giving his children so much as a sou. Their relationship resembled a real journeyman baker's marriage far more than Marie-Louise Gobelet suspected.[12]

It would be wrong, however, to leave an entirely negative impression of journeyman family life. Many marriages did succeed. These strong marriages were the upwardly mobile ones, the ones in which the couple could realistically envision a transformation of their material and moral circumstances. These were the marriages in which the journeyman was able to attenuate or escape gradually from the terrible incubus of daily labor in the bakerooms by himself becoming an employer. But these were the best marriages, made by only a few, by the elite of the journeyman world, sons of masters and a handful of others who had the means to break free. There was also, perhaps, a certain stability in those journeyman marriages which were organized around the bakery and in which the profession became a source of fusion rather than fissure. One thinks, for example, of the Campé family. Anton worked as geindre and his wife as porteuse in the same shop, while their six children toiled for various faubourg

Saint-Antoine bakers. Or François Lefort, who worked as a journeyman in the same shop where his wife was porteuse. Journeyman Pierre Lefebre's wife was the countergirl in his master's shop. They married two of their daughters to bakers.[13]

Journeyman Fortune

With relatively few exceptions, journeyman bakers remained poor all their lives. The exceptions were some of those who rose to masterships—some because there were, after all, a substantial number of masters who were hardly better off in purely material terms than many journeymen. Of those bakers who obtained mastership, the most favored economically were the sons or nephews of masters or those who married masters' daughters or widows. Endogamous marriages of masters' sons to masters' daughters, as we have seen, produced the most auspicious takeoff capital. But most of the marriage contracts we considered suggest humble beginnings and many obstructions on the road to upward mobility.

The other broad category of journeymen who escaped poverty were those who made their fortune outside the bakeroom and outside the law, such as Pierre Couppe, who conspired with his wife to prostitute his stunning daughter Edme, or François Noël, who was a forger and a con man. Finally there are anomalous cases that we cannot explain for lack of information, such as the widow of journeyman Pierre LeFebre, who bequeathed substantial amounts to her two daughters (both married to bakers), to a niece, and to the priest who served as testamentary executor.[14]

Occasionally journeymen struck it rich (in their terms) by accident of succession. François Tessier, a remarkably stable worker who had been with the same master for years and who had saved from his wages with a view toward his future, inherited nineteen hundred livres from his father, whose profession, alas, we do not know. But much less opulent bequests were the common lot of journeymen. Charles Delyon, a twenty-four-year-old baker boy who could not sign his name, received ninety-six livres, one-half of the entire estate of his vintner father. Antoine Tessier sold his part of a family legacy for ninety livres.[15]

The little data available on the possessions of the journeyman reinforce the impression of poverty. To be sure, René Alexander and Vincent Lelong vaunted wigs and canes, the former for courting and the latter for brawling. But one finds little else in their case of belongings (*cassette*) to sustain this swagger: a suit of Elboeuf cloth lined in serge, two shirts decorated with muslin, three vests (one in satin, one in cotton, one printed calico), a handkerchief, and two white cotton caps. The trunk containing all the effects of the late journeyman John-Baptiste Delatre told the same story: one Elboeuf suit, three old jackets, two pairs of pants, two sets of stockings, a pair of shoes, two vests, an old hat, a shirt and two collars, eighteen livres in cash, and three volumes of "books of piety." Another journeyman had amid his treasure a three-volume *History of France* and a Corneille play as well as several books of spirituality. But he had spent many years abroad in the army, which made him more cosmopolitan than his peers.[16]

The Journeyman's Home

As far as we can tell, a substantial majority of working journeyman bakers lived where they worked. As we have seen, the conditions in which many of these journeymen lived were hardly better than the conditions in which they worked. Indeed, even as work time and rest time seem intermeshed and inseparable, so there was often no spatial or physical discontinuity between where the journeyman worked and where he slept. We have seen the journeyman asleep "atop the oven," in flour-sack sleeping bags, in makeshift beds next to the kneading trough, or behind the bolter.[17] For the most part journeymen had no room of their own, no privacy, no real respite from the odors and tools and the space of work. They folded up their mattress when it was time to work and spread it out when it was time to rest. They took their meals on the table in the bakeroom where they cut, weighed, and fashioned the dough, and they sometimes washed in the kneading trough or the water pail. On occasion they left their valued belongings in an inn with a friend or a trusted innkeeper.[18] Within the master's house the journeymen were segregated. In fact, they were not a part of his household; they were relegated exclusively to the "cave" and they surfaced only to go to the shop or to go outside. In these conditions it is hard to imagine many strong voluntary bonds forming between masters and journeymen.

A few masters did try to make their journeymen more comfortable. Mistress baker Fayel gave her three baker boys one large bed to share and a toilet-commode (*chaise de commodité*) in an independent room that served to store sacks of bread crumbs and other merchandise. At master Lepage's the garçons had a room looking onto the court equipped with two old beds with wooden frames and feather mattresses evaluated together at eight livres. A small minority of journeymen somehow managed—at least nominally—to maintain residences outside the shop.[19] Julien Quissier stayed in the bakeroom only until it was "time to go home." Jacques Fortin worked for Dame Joly, but he lodged in a rooming house, perhaps because at fifty-three years of age he had earned the right to a little intermittent tranquility. Louis Peret worked for master Tugnot, but after a month of boarding in the shop he moved into a room at an inn with several other journeymen.[20]

Even when they lived in a shop many journeymen kept contact in one way or another with innkeepers, who could be very useful to them when or if they "left condition," that is to say, left or lost their jobs. Antoine Vache recruited tenants for a baker lodger inauspiciously named Dame La Cruelle in the faubourg Saint-Antoine. Though he slept in the bakeroom, Jean Mezure contrived to keep his room at Dame LeFebvre's, perhaps by subletting it to his transient comrades. Many journeymen left their trunks and other possessions with their preferred lodger. The working journeymen were joined in the inns and rooming houses by a mass of unemployed, runaway, and rootless baker boys.[21]

Journeyman Inns

There were thousands of places to lodge in Paris—private houses, garnis, inns, boardinghouses—but the journeyman bakers, like their comrades in other professions, tended to favor a certain number of establishments which came to specialize in receiving them and which they came to regard as their own. Dame Brocq of the rue neuve Saint-Denis, Sieur Buttin of the colorfully named rue du Bout-du-Monde, Vautier of the rue neuve d'Orléans, Martin of the rue et faubourg Saint-Jacques du Haut Pas, Dame Joly of the rue neuve Saint-Sauveur, Sieur Mener of the rue Saint-Martin, widow Roussel of the cour des Miracles, and widow Gaillard of la grosse Tête were all "lodgers of journeyman bakers." Simultaneously some of them took the title of *aubergiste* (innkeeper) or *cabaretier* (tavern keeper). Another lodger-specialist, Deslombert, rue Saint-Lazare, inspired a special confidence among the journeymen, for he had been until recently a journeyman baker himself. The baker boys also held in high esteem the inn kept by Claudine Bichet, rue du faubourg Saint-Martin, because she was the wife of Vincent Ragonneau, who was one of them.[22]

The baker's inn was an enormous source of consolation and encouragement to the journeymen. It was a social institution that provided invaluable service to the journeymen who were occasionally fugitives, frequently without family or nearby kin, sometimes recently arrived from the provinces, commonly without resources. Frightened, lonely, jobless, hunted, down-and-out, or simply sick and tired of working hard, the journeymen turned to the baker's inn, not only for lodging but for credit, counsel and protection, sociability and sometimes even a family ambience, as well as job placement. The journeymen were their own masters in the inn. There was neither a protocol of deference nor a confining hierarchy to respect. They felt at ease; they drew strength and confidence from one another. They had a taste of a kind of sovereignty or self-determination in which they (fleetingly) delighted. They had no illusions; they knew it was only a momentary respite. But it energized them powerfully.

At the inn the baker boys commonly slept two or three to a bed and ate out of a common bowl. They played a betting game called "à la Mouche" and all manner of card games. They drank heartily, the circumstances and consequences of which we shall examine presently. In some of the inns, they were tended to materially, morally, and even medically by the innkeeper or by the *hôtesse,* analogue of the "mother" of the clandestine journeymen's associations called *compagnonnages.* Benoist Desplat, a lodger and beer merchant, helped to get his tenant Louis Jannard, a journeyman baker, out of jail where he was sent for brawling at Les Porcherons. Philippe Laurent, a twenty-nine-year-old baker boy without a job, was tempted by a proposition made by another journeyman to transport some flour at night in return for a new pair of shoes and perhaps some stockings. Before replying, Laurent went to consult his "hôtesse," Dame Brocq. She warned him that he courted serious risk, for he could be charged with dealing in contraband and sent to jail. Laurent accepted her advice and turned down the tainted offer. Within a few days, Dame Brocq found him a place with a master baker in the parish of Saint-Germain-de-l'Auxerrois.[23]

"Foreign" journeymen, such as Claude Richet from Sully-en-Loire, who came to Paris in search of work, went not to the guild bureau but instead to Sieur Dubois's inn in the rue Verderet.[24] How did Richet know where to go? He knew from the recommendations of friends or from the journeymen's "wife," a system of rapid oral communication that linked and integrated the members of the group. The innkeepers had their own wire, one that plugged them into such diverse institutions as the labor market and the market in brief amorous encounters, into both of which the garçons sought entry.

Bonds and Tensions

Inn life nurtured a certain fraternal solidarity, a solidarity that was at once the cause and the effect of the journeymen's collective consciousness that their interests differed in many ways from those of the masters. This solidarity took many forms. In professional terms, it expressed itself in hostile actions such as the blackballing or boycotting (some called it the "damning") of especially odious or tyrannical masters. Journeyman solidarity also took more diffuse forms of mutual assistance that were not strictly speaking in the domain of labor relations. When a journeyman baker named Lalemand was arrested for having beaten the wife of a master baker, he pronounced this warning: "the baker boys of the inn" were ready to avenge him. Journeyman sentinels posted outside the baker inn on the rue neuve d'Orléans rescued baker boy Joseph Bonnefoy from two enraged quarrymen who denounced him for theft. When one journeyman swore his *"foy de garçon"*—his journeyman's pledge—all of the others believed him, for there was no loftier engagement that a journeyman could make.[25]

Journeymen had many ties that bound them closely, but it would be wrong to picture them inveterately in fraternal embrace. They helped each other, but they also hurt each other. They fought together, but they also quarreled among themselves. Few of them viewed the world in terms of constant higher solidarities. If indeed they often joined one another on matters of common concern, they did not for a moment infer the need for a collective discipline for everyday life. On the contrary, each felt constrained by the imperative need to survive as best he could. Their outbursts of solidarity were counterbalanced and in a way complemented by eruptions of a rough, street-level individualism.

Not even the inns were sacrosanct. Journeymen brazenly stole from their roommates and bedmates. The incidence of theft was astonishing. Hubert Lelvier suspected his comrade Coquard of filching two coats, a shirt, a silver buckle, stockings, and several bonnets. Jacques Leger was arrested for stealing four livres four sous from Guillaume Bory, with whom he lived. Charles Jannard had his chest broken into, presumably by a fellow guest, in "the hotel of the garçon bakers" of the rue Guénégaud. Christophe Ambress had the same experience at Wolf's journeyman bakers' boardinghouse. Antoine Bertrand and Louis Peret, called Dauphiné, both had their trunks stolen in their respective inns. Nicolas Burnot was sure that his wallet was stolen by his bedmate, a journeyman whose name he did not know, despite the intimacy of their contact. Pierre Rutard and François Lanoue had clothes stolen from them, while another colleague in yet another inn lost jewelry to a thief. Of course, there were instances of renascent solidarity in these times of personal crisis. Noel Savary, called Le Contois, returned to Leroy's inn to find his room ransacked and most of his valuables stolen. Joined by a half-dozen baker-boy colleagues, he searched and questioned all the residents and examined every room. Suspicion pointed to a forty-year-old journeyman named Denis Morlet, called Montargis, who claimed to have been in the constabulary.[26]

Journeymen stole from each other in the bakerooms as well as in the inns, at work as well as at rest. The most valuable possession of many baker boys was their watch, often set in a silver or gold case and embellished with a fancy fob. A luxury investment, the watch was also a work instrument, for the journeymen had to be aware constantly of the time. While at work, the baker boy ordinarily hung his watch on the bakeroom wall near the oven. Jean-Baptiste Pradant, Guillaume Fortin, André Degas, Edme Cotel, and Pierre Aigrain were among the journeymen whose highly prized watches were stolen from the bakeroom during their momentary absences. In each case another journeyman was suspected. Nicolas Martin told the police that his cache of clothing and jewelry, including diamond studs hidden "in a little redoubt in the bakeroom," was ransacked while he was sick in bed. He accused his coworker, François Jardin, "a bad subject," who had not returned to work since the theft was committed. Similarly, François Dubois reported the theft of a pouch containing a gold ring and twenty to twenty-five livres in cash that he kept in a dresser over the oven. He suspected Joseph Morisset, a journeyman next to whom he had labored for over a year and a half.[27]

Nor did a sense of common identity inhibit journeymen from brutalizing one another. Georges Pacquemare accosted his colleague Laurent Leveque at ten o'clock at night and threatened to beat him if he did not lend him money. When Leveque insisted that he had none, Pacquemare hurled him to the ground and stole his hat. For reasons that are unclear, mistress baker Colombel fired baker boy Fourcy after two years of service and replaced him with Martin Macadrez. Bitter and jealous of his replacement, Fourcy attacked him with a broomstick while he was asleep atop the oven. Seriously wounded and fearing for his life, Macadrez asked the police for protection. Jean-Baptiste Anquetin settled a score with Claude Desmart, a journeyman with whom he had frequently worked, by assaulting him while he was asleep in the bakeroom. Jealousy of a nonprofessional sort also caused friction. Edme Sellier jumped Pierre Legrand, pummeled him, and (here the crime acquires larger significance) broke his *tailles* because Legrand allegedly called Sellier's girlfriend a whore. Journeymen from the same *pays* banded together against others who did not enjoy the merit of sharing their roots: "Germans" bickered with Burgundians, Auvergnats with Picards, Normans with Flemish.[28]

In other clashes it is more difficult to discern a motive. Pierre Fraussard and Antoine Marcel worked side by side, but when the latter got drunk he beat the former's head against the kneading trough until it split open. Ostensibly, it was also drink that clouded the mind of Fernand Dumat and caused him to assault his friend and coworker Blaise Dutour. The latter was so badly wounded that he had to find a replacement to work in his stead. Two journeymen in Moraze's shop plotted to compromise and get rid of a third, whom they found uncongenial. They planted a gun in his pocket and called it to the attention of the master, along with denunciations of poor workmanship.[29]

A Violent Temperament

If the journeyman bakers had a reputation for violence, however, it was earned largely from the violence they committed against others rather than themselves. There was no specifically "baker" criminality, nor can we pretend to show that baker journeymen had a greater propensity to crime and/or violence than journeymen in other professions. There is no doubt, however, that bakers were reputed to be the edgiest and the most tempestuous Parisian journeymen, and there is considerable evidence, albeit of a scattered and impressionistic kind, to make this persuasion credible if not absolutely convincing.

Nor were violence and crime the appanage of the unemployed and the fugitive, though their behavior, as we have seen, was sufficiently deviant and tumultuous to account for the stereotype. Working journeymen got into trouble frequently enough to call attention to themselves as a group. Fremin *fils* was a neighborhood bully. Jean-Edme Oville was involved in the murder of his wealthy brother-in-law, a crime that caused a sensation in the community. On separate occasions, Jean Imbert and Baptiste Lorivet tried to stir a "rebellion" among the people against the guard. A journeyman with a pockmarked face named Bourguignon raped a girl aged five years and eight months, wounding her seriously "à la matrice," the term the police used to name her reproductive organs. Other journeymen were involved in knife fights, muggings, illicit fireworks, brawling, exhibitionism, and harassing neighbors with insults.[30]

Jacques Jemelet and Pierre Legrand were sentenced to be strangled and hung for breaking and entering in order to steal.[31] Scores of other journeymen were either branded with GAL (*galères*) and sent to the galleys or branded with V (*vol* or *voleur*) and exiled from Paris for various degrees of theft.[32] The fact that Jean Geoffroy, nicknamed Petit Pain, was in prison already did not inhibit his sense of enterprise. He stole a silver tobacco box from a visitor who was distributing alms in the courtyard of the Grand Châtelet.[33]

Police officials and other commentators thought there was a link between the nature of the baking profession and the temperament of the bakery workers. Bakers, they suggested, had an occupational propensity toward contentiousness, irascibility, aggressiveness. The journeymen were not like this when they first began working in the bakeroom. Their character and comportment were changed on the job. The social and moral segregation imposed by their work, the exhaustion, the nighttime labor, the lack of sleep, the tremendous pressure to perform rapidly and well, the yoke of routine, the unhealthy work environment, and the debilitating illnesses they contracted—these were the factors, observers alleged, that collectively disposed the bakery worker to violence and crime. Work barbarized the workers instead of civilizing them. It eroded their sense of community, cankered their values, and weakened their ability to control themselves.[34] The interesting thing about these remarks is that they betrayed neither surprise nor genuine consternation. Given the brutalization of work, bakery workers were more or less expected to exact a certain price from the society they served. It should be added that according to this argument, the worker was broken but not alienated by his work. His reactions were instinctive rather than self-conscious; he did not really know why he acted the way he did.

Drinking: From Sociability to Subversion

This argument suffers in want of some pieces. Work obviously had a profoundly disorganizing impact on the bakery worker, but the causal relationship between work and antisocial comportment remains to be demonstrated. One of the habits that work encouraged journeymen to acquire was the practice of drinking—too often and too much. Drink could be one of the missing links in the argument connecting work and behavior, for there is little doubt that intemperance fostered conduct generally deemed to be deviant. Work, it might be contended, wore the worker down and made him vulnerable, and then drink—wine and spirits—did the rest.

Journeymen drank for many different reasons. They drank not to become antisocial but, on the contrary, because they intensely craved sociability. (Indeed, some of them may have drunk largely because it was a condition of enjoying sociability). Drink liberated, in wonderfully sudden and powerful ways, and it liberated each man according to his own fantasy. But drink was not simply a matter of escape. It would be wrong to overlook the "positive" contribution of drink, a contribution that the authorities tacitly acknowledged by continuing to permit workers to imbibe even as they moralized about the horrors of intoxication. It was an analgesic as well as a hallucinogen. Drink aided many workers who needed or wanted to run away, as we have seen in our discussion of journeymen without work and without papers. But drink also helped working journeymen to stay on the job. Drink was a surrogate for rest and recreation. It helped the worker to keep going, to sustain the pace, to discharge his obligations. Thus drink could be, from a short-term social perspective, either productive or deleterious. Everything depended on the dose.[35]

Given the nature of the archives (or rather of the administration that created them), we know much more about overdoses than about anodyne rations. Jean Urbain, age thirty-one, and Nicolas Chapiteau, age twenty-four, drank too much wine and got into a brawl in a tavern. Windows and furniture were broken and many blows were struck. In the fracas, Chapiteau hurled a stone that killed one of the clients. Chapiteau was sentenced to be strangled and hung and Urbain to perpetual exile from Paris. At different times the guard arrested Alexis Cresson, Guillaume Gaudripon, Jean Daniel, Georges Arbormisch, Nicholas Porcaboeuf, Simon Langlis, Jean-Baptiste Anquetin, Louis Jannard, Pierre Prot, and Louis Nignet for drunkenness and fighting in taverns. "Overheated with wine," Simon Cocquart tried to resist arrest by inciting the populace to rise against the guard. Inebriated, Maurice Bignon, called La Plante, and Michel Guillaume tried to force a tavern keeper to serve them after he had closed his shop by beating him and his servants. Siphorien Perot and Jean Lesueur were jailed for a similar assault on another reluctant wineshop keeper. René Fort was badly beaten in a brawl with some strangers in a tavern, while another journeyman baker was murdered in the bakeroom by an irate domestic who followed him home to avenge a drunken quarrel in a tavern. Other journeymen either committed thefts or were themselves robbed while under the influence of alcohol.[36]

As far as we can tell, journeyman bakers visited taverns (and the drinking places in the inns and in the wine merchants' shops) quite frequently. They were habitués of neighborhood bars near their shops. One found them there most frequently in the late afternoon after they finished working and before they went to sleep. On Sundays they renounced the confines of the home quarter for a more exotic itinerary: the *guinguettes* of Les Porcherons and the converted mills and other drinking houses outside the Parisian fiscal barriers, where the wine was cheaper than in the city proper. Moralists exhorted workers to go to church instead of to the tavern. Many baker journeymen, like Philippe Laurent and Michel Mulin, did both, "heading directly after mass to drink in a tavern across the street from the church."[37]

For these journeymen, the tavern was a social institution, no less fundamental to their well-being and to their daily life than the church. It was their turf, the social ground that they had staked out. Like the inn (with which it sometimes overlapped), the tavern was free of the constraint of subordination. Here the workers wrote the scenario; they did not have to play the role imposed by their work and their rank. The ambience, the sociability, and the wine—"a venom as dangerous as that of hemlock," warned Delamare—gave the journeymen a sense of comfort that they rarely experienced elsewhere.[38] In the tavern they talked about things they dared not discuss elsewhere. The tavern facilitated and symbolized the link between wine on the one hand and crime and sedition on the other, between individual behavior and collective action, between libertinage and liberty, between morality and politics.

The police and the moralists abominated these drinking houses. "Taverns of debauchery," observed Fréminville; "haunt of scoundrels, meeting place of rogues, fences, pimps, and gens sans aveu," wrote Des Essarts. Tavern keepers, like their clients, were recruited from a dubious milieu, "persons without education, without any principle other than their self-interest." Tavern keepers and their clients together seemed to mock the idea of good order and those responsible for assuring it. The situation was dramatic, according to these critics, because the number of taverns had grown rapidly throughout the century. In their view, this increase helped explain the mounting incidence of theft, of public besottedness, of unlawful assemblies of workers, of ruined worker families. If the number of taverns were severely limited, the *Encyclopédie méthodique* predicted, there would result "the prompt reestablishment of good mores." The police preferred journeymen like Louis Delalande, who boasted, as a gauge of good conduct, that he never lodged in the inns, and like Symphorien Lorient, who told the police that he resisted pressure every day from his comrades to go drinking. Yet if taverns not only survived but continued to multiply, it was in part because the police recognized that they served certain social needs even as they degraded the society that tolerated them.[39]

Places of debauchery and nurseries of common crime, the taverns were also posts of subversion. Not only did the journeymen get drunk in the taverns, but they also plotted there. The tavern, like the inn, served as a base for organization for journeymen of virtually all professions. The confréries of the stocking makers and the painters, as well as secret associations of bakers, frequently met in the taverns. One of the tavern keepers who welcomed stocking makers was Neureux, himself a former journeyman in that profession. Lidy, an innkeeper on the rue du Mont-Hillaire, probably a former bookbinder, was "the guiding force" of the journeyman bookbinders' strike of 1776. Numerous "cabals" and strikes were born in taverns and inns. The drinking houses served as meeting halls, centers of propaganda, hideouts, welfare centers for indigent journeymen providing lodging and meals on credit. Ultimately the authorities were far more alarmed by tavern-linked sedition than by tavern-associated crime. Both were antisocial in nature, but whereas the latter usually involved individuals and had a very limited social impact, the former generally mobilized large numbers of workers engaged in a collective action that could easily spread to other sectors and that could result in the emergence of quasi-permanent worker associations.[40]

Given the extreme paucity of evidence, it is virtually impossible to determine the extent to which journeyman bakers were organized in the eighteenth century.[41] There are some indications of inn- and/or cabaret-based associations, but it is clear neither what the criteria of membership were nor whether the individual organizations had ties among themselves. Journeymen from the same *pays* tended to frequent one another, but there is no hard evidence to show that they actually imposed a structure on their relationships. In the course of the eighteenth century, bakery workers repeatedly formed "cabals" to try to "dictate the law" on various issues including wages and placement. There are unmistakable signs of organization in each instance (e.g., 1704, 1738, 1749, 1756, 1776), but one cannot tell whether these were transient crisis-wrought organizations (the particular crisis generating the organization necessary to deal with it) or whether they were more or less permanent, clandestine organizations (the organizations provoking and/or exploiting the crisis). The journeymen revealed a remarkably sophisticated division of labor in a number of these "cabals" and strike movements. They had leaders who planned the overall strategy, treasurers who collected a strike fund, field agents who enforced collective discipline (in part by encouraging or intimidating reluctant workers to join the movement), messengers who spread the word from shop to shop, and fixed assembly points in different quarters (including various boardinghouses and taverns).

Compagnonnages and Confréries

I have found no indication of the presence of a *compagnonnage* organization among the baker boys. It must be remembered, however, that the compagnonnages were sub-rosa associations that, if they operated according to plan, should not have left any trace in the public record. In fact, as a result of carelessness, espionage, or the pressure of interrogation, one finds evidence here and there of compagnonnages in other Parisian professions. Still, it is quite possible that the compagnonnage functioned surreptitiously and discreetly in the bakery. In a number of trades an association called a confrérie (or *confrairie*) played virtually the same role as a compagnonnage, and it may indeed have served to mask the compagnonnage.[42]

Though they were by and large proscribed by the eighteenth century, the confréries had a more nuanced ancestry than the compagnonnages. Initially they were highly respected associations of piety rather than outlaw clandestine organizations. Over time some of them remained faithful to their original mission while others were transformed into sociopolitical pressure groups—cells of subversion in the eyes of the police. In order to protect themselves they sought to maintain the illusion that they were exclusively devoted to works of charity and spirituality. Since the authorities often had trouble distinguishing between good and bad brotherhoods, after the end of the fourteenth century they tended to favor a general suppression.

As usual there were many exceptions allowed in deference to groups that enjoyed influential patronage or demonstrated convincingly their innocuousness. Predictably, there was far greater toleration for guild-sponsored confréries of master-employers than for worker brotherhoods composed of journeymen. Some worker confréries, such as those of the carpenters, hatmakers, polishers, and secondhand clothes dealers, tried to obtain formal legitimation in the eighteenth century. Several Paris police chiefs, especially Lenoir, were impressed with the utility of certain confréries as mutual-assistance organizations. Other brotherhoods, those, for example, of the stocking makers and the painters, remained underground. These were the confréries that, in the eyes of the police, discredited all the others, for they taught insubordination, organized strikes, and generally troubled the public order. In function and organization these brotherhoods bore a striking resemblance to compagnonnages.

An episode in 1719 reveals that the baker boys, too, had a secret organization called a confrérie. It was denounced by wrathful masters who blamed it for drawing the journeymen away from the bakerooms to illicit meetings in the inns and taverns. There the workers plotted against the masters and encouraged one another to stay away from work. "The disorders and absences" occasioned by "the pretext of a *confrairie*," according to a guild petition, jeopardized the regular supply of bread for the capital. In response to this complaint, the lieutenant general promulgated a sentence forbidding journeymen to assemble and cabal in the cadre of a confrérie or for any other reason.[43] It is very likely that the brotherhood continued to exist, despite the prohibition and the threats of vigorous punishment, including prison. But we know nothing either about its recruitment or its activities in the course of the century. Given the large number of baker boys and their dispersion throughout the city, it is quite possible that there were several confrérie-type associations.

Outside of work, the compagnon sought companionship of a different nature than he found in the bakeroom. He looked persistently for love, with a prostitute or with an acquaintance. Those with indomitable will or means managed to form more enduring unions, usually of the formal matrimonial sort. Obviously it was extremely hard for the journeyman to nourish a conventional family life, and there is considerable evidence that work placed terrible strains on marriage.

The garçon found other forms of sociability in the taverns and the inns, and probably in the confréries as well. All three venues constituted spaces of relative freedom and shelter where the journeymen could affirm and develop their sense of identity and personhood. Yet none became a veritable institutional pole around which the garçon could build coherently and serenely his life beyond work. Each situation placed the bakery worker under some form of insidious pressure that made it hard to find sustained fulfillment. He had to hurry or act surreptitiously or engage in sundry ritual challenges that tested his honor, his strength, or his prudence. His very clothes heralded a temperament that was quite often not at all his own. One has the impression that journeyman bakers frequently acted out roles that they really did not want to play. It is a shame that we cannot follow substantial numbers of permanent garçons through the life cycle (public and private, to the extent that they were not totally convergent) in order to see how they dealt with their (relative) poverty, their professional instability, their physical burdens, and their moral incubus.

Chapter 10

Establishment

Upward mobility for a journeyman baker meant some sort of independent status in which he would operate as his own employer for his own account. There were several ways—professional and extraprofessional, legitimate and illicit—to achieve this autonomous establishment. The classical path of ascensions in the corporate world culminated in the attainment of mastership (though it is well to note at the outset that mastership itself did not provide independent establishment; it merely permitted the aspirant to set himself up within the corporate jurisdiction—if he could). Given the power and presence of the guild, it is easy to forget that there were certain opportunities for establishment outside corporate dominion. A journeyman could attempt to launch a baking enterprise in certain of the so-called privileged places, including the vast expanse of the faubourg Saint-Antoine. If he preferred to operate outside the city, he could become a forain, in the suburbs or the hinterland. In complicity with a master or a mistress (widowed) baker or some other patron with means, or, if he was especially bold and resourceful, on his own, the journeymen could install himself as a baker illicitly, brandishing a false title or a borrowed name (but it could be argued that the newly risen journeyman's relationship with silent partners or sponsors diminished his independence). Another road to independence passed through a well-calculated marriage that could lead to a mastership or extracorporate establishment. Each of these channels of upward mobility within the bakery had its own peculiarities, but from the perspectives of many baker boys they may have seemed to be pretty much the same, for they all required money in more or less substantial amounts.

The achievement of mastership was the surest, the most prestigious, and the best-known mode of socioprofessional ascension. Mastership was a kind of moral and political regeneration. After a more or less long probationary period, the worker redeemed himself from abjection and domesticity by joining his former superiors on terms of status equality. Mastership was sovereignty even as journeymanship had been subordination. It represented an exhilarating psychological emancipation. The attainment of mastership was a vindication not only of the journeyman but of the whole system. It confirmed the system's validity and the elaborate division of society on which the system rested. The sense of justice (and pride) that his own ascension inspired helped the ex-journeyman to bridge the enormous chasm between his new status and his old and to become a true-believing master much more conscious of his new responsibilities than mindful of his past sufferings. Invested with a new professional role and located in a new social space, the recent master saw himself and his world in a new light. And for the first time he envisioned the prospect (which he should not have mistaken for the *promise*) of a far more comfortable style of life and work.

According to corporate theory, guild mobility was a natural, organic process of selection and assimilation. In fact, it was a rather arbitrary, discriminatory, and often treacherous system of recruitment and containment (at least it was so by the eighteenth century, even if that is not always the way it had been). All journeymen were not equal in the eyes of the guild, and the rules governing mobility reflected the guild's Manichean vision. There were the sons of masters, the anointed journeymen whose success was desired and promoted, and there were the others—outsiders, *étrangers,* corporate orphans without any kinship link to the bakery, whose future mattered very little to the guild fathers. The primary goal of the corporate leaders was to preserve their monopoly. They wanted to keep tight control of the number of master bakers in order to forestall excessive competition and thus protect their collective prosperity. Since places were to be limited, the masters endeavored to advance their own familial interests. As a consequence, two different regimes for mobility were written into the guild bylaws.

In the eighteenth century, until 1776, a nonson who aspired to mastership had to complete a formal three-year apprenticeship, evidenced by the award of a license, or *brevet,* and at least a three-year journeymanship (it goes without saying that the outsider as well as the son had to be Roman Catholic, "of good conduct and mores," and free of any contagious disease).[1] If a candidate was at least twenty-two years of age and, in the words of the statutes of 1746/57, "if there appears to be no defect either in the person or in the titles of the aspirant," then he could prepare the *chef-d'oeuvre,* or masterpiece, by which the guild leaders would judge his professional capacity. The corporate panel that reviewed his case consisted of the jurés, the anciens (those who chose to concern themselves), four *modernes,* and four *jeunes.* The jurés named a "meneur," or tutor, to supervise the aspirant's project. It was symptomatic of the corporate weltanschauung that the meneur was expected to protect and represent simultaneously the interests of the guild and those of the outsider-candidate. The meneurs seemed to have been chosen as a rule from the corporate aristocracy in order to underline the seriousness with which the guild regarded the master-piece test. For example, master Nicolas Driancourt, a former chief financial officer (*juré comptable*), was named as meneur to Nicholas Michel Edhieres, an illiterate, unconnected journeyman who counted heavily on his tutor's advice.[2]

In the seventeenth century the masterpiece task consisted of converting three setiers (Paris measure) of "good flour" into one-pound loaves of a white bread called *pain broyé*. *Broyer* means to pound; this bread, composed of a hard or firm dough (the antithesis, say, of the light *mollet* bread so coveted by wealthy Parisians), had to be pounded with so much force that the bakers kneaded it with their feet or used a paddle.[3] In the eighteenth century, the aspirant had to demonstrate a broader range of skills, none of which, however, had to be mastered with the level of command considered mandatory a hun-dred years earlier. He had to transform the same three setiers into "diverse kinds of doughs and breads" as prescribed by the panel of examiners. If the masterpiece was accepted, the aspirant had only to pay the necessary fees and take his corporate oath of fidelity before the royal procurator of the Châtelet. If the panel judged the masterpiece to be "defective or unacceptable," it in-structed the journeyman either to take a new test or "return to serve the Mas-ters for an appropriate period of time in order to become more competent."

The baker's masterpiece was not a complex or esoteric test contrived to frustrate the ambitions of the candidates. On the contrary, it was a straightfor-ward and fair-minded trial of skill that should not have posed any problem for an experienced journeyman. Yet it was not a test that could be judged by standardized, objective criteria. As a writer in the *Encyclopédie* remarked, if the guild fathers found the candidate "odious," there is no way he could pass the test, regardless of the apparent quality of his masterpiece.[4] Unfortunately, we have no data concerning the rate of rejection or the deliberations of the exam-ining panel. It is likely that we would have detected the echo of complaints of prejudicial evaluation had this blackballing practice been common. We cannot be absolutely sure, however, that the jurés did not use the masterpiece as a filter to control the number of admissions or that certain masters did not influence the panel decisions in order to settle scores with individual journeymen (or their sponsors).

A master's son had a much easier time of it, and he was certain of success provided he was born *after* his father had become a guild master. Corporate bylaws, seconded by police sentences, stipulated that sons born prior to their father's ascension to mastership had to pass through the outsiders' regime.[5] There was no doubt that these "premature" sons would be received, but they would be obliged to go through all the extra time, expense, and trouble of outsiders. Nor was this rule merely systematically violated or ignored by the jurés in deference to the importunities of their fellow masters. The most striking illustration of corporate staunchness on this matter occurred in 1762, when the jurés instituted and won a suit in the Châtelet against one of their distinguished colleagues, a former juré—an ancien—named Clément Grégoire.

Grégoire was found guilty of deceiving the guild by presenting a baptismal certificate for a son who aspired to mastership that in fact belonged to another son, now dead, who had been born after the father's rise to mastership. As a consequence of his manipulation, the father lost his seniority and was struck from the ranks of the anciens and placed in the category of the modernes, punishment for having "abused the confidence of cojurés and anciens" and perjured himself before the royal procurator in the Châtelet. Grégoire's son was ordered to close his shop immediately and cease exercising a profession for which he lacked the official status, or "quality."[6] That a former guild official like Grégoire turned to fraud suggests how grinding were the combined burdens of apprenticeship, journeymanship, the masterpiece, and outsider fee schedules.

A legitimate son (in corporate terms) was dispensed both from apprenticeship and the journeyman's training period. In place of a masterpiece he had only to perform what was called a "légère expérience," a superficial exercise involving a *mine* of flour, a mere half a setier. If his father was an ancien or a juré, the candidate could pass the test in his own bakeroom at home. The guild celebrated the *expérience* as proof that it subjected its sons, too, to professional scrutiny. But no one took the *expérience* seriously. It seemed to be a vapid and hypocritical gesture, and it tended to discredit the corporation rather than enhance its reputation for rigor and honesty.

There was yet one more requirement—the last in addition to the assessment of fees—which also raised questions about the guild's conception of mastership for sons. Whereas the corporate bylaws of 1719 considered any age below fourteen years to be too young to begin apprenticeship, they accorded mastership to sons as of age twelve. To the credit of the guild fathers, they were as firm about this frontier as they were about distinguishing legitimate from illegitimate sons. When the widow of master Jean Boucheu presented her ten-year-old boy for reception, he was refused on the grounds that he was not yet mature enough to assume mastership responsibilities. But did the guild leaders really believe that her son would be ready to run a shop and participate in guild affairs two years later? The guild fathers themselves seemed to have been embarrassed by their position, for they raised the minimal age to eighteen when they next revised their statutes in the mid-forties. As early as 1720 the lieutenant general of police complained in a public sentence that too many sons of masters without any proficiency in the craft were receiving masterships. These unprepared and unaccomplished masters either served the public badly or lent their names to others who practiced their trade illicitly.[7]

Guild Admission Fees

The two mobility regimes differed in one more crucial respect: sons paid much smaller sums in reception fees. First of all, let us remember that sons were already spared the heavy cost in time and money of the six years combined apprenticeship-journeymanship. Outsiders had to buy three setiers of high quality flour—fifty livres more or less in a year of subsistence ease—and had to pay the meneur (30 sous in 1719, 60 sous in 1757), another master of the candidate's choice (30 sous), and the six jurés on the examining board (30 sous in 1719, 60 sous in 1757). The son had to buy a *mine* of flour, and apparently his *expérience* was administered free of charge. The outsider had to pay additional *droits* to the jurés (5 livres each in 1719, 6 livres in 1757) to the anciens (50 sous in 1719, 60 sous in 1757), to the four modernes and four jeunes who sat on the board of reception (30 sous each in 1719 and 1757), and to the clerks (3 livres each in 1719 and 1757). Officially, the son paid only half of these charges; in practice, sometimes even less. In addition, both son and outsider paid twelve livres each to the confrérie and three livres for Saint-Lazare. The total of these miscellaneous fees and expenses for an outsider was approximately 150 livres in 1719 and 170 livres in 1757. For sons it varied between fifty-seven livres and sixty-three livres.

The largest single charge was the droit, or right of mastership. Because of gaps and contradictions in the documentation, it is not always easy to determine the amount at which it was set. The bylaws of 1719 fix the payment at four hundred livres for outsiders and one hundred livres for sons. Now, the bulk of these fees were assigned, by royal order, to a debt amortization fund. By 1740 the major part if not all of the debt was fully paid. As a result, by an arrêt du conseil the government reduced the mastership fee to two hundred livres for outsiders and fifty livres for sons (and the apprenticeship registration fee to five livres, down from twenty livres). But the guild must have contested this decision vigorously and persuaded the ministry to annul it, for the revised bylaws of 1746/57, approved by the royal administration, put the fees at four hundred and one hundred livres.[8] According to "tables of comparison" appended to the edict of August 1776 reestablishing the corporate system, the droit in the baker's guild had risen to nine hundred livres by the late sixties or early seventies. There is insufficient evidence to corroborate this claim.[9] It is not impossible that the royal officials exaggerated the amount of the so-called old fees in order to make the changes introduced in August 1776 appear more reasonable and necessary.

Nor can we be certain that the fees officially established corresponded to the amounts collected by the guild for the right of mastership. Nicolas Bourlon and Jean-Baptiste Grimbert, both sons of masters, were received by the guild at approximately the same time. But Grimbert's family evaluated the price of his mastership at 150 livres, while Bourlon's fixed his at six hundred livres. Though master baker Simon Poton should only have paid one hundred livres for his son's mastership, he claimed to have expended three hundred livres. An administrative source records that masters' sons remitted 117 livres around midcentury. Gentien Aubert was apparently a first-generation aspirant, yet he paid three hundred livres instead of four hundred livres in 1735. The widow of master baker Sébastien Lapareillé testified that it cost her 238 livres 19 sous and 9 deniers to get each of her sons admitted to the guilds in the early forties. Another mistress baker, widow Constant, put her costs at four hundred livres, as did master baker Guy Leguay for his son. On the request of the jurés, in the fall of 1775 the procurator's court of the Châtelet convicted a freshly arrived master for failing to pay his fee of 241 livres.[10]

How does one account for these discrepancies and inconsistencies? It may be that in some instances the baker families confused, by error or by design, total costs of reception (which included the miscellaneous fees discussed above along with other unofficial impositions, which we shall consider below) with the right of mastership. Those of our estimates which come from marriage contracts may have been inflated by parents eager to constitute an impressive paper dowry for their sons. Figures from the division of estates may similarly have been swelled in order to make a son's inheritance seem larger than it actually was. One encounters both the financial incentives to exaggerate and the amalgamation of total reception costs in the affidavits of masters filing for bankruptcy or in serious economic difficulties.

René Morin insisted that he paid a thousand livres cash for his mastership, money that he borrowed and later had trouble repaying. André Devaux also claimed that he paid this amount. Jean-François Pigeot entered in the "losses" column of his balance sheet eleven hundred livres "for his mastership of master baker." Seeking police assistance to continue his business, Pierre Tauvin griped to Commissaire Grimperel that he "had ruined himself (in part) by his reception to mastership, which he maintained had cost him fifteen hundred livres."

Even if all four of these masters were outsiders, their evaluations seem extravagant on the face of it, unless the nine-hundred-livre figure put forth in the edict of 1776 was correct (at least for the sixties and seventies), or unless the under-the-table charges were considerable. But even so Noirot the elder, himself a master baker, should not have had to pay fifteen hundred livres for his son's admission.[11]

In earlier days, a fairly elaborate ceremony marked the ascension of a new master. In an act similar to certain feudal rites, the candidate used to announce that he had completed his preparation, and, after the authorities agreed to receive him, he smashed a pot of nuts against the wall as a reaffirmation of his oath of fidelity.[12] Then the new master shared his happiness and expressed his gratitude by offering a more or less sumptuous repast to the guild fathers and sometimes also by presenting them with *jetons* or *pourboires* or other monetary rewards. In the eighteenth century the reception ceremony was stripped of its colorful ritual. It was a businesslike affair that may or may not have been followed by drinking or eating at the new master's expense.

There is little doubt that the candidate was expected to offer gifts of appreciation, but it appears that they were always solicited (or offered) *before* admission to the guild was a foregone conclusion—when there was still some chance for the play of influence. It is impossible to determine how widespread these practices were, whether they pertained exclusively to outsiders, and how much money they involved. If one is to give any credence to the blast of accusations made against exorbitant and extortionate corporate demands on the admission of outsiders in the second half of the eighteenth century, then one is tempted to think that these sub-rosa exactions were substantial, for the official fees by themselves, albeit considerable, were not as onerous as this criticism indicated. Moreover, each time it treated guild mobility in a piece of royal legislation, the government had to remind the masters that they were forbidden to demand any fees or gifts beyond those stipulated in their statutes or in royal law.[13]

The testimony of a single witness is never convincing in the historian's court, but one is loathe to ignore it when it is particularly suggestive. Let us see what the case of François Royon leads us to think about corporate extortion and about the gaping incongruity that often existed between the formal rules and the way the game was actually played. First garçon for the Jesuit bakery on the rue Saint-Jacques at the time of his application for mastership, Royon had his apprenticeship license in hand and ten years of work as a journeyman behind him. During all this time, he boasted, "he had discharged his duties with honor and distinction and without any complaint against him being addressed to the guild." With this promising record, he asked to stand for mastership. But,

Royon complained, the jurés exacted a fee of seven hundred livres above the amount stipulated in the corporate bylaws. Royon filed a protest with the royal commissioners who were charged with settling corporate disputes concerning advancement. The commission found in Royon's favor and ordered that he be admitted immediately as master upon payment of the prescribed four-hundred-livre fee.[14]

One can only wonder how many times the jurés demanded a supplementary payment and received it, how many times they demanded it not for venal reasons but in order to prevent a candidate from attaining mastership, and how often victims of this extortion lacked the tenacity and knowledge to raise their voices in protest and seek redress in Royon's fashion. Given all the contradictions, uncertainties, and gaps in our information, it is impossible to determine the exact cost of an outsider's mastership. Passing over all the indirect expenses associated with apprenticeship and journeymanship, the official surface price of mastership, including the droit itself and the miscellaneous legitimate fees attendant upon it, seems to have ranged during the reign of Louis XV from roughly six hundred to eleven hundred livres. If one allows for hidden exactions and gratuities, it is probably reasonable to augment this base by 25 to 100 percent. Was the price of mastership prohibitively expensive for the outsiders? Did it effectively limit recruitment to master families, thereby reinforcing the dynastic character of the monopoly, the same way that other "constituted bodies" of the eighteenth century barred *roturiers*—commoners—in order to preserve their purity and exclusivity?

The Liberal Critique of the Closed Corporate World

If one is to take the critics of the corporate system at their word, there is no doubt that this is what happened. Instituted originally "to maintain good order among workers and merchants," mastership had degenerated, wrote Faiguet de Villeneuve in the *Encyclopédie*, into "cabal, drunkenness, and monopoly." Apprenticeship and journeymanship were all long, oppressive, and fruitless, a form of institutionalized "brigandage" for the master and "servitude" for the worker. The masterpiece was a cynical invention meant to keep out even the frugal and industrious outsiders. The recruitment system was a major source of misery, mendicancy, and ultimately criminality, for it denied the little people a chance to work and become socially integrated and useful. Jealous of their authority and avid for more wealth and power, the corporate fathers had transformed the mastership from a system of renewal into a system of exclusion by means of which "the richest and the strongest ordinarily succeeded in keeping out the weakest."[15]

The most stringent criticism of alleged Old Regime abuses appeared not in the tracts of the philosophes but in the preambles of the legislation drafted by certain controllers general. Turgot's denunciation of the corporate system is a striking example. The guild system kept prices high and perpetuated inefficiency by restraining competition, Turgot maintained. It inhibited technological innovation. It encouraged waste, corruption, contentiousness, litigiousness. The corporate system harmed all of society, but its most wretched victims in Turgot's view, were the laboring poor. The guild structure deprived the simple citizens of the right to work, "an inalienable right of humanity." In most cases the work they were refused the right to perform was the only property that these poor people possessed. The guild bylaws were frank statements of the vicious goals of the system:

Their general bent is to limit, as far as possible, the number of masters, to make the acquisition of a mastership almost insurmountably difficult for anyone other than the children of current masters. It is toward this end that are directed the multiplication of expenses and formalities of admission, the difficulties of mastership, which is always arbitrarily judged, especially the high cost and the useless length of apprenticeship and the prolonged servitude of journeymanship, institutions whose object is to enable the masters to enjoy at no cost the labor of the aspirants for several years.[16]

Turgot's indictment embraced the whole corporate system. Abbé Coyer, a philosophe interested in problems of social organization, education, and civic spirit, directly addressed the issue of advancement in the bakers' guild in his sprightly tale of Chinki, an Indochinese laboureur whose Edenic happiness and prosperity were undermined by the Europeanization of the state and the social order. Convinced that the land could no longer support his family of twelve children (born of two wives in a space of six years), Chinki decided to take them to the capital where the arts and crafts flourished and find a position with a future for each of them as they came of age. The city experience was traumatic for the honest country bumpkin, for he could detect neither justice nor good sense nor economy in any of the customs he observed. Chapter 9, entitled "Why Chinki Could Not Succeed in Placing His Son with a Baker," reads as follows:

"Master," said Chinki to a Baker, "I bring you an apprentice, if you wish to receive him." "Is he the son of a Master?" "Yes, of a Master Farmer, yours truly his father." "My good man," replied the Baker, "understand that your son after his apprenticeship, even if he were as skilled as I, would not be received as a Master, not being the son of a Master Baker. . . ." "I thought," said Chinki, "that the worker was judged by his work, and not by his birth. Does the son of a Master inherit his father's skill? Mine will not be a baker."[17]

Was guild recruitment in fact restricted to masters' sons in the eighteenth century, as Turgot, Coyer, and other critics charged? What was the real effect of corporate admissions policy on baker recruitment until the abolition and reestablishment of the guild system in 1776? There are three categories of admission in the reception records that begin in 1736. Two of them are familiar to us: those who become masters "as sons of masters" and those who acquire the rank "by apprenticeship and masterpiece." The third rubric consists of those who reach mastership through extraordinary channels: either as a consequence of royal nomination in an arrêt du conseil or as a result of service in the hospitals of Paris as apprentice and journeyman.[18]

Broadening of Recruitment

In the four decades between 1736 and 1775, the guild admitted 826 new masters at an average rate of 20.66 a year (ranging from a low of 11 in 1751 to a maximum of 40 in 1764). During thirty-three of the forty years more outsiders were admitted than sons. (If one adds the third category to the outsiders, then the sons were more numerous than the outsiders only six times during this period.) Of the 826 admissions, 32.1 percent or 265 were sons, while 60.4 percent (499) were outsiders and 7.5 percent (62) rose through extraordinary channels. On average each year the guild admitted almost twice as many outsiders as sons, 12.48 over against 6.63 (only 1.55 new masters came from the third category). Thus it is not at all true that the baker corporation blocked admission to outsiders. Between 1736 and 1775, two of three new masters were *not* sons of masters. If lineage was the only barrier, then Chinki's son had a better chance than Coyer led us to believe.

In fact, over the course of the century, there was a clear trend toward a liberalization of admissions policy. Even as the critics of the guild system became more strident—perhaps because of this crescendo of strictures—it became easier for an outsider to advance to mastership. Let us compare recruitment in the period 1736–57 with admissions in the interval between 1758 and the Turgot reforms, when the guild system came under intense and unremitting attack. Four hundred eight aspirants attained mastership during the earlier period at an average rate of 19 a year. There were 8.3 sons a year as opposed to 9.2 apprenticeship-masterpiece outsiders and 1.5 extraordinary outsiders. Sons represented almost 44 percent of total recruitment and outsiders a little more than 56 percent (48.3 percent by ordinary and 7.9 percent by extraordinary channels). Thus between 1736 and 1757, almost one of two new masters was the son of a master. This is hardly a policy of exclusion, though it is obvious that the relative advantage of sons was considerably greater in the early period than for the whole run of years for which we have records.

The later period presents the dramatic contrast. The size of the class was only slightly larger: 22.67 a year. But of the 408 new masters admitted during these eighteen years, only 82 (20.1 percent) were sons of masters, while 297 (72.8 percent) did apprenticeship and masterpieces and 29 (7.1 percent) rose through the extracorporate arteries. Every year on average 4.6 sons became masters over against 18.1 outsiders. Thus four out of five new masters were drawn from outside the hereditary guild structure. By any reasonable measure, it is impossible to construe these results as the fruit of a rigorously clannish, self-regarding, and self-perpetuating recruitment strategy.

It is far more difficult to account for this striking broadening of the base of recruitment than it is to demonstrate it. It is unclear whether it was the product of a policy decision made by the guild to liberalize admissions in order to allay criticism and strengthen its position (financial as well as political, for the outsiders brought in more money) or whether it merely bespoke a shortage of masters' sons needed to replace dying or retiring or failing masters or to meet a slowly expanding demand. It is hard to see why there should have been a demographic crisis within the bakery around midcentury or a sudden hemorrhaging of bakers sons into other professions. Nor is there any evidence to suggest that a larger number of outsiders reaching the age of aspiration after midcentury had substantially more money to spend on upward mobility than their predecessors (though it could be argued that changes in guild policy and practice increased the corporate purchasing power of the outsiders). Finally, the relatively small increase in the annual size of the classes suggests that the guild did not have a lucid appreciation of subsistence demand and that it was still very much concerned about protecting its members from "excessive" competition.

During the decade and a half preceding the Turgot reform, the corporations were sharply criticized for charging excessively high admission fees not only by a handful of liberal propagandist-philosophes but also by the royal government itself. In an edict of 1767 Louis XV deplored the bloated "costs of admission" that made it very difficult for humble journeymen to rise in the guild's hierarchy. Without regard to the requirements peculiar to each corporation, the king announced that he would name as many as twelve deserving journeymen to mastership in each of the guilds. Unlike earlier "creations" of royal masterships, this one does not seem to have been primarily motivated by fiscal needs.[19] The government appeared genuinely interested in calling attention to the plight of poor journeymen.[20] One of the goals of the edict reestablishing the guild system after the fall of Turgot in 1776, according to its preamble, was to facilitate accession to mastership. Louis XVI promised that the right of mastership as well as the sundry costs of admission to the guild would henceforth be "reduced to a very moderate level" and would no longer be "an obstacle" to corporate mobility.[21]

The Reforms of 1776

Beginning in August 1776, the fee for baker mastership was fixed at five hundred livres, a four-hundred-livre reduction from what it had been, according to the edict. The new administrators of the corporation—syndics and *adjoints*—were forbidden to demand a greater droit or to exact any meals, gifts, or money. If one is to judge by the number of new masters admitted in the following years, as we shall shortly see, then clearly it did become easier to penetrate the guild. One hesitates to claim that extortion and corruption disappeared. They seem, however, to be far less common in incidence, in part because the guild was more rigorously policed by the Châtelet, in part because there was a genuine change in the moral and political climate after the convulsive passage of Turgot through the *contrôle général*, the lead ministry. Jean Lanier and Jean-Baptiste Vosniez each reported to have paid six hundred livres to become a master in the 1780s. In these two cases there seems to be no disproportion between the official fee published in the texts and the amount actually paid, for, by an edict of 1782, the king authorized each guild to raise its mastership charge by one hundred livres, which thus made the baker droit worth six hundred livres.[22] In addition to the mastership fees, a candidate had theoretically only two other payments to make in all: twenty livres to the admissions panel presided over by the syndics and twenty-four livres to the royal procurator who received his oath.

If aspirants succeeded in obtaining masterships more easily after 1776, it was not only because it was cheaper. The rules were fundamentally changed. The Manichean advancement system was to a large extent abandoned. In formal terms, the only advantage that the son retained was that he could receive mastership as of age eighteen, two years before the outsider who had completed a three-year apprenticeship and seven years before the journeyman without an apprentice's license. The latter possibility implicitly rendered contract apprenticeship optional.[23] Both the masterpiece and the son's *expérience* were abolished. An aspirant now had to submit to an oral examination of at least two hours on all matters relating to the profession conducted by a five-member admissions panel that reached its decision by a plurality vote. Finally, to qualify for promotion, the aspirant had to present the panel with testimony of his "good conduct and mores" by two master bakers and two "bourgeois notables" from outside the corporation. Moreover, women could now be admitted to the guild on their own, not exclusively as widows of masters, provided they were at least eighteen years old and satisfied all standard requirements. Widows could freely exercise their husband's mastership for one year following his death, after which they had to apply for mastership on their own account through regular channels.[24]

It became easier and cheaper for outsiders to obtain masterships after 1776, though I think the stunning increase in admissions was due less to fee reduction per se than to a major internal reorganization of guild procedures and to the intense publicity to which corporate life was henceforth subjected. In 1776 the era of open admissions begins. In principle there are no more distinctions between outsiders and sons, though it is still likely that corporate leaders made it easier for sons to pass the examination and obtain certification rapidly. In any event, the records of the Châtelet no longer recognize or note any difference in the socioprofessional origins of candidates or in their modes of ascension. Between 1776 and the Revolution, 615 new masters joined the baker community at an average rate of 43.9 a year, more than twice as many as in the period from 1736 to 1775. Twenty-four women (4 percent of total admissions) became masters after 1776 at the rate of 1.7 a year. Almost two-thirds of them were widows of masters.

It is difficult to assess precisely the significance of this surge of admissions. It testifies to the frustration of many journeymen who yearned for advancement but lacked the courage or the means to envisage it under the old system. These were the journeymen who celebrated Turgot's *coup de force* with a mixture of unbridled joy and violence. Surely many of them looked beyond settling scores with masters to the prospect of working for themselves or working for employers who could not rely on corporate structure as a means of subordinating and domesticating them. To many of these journeymen the reestablishment of the guilds only six months after their suppression must have been a terrible blow.[25]

Establishment: Acquiring a Shop

For some, however, there was consolation in the new opportunities that the reorganized system offered them. The number of admissions doubled after 1776 because many journeymen managed to take advantage of these opportunities. Nor did this enthusiasm spend itself quickly. The number of new admissions remained quite high—above forty-eight a year—through 1782 (the peak year: sixty-three new masters). But how many of the new masters actually established themselves in shops? How many survived the first year or two? These are the critical questions beyond the raw mobility statistics that we cannot answer.

For a young baker who wished to establish himself within the corporate jurisdiction of the capital, mastership was a necessary but not a sufficient accomplishment. He needed a bakeroom with its equipment, a shop, clients, an attractive business opportunity. He also needed circulating capital or a significant dose of credit in order to acquire raw materials. Vincent Barbacois, master baker, waited for years until a small inheritance enabled him to rent and equip a shop. Even then he desperately searched for a hundred livres in cash to start up.[26] In the eighteenth century very few new masters started out from scratch. All the best spots in the city had been staked out for years and years, consumer buying habits had long been fixed, and the guild discountenanced new installations that would increase competition and thereby menace the well-being of senior masters. Instead of starting up their own enterprises, young masters characteristically purchased the businesses (*fonds de boutique*) of older bakers who ceded their places in order to retire or to move to a different area. These "successions" led to a remarkable continuity in the pattern of implantation and operation of the shops.

To track down and purchase a going business, one needed good contacts in the profession and a substantial amount of money. One could obtain a succession for as little as 350 or 400 livres, but it would most likely be a business in shambles, utterly lacking in fixed capital investment and in community following, or a lethargic business operated in a marginal location. Certain flourishing enterprises, by contrast, commanded more than six thousand livres. For forty-one cases in which the actual contract price is known, the average cost of a going concern is 1,661.6 livres, the median 1,330 livres. If one adds prices culled from sixty bankruptcy petitions and one letter of respite—prices that may have been somewhat bloated to impress creditors with the underlying solidity of the bakers' economic situation—then the average rises to 1,934.3 and the median to 1,500 livres.[27] While there was a genuine market for fonds, we ought to note that the purchase price (apart from ex post facto distortions) did not always neatly reflect supply and demand factors. Transfers often took place within families or between members of closely associated groups, and the price in these instances seems to be as much a function of kinship exigencies or mutually beneficial financial and commercial arrangements as of so-called market realities.[28]

Normally the price of a fonds consisted of two elements. The first represented the tools, equipment, and all other objects relating to baking (save merchandise in flour and grain, which was always considered separately). The second component was called the *pot de vin* (literally "jug of wine"). In the modern sense the term pot de vin generally signifies a bribe or hush money. But in our context it stood for the commercial value of the business apart from its fixed capital investment. The pot de vin was meant to be compensation for the years of hard work of the baker-seller, his good reputation, and his more or less elevated level of earnings. Or, as it was often defined in the contracts, the pot de vin was "the price of the clientele" that the seller passed on to the buyer.

Most contracts do not distinguish the equipment component from the pot de vin, although they usually hint that the latter is more important than the former. The average price of equipment for six cases for which the data is available is 1,044 livres, or 54 percent of the average price for forty-eight business transfers.[29] My guess is that for all forty-eight cases the average part for equipment would be considerably lower, probably in the range of seven to eight hundred livres. The high prices cited for equipment are sometimes quite inflated. Upon examination of the summary inventory (and in comparison with others), for example, the three-thousand-evaluation of the utensils of Françoise Constant, the widow of a master baker, appears overblown.[30] Now, widow Constant sold her business to her son, to whom she owed money coming from his father's estate. In order to reduce her liability to her son she may have purposely exaggerated the value of the equipment in the business. The widow of master baker Pierre Lepage, who also owed her son a share of his father's estate, seems to have done exactly the same thing in setting the price of the bakery matériel she sold to him at fifty-seven hundred livres.[31]

Though the "utensils" included in the sales varied from contract to contract, a number of items were common to most of them. Virtually every buyer acquired at least one kneading trough, and sometimes two or three (the best were made of oak); scales and weights to weigh the dough; dough cutters, scrapers, and knives; between twenty and one hundred *bannetons*, or wicker baskets, long or round in shape, in which the dough for certain loaves was placed to rise; a dozen or more *couches* (the word evokes both the birth of a baby and the baby's diapers), wooden planks or drawers (or even tables) on or in which the dough was placed and sometimes covered with a cloth either to proof or to await baking; long- and short-handled flat wooden and metal shovels for placing dough in the oven and removing bread, for removing embers, and for conditioning flour; storage cabinets or shelves; a flour-storage trunk; baskets for delivering bread and transporting flour; vats and pails for water storage and transfer; a metal pail for extinguishing embers and a trunk for storing them; a number of brooms; a set of lamps; a yeast box. Many contracts included an oven; several mentioned a second oven, which in some cases was located outside in the courtyard and destined for warm-weather baking.

Some bakers were fortunate enough to acquire shop furnishings as well as bakeroom tools: a counter and drawer; an iron grill for the facade; a screen to separate the shop from the anteroom; a bread cutter, shelves, and storage boxes for retail sales of flour, bread crumbs, crusts, and embers; a measure for the flour. Finally, a number of sales included rights of well usage and a well cover, mattresses for baker boys, an encased clock for the shop counter and an alarm clock for the bakeroom, sacks for flour purchase, flour-bolting apparatuses, and instruments for cleaning wheat such as vanning machines.[32]

Ceding Clients

In return for the pot de vin the seller made two promises that were of paramount importance to the buyer. With the first the seller passed on his clients to the buyer. In most contracts this transfer was referred to as an actual "sale" of clients with the stipulation, however, that the sale was made "without any guarantee." That is, once the seller presented the buyer to his clients and introduced them to him, he had discharged his obligations. It was up to the successor "to get himself accepted as bread supplier by the customers at his personal risk and fortune." There was no formal "trial period" prior to purchase until the nineteenth century, but the serious purchaser studied the business carefully before making an offer, visited periodically to survey the level of business, and consulted with neighbors to learn about local preferences and habits.[33]

Usually the sale of the fonds involved the cession of a shop-based clientele. In two of our contracts, however, the sellers transferred their stalls in the bread market, along with their regular market customers, to the successors. The stalls were concessions that the lieutenant general of police was supposed to award to "meritorious" bakers of his choice rather than alienable property. In practice, however, bakers treated the market assignments as conveyable possessions. In the first of our two cases, the market transfer was merely accessory to the sale of the shop and its clientele. But in the second contract, concluded between faubourg Saint-Antoine bakers, the market clientele and commerce were probably more important than the shop business. In light of the market stall's significance, the seller pledged to introduce the buyer to Trudon, the police commissaire in charge of the Saint-Paul market, in order to assure that he would succeed to the stall.[34]

The second promise exchanged against the pot de vin complemented the first. With the first the seller ceded his clientele to the baker without any guarantee. With the second, however, he vowed that he would not attempt to retain or win back any of those customers. The seller solemnly renounced the right to engage in the bakery business for a period that ranged from six to ten years ("forever" in one contract) in an area near his former shop. The spatial taboo varied substantially from contract to contract; its breadth seems to have been roughly inversely proportional to the amount of confidence the seller inspired in the buyer. One contract vaguely puts off-limits "neighboring streets"; another, concerning a shop on the rue Dauphine, prohibits installation in the parishes of Saint-André, Saint-Sulpice, and Saint-Germain-de-l'Auxerrois; a third, in support of the shop in the Place Maubert, declares a ban in the parishes of Saint-Etienne-du-Mont, Saint-Séverin, Saint-Nicolas-du-Chardonnet, Saint-Yves, Saint-Benoist, and "other adjacent parishes."[35]

Presumably, the buyer would resort to the courts against the seller who abjured his renunciation, though the precise remedy is not specified. But recourse to such a remedy was time-consuming and costly. As master baker Louis Fairet learned, one could lose everything while engaged in legal proceedings. The former owner of the shop that Fairet had purchased on the rue Montmartre, "against his word of honor," established a new business within the territory from which he was banned by the contract. Fairet complained that this treacherous competition from his predecessor, who retained the fealty of his old clients, caused his "commerce to fall into ruin." I suspect that there were many other cases of alleged betrayal and quarrels over territorial rights. The guild probably intervened to resolve some of them out of court. Had the contract specified in advance the penalties for violation, the incidence of such conflicts might have been substantially reduced. In only two of our contracts was there such a concrete stipulation and warning. In the first, the seller was forbidden to set up commerce either in the faubourg Saint-Germain or in the bread market of the quai des Augustins on pain of five hundred livres in damages, a surprisingly paltry level of sanction. In the second, the seller pledged not to do business in a multiparish area on penalty of restitution of the entire sale price of his fonds, a more credible deterrent.[36]

Nor were these arrangements always devoid of ambiguity. In the contract ceding his shop on the rue du Four, Saint-Sulpice parish, to master baker André François Leroux, Pierre Brot "renounces in good faith the right to supply bread in the future to any other customers attached to the aforesaid shop and the right to open a bakeshop for a period of ten years beginning next 1 July in the entire expanse of the Saint-Sulpice parish." Nevertheless, Brot had no plans to retire, and he made it plain that he meant to keep all those clients and "to conquer new ones" within the area he esteemed to fall in the sphere of commercial influence of his new shop. Leroux beware![37]

This same contract also reveals how complicated transfer arrangements could become. André François Leroux continued to work for and with his father, a master baker on the rue de la Huchette, for several years after he attained his mastership. Now twenty-three years of age, he wanted to establish himself independently. He coveted Pierre Brot's business, but even with his father's help, he lacked the money necessary to make the purchase. Meanwhile, another master baker named Pierre Bonjean, who had a considerable fortune, was ready to retire. He found a buyer for his shop in the person of Pierre Brot, but Brot probably made his offer contingent on his ability to sell his own fonds and raise the extra money needed to buy Bonjean's (Brot was asking twenty-seven hundred livres and Bonjean four thousand livres). Ultimately, Bonjean solved everybody's problems by financing both purchases. He lent twenty-five hundred livres to Leroux (whose father pledged the shares he had in two mills as collateral) and four thousand livres to Brot. The loans were masked as rentes in the amount of 125 and 200 livres respectively, a nice annuity for Bonjean in his retirement.[38]

Acquiring a shop was for most young masters an enormously trying and costly operation. Save for a favored few, the initial expenses were weighty and the first years of independent establishment were rocky times. Jean Galland evoked a situation common to many fellow bakers when he recalled "the losses that are the infallible consequence of the beginnings of the profession of Baker." Louis Martin Meusnier had a painful memory of his early years. He began without a familial or financial cushion. He suffered heavy losses because he lacked "starting capital" and because he "was engaged in a profession that he did not yet know." Some young bakers who lacked funds of their own had the good fortune to obtain advances from their families. Others had to contract loans that they had difficulty repaying. Master Nicolas Eloy had to borrow 540 livres from a valet de chambre to set up his shop. Master René Morin had to borrow a thousand livres for his mastership and another thousand livres to "set up a shop on the rue de la Lingerie and pay the pot de vin."[39]

Nor was the problem of establishment a narrowly financial one. Paying for the fonds was only part of the job. The young baker then had to profit from the franchise that the purchase of the fonds awarded him. He had to convince the customers to give him their allegiance, he had to run a complicated business with both manufacturing and commercial components, and he had to be prepared to face what Meusnier called "the misfortunes of the moment."

Changing Fonds

A majority of the bakers appear to have remained in the fonds in which they first planted their roots. Yet, despite the heavy initial investment in money, time, and passion, a considerable number of bakers moved about the city. Some of this mobility was the fruit of ambition, but most of it seems to have been the result of failure: hard times, heavy debts, bad management. Master Jean Lanier abandoned his first fonds in the Marais, for which he paid two thousand livres. He withdrew to the outskirts to reestablish himself, purchasing a business at La Courtille for six hundred livres. Then he invested fifteen hundred livres in a fonds on the rue de la Calandre. Master Pierre Dufour's first shop on the rue de Buci, appraised for 2,874, burned. He invested fifteen hundred livres in order to rebuild it. Then he moved to the rue des Orties, where he paid twenty-two hundred livres for the installation and two hundred livres in repairs. He ended up in the rue Fromenteaux in a business that cost around three thousand livres. Master Claude Gerard went through three fonds in ten years, all in the city center. Burdened with debts from his first shop location, René Morin moved three more times in search of an elusive stability and success. Masters Jean-Baptiste Chenart and Joseph Françoise each changed fonds in hopes of doing better. Jean Michel Thevenon abandoned his shop on the rue de la Ferronnerie after a three-year effort to make it profitable. "In search of his fortune," he set up shop on the rue Sainte-Margueritte, where his efforts proved equally abortive.[40]

Leases and Rents

By itself the fonds had no value even if the contract was scrupulously respected by both parties. It had to be accompanied by the transfer of the lease that the seller held on the shop and its dependencies (or the letting of the shop if the seller of the fonds was the owner), for the fonds had to have a physical emplacement as well as a commercial and moral content. Virtually every sale of a business was predicated on the buyer's ability to obtain a lease of the same property from which the seller had exercised his business. In numerous instances, the owner of the property who agreed to the lease transfer from the seller of the business to the buyer (or the seller himself if he was proprietor) demanded a whole year's or six months' rent in advance. Thus the cost of acquiring the fonds was not the last major expense that the young baker faced prior to the burdens of actually beginning his commerce.[41] In addition, in many cases the baker also had to contract to have his oven built at a cost ranging between 100 and 225 livres and/or engage in diverse repairs in the bakeroom.

The average baker's rent in fifty-six leases signed during the eighteenth century was 763.9 livres (the median stood at 618 livres). The rents ranged from 140 livres in the rue du faubourg Saint-Lazare, 170 livres in the rue du faubourg du Temple, and 192 livres on the rue de Charenton in the faubourg Saint-Antoine on the low side of the spectrum to 3,200 livres on the rue Aumaire, 1,800 livres on the rue de Beaune, and 1,465 livres on the rue Dauphine on the other.[42]

On the surface, it appears that properties in the northern and eastern faubourgs were cheaper than those located in central Paris on either bank or in the west. But it is not an easy matter to plot a geography of rents, for one finds leases of three hundred livres on the rue Saint-André-des-Arts, the rue Mouffetard, and in the Marais, while one encounters leases amounting to twelve hundred livres in the faubourg Saint-Antoine. Moreover, one must compare space as well as place. A cheap lease may involve merely a shop, bakeroom, and one or two other rooms for living and/or storage, while a dear rent may represent an entire house. The lease for eighteen hundred livres on the rue Aumaire, for example, consisted of a huge house comprising two main buildings (*corps de logis*) connected by a passage, four stories of rooms and an attic of storage rooms, four shops and other rooms on the ground floor, a bakeroom, a latrine in the courtyard, and so on. Baker Jean Go paid only five hundred livres, by contrast, for his lease on the rue du Vieux-Colomnier, but he had exiguous quarters (a shop, anteroom, bakeroom, and one room) and was not even principal tenant. Similarly, Michel Desbordes paid only 280 livres a year for a cramped installation on the rue du Bac.[43]

The widowed mistress baker Jeanne Simon rented her entire house on the rue Bodet, consisting of a shop, bakeroom, three storage rooms, three stories of living quarters, granaries, and latrines, for only 350 livres a year. But she may have asked for less than she could have obtained on the open market, for she rented to her son and daughter-in-law. The conditions of the quarters also mattered. If Jean Legrand paid only 170 livres on the rue du faubourg du Temple, it was in part because the owner was rebuilding the house, resulting in enormous inconvenience for the baker, who was obliged to move his bolters into another house in order to keep debris out of the flour.[44]

The general terms of a contract were usually the same. The lessee was required to furnish his quarters with possessions (including merchandise in grain and flour, some contracts allowed) of sufficient value to serve as collateral for the regular payment of the rent. He was responsible for making repairs (major as well as minor, a number of leases stipulated) and for discharging all fiscal and civic obligations such as the poor tax (*taxe des pauvres*). A number of contracts specifically forbade the lessee to engage in or sublet to persons who engaged in certain kinds of activities (e.g., artisans whose hammering made loud noise or whose products caused pollution of the air or water or damage to the property). Several contracts expressly authorized the tenants to build ovens in the basement or bakeroom. But, since the lessees were also required to return the property at the expiration of the lease time in exactly the same state in which they received it, those bakers who planned to build ovens had to demolish them before leaving the premises definitively. If the owner did not know the tenant or entertained some doubt about his capacity or his good faith, he demanded that the tenant present cosigners, or *cautions,* to guarantee payment.

Leases were signed for between three and nine years, the latter term being by far the most common. Several leases were for five and six years; the two shortest terms were set at three years each. The average length of our fifty-six contracts was 7.3 years. Owners were generally willing to liberate bakers in difficulty from their leases provided they gave at least six months' notice. It was not surprising that Jean Delu, bourgeois de Paris, refused to discharge a baker named Becheret who wanted immediate freedom from a lease that had seven years to run. Infuriated at his landlord's obstinacy, Becheret threatened him physical harm and called him a "thief, son of a bitch, and other atrocious and scandalous insults against his honor and reputation."[45]

Establishment by Inheritance

Launching a master's career was quite often a family affair. Sons frequently succeeded their fathers or widowed mothers, either as a result of the father's death followed by an estate settlement geared to launch the son's career or as a consequence of the sale of the business by the father or his widow to the son. The Jombert estate division (*partage*) illustrates the first case. When master baker Antoine Jombert died, his son Jean-Baptiste, also a master baker, took over the business at the request of his four siblings in order to operate it for their "common profit." When the estate was apportioned, Jean-Baptiste received all the bakery equipment, some grain and flour, and a large number of his father's accounts receivable for bread supply. As a reward for the pain he took managing a shop before it actually became his, his brothers and sisters awarded him the fonds ("with all the customers attached to the shop") as a bonus over and above his share of the estate.[46]

Numerous masters passed on their fonds to sons before they died. Given "the resolution of Sieur Guy Leguay to retire from a trade that an advanced age no longer allows him to practice with the same energy of which he used to dispose and on the proposition made by his son Sieur Pierre Leguay to replace him in the continuation of his business in order thereby to enable him to set himself up in a situation commensurate with his age [twenty-five years] and with his status as master baker of this city," the father agreed to pass on not only his fonds but also merchandise and accounts receivable. About half the price was covered by Pierre's mother's estate (4,158 livres). The father, who had a fortune ranging between thirty and forty thousand livres, financed the remainder by making his son a loan to be repaid in the form of a rente.[47]

Though it is not clear that master Jacques Chartie's business had a substantially lower market value than Leguay's, Chartie made it easier for his son to take over by fixing the price at seven hundred livres (419 livres for the equipment and 281 livres for his customers both at the shop and at a stall in one of the bread markets).[48] Nor did the son have to pay cash for the purchase: it would eventually be deducted from his father's estate. One reason the father may have been so tender with his son is that the latter worked for him for five years after attaining mastership without taking any wages. Widow Constant also set up a high price on the business she passed on to her son Jean. Of the 6,274 livres she demanded, more than half was absorbed by the share mother owed son in her late husband's estate. Jean had a year to pay the balance at 5 percent interest. Jean Estienne Dupuis sold his business to his son for 1,737 livres, but he required only a down payment of fifty livres and set no timetable for acquitting the balance.[49]

Some of the parents—especially those who were left alone by the death of their spouse—used the occasion of the business succession to obtain a certain security for themselves in old age. In a leasing arrangement appended to the fonds contract, Guy Leguay demanded that his son reserve for him three rooms of his choice for himself, his servant, and his property. Widow Constant retained a room for herself as long as she lived. Widow Guerin imposed the same condition in the cession of her business to her daughter and son-in-law. Shortly after his wife's death, Alexander Tribouillet passed his business to his nephew, Pierre Nezot, who lived with him. Instead of extracting a fixed price, Tribouillet imposed a much longer term transaction. In return for the business, Nezot was to lodge his uncle, provide him with heat and light and medical care, do his laundry, and nourish him at the family table.[50]

These cases suggest the existence of three stages of family organization in certain baker families in which the son eventually takes over the business. In the first stage, the baker parents train the son in their profession and profit from his labor. This process begins in childhood and can last until the mid-twenties (when the son incidentally reaches legal majority). Phase two overlaps with the first, for it begins with the son's marriage and the entry of a wife into the parental household. Marriage frequently heralds the son's succession to the business, though the parents can delay the transfer (by making promises of eventual succession or by proposing accords of commercial association) for a number of years. While he is awaiting the opportunity to force his parents (or parent) out of the business, the son founds his family and begins to have children. The household takes on an extended physiognomy for a while.[51] In phase three the son displaces his father (or mother) as head of the business and the household. But the family retains a stem character as the parents (parent) remain until death as quasi-pensioners in the household. Subordinate in phase one, the son lives in a certain tension with his parents in phase two and eclipses them in phase three. The parents dominate and exploit the son in phase one, humor him and negotiate with him in phase two, and defer to him (but not unconditionally) in phase three.

Establishment by Marriage

Marriage is often the signal and occasion for the transfer of a business and the launching of a career. In Jean Charles Delaitre's marriage contract, Delaitre's parents pledged to cede him "their clients and commerce" at the moment of his wedding. "In consideration" of his niece and ward's marriage, master baker René François Saunier gave the newlyweds the right to exercise his commerce at their profit. In return they promised to provide Saunier with a four-pound cake every year for Epiphany.[52]

As part of her dowry, Marie Françoise Bassille received her father's bakery business, which she planned to operate with her husband, Guillaume Rousseau, a master baker. Fonds constituted marriage portions for brides or grooms in ten other marriages that are part of our sample of 160 contracts. Remarriage also provided an opportunity to redeem or relaunch an ill-starred career. For example, Nicolas Bourlon used the estate settlement of his second wife along with the dowry of his third to purchase a new fonds and begin all over again.[53]

Master Martin Nivet invited his son-in-law, master Etienne Garin, to live with him and take over his business. As an inducement, he dowered his daughter with his baking equipment. But Nivet imposed a condition: he obliged Garin (in contravention of the corporate rules against association with persons lacking "quality") to give his other daughter, who was unmarried and lived at home, a one-third share in his commerce.[54]

I suspect that a number of parents and parents-in-law covertly insinuated themselves into the business operation that they theoretically sold and transferred to their children. This, too, was a form of old-age insurance. In some cases it might help explain the remarkably moderate face value of the fonds contracts. One contract frankly defines a "society" between mother-in-law and newlywed couple and illustrates the way in which the collaboration took shape. Marie Anne Legrand, twice the widow of master bakers, married her daughter Anne Perrette to Jean Blezimas, a Paris flour merchant whose father had also been a flour merchant. Written into the marriage contract was an agreement by which, beginning on the day of the nuptial celebration, the couple and the widow joined together to exercise "the commerce of bakery under the name of the said dame widow Legrand." Each of the two parties was to subscribe to one half of the capital of the partnership (in cash, merchandise, or tools) and they were to share equally in the profits and losses. The "society" was to exist until the death of widow Legrand, after which her daughter and son-in-law would succeed to the fonds and operate it as their own (illicitly, however, for she could not bequeath her mastership status to her daughter or her son-in-law).[55]

Extracorporate Establishment

An ambitious journeyman who disposed neither of substantial resources nor of the patronage of family or friends could envision an entirely different strategy of mobility. Instead of opting for the arduous pursuit of mastership and then attempting to acquire a business in the corporate jurisdiction of the city, a journeyman could seek to establish himself as a baker outside the corporate frontiers, either in one of the open districts (Saint-Antoine was by far the most important) or in the suburbs or the nearby hinterland.[56] To set up a business in one of these places, a journeyman had to satisfy a number of police regulations regarding the manufacture and sale of bread, but he did not have to serve an apprenticeship or probationary term, pass an examination, or pay any special fee (beyond local taxes to which all citizens or tradesmen were subject). In theory, the young baker was *free* to establish himself wherever he wished without regard to any authority other than the police; as a rule, to promote the cause of "abundance," the police encouraged the implantation of "foreign" (forain) bakers on the outskirts of the capital or in the countryside.

This sort of establishment was naturally less trying than a corporate installation. The beginning baker saved the heavy cost of formal training and especially of mastership. Though some of the shop-house rents in the faubourg Saint-Antoine were competitive with those in the corporate territory, as a rule they were cheaper outside the city proper, especially in the suburbs and the hinterland. For a whole house including the shop, a bakehouse, granaries, a bolting room, stables, a courtyard, and four rooms, Jacques Vavesseur, a baker at Issy, paid one hundred livres a year. Baker Denis Poisson of Ivry, paid 220 livres for an even larger house (which boasted a mechanical system for loading grain and flour into the storeroom). Since the shop was generally much less important to the forain (and in a few instances to the faubourg baker as well), on average the cost of purchasing a fonds was substantially lower than in the corporate jurisdiction.[57]

Indeed, in some cases the young baker did not even bother to acquire an ongoing business (though in most instances he had to procure a space at the Paris markets). He merely staked out his ground in a spot he judged promising and he commenced his trade. The forain baker had fewer obligations and suffered fewer constraints, both moral and material, than his corporate counterpart. His chances of constituting a considerable fortune were perhaps not as great, but he had a calmer life. Thus it would be wrong to view the forain as a *raté*, a failure who could not make it in Paris. Among the forains were not only journeymen who could not achieve mastership but also sons of masters who preferred to renounce the guild and the city (fleeing the type A baker personality?).[58]

Yet this sort of commerce outside the corporate sphere was not as easy to launch as it appears on the surface. Theoretically, the arena was "free." But few young bakers had the audacity to venture into strange territory without the support of family or friends. Indeed, much of the best terrain was already claimed by forain families with the same dynastic pretensions and the same ferocity of exclusiveness that characterized some of the guild families. The "foreign" bakers of Gonesse, Bonneuil, Versailles did not like outsiders. One had to marshal influence as well to acquire a market spot. One needed support to obtain credit from the local merchants, and while it was clearly less expensive to set up beyond the corporate jurisdiction, one still required funds for the basic tools of the trade and for living quarters. Thus it would be wrong to depict the faubourg Saint-Antoine or Vincennes or Gonesse as El Dorados for young bakers hungry for upward mobility.

Illicit Establishment

There was still another avenue (or, more precisely, a bypass) that could lead the young baker to a more or less independent establishment *within* corporate territory. What was peculiar about this alternative path of mobility was that it was decidedly illegal according to guild regulations—regulations that, because they were ratified by the Châtelet and the parlement, had the force of law. The statutes specified that only a baker with a proper "title," that is, with a rightfully acquired mastership, could exercise the profession of baker in a business of his own within the guild bounds of the capital. They also prohibited masters to "lend their names" or "let out their mastership directly or indirectly" to anyone or to "associate with another person who is not a Master." Implicitly masters were also required to reside in the same place where they baked their bread, to operate their own oven, to sell only the bread that they themselves baked.[59]

The journeyman bakers who opted for this illicit enterprise and their accomplices violated guild rules in one way or another. These spurious establishments took three main forms. First there was the premature transfer in which the owner of a master's fonds sold it to a journeyman who was "awaiting mastership." Because legitimation was anticipated, the gravity of the transgression was considerably mitigated. The second type of underground establishment, a prevalent one, was the fruit of an actual contract by means of which the holder of a fonds and/or "right" or "privilege" transferred one or both of these possessions to a young baker for a limited period or forever in return for a fixed price or a term lease. The third and least common type was in some ways the most audacious of all. The journeyman simply ventured to set up shop in corporate territory on his own with practically no cover.

Jean Leblanc and Gabriel Lepileur were the principals in one of the most striking cases of premature transfer. Leblanc, a master baker on the rue du Four, sold his business (for 1,500 livres) and sublet his quarters for five years (for 600 livres) to Lepileur, a mere journeyman baker who still had almost two years of probationary time to serve before he could pose his candidacy for mastership. Each man tried to protect himself carefully, though neither seemed to have perceived himself as especially vulnerable. Leblanc and his wife promised to continue living in the house now rented by Lepileur "and to protect the said Lepileur for two years." In fact Leblanc was to serve as a front for Lepileur: the latter was to operate the commerce at his own risk and with his own money, to stand responsible for all penalties and to pay all charges, but because he lacked the title authorizing him to function professionally he would run the business (until he obtained mastership) "under the name of the said Sieur Leblanc." This transaction was unequivocally illegal, but the fact that both parties signed an open contract hiding nothing before a notary suggests that the premature transfer may have been practiced fairly commonly.[60] It is interesting to note that although Lepileur became a master at the end of 1760, for reasons that remain obscure, he retroceded the fonds and the lease to Leblanc.[61]

Vaguely uneasy, Leblanc had insisted that Lepileur find someone of substance to stand surety for him, not only to guarantee payment but also to take responsibility for "all events of any sort [that might arise] concerning the said business." Similarly, when François Noël, a journeyman baker awaiting mastership, sought to borrow fourteen hundred livres from an oven maker named Nicolas Hautelin in order to purchase a master's fonds, Hautelin hesitated. A simple note would not be sufficient; he wanted Noël to find a person to stand surety. "I would like to deliver this shop to you," Hautelin told Noël, but "you must furnish me good security because you are not a master and your property could be legally seized at any moment." The jurés began proceedings for the confiscation of the tools and merchandise of Gentien Aubert, a journeyman baker who had purchased a master's business on the rue Montmartre. At the eleventh hour Aubert was saved by his uncle, a "cavalier of his eminence Monseigneur Cardinal de Fleury," who paid three hundred livres cash to the guild for his nephew's mastership.[62] Journeyman Gilles Pasquier courted fewer risks than the others, for he was only three and one-half months away from receiving his letters of mastership when he purchased his widowed aunt's business.[63]

There were numerous instances of the second type of illicit establishment resulting from the acquisition of a business by a journeyman baker *without* any real prospect of attaining mastership. Widows of master bakers, who lacked the means or the desire to carry on the business, were frequently the other party in these transactions. Claude Gillain was only seventeen years old and had barely finished his apprenticeship, but he called himself a master baker and operated out of a shop belonging to widow Noiseux of the grande rue du faubourg Saint-Laurent. Challenged to justify his professional authenticity by the jurés, Gillain took refuge in the convenient (and sometimes difficult to verify) fiction that he was merely working for Noiseux as her first garçon. But the registers of bread sales were in Gillain's name and all the flour sacks bore his mark as well. More devastating was the unvarnished testimony of widow Noiseux, who said that Gillain was not her employee but, rather, that "he borrowed her name," under which he ran his own business. Another widow lent her bakeroom and her credentials to a journeyman named Pierre Descolas who used them to launch his own business. In both these cases the jurés seized all the equipment, merchandise, and bread.[64]

Like many mistresses (and masters), widow Teissier must never have read the guild statutes closely, for she staunchly believed that the mastership that devolved upon her after her husband's death was a marketable license or right that she could dispose of as she pleased. On the occasion of her daughter's marriage to a journeyman baker named Jean Blin, she "ceded and abandoned the rights and privileges of mistress baker" to the newlyweds on condition that they pay all of her guild obligations, including her share of the capitation tax, support her until her death, and provide her with a pension of fifty livres a year. Widow Teissier reserved the right, if she was dissatisfied with her son-in-law and daughter, to break the contract and "to rent out her said privilege to whomever she pleases."[65]

Like the widows, albeit far less numerous, the "privileged bakers" had an interest in providing fronts for journeymen seeking establishment. Appointed to "follow" the royal court and cater to its needs, the privileged bakers, theoretically a dozen in number, maintained their primary residence in Paris.[66] Frequently they opened shops in the capital, a practice the authorities appear to have tolerated. But a number of them tried to profit from their privilege by renting it out; some even attempted to parlay it by letting it out simultaneously—franchising it—to several journeyman bakers. The latter had a better chance to escape unnoticed by contracting with widows, for the jurés kept the privileged bakers under close surveillance in the hope of repressing what they considered to be abusive and unlawful competition. When the jurés challenged Claude Frion to justify his title as baker, all he could produce was the "lease of a privilege" let by a court baker for three years. The price was 130 livres a year, cheap for an establishment, yet obviously of precarious character. Another journeyman baker, Louis Loiselle, rented a similar privilege for 145 livres a year. Claude Magnier, a "baker without quality" on the rue des Vieux-Augustins, purchased his concession for 150 livres a year, as did journeyman Pasquier of the rue Saint-Honoré. Journeyman Jean Philippe Mutel, "without either quality or right," ran a shop in the name of Rollin Villemesse, a "boulanger de la garde du Roy." In all these cases the jurés, with authorization from the lieutenant general of police, closed down the shops and seized the matériel and merchandise.[67]

For lack of widows or privileged bakers, dauntless journeymen turned to the ample pool of master bakers who, for one reason or another, were willing to risk connivance or collaboration with them. Usually they rented or purchased a business from a master who was willing to lend his name as a cover. In some instances the master was too old to work himself and found no better way to support himself than to rent out his name one or more times. In other circumstances, the master continued to run his own business and regarded the journeyman's enterprise as a sort of branch operation. Prevost, a journeyman keeping shop on the rue du faubourg Montmartre, told the jurés that he merely worked for a master baker named Etienne Mongueret. The jurés knew that Mongueret had retired and that he lived elsewhere, so they seized all Prevost's possessions.[68]

Nicolas Frambois admitted "that he has no quality by his own right," but he insisted that he "only works under the authority given to him by Sieur Pierre François, master baker." But Frambois was not the simple journeyman he suddenly pretended to be. Pierre François himself confessed that "he had lent his name to Sieur Frambois for the past three years." After marrying his last daughter and selling his fonds six years earlier, François had fallen on hard times. Without the "rent" that Frambois paid him for the use of his name, he feared that "he would be forced to retire to the general hospital," a prospect that, like most Parisians, he deeply dreaded. Master Edhieres and a journeyman named Baron inverted their world. Baron purchased a fonds in Edhieres's name and promised, in a formula redolent of the apprenticeship contract, "to feed him, to keep him, and to make sure that he lacked nothing, and to give him a wage when he learned how to work better than he now knows." Master Jean Thevenot associated himself with a journeyman in order to obtain an interest in a second shop.[69]

Denounced by the jurés as an imposter for operating a shop on the rue Poissonnerie, Jean Aniballe swore that he was merely the geindre of master Charles Blanvilain. Blanvilain backed him up vigorously, but the room in which he claimed to live behind the shop was found to have no furniture or clothing in it. The jurés alleged that they had information according to which

Blanvillain lived near Les Porcherons and had sold his business to the ex-journeyman. Master Cousain and journeyman Fery had secretly exchanged roles, but when the guild offices accused Fery of running the business by himself and for himself ("it is Fery who without credentials . . . makes and sells bread, buys grain and flour"), Cousain stubbornly maintained that "it is he who exercised the trade of baker . . . and that Fery is only his journeyman." Charged with the same usurpation and fraud, journeymen Etienne Dubois, Jacques Gauguin, Rene Noury, Louis Peret, and Laurent Vivier, and the masters with whom they were associated, all told the same story.[70]

The jurés hunted down irregularities with a certain fury, but the fines pronounced against masters who "lent" their names surreptitiously to journeymen were rarely oppressive. The jurés asked for one thousand livres in damages against Blanvilain, a troublesome master with an abominable reputation. The procurator's chamber in the Châtelet ordered him to pay thirty livres plus court expenses. Another master suffered a three-livre fine for the king and a six-livre judgment in damages for the guild and in addition had his illegal oven dismantled and his shop closed.[71]

Finally, there were those enterprising journeymen who managed to set up shop without the cover and apparently without the assistance of a master. Louis Landrieux launched a small business in the faubourg de Gloire. He had enough capital to keep a mammoth stock of twenty-eight muids of flour, and he apparently employed two garçons. Yet Landrieux confessed, when tracked down by the jurés, that "he was only an apprentice in the city and that he had not yet completed his time." Michel Leblanc also admitted that he lacked "title and quality" but stubbornly resisted the guild's effort to close him down, for he had a thriving business on the rue des Fossez-de-l'Ancien-Estrapade. After a protracted courtroom battle, the jurés triumphed and wasted no time in walling up Leblanc's shop and dismantling his oven. Noel Grain and Pierre Cochois faced similar sentences for having undertaken to exercise the profession in the commercial underground.[72]

Was this alternative road to establishment a dead end? Surely all the journeymen who tried it were not exposed and convicted; our tainted evidence paints too bleak a picture. The fact that young bakers kept trying to rise this way throughout the century suggests that it was not a path of despair. Many of the covers were convincing; they afforded the upstart bakers protection if not legitimacy. For obvious reasons the guild was deeply concerned about this illegal activity. But the jurés by themselves could not root it all out. And they were not effectively seconded by the police. For the police, despite their general hostility to sub-rosa or *chambrelan* work, were not averse to the prospect of seeing the bread supply of the capital increased—even by these scandalous means.

Multiple roads led to establishment. The most prestigious arrived at mastership. Despite the vociferous allegations of the liberal critics, the guild was a relatively open institution that became even more hospitable to outsiders in the second half of the century (in part in response to anticorporative strictures and menaces). While it is true that sons had it easier in every way, aspirants who passed through the costly and lengthy channels of apprenticeship and journeymanship held a clear majority. Politically, the family connection favored the influence of the sons in the guild, yet genealogy by itself did not constitute a principal line of cleavage in corporate politics. Economically, the family connection facilitated establishment.

Establishment could be realized in other ways, legitimately in the territories juridically endowed with "privilege," and illegally in the corporate space itself, often in complicity with various guild actors. Setting up for masters and bakers "without quality" presupposed a certain capital investment and a marketing strategy. By and large, it seemed imperative to bakers of all origins in the city and faubourgs to have a shop as a base of operations in order to reach a regular clientele (though a minority confined themselves to the marketplaces or to street solutions). Given the density of existing baker implantation, it was difficult, albeit by no means impossible, to create a shop ex nihilo. The bulk of newcomers preferred to take over going concerns, thus stimulating a brisk market in bakery businesses (fonds). Inheritance, marriage, and kin, and friendship networks enabled many beginners to bypass or manipulate this market and launch their professional lives with a running start. One's social capital carried almost as much weight as one's money in the takeoff process. The trajectories of mobility remind us both of the significance of the guild in framing careers in the public sphere and of the limits of its influence in organizing economic activity.

The transmission of business played an important role in regulating the professional life cycle. Retiring masters and widows used their leverage to obtain favorable arrangements for their care in old age. As brokers of establishment, parents and patrons also retained considerable influence over the matrimonial decisions and the commercial choices of the young baker. The costs of more or less subtle dependence on others proved to be exorbitant in the experience of numerous bakers. In the end, they returned for a second (or third) chance to the marketplace for businesses, preferring the uncertainties of impersonal exchange to the choking familiarities of the kinship nexus.

Chapter 11

Marriage Strategies and Family Life

No decision a baker ever faced was more critical than the choice of a wife. In many instances, his professional establishment and his social destiny depended on it. He needed the liquid capital or the *fonds de boutique* (the baker's shop and practice) or the market access that his bride contributed through her matrimonial portion. Or he required the help that members of the bride's family and its allied networks could provide: technical professional assistance, commercial connections, credit, and so forth. Or he craved the prestige in the community that a certain union could afford him. The intrinsic merits of the wife also mattered to him, not only in terms of affective bond but as an occupational concern. In most baker households the wife played a key role in the affairs of the shop. An adroit woman could make the difference between success and failure by way of the manner in which she dealt with customers as well as suppliers or managed accounts receivable and payable.

Marriage was the most serious business affair of the baker's life. Given the importance of the affair, it was rarely arranged by the groom on his own, at least not in most first marriages. This business affair engaged the honor, the reputation, and to some extent the fortune of the whole family. Thus it was a family matter, carefully discussed in a family council presided over by the father, in his absence by the mother, or in the case of a minor, often by the guardian(s). Marriages were prospected, negotiated, and then contracted before a notary. It is not too much to say that certain families had veritable marriage strategies, meant not merely to further the interest of individuals in the family but to serve the general or collective interest, in the short and long term, as well.

Endogamy

Of 154 marriages for which we have full information, bakers married daughters of bakers in a little less than half the cases (46 percent). Thus, if we are correct in insisting on the primacy of business and family interest in contracting marriage, endogamy must not have been viewed as the sole criterion for matrimonial success. Nevertheless, endogamous marriage was attractive to a large number of bakers. The endogamous act flattered the baker's sense of belonging to an exclusive group. In addition to enhancing the couple's self-esteem and (perhaps) their status, the endogamous marriage was obviously of great professional value, for it brought the baker a wife already familiar with the trade and a family of in-laws with experience and connections in the business. The psychological value of endogamy may have been greatest among the masters, who had a clearly defined sense of group, of belonging to a corps with privileges and prerogatives of which its members were deeply jealous.[1]

Yet only a little over a third of the masters married endogamously. Of eighty-one masters in our sample (see table 11.1), twenty-nine married daughters of master bakers. Five masters married women below their socioprofessional status, the daughters of plain bakers. The dowries of these women varied slightly, above and below the average for all dowries in this group, so material incentives do not seem to account for these hypergamous unions. More than half of the master contingent, 58 percent, or forty-seven grooms, married entirely outside the bakery. The role of endogamy was slightly higher among masters who were sons of master bakers (thus at least second-generation masters) than among the whole master population. Fourteen, or 40 percent, of the thirty-five sons of masters married daughters of masters.[2] It may be that the second-generation masters had a keener sense of group identity and solidarity than newly arrived masters. In theory, however, it could be argued that first-generation bakers would be highly motivated to consecrate their upward mobility by marrying into master families. Twelve of the thirty-nine masters who were the sons of nonbakers did just that. Yet twenty-five others, 64 percent of the group of masters whose fathers were not bakers, married daughters of nonbakers, which suggests that the first-generation masters were in a sense freer than more deeply rooted masters to choose a wife from whatever milieu they wished. The sons of simple bakers illustrated both tendencies: three married daughters of masters and three married outside the bakery.

Endogamy was slightly more pronounced among nonincorporated (or plain) bakers than among masters. Of plain bakers, twenty-nine, or 40 percent, married women whose fathers held the same socioprofessional rank. Fewer than half of the plain bakers married outside the group. Eleven percent of the bakers married "up" into the families of masters. In three of these cases they were sons of masters who were already introduced to the profession and who had some prospect of attaining mastership. In the remaining instances these

Table 11.1 Marriage Patterns of Master Bakers

Profession of Groom's Father	Profession of Bride's Father (N = 81)			
	Master Baker	Baker	Nonbaker	TOTAL
Master Baker	14	2	19	35
Baker	3	1	3	7
Nonbaker	12	2	25	39
TOTAL	29	5	47	

upwardly mobile bakers were the sons of plain bakers and they had larger dowries than their wives, whose means were not at the level a master's daughter ordinarily vaunted.

Of the seventy-three plain bakers, twenty-nine were sons of bakers, four were sons of masters, and the remainder were sons of nonbakers (see table 11.2). The sons of bakers had a much higher rate of endogamy than the sons of nonbakers. Fifty-two percent of the bakers' sons married women whose fathers were bakers, but only about a third of the nonbaker sons married into the bakery. While only 31 percent of the baker sons took wives outside the bakery, more than two-thirds of the nonbaker sons married strangers to the profession.

Table 11.2 Marriage Patterns of Bakers

Profession of Groom's Father	Profession of Bride's Father (N = 73)			
	Master Baker	Baker	Nonbaker	TOTAL
Master Baker	3	1	0	4
Baker	5	15	9	29
Nonbaker	0	13	27	40
TOTAL	8	29	36	

Endogamous marriage, often repeated across generations, created a number of interlocking family dynasties. A striking example was the merger of the Delamare and Félix families, both implanted in the faubourg Saint-Antoine with common ancestral and professional roots in the Gonesse area. Simeon Delamare, master baker and son of a master, married Marie-Jeanne Félix, daughter of a master. On the same day Marie-Jeanne's brother Gilles married Simeon's sister Marie-Antoinette. Already linked on each side to a half-dozen prominent baker families, in the next two decades the Delamare-Félix clan reproduced prolifically and married their children into other solid baker families, such as the Leroux and the Lapareillés.[3]

Master baker Pierre Paul Leroux, son of an ancien juré, took as his first wife another Marie-Jeanne Félix, the stepdaughter of an ancien juré named Bellange whose wife was a Delamare. Leroux had two brothers and three uncles who were master bakers and a sister who married one. Three of his wife's uncles were masters as was her sister's husband. Three of Leroux's mother's sisters married the Desfousse brothers, all three of whom were masters.[4]

Now, Claude Lapareillé, himself the son of a master baker, had married Jeanne Hutinet, daughter of a master and widow of master Vavasseur. Two of Claude's sisters married bakers and two of his brothers were masters. One of the latter, Toussaint, eventually married Marie-Jeanne Vavasseur, his brother's stepdaughter. Two of Vavasseur's brothers were bakers, one a master, and one of his sisters married a master. One of Claude Lapareillé's sisters became the wife of Joseph François Fourcy, master and son of a master. Fourcy's sister, Marie-Margueritte, took as her husband a master named Lazare Garnier, one of whose brothers was a master. Another Fourcy brother was a master and a sister also married a master.[5]

The Destors had similarly tentacular ties with a dozen different baker families located in the city, in the faubourg Saint-Antoine, and in the Gonesse area. The daughter of a Gonesse baker, Marie-Jeanne Chapon, whose mother was a Destors, married Jacques-Nicolas Félix, scion of a Bonneuil-Gonesse dynasty. Two of the leading Gonesse farmer-baker households, the Gouffes and the Michels, united at the end of the seventeenth century. Two Gouffe sons, both bakers, married the Michel sisters. Another Michel sister married a baker at Bonneuil, where her brother had set up as a baker. Two other Gouffe sons were bakers, while their sister Noelle married a baker named Denis Delamare.[6]

Brothers Nicolas and Antoine Berson, masters and sons of a master, married the Cousin sisters, daughters of a Saint-Antoine baker whose uncle, François Cousin, was a master and a friend of Berson's father. The two brothers of the Cousin girls were bakers. Another Berson brother, also a baker, married a baker's daughter, Madeleine Tournade, whose sister Nicolle was the wife of a faubourg Saint-Antoine baker. The Tresse sisters, daughters of a baker, married the Bardy brothers, bakers and sons of a baker. The Nezot sisters, whose father and brothers were all master bakers, married masters Geoffroy and Tribouillet, themselves sons and brothers of masters. Baker Sébastien Tourny had two sons who became bakers and two daughters one of whom married a baker and the other a master baker. It is possible there was a third son who was a flour dealer. Similarly, Nicolas Herbert's two sons followed in his footsteps as Saint-Antoine bakers, while his two daughters married bakers. Master Jean Petit's son took up the same profession and his two daughters married masters.[7]

Remarriage by widowed baker wives forged interfamilial bonds. Jeanne Personne married successively two master bakers, Savalle and Jobidon. Two of her sons from each marriage became master bakers. Jeanne Victoire Le Sueur had a daughter from her marriage to master Etienne Huin who married a baker and two sons from her marriage to baker François Livry who were bakers, as well as two daughters from the second marriage who married bakers. From her marriage to master baker Pierre Hervi, Marie-Michelle Chavirot had two sons who became bakers, and from her marriage with faubourg Saint-Antoine baker Edme Pollard she had a son who entered the bakery. Catherine Boucher raised a master-baker son in each of her two marriages.[8]

A similar dynastic phenomenon occurred when baker widowers remarried. Baker Charles Plé from the Gonesse area married three times (the last time to a Gouffe) and had sons and daughters involved in baking in at least two of the marriages. Beset with the same matrimonial misfortunes, Lubin Hetard, a master baker, and one of his sons, Pierre Lubin Hetard, also a master baker, each married four times and each left at least half a dozen baker sons and daughters wedded to bakers.[9]

Marriage-contract data do not always permit us to reconstitute professional and familial genealogies, but information from our samples suggests the same sort of quasi-dynastic trend.[10] Twenty-two bakers of all sorts had at least one sibling in the bakery, one had two, one had four, and one had five. Masters had more baker siblings than nonmasters (forains had the fewest), and sons of bakers predictably had more than nonbaker sons. Eighteen wives had a baker sibling, three had two, three had three, and one had four. Like baker sons, daughters of bakers had more baker siblings than nonbaker daughters. Both grooms and brides had three times as many baker siblings in the last two decades of the Old Regime as in the first quarter of the eighteenth century. Among the witnesses to these marriages were seventy-one baker uncles, forty-three for the brides and twenty-eight for the grooms.

Marriage Capital

For the baker, marriage was a transaction in human capital: the professional skills of the wife and the nexus of contacts and relations into which she introduced her husband strengthened his business considerably. At the same time marriage was a financial transaction. The young baker counted on his bride's dowry to help him launch his career or to fortify the ongoing business and set up his household. (An older baker entering a second marriage viewed the dowry as a means of liberating himself from debt or expanding his operations.) If the baker groom was fortunate, then the marriage meant not only a dowry from his wife but also a dowry for him constituted by his family, for the convention in Paris was for both parties to the marriage to contribute to the "community" established between them.[11] A baker without familial resources would normally have waited long enough before proposing marriage to have put together his own dowry from savings, from his personal effects, and from his commercial investments in fixed capital or stock. The combined dowries made up the marriage fortune, which helps us sort out levels of well-being within the bakery and take a measure of the baker's prospects for commercial and social success.

We will examine successively the marriage contributions of the bride and the groom, combined marriage fortunes, the components of the dowry on each side, and other financial arrangements (see figures 11.1 and 11.2). But first we should note certain limitations and peculiarities of the source material. We do not know what proportion of bakers of all types had recourse to a notarial contract. Yet, given the substantial amounts of wealth often involved and the familiarity of the bakers with notaries for business reasons, it's likely that the vast majority signed marriage contracts.[12]

Diplomatic conventions governing the drafting of the contracts were relatively loose. Thus there was considerable variation in style and procedure from notary to notary. One must be careful to make necessary adjustments in order to render all the data comparable. Notarial silence must not normally be interpreted as positively significant (for instance, as an implicit avowal of poverty). Frequently it resulted from a genuine lack of available information, from negli-

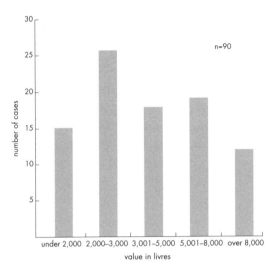

Figure 11.1 Distribution of Baker Marriage Contributions: Husbands and Wives Combined *(Source: marriage contracts)*

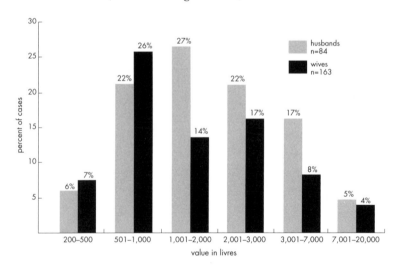

Figure 11.2 Percentage Distribution of Baker Marriage Contributions: Husbands and Wives *(Source: marriage contracts)*

gence, or from a particular notarial custom. Unless a contract specified that there was no marriage contribution from one side or the other or both, I treated the absence as missing data rather than as zero dowry. When a contract stipulated that a dowry was to be drawn from an after-death inventory not yet processed or an estate not yet divided up, I considered it as missing data with the mention that a dowry was very likely rather than as an effort to camouflage impecuniousness.[13]

The calculations for marriage fortunes are based exclusively on those cases for which clear-cut information from both bride and groom was available. Whenever possible, both marriage contributions were broken down into their component parts in order to see how the dowry could be used and how "real" it was. Some notaries did not disaggregate marriage contributions, while others used a wholly formulaic allocation procedure that must be rejected because it obfuscates the actual composition. Revenue from annuities is reasonably easy to capitalize, but in many instances it is impossible to assign a value to real property cited descriptively or topographically. Allowance must be made for inflated dowries, betrayed by the mention of suspect components. "Mastership costs" or "business losses," for example, may be fictions designed to puff a meager dowry or to equilibrate the marriage contributions. As a result of documentary omissions or analytical corrections, most of the following studies are based on fewer than the 175 marriage contracts that make up our core sample.

In 165 cases, the bride's portion ranged from nothing (one case) to twenty thousand livres (one case), with an average of 2,182, a median of 1,500, and a mode of 1,000 livres (see figure 11.3). Almost two-thirds of the dowries were under 2,000 livres. More than half were concentrated between 1,000 and 3,000 livres. Master bakers (n = 90) attracted substantially larger dowries than other types of bakers. The average master dowry was 2,903 livres, the median 1,983, and the mode 3,000. More than half were between 1,500 and 3,000 livres and one-fifth were over 3,000 livres. Three-quarters of the contributions of the brides of unincorporated Parisians bakers (n = 63) were located between 500 and 1,500 livres, while less than 16 percent were over 1,500 livres. For these bakers the average dowry was 1,372, the median 1,010, and the mode 500. Half of the twelve forain contracts were situated between 1,000 and 1,200 livres, with the average at 1,115, the median nearly 1,025, and the mode at 1,000. The average size of a dowry grew till midcentury, receded sharply during the next twenty years, and rose again to midcentury levels during the last two decades of the Old Regime (see table 11.3).[14]

The grooms' contributions were larger than the brides' (see figure 11.4). For ninety-four contracts of bakers of all kinds, the average portion was 2,609 livres, the median 1,900, and the mode 1,000, within a range from nothing (one case) to 13,000 (one case). More than one-fifth of these contributions were over 3,000 livres while 43 percent were over 2,000 livres. Almost half fell between 1,001 and 5,000 livres. Even as the masters attracted a larger portion than the

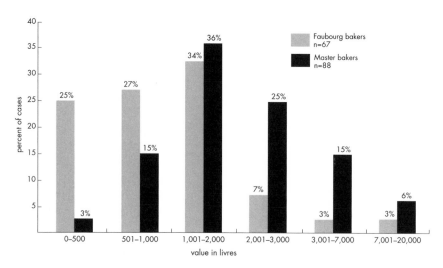

Figure 11.3 Percentage Distribution of Baker Wives' Marriage Contributions
(Source: marriage contracts)

Table 11.3 Evaluation of Marriage Contributions, 1711–90 *(Averages in Livres)*

Dates	Bride	Groom	Douaire (Jointure)	Préciput (Death Benefit)
1711–25	1,859	2,337	709	319
1726–40	2,442	2,673	958	416
1741–50	3,124	2,188	1,131	441
1751–60	2,705	3,267	1,089	539
1761–70	2,148	2,712	1,215	500
1771–90	3,118	3,783	1,557	714

nonguildsmen, so they tendered a larger one to the marriages. The average master's contribution (n = 59) was 3,169 livres, the median 2,413, and the mode 3,000. Only 10 percent were under 1,000 livres, while 20 percent were over 4,000. More than half the masters furnished between 1,501 and 4,000 livres. Over three-quarters of the Parisian nonmasters (n = 26) meted out 1,500 livres and under. The average size of their portion was 1,989, the median 1,200, and the mode 1,000. Four of the nine forain contributions were under 1,000 livres. The forain average was 1,322 livres, the median 975, and the mode 1,000.

We have marriage fortunes or combined portions for ninety couples, ranging from 400 to 31,950 livres.[15] The average for bakers of all types was 5,036 livres and the median 3,600. Almost half the fortunes were concentrated between 2,000 and 5,000 livres. More than one-third were over 5,000 livres, while only

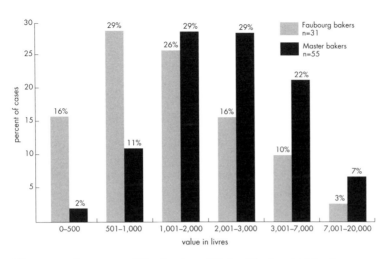

Figure 11.4 Percentage Distribution of Baker Husbands' Marriage
Contributions *(Source: marriage contracts)*

one-sixth were under 2,000. How do the bakers compare in marriage-fortune
levels with other socioprofessional groups? The bakers must be measured
against the somewhat incoherent sector that Adeline Daumard and François
Furet labeled "masters and merchants" in their survey of Parisian marriages in
1749. Almost 28 percent of the master-merchants fell into the dominant baker
range of 2,000 to 5,000 livres. There were fewer poor newlyweds among the
bakers than among the master-merchants, but there were also fewer truly
wealthy ones (38.4 percent of the master-merchants had fortunes ranging from
5,000 to 50,000 livres). The bakers were entrenched among those artisan-
merchants of relatively modest fortune.[16] But the bulk of the marriage fortunes
were sufficiently ample to give the couple a good start.

Master couples were decidedly better off at the outset of their marriage
career than the households of unincorporated bakers. Only 3 percent had
fortunes under 2,000 livres while almost one-third had fortunes over 6,000.
The average master fortune was 6,192 livres, the median 5,000, and the mode
6,000. Over two-fifths of the Parisian nonmasters had fortunes under 2,000
livres; half clustered between 1,501 and 3,000 livres. The average baker marriage
fortune was 3,386 livres, the median 2,050, and the mode 2,000. Whereas
almost a fifth of the Paris bakers had fortunes over 5,000 livres, the largest
forain fortune was 2,900 livres. A third of the forains were located between
1,501 and 2,400. Their average fortune was 1,958 livres, and the median as well
as the mode was 2,000 livres.

Neither the raw figures in the marriage contracts nor the measures of central tendency convey the full significance of many unions. A number of contracts, for example, made allusion to the acquisition of business properties or franchises, without appraising their value or subsuming them under the rubric of marriage contributions. Gilles Pille, the son of a deceased master baker, had served as his mother's geindre till just before his betrothal. His wife brought not only a large dowry (5,800 livres, three-quarters of it in cash) but also her mother's pledge to sell the couple the highly successful bakery business she had run since her husband's death on the grande rue du faubourg Saint-Martin. In return for an annual pension of three hundred livres, the father of master baker Jean Charles Delaitre promised to sell his business to his newlywed son and thereby launch his career. The future father-in-law of master Rene Delahogue pledged to cede his shop to his daughter and her husband. The groom's mother provided forain Germain Herbert and his wife with a fonds on the occasion of their marriage. Guillaume Rousseau obtained his father-in-law's baking practice as part of his marriage accord, but the object of the transaction was less to enhance the chances of the new couple than to extricate the bride's floundering father from an impossible financial and commercial situation. Certain families donated spaces at the bread markets in order to celebrate the weddings of their children.[17]

As a rule each groom promised in the marriage contract to provide a certain sum as a household pension for his wife's use after his death. This *douaire,* analogous to the English jointure, can be used as an indicator of wealth and as a proxy for an unspecified but acknowledged marriage contribution. The douaire may also reflect the esteem a man felt for his wife-to-be, but in some instances it was a sort of inflationary device the future husband used as a way of compensating for the meager marriage portion he pledged. Although the idea of the douaire was not without ambiguities, commonly it was conceived of as half of the groom's declared wealth at the time of marriage. The average douaire among bakers of all types (n = 92) was 54 percent of the male marriage portion and the median 42 percent. In four cases it was equal to the groom's contribution and in nine other instances greater than it.[18] The bakers with more modest fortunes subscribed to proportionately larger douaires than the wealthier bakers. The average douaire of masters amounted to 50 percent of their marriage contribution and the median 41 percent; of Paris bakers 52 percent and 43 percent; of forains 80 percent and 63 percent.

In addition to the douaire, most contracts (n = 170) specified a *préciput,* a sort of life insurance policy automatically assigned to the surviving spouse. Unlike the douaire, it was not imposed by customary law; if it was not stipulated in the contract by express convention, it did not exist. The préciput was to be drawn prior to the division of the estate only from wealth held in community. Among bakers of all sorts it represented on average 26 percent (median of 20 percent) of the husband's marriage contribution (n = 92) and 49 percent (median of 50 percent) of the douaire (n = 167).[19]

Family Origins and Lucrative Marriage

It is interesting to compare marriage fortunes in terms of the profession of the fathers as well as the baker category of the future husband. The men with the greatest wealth at marriage were the sons of masters (3,686 livres on average), followed by sons of bourgeois de Paris who were often rentiers of various sorts (3,340). The hierarchy steepened with the sons of men in urban professions outside Paris (2,400), followed rather closely by the sons of artisans or merchants of different kinds (2,191), the sons of laboureurs (1,909), and lastly the sons of plain bakers (1,392). The sons of masters drew the largest dowries (3,418) livres), followed curiously enough by the sons of laboureurs (3,024), who seem to compensate for their modest wealth by promising huge douaires (76 percent of their contributions versus 48 for the masters).

The women with the biggest dowries were the daughters of master bakers (5,794 livres), tailed closely by the daughters of bourgeois de Paris (5,583). A substantial gap separates these two groups from the others: daughters of diverse artisans and merchants (3,686 livres), daughters of men in urban professions outside Paris (3,393), daughters of laboureurs (2,390), daughters of simple bakers (2,000), and daughters of men exercising other rural professions (1,209). The "bourgeois" brides attracted slightly wealthier husbands (4,100 livres) than the master-baker wives (3,945). The ranking of the others precisely imitated the hierarchy of dowries.

The most lucrative marriages united master bakers who were the sons of master bakers to the daughters of master bakers. Their joint marriage capital amounted to almost ten thousand livres, a sum that augured brilliantly for commercial success, household comfort, social competitiveness, and psychological serenity. The next largest fortune, a little more than half as large, belonged to masters, themselves sons of masters, who took wives outside the profession, perhaps because they could not find families within the master milieu willing to help them *redorer leur blason* (the dowries attracted by these men were larger than their own matrimonial contributions). Close behind were a handful of masters, the sons of masters, who married plain baker daughters. Then came a group of masters whose achievement was remarkable because they were "new" men: sons of men of nonbaking professions who married the daughters of men whose occupations were also outside the bakery. There followed the sons of masters who were not themselves masters but who managed to marry into master families (perhaps on the strength of their father's status and perhaps with the prospect of attaining mastership).

Sixth in the order of marriage fortunes was another group of first-generation masters who married conservatively and profitably into master families. They barely outdistanced the master bakers, upwardly mobile sons of simple bakers, who married outside the bakery. It is surprising that the masters who were sons of plain bakers marrying into master families ranked only eighth. It might be that these bakers started out slowly because they had to recover from the costs of achieving mastership as outsiders. But the trouble with this argument is that they were not the only first-generation masters. The fortunes of the eight remaining couple types, all of whom, save one, involved a plain baker as husband, were significantly below the level of the first eight.[20] Bakers who were sons of bakers marrying baker daughters, together amassing an average marriage fortune of about twenty-seven hundred livres, were the wealthiest in the bottom half of the hierarchy.

Residential Pattern

A remarkably large number of bakers, almost two-fifths of the whole population, chose wives who lived in the same quarter as they. In fairy-tale fashion, 17 percent of baker husbands of all types married the girl next door, or, more precisely, women who lived on the same street as they. The neighborhood in these cases was a sort of village, a symbolically bounded space of relative social intimacy, and it was natural for a prospective groom to look first for his bride locally.[21] In some cases the couple may have grown up together and their fathers or mothers may have been colleagues or friends. In other instances the young baker may have purchased a business on the street prior to beginning his search for a wife. Marriages between men and women from the same street yielded the largest fortunes, over fifty-seven hundred livres on average. Another twenty-two percent of the newlyweds came from the same parish. Their marriage fortunes were about a thousand livres lower than those of the couples from the same street.

Almost half of the grooms and brides came from different parts of Paris, and their fortunes were only a few hundred livres below those of the same-street couples.[22] A little over 5 percent of the marriages (seven) joined provincial women to Parisian bakers. Their fortunes were about the same as those of the Parisian couples from different parts of the city. The forain households were the poorest, though when a forain married a Parisian woman (only 3 percent of the cases), the couple constituted a fortune more than one-third again as large as that of the forains who took country wives.

Literacy

Most grooms and brides knew how to write, though the level of literacy was higher among masters than among plain bakers. Ninety-one percent of the master grooms signed and 82 percent of the master bakers who served as witnesses signed their names. Eighty-four percent of the unincorporated Parisian bakers signed, but a surprisingly high proportion of their future wives (39 percent) could not do so. Of the bakers in this category who served as witnesses, 79 percent signed. Among the forains, 71 percent of the grooms and 80 percent of the brides signed. Couples composed of literate brides and grooms (70 percent of the population) had marriage fortunes (averaging over five thousand livres) almost two-thirds over and above those of households in which one or the other partner could not sign (29 percent). There were only two marriages in which both husband and wife were illiterate; they possessed a little more than one-fifth the wealth of literate households.[23]

Remarriage

In forty instances we meet individuals who were not marrying for the first time: in four cases both parties had been previously married (in three of those, the women had taken bakers as their first husbands); fifteen of the sixteen previously married women who married previously unmarried bakers were widows of bakers; in thirteen cases, bakers who had been married once before took previously unmarried women as wives, while seven bakers who had been married twice before married women who had never been married. Because of pending estate settlements, the data on marriage contributions in cases of remarriage are extremely incomplete. We do not know the magnitude of the wealth of the widower husbands. The average dowry of the baker widows who took as husbands previously unmarried bakers proved quite stunning, 6,337 livres, especially in light of the mean size of their husbands' contributions, 1,842 livres, a figure substantially below the average husband's portion for the whole universe. These men made amends for their quite modest circumstances with their ostensible vitality and with douaires almost as large as their marriage contributions (1,616 livres on average).

Marriage to a well-established widow often represented a singular chance for a young baker to make good. Had he not married Marie-Madeleine Gouhier, the widow of master baker Thomas Ligon, it is quite likely that Amable Guillaume Jubin, a journeyman baker, would not have been received as a master. Similarly, Tournay, also a journeyman, rose to mastership directly as a result of his marriage to the widow of a master.[24]

In 1700, Denis Larcher, the son of a deceased laboureur, married the woman whom he served as geindre, herself twice over the widow of master bakers. This marriage permitted Larcher to attain mastership and take over his wife's flourishing business. Upon her death in 1717, Larcher married another widow, whose first husband had been a mason and probably a part-time baker. Denuded of any resources in his first marriage, Larcher now contributed a portion of five thousand livres. With the help of his wife's dowry (2,766 livres) and the savings accumulated from the bakery and from his first wife's estate, Larcher was able to purchase two half-houses in Paris (in addition to another he received from his first wife's estate) and a house in Villejuif. In 1730, after his second wife's death, he married the daughter of a tapestry merchant who brought him a portion of four thousand livres, and he moved his business from the faubourg Saint-Marceau to the more fashionable faubourg Saint-Honoré neighborhood where his wife's family lived.[25]

Twice-married baker widowers placed a very high value on maiden brides as their third wives. The latter averaged meager dowries of 1,516 livres while their husbands' average fortune was an impressive 5,078 livres. It is interesting to speculate why these bakers chose inexperienced and relatively poor wives rather than widows—especially baker widows—whose wealth and professional skills could have helped them consolidate or reinvigorate their businesses. Baker widows usually had children as well as property for adoption. Already blessed with children from one or both of his earlier marriages, it is possible that the baker widower did not esteem that the latter compensated for the former.[26]

Moreover, it is likely that the putatively virgin bride would give the baker children in short order. Bakers who were married once before also paid a premium for maiden wives, though not as high a one as their twice-married confreres. Their average marriage contribution was 3,369 livres; the mean dowry of their wives, 2,108 livres, closely approached the average for the whole universe. It is possible that the baker widow preferred beginners to more seasoned bakers because she could exercise more influence over the former than the latter, especially since the new husband almost always joined the business that she had operated with her previous husband.

Composition of Marriage Contributions

The composition of the average male and female marriage contributions varied considerably (see figures 11.5 and 11.6). On the one hand, almost half the bridal dowry was in cash, while less than a third of the husband's assets were liquid. On the other hand, 36.4 percent of his portion consisted of commercial goods and tools, while business investment represented less than 4 percent of the female dowries. The other major component in the bridal dowry was clothing and furnishings, basically the trousseau, which represented a little over two-fifths. This category accounted for only 10 percent of the husband's wealth. The male and female contributions complemented each other. The husband provided the infrastructure for the trade along with a dose of liquid capital for purchasing stock and amortizing debt, while the wife supplied the infrastructure for the household as well as large amounts of cash for both domestic and commercial needs. The bride had almost twice as much real property as the groom, but the amount involved was extremely modest: 4 percent of the dowry. Almost one-fifth of the husband's contribution consisted of rentes, a sort of mortgage guarantee for the eventual payment of the douaire.[27]

The component distribution did not change substantially in second or third marriages. For the women, the trousseau remained a dominant element. Even in the cases where the wife was a baker widow, the commercial category represented only 23 percent of the dowry.[28] Baker widows contributed a higher proportion of cash and real property than maiden wives.

The relative importance of constituent parts of the marriage contributions varied from decade to decade (see figures 11.7 and 11.8). The cash component of the groom's fortune dipped brusquely below 20 percent between 1751 and 1760 and then recovered. Goods and tools also receded in the same decade and then rose to almost two-thirds of the husband's contribution in the last twenty years of the Old Regime. To compensate for the erosion of cash and commercial investment, clothing and furnishings swelled between 1751 and 1760 and then fell to under 10 percent in the following decade before utterly disappearing between 1771 and 1790. The cash portion of the bridal dowry dropped from 50 percent in 1726–40 to 35 percent in 1761–70 and then soared to 63 percent in 1771–90. The commercial component stagnated at very low levels throughout the century (0 in 1761–70) and then surged to over a fifth of the dowry in the

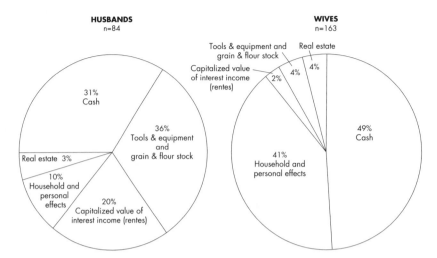

Figure 11.5 Percentage Distribution of Components of Baker Marriage
Contributions *(Source: marriage contracts)*

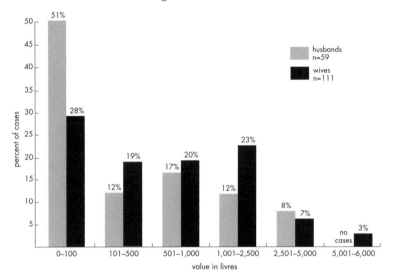

Figure 11.6 Percentage Distribution of Components of Baker Marriage
Contributions: Cash *(Source: marriage contracts)*

last twenty years of the Old Regime. The importance of the trousseau increased from 40 percent in 1726–40 to over 56 percent in 1761–70 before plunging to under a quarter of the dowry in 1771–90.

One might expect that sons raised in a baker household would have a larger
proportion of baker's goods and equipment in their marriage contributions
than others. At first sight this seems to be so: almost half of the wealth of sons

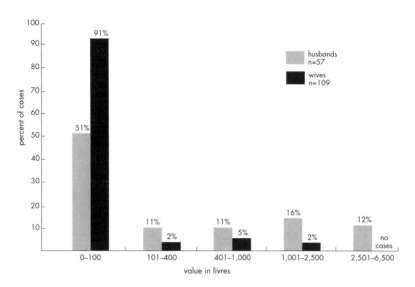

Figure 11.7 Percentage Distribution of Components of Baker Marriage
Contributions: Tools & Equipment and Grain & Flour Stock
(Source: marriage contracts)

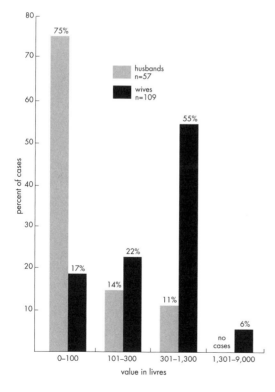

Figure 11.8
Percentage Distribution
of Components of Baker
Marriage Contributions:
Household and Personal
Effects *(Source: marriage
contracts)*

of master bakers is in commercial investment. Yet less than a third of the contributions of bakers' sons was in goods and tools, a portion well below the average for the total population. In contrast, 50 percent of the fortunes of sons of laboureurs and 44 percent of the fortune of sons of various artisans and merchants consisted of professional items. Obviously, in many cases the sons constitute all or part of their marriage contributions from their own labor and savings. Fittingly, the sons of bourgeois de Paris had by far the highest proportion or rentes (62 percent). The sons of men working in urban occupations outside the capital had the largest percentage of cash (62 percent).

The daughters of bakers, whether masters or plain bakers, were the only women with goods and tools in their dowries, and neither category accounted for substantial amounts: 6 and 7.4 percent respectively. Cash and the trousseau represented the paramount elements in the dowries of the daughters of men in all professions. The daughters of bakers and other merchants and artisans had proportionately more clothing and furnishings, and the daughters of bourgeois de Paris and laboureurs had more cash. The marriage couplings with the largest cash component in their combined fortune involved masters, themselves the sons of masters, marrying daughters of masters, followed by masters who were sons of masters marrying outside the bakery, and then by masters, the sons of nonbakers, marrying daughters of nonbakers. The household in which commercial investment bulked largest consisted of masters who were sons of bakers marrying outside the bakery, followed by masters, themselves sons of masters, marrying outside the bakery, and finally master-baker sons of masters taking daughters of masters as wives.

Commercial investment represented a much greater part of the fortunes of master bakers than of nonmasters, perhaps in part because the masters had a more elaborate bakery establishment. Equipment and goods represented half of the average master marriage contribution, whereas it was less than one-fifth for country and faubourg bakers. Dowries bestowed on forains contained no commercial component, while the value of tools and merchandise was approximately the same in dowries received by masters and plain bakers. Surprisingly, the forains had the largest cash component, both absolutely and in relation to the other parts of their marriage contribution. The bakers had the largest share of clothing, furnishings, and real property, while the masters had a far greater investment in rentes than the others. The wives of the masters had more cash, more property, and more rentes than the other brides. The master wives also had a fancier trousseau than the others, while the clothing and furnishings category of the forain wives (80 percent) and the baker wives (51 percent) loomed much larger in their dowries than it did in the dowries of the wives of masters (31 percent).

The Wife's Special Role in the Bakery

Given the juridical, social, and professional preeminence of the man in Old Regime society, most of the records concerning the bakery focus on him. It would be a serious error, however, to fail to insist upon the cardinal role played by the baker's wife not merely in the household, in which her moral and material sovereignty was rarely contested, but also in the business. Habitually, the baker family lived behind and/or above the shop. The wife seems to have spent more time in the shop than in the apartment. She slept at night, but she was in the shop by 6:00 A.M. to organize the distribution to clients at home, to display bread behind the grill, and to prepare bread crumbs, crusts, and charcoal for retail sale. On market days she was up even earlier, for it was very often the wife who ran the entire sales operation.[29]

As a rule, the wife was in charge of relations with clients, even as she is today in Parisian bakeshops. Installed (in the more opulent shops practically enthroned) behind the counter, she chatted with the customers in order to keep them well disposed and faithful to the shop. She ordered an apprentice or journeyman or a delivery girl or a domestic to serve the client, or, in the more modest shops, she herself got the bread and the other articles desired by the consumers. The wife took payment in cash, which she placed in a money box in the counter drawer, or she extended credit, which she either recorded in a register or marked on a set of *tailles* (or she did both, a useful control system). The *taille* was retained in the eighteenth century not because the wife could not read or write—we have seen that the contrary was true in most cases—but because it was still demanded by many customers; it was considered the surest way for delivery girls and boys to keep note of sales, and it was a simple way to keep records during periods when the wife was absent.[30]

Credit management was an extremely complicated and critical affair, especially in light of the enormous part that sales on credit played in the bakery business. The wife had to keep track of scores of accounts at a time. She had to decide, frequently on her own, how much credit to extend to a person or if any credit at all should be allowed. It was she who asked for *billets*, or IOUs (commonly addressed to "ma boulangère"), to supplement and reinforce the primitive notations in her books, or full-fledged avowals of indebtedness and pledges of repayment sworn to before notaries when the debts amounted to excessive levels or when the client's credit rating came into question.[31] The wife had the highly sensitive mission of demanding payment without alienating the client. She had to make sure that the business was sufficiently liquid at all times to enable her and her husband to meet their professional, household, and civic obligations. Debt collection was difficult and was sometimes dirty work that the husband gladly left to the wife. Wives were often verbally abused and on occasion severely beaten while attempting to collect bread debts.[32]

Bookkeeping included purchases as well as sales. The wife was often in charge of relations with suppliers, at least as far as those relations concerned acknowledging receipt of merchandise, negotiating credit terms, and meeting installment payments. In the event of quarrels over contractual obligations, quality of merchandise, price, or amounts due, the wife frequently represented her husband and their business in the consular courts and before other jurisdictions. Not only did the wives keep the books on grain and flour purchases, but according to a brief of the Paris grain merchants of the early thirties, "it was they who most often bought grain at the ports and at the Halles." This observation echoed another made by Parisian grain dealers eighty years earlier. In 1653 the confraternity of grain merchants petitioned the courts to permit them to hold baker wives jointly responsible juridically for the payment of merchandise, "since it is the woman who runs the business in the markets, [who] receives and handles the monies coming from the [utilization of] the grain."[33]

The women's part in buying may not have been as great in the seventeenth century as the merchants contended, but in the eighteenth century the evidence is overwhelming that women commonly purchased grain at the ports and, especially, flour at the Halles. Françoise Picard, the wife of a Porcherons baker, told the police in 1760 that she had been doing all the flour buying for the previous six years. Dame Georgois of the faubourg Saint-Antoine regularly bought flour from factor Bazin, just as Dame Papillon from La Courtille frequented factor Chicheret. The wife of master Louis Chevau of the porte Saint-Denis had her purse stolen while buying flour at the Halles. Invited by an official to fetch her husband after she fell into a quarrel at the Halles over twelve sacks of flour, the wife of master Gilles Nerodeau replied indignantly "that she had no need of him [to settle the affair]." There seems to have been a common division of labor regarding grain and flour procurement: the husbands took charge of buying in the countryside and the wives assumed the task of internal purchasing. Moreover, perhaps in response to the complaints of the merchants, the wives frequently cosigned their husbands' notes and jointly guaranteed payment with them.[34]

Given the wife's major role, it is no wonder that baker Jacques Betemont affirmed that he simply could not conduct his business without his wife's constant participation. Master Joseph Verdet gave his wife his power of attorney and legally authorized her to "fix accounts and settle with all clients, . . . make all purchases of merchandise, . . . take care of all legal matters, . . . and generally undertake any steps in management of the business that his wife considers most effective." Now, it may be that Verdet was ill or that some other special circumstance impelled him to invest his wife explicitly with this authority.[35] It would not be truly surprising, however, if he did it as a matter of course and convenience, rendering de jure what was already de facto. Nor should his creditors have been surprised when master Denis Larcher insisted that he did not know most of the persons who had signed notes for the bread they bought on credit. It was his wife, he explained, who handled customer business. Catherine Marois, wife of a master baker of the rue Saint-Denis, did not hesitate to refer to "my shop." Many other baker wives must have felt the same way.[36]

Widows

Years of shop experience—in the case of daughters of bakers, the training began long before marriage—prepared the wife to carry on in the event of her husband's demise. Though we cannot measure the relative frequency and the age-specific incidence of baker mortality (especially death that is sociodemographically "premature"), there seem to have been a significant number of baker widows young enough for remarriage and/or active professional lives at any given moment. Baker families could not allow death, however anguishing, to cause too much disarray. In the midst of the trauma and grief, the widow had to think first of continuing business as usual. Competition for customers was keen; any interruption of service threatened to damage the business durably. On the bottom of the page of certain shop registers one notes an unadorned, pithy entry recording the baker's death. It is also striking to find that not a single day of baking or selling was missed, even in a small shop with only one journeyman and an apprentice to assume the burden along with the widow.[37]

One widow borrowed 450 livres from her daughter in order to buy flour and continue supplying her customers. Another sold a horse and pressed the collection of sums due for bread sold on credit for the same purpose. Widow Motot planned to sell her shop, but she was careful to keep it running smoothly after her husband's death "so that it would not lose value." "If our customers are allowed to disperse," wrote widow Moto, "our business will be annihilated."

The widow of master Pierre Coussin petitioned the courts to permit her to free the merchandise that was placed under seal with all the other property in the estate in order to allow her to "continue her commerce" and thus avoid "damaging the inheritance."[38]

Even as marriage was in part a business deal, so the rupture of matrimonial ties entailed if not the dissolution, then the reconfiguration of the business. The family of the deceased was interested in protecting their "investment," either for children of the marriage or for collateral claimants. Instead of bringing families close together, the tragedy of death sometimes resulted in ugly brawling between contestants from the two lines. Yet, whatever their differences, usually they all understood the urgency of keeping the business going round the clock.

Bruising strains marked the efforts to settle the succession involved in the (re)marriage of a widow and a widower baker. In 1732 Marie Brion had married master baker Jombert, bringing a modest dowry of one thousand livres to the union. Upon his death two and a half years later, she formally (and quite legally) renounced the succession of the matrimonial community because it was laden with debt. After a brief widowhood of six months, Brion married master baker Guy Leguay, himself a recently bereaved widower. Upon Jombert's death, Brion had not demanded the reimbursement of her dowry, for it turns out that it had never been paid. Nor had she pressed claims for the ordinary insurance sums that should have accrued to her, apparently for want of a familiarity with her rights. Leguay then took matters into his own hands, filing suit against the trustee of the vacant succession for recovery of the douaire of three hundred livres and the préciput of two hundred livres, as well as compensation for community expenses such as the funeral (50 livres). This offensive gave rise to a flurry of protracted contestations, and expressions of indignation and bitterness from Jombert's close relatives, for whom Brion now loomed as a traitor and a harlot. Though Marie Brion died in 1749, her second husband stubbornly pursued the case against her dead first spouse until he passed away in 1765.[39]

Brion's choice of a marriage partner made good sense: in many ways the best solution for a baker widow was to remarry a baker. Many women were of remarkable professional competence. It was extremely difficult, however, to run a shop on one's own, whether one was a man or a woman. And there were certain aspects of the business—country buying and especially bakeroom operations—that women were poorly equipped or not normally allowed to undertake. Moreover, it goes without saying that needs other than professional collaboration drove widows to seek new husbands. Eighteen of the wives in our marriage study were baker widows, and we have found evidence elsewhere of twenty-eight women who were married at least twice to bakers.

If her young first-marriage children served as a disincentive to prospective husbands, her experience, her resources, and her fonds were often powerful lures. Baker-widow marriages usually led to a buttressing of the business, at least in the short term. About two-thirds of the remarriages were endogamous, while a handful were hypergamous. Some resulted in the fusion of rugged dynasties, such as the union between the widow of Nicolas Lecoq and Nicolas Jacob. In a few cases master widows married "below" their station, either becoming the wives of faubourg Saint-Antoine bakers or marrying their first journeyman.[40]

The relationship between a widow and her journeyman was a very delicate matter, for it could generate rumors of scandal and cast opprobrium on both the business and the widow. Writing during the first empire, Grimod de la Reynière bespoke the conventional wisdom about the solitary baker widow. Referring to the widow of a well-known baker, Madame Gaumard, he hailed her decision to remarry with an equally successful baker. "The profession of baker is one of the large number of those that a woman cannot handle on her own," he observed, "she would find herself at the mercy of her garçons, and her shop would degenerate if she could not find a master to govern them."

Surely there were widows, born of baking families, steeped in knowledge of the craft, aware of precisely what they could not do themselves as well as what they could handle, who had the temperament and authority to govern a shop. Given the normative views that held sway, however, a "marriageable" widow was expected to marry, precisely to preempt suspicions and innuendos. A widow such as Dame Leguay risked her reputation by making one of her journeyman legal guardian of her children but formally keeping him out of the family.[41]

The widows remarrying "below" probably hoped to transmit their masterships to their new husbands. But this ambition rarely was fulfilled, for the corporate statutes forbade such transfers. Once she remarried, a widow no longer had the right to exercise her former husband's profession as a master (though the guilds usually allowed a period of grace of up to a year for the new household to regularize its status). Françoise Aubry, a master widow with two minor children, married a journeyman baker who brought merely two hundred livres to the community. She counted on her good relations with the jurés and the influence of her brothers, both masters in the guild, to hasten the integration of her new husband.[42]

Married to master widow Beaudin for fifteen months, Jean-François Chevriot was accused by the jurés of practicing baking "without title or quality" because he personally had not earned mastership. The jurés seized all of the couple's tools and merchandise, effectively putting an end to their business. If the guild rules induced certain master widows to marry their own kind, they impelled others not to marry at all, at least not legally, in order to retain their masterships and legitimize their businesses. Thus Barbe Bureau, the widow of a master baker, lived with a cook named Marin for seven years after her husband's death and only broke with him when he started to proclaim publicly that he was her husband.[43]

These remarriages were not without complications, both professional and personal. Faced with a legal battle over her husband's estate mounted by his family, master widow Delaitre shrewdly protected her interest by agreeing to the sale of the business to master Deline . . . and then several months later marrying him! Though she renounced her community with her dead first husband, widow Coussin's new marriage to Jean-Baptiste Robert was marred by the persistent effort of her first husband's creditors to get Robert to assume responsibility for the debts.[44]

The remarriage of a widow who had a baker son sometimes generated violent Oedipal tensions in the family. The son felt that he should succeed his father as head of the business and bitterly resented his displacement by an outsider who took command of both shop and household. Besides the deep psychological wound, in practical terms the son might now find his career blocked. Without the means (and perhaps the initiative) to set off on his own, he would have no place to go. This seems to be at least in part what caused problems in the Fontaine home. After his mother remarried, Fontaine Junior began to feel a profound antipathy for her, so he avowed. He called her names, beat her, and stole goods from the house.[45]

Jean Langlois felt a similar anger and frustration. The son of a master baker whose name he bore, he watched his mother marry master Louis Dugland very shortly after his father's death. Langlois quarreled wrathfully with his mother and stepfather, whom he continued to serve as a journeyman, next to Dugland's young son with whom he also had strained relations. In the first three years of the new household, the guard had to be called several times to contain Langlois's rages. In 1739 Langlois was still a journeyman and he could not write, though he was almost an adult. Ten years later he could scratch his name and he was finally a master, but he apparently still worked for his mother and stepfather. In November 1749 he filed a complaint with the police against his stepfather for brutalizing him. He told the commissaire that his mother had had "nothing but misfortunes" since marrying Dugland. The son charged that

Dugland had spent "all their money in debaucheries" and that he jeopardized their business. We do not know what Langlois's mother had to say about any of these accusations.[46]

Some of the widows who did not remarry kept their businesses going by associating themselves more or less clandestinely with persons who did not always have the credentials to engage in the profession. Geneviève Villemain, twice married to masters, now "exercised the profession of baker as widow mistress." In 1770 she married her seventeen-year-old daughter to her twenty-five-year-old journeyman named Joseph-François Thevenin, the son of a baker from Champagne. Thevenin's fortune consisted of four hundred livres in personal items and six hundred livres in savings that he had loaned to the widow Villemain "for use in her business." The bride's mother promised a dowry of three thousand livres provided that this money not be requested until her death. Thus the newlyweds had virtually no marriage fortune with which to establish themselves. Nor was the widow Villemain's intention that they should launch an independent career. Indeed, she proposed to take the newlyweds as 50 percent partners in the business, which she "intends to continue to operate under her name." They were to share expenses and profits equally, though widow Villemain alone was to keep the books, handle purchases, receive monies due, and make disbursements. Only she could dissolve the partnership, at which time she would be obliged to pay the couple their dowry.

Since masters could associate themselves only with other masters, this partnership was illegal. Moreover, it did not augur well for Thevenin's future, for his mother-in-law did not commit herself to sponsor his admission to the guild. Fifteen years earlier widow Legrand, also twice married to masters, created a similar "society" between herself and her son-in-law and daughter as part of the dowry agreement. Widow Legrand would have had more difficulty than widow Villemain in justifying the arrangement under her name, for her son-in-law was not even a journeyman baker but a mealman.[47]

To help defray operating costs and to ease the management burdens, mistress widow Roizeux "shared" her shop with a man who had no professional status. Each of them had her/his own clients and kept her/his own books. Her associate did all the actual baking and in return "she lent him credit in the profession." Widow Thevenet asked her cousin to join her business, while widow Vie brought her brother into her establishment. Master widow Bruché had her business run "for her account and under her name" by her son, a journeyman. Another master widow operated a "society" with a mistress dressmaker who brought capital rather than professional knowledge to the arrangement. Instead of actively remaining involved in commercial affairs, a number of widows rented out their names and in some instances their shops and equipment as well. An apprentice named Claude Gillain operated a shop under the name of widow Denise Noiseux, mistress baker. A geindre named Jacques Gauguin practiced as a master under the name of widow Nuzé.[48]

Finally there were widows, usually bound by family ties, who pooled their resources in a collective business venture. The Chenais sisters, both widows of masters, seconded by a spinster sister, operated a shop near the Saint-Michel market. Four widows ran a shop together in the faubourg Saint-Antoine, supplying at the same time several different markets on Wednesdays and Saturdays. The titular head was the widow of baker Nicolas Herbert, mother of baker widows Goix and Hatié and mother-in-law of baker widow Denise Herbert. When the matriarch died, the other three women kept the business going.[49]

The widows we know least about were the ones who were too sick or too old or too poor to pursue their late husbands' profession. A substantial number, like mistress widow Cretté, boarded with their children, either paying an annual fee (Cretté paid 150 livres a year) or living off capital invested in their children's business or marriage. When mistress widow Moriceau's one-room flat was robbed, she lost her life savings, seven hundred livres kept in a pottery piggybank. Sixty-five years old and without other resources, she was desperate. Another widow of a master, Jeanne Pignon, lived in a one-room flat and owned possessions worth a total of 450 livres. Similarly, mistress widow Jeanne Caquet lived meagerly in a tiny fourth-floor room. Styling herself a bourgeoise de Paris, by contrast, Françoise Lallemand, a master's widow, appeared to have a comfortable retirement.[50] In very few cases did baker families appear to plan for tragedy or for retirement, aside from the douaire and the préciput built into the marriage contract, sums, alas, that did not always materialize or that were quickly eaten up.

Marriage Strains and Separations

Baker marriages, like other marriages, sometimes did not work. Couples with matrimonial difficulties faced a serious dilemma, for they did not usually have a legally and morally legitimate alternative to weathering the storm *en famille*. Theoretically the marriage was indissoluble. "The vaults of the Temple of Justice," wrote Mercier, "reverberate with the wailing of married persons tired of one another." He ironized over "these sacred links" that "battered and rent." Inflexible and insensitive, the law made no provision for derelict unions, and the result, according to Mercier, was that many couples in Paris separated voluntarily. Most of the separations to which Mercier referred were informal, de facto arrangements, for it was not easy to obtain a legally sanctioned separation.

Mercier claimed that husbands and wives seeking separations "had only to stage a fight before two witnesses, the courts would separate them immediately." The evidence suggests, however, that Mercier exaggerated the facility with which separation could be obtained, especially in the milieu of relatively modest artisans and merchants. The Goncourts maintained that there was an "enormous" number of demands for separation in the eighteenth century, but they did not document their contentions. Though technically in violation of civil and/or ecclesiastical law, voluntary separations were generally tolerated provided neither party complained. Yet they were viewed as "contrary to good morals" and they sullied the reputation of both the husband and the wife, who were still considered married.[51] I suspect that the pressure of public opinion along with the desire to safeguard the business and protect the children probably impelled many troubled baker couples to remain together.

A separation could be sought from a civil court when a husband dissipated the couple's fortune or when he exercised violence of some sort upon his wife. In the first case the court pronounced a separation of *biens*, or property, and in the second a separation of *corps*, or domiciles. The first could occur without the second, but a physical separation perforce implied a financial separation. Dissipation meant unmitigated prodigality or simply bad management (including "madness") "that led the couple to the brink of impoverishment." Before pronouncing a verdict, the court conducted an investigation of the charges against the husband. It wanted to be sure that this request for separation was not merely a ploy to deprive creditors of their claims on the community that the wife now renounced. A separation of property could only be sought by a wife, because the husband was "the master of the community" and thus "however considerable his wife's dissipation might be, . . . it is his own fault if he does not take measures to remedy the situation."[52]

A woman could seek physical separation if her husband subjected her to beatings, placed her life in danger, or "conceived against her a deadly hatred." This separation had to have "grave causes." It could not be granted for "petty altercations" or "temperamental differences." Even insults and threats were "ordinarily" insufficient—a prudent limitation in the baking milieu where foulmouthed irascibility and short tempers were rife.[53] Physical separation was almost always sought by the wife rather than the husband, for a husband was supposed to be able to impose his authority over a wife no matter how violently she behaved. If the situation became truly unbearable, he could always have her incarcerated in a convent. As in the separation of property, the court did not pronounce until it carefully scrutinized the evidence, including the testimony of witnesses. When a wife obtained a separation, her husband could not force her ever to return to live with him. If her suit was dismissed, however, the court ordered her to return home and enjoined her husband "to treat her as a husband should."

Physical violence led baker wives to seek *séparations de corps*. Usually it was a compartmentalized kind of violence. The husband brutalized the wife, but he did not mismanage the business affairs. The fourteen years of her marriage, declared the long-suffering wife of master Jean Fontaine, were "the saddest days of her life." Her spouse beat her continually. What prompted her finally to seek estrangement was the physical and verbal abuse she suffered at the hands of her eighteen-year-old stepson, Fontaine's child from a previous marriage. Marie-Jeanne Bounnenie left her husband because she was "constantly black and blue at his hands," but she returned when he promised her and her parents to treat her well. Now she asked for juridical leave because "he broke all his promises" by continuing to beat and curse her. Marie Tranchard "feared for her life," for her baker husband "had neither faith nor religion," drank hard and gambled (but apparently after work only), and beat her regularly even when she was pregnant.[54]

A forain's wife had the good fortune of convincing the authorities that her husband beat her because he was deranged. Thus he was peremptorily locked up in Bicêtre. Married for only fifteen months, master Etienne Garin and his wife Agnes fought every single day. His wife claimed that he did not love her, that he married her "only for her money," and that he caused her to have a miscarriage by beating her. Jeanne Mantion trembled for her safety, for her husband of thirteen years suffered volcanic eruptions of fury that led him to beat her till she bled, to try to strangle her, to cut her with a knife, and finally to attempt to throw her into the bread oven, which he would have succeeded in doing had he not been stopped by the journeyman.[55]

Dissipation seems to have been a more common source of household turmoil than brutality, and thus one encounters more requests for financial than for physical separation. The wife of master baker Le Cocq sought a separation on the grounds that her husband had "totally ruined" his business, "preferring gambling and wine to commerce." One of his confreres who had known him for twenty-seven years testified that even before wine addled Le Cocq "he had never mastered the fundamentals of the baking profession, for he often bought grain above its real value." Similarly, the husbands of Jeanne Calamel and

Marie Bisset drank excessively, devouring their marriage capital and incurring heavy debts. According to his wife, master Jean Mathes, drank so much "that it was impossible for him even to supervise work in the bakerooms, so that his journeymen made a poor quality bread that induced his customers to buy their bread elsewhere." Master Joseph de Caut's willingness to give credit to "insolvent persons" resulted in heavy indebtedness that threatened to eat up his wife's property. The wives of masters Louis Suire, Jean Geoffroy, and Louis Buteaux accused their husbands of sacrificing their businesses entirely "to their pleasures," contracting heavy debts, and compromising their wives' financial future.[56]

In some cases the wives complained simultaneously of profligacy and violence. Already seven thousand livres in debt when he married widow Sergent—one wonders why this experienced master's widow accepted such an inauspicious union—master Claude Roger added three thousand livres more from drinking and whoring, sold his shop to support his excesses, gave away his wife's clothing to his "girls of debauchery," denied his wife any support yet refused to authorize her to work, beat her, and then tried to have her locked up in a hospital for bad conduct. Master Pierre Dubreuil behaved very much the same way: after squandering a three-thousand-livre dowry, he began to sell family furnishings and clothing piece by piece and apparently also sold the business. His wife charged that he beat her habitually, triggering a miscarriage in one attack.[57]

Marie-Elisabeth Mandon was disenchanted to discover that her master husband, who had boasted when they were married two and one-half years earlier that he was "very rich," in fact languished in debt. Moreover, as a result of his "bad conduct and his unsettled state of mind," Mandon not only frittered away her four-thousand-livre dowry and twenty-five-hundred-livre inheritance but in addition mired them even more deeply in debt, leading to the seizure of their furniture and expulsion from their house. Marie-Elisabeth claimed that she might have been able to stand for all this if her husband had not compounded his error by beating her savagely, usually when he was drunk. It was rumored that his first wife owed her premature demise to "the turbulence and brutality of Mandon's character."[58]

The most extraordinary story of marital conflict concerns the Blanvilain family, whose saga one can piece together from thirty years of fragmentary police archives. As early as 1726, Ursule Evrard, the wife of Julien Charles Blanvilain, a master baker, asked the lieutenant general of police to incarcerate her husband on the grounds that he spent all their money on "woman and girl prostitutes." The next trace of the Blanvilains occurs in August 1732 when

Ursule denounced Marie-Jeanne Millet, the young daughter of a gypsum dealer, as "my husband's whore." The following month Ursule apparently left home, taking refuge with her brother, another master baker. Blanvilain quarreled violently with him for sheltering his wife. When she returned home later, he beat her so badly that she had to leave again. In October she formally petitioned for physical separation.[59]

For the first of several times Ursule abandoned her petition, perhaps both because she was still attached to her husband in certain ways and because she was susceptible to all the social and psychological pressures that favored a wife's subjection. In February 1734, he beat her bloody again because he found her conversing with her sister, whom he suspected of plotting against him. He threatened to kill her if he found her in her sister's shop again (for the third time, according to witnesses). Bitterly jealous of her spouse, Ursule continued to insult Marie-Jeanne Millet wherever she encountered her, calling her "a whore and a bitch" and charging that she had had three children with Blanvilain. Meanwhile Blanvilain got in trouble with the police for brawling with another baker in a tavern.[60]

At the end of August 1734 Ursule again sought a physical separation, alleging that, while she was "totally devoted to her household and business," her husband was drunk every day, beat her inveterately, and threatened her life. After another vicious thrashing in October—Blanvilain had been intoxicated from early morning till almost midnight—Ursule renewed her demand urgently. Then she again changed her mind, "in the hope that her husband would come to his senses, and would have more respect for her." But she soon pronounced her hopes "vain" because in late November 1734, after cudgeling her again, Blanvilain locked her out of the house.[61]

This time the request for separation precipitated the standard judicial investigation, including the testimony of witnesses to the alleged brutalities. Meanwhile Blanvilain battered her again and continued to frequent Marie-Jeanne Millet, whom Ursule again reviled for breaking up her family and for taking money to have sexual relations with her husband. In January 1736 Millet and other witnesses charged that Ursule, while drunk, insulted her publicly and struck her with a wine bottle.[62]

Once more Ursule abandoned her suit for separation when her husband solemnly swore "to be assiduous in his commerce and to treat her as an *honnête homme* should treat his wife." Within days, however, Blanvilain returned to his old ways: drinking all day, returning late at night, pummeling his wife. Some sort of truce or transaction seems to have intervened, for Ursule remained at home for the next six years, save for periodic and short-lived flights to safety.[63]

During this period, apparently as a result of his negligence and dissipation, Blanvilain was thrown in debtor's jail. He importuned Ursule for her help, pledging to turn over a new leaf. With the hope of "drawing her husband closer to the household," she sold a valuable necklace, paid his creditors, and freed him from jail. Within days, however, Blanvilain revealed that he had no sense of gratitude, for he assaulted her in the street, tearing off her clothes and threatening to take her life.[64]

In June and December 1743 and again in July 1744, Ursule cried out against her husband's "violent conduct" to the commissaire. Then in October, while he was out of town, she moved out of the house. This time it was not a false alarm, for she moved almost all the household furnishings and linen as well as her personal effects. In May 1745 she again filed for separation, documenting her request with a surgeon's certificate testifying to recent brutalities to which her husband had subjected her. There was absolutely no hope of saving their marriage and, in addition, their business was in serious jeopardy (the fact that it had survived this relentless turbulence borders on the miraculous). Blanvilain used all the cash generated by the shop to pay for his philandering and drinking. Without money for merchandise or even for subsistence, Ursule was unable to supply their stall at the market or meet the needs of the shop clients. Submerged in debt and threatened with the confiscation of the family's remaining property, Ursule feared that the family would be reduced to beggary.[65]

Astonishingly, Ursule appears to have returned home again. In April 1749, reporting yet another beating to the police, she complained that "her state of poverty" had forced her to give up the physical separation suit that she had launched (or resuscitated) in 1745. It is true that the separation procedure demanded legal fees, and the burden may indeed have been onerous and dissuasive to many battered wives among the common people. Yet it seems inconceivable that Ursule could have found no recourse had she really been tenacious and lucid. Now she implored the police for protection, for she could not stand any further beatings, she claimed. The business seems to have recovered, for we again find Ursule selling bread at the Halles market, where on two occasions in April her husband struck her and called her names before scores of witnesses.[66]

We know that in the next months Blanvilain threw his wife out and replaced her with "a woman of confidence whom he paid to sell his bread at the market and to take charge of the other aspects of the business." We know this because one of the embattled couple's sons, Jacques Blanvilain, a journeyman baker, came to the Halles and exploded in a rage against the woman who had usurped his mother's role. He denounced her as "his Father's bitch and whore" and execrated his father for having eaten up all the family's fortune.[67]

Once he made up his mind, Blanvilain proved far more resolute than his wife. One has the sense that she expected to return to her home and her post in the shop once tempers cooled as she had so many times before. But her husband refused to take her back under any conditions: this was *enfin* the definitive rupture. There now occurred a strange and sad role inversion: she began to drink very heavily and frequent various men, while he appeared to pull himself together and to focus his energies on the business. Forlorn and disoriented, Ursule was also reduced, in her own words, "to great misery." She had no funds left at all once she sold the furnishings and clothes that she no longer used. Blanvilain accorded her "a pension" of two four-pound loaves a week, which she quite rightly judged insufficient. Perfectly familiar with her husband's customers, Ursule tried to collect money from those who had purchased bread on credit. But he had already warned them not to pay, for he would not recognize any sums remitted to her as liquidation of debts.[68]

To avenge herself against her husband, Ursule resorted to more or less violent measures. Accompanied by a group of her woman friends—a band of furies in her husband's eyes—she descended several times on his stall at the market, screaming insults, "raising the crowd against him," and forcing him to abandon sales. On several other occasions, Blanvilain complained, she and her friends "pillaged" his shop in his absence. They allegedly gave bread and money away, broke windows, and tried to set fire to the shop. Just as his wife had complained of his violence and insisted on her sobriety years before, so now Blanvilain presented himself as an earnest businessman whose commerce was threatened with "destruction" as a result of the rampages and the calumnies of his wife. He even claimed that he feared for his life as a result of several beatings administered to him by his wife's accomplices. The once notoriously unstable Blanvilain now characterized his wife as "a madwoman."[69]

As late as 1756, the tragic and tumultuous conflict between husband and wife remained unresolved. Blanvilain continued to file complaints with the police against his wife with the same frequency and intensity with which his wife used to denounce him to the authorities in the thirties and forties. Police Inspector Poussot had considerable sympathy for Ursule. Blanvilain, he wrote, was clearly "a very bad subject" whose "evil conduct" had come to the personal attention of several lieutenants of police. Despite their threats of punishment, he never "reformed." Ultimately, according to Poussot, he drove his spouse, once a model wife and mother, to debauchery and ruin. Morally, all was lost in Poussot's eyes, but he proposed a solution to Ursule's material problems. Blanvilain should be forced either to pay her five livres a week or to allow her to operate for her profit his space in the Halles market.[70] The subsequent silence on both sides suggests that Poussot's recommendations may have been adopted.

Property separations were usually sought in extremis by women who were alienated by their husbands' dissipation, negligence, or incompetence, but there were some instances in which women requested separation reluctantly and respectfully, without any trace of ill will or any desire to disrupt the household. For example, Marie-Jeanne Desauge asked for a separation to protect her marriage fortune. All the witnesses interviewed concurred that Desauge was "a very fine man" who worked hard but whose business had been undermined by the serious dearth that afflicted the kingdom in the late sixties. Deeply in debt to his suppliers, he was unable to extract payment from customers who owed him money, and as a result he faced business failure.[71] In a situation such as this, the court may very well have refused to grant separation, on the grounds that the wife had to share in the husband's losses even as she profited from his gains, provided that he was not guilty of misconduct.

Since in practice only women could request separation, our study has focused on cases of husbands' dissipation or violence. It ought to be noted, however, that in baker families men were sometimes the victims of their wives' "bad conduct." Apparently frustrated by her husband's inability to succeed in business, the wife of master Louis Tayret abandoned him, taking their two children and most of their valuables. Profoundly shaken, Tayret appealed for their return. A month after his second marriage (his first wife having died six months earlier), master Louis Suire suffered another spousal loss. His new wife, aided by her mother, ran away with over three thousand livres in cash. Louis de La Plaine of the faubourg Saint-Antoine had virtually the same experience. The wife of another baker left him in order to live with a butcher, with whom she had a child. Master Jean Duboisson was cuckolded by his grain supplier, a merchant from the rue de la Mortellerie. His wife abused him physically and verbally, "lied to him," and "caused him despair." Finally she absconded with her lover and the household furnishings.[72]

Emotionally, professionally, and socially, marriage represented a crucial act in the life of the baker. In many instances, it was not merely the fruit of individual choices but the considered result of family calculation. Widespread endogamy testifies to the value placed on family networks. Tentacular dynasties implanted themselves throughout the bakery along with more modest family alliances. In most cases, the parties carefully measured the contribution each side would make to the new entity, though they were generally too shrewd to reduce everything to cash value. Female and male dowries took diverse forms, strengthening the new couple in different ways. In gross terms, men brought more than women, masters invested and attracted more than nonmasters. Combined matrimonial fortune situated bakers in the middle to lower ranks of the artisan-merchant population, too vague and broad a context for useful comparison. In concrete terms, the average couple started off with a promising launch fund. Disaggregated, marriage fortunes revealed important disparities in composition, according to the socioprofessional makeup of the union. By and large, "master" marriages had a more useful mix of elements as well as a higher total worth. Faubourg matches were closer to the master type than were forains. Variation across time in quality and quantity of wealth are hard to explain in terms of business cycles, though the second half of the century, a blend of galloping prosperity and recurrent socioeconomic crisis, seemed to favor more flexibility and easier adjustment than the first half, an era marked by slower economic growth and more rigidities.

Wives played an absolutely critical role in the management of the business as well as in the governance of the family and household. Though not an acknowledged "quality," baker-wife was a virtual profession. A woman without a certain aptitude for business and hard work would have grave difficulty in acclimating to the position. Work and home were wholly imbricated in the bakery. With different work schedules, at least at the beginning of the career, husband and wife assured a virtual round-the-clock operation, incidentally creating matrimonial contretemps that sapped certain unions while solidifying others.

Second (or third) marriages were common and were sometimes contracted with striking celerity. Widowers seem to have had as hard a time as widows in making it alone, though not necessarily for the same reasons. Socially, economically, and psychologically, widows were far more constrained in the range of accommodations they could make than their male counterparts. When remarriage seemed beyond reach, widows fashioned survival strategies against sometimes harsh odds. Figuring prominently in all strata of French society, widows need to have their story told.

Failed marriages, as we might understand the notion, rarely came to the surface. Powerful juridical and social incentives disinclined men and women to think in terms of insurmountable incompatibilities. Everything induced couples to adjust, and to repress, in some fashion or other. Under special circumstances, separation could be envisaged, if women could demonstrate the imperative need to shelter their wealth (and that of their children) against dissipation or to protect their safety from cruel aggression. It is not surprising that the latter case was harder to make than the former, though further research ought to be done before this impression hardens into generalization.

Chapter 12

Fortune

Just as the marriage contract offers a picture of the wealth of a baker couple at the outset of their career, so the after-death inventory (*inventaire après décès*) conducted by a notary is the key to forging a notion of the baker couple's fortune later in their professional life, if not at its end.[1] An inventory was not automatically conducted upon every death. A costly and often complicated procedure, it had to be "requested," in the juridical sense, by an heir worried about the solidity of the estate or suspicious of other heirs, by a guardian representing the interests of minor children, or by creditors seeking to protect their rights. Sample surveys conducted in the Parisian notarial archives, like similar studies at Lyon, suggest that only about one in ten deaths triggered inventories (though in Paris, unlike in Lyon, the more affluent families were the ones that demanded inventories).[2]

Among bakers, recourse to inventories appears to have been far more common, perhaps because of the need to take stock of accounts payable and receivable and of merchandise on hand in order to permit survivors to continue the business or dispose of it on favorable terms. The high costs of the inventory must have deterred certain families from requesting that one be performed. The family of master Firmin claimed to have paid 283 livres 12 sous for the drafting of the inventory and another 386 livres 11 sous for the evaluation and sale of furniture, not to mention 514 livres 14 sous for the "cost of seals," the application of which by a police official frequently preceded the notarial assessment. Their case was surely an exception, given the large estate in question, but it does suggest how heavy the burden could be. All the members of another master baker family with considerable wealth—but without creditors to annoy them—agreed to "not have an inventory done of property and effects in order to avoid expenses." Instead they had an informal evaluation made "by common friends."[3] Given the extraordinary range of fortunes I have encountered, it is my impression that most Parisian bakers with going concerns had inventories conducted.

In most cases a police commissaire placed seals on the property and papers of the deceased shortly after his (or her) death at the request of one of the "interested parties." Cumbersome and constrictive, the seals hindered both household and business operations, though members of the immediate family and employees were permitted to make use of common effects necessary for everyday activities. The seals were not usually removed until an inventory was completed. Thus, while they had up to three months to have the inventory performed, most families preferred to have it done expeditiously. The purpose of the inventory, according to a leading eighteenth-century jurist, was "to determine in a detailed manner the number, the quantity, and the type of possessions that compose the estate."[4]

The inventory was habitually conducted by one or two notaries and a licensed appraiser (*huissier-priseur*), occasionally assisted by experts called upon to evaluate items that required special knowledge (such as baking tools and wheat and flour stock).[5] Usually they began with household furnishings and other personal items and then dealt with producer goods and stocks in the bakery. More often than not, the inventory was exhaustive and meticulous. The notaries put a price on every item from frying pans and soup pots to corsets and shoe buckles to silver decanters and ivory crosses to handkerchieves and portable toilets to chairs and dressers to bolters and kneading troughs to bread molds and dough blades.

The appraisal was followed by an exacting examination of all titles, deeds, and papers of the deceased. The notaries sought to uncover all assets and liabilities: the amounts due to the estate for bread and other goods furnished (calculations based on business registers, *tailles*, and notes) or as a result of promissory notes of sundry provenance; the real property, offices, annuities, and other investments possessed by the deceased; and the amounts due on the part of the estate for supplies, services (a remade oven, milling charges, legal fees), loans and interest, and so on. The notaries sometimes reported tax returns (capitation and "industrie"), guild charges (mastership, confirmation of mastership, confraternity dues), rent receipts, leases, medical and funeral bills, lawsuits and judgments for and against the deceased, the marriage contract(s) of the deceased, vouchers for dowries accorded children or for advances made on anticipated inheritance, legacies received, and property or rights transferred. Finally, members of the family were invited to add any further information pertinent to the estate. Heirs sometimes called attention to properties that had not been mentioned or had allegedly been concealed; they challenged the enumeration of the accounts due on the grounds that the debtors were unknown to them or were of questionable solvency, and they produced receipts to prove that the claims of certain creditors were specious.

Was the portrait of a baker's fortune fashioned by the inventory as reliable as it was thorough? Surely there were errors, instances of missing data, and cases

The Baker's Outfit, a caricature portraying numerous bread-making operations, dominated by the task of bolting flour, a product that the baker might obtain in rough form from the water mill in the background. *Source:* © *cliché Bibliothèque nationale de France, Paris.*

of negligence and fraud. Historians can rectify some of the internal mistakes by subjecting each inventory to rigorous critical scrutiny, purging it of double entries, restoring items to the category in which they belong, recomputing *all* notarial calculations from the simplest kind of arithmetic to more complicated operations involving debt amortization, the derivation of capitalized annuity values, and the discounting of notes. Occasionally one can even supply ostensibly missing information by inferring it from internal evidence or by locating it elsewhere in the notarial or police archives. None of the notarial decisions was rendered arbitrarily. The "interested parties" had the right throughout the process to contest evaluations. Each heir was represented and a "contradictory" climate was fostered in the hope that an eventual consensus approximating the truth would emerge from the adversarial give-and-take. Fraud is not easy for us to discern, but it was not always easy to perform it either. To be sure, there must have been opportunities for the surviving spouse to siphon off certain effects (cash, jewels, silver, various titles) during a terminal illness preceding death or immediately upon death. Or the major heirs, in order to thwart creditors or frustrate collateral claimants, might have "skimmed" the estate by joint agreement. One cannot take at face value the oath sworn by members of the deceased's household that nothing has been diverted from the estate.

Yet there is no reason to assume that fraud was regularly practiced. Apart from the possibility that many of the bakers were more or less honest, one must remember that dying, like marrying, was a business transaction. Death, in a sense, marks a brutal reversion to the more or less Hobbesian premarital state of tension. Even as an alliance disintegrates into its constituent parts in certain circumstances, so the family cleaves and organizes along lines of affiliation to the husband and wife, with the children theoretically represented by a third-party standing between the suddenly resurgent family of the deceased and the survivor, who turns to his/her own family for support. Mercier's depiction of vulturous kin calculating their shares instead of grieving over the death of a loved one is a caricature, but one, like most of his vignettes, that takes off from a bit of observed reality.[6] The stories about the relative standing impatient vigil over the last agonies of a dying family member in order to be in position to call on "his" (or "her") commissaire to place seals at death before his rivals could alert "their" commissaire were not all apocryphal.

Each party to the death wanted to make sure it received fair if not privileged treatment. Even as they served as checks on the notaries and appraisers, so the contending "interested parties" served as checks against each other's infidelity. Like the father of the deceased wife of master baker Firmin Delamare, who accused his son-in-law of cheating the heirs by "dissipating" merchandise and effects prior to the inventory, the heirs in many situations were highly alert to the possibility of fraud.[7] Their watchfulness acted as a general deterrent even if it did not prevent knavery. Indeed, the "interested parties," even in families whose solidarities did not erode under pressure of death, had recourse to an inventory precisely to avoid or attenuate mutual suspicion and recrimination. Moreover, in many cases, there were other keen-eyed checks against misrepresentation in the person of the creditors, who were usually neutral parties outside the circle of family passions, whose only concern was to recover monies due them.

On balance, the inventory seems to me to present a reliable notion of baker fortune. There was probably a bias toward underestimation of merchandise and of relatively liquid items such as silver and jewels, which were more often than not appraised below their market value. (But then again, what criteria does one apply to used sheets or socks or pots, for which "market" value makes little sense?) The value of real property may have been underestimated, for notaries often cite the original purchase price without allowing for unearned increment or future capital gains. The same may be true of some offices, but not of others, which stagnated or declined in face value. Certain investments and some heirlooms may not have been enumerated at all in the inventory. In a number of cases the accounts receivable category is incomplete, but in others it may be excessively optimistic in its elaboration of scores of more or less old obligations. Similarly, it is likely that all the claims against the estate were not registered, especially if the inventory took place immediately upon death (after the mandatory three-day grace).

I do not believe that these distortions amounted to very much. In my view the inventories reflect real orders of magnitude of baker fortune and they can be considered reliable not only for internal comparison (where "errors" would ostensibly cancel each other out) but also for comparison with the fortunes of other social and professional groups. A more serious problem of comparability arises from the fact that the length of the "career" of the bakers varied significantly (measured, say, from the marriage contract to the death of one of the spouses).[8] It is important to add that the significance of the inventory is not merely to be measured quantitatively in terms of the total fortune it presents. Even if it were not a staunch guide to measurable wealth, the inventory would be precious for all it reveals about family relations, business practices, consumption patterns, taste and fashions, lifestyle, the use of space, the role of credit, and affective bonds.

Baker Wealth

The total assets of the 101 bakers of all types for whom we have inventories or surrogate inventories (*partages, comptes*) range from 184 to 99,988 livres (see figure 12.1). The average was 14,100 livres, but the median, nullifying the impact of the largest fortunes, stood at 5,219 livres. Over 37 percent had assets over 10,000 livres and 22 percent over 20,000 livres. The total liabilities of the bakers ranged from zero to 24,536 livres, with the average at 1,784 and the median at 721 livres. Less than one-third of the bakers had liabilities over 1,500 livres, and fewer than 8 percent had claims against them of over 5,000 livres. I consider the baker's total fortune to be his net worth, or his assets less his liabilities. Net worth ranged from *minus* 6,422 to 99,988 livres. The average was 12,422 and the median 4,228 livres. Ten percent of the bakers had a negative net worth. Over a third were worth under 2,000 livres (including the negative balances), while about the same number had a fortune of over 10,000 livres. Almost a fifth of the bakers had fortunes over 20,000 livres; a little more than one-eighth had a net worth above 30,000 livres.[9]

During this century of considerable economic expansion and rising prices, baker fortunes grew substantially. As table 12.1 indicates, the assets and net worth of bakers after 1745 were more than twice as high as before that year (liabilities were three times as high).[10]

A decade-by-decade study, however, reveals that fortunes did not increase continuously from the beginning of the century to the end (see figure 12.2). After an upward bound between 1711–25 and 1726–40, net worth stagnated during the next ten years and dropped slightly in the following decade. Unlike net worth, assets increased substantially between 1741–50 and 1751–60, but so did liabilities. The period between midcentury and 1775 constituted the years of most stunning development in bakers' wealth. Liabilities declined sharply, as bakers freed themselves from debts and depended less heavily than usual on credit for the purchase of grain and flour. Simultaneously, assets soared, followed closely by net worth, which almost reached the twenty-five-thousand-livre level. Baker fortunes suffered a brutal reversal between the mid-seventies and 1792 as assets dropped below twenty thousand livres and liabilities mounted dramatically from under two thousand to almost thirteen thousand livres.

It is difficult to account for the movement of fortune from midcentury onward, especially the phase of spectacular expansion. It appears that many bakers must have profited substantially between 1751 and 1765, though grain and bread prices tended to stagnate despite the Seven Years War. At least on the surface, bakers seem to have defied the popular adage that only dearths en-

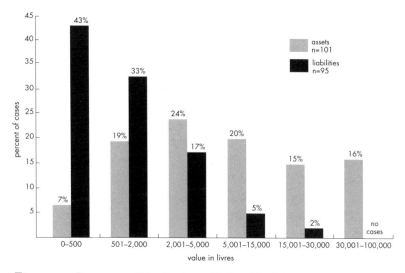

Figure 12.1 Percentage Distribution of Baker Total Assets and Liabilities
(Source: after-death inventories)

Table 12.1 Baker Net Worth *(in Livres)*

| | Before 1745 | | After 1745 | |
	Mean	Median	Mean	Median
Assets	9,903	3,016	21,058	14,357
Liabilities	1,003	454	3,185	1,323
Net worth	8,932	2,592	18,208	11,591

gender affluence. The following decade was marked by a hectic series of subsistence crises, the longest and most intense of the century. Acute protracted scarcity issued in a general economic crisis as well as in political upheaval.[11] Some bakers skillfully took advantage of the uncertainties, but others suffered from scarcity, market disorganization, price controls, inadequate cash flow, and tight credit. Is it possible that the sharp rise in net worth in the periods 1751–60 and 1761–65 reflected an achievement substantially completed by 1765? Normally, there ought to be a certain lag between the accumulation of wealth and expenditures in consumer goods and investments. Similarly, skyrocketing liabilities and declining assets toward the end of the Old Regime may bespeak the results of the crises between 1765 and 1775, though it is true that the Parisian (and "national") economy suffered a severe general setback in the decade before the Revolution that undoubtedly had an adverse effect on the bakery.[12] We

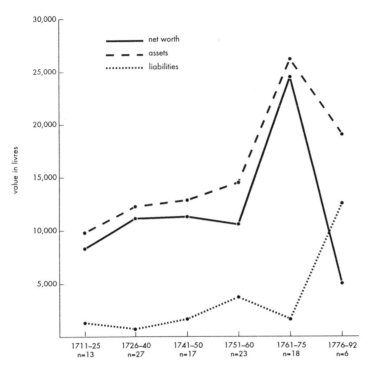

Figure 12.2 Mean Baker Fortunes, 1711–92 *(Source: after-death inventories)*

Table 12.2 Baker Fortunes *(in Livres)*

| Status | Number | Assets | | Liabilities | | Net Worth | |
		Mean	Median	Mean	Median	Mean	Median
Master	63	16,224	8,620	2,296	1,008	14,074	7,379
Faubourg	34	10,205	2,835	712	320	9,535	2,684
Forain	4	13,757	10,467	2,808	2,290	10,950	8,671

have too few cases, especially in the last years of the Old Regime, and too little detailed knowledge spanning the century, to conclude that the baker corps was gravely and cumulatively damaged as the Revolution erupted. The crisis of 1788–89 put scores of bakers in serious difficulty, but it is impossible to say how many were structurally debilitated as opposed to cyclically/transiently wounded. Despite periodic shakeouts, taken together the bakers seem to have resisted in terms of net worth and the relation of net worth to operating capacity.

Across the whole period, the masters, as we might expect, were richer than the other bakers, as table 12.2 indicates. Only a fifth of the masters had assets

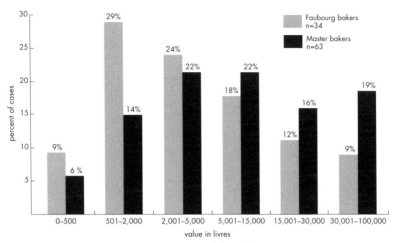

Figure 12.3 Percentage Distribution of Total Assets of Master and Faubourg Bakers *(Source: after-death inventories)*

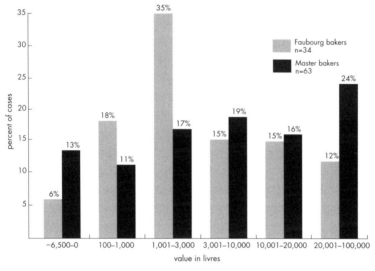

Figure 12.4 Percentage Distribution of Net Worth of Master and Faubourg Bakers *(Source: after-death inventories)*

under two thousand livres, while 40 percent had more than twelve thousand. Half of them had liabilities under a thousand livres, a third under 375. Thirteen percent of the masters had negative balances, while about one-quarter were worth over twenty thousand livres. Though the faubourg bakers were substantially less indebted than their guild counterparts, only a little more than a quarter of them had assets over twenty thousand livres and only 29 percent had a net worth of eight thousand livres (see figures 12.3 and 12.4).

Yet the superiority of the masters erodes strikingly if one compares them not with all the faubourg bakers but with those whose well-being, self-image, and ambition prompted them to appropriate the wholly unofficial dignity of "merchant baker."[13] The merchant title conferred and represented a significant social capital, though if it were to prove useful to the actors it had to command real economic credibility. Merchant became a response to master, a liberal alternative to the corporatist pole. In mean terms, the assets of the merchant class, numbering eighteen, amounted to 15,807 livres, a little more than a thousand livres under the masters' average. Their liabilities were under five hundred livres, barely more than one-fifth of the masters'. At 14,593 livres, they outdistanced the masters in net worth by almost five hundred livres. Thus substantial wealth was not the appanage of the guild. The "merchants" constituted a noncorporate elite with profitable businesses, comfortable homes, and solid investments.

The relative opulence of the country forains is surely misleading. It is very likely that only the very richest requested Parisian inventories. The others either did not bother with inventories or had them done locally. The key to country forain wealth was the valuable parcels of property that the richest bakers (probably laboureurs-bakers) owned. It is likely that the level of wealth of forains without lands was clearly below that of the Parisians, be they of faubourg or guild.

Of what did baker assets consist? They can be divided into three broad and sometimes overlapping categories that we shall successively examine: household; business; and property and investments, including paper and titles not directly related to the baking business.[14]

The Household

Although the headings (*intitulés*) of the inventories are supposed to introduce us to all the heirs, they do not always give a complete or precise idea of the size and composition of the family. Moreover, we rarely know where the family is in the normal cycle of development. On average each family had about three children, surely an underestimation even if we allow for predeceased offspring. This "fertility" did not vary according to baker type. Twenty-six of our bakers had children in the baking business, an average of almost two per family. There is little doubt that a far higher proportion of families had at least one child in the bakery.[15]

Thirteen bakers owned the houses in which they lived (others owned houses

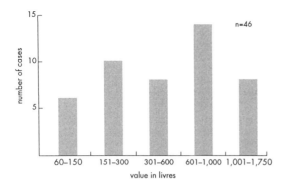

Figure 12.5
Rents Paid by Bakers
(Shops with Apartment)
(*Source: leases*)

that they rented out while they lived as tenants elsewhere), fifteen were "principal tenants who exercised a moral and economic jurisdiction over the other tenants," fifty-five were plain tenants, and the status of the remaining bakers is unknown.[16] The house owners and principal tenants had assets of roughly equal size, but the latter's liabilities were more than three times the former's. Thus while the average net worth of the owners amounted to almost fourteen thousand livres, that of the principal tenants was barely over ten thousand livres, only about fifteen hundred livres more than the net worth of the simple tenants.

Rent for tenants of both types varied from sixty to one thousand seven hundred fifty livres a year. Almost a third paid under three hundred and more than three-quarters paid under a thousand. The average rent was 595 livres, twenty livres above the median. Predictably the masters, lodged in the city center, paid a much higher average rent than the faubourg bakers: 718 versus 246 livres. Nor is it surprising to note that the principal tenants, who took responsibility for the entire house as a rule, paid almost twice as much rent on average as the simple tenants: 929 versus 469 livres. As one might expect given the rise in prices and the evolution of baker wealth, average rents after 1745 were almost twice the level prior to midcentury (see figure 12.5).

Excluding the kitchen (which sometimes served as bakery anteroom), the shop, and the bakeroom (most often underground), baker families did not dispose of a great deal of space in which to live. They had a little over two rooms on average (median = 1.74). Forty-one percent possessed only one room, 35 percent had two rooms, and only 24 percent disposed of three rooms or more. A study of fifty baker *scellés* (preinventories conducted by police officials who placed seals on all the catalogued items in the estate) reveals a more favorable pattern of available space. Bakers averaged 2.6 rooms apiece and almost half had three rooms or more. These figures adumbrate the boundaries

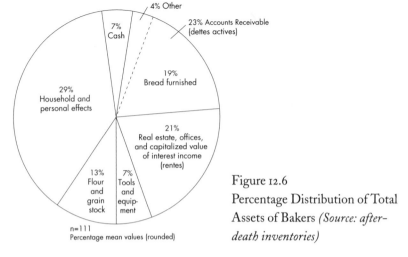

Figure 12.6
Percentage Distribution of Total
Assets of Bakers *(Source: after-
death inventories)*

of the world in which the baker families moved, but they are of limited comparative value because we do not know the size of the rooms in question.
Though they often doubled as living rooms, virtually all of the rooms were
designated and furnished as bedrooms (the kitchen of course also played the
role of salon).

Bakers had on average half a room more in which to live in the second half
of the century than in the first. Generally the amount of space available varied
more or less directly with the amount of the rent: bakers in the highest quartile
of rent payers had an average of two and one-half as many rooms as those in the
lowest quartile (though the faubourg bakers, who paid less rent than the masters, disposed of slightly more space: 2.19 versus 2.08 rooms). There is a weaker
relationship between rent and net worth (r = .35 at .008), for the bakers in the
third quartile of rent payers had a substantially higher fortune than those in the
higher quartile. Paying a high rent, then, was not necessarily a sign of either
solid fortune or good resource management.[17]

The inventories indicate that only eighteen families had a servant, but the
incidence of domestic help is almost certainly underrepresented. The bakers in
the scellés had on average one servant each. It is safe to assume that all of these
domestics worked both in the household and in the business, there being no
fine line between house and "work" domesticity as there was in bourgeois and
aristocratic households.

Household assets, including clothing, linens, furniture, jewels, silverware,
and wine holdings, along with cash (cash straddles the line separating household and business effects), represented on average over a third of all total assets
(see figures 12.6 and 12.7). Thus defined, the baker household ranged in value

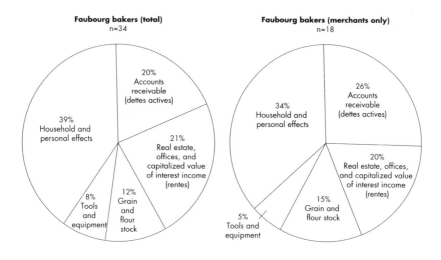

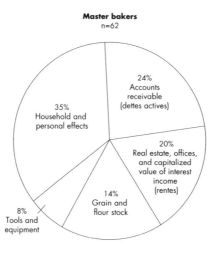

Percentage mean values (rounded)

Figure 12.7 Percentage Distribution of Assets of Master, Faubourg, and Merchant Bakers *(Source: after-death inventories)*

from thirty-five to 23,400 livres and was worth on average 2,760 livres (median = 1,237). The masters averaged about four hundred livres more and the faubourg and forain bakers between seven and eight hundred livres less. Two-fifths of the whole baker population had household assets evaluated at more than two thousand, 27 percent over three thousand, and 6 percent over six thousand livres. The larger the household fortune, the greater the value of the family fortune.[18]

The single most important component of household assets was cash, the mean value of which amounted to 8.7 percent of total assets (median = 2.5 percent). Cash holdings ranged from zero to 18,360 livres, with an average of 1,348. The median, 170 livres, suggests how uneven the distribution was: almost three-quarters of all bakers had less than one thousand livres in cash (84 percent of the faubourg bakers possessed less than 425 livres in cash). The average amount of cash held in the fifty scellés was 196 livres, but twenty-eight of the families had none at all. The masters had 207 livres and the faubourg bakers 172 livres on average. Cash holdings in both the inventories and preinventories reached a peak in the period 1751–60.

More widely shared, holdings in furniture loomed larger than cash in most baker families: 527 livres per household (median = 424). Furniture capital was a good predictor of total assets.[19] Though every family had some frills, most of the furniture was of a highly practical sort: beds—over two in each household—represented on average more than one-third of total furniture value.[20] The value of furniture in each household rose steadily from the beginning of the century to the Revolution.

Clothing amounted to 6 percent of total assets on average (median = 3.9 percent). The wife's wardrobe, which usually included several dresses and many accessories, was worth more than the relatively austere garb of her husband: 248 versus 152 livres. In the countryside this ratio was reversed, but the calculations rest on too few cases to permit any generalization about the roles of rural versus urban men and women. Guild members evidently esteemed clothing and appearance far more than their faubourg colleagues, perhaps because they socialized more frequently in public and semipublic gatherings (corporation and confraternity meetings, for example). Masters had 197 livres in clothing and their wives 354 livres, while faubourg bakers could vaunt only seventy-five and eighty-five livres' worth (see figure 12.8).[21]

Another symbol of status and self-esteem was the wig. Bakers who sported it could pass for bourgeois de Paris or for successful merchants, and a handful cultivated this image. (Perhaps to enhance both his status and his self-confidence, Jean-Baptiste Hornet wore a wig on his rounds to collect money due for bread).[22] The record holder, a merchant baker, had five wigs of different hues. One baker had four wigs, seven bakers had two wigs each, and fourteen others had one apiece. Unknown in the country lifestyle of the forains, wigs were almost as common among faubouriens as among masters. If one judges by their holdings in clothes and jewelry, the baker "wigs" were vain. The wigs may have covered eggheads, for bakers with wigs owned twice as many books as the rest of the population. Their household fortune was below average. But they possessed grain and flour worth 40 percent more than the others, real property worth 37 percent more, and they paid a higher capitation. The total assets of the wig wearers were 43 percent above the universal average and their net worth 44 percent above it. So bewigged bakers were not merely swaggerers; they were men of substance to be reckoned with.

Household linen was rather evenly distributed (186 livres on average), but only 28 percent of the families possessed jewelry, worth an average of sixty-three livres. Silverware of all sorts was a much more important luxury investment item. More than four-fifths of the bakers owned some silver with average holdings at 325 livres (median = 169). Masters boasted barely more jewelry than the average, but they had almost 30 percent more silver.[23] Together, silver, jewelry, and cash constituted a sort of liquidity factor that represented on average 13 percent of total assets (median = 7.8 percent). Wine was another (extremely liquid) luxury item that was kept by only 27.5 percent of the baker group. The average cellar was worth a derisory forty-four livres, but the amount increases to 161 livres if one excludes those who owned no wine from the calculation. Two-thirds of the bakers kept a modest stock of wood worth thirty livres on average, but it is not clear whether it was heating or cooking fuel.

The "household" portrait of the baker family reveals a simplicity of lifestyle. Since the members spent most of their time in the shop or the bakeroom, there was little need for living space other than the bedroom and the kitchen. Rooms were sparsely furnished, with a bed, a chair, a dresser of sorts, and sometimes a wardrobe. There was little ostentation, save perhaps for a costume of Sunday elegance. Silver, the most solid and traditional domestic investment, absorbed whatever surplus capital could be drained off into the house. Most families probably drank beer rather than wine at home.

A number of bakers had other belongings that throw some light on their attitudes and values. Fifty households contained religious objects such as crosses (wooden, ivory, and occasionally bronze alloy), prints of devotional scenes, and icons, while forty apparently had devotional monuments or mementos.[24] Twenty-six families had other cultural and aesthetic investments such as family portraits, still life and landscape paintings, and pieces of decorative sculpture. Only a minority seemed interested in reading if one can infer real utilization from ownership, though most were literate if one can infer the ability to read and comprehend from the ability to sign a contract.[25] The readers are interested primarily in subjects that we call traditional. The entire baker population averaged nine books per family, but if one counts only the 20 percent who actually owned books, the average holding increases to forty-four

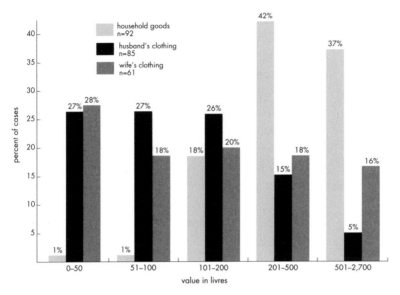

Figure 12.8 Percentage Distribution of Baker Household Goods and Clothing
(Source: after-death inventories)

volumes. Merchant bakers possessed more than four times as many books as country forains and more than twice as many as masters. Though the number of books a baker disposed of appears to vary with his wealth, the correlation coefficient (r = .53 at .001) does not suggest a very potent association. Devotional themes were the subjects of the majority of the thirteen cases for which we have information. In five libraries history and folklore joined confessional works. Religion appears to be their major "cultural" preoccupation, given expression by the physical symbols of worship with which a large number adorned their houses as well as by the choice of books made by the baker intellectual elite.

It will be comforting to those who believe that there is a connection between piety and worldly success to note that baker families who kept religious shrines of diverse sorts had a net worth almost twice that of the undemonstrative families. Their household fortune was more than twice as great, they had more than three times as many books, and they spent almost a third more on the funerals of their departed members. Of course it is quite possible that these families celebrated their religiosity because they were relatively well-to-do and not vice versa. The same might be true concerning artistic capital, whose owners had a net worth almost 70 percent higher, six times as many books, and far more real property and household assets than bakers who did not possess "cultural" goods.

Shop and tools, the first items in the "business" rubric of the fortune, represented on average only 7.4 percent (median = 4.4 percent) of total assets (see figure 12.9). These investments ranged in value from zero (in these three cases, the bakers must either have just retired and disposed of their equipment, or for some reason, the appraisers did not include it in the evaluation) to 1,297 livres with an average of 287 livres (median = 222) per baker. More than two-thirds of the bakers had tools estimated at under three hundred livres. In many instances the baker equipment was old and badly worn; in other cases it was undervalued by extremely conservative experts. The relative cost of the equipment seems to have been infinitely greater from the perspective of a newly launched baker at the time of his marriage or purchase of a fonds than it later appeared from the vantage point of mid- or late career.[26]

Investment in equipment remained remarkably stable across the century, varying on average between 235 and 284 livres, save for the period 1751–60, when it swelled to 544 livres. Masters had more elaborate shops and bakerooms (average = 341, median 270 livres) than faubourg bakers (214 and 171 livres) and forains (174 and 177 livres). The higher the value of one's capital equipment, the more grain and flour one was inclined to stock and the more journeymen one tended to employ—in a word, the more vigorous, if not always the more profitable, one's business appeared to be.[27]

Among the elements included under the equipment rubric were the furnishings necessary for the shop itself: counter, scales and weights, bread knife, shelves, bins, window grill. While the average value was fifty-seven livres, fifty-five percent of the baker contingent had a shop worth under fifty livres. The value of shop investment rose steadily throughout the century, reflecting not only increasing prices (and in some cases burgeoning prosperity) but perhaps a desire to make the shop more attractive and efficient. The masters had the highest shop value (69 livres) followed by the merchants (43 livres). The shops of faubourg bakers were often closet-sized and summarily furnished. Some of the forains appear to have maintained tiny retail outlets for local customers and as a result baked more than twice a week.

The kneading trough was probably the most important single tool the baker had (besides the oven, which was never evaluated in the inventory, perhaps because it was considered to be part of the fonds or of the leased property in which the baker lived). The trough occupied the same place in the business as the bed in the household: it was an instrument of necessity and an emblem of relative status. On average bakers kept two kneading troughs each (mean value

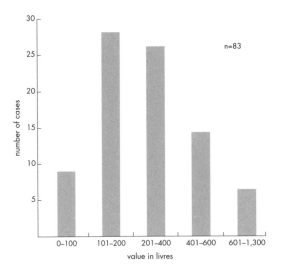

Figure 12.9
Baker Investment in Shop
Fittings and Tools *(Source:
after-death inventories)*

43 livres) and country forains between two and three each (mean value 50 livres). In the preinventories, the bakers averaged a little under two troughs per shop (the masters a bit less, the unincorporated bakers a little more).

Traditionally bakers did their own bolting, in part because millers were unequipped or unwilling to do it, and in part because the bakers preferred to take personal responsibility for fashioning the different grades of wheaten flour according to their own standards and their customers' taste. Paris bakers had an average of 2.4 bolters per shop (worth 77 livres), while the country forains had only one (because they consumed less flour? because they did not produce as many different types of flour as their Parisian counterparts?). The average value of bolters dropped by almost 60 percent and the average number per bakeroom by one full machine between the periods 1751–60 and 1761–65. Did this change—the reversal of a secular trend—reflect transformations in milling technology and flour production?

Too slim to be conclusive, the evidence is nevertheless suggestive. A growing number of millers, especially around midcentury, began to do their own bolting at the mill, with machines often powered by the mill force. These were the same millers who did less and less contract milling for individuals and individual bakers and more and more speculative grain buying and conversion on their own account for wholesale distribution to bakers, either directly or through brokers at the Halles. Logistically these "flour merchants," as the millers liked to call themselves, made things easier for the bakers, and by eliminating the grain merchant they tended to make things a bit cheaper as well. They also saved the baker time and labor (and therefore money) by sparing the baker the job of bolting, which required not merely the services of a more or less skilled journeyman and a machine but also considerable space. Moreover, the flour product of these mealmen was considered more often than not to be of excellent quality.

Home bolting did not disappear for a long time, but the fundamental technological and commercial shifts of the future were already visible around mid-century. Before 1745, the average bolted stock held by a baker at the time of inventory amounted to a value of 107 livres; after that date it was worth 1,141 livres, an enormous difference even allowing for the short- and long-term price movements. Before 1745, unbolted flour represented 73 percent of stock; after, 65 percent. In the preinventory study, whereas unbolted flour represented 92 percent of the stock in 1751–60, it fell precipitously to only 18 percent in 1761–75 (at the same time average bolters per baker dropped from 2.5 to 1.67). Finally, it is crucial to note that economic milling was introduced in the sixties. By this technique bolting was fully integrated into the milling process. Before this system became widely available, certain bakers kept bolters primarily to extract middlings that were sent back for regrinding. One finds significant amounts of middling flour—an elite meal—in baker stocks as early as 1725.[28] Economic milling extracted and ground the middlings in a single graduated process at the same time that it differentiated several other flour types. It made home bolting not merely gratuitous but highly uneconomic.

The average value of all utensils assumed under the equipment rubric was 117 livres. The masters had about half again as much in tool value as the faubourg bakers, and bakers in the second half of the century had more than twice the utensils investment as their counterparts in the first half. Bakers were supposed to have their own cloth sacks, marked with their name, for the purchase of flour and grain (though in fact we often find them sued by suppliers for failing to return sacks they borrowed). Relatively speaking, bakers disbursed substantial sums on sacks—sixty-two livres on average, more for masters, less for others. Bakers needed transportation in order to scour the markets and the countryside for grain and flour, to deliver and pick up their flour, and to send bread to the market. Most bakers apparently borrowed or rented horses and wagons (with or without drivers) when they needed them, for only 11 percent of the bakers in our inventory owned horses (the incidence of ownership being considerably greater among unincorporated bakers than among guildsmen). Similarly, only 12 percent of the preinventory population possessed horses (mostly faubourg bakers), and only 10 percent owned their own wagons.[29]

The average and median number of journeymen per shop among the inventoried bakers, a figure that seems to correspond to the magnitude of capital investment, was 1.9.[30] Fewer than a third of the bakers had only one journeyman, while 16 percent had three and 3.2 percent had four. Master and faubourg bakers had virtually the same number of workers (though the guild bakers had apprentices, who were theoretically not included in the inventory listing). The average number of journeymen per shop increased from 1.3 in 1701–10 to 2.7 in 1751–60, which suggests that the increase in bakeroom capital investment reflected a growth in the average size of baker businesses rather than merely a rise in prices. Given the fact that the global baker corps did not expand across the century (at least not before 1776, and even then nonmaster contraction may have compensated for guild growth), the larger work force also reflects the need to meet the demands of a vigorously increasing population. Each of the fifty bakers in the preinventories employed 1.7 bakers on average (median = 1.8). Masters had about two workers while faubourg bakers averaged 1.5. In this group as in the inventory sample, the bakers with the most workers maintained the largest stocks. Preinventory bakers with the largest number of journeymen also appear to have produced the most bread per day.[31]

The value of baker grain and flour stocks is of interest not only because it represented on average 13 percent of total assets (median = 9 percent) but also because it will enhance our understanding of the relative vulnerability of Paris to grain and flour dearth.[32] Paris, it should be recalled, had no formal organization of abundance on the model of Lyon, Marseille, or Geneva. Starting around 1725, the authorities put together an emergency reserve system located in the religious communities, colleges, and public-assistance institutions, but it was of only limited potential aid. Beginning in the second half of the century the government maintained a relatively modest reserve in a river town and milling center not far from the capital. Both the "community" and the royal granaries were of short-term tactical significance. Neither singly nor together were they substantial enough to avert disaster in a time of prolonged crisis. A number of subsistence theorists and administrators argued that the bakeries were the best granaries and proposed that all bakers be required to store at all times an amount of grain or flour (preferably the latter) equal to a certain percentage of their usual annual consumption. It was not until the first empire, however, that a version of this plan was put into effect. It proved to be highly successful, at least in psychological terms, the psychological factor having always been more than half of the subsistence battle.[33]

Did the bakers serving Paris ordinarily stock a substantial amount of grain and/or flour? Can one speak of a more or less concealed commercial reserve that afforded Paris some protection against the vicissitudes of distribution? Each of our 101 inventoried bakers had an average of 8.5 setiers (median = 0) of wheat and forty-nine setiers (median = 18) of flour. This stock represented approximately three weeks' production for the average baker—an inadequate reserve in absolute terms, but a precious margin when joined with other emergency resources and a larger buffer than most police officials thought was available. This hidden, built-in reserve grew steadily and significantly across the century, rising from the equivalent of 1.5 muids expressed in terms of wheat toward the beginning of the century to over seven muids in 1761–92. In the last three decades of the Old Regime, then, the median holding amounted to over a month's worth of the average baker's requirements. During this period all stock was held in the form of flour.[34]

If one shifts attention from central tendencies in the whole baker universe to just those bakers who actually held reserves, the figures of course project a different picture. Over 80 percent of the inventoried bakers held some form of flour stock. The average holding for the sixty among them whose stock could be counted was a very hefty 82.6 setiers (median = 52). Only eleven inventoried bakers kept wheat, amounting to an average of seventy-eight setiers a shop (median = 63). Guild bakers held almost twice as much flour as unincorporated bakers. For reasons that are not clear to me, masters residing in the Right Bank faubourgs of Saint-Denis, Saint-Laurent, Saint-Martin, and Montmartre had one-third again heftier stocks than masters in the old city nucleus and five times larger holdings than masters in the Left Bank faubourgs of Saint-Marcel and Saint-Jacques.

For the fifty preinventories, the average wheat holding was 1.58 setiers, but this figure is of no real significance since forty-eight out of fifty bakers kept no stock in wheat at all. The average stock in flour amounted to sixty-one setiers, or a little over five muids (median = 30 setiers, or 2.5 muids), a quantity larger than we find in the inventories. Nor does the evolution in holdings follow the inventory pattern. Between 1727–30 and 1741–50, median stock dropped from about 2.6 muids to 1.6 muids.[35] After skyrocketing during the relatively easy subsistence years of 1751–60 to 5.6 muids, the median stock regressed to approximately two muids—still a substantial reserve buffer—between 1761 and 1775, a period of extreme disorder in the provisioning trade. Though the discrepancy is less great in the preinventories than in the inventories, masters still have a large edge over faubourg bakers in holdings: an average of seventy setiers (median = 33) versus forty-seven setiers (median = 27).

On average, in the inventories there was almost twice as much unbolted (gross) flour as bolted flour, hardly encouraging evidence for the precocious dissemination of the new economic milling technology. Almost three-quarters of the bakers had stock in unbolted flour, whereas only 45 percent of the population stored some form of bolted flour. In the preinventories unbolted flour made up only one-third of the average stock, though another third, consisting of flours of unidentified types, might have contained a substantial portion of the gross meal. White represents 23 percent of average holdings, middlings 6 percent, dark 3 percent, and mid-white 2 percent. The faubourg bakers had as much white as the masters, but almost twice as much gross. The comparison is largely vitiated, however, by the fact that 53 percent of the master holdings are of unidentified flour types.

Virtually every baker (98 percent) allocated a special area for grain and flour storage. Among the inventoried bakers, on average a room and a half (median = 1.4) were reserved for stocks. The preinventory bakers each had 1.67 (mean and median) granaries on average, and in addition they each maintained a flour room for bolting, mixing flours, and preparing flour for kneading (sometimes a pipe or conduit led directly from the flour room to the kneading trough).[36] The storage areas had to be dry places with good ventilation. Over the course of the eighteenth century, bakers paid more and more attention to storage facilities, for they understood that well-kept grain and flour produced not only better but also more bread. They needed space in the granaries for turning and aerating both wheat and flour, for cleaning wheat, and for allowing flour to cool and "recover" after milling.

In addition to selling bread, bakers retailed flour in small amounts for household purposes and the fuel residue (charcoal) left in the ovens after baking for use in domestic cooking and heating. Small amounts of charcoal averaging under ten livres in value were found in the shops of one-quarter of the inventory pool.

Paper Assets: Accounts Receivable

Over a fifth of total assets of the inventoried bakers consisted of amounts due to bakers for various reasons (what the French call *dettes actives,* or what we call accounts receivable, in contradistinction to *dettes passives,* or accounts payable) (see figure 12.10). The average amount of money due per baker was 3,983 livres (median = 1,241). A quarter of the bakers had under three hundred livres' worth of accounts receivable and almost two-thirds held paper valued at under three thousand livres, while 20 percent of the bakers claimed sums over six thousand livres. The average amount of money due per baker increased steadily across the century, mounting from 2,472 livres in 1711–25 to 5,569 livres in 1761–65. The masters, with an average of 4,139 (median = 1,435), had larger debts due than the

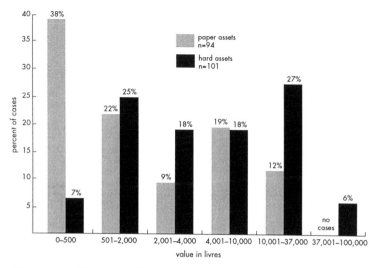

Figure 12.10 Percentage Distribution of Baker Hard and Paper Assets
(Source: after-death inventories)

faubourg bakers (3,739), but it was the merchant baker who held the record for accounts receivable (5,765). In most cases, these debts were well founded; they do not represent paper dreams, even if in certain cases the solvency of the debtors cannot be guaranteed. Accounts receivable are a good predictor of total assets (r = .84 at .001) and of net worth (r = .79 at .001); the richest bakers had the largest debts-due component.

The lion's share of accounts receivable consisted of money due for the supply of bread on credit. On average the bread part was 78.6 percent, but the median, 91.5 percent, indicates the character of the distribution. In 39 percent of the cases bread debts equaled total debts due, and in 60 percent of the cases bread represented over 85 percent of all accounts receivable. Masters (mean = 82 percent, median = 92 percent) and faubourg bakers (76.6 percent and 94.7 percent) evidenced substantially the same proportion of bread debts (though in absolute terms the holdings of the guildsmen in accounts receivable for bread was substantially higher), while the forains' part was less than half the universal average (41.4 percent and 42.4 percent), most likely because these itinerant bakers sold a much smaller percentage of their volume on credit.

Accounts receivable for bread for the whole population ranged from zero (only 12.5 percent of all cases) to 33,245 livres, a record debt portfolio belonging to a faubourg baker. Most bread debts were relatively small and scattered among a large number of persons (and a few collectivities). Bakers had from six to 153 bread debtors, the average being thirty-seven (median = 28). Masters had more customers in debt (mean = 40, median = 31) than faubourg bakers (30 and 20) and country forains (23 and 17). Still, there was a handful of extremely large individual debts, usually emanating from institutions, from operators of inns and pensions, or from influential public figures, for these were the only clients to whom our bakers would ordinarily show more or less unbounded indulgence.

Highest individual bread debts ranged from forty-three livres to 9,990 livres, with the average at 1,799 and the median at 481 livres. Only 7 percent of the inventories enumerated "bad" or "extremely doubtful" debts; they have not been included in the figures either for total or for bread accounts receivable. Frankly avowed write-offs ranged from eighty-nine livres to 2,574 livres (a master took the greatest loss), with an average of 1,060 livres (median = 873 livres). Though we were able to find concrete mention of the practice of the *taille* in only nine cases in the inventories and twenty-eight of 144 instances in the business-failures study, it is certain that many more bakers utilized the *taille* as a primary or supplementary form of bread-debt accounting. The nonbread accounts receivable of the bakers resulted from loans and advances of various kinds (sometimes to bakery customers). They are far too atomized and personal to claim analysis here.

Investments

The last category of assets covers real estate, annuities (*rentes*), and venal offices. Together they represent a substantial portion of total assets—over 20 percent on average. But average figures (Paris real estate = 1,974 livres; rural real estate = 527 livres; rentes = 311 livres; offices = 732 livres) are misleading, for the distribution of these properties was extremely unbalanced. Only 21 percent of the bakers owned real estate in Paris, holdings that varied in value from 250 to 28,025 livres. Five and a half percent owned Paris property worth under five thousand livres and 7.4 percent had real estate valued at more than ten thousand livres. About the same proportion of masters and faubourg bakers owned Parisian real estate, but the former possessed 75 percent more than the latter, while the forains boasted none. If one scrutinizes only those who owned Paris property rather than the whole baker universe, the average value was 9,155 livres, the median 8,153. Thus the twenty bakers who possessed real estate in the capital—thirteen masters and seven faubourg bakers—had quite substantial holdings.

Only 12 percent of our bakers—five masters, four faubouriens, two forains—

owned property outside Paris, for the most part in the countryside, ranging in value from three hundred to 16,500 livres. If one examines only those who held real estate rather than the whole population, the average holding was worth 4,452 livres, the median 1,783 livres. Over a quarter of the bakers had real property worth under six hundred livres and only a quarter had possessions valued at more than six thousand livres. Not surprisingly, the country forains had impressive rural holdings: parcels worth 8,379 and 11,000 livres for the two cases we have. Master-owned real estate outside the capital was greater in value than that held by the faubourg bakers. Both Parisian and rural real property ownership fluctuated throughout the century, permitting one to discern no pattern across time.

Thirty-one percent of the bakers possessed annuities, slightly more than the number who owned real estate. These rentes were far more often privately arranged than publicly/institutionally grounded. Like the owners of Parisian real estate, these rentiers—nineteen masters, eight faubouriens, two forains—had very substantial portfolios. The average holding amounted to 10,107 livres and the median 3,781, with a range from a piddling 162 to a very bourgeois 47,390 livres. Masters had almost three times the rentes as the forains and half again as much as the faubourg bakers. Annuity holding declined substantially between 1740 and 1750 but recovered to reach its eighteenth-century zenith in the largely turbulent period between 1761 and 1775.[37]

The distribution is also the key to understanding the significance of office holding in baker assets. Only ten bakers—eight masters and two faubouriens—owned venal posts, for the most part market or municipal offices. They varied in value from 660 to 20,000 livres, with an average of 6,881 livres (median = 5,200).

There are two other indicators of wealth, both extremely imprecise, which do not as a result enter into our quantitative estimates. Three bakers owned mills, two masters and one faubourien, but we do not know their value (which, depending on type, location, buildings, and land, could range from a few thousand to several tens of thousands of livres). The inventories report the capitation tax paid by fifteen bakers.[38] It ranged from six livres (a faubourg baker), a very modest impost not far from minimal assessment, to ninety-eight livres (a guildsman), a levy that surely denotes real well-being. A third of the capitation payments were under twenty livres, while 13 percent were over fifty livres. The average tax for all bakers was thirty livres (median = 23). Masters were assessed more highly (mean = 39, median = 28 livres) than faubourg bakers (mean = 13, median = 9). The capitation appears to be a good predictor of total assets. Bakers in the highest quartile of capitation payers had twice the assets of those in the second quartile and over nine times the assets of those in the lowest quartile.[39] Obviously the capitation figures will make far more sense when we have more of them and when we have analogous data about other socioprofessional groups with which to compare the bakers.

We have already noted that "merchant" bakers were as wealthy as the guild bakers. The pie diagrams in figure 12.7 indicate not only that they had fortunes quantitatively comparable to the masters but also that their fortunes were strikingly similar in composition. Like the masters, the merchants were not afraid of extending credit and they maintained substantial stocks of merchandise. The difference in capital investment was probably located in the shop—better decorated, larger, and generally more attractive in the corporate sites than in the faubourgs—rather than in the bakeroom. Though the merchant fortune pattern is closer to that of the masters than to that of the merchants' own colleagues in the faubourg Saint-Antoine, it should be noted that the composition of Saint-Antoine assets was not radically at variance with the master-merchant model. Four-fifths of the faubourien's assets were "hard"—that is, nonpaper assets, a higher proportion than we find in the case of the masters or merchants. Hard assets obviously were more compelling because they were already in existence and not dependent on the vagaries or costs of collection. The nonmerchant faubourg bakers were far more niggardly in extending credit, and they did not allow debts to swell very large before collection (in part because they had fewer institutional or influential individual clients). The faubouriens as a whole invested more in equipment than the merchants, proportionately speaking, not in absolute terms. They had smaller flour inventories, a larger proportion of family possessions, and about the same share of noncommercial investments.

Marriage and Inventory Fortune

The marriage contributions of the inventory bakers were on the whole smaller than those of the large baker population that we have examined (see table 12.3).[40] There is, however, no strong association between marriage fortunes and either total assets or net worth. Correlation analysis reveals that changes in neither the wife's nor the husband's marriage share account for more than 13 percent of the variation in total assets and 10 percent of the variation in net worth.[41] The lack of a strong association between marriage fortune—the couple's resources at the beginning of their career—and inventory net worth—their fortune considerably later in their career, if not at the end (eighteen and one-

Table 12.3 Marriage Contributions in Marriage Contracts and Inventories
(in Livres)

Population	Groom		Bride	
	Mean	Median	Mean	Median
Universe in marriage contracts	2,609	1,900	2,182	1,500
Universe in inventories	2,541	1,650	1,637	1,221
Inventory masters	2,971	2,325	1,859	1,488
Inventory faubouriens	1,784	1,025	1,099	919
Inventory forains	1,000		804	852

half years is the average interval, sixteen years is the median)—suggests that auspicious "takeoff" conditions, measured in material terms, were not absolutely crucial to long-term success.[42] It would be absurd to assert that well-launched bakers did not have it much easier than others who started with fewer resources. But a "good" marriage, so styled according to quantitative criteria, neither predicted nor predetermined "success," measured equally in quantitative terms. Clearly bakers who started slowly had a chance to catch up.

Liabilities

Average total liabilities for all bakers came to 1,784 livres, but the median, 721 livres, makes plain how extensive the dispersion was (see figure 12.11).[43] Liabilities ranged from none (13.7 percent of the cases) to 24,536 livres. Not quite a quarter of the population had liabilities over two thousand livres, and only 7.4 percent owed more than five thousand livres. Before midcentury, accounts payable were considerably lower, averaging 1,003 livres (median = 454). For reasons that are not clear, after 1745–50 they rose substantially, to 3,185 livres (median = 1,323). Mean liabilities climbed from 854 livres in 1725–40 to 3,896 in 1751–60 (bespeaking a postcrisis expansion in a climate of recaptured confidence?), dropped to 1,676 livres in 1761–75 (an unlikely outcome, given the terrible burdens of prolonged dearths and dislocations), and then sprinted to a heady 12,821 livres in 1776–92 (perhaps reflecting in part a delayed effect of earlier crises, compounded by the economic traumas of the early Revolution). Masters, with accounts payable averaging 2,296 livres (median = 1,008) were far more heavily indebted than the faubourg bakers (712 and 320 livres). With mean debts of 2,808 livres (median = 2,290) the forains were more heavily

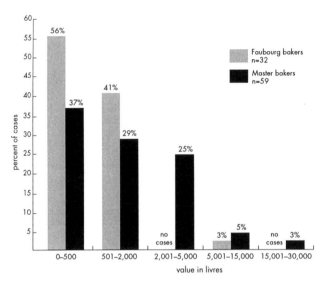

Figure 12.11 Percentage Distribution of Total Liabilities of
Master and Faubourg Bakers *(Source: after-death inventories)*

freighted than either of the Paris baker contingents, but their number is too
limited in the sample to give the result great significance.

Commercial liabilities (stock plus mill fees) represented about half of accounts payable (mean = 48.8 percent, median = 53.7 percent). Flour purchases
were the most important element, accounting for an average 711 livres. Yet
fewer than a fifth of the bakers had flour debts surpassing a thousand livres and
almost 56 percent had none at all. If one counts only those bakers who actually
owed money for flour rather than the whole group, the average debt was 1,604
livres (median = 794). Thirty-two of fifty-six masters were indebted to flour
dealers and millers for an average of 1,081 livres (median = 305). One master
owed his suppliers a total of 12,848 livres—a figure well beyond the accounts-
payable threshold of many business failures. Mean faubourien flour debts were
very low (56 livres), because almost 80 percent of them owed nothing for this
merchandise. Forains averaged 142 livres each in flour debts. Under 220 livres
till the forties, mean flour indebtedness for the entire group climbed sharply to
an average of 1,698 livres in the fifties, and then fell back to 788 livres in the
period 1761–75.

Grain liabilities for the whole population average 261 livres, but this figure
reflects above all the 73 percent of the bakers who owed nothing for grain. If
one excludes them from the calculation, the average grain obligation of those
actually indebted was 957 livres (median = 618). One master owed five thousand

livres to grain dealers, but only 7 percent of the population had grain liabilities over one thousand livres. Mean grain debts for masters amounted to 339 livres, for forains 240 livres, and for faubourg bakers 113 livres. Stagnating around 130 livres from 1711 through 1740, grain debts rose to 580 livres in the forties and fifties before returning, countercyclically, to the earlier levels in 1761–75.[44]

Virtually all the bakers depended on credit. Many of them, as we shall see, encountered serious problems meeting their obligations. They juggled, groped, evaded, procrastinated, faltered, and sometimes succumbed. The inventory bakers show us the positive side of the spectrum, their business practices being marked on the whole by prudent management and swift and regular reimbursement.

Only 18 percent of the inventory bakers owed money to millers specifically for grinding wheat on demand (many millers were among baker creditors, but for supplying flour rather than for merely converting wheat that the baker himself purchased). The average mill-fee debt for the whole population was twenty livres, but the mean debt just for the sixteen bakers who owed money to millers amounted to 112 livres (median = 95). Almost a third of these liabilities were under fifty livres, while 7 percent were over one hundred livres (see figures 12.12 and 12.13). Faubouriens owed two and one-half times more for mill fees than masters, while forains owed nothing.

There is no significant relationship between the magnitude of the baker's merchandise liabilities and the size of his flour and grain stocks.[45] The fact that a baker was more or less seriously indebted to his suppliers did not imply that he kept a substantial inventory of merchandise. Nor did the fact that a baker had large stock imply that he had heavy liabilities. Rather, the bakers who stored grain seem to have been better managers and shrewder traders than those who were mired in accounts payable. Finally, neither granary space, available sacks, house ownership, or bakery capital investment appear to have influenced the constitution of stock liabilities.[46]

Although they were not included in the rubric of commercial liabilities, two other accounts payable were directly linked to the business. The first was money owed to wood merchants for supplies for baking. Surely the inventory does not reveal the true state of wood indebtedness, for only five bakers (four masters and one merchant) owed sums for wood, sums ranging from eighteen to 850 livres, or an average of 230 livres each (median = 88). Wood indebtedness did not vary with the amount of wood stocked.[47] Fourteen masters, seven faubouriens, and one forain owed an average of 62.6 livres (median = 28) in wages to their journeymen. Though one master owed 258 livres, most of the sums were relatively modest, at least from the employer's point of view: only 18

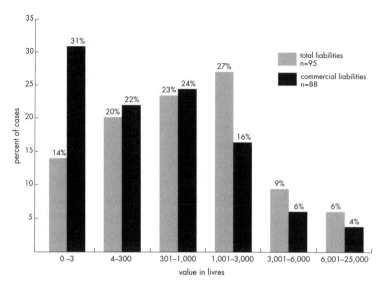

Figure 12.12 Percentage Distribution of Baker Total Liabilities and Commercial Liabilities (Grain, Flour, and Mill Fees) *(Source: after-death inventories)*

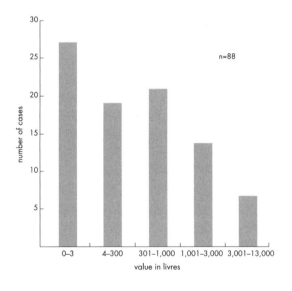

Figure 12.13 Baker Commercial Liabilities: Grain, Flour, and Mill Fees *(Source: after-death inventories)*

percent of debts exceeded eighty livres. The size of one's wage liabilities was not dependent on the number of journeymen one employed.[48] Wage indebtedness remained constant until the last decade before the Revolution, when it jumped to four times the long-run average. As we have seen, journeymen often requested their employers to "bank" their wages for them, though there were cases in which the bosses withheld wages in order to exercise pressure on their workers for one reason or the other, or because they were very short on liquidity. Revolutionary teleologists might be tempted to infer a chain of burgeoning tensions in labor relations on the eve of 1789: journeyman indiscipline either exacerbated by dilatory payment of wages or resulting in punitive wage retention leading to further acts of insubordination helping to create a revolutionary disposition within an increasingly favorable political opportunity structure. The script, however, cannot be sustained by the far less linear and more equivocal realities of the day.

Straddling the boundary between business and personal liabilities was indebtedness for arrears in house or apartment rent. The average for all bakers was merely eighty-two livres, but if one includes only the thirty-three bakers who actually owed rent, the average swells to 223 livres (median = 154). Rent liabilities ranged from twenty-five to 998 livres. The nineteen masters with rent debts owed almost twice as much on average as the thirteen faubouriens (though less than the one forain with rent due). Rent indebtedness did not vary with the size of one's rent.[49] Save for a fleeting surge forward in the period 1751–60, rent indebtedness remained at about the same level throughout the century.

The remaining liabilities were of a "familial" (in the sense that we have understood it) or personal character. The inventories of fifteen bakers (twelve masters, two faubouriens, one forain) enumerated debts for medical or nursing care administered to the soon-to-be deceased on his (her) sickbed, averaging 108 livres (median = 82). Ranging from six to 384 livres, these sickbed costs reached above two hundred livres for 13.3 percent of the bakers. The masters spent nearly twice as much on average as the faubouriens. Bakers as a whole spent twice as much on sick- or deathbed care in the second half of the century as in the first.

The inventories specified funeral expenses in only thirty-eight cases, though there is no doubt that such costs were borne by all the estates. Although the average was 148 livres, 47 percent of the families spent under one hundred livres (median = 102.5). While almost one-fifth of the funerals cost over three hundred livres, there was only one instance of extravagance: the master whose baroque passage cost 654 livres. The twenty-seven masters averaged a little less than twice the funeral expenses of the faubouriens and almost three times the costs of the country forains. Burial expenses were more than one-third higher in the second half of the century than in the first, but they rose quite slowly between 1711–25 and 1776–92. The greater the assets of a family, the more they tended to expend on funeral ceremonies, an investment in their community standing and prestige as well as a solemn rite.[50]

Let us leave the domain of collective statistical description in order to meet a number of bakers whose individual experiences were at once registered or reflected and obfuscated by our means and medians. Though he married a master's daughter and employed two garçons in what must have been a fairly active commerce, Jean Denise of the faubourg Saint-Antoine died a relatively poor man. He lived with his wife and three minor children in a one-bedroom apartment for which he paid 170 livres a year. The family owned remarkably little: the apartment was meagerly furnished, with a table, some chairs, two beds, a wardrobe, a mirror, and some wall hangings worth a total of 139 livres. A bone cross, a three-volume *Life of the Saints,* and thirty small glass-encased engravings do not tell us in themselves that the Denises prayed or read or appreciated decoration, but the fact that they were among the few possessions that they kept suggests that they valued them. The widow had more clothes than one would have expected: nineteen blouses, two skirts, a jacket, a corset, four petticoats, an apron, ten pairs of woolen stockings, and two pairs of shoes. Denise had eighteen shirts, some collars and cuffs, and an old jerkin.

The bakeroom contained two bolters, two oak kneading troughs, two charcoal boxes, scales, thirty-six empty sacks, some canvas, and various pots, pails, peels, and shovels. Denise may not have had a shop at all, for there is no evaluation of shop materials by the expert baker who appraised the bakeroom. Denise's journeymen, to whom he owed eighteen and eight livres in back wages, could continue baking without interruption, for there was an equivalent of almost three muids of merchandise on hand. Though he had two old horses and a wagon for country buying, the merchant to whom Denise owed the most money was a grain dealer from the rue de la Mortellerie, the home base of the leading port traders. If one does not count the money due him for bread furnished (400 livres), Denise's net worth was thirty-eight livres.[51]

Jean Dubuisson lived nearby and had much in common with Denise besides faubourien status. He paid 230 livres for a one-room apartment apparently attached to a small shop. He had three minor children after a thirteen-year marriage. The family owned two beds, a wardrobe, some chairs, and a mirror, together worth 128 livres. Clothing and linen amounted to 194 livres. Dubuisson left ten setiers' worth of flour in the bakeroom: given the primitive nature of his matériel, I doubt that he could bake much more than that in a week. Fourteen customers owed him 208 livres, while he had to reimburse a grain factor 277 livres. Dubuisson's net worth amounted to 562 livres, including accounts receivable. His wife's illiteracy may have hampered her ability to take over the business with confidence.[52]

Nor did guild membership, as we have noted, offer an immunity from straitened circumstances. A master on the rue Saint-Jacques, Jean Chocarne left his wife of nineteen years and five children ranging in age from five to fifteen a fortune of 614 livres, only 114 livres more than his wife's dowry. They lived in a two-and-one-half-room apartment above the shop. Three beds represented almost a third of the total value of their furniture and personal possessions, which included six small engravings and five pistols. Their real extravagance, however, was 319 livres in jewelry and silverware. The shop was sparsely equipped: it had a counter and a grill. The bakeroom had two kneading troughs, twelve bread-proofing baskets, a dozen proofing beds, shelves, shovels, scales, a charcoal box, a kettle, and a pot. Chocarne kept a substantial reserve of wood and he had twenty-six sacks of white flour. It was flour indebtedness, however, that sapped his position. He owed 2,544 livres to his suppliers—the bulk of it to a single flour broker.[53]

One of the poorest masters, Pierre Anquetin, died on 22 May 1741 in the house on the rue de Bourbon of which he was principal tenant for an annual rent of eight hundred livres, a price drastically incommensurate with his end-of-life fortune. He left six minor children ranging in age from twenty-year-old Pierre to two-year-old Anne Gabrielle, all of whom were juridically under the guardianship of the deceased's brother, also a master baker. Married in 1719, Anquetin's wife had brought him a modest dowry of fifteen hundred livres. On the basis of his douaire, one may conjecture that his contribution was about the same. He died at a relatively young age, mired in debt with a negative net worth amounting to 4,202 livres.

Anquetin and his large family lived in two bedrooms over his shop and a kitchen, also serving as the dining room, located in back of the shop. They did not possess a great deal of furniture: an oak buffet, a small cupboard containing a few pieces of faience, a copper standing sink, a table with eight chairs, another table with four chairs, two straw armchairs, two framed mirrors, some wall tapestry, and six beds. The furniture was evaluated at 517 livres, 43 percent of which represented the beds. Anquetin and his wife had predictably modest garments. Among his clothes were five shirts, six muslin collars, two pairs of woolen stockings, one pair of shoes, two overcoats, three baker's caps, a hat, and a wig. His wife's clothing included four blouses, four pairs of socks, four pairs of cotton stockings, several petticoats, an apron, a lined dress of Siamese calico, and a lined dress of serge. All their clothing plus household linen (two dozen towels, six tablecloths, four pillow cases) was worth a little over two hundred livres.

Two fellow master bakers appraised Anquetin's bakery for matériel and merchandise. It contained three bolters (which, along with the two sacks of gross flour also found in his bakeroom, suggests that he habitually did his own sifting and mixing), two scales with weights, nine wooden bread shovels, two kneading troughs in bad condition, baskets, shelves, sacks and canvas, bread molds, charcoal boxes, a charcoal rake, a kettle, an oven door, a large pot, and a salting tub (for a total value of 224 livres). The shop was small and rather austere, boasting a traditional counter with a drawer and a bread knife, scales, an oak bench, an iron grill, boxes for flour, bran, bread crumbs, and charcoal, various measures, shelves for bread, a lamp, and a clock (worth a total of 77 livres).

Anquetin was a relatively poor but shrewd baker. Short on cash, he scattered his purchases among rival merchants (he owed large grain debts to Derain, a Mortellerie baron, and to Nicolas-Louis Martin, one of the leading dealers at the Halles), and he bought flour from a broker at the Halles as well as from an independent mealman. He owed his suppliers 5,573 livres (including one hundred livres to the wood merchants) at the time of his death. To balance out his liabilities, he had only 458 livres ("en petittes sommes") in accounts receivable for bread furnished. Prospects could not have been bright for his widow, with heavy debts and five children under the age of ten. She probably renounced their financial community, thereby writing off some of the debts, though such a gesture might have made it difficult for her to continue her husband's business (seconded by her firstborn son).[54]

Anquetin's guild colleague Noel Bossu also died in debt, leaving a net worth of minus 878 livres. The principal tenant of a shop and two-room apartment on the rue de la Calandre, Bossu had two minor children. He launched his career on a far more sanguine note, for his wife had brought him 4,187 livres in dowry nineteen years earlier (his contribution had been only a thousand livres). The family possessed three beds, seven chairs, a table, two wardrobes, and a mirror, all of which added up to 307 livres. The beds represented almost a quarter of this value. Bossu's shop was as bare as Anquetin's. Though he had about one thousand livres in flour on hand (his wife would have no trouble baking for two weeks at least), he owed almost two thousand livres to a Versailles mealman. The fact that his customers owed him only 224 livres suggests he was extremely exigent in collection or that his clientele was quite small. Making use of the guild's pall and silver enabled the widow to keep funeral costs to ninety livres.[55]

Chances are, however, that Noël's brother Jacques, also a master baker, helped out. Judged by his fortune at death thirteen years later, his advice was to be prized. Jacques had married less auspiciously than Noël, for his wife, the daughter of a wine merchant, brought him only two thousand livres. At his death, his widow and six children (five minors) shared his net worth of over thirty-six thousand livres. Less than a quarter of his assets were paper (half for bread debts, much of which the family succeeded in collecting, according to the estate settlement that followed the inventory), while 30 percent consisted of real property and 11 percent represented the capitalized value of annuities. The Bossu owned jewelry and silver worth over two thousand two hundred livres and they kept a wine cellar. The fact that Jacques owed no money to anyone underlines how well he managed his business. No wonder his family was willing to spend over three hundred livres on his funeral.[56]

Jacques Bossu was well-off, but he was not among the most affluent masters. His Left Bank neighbor, Rene Delahogue, began with a combined marriage pool of only forty-four hundred livres (but including the option to buy an established fonds) and amassed a net worth of 62,973 livres in fifteen years of work. To be sure, a third of his assets were paper, the bulk for bread furnished. But Delahogue was optimistic and aggressive about collection.[57] Most important, he had only 665 livres in accounts payable, a bagatelle, so he was not pressed. Delahogue had an astonishingly large flour stock of 350 sacks weighing 325 pounds each and collectively worth 11,574 livres—a stock that might have qualified him for arrest as a hoarder if the police had not been favorably disposed to individual baker granaries. The baker planned for his future with annuities worth twenty-two thousand livres. A principal tenant, he paid one thousand livres a year for spacious quarters consisting of a kitchen, a master bedroom, two small bedrooms, a sitting room, and an antechamber in addition to his shop, bakeroom, storage room, and wine cellar (where he kept 160 bottles of red burgundy).

A servant helped to care for three minor children and three journeymen, who slept in a bakeroom on individual mattresses and virtually liberated the master from baking. The bakeroom boasted the latest technology, including an automatic feed line from the flour room to the single, large kneading trough. It also contained two scales, seventy-eight bread-proofing baskets, shelves, two oven doors, a sink, eight shovels, a lamp, and miscellaneous tools. An oak counter mounted on a sort of dais dominated the shop, which had two grills looking out on the street, shelves for displaying bread, sacks for bread crumbs, a retail flour box with measures and shovels, scales with weights, and baskets for transporting bread.

The master bedroom was the centerpiece of this well-furnished apartment.

It featured two beds, a wardrobe, a commode, a storage chest, a standing clock, four mirrors, curtains, six chairs carved in wood, two armchairs, a carpet and wall hangings, eight engravings, a plaster Christ on a wooden cross, and a bronzed Christ. Delahogue's late wife was a likely candidate for best-dressed *boulangère* with twenty-five hundred livres' worth of silks, satins, calicoes, laces, brocades, an elegant assortment of hats, and thirty-six blouses, among other garments, common and elegant. On special occasions she wore her necklace of fifty-nine "fine pearls," her diamond earrings, and her decorative cross composed of five large and eight small diamonds.

Nor was Delahogue indifferent to sartorial matters. He vaunted a black cloth suit, a pair of black velvet trousers, two jackets of black satin, a jacket of brown silk and two of black silk, nineteen shirts (eight extrafine) and accessories, several coats, two pairs of suspenders with silver buckles, numerous stockings, and several pairs of shoes. To complete this bourgeois mien he flourished a fancy gold-tipped and leather-bound cane, two grayish wigs (one of which was recently returned for "adjustment"), and a fur hat. It is not surprising that Delahogue spent 135 livres on mourning clothes tailored for himself and his children. Other funeral expenses for his wife amounted to 403 livres, including fifty-nine livres for music.[58]

After forty years in the bakery, including a period as guild juré, François Brivot retired with a fortune nominally the same as Delahogue's but sounder because it rested on more hard assets. Married to the daughter of a master cobbler who had brought him only fourteen hundred livres in dowry, Brivot took pride in being able to settle a sumptuous portion of eight thousand livres on his daughter in the first of her two marriages to master bakers. Brivot lived in a comfortable two-and-a-half-room apartment in Paris that he rented and also in a large house with a garden in the suburb of Sceaux that he partly owned. What he lacked in taste for wine (Orléanais red), his wife made up in diamonds, gold, and pearl jewelry. He did not invest a great deal in furniture, though he owned two superb hand-carved standing clocks. He put his money in annuities (over 30,000 livres in principal) and in a venal office (garde bateaux for 20,000 livres) that brought a regular return as well as a dividend of prestige. Brivot kept almost a thousand livres in cash on hand for current expenses. Paying cash for everything he bought was obviously part of his serenity. His sole debt was the sixty livres that he "saved" for his servant-girl.[59]

One of the richest bakers in all of Paris with a net worth of 82,818 livres was not a master but a faubourien named Pierre Lepage who was one of at least twenty-four bakers on the bustling rue de Reuilly.[60] He was the owner of the house in which he lived, number 3 on the right, which was worth thirteen thousand livres. The family enjoyed capacious quarters: four and one-half bedrooms, a living room, a dining room, and a kitchen. In back of the dwelling stood a barn and a pigeon house, surrounded by a garden. A basement served for storage of a large amount of apparently ordinary wine along with wood, wax, and olive oil. Lepage owned another house elsewhere, as well as one-third and one-half portions of two mills, but for some reason they were not evaluated in the inventory. In thirty-four years of marriage the baker had nine children, two of whom died in childhood. The other seven were grown up by the time of his death. At least two of his sons became bakers and one of his daughters married a "merchant baker" like her father; the guild obviously did not turn the heads of the Lepages.[61]

The most striking feature of the furnishings aside from several ornate wardrobes and commodes manufactured by fellow Saint-Antoine artisans, was the large number of paintings. On the dining room walls hung two flower and fruit still lifes, two battle scenes, a country festival, and a "Moses found in the bullrushes." One bedroom was decorated with a portrait of a man and a woman, a "petits amours," depicting lovers, and four small landscapes, as well as three marble busts and some lesser figurines. Portraits of family members—four in all, two in pastels—adorned the walls of another bedroom, along with eight unidentified paintings. Two other tableaux hung in another small bedroom.[62]

Equally notable was the unusually large and rich library of more than two hundred books containing, inter alia, four volumes in folio of the Sacy Bible of 1717, three volumes of the Frizon Bible of 1721, a manual of sacraments, twenty volumes of Fleury's *Histoire ecclésiastique,* twelve volumes of Letourneux's *Année chrétienne,* a *Catéchisme de Montpellier,* seven volumes of public prayers, a three-volume work on the commandments of God, several books of sermons, a number of Jansenist tracts, seven volumes of Boileau, Moréri's *Dictionnaire,* a history of the Jews, a history of France, and eight volumes of travels in Italy.[63]

Neither Lepage nor his wife seems to have been infatuated with clothing, though both dressed *bourgeoisement.* The baker had several sets of dark suits, four auburn-colored wigs, two fur hats, a decorated cane, and a silver sword. The family had no jewelry and only a modest hoard of silverware (worth 1,942 livres). Lepage's major investments outside real estate were in annuities, capitalized at 24,604 livres. He kept on hand 1,224 livres in cash, a derisory amount compared to the startling cache of 16,219 livres left by his wife eight years later.

Though he had only two journeymen, Lepage had a very large business (no doubt undertaken with the help of some of his children as well). Unlike certain faubouriens, he maintained a shop, equipped rather starkly with an oak counter and shelves for displaying bread. Unlike most bakers in the city or faubourgs, Lepage operated two bakerooms with at least one oven in each. (Both bakerooms, incidentally, were on the ground floor, another unusual arrangement in a world of subterranean bakeries.) Each was fitted with two long kneading troughs and the usual array of shovels, kettles, baskets, and tools. With a very large stock of thirteen muids in unbolted flour worth 2,355 livres, it is not surprising to find a flour room with three bolters, one for white, another for mid-white, and a third for dark flour.

Lepage baked for a wide range of clients, and he offered credit liberally—perhaps with exorbitant forbearance and/or lack of discrimination—for he left over twenty-five thousand livres in accounts receivable for bread. Nor does his wife seem to have succeeded in collecting, for when she died eight years later, over twenty-eight thousand livres were due. Lepage's clientele was extremely diverse—dayworkers (who obtained credit without a problem and did not restrict themselves to dark or even mid-white wheaten loaves), journeymen, master artisans, religious communities, and a sprinkling of *le tout Paris*, including the cardinal de Noailles, the prince de Soubise, the prince de Rohan, Prince Charles, the duc de La Rochefoucault, the duc de Sully, the présidentes de Brosse and Langlois (wives of eminent parlementaires), and Advocate General Daguesseau. Lepage shared with master Delahogue the burden of supplying the princesse de Conty and of trying to pry payment from her: she owed him 3,568 livres.

Though he called himself a master baker, Louis Reverard was a forain from Saint-Germain-en-Laye. It is instructive to glance at his fortune if only because data on forains are relatively rare. Little, in fact, distinguished him from the middling level of Parisian counterparts. Save for a pile of manure (*et encore!* in the faubourgs) evaluated at three livres and perhaps his stock of hay and oats (but this, too, is found in outer Paris), Reverard's inventory looks much like the others.

Married to the daughter of a Saint-Germain baker who matched his matrimonial contribution with a dowry of one thousand livres, Reverard built up a moderate but solid fortune of 13,685 livres in twenty-eight years. His security reposed on the ownership of a house worth eleven thousand livres. Though there were five children, the family did not appear to have occupied all of it, since there is some evidence of a rental income. The Reverards' furniture, almost half of the value of which represented beds, was quite simple. Between them the husband and wife owned only 155 livres in clothes, a penury partly compensated for by the latter after the death of the former.[64] They possessed no jewels and a trifle of silver worth 164 livres. Two paintings and four engravings beautified their home, while fourteen devotional books may have illuminated it. The bulk of Reverard's debts consisted of an annuity that he owed to one of his sons—not a pressing matter of liquidity.

Reverard had a bakeroom in which most Parisian bakers would have been comfortable. He worked at two kneading troughs, proofed in baskets and on shelves or in drawers, and utilized the familiar spectrum of shovels, dough cutters, charcoal collectors, scales, and so on. He had a reassuring stock of flour valued at one thousand livres, all of it white, which explains why he had no bolters on hand. Reverard also had a shop; he was not uniquely a market baker. Moreover, many of the breads that he had on sale there were of the sort one would expect to find in a shop rather than in an open-air stall, including one hundred round and long four-pound loaves, all of assorted white and mid-white qualities—no twelve-pounders and no dark bread.

Where Reverard differed from the mass of Paris bakers and where he betrayed his forain identity was in his exercise of an auxiliary profession. In addition to baking for the capital and for his hometown, he functioned as a carter, transporting such items as leather from a Saint-Germain factory to Trianon. Because he had to maintain a cart and horses, worth together a thousand livres, for buying grain and bringing bread to Paris, it made sense for him to extract optimal use from his capital investment. Since three of his sons became bakers (two eventually establishing themselves as Paris masters), it was easy for him to leave the bakeroom and shop.[65]

Quantitatively and qualitatively, the evidence belies Marat's affirmation that "the trade of baker is the best of them all, it's the only one that always goes well, and always on the same level."[66] The Revolutionary journalist tended to view the world from a consumer's perspective, and from that vantage point (one of more or less agonizing dependence), what could seem more intuitively true? Operating virtually every day and dealing in a commodity that everyone urgently needed, the baker seemed like a tradesperson destined to prosper.

Despite in-built inelasticities, economic and political, the bakery business did not "always go well, and always on the same level." Though we have insufficient data to map out and then account for the movement of the bakery across the century, bakers varied considerably in how they practiced their craft and commerce and in how they fared. Crises were supposed to raise bakers' daughters to a comfortable marriage portion, but in fact they frequently ensnared the bakers in onerous if not inextricable difficulties. Crisis favored a relatively small number of bakers; it was rarely a time for middling let alone poor bakers to make it speculatively. Still, on the basis of our inventory figures, it is hard to measure the precise impact of stress periods; the time, cost, and likelihood of recovery; and the time lags involved in adjusting. While numerous bakers could not overcome indebtedness (as we shall imminently see), others showed a remarkable resilience, in some cases as a result of systematic risk aversion and in others as a consequence of audacious risk taking. The strong survived, through a sort of Darwinian winnowing out, but so did many of the feeble, despite or perhaps because of their very frailty and marginality. In any event, the bakery as a whole does not seem to be weaker at the end of the century than at the beginning.

As we have noted, only a handful of bakers were truly rich. While a substantial number were well-off, an equal proportion lived in quite modest conditions, many of them frankly poor.[67] As a rule, bakers lived without extravagance. Their homes were simply appointed and they indulged themselves in little decorative luxury, though many amassed social capital and practiced the sort of measured ostentation that bespoke power and prestige in the community. Save for some significant (and aberrant) property owners, they all had the same kinds of assets and liabilities, albeit obviously in varying proportions. Paper claims and obligations loomed very large in almost all cases. Those who started off auspiciously in marriage were not necessarily the ones who ended up most comfortably. The bakers' world was as unstable as everyone else's, including Marat's.

Chapter 13

Bakers as Debtors

In order to bake bread, the baker had to acquire the raw materials. In some fashion or other, passively or actively, modestly or extravagantly, in the city or in the countryside, he (more rarely she) was involved in the grain and flour trade. I have told the story of this critically important and complex commerce elsewhere. In this chapter I focus narrowly on the commercial relations between bakers and their various furnishers: grain traders, mealmen, millers, brokers, and laboureurs. Without obtaining cooperation from the suppliers, it would have been impossible for most bakers to operate. The precise nature of that cooperation varied considerably.[1]

The Great Chain of Credit

The entire provisioning nexus was built on a cascading foundation of private credit, and here there were no institutional guarantees against collapse (of the sort, say, that the butchers enjoyed).[2] Without the constant flow of credit, the supply system could not have functioned; credit was the source at once of its dynamism and its fragility. The great chain of credit covered the entire distance from the fields (though the smaller producers tended to demand cash) to the tables of consumers: consumers bought bread on credit from bakers; bakers bought grain on credit from traders who obtained the grain either on credit or for cash from still other intermediaries or from producers; millers intervened, speculatively or on request, to perform conversion operations or provide flour on credit. Indebtedness was rarely so great that the failure of one link could generate a chain reaction of disaster, as, say, in the banking milieu. But given the narrow margin on which any of these artisans and merchants operated, the inability to collect could mean trouble.

    The purpose of credit was not just to capitalize trade but also to stabilize ties between buyers and sellers.[3] Credit was often the expression of a broader prestatory relationship that aimed to give form and persistence to exchange. This relationship, called *pratik* in Haiti and *suki* in the Philippines, linked a seller and a buyer in a most-favored-client connection. Based on personal engagement on each side, the client bond lowered the costs of search in the trading arena, attenuated the imperfections of the information system, increased the security of transactions (not only by providing guarantees of product but also by serving as a sort of social analogue to contract law), and generally helped bring order to the market.[4] In return for a grain or flour buyer's fidelity, the seller granted certain concessions, for example, a price rebate or a bonus in kind (known as *brawta* in Jamaica, *yapa* in the central Andes, *dash* in Africa, and *paaman* in the Philippines, it was called the thirteenth bushel or the "good measure" or "reward" in the grain and flour markets of the Paris region) or credit. Without a pratik arrangement, the baker who obtained open-market credit risked having to pay a higher price for the merchandise or having to pay interest on the debt from the moment of purchase.[5]

Nor was the pratik relationship without ambiguity and asymmetry, at least in the context of the eighteenth-century grain and flour trade. At a certain level, the parties remained adversaries even as they pledged partnership, each seeking an advantage over the other. Given the intensity of the competition for custom, the sellers had to court the buyers with more fervor than they would have liked. And given the multiplicity of sellers, the buyers felt less constrained to be faithful than they should have been.

Of all the pratik concessions, credit was the most significant and the most problematic. The seller extended credit in the hope of binding the buyer to him. Sometimes the seller even refused cash, as a gesture of trust and an effort to establish a long-term relationship.[6] It was relatively easy, however, for the seller to become snared in his own trap, for the debtor-creditor relationship generated mutual dependence and mutual exploitation. The lender provided the borrower with the wherewithal to do business, but the borrower could threaten to take his custom elsewhere if the lender pressed him too hard or cut him off. The smaller the amount the seller advanced, the stronger his position (the safer and more effective his "investment"). Yet if the balance became too small, the debtor might be tempted to turn elsewhere. The debtor had little incentive to make payments unless he received further advances. Thus if he refused further credit, the creditor risked losing not only the debtor's business but also the outstanding accounts receivable. The larger the debt grew, the stronger the position of the debtor (following Keynes's principle that if you owe your banker a thousand dollars, you are in his power but if you owe him a million, he is in your thrall).

In this highly personalized system credit was usually based on personal reputation and trust rather than on the provision of direct collateral. The lender had a theoretical mortgage on the borrower's property, but the debtor enjoyed full usage and several layers of protection against peremptory seizure. In most cases the chief collateral was the grain and flour he bought on credit.[7]

Pratik reciprocity formed many lasting bonds between bakers and their suppliers (even as it did, on a different scale, between bakers and their customers). The Mortellerie barons feted "their" bakers at the Hôtellerie du Barillet d'Or, whence they could watch the arrival of grain boats. Nicolas-Louis Martin, one of the foremost grain dealers at the Halles in the thirties and forties, financed purchases and served as surety for bakers, to whom he also advanced cash. In addition, he sold them larger lots than legally permitted and provided storage facilities and delivery service on occasion. Flour merchants greeted bakers at the Epée de Bois, the Croix Dorée, and other taverns, not only to do business but to gossip and relax. They served as witnesses at weddings of each other's children and as godparents on the occasion of baptisms. While the mealmen extended credit to the bakers in the normal course of events, sometimes the bakers lent the mealmen money to help them out at a difficult juncture. They represented each other in the consular court, even at the risk of alienating colleagues who were parties to the suit. Together they contrived schemes to bypass the market by sneaking flour, sometimes made from illicitly acquired wheat, past the measurers, Arguses of the physical marketplace, directly to the bakeries.[8]

The licensed brokers (*facteurs*) wooed the bakers in myriad ways. They furnished the type and quality of flour favored by a client (perhaps a middling meal rich in gluten or a mid-white flour more golden in color than the competitors'). They promised free delivery or priority measuring (through more or less corrupt arrangements with the officials). The brokers lured the bakers with various kinds of rebates, bonuses, and price differentials. Most common was the "good measure" premium, from three to five bushels per *voie,* or wagonload, that each baker was supposed to believe was reserved especially for him. Brokers Chicheret and Pilloy sealed their sales with a collation at the taverns near the Halles.[9]

The credit the brokers extended, however, was at the core of their relationship with the bakers. Often the broker was caught in a double pratik bind, torn between the fear of alienating a baker client by pressing too hard for payment and the need to appease his merchant clients. The plaintive note sent by one broker to master baker Albert bears witness to this dilemma: "Monsieur, here are the twelve sacks of flour that we settled on at the price of forty livres per sack. I beg you to give as much money on account as you possibly can to my driver, as well as your empty sacks."[10]

Most bakers could not have done without credit. The police were sensitive to their need. On the one hand, the proliferation of credit provoked endless quarrels and litigation, and it multiplied the risks of serious dislocation through failure. On the other hand, without credit, the provisioning trade would have contracted dramatically and gradually died of inanition. For this reason Commissaire Delamare gave his tremulous blessing to the practice of according credit to the bakers: there was "nothing [inherently] vicious" in it and it could be "of great utility."[11] While the pratik union could degenerate into collusion and monopoly, when it operated openly the police appreciated the way it smoothed the flow of goods and services.

Bakers in Arrears

The most serious strain on the pratik liaison was the failure of the bakers to make regular payments, or at least to account convincingly for their (moral and financial) default. A baker would first try to use the pratik lever to extract more credit, more time, more concessions. Implicit in his démarche was the message/threat that he could go elsewhere. Competition for clients was intense. That is why the suppliers could ill afford to abandon their courtship and why the most bitter altercations in the markets involved recriminations over pratik seduction or betrayal. At a certain point, however, the supplier decided that he (or she) could go no further.[12] His bluff called, the baker had to pay enough to restore pratik confidence or turn elsewhere in the hope that news of his commercial problems would not deter other suppliers from linking up with him.

Broker Bazin cut off credit to baker Dame Georgois because she no longer seemed a reasonable risk, but at least one broker-colleague did not share this pessimism. Master baker Rousseau was serenely confident that obstinacy and abuse would stir his supplier, Joachim Letellier of Versailles, to continue his flour shipments. So he rebuffed several requests "to count" and he called his merchant the most atrocious names. Though Letellier complained to the police, he was still half willing to resume his deliveries to Rousseau. Saint-Antoine baker Jean-Baptiste Guillaume learned that it was not always easy to rebound from one broken pratik relationship to a new one. The wife of Halles flour merchant Meunier refused credit to Guillaume's wife on the grounds that "her husband owed money to all the merchants of the port and that he had no credit any more."[13]

How deeply in debt were eighteenth-century Paris bakers to their suppliers, and how significant was this indebtedness? As a rule bakers do not seem to have been deeply in debt to grain merchants. To judge from the situation of some 120 bakers whose after-death inventories (or whose wives' inventories) I have studied, bakers characteristically paid swiftly for their merchandise. Almost three-quarters of these bakers owed nothing to any grain merchant (from the ports, the Halles, or the country). Only 7 percent owed more than a thousand livres. Mean accounts payable for grain amounted to only 261 livres, less than one week's provisions.[14] The mean obligation of masters (339 livres) was 41 percent higher than the grain debt of forains and three times above that of faubourg bakers.

Flour debts were more significant. Forty-four percent of the bakers owed millers, mealmen, or brokers something, and almost 10 percent owed over two thousand livres. Mean accounts payable for flour rose to 711 livres.[15] Masters were far more heavily burdened for flour purchases than the other bakers. These "inventory" bakers were not the bakers who posed problems for the grain and flour traders, though it is plain that if a merchant found himself saddled with a dozen baker debtors, each of whom was in arrears a week, he would suffer serious cash-flow difficulties.

Failures

The bakers who threatened the grain merchants were those who failed. Of the bakers in our pool of 144 *faillites,* or business failures, each on average owed grain merchants of all types 2,186 livres (median = 0) and flour dealers 6,645 livres (median = 4,735). If one focuses exclusively on the bakers who were actually in debt for merchandise, the figures rise. Average grain accounts payable for forty-three bakers amounted to 3,972 livres (median = 2,201), while 128 bakers owed an average of 7,251 livres (median = 5,570) for flour. Nonmasters owed more than masters for flour, but they had about the same mean liabilities for grain. One can see how baker insolvency could jeopardize a merchant's situation.

The average number of supplier creditors in the faillite universe was seven (median = 6), and it did not vary much across the century. Masters, with over seven suppliers, had more commercial contacts than nonmasters, who barely averaged five. Bakers did not allow pratik engagements to confine them too narrowly. Their suppliers tended to be based in different places, thus sparing most-favored baker clients some embarrassment. One master baker had twenty-four suppliers and a merchant baker had seventeen: an adroit practice of diffusing orders widely in order to keep afloat on credit a long time.

Each of the eight Mortellerie grain merchants who failed had mean claims of almost eight thousand livres on an average cohort of seven bakers. One port merchant, Charles Carlier, alleged that he lost almost ten thousand livres in grain sales to insolvent bakers. In addition he still hoped to collect thousands of livres in accounts receivable from other bakers. Based inland at the Halles, grain trader Nicolas-Louis Martin had remarkably symmetrical entries in his balance sheet: 11,600 livres in losses to insolvent bakers and over five thousand livres due from eleven bakers presumed to have means. The other failed merchants at the Halles counted on collecting almost as much from their baker debtors. Like the Parisian grain merchants, forain dealers at the central market extended pratik credit liberally—about two-fifths of their total assets consisted of accounts receivable owed by bakers. The flour merchant-millers also filed large paper claims on delinquent bakers, an average of five of whom appeared on each failed baker's balance sheet. Fifteen bakers owed Piedelu of L'Isle-Adam a total of 24,740 livres. Since this flour merchant had virtually no hard assets, his creditors, mostly grain dealers to whom he owed 10,274 livres, were alarmed.[16]

The business failures of the brokers, more than those of any other actors in the provisioning theatre, engaged the public interest because they affected such a significant part of the supply structure. Virtually all the assets of seven failed brokers, about twenty-five thousand livres on average, represented accounts receivable owed by an average of thirty-seven bakers per broker (median = 31) for flour they purchased on credit. In turn the brokers could not pay their flour dealers. On the one side the merchant-millers were probably being pressed by their grain suppliers, and on the other side the bakers had trouble calling in debts from their customers.

Casting the precariousness of the credit structure into stark relief, the seven broker failures in the seventies led the police to attempt to reform the entire brokerage system with the aim of drastically reducing credit abuse and collection problems.[17] Almost a century and a half earlier, the port merchants, then the dominant figures in the provisioning trade, had complained in sharp language that the bakers were abusing their terms of exchange. "We buy at great expense and risk, borrowing large sums for this purpose," they claimed. "Instead of meeting their debts as they are obliged to," their baker clients used delaying tactics, especially "letters of respite," which the courts willingly granted them. The public might suffer unless the merchants were seconded in their efforts to recover debts, they hinted darkly. In fact, the king responded favorably by revoking the baker letters and allowing the merchants to pursue the bakers for collection. Periodically, the merchants asked that the royal order prohibiting respite be renewed.[18]

Some eighty years later the merchants again protested vehemently. This time, instead of veiled threats, twenty-seven of them, including all the leading Mortellerie dealers, issued an open warning: the provisioning of the capital would be disrupted and "abundance would be compromised" if they were not given greater facility in debt recovery. They attributed the erosion of the number of port traders in part to the inability of merchants to stay solvent. They did their country buying in cash, they (falsely) alleged, and they could not go on much longer because "all their capital is scattered about among most of the bakers through the sale to them of wheat on credit." The merchants depicted the bakers as sly and deceitful, remaining at home where they could evade sentences and summonses. They hid away their most valuable possessions, depending on their wives to buy supplies and run their businesses "with the grain traders' money."

Pursuing the Bakers

The merchants demanded a number of changes in the regulations governing pursuits. They sought the right to have bakers arrested in their homes; they asked that the baker debtors be declared *banqueroutiers*—fraudulent failures—making them liable to criminal as well as to more stringent civil action; they sought the power to keep them in jail once they were arrested until they acquitted their entire debt; and they pressed to have their wives, who often were able to siphon off family resources by means of a juridical "separation of property," held jointly responsible for all grain debts. "The security of the grain trade," the merchants esteemed, "is directly contingent on the capacity of the merchants to extend credit to the bakers without risk and with the assurance of getting their money back."[19]

The Assembly of Police, to whom the merchants addressed their petition, bristled at its threatening tone and regarded its complaints as exaggerated. The goal of the authorities was to sustain as many bakers as possible through the great chain of credit without, however, appearing indifferent to the commercial situation of the merchants. The Assembly denied most of the merchants' demands on the grounds that they contravened time-tested law and customs. Bakers said to be in hiding could not be declared fraudulent bankrupts unless they had committed and were convicted of fraud. Bakers, like others, had the right to petition for release from debtor's jail once they remitted an initial payment (usually one-fourth) and pledged to pay the rest in installments. Grain dealers in principle could not arrogate precedence over other creditors, even as "special cases" could not be allowed to supersede "general rules." Nor could creditors obstruct or undo legitimate property separation in marriage. But the Assembly made one significant concession: it agreed to recommend that the parlement promulgate an order permitting arrest in the home for debt.[20]

Creditors pursuing bakers in the courts had a choice of three primary jurisdictions: one of the Châtelet's civil chambers, the municipal court, or the consular jurisdiction. The lieutenant civil of the Châtelet and the prévôt des marchands worked hard, seconded by a large number of parlementaires, to challenge the overwhelming hegemony of the consular jurisdiction in commercial litigation. The Châtelet promoted itself as a court of convenience and expeditiousness, in part as a result of its close connection with the police, which linked it to precious sources of information and facilitated the execution of its decisions. The municipality sought to build its clientele around its special competence in all matters pertaining to waterborne trade. But the consular jurisdiction, ardently supported by the corporate business establishment, proved extremely resistant. Boasting judges drawn from the guilds with years of practical commercial experience, it focused unrivaled expertise on the cases it examined. It worked industriously and rapidly, hearing parties three times a week, sometimes till past midnight. Adjudication was not without costs, but, as the exponents of the consular court boasted, expenses were quite modest relative to the price of litigation in other jurisdictions. Consular judges spoke the everyday language of business (no "vain subtleties") and evinced no disdain for artisans and merchants who eschewed counsel and spoke for themselves. The consular jurisdiction lost some ground in the eighteenth century, but not much.[21]

Grain merchants pursuing bakers for recovery of accounts receivable addressed the municipal and consular courts, and seem to have obtained similar results in the two jurisdictions. The port merchants pled before their municipal protectors with a virtual guarantee of success as far as the judgment was concerned.[22] Market and country-based grain merchants preferred the consular court, which appeared to worry more than the other jurisdictions about the problem of collection beyond the promulgation of the sentence. It invited convicted bakers to spare themselves the risk of incarceration or seizure by having a family member or friend stand as surety.[23] It encouraged plaintiffs and defendants to agree on installment terms since it esteemed unlikely that bakers could or would agree to acquit their obligations in entirety immediately.

Thus master baker Richer (the elder) was directed to pay 150 livres 10 sous over six months. After a down payment of fifty-two livres on the spot, merchant baker Guichoux had two months to pay the remaining two hundred livres. Forain baker Louis Félix was enjoined to pay twenty-two of 162 livres within two weeks and the remainder in equal installments over three months. Baker Emard had to pay only nine livres a month for half a year, while forain Lionnet had to come up with 650 livres during the same elapse of time. Certain bakers, perhaps because of their reputation or the specific character of their debts, were ordered to make immediate restitution.[24] Bakers felt comfortable in the consular court because they knew that any incriminating information that incidentally surfaced regarding illicit grain- or flour-trading activities would not be used against them as it might be in a court attached to a police or administrative jurisdiction.[25]

Usually loathe to extend credit to bakers, laboureurs who had debts to collect for grain turned to the consular jurisdiction despite the distance that separated their residences from the capital. Indeed, very often they sued forain bakers, who were also from outside Paris. The disposition of their suits was very similar to the results of the proceedings in which grain merchants engaged. Occasionally the laboureurs won sentences demanding immediate remittance. For example, Denis Pillée, a master baker, was condemned to pay a Brie laboureur the 336 livres remaining from the purchase of forty-eight setiers of wheat, delivered by the seller to the baker's mill, more than a year and a half earlier. More commonly, installment terms were arranged, either at the suggestion of the court or at the request of the baker. Chatelain, a Gonesse forain, was found to be remiss in his dealings with three laboureurs on the same day (for 140, 180, and 150 livres). He sought six months during which to pay and was granted these terms in two of the three cases.[26]

Flour merchants, like their counterparts in grain, had to devote an inordinate amount of time, energy, and money to the task of recovering accounts receivable from bakers. Generally, flour transactions were sealed by a down payment that ranged from 25 to 50 percent of the total price. Bakers sometimes negotiated longer terms, but as a rule full payment was due a month after delivery, even in the warmest pratik relations. After fruitlessly summoning delinquent bakers to pay, miller-merchants usually sued in the consular jurisdiction. Some dealers seem to have been unnecessarily precipitate in litigating for collection, but they knew more about their clients' characters than we do. Etienne Denise of Beaumont waited barely three weeks before suing widow baker Loriset for 2,003 livres. Nicolas Noret of Charenton similarly showed less than a month's patience toward his neighbor, baker Guiton of the faubourg Saint-Antoine, who owed him almost a thousand livres.[27]

On average, however, the miller-merchants gave their clients up to seven and a half months before filing suit. They were sometimes compelled to take rather impetuous action by their own suppliers, who took them to court. Three-way suits were not unusual: miller Augustin Vieuxbled, pursued by laboureur Jacques Lecler for payment of 345 livres in wheat, sued baker Vassou of the faubourg Saint-Antoine (and by doing so avowed that he was the baker's buying agent, a commission that was an infraction of the law). Sued by a Brie laboureur named Gilbert, miller Touroux had baker Claude Ver sentenced to pay the 232 livres remaining on a huge transaction of sixty-four muids. The aggressive Etienne Denise went a step beyond most of his confreres by filing suit not only against baker widow Loriset but against the consumers who owed her money. He agreed to drop the suit—which caused Loriset considerable embarrassment as a result of her inability to protect her customers from harassment—only after she signed a promise, backed by her son's bond, to reimburse him at the rate of one hundred livres a month.[28]

The flour brokers also frequented the consular jurisdiction for recovery. Though justice was economical here, the brokers still complained of the burden of hiring lawyers to speak for them (because they were too busy to attend court themselves) and huissiers to serve summonses. Their claims varied enormously, from twenty-six livres that a forain owed broker Marguerite Lepetit to 1,464 that broker Ancelin sought from master baker Picard.[29] Brokers seem to have accepted more liberal installment terms than other plaintiffs. Ancelin gave baker Cousin a year to repay eighteen hundred livres, the same grace that Lepetit accorded widow baker Austry to remit 367 livres. Broker Delaistre gave master baker Bire ten months to reimburse 1,268 livres.[30] To escape seizure, baker defendants generally were required to find persons to stand surety for them. When one of his debtors, baker widow Viery, died "insolvent and without property or effects" still owing him 2,983, Ancelin sued her "bond" for payment, an unsuspecting master baker named Roussel.[31]

Because the bakers were elusive and unhurried, the brokers had to be tenacious and clamorous. When master baker Jacques Courtois found a hundred reasons to question one finding against him for 843 livres, broker Ancelin sought a second. In 1742 broker Louis Delaroche asked the consular court to compel master baker Berton fils to restitute the same 331 livres for which his father, also a broker, had obtained a payment order six years earlier. It was one thing to win a judgment and quite another actually to collect. Broker Jacques Frémont had baker Dame Duhamel condemned to pay 557 livres for flour he supplied her. She "transported" the obligation to her mother, the widow of master baker Petit, who assigned Frémont to collect the 557 livres from one of her debtors, himself in the avid hands of a consortium of creditors (*union des créanciers*)—a hornet's nest that Petit was happy to flee in favor of the hapless broker.[32]

As an expression of defiance and disdain, a great many baker defendants did not show up for their day in court. These were the bakers who knew they would lose. Yet by defaulting, they may have sacrificed opportunities to improve their situation. Bakers present at the sentencing could appeal in person for specific terms. The judges usually pressured the plaintiffs to grant a reasonable amount of time. Bakers in attendance were also able to petition for remission of interest payments and occasionally for court expenses as well. By agreeing to reimburse in two installments, widow baker Duclos won forgiveness of penalty from miller Horn. By arguing his case instead of boycotting the session, a baker had the opportunity to contest the figures adduced by his suppliers. Master baker Martin Chantard swore that he had already paid in full the 192 livres sought by merchant-miller Jacques Denis of Eperon. While the court did not believe Chantard, it supported Mosny, a forain from Vincennes, who insisted that he only owed miller Horet 237 of the 318 livres demanded by the latter.[33]

Mediation

Finally, bakers who stood up for their cause, however intrinsically weak it was, were sometimes able to persuade the judges to refer the case to an *arbitre*, either an expert in the trade or a person believed to be of special sagacity, who would seek to reconcile the parties in an amicable accord or would make a recommendation to the judges on the appropriate action to take.[34] It was not impossible that an arbitre would revise the debt downward or seek some other way to temper the blow against the debtor even as he sought justice for the creditor. The style of the arbitre depended on his temperament, his experience, his profession, and his intuition about the case at hand. Some arbitres remained largely on the periphery; others were activists who aggressively sought to get to the bottom of things. From the consular court's perspective, the best arbitres were those who earnestly sought to hammer out "arrangement" between the parties.

The effort to "raccommoder," as the arbitres put it, favored a priori the baker-debtor defendants. Flour broker Barbier was not unsympathetic to baker Eloy, sued by Hacquin, a flour merchant from the Dammartin area, for 2,080 livres. Hacquin contended that he had "totally proved his case" by submitting an impressively ordered register of sales and receipts. Account books were the first thing arbitres looked for; Barbier acknowledged the strength of Hacquin's documentation. Still, the broker pressed Eloy to discuss the matter, though he could not persuade him "to face the plaintiff for a full, mutual explanation." From the baker's wife, who conceded the validity of Hacquin's claim, at least on the surface, Barbier learned why Eloy was so recalcitrant and so bitter: "his discontent was with the merchandise, which made bad bread, and it was this that completely tore him apart." After much persistence, Barbier finally got the flour merchant and the baker together, but "Eloy refuses any arrangement." The broker-arbitre had no choice but to recommend that Eloy be directed to remit the full 2,080 livres sought by Hacquin. Committed to "reconciling [the parties] if I could," an official at the Versailles flour market named as arbitre could not get master baker Binet to meet with him and thus ended up endorsing flour merchant Simon Serinam's claim of 360 livres.[35]

Dolivier, a grain merchant at Etampes, pursued merchant baker Cottin for 1,642 livres' worth of flour. After the first arbitre, a master baker, withdrew from the case, the consuls asked the baker-jurés to serve. Cottin told the jurés that he could not find the "little book" in which lay proof that he had received only fifty-five of the seventy-five sacks Dolivier alleged that he had delivered. Nor could the jurés verify Cottin's claims that the twenty contested sacks had been sent instead to his colleague, baker Dupuis, because Dupuis had since died. Indeed, the jurés impugned the validity of the affidavit that Cottin produced with Dupuis's signature on the grounds that it was composed by several different hands. Finally, Cottin called attention to evidence that his wife had already paid four hundred livres to Dolivier, proof that he had curiously failed to mention before, and that his wife could not verify, she, too, being dead. Despite the highly dubious assertions of the baker, "nevertheless, in order to conciliate the interests of the parties and by way of accommodation," the jurés recommended that Cottin pay only twelve hundred livres, less than three-quarters of the amount demanded by the merchant.[36]

In the Cottin case the baker-jurés favored a fellow baker (though not, incidentally, a member of the guild). So did a former juré, Pierre de Saint-Martin, the arbitre in a case between a Brie grain merchant, Mondollet, and Delavigne, a master baker. The dispute turned on sixteen setiers of wheat that the baker asserted he had never received. Since Mondollet had sent the grain to Delavigne's miller, not directly to the baker, the arbitre advised that the latter be discharged and that the suit be redirected against the miller. The appointment of a baker as arbitre, however, should not have automatically been cause for despair on the part of the plaintiff and jubilation on that of the defendant. After hearing the parties separately and then "confronting them," the baker-jurés found in favor of a miller against a baker whose documentation "was not at all in proper order." To help determine the truthfulness of baker Thouin's contention that he had paid in cash for each shipment of flour delivered by merchant-broker Chicheret, the jurés convoked seven bakers who knew Thouin and were familiar with his practices. Their testimony convinced the arbitres to support the plaintiff. Jean Cochu, a Paris miller suing a Saint-Antoine baker, widow Henriette, considered one of the two master bakers named by the consuls as arbitres to be "his arbitre" and the other to be the defendant's, as if they were serving as counsel. The baker's arbitre could not persuade Henriette and her son to desist from insulting and assaulting their adversary.[37]

Along with bakers and brokers, the market officials called measurers were the arbitres most frequently assigned to baker cases. In many ways, the measurer was an excellent choice. He was equally suspect in the eyes of both buyer and seller because of his interventionist role in grain and flour transactions and his often bumptious manner. He was also intimately familiar with the business habits of all the parties. Measurer Marin succeeded in getting master baker Cheroux and flour merchant Houdun of Persan to agree on a settlement and a reimbursement schedule. In another case involving the same arbitre, widow Fasquel, a flour merchant from Senlis, sued former baker-juré Dorigny. According to Marin, the business books (*livres-portatifs*) of both parties "were very much out of order." It looked as if Dorigny owed 452 livres (from a total transaction of 3,729 livres), but the baker claimed to have furnished "certain objects" to Fasquel worth 133 livres. Unable "to penetrate to the light," Marin left the decision "to the wisdom of the consuls."[38]

The critique of baker bookkeeping reads like a refrain in the arbitres' reports. Aubry, a miller near Meaux, sued merchant baker Guillement for over a thousand livres for grain purchases, flour production, and delivery fees. The baker maintained that he owed nothing, that he had paid for everything in cash. The arbitre accused him of "bad faith" on the grounds that he kept no records whatsoever of his transactions. With her "little piece of paper on which we scribbled all sorts of notes," how could the wife of forain Guilleman of Vaugirard overcome the authority of the broker Dame Guillain, who brandished a register-journal "in excellent order"? The arbitre urged that the baker be compelled to pay the broker the 705 livres she sought. Similarly, since baker Landet could produce no books to contest the claim of Frenel, a flour merchant, that he owed 1,473 livres, the arbitre found in favor of Frenel.[39]

Another arbitre was helpless to decide between baker Courtin, a forain at Chaillot ("I owe only fifty-five livres for all remaining obligations"), and Louis

Lalande, a flour merchant from Maintenon ("Courtin owes me the sum of 1,045 livres"), because neither defendant nor plaintiff kept "any sort of register or accounts." With the installation of the flour-weighing station at the Halles, another arbitre found it possible to get at the truth without business records. Merchants were required to declare the amount of flour sold and the price at the station. By obtaining a copy of that declaration, the arbitre, a flour merchant himself, was able to show that the baker was lying.[40]

If a baker felt very strongly that a given arbitre was prejudiced against him, he could appeal to the consuls to appoint a new one. Baker Jubert protested sharply that the broker assigned as arbitre in his case had commercial and perhaps familial ties with his adversary, a grain and flour dealer. The court replaced him with a measurer. Baker Deshayes denounced arbitre-broker Barbier for basing his evaluation in favor of the plaintiff on extraneous and irrelevant evidence. In the hope of clarifying the dispute, the judges turned to a secondary arbitre of authority and experience in the subsistence sector: Commissaire Courcy, the senior police official in the "department" of the Halles.[41]

Certain creditors tried to bypass the usual channels of recovery in the hope of persuading a powerful official, the lieutenant general of police, to effect an extrajudicial solution. They turned to him because he was responsible for the conduct of the bakers as part of his mission to assure the provisioning of Paris. As a rule, the lieutenant general scrupulously avoided becoming embroiled in litigious matters. If, however, they engaged the public interest, or involved persons of influence, he found it hard to remain aloof. In 1725 one of the foremost flour merchants of Pontoise threatened to boycott the capital because a baker refused to pay his account of eight thousand livres. Pontoise was one of the chief sources of Paris flour, the capital was in the throes of a serious dearth, and other merchants would surely follow the lead of this "big mealman [gros farinier]." In these circumstances, Lieutenant General Hérault vowed to extract payment from the debtor baker and any others who gave the flour dealers a pretext to abandon the supply of the capital.[42]

Nor could Hérault spurn the request of chevalier Bernard that he compel baker Estienne Meusnier to pay 1,368 livres for wheat purchased. At the king's demand, Bernard, an international banker, had entered the worldwide grain trade massively in 1725 to help France deal with the subsistence crisis. It was not in the interest of provisioning Paris to deny him. Less compelling was the plea from Madame Bignon that the lieutenant general goad three forain bakers, corenters of her mill, to pay the overdue rent. She traded on the high place that her husband had exercised in the royal administration. The creditors of baker Felize addressed the lieutenant general because Felize was the bread supplier to the prisons over which the lieutenant general had (partial) jurisdiction. The creditors of baker Fremin appealed to the lieutenant general because Fremin had taken shelter from pursuit in the household of one of the major honorific figures in the government of the generality of Paris.[43]

Resistance

Once condemned by judicial sentence to pay, what options remained available to the baker disinclined or unable to obey? One course of resistance was appeal. There were two different appeal strategies. A "horizontal" route focused on procedural irregularities, kept the parties in the same court, and minimized costs. Thus widow baker Chevriot filed an "opposition" within the consular jurisdiction to the execution of a judgment it had awarded broker Pierre Girard. The latter countersued, asking for and obtaining the quashing of this elliptical appeal and dilatory action. A "vertical" appeal took the defendant to one or more higher arenas and involved considerable expense in court and counsel fees and in time. Condemned by the municipal court to reimburse Vincent Pelletier, a grain merchant of the rue de la Mortellerie, baker Antoine Gouffe contested the verdict for three years, all the way to the parlement. Baker Jean-Baptiste Guillaume vainly asked the parlement to quash a sentence of seizure that the consuls pronounced in favor of a flour merchant.[44]

Another option was to hide or flee. Condemned to pay flour merchant Jean Gilbert 694 livres, baker Benoist went underground, appearing only on Sundays and holidays when processes could not be served. Convinced that Benoist was secluded in his own house, Gilbert petitioned parlement for authorization to have him arrested in the sanctity of his domicile. Under pressure from at least four flour dealers for thousands of livres in flour sales, master baker Seurre took more dramatic evasive action. "Furtively," in the middle of the night, he left his house "with the major part" of his furnishings, effects, and merchandise. He allegedly sold several beds, other furniture, paintings, and clothing to a journeyman surgeon dealing on the side in secondhand goods who paid him nineteen hundred livres (four hundred livres of which represented reimbursement of a loan he had previously made to the baker). Seurre left other property with an innkeeper and his baker equipment with two bakers, one of whom had taken on one of his journeymen. He sold a considerable stock of flour to one of these bakers, and he deposited some silver in the safekeeping of a tailor. Nor did he forget to send a journeyman to collect several hundred livres in sums due from his customers for bread. Pressed by the creditors, one of whom was a particularly influential Halles-based flour dealer, the police scoured the city for Seurre, arresting in the process the surgeon–used articles dealer, the baker's wife, and his stepdaughter, each of whom denied any knowledge of the fugitive's whereabouts.[45]

Bakers who could not pay their rent frequently absconded at night, with as much as they could transport with them. With the aid of several off-duty soldiers from the guard, baker Milan moved his possessions out of the house of widow Leblanc to whom he owed 105 livres in rent. Leaving behind a pile of junk and a rent bill of 334 livres, master baker Thomas Barbier set up shop in a different neighborhood. Master baker Michel Desbordes repudiated five years on a lease on which he was nine months behind in rent, taking with him all his furniture and equipment. One wonders exactly what strategy master baker Simon Guiton had in mind. He owed his landlord, another master baker, 366 livres in rent, for which claim the latter had instituted a request for a sentence of seizure. Guiton called in masons, carpenters, and other artisans in the middle of the night to dismantle his entire bakery and reassemble it in a house across the street, a move that he could hardly keep secret for very long.[46]

Borrowing as an interim solution was not an option available to most debtor bakers simply because they were such poor risks. To borrow, a baker had to have substantial collateral that he could cogently argue was safe from seizure or some friend or relative who agreed to help. Desperate for liquidity, Saint-Antoine widow Hedé borrowed five hundred livres against a half-house—lien-free—that she owned. Master baker Marin Laurence borrowed against an annuity he had inherited. Pursued by a swarm of creditors, master baker François Sauvegrain obtained loans of first twenty-five hundred livres and then five hundred livres from a relative at Arpajon (against which the baker mortgaged all his property). There were success stories as well as tales of woe: master baker Joseph Barre repaid the one thousand livres that he borrowed from a master shoemaker in eight months.[47]

A baker could stop, or at least slow down, the pursuit of creditors by declaring *faillite,* a business failure defined in eighteenth-century terms as "innocent bankruptcy." But faillite imposed a number of unpleasant strictures. It subjected one's resources and one's business practices to probing scrutiny. It meant the end of one's commerce and the forced sale of one's assets if creditors refused to tender a second chance. It blemished one's reputation, regardless of the outcome. Just short of faillite, a baker besieged by creditors could seek terms at the eleventh hour. Often these terms, granted collectively by creditors organized in a "union" to enhance their pressure, proved to be more attractive than those granted in court. The baker wanted time to get back on his feet. The creditors needed to weigh the risks of postponing immediate liquidation against the prospects for earning more sous per livre from a revival of their debtor's enterprise.

Two Mortellerie grain merchants joined several other creditors in giving merchant baker Pierre Sacre a delay of two years in light of the "misfortunes" and "losses" he had suffered. Another baker agreed to repay the two thousand livres he owed Mortellerie dealer Armet in twenty-seven consecutive weekly installments. Master baker Nicolas Sauvegrain implored his creditors (who included two grain merchants, three flour dealers, and a miller) to join together in a united bargaining front. In the red by forty-seven thousand livres, he found it impossible to deal rationally with his situation while his creditors squabbled among themselves for priority. He offered to abandon to the union three houses he owned as well as the right to attempt to collect his bread accounts receivable, provided the creditors agreed to renounce all other claims and to endow a small "subsistence fund" from the sale of the houses to afford Sauvegrain some old-age insurance. The union accepted these terms, but four years later at least one of the houses had yet to be sold.[48]

Master baker Jacques Michel Guenée tried to persuade four millers and three flour merchants, to whom he owed twenty-three thousand livres, that his rehabilitation was a good bet. During its first fifteen years, his business had flourished: "He always fulfilled his commitments with honor." In recent years heavy domestic expenses (raising six children) and "difficult times" (high prices and scarce supplies) had caused him "big losses." Given "the good name of his establishment, the number and nature of his customers," he was "optimistic that he could repay his creditors" if they were willing to allow him eight years in which to amortize his debts and to forgive him interest and penalties. Convinced that "there was no bad faith" on Guenée's part after examining his balance sheet, the creditors embraced his proposition, which also gave them a privileged claim on all his assets if he missed an installment.[49]

A victim of the same "hard times," merchant baker Adrien Loiselle asked not only for the usual grace on interest and recovery costs, but also for a 50 percent reduction on the principal and six years' time to repay the remainder. Eager to "maintain him and help him as far as they could," the creditors, including two flour dealers and a broker, drew the line at taking a 50 percent loss at the very outset. They gave him a year's freedom from any restitution and seven years of biannual installments during which to pay off the full principal, with interest and expenses forgiven.[50]

Bakers who wanted immediate closure—that is, an absolute guarantee against any further pursuits, seizure, or arrest—could cede all their effects to their creditors. Cession usually implied the renunciation of business. In most cases, the baker had precious little to abandon and the creditors had to resign themselves to significant losses.[51] Master René Morin is the only baker I found who survived one cession only to surrender all his property to a second legion of creditors fourteen years later.[52]

If a creditor who had already won a sentence *par corps* believed that a baker had adequate assets to cover his obligations, or if he had no confidence in the probity or the competence of the baker, or if he wanted to attack the baker's resources before other creditors arrived on the scene, it was quite likely that he would attempt a confiscation of his property (*saisie*). Marin Denis of the rue de la Mortellerie had the house of baker Charles Lelièvre seized; the two fought over its disposition for thirty-seven years. Another Mortellerie trader, Joachim Armet, effectively incapacitated master baker Denis Petitfils by securing liens on all his property. The widow of another port merchant, Charles Vassou, won seizure of two houses and some land belonging to the widow of a forain. Master baker Jean Legrand's landlord was disgusted to find only paltry effects when he had authorities "open his tenant's doors" as part of a saisie operation. The landlady of baker Pierre Hebert did considerably better: the baking equipment, flour stock, and furnishings easily covered his three hundred livres in rent arrears. Master baker Jean Go repelled the efforts of a creditor, who had rebuffed his plea for a year's grace, to seize his possessions on the grounds that his representative lacked the proper judicial authorization.[53]

As a rule it was in the creditors' interest to see the baker remain in business. When they were sufficiently exasperated, however, creditors became pitiless. They turned to punishment both as a sort of moral compensation and as an instrument of leverage. Following the letter of the law prescribed in *par corps* convictions, they had their baker debtors arrested. In jail they were not segregated from the criminal population. They slept on piles of straw that were often humid and littered with filth. Their cells were exiguous, unheated, dark.[54] The debtors languished in prison (usually the Grand Châlelet, the Conciergerie, or La Force) either until their families raised sufficient funds to warrant release or until the creditors, legally responsible for their upkeep, stopped paying for their food.

The eighteenth century witnessed a lively debate concerning the treatment of debtors. Mercier was ferocious: "there is not enough severity [against the debtor]." He should be "pressed hard"; the faithless debtor ought "to be stained with infamy." Early in the Enlightenment the abbé de Saint-Pierre called for a reform on the British model, which freed thousands of debtors from jail. Dupont, a physiocrat and a passionate defender of property rights, also found the English system to be superior to the French. "According to the natural order," proclaimed Dupont, "prison must not be decreed against a nonfraudulent debtor." Citing Beccaria, he denounced the penalty as incommensurate with the crime. It was "barbarous" to lock up poor tradesmen in "horrible" dungeon-prisons. The only sensible solution, concluded Dupont, was to allow the debtor the freedom to try to redeem himself and to restrict creditors' liens on his property and effects, sheltering his person. On this point, the vehement antiphysiocrat Simon Linguet was entirely in accord with Dupont. Prison brutalized and alienated the debtor, reducing him to forced indolence and enveloping him in an environment of evil and corruption. Debt imprisonment was useful neither to the creditor, who paid the expenses and deprived the debtor of his power to repay, nor to society, which ultimately had to reintegrate the detainee.[55]

Jail

Bakers went to debtor's prison in substantial numbers throughout the eighteenth century. Despite the withering critique of reformers, they were still being incarcerated on the eve of the Terror. For the sake of discussion, let us divide the baker prison sojourns into two categories: short stays (under a week) and long stays (ranging from three weeks to thirteen months). The mean short stay for nine bakers was 3.2 days (median = 3). These bakers owed between 120 and 477 livres (mean = 325, median = 400). Twenty bakers averaged a long stay of 6.5 months in jail (median = 6) for debts ranging from 210 to 15,500 livres. The mean debt of these detainees amounted to 2,910 livres (median = 1,478). There was a partial correlation between the length of the stay and the magnitude of the debt: thirteen months and 9,110 livres, twelve months and 5,099 livres, ten months and 15,500. Yet there were quite a few exceptions: 330 livres in debt issued in a seven-month stay; 2,738 livres in debt kept a baker behind bars for thirteen months.

The short-stay prison experience reveals the efficacy of scare tactics. Unable to move the bakers any other way, the creditors wagered that the prod of jail would jolt them into action. Arrested on 11 March 1736 at the request of two creditors (240 and 54 livres in claims), master baker Pierre Lesueur left the Grand Châtelet two days later after paying one-third of his obligation and pledging to remit the remainder in two installments over two months. Master Nicolas Tourneville barely spent a day in jail, reimbursing his full debt of 150 livres almost immediately. Another Tourneville, perhaps the same, spent two days in the Grand Châtelet in 1730 as a consequence of a 409-livre debt to a broker. A one-quarter down payment won his release. Merchant baker Jacques Porcheon got out after five days in prison with a payment of two hundred livres toward a principal of 424 livres. Flour merchant Quentin Dubourg earns the title of creditor with the meanest spirit. He had baker Pierre Laurent jailed on Christmas eve 1725 and not released till he raised one-third of his 159-livre debt on 26 December.[56]

Master baker Genard seems to have been freed after nineteen days behind bars largely because a reputable master glazier agreed to stand surety. It took master baker François Provendier over two and a half months to obtain his freedom because he had to deal with sixteen different creditors: nine flour merchants, four grain dealers, and three brokers, each of whom accepted a payment averaging 164 livres (over against an average claim of 610 livres). Bakers raised money in diverse ways. Their wives importuned delinquent customers to pay, or they sold some of their flour or wheat stock, or they got a supplier to whom they vowed fidelity to bail them out. Master Jean Devaux, jailed by the relentless Quentin Dubourg, borrowed sixteen hundred livres from several friends. The wife of master Julien-Charles Blanvilain sold a precious necklace in order to pay the price of her husband's liberty.[57]

Twelve creditors with claims totalling 15,500 livres kept master baker Martin Hetrin in the Grand Châtelet for ten months. His colleague Jean Devaux endured incarceration for a year until Quentin Dubourg yielded. After holding baker Angelbus in jail for ten months, a flour merchant from Persan finally granted him three years in which to make good his debt of 2,072 livres. A wine merchant refused to release baker Philbert Bourgeois till he paid his entire debt of 1,209 livres. It took him a year and a day in jail. Master baker Dutoq thought he was home free once he paid Mortellerie merchant Jean-Pierre Armet one-third of a 1,152-livre debt. But another Mortellerie dealer, Jean-Baptiste Jauvin, nabbed him on the way out for an obligation of eleven hundred livres, and he apparently pined for another eleven months in the dungeon of the Grand Châtelet. It made no difference that the Revolution had changed almost everything else: baker François (the François who was not strung up in 1789!) spent half a year in La Force in 1791–92 for 914 livres in debts.[58]

If the creditor failed to pay for his prisoner's bread one month in advance, the jailor was authorized to release the prisoner. In practice, the creditors were given considerable leeway. This was a test of the creditor's determination and a constant invitation to him to reassess his strategy. Master baker Charles Gouy's creditors, among whom were a flour merchant and a bourgeois de Paris, could not agree on how to share the burden of support. After a year in the Grand

Châtelet, the baker was freed "at 5:00 A.M. on 8 March 1721 . . . for lack of provision of food." In debt to the royal government for the purchase of the king's grain, to a Mortellerie dealer, a flour merchant, a broker, a miller, a flour merchant, and a receiver of fines (for the one hundred livres remaining of a three-hundred-livre fine for a dozen short-weight four-pound loaves), Claude Trudon, another master baker, spent six months in jail. He, too, walked out "for want of food." In a spasm of rancor, a failing flour broker named Viollet had eight of his intractable baker clients imprisoned. For lack of liquidity, he was obliged to allow five to go free; given the creditor's precarious situation, the others were probably not long for Bicêtre.[59]

Debt-Collectors and Devils

When they looked up to see the huissier entering to arrest them or confiscate their property, most bakers understood there was no longer any way to avert their downfall. (After the creation of ten *officiers-gardes du commerce* in November 1772 to execute sentences against debtors, it became even more difficult than ever before to avoid civil imprisonment, called a *contrainte par corps*.)[60] A handful of bakers resisted violently. A huissier visited Saint-Antoine baker Desmures for the purpose of collecting the sum of slightly more than a thousand livres that he owed a miller from Vaux or seizing his furniture. After showering the huissier with curses, Desmures chased him out of the shop and down the street swinging a four-foot axe. Supported by his two journeymen, master Goguin met the efforts of a huissier to collect 133 livres 12 sous with "a violent rebellion."[61]

Waving his *baguette distinctive*, Jean Monglas, one of the original ten gardes du commerce, ordered baker Levêque to pay a judgment of 3,309 livres or accompany him to jail. Levêque angrily refused and fled, assisted by his journeymen, who attacked the intruder with bread shovels and oven rakes. One of Monglas's colleagues, Jean Lucas, encountered a similar reception a number of years later. Baker Colle assaulted him with a pitchfork when he served a sentence for 1,030 livres or imprisonment.[62]

A sixty-year-old baker from the rue des Noyers unleashed another act of violence—against himself. To escape his creditors he tried to cut his throat with the tool bakers used to cut and inscribe the dough. Discovered in time, he was saved at the Hôtel-Dieu, where "he was treated as if he were a madman." Death gave the bakers a macabre last laugh. Eight of broker Viollet's baker debtors died before he could collect. Nor could he or other creditors easily pursue their widows, for the women kept them at bay by "renouncing" their husbands' estates because they were insolvent. Yet, to the exasperation of the creditors, the widows usually continued their husbands' businesses (in which they had been full-time collaborators all their married lives). Widow baker Deline denied any responsibility for the price of eighty setiers of wheat (1,331 livres) that her husband had purchased from grain and flour merchant Jean Guibert of Coulommiers. Widow Constant scoffed at mealman's Desrues's claim of 392 livres on the same grounds: it concerned her late husband, not her.[63]

In the hope of finding a way to appease his creditors and thus escape jail or confiscation, a forain from near Gonesse turned to the occult. Despondent, Jean Tremblay succumbed easily to the lure of "a treasure" offered by a man he knew vaguely from the bread market. Tremblay was not greedy: he insisted that he wanted only enough to pay his creditors. His friend introduced him to two strangers who expressed confidence that they could find "a million." To seal the "contract," they induced the cash-poor Tremblay to give them eighty-four livres immediately. At the appointed time, close to midnight, the baker met them under a tree on the Champs-Elysées. One of them drew a circle around Tremblay and called to the devil in an exotic language.

Subsequently, one of his mysterious guides brought Tremblay a paper indicating that the devil required a little more than eighty-five livres for registering his act, as if it required official documentation. He borrowed the money and paid them. Then, at a second ceremony under the same tree at midnight, Tremblay heard a voice call urgently for another seventy-three livres. Once again, he gave it to the devil's advocates. Yet another night the devil bid him to remit a further fifty-two livres. When the pitiful Tremblay broke down and avowed that he could raise no more money, the treasure hunters told him "to go to hell and never ask them for anything again."[64]

There is no doubt that credit had a diabolical aspect. Yet in this highly stratified society, credit (or debt) was the great leveler. In diverse ways, it touched virtually everyone at one moment or another. Without credit, it would have been impossible for the grain and flour trade to function autonomously. Indebtedness per se was normal; the story of baker debt is not characteristically a lachrymose tale of the decline of a business and the unraveling of a household. Rather, it is the narrative of management: those who thrived were the ones who kept a viable proportion between what they owed and what was owed them and who avoided becoming excessively dependent on either a few of their creditors or a few of their debtors.

Serious breakdowns in the credit-debt balance sometimes issued in wrenching human dramas. Fail-safe devices did not lack, but they were not always efficacious. A commissaire or a broker or a common friend, usually in the trade, might try to reconcile antagonists. The baker-jurés intervened occasionally. The business court designated mediators whose function was precisely to find common ground. The filing of failure (what we call today bankruptcy) through this jurisdiction frequently resulted in debt consolidation and revitalization. The working assumption of all these agents was that it would be cheaper for everyone, including the public, if a solution could be found and a business (or several) salvaged.

When the institutional and informal alternatives were exhausted—or rather when the will to exploit them optimally was—the parties reverted to a sort of commercial state of nature. In the worst case, bakers lost everything and lapsed into jobless and/or homeless misery; or they languished in jail (which suggests that their creditors believed that they had hidden or sheltered resources); or the bakers resisted, through violence or flight, the one ultimately no less vain than the other. The shattered baker, mired in debt and doubt, could hope only for posthumous revenge: his widow's renunciation of his onerous succession, the most devilish and desperate strategy of family survival.

Chapter 14

Failure

Another source can be used for the study of the relative haleness of bakers: the *fonds de faillites,* or business-failure records. Like police records, the faillites are a pathological source and they must be used with critical discretion. Whereas a strong case can be made for the representativeness (within the limits previously indicated) of randomly chosen inventories, preinventories, and marriage contracts, it is obvious, from the very nature of the document, that business failures reflect an unusual and unhealthy state of affairs. By definition, a "failed" baker was one who could not meet his obligations as a result of some sort of "accident" or combination of circumstances judged to be "not of his fault" (for example, "the misery of the times," "real losses on merchandise," fire, war, bad weather, and so on).

In fact, "failure" was a type of bankruptcy, but the word bankruptcy in eighteenth-century France necessarily implied fraud and bad faith, whereas the insolvency of the genuine *failli,* or failure victim, was said to be "innocent." As a journalist put it in 1791, failure was "a misfortune" rather than "a crime." Provided he was guilty of no "artifice to deceive," and provided he submitted a cogent written dossier (good bookkeeping was esteemed to be the best test of "sincerity"), a failed artisan or merchant could not be subjected to any criminal penalty. I have used the word faillite instead of bankruptcy to avoid confusion with the criminal maneuvers inherent in the eighteenth-century French notion of bankruptcy.[1]

Despite the formal distinction between fraudulent and simple, or innocent, bankruptcy, in business circles and in the community at large the failli suffered opprobrium. He could purge the stain by satisfying his creditors according to a mutually devised "arrangement"; ultimately, he could seek formal rehabilitation by soliciting letters from the Keeper of the Seals. A number of writers, including Letrosne, Saint-Péravi, and Mercier, sharply assailed the legislation governing failures. They considered it far too permissive, and they cited (without figures) the ever-increasing incidence of failures as evidence of abuse and deception and the impunity with which this "crime" could be committed as the reason for its proliferation. Bankruptcy, they complained, had become a "lucrative" enterprise, "a game" by which cunning and dishonest men made fortunes at the expense of reticent or busy creditors who were deterred by the cost and time of litigation from pursuing their rights. These writers demanded a more rigorous examination of the bankrupt's business records; stiffer penalties, including jail; and "a more or less great mark of dishonor," depending on the gravity of the case.[2]

While the critics conceded that some faillis were honest businessmen afflicted with bad luck, they gave the impression that the failure petition, or *bilan*, was more often than not an elaborate counterfeit, a series of figures ("exaggerated losses" and "imaginary outstanding debts") orchestrated to make credible the failli's stand in the negotiations he would have to conduct with his creditors. To what extent do these accusations seem to be true, particularly in regard to the bakers, and what are the implications for the historian who wants to use failure papers to study not only failure itself, and business practices, but also fortune and family? No one doubts that there was some fraud in the failure procedure. Yet the critics are probably as guilty of hyperbole as the alleged businessman swindlers.

It was no secret that the filing of a *bilan* was an opportunity, if not an invitation, for cheating. Precisely because every failli was prima facie suspect, it would be absurd to imagine, save in instances of pervasive collusion, that each case was not closely examined. Even if the authorities were on occasion lax, there were the creditors to be reckoned with. Mercier and the others underestimated their tenacity, less intense perhaps on the part of the big bankers than on that of the relatively small merchants who felt personally involved in their client's failure and who badly needed the money that was due to them. If the creditors accepted "arrangements" with the petitioner, it was not because they were discouraged or forlorn about the prospects of collecting their debts but rather because they felt this was the surest way to obtain payment, or at least some payment. Where it was in their interest, they showed indulgence (though they were not without compassion for solid clients and/or neighbors who fell unexpectedly into hard times). The creditors generally had a good idea of the economic (and moral) situation of the failli. They or their procurators scrutinized his books and his balance sheets.[3]

Claims of losses were probably systematically inflated because one did not have to document them. But it was much less easy to lie about accounts receivable, for which registers, notes, or *tailles* had to be produced (though the solvency of the baker's debtors was a matter that could not be verified), or about real estate or annuities (rentes), for which titles had to be shown, or even about household possessions, for a failli's manner of living and his *reputed* socioeconomic standing could easily be ascertained. Though less precise than inventory data, the failure figures for stocks or stock value, property and annuities, and even bread debts due warrant confidence, in my estimation. On average, household assets were probably underestimated, less in the largely vain hope of concealing assets than because many bakers did not seem to have understood that they could or should include nonbusiness or strictly domestic assets in the balance sheets. As for the liabilities, the balance sheets present them as one might imagine, with a fanfare that leaves us no reason to regret the inventories.

Finally, we ought to take note of the specific characteristics of the baking profession in relation to what Mercier called the "game" of bankruptcy. Bakers had little incentive and little opportunity to play this game. Their profession was one of the most closely regulated; given its sociopolitical importance, it was under constant surveillance, not only by the police but by the consumer-people as well. Ultimately, the baker's success depended on the fidelity of his clientele. He had to retain their confidence, he required a good reputation, he had to produce and sell bread of good caliber every day without fail. The baker also needed to cultivate a positive rapport with his suppliers. He counted on obtaining credit from them, for he was obliged to extend credit to his customers and thus suffered a chronic liquidity problem. Dealing with a single, simple product, the baker had little to juggle. If he got into difficulty, his first concern was to extricate himself as expeditiously as possible: to get seals removed in order to continue his baking and to purge the stigma that marked him in the neighborhood.

A few faillis managed to change fonds, but it was a doubly difficult operation. First, it was hard to sell a business tainted by failure, with the inexorable erosion in customer fidelity. Secondly, even if the baker did not wear the humiliating green bonnet of fraud, it was difficult to keep his failure a secret and to win the confidence of customers in his new neighborhood. He could not afford protracted discussions or the risk of jail. He needed to persuade his creditors that he could turn his business affairs around if given the chance. In most instances, he found that candor served him best. The baker who postured or embroidered excessively or whose story appeared devoid of verisimilitude did not get "the terms and delays" he sought.[4]

The failli who did not win the confidence of his creditors could be forced to cede all of his assets immediately in order to satisfy partially his debts. Such a fate virtually constrained the failli to renounce his business. That is what happened to master baker Michel Danhuser, whose chief flour creditors threatened to have him jailed if he did not raise cash immediately by selling his fonds, his tools, and his household goods—despite the fact that his debts were not astronomical (8,694 livres) and that the discrepancy between them and his claimed assets was a trifling 440 livres.[5]

The overwhelming majority of baker failures were juridically "innocent" bankruptcies. Some resulted from mismanagement and others (fewer) from dissipation, a moral rather than a technical defect. Still others were the fruit of bad luck. It is significant, for example, that twenty-seven bakers petitioned for failure between 1772 and 1774, after several years of grave disorders in the provisioning trade that badly buffeted dealers and bakers as well as consumers, whereas bakers had filed only six petitions in 1760–63, four in 1756–59, and six in 1752–54. Twenty other bakers failed in 1775–77, most of them also hurt by the widespread economic crisis that prevented them from collecting money due or from buying on credit and by the squeeze caused by soaring prices on the one hand and the unwillingness of the police to allow them to pass the full brunt on to consumers on the other. A handful suffered losses as a result of the Flour War riots in 1775.[6] There was a speculative element in the baking business, as in the grain trade, whose implications the baker could not always foresee or control. Many bakers were small businessmen, working on a thin margin, so it did not take very much to put them into disequilibrium.

Assets in Failure

The total assets in 144 failures averaged 11,073 livres (median = 6,611), with a range from 173 to 72,400 livres. On average, they were three thousand livres less than the inventory assets, though the median value of the failures was about fourteen hundred livres above that of the inventories. Assets fluctuated quite erratically between 1740 and 1790, the half-century for which we have failure records. Averaging 15,515 livres in 1740–45 despite the catastrophic dearth of 1739–41, assets subsequently dropped to 3,315 in 1756–59 before climbing back to the older levels in 1764–66 (16,228 livres). They fell again to 6,447 livres in the crisis period of 1772–74 and subsequently recovered, never dropping below ten thousand livres after 1778–81. Assets reached their apex in the period 1785–88 with an average of 20,143 livres (see figure 14.1).[7]

It is striking that thirty-eight nonmasters in the group had average assets of 11,932 livres, thereby surpassing the ninety-seven masters, whose mean holdings amounted to 11,195 livres (but the masters had a higher median value: 7,522 v. 6,863 livres). The vanguard of the nonmasters was the "merchant" group with average assets of 20,188 livres, followed by the forains whose 13,262 livres in mean assets also put them substantially ahead of the guild bakers. Nonmas-

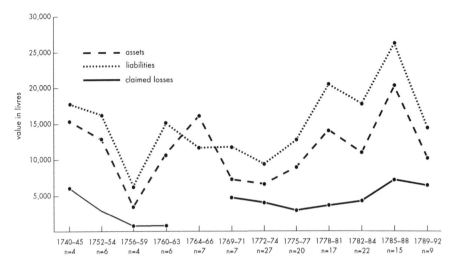

Figure 14.1 Baker Failures: Mean Total Assets, Liabilities, and Claimed Losses, 1740–92 *(Source: failure records)*

ter assets remained below five thousand livres on average until the late seventies, after which they never fell below sixteen thousand livres, reaching a zenith of almost twenty-nine thousand livres in 1785–88. The masters fared better much earlier, enjoying their highest assets in 1740–45 (26,425 livres) and never again reaching sixteen thousand livres after 1764–66. Over the course of the century, it appears that the wealthier (and thus the stronger) guild members became less and less likely to sink into failure. They may not necessarily have been better managers than their less successful confreres, but they had a much greater economic cushion to insulate them from trouble. The nonmasters enjoyed a marked development of their wealth, but their fortunes appear to have been built on more fragile foundations than those of the masters. The nonmasters buckled under the same kind of pressures that the guildsmen with assets of comparable magnitude were able to resist.

The single most important component in the failure assets was accounts receivable for bread sold on credit. Ineffective debt collection was the ostensible cause of many failures, for in a large number of cases bread debts more than covered the disparity between assets and liabilities. For the 122 bakers who claimed bread accounts receivable, those accounts represented on average over half of total assets.[8] Mean bread debts amounted to 4,511 livres (median = 3,301), with a range from 170 to 19,311 livres. Bread accounts receivable for nonmasters amounted to 5,176 livres (median = 3,720), a figure about 16 percent higher than the bread debts due masters. Merchant-bakers alone had almost twice the bread accounts receivable of the same magnitude as the masters, thus revealing a far greater propensity to grant credit than the inventories would have enabled us to predict. Bread accounts receivable for both guildsmen and unincorporated bakers were twice as high in 1785–88 as they had been in 1742–54. Only further study will determine whether this development (controlling for demographic change) reflects a gradual impoverishment of the laboring poor as real wages lagged in a century of brisk inflation and they were compelled to rely more and more heavily on credit.

Mean bakery investment stood at 2,642 livres (median = 1,900). For the seventy-eight cases for which we have information, it accounted for 30 percent (median = 21 percent) of total assets. Average bakery investment was more than nine times the inventory figure because, more often than not, it included an estimated price of the fonds. This is a somewhat diffuse rubric, for it is not always possible to disaggregate its contents in order to determine the respective claims of tools, shop installation, clientele, merchandise, and so on. On the surface, it seems surprising that nonmasters (2,735 livres) had slightly more money invested in their businesses on average than masters (2,678) and that merchant bakers (6,333 livres) had more than twice the investment of masters.[9] It is likely that the nonmasters, especially the merchants, exaggerated the worth of their businesses, but it should also be kept in mind that we are comparing, to a considerable extent, some of the poorest masters with some of the wealthier faubouriens. Nonmasters had a much greater bakery investment toward the end of our period than at the beginning, whereas there was no clear pattern of evolution for the masters.

Since one of the chief reasons that bakers failed was that they had no more grain or flour to convert into bread and no means of obtaining any, it is understandable that only a small segment of our population had stocks listed in their balance sheets. Yet the twenty-nine bakers who held stocks—twenty-three masters, four forains, and two faubouriens—had very substantial average holdings, valued higher than mean inventory stocks.[10] For these twenty-nine bakers, raw materials represented on average 17 percent (median = 12 percent) of their assets, a higher proportion than the inventory mean. One master with stocks worth twelve thousand livres, representing over forty-eight muids, had a larger holding than the inventory maximum.[11] Stock holdings were more significant in the last two decades of our period than earlier; the faillis claimed virtually no grain or flour storage prior to the seventies.

Though all bakers had household assets of one sort or another, including furniture, linens, clothing, and utensils, for reasons that are not clear, fewer than a third of our population enumerated them.[12] Mean household assets amounted to 1,946 livres (median = 965), representing 17.5 percent (median = 13 percent) of the total assets of the forty-five bakers involved. Masters enjoyed slightly more than nonmasters as a group, but merchants alone averaged almost three thousand livres. The mean value for the whole population was significantly lower than the average total family fortune in the inventories. Not surprisingly, the faillis had no cash to speak of, and they acknowledged no items such as jewelry and silver that were easy to turn into cash.

In addition to the mottled category of miscellaneous assets, which we shall not examine, there are two other components in the failli fortune: real estate and annuities. The latter can be disposed of rapidly, for only four bakers—three masters and one merchant—possessed rentes. Ranging from a petty 325 livres to an impressive portfolio of 21,700 livres, these annuities represented on average 23 percent of the total assets of their owners (median = 13 percent). Nineteen bakers owned an average of 15,798 livres worth of real estate (median = 8,850) ranging in size from 1,021 to 48,500 livres.[13] Ten masters averaged 14,472 livres, but the richest proprietors were four country forains with houses and arable land averaging 24,125 livres. Eight holdings, including the properties of the forains, were rural and averaged 17,306 livres in value, while the other real estate was Parisian and represented a mean value of 13,352 livres.

Liabilities in Failure

The most sacred entry in any balance sheet is liabilities, or accounts payable. Mean liabilities amounted to 15,258 livres (median = 11,110), almost nine times average accounts payable in the inventories (see figure 14.2). Only 10 percent of the failli population had liabilities under four thousand livres and only 30 percent had liabilities under seven thousand livres. Almost one-fifth owed over twenty-five thousand livres and 15 percent were indebted by more than thirty thousand livres. Masters and nonmasters had virtually the same mean accounts payable—15,866 and 15,729 livres respectively—but merchants, with 21,579, and forains, with 19,033, were the leaders in this category. There is no clear pattern for the evolution of the liabilities of the whole population over time. They tended to be higher beginning in 1778–81 than before, largely as a result of the enormous growth in nonmaster debts in these years, presumably the delayed consequence of the terribly stormy decade of 1765–75.

It is hard, however, to establish unequivocal and unmediated connections between subsistence or economic crises on the one hand and the magnitude of debts on the other. In the short term, for instance, it is reasonable to infer that

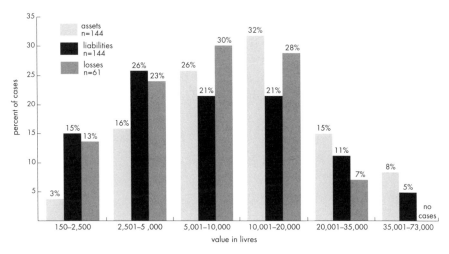

Figure 14.2 Percentage Distribution of Assets, Liabilities, and Claimed
Losses of Bakers Who Failed *(Source: failure records)*

the masters of 1740–45, with average liabilities of 29,084 livres, suffered more or less directly as a result of the long, deep dearth of 1738–41 and that, similarly, the troubles of the nonmasters of 1789–92, with mean liabilities of 34,990 livres, derived to some degree from the dramatic price rise and widespread scarcity accompanying the outbreak of the Revolution. Yet one would have expected an earlier trace of the seismic dislocation of 1765–70, a period that registered an implausible erosion in mean liabilities (but also in average assets). One might distinguish between precariously balanced bakers cast into the abyss by the slightest pressures and more resistant yet still vulnerable bakers who hold out for some time before succumbing. Even allowing for a time-lag effort of accumulated tribulation/attrition, it seems surprising that the failure level for 1766 through 1771 remained so low.

Stock liabilities accounted for about three-fifths of all accounts payable. Forty-two bakers—twenty-nine masters, eleven faubouriens, and two forains—owed an average of 3,972 livres worth of grain (median = 2,201), an extremely large amount, especially when juxtaposed to the practically infinitesimal grain liabilities in the inventories. Of course, the two items are not really comparable, for whereas the inventory entries may indeed reflect the relative importance of grain purchases on a day-to-day basis, the balance-sheet entries might very well represent accumulated debts for many months or even years. The nonmasters, propelled by the merchants, who averaged over seven thousand livres in grain debts, owed about 27 percent more on average than the masters for grain. Mean flour liabilities for the one hundred twenty-eight bakers who acknowledged such debts came to 7,251 livres (median = 5,570), ranging in value from 154 livres to 52,689 livres (a merchant baker). Ninety-two masters averaged 7,485 livres (median = 6,025) while twenty-nine nonmasters averaged 7,472 livres (median = 4,571 livres). There appears to have been a weak association between commercial liabilities and stock holding.[14] The average number of grain and flour creditors, which could serve as a reasonable proxy for the number of suppliers, was seven (median = 6). On the whole, there appears to be only a moderately strong association between total commercial liabilities and the number of supplier creditors.[15]

Net Worth

Average net worth is, of course, deficitary: *minus* 4,185 on average (median = *minus* 3,014) (see figure 14.3). Five percent of the population was more than twenty thousand livres in the red, 15 percent more than ten thousand livres, 40 percent more than four thousand livres, and almost three-quarters had a negative net worth. Almost 15 percent claimed to have between three thousand and thirty-one thousand livres after liabilities were deducted from assets, but they were on the whole a marginal element of the faillis. The deficit tended to be larger after 1778–81 than before, but there were considerable fluctuations in net worth across time that defy any global explanation. Masters had a lower net worth than nonmasters. The forains had the highest mean negative net worth, 5,772, and the merchants the lowest, 1,391 livres.

Claimed Losses

Sixty-one of our bakers claimed between three hundred and 34,700 livres in what they called losses. They attributed them to a wide variety of causes: speculative losses on grain or flour trade, spoilage or other physical loss of merchandise, bad debts, legal fees related to debt collection and to contested claims, illness, individual or collective criminal action against them, hard times, marriage settlements, childbirth, mastership costs, the bankruptcies of other persons, the death of horses, police fines, the discount of notes in order to raise cash, shop and oven repairs, fire damage, key money for new leases, marriage breakup and separation of property, the wastefulness and incompetence of journeymen, a change in fonds, and setting up a new shop. Obviously, many claimed "losses" were in fact ordinary business costs, depreciation, and

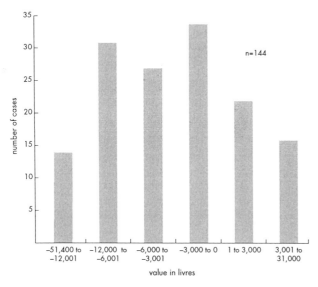

Figure 14.3 Distribution of Net Worth of Bakers Who Failed
(Source: failure records)

familial or household expenses that had nothing to do with the business (in a modern accounting logic). Bakers with significant disparities between total assets and liabilities liked to amalgamate losses (generally undocumented and frequently puffed) and assets to create a sanguine impression of a sort of long-run or nostalgic positive net worth.

For the sixty-one faillis with loss claims, losses on average amounted to over three-quarters of total liabilities (median = 58 percent).[16] Mean losses per baker were 9,027 livres (median = 7,450 livres). A little more than a third of the bakers claimed losses under five thousand livres, while about the same number indicated losses over ten thousand livres. Mean losses appear to have been heaviest in the aftermath of dearth (1740–45 and 1769–74) as well as during periods of economic contraction (1785–89). Masters, merchants, and forains all had about the same amount of mean losses (between 9,300 and 9,900 livres). The non-merchant faubouriens, with only 3,690 livres in average losses, brought the nonmasters' average down. In some of our cases, we can identify the nature of the losses claimed. Thirty-seven bakers claimed an average of 5,608 livres (median = 3,106) in losses directly related to their commercial activities (e.g., bad quality merchandise, unexpected price shifts). Thirty-two others claimed an average of 3,118 livres (median = 2,165) for losses stemming from their inability to collect money due for bread sold.[17]

Recidivism

It is plausible to assume that the bulk of bakers filing for business failure for the first time reached some kind of accommodation with their creditors and managed to return to their commerce. In most cases, it was in the mutual interest of the faillis and the creditors for the baker to get back to work. The prospects of bakers who were obliged to file for failure a second time, however, were much less propitious. Fourteen bakers in our group filed for failure on two separate occasions. Predictably, second failures revealed a general erosion in all categories of assets. Mean total assets fell from 11,073 to 9,643 livres. Nevertheless, there is some indication that an effort was made after the first failure to revive and cleanse the business. Liabilities dropped from 15,258 to 12,839 livres.[18] Net worth remains negative, but it is slightly lower: from minus 4,185 to minus 3,196. On average, there are two more creditors in the second failure than the first: the bakers managed to find new dealers, probably familiar with their new clients' backgrounds, willing to risk advances in grain and flour. There is one case of a triple failure. On the third balance sheet, the baker's liabilities amounted to 52,679, far outstripping his assets. One wonders what this baker could have said to his seventeen creditors to convince them to finance his dreams of recovery.

Case Studies

While there are no "typical" faillis, a few case studies may help to illuminate the experience of failure. Let us look first at bakers who fall into the category of ordinary or common failures according to our parameters. Master Jean-François Pigeot of the rue de la Verrerie in the center of Paris reported assets of 11,452 livres. They consisted of his fonds, evaluated more or less plausibly at three thousand livres; his household furnishings, clothing, and professional equipment estimated at twenty-four hundred livres; flour stock worth five thousand livres; and thirty-four accounts receivable, mostly quite small and presumably representing bread furnished, that amounted to 1,052 livres. Certainly not affluent, Pigeot nevertheless did not appear to be a marginal baker; nor had he dispersed his energies in too many directions. His liabilities, even more than his very conventional assets, were extremely compact. Pigeot was guilty of no ill-conceived speculations or frivolous adventures. All his liabilities, adding up to 10,452 livres, were for the merchandise supplied to him. To maximize credit opportunities and multiply his options, Pigeot dealt with nine flour merchants (three from Versailles, two from Pontoise, two from Melun, one from Etampes, and one from Paris), one flour broker (at the Halles), and two grain merchants (one from the rue de la Mortellerie, the Paris connection, and one from Méry-sur-Seine, deep in the supply zone). By 1770, the year of Pigeot's failure, it is no surprise to find that over 89 percent of his purchases are in flour.

Pigeot's balance sheet sketched the portrait of a businessman without any grave flaws who seemed unable to generate enough business, perhaps because of the hard times of the previous few years. In this light, he had no difficulty in obtaining terms from his creditors and refloating his enterprise. Doubtless no one expected that Pigeot would return to the consular jurisdiction to file again two years later. In a year and a half his assets frittered away to a mere 6,585 livres. He reckoned his fonds at only two thousand livres and his household goods at nine hundred livres, perhaps as a result of a fire to which he vaguely alluded and that required him to "reestablish." Accounts receivable tripled with bread debts owed by fifty-eight clients. Why didn't this hard-pressed baker insist on payment? One suspects a double bind: reluctance to squeeze for fear of alienating the customers. But it is also likely that the acute economic crisis of 1770 (1768–72) crippled many consumers, who postponed reimbursement or repudiated their debts. Pigeot no longer had any stock on hand. He owed 16,140 livres, all save eight hundred livres to suppliers.

While Pigeot had managed to reimburse fully only the two grain merchants and the flour factor (a total of 1,300 livres), he had reduced eight other old debts by an average of 49 percent—a very promising beginning, all things considered. The problem was that to stay in business, the baker was obliged to contract new debts: 11,540 livres' worth, an enormous weight on his precarious financial foundation. Pigeot merits commendation for his enterprise in ferreting out new suppliers, all in flour: three from Pontoise, where his previous difficulties should have been known, the others astutely scattered in many directions (two in Saint-Denis, two in Coulommiers, one in Dammartin, one in Crépy, one in Chartres, and one in Paris—another broker).

Pigeot tried to account for his trouble by citing losses this time. In addition to eight hundred livres for fire damage and twenty-four hundred livres for a new shop, apparently consequent upon the fire (but no dates or details are given), the baker indicated fifteen hundred livres in losses as a result of price fluctuation; five hundred livres in bad debts for bread furnished; eighteen hundred livres in expenses occasioned by his creditors (legal fees? damage and interest payments?); nine hundred livres in illness suffered by himself, his wife, and his five children (the familiar *drame larmoyant* rubric: part ploy and/or hyperbole of dubious relevance, part genuine tale of woe with real commercial implications); and eleven hundred livres for his mastership (hardly a commercial loss but surely a capital drain). It is doubtful that this crestfallen baker obtained a third chance.[19]

Another master baker, Antoine Riot of the rue Mouffetard, pleaded for terms and seems to have received some sort of reprieve, though not the eight-year total moratorium that he had the nerve to request. His position was highly unfavorable, but his earnestness, seconded by his wife's remorse, impressed the couple's creditors. Riot had only 422 livres in household furnishings and apparently no other hard assets (he mentioned neither bakery matériel nor a fonds). To raise cash, he had already pawned a diamond ring for 121 livres. He was owed 5,204 livres, most of it for bread furnished, from a long list of customers including five carpenters, two furniture makers, two shoemakers, two soldiers, two potters, two match makers, a jeweler, an innkeeper, and several nobles. To account for his limited resources, he cited a theft of four thousand livres in silver and other effects committed against him by his domestic while he was in the army six years earlier. He also mentioned that his expenses for the maintenance of his wife, five children, and a journeyman amounted to three thousand livres a year (not including lodging, worth another 636 livres)—a far larger sum than might be expected (cause—or discursive consequence—of his difficulties?).

Riot avowed liabilities of 12,647 livres, more than half that amount for supplies (mostly flour) and the rest for reasons not stated. He owed money to ten flour merchants, two flour brokers, a miller, and four grain dealers. At least five suppliers obtained judgments against Riot at the consular court while Greban, a Mortellerie-Grève dealer, sued him before the Paris municipal court. The baker was also indebted to two domestics for wages and/or loans and to a priest from the faubourg Saint-Antoine.[20]

The leading Paris port grain traders, the Mortellerie clan, complained bitterly that their inability to collect from Paris bakers for merchandise furnished jeopardized their business.[21] The failure of master baker Antoine Tayret lends credence to their contention. He owed five Mortellerie merchants an average of 1,189 livres each and was also indebted to two millers (500 and 400 livres) and a grain dealer from La Ferté-Milon (2,262 livres). His total accounts payable amounted to 11,312 livres. Apparently because his second wife had recently died, and thus their estate was in the process of being inventoried, Tayret listed no hard assets. He enumerated in great detail the sums owed him by one hundred eighty customers, ranging from eight sous to 2,190 livres. Tayret seems to have lent money as well as having extended credit to certain clients.[22]

The faubouriens do not modify the image we have formed of the failli. Gilbert Paris of the rue de Montreuil owed 4,330 livres, all of it for grain or flour, save eighty livres due to his wood merchant. He was in debt to two Paris flour brokers (2,260 livres) and a miller (500 livres). But he also brought grain from laboureurs, perhaps illicitly at their farms (1,390 livres). Without specifying, Paris claimed two thousand livres in matériel. Nineteen individuals owed him a total of 3,093 livres, at least a third of which represented bread sold on credit. Like many other bakers, Paris cited recent illnesses and losses (1,500 livres in three years) as one major source of his trouble.[23]

Antoine Battu of the rue de Reuilly broke down his assets as follows: 800 livres in tools and equipment, 300 livres in merchandise, 260 livres in household furnishings, and 200 livres in clothing and linen. He had lost a horse worth 300 livres, suffered illness during the previous years estimated to cost 450 livres, and absorbed losses of 2,000 livres, apparently over several years, from bad debts and the business failures of others. Battu owed six flour dealers a total of 2,870 livres and he was indebted to a surgeon for 700 livres (which, if due for professional reasons, perhaps justified his losses ascribed to illness).[24]

Another faubourien, named Deshayes, was in worse financial health. Two laboureurs and five grain merchants demanded 3,187 livres for grain, and five millers and one flour merchant posted claims of 1,330 for flour. The 2,471 livres that Deshayes owed to ten fellow bakers may also have represented loans for merchandise or advances in kind. He owed his journeymen two hundred livres for wages and/or loans. Juxtaposed to total accounts payable of 9,602 livres were seventy-one accounts receivable amounting to 3,187 livres, for the most part in small sums. The Deshayes affirmed that they owned no furniture at all. They implored permission to keep their "professional utensils" without which they would be "reduced to beggary."[25]

Faced with 9,240 livres in accounts payable, Deshayes's colleague, Pierre Druet, offered to cede all his tools and his fonds to his creditors, among whom were nine flour merchants scattered from Beaumont to Saint-Denis to Meaux to Etampes. Druet was in far less desperate shape than Deshayes, for in addition to his bakeshop (for which he demanded three thousand livres), he had a considerable amount of furniture.[26]

The forains seemed to project more confidence in themselves than the faubouriens. Louis Vaugeois was sure that he would win favorable terms. After all, he was a hard-working man of impeccable reputation who had produced sixteen children in twenty-five years of marriage (along with eighteen dead horses and some dead cows, he calculated that the family had cost him twenty-two thousand livres in "losses"). He owed 10,521 livres to laboureurs, millers, and bakers, most of which probably concerned merchandise. But his assets, a crazy quilt of small parcels of land, a house, horses, cows, and chickens, household furniture and bakery equipment, and accounts receivable for bread furnished, were greater than his accounts payable, totaling 11,116 livres.[27]

Widow Joly of Charenton had very few assets: 250 livres in matériel, two sacks of flour worth ninety-six livres, and two small carts and a very old horse who could no longer work evaluated at 120 livres. The widow claimed to have suffered 120 livres in losses caused by "the brawl over bread in Paris," perhaps an allusion to the Flour War riots. She cited other losses of 2,750 livres, including the death of five horses in three years (two to glanders and three to colic) and four serious illnesses, two of which "reduced her to agony." The widow faced claims of 9,476 livres, at least half of which were for merchandise.[28]

The dossier of forain Louis Hiest reads like the saga of an ill-starred life. The balance sheet itself was pedestrian: 7,630 livres in assets, mostly paper, and 17,042 livres in liabilities, of which a third to a half seem to represent merchandise. The "losses" rubric provides the vehicle for Hiest's tale. Shortly after establishing himself at La Courtille (it is unclear when), the baker had to tend to business in his wife's home province. His journeyman and his deliverywoman took advantage of his absence to collect four hundred livres from his customers and abscond. As a result of their indifference or incapacity, flour worth 340 livres also spoiled while the boss was away. Sometime later, he lost the use of one hand and had to abandon his business at a loss of six hundred livres, a difficulty compounded by the illness that befell his wife, which cost another twelve hundred livres.

Once he recovered the use of his hand, Hiest negotiated for a lease on a house at Corbeil with the miller and entrepreneur Malisset, celebrated in the famine plot scandals.[29] But the baker imprudently engaged himself in the expectation of obtaining six thousand livres from an inheritance due his wife. His wife's estate trustee died, however, before settling. To cover himself, Hiest had to borrow on a rente and, in addition, pay a *pot de vin* to Malisset for the lease. The house, it turned out, was in poor condition and parts of it began to collapse when he built his oven. Construction and repairs resulted in losses of up to thirty-eight hundred livres.

Hiest suffered either bad judgment or bad luck in his choice of personnel, for once again during an absence from his new bakery for the purpose of picking up the scent of his wife's inheritance, Hiest's workers stole merchandise worth six hundred livres. During yet another absence—what an object lesson for masters who did not like the bakeroom! what ringing confirmation of Parmentier's warnings to those masters! (or what self-serving scapegoating!)—

his journeymen accepted shipment of rotten flour, made five hundred eight-pound loaves out of it, and sold much of this bread to a group of workers who found it unfit to eat. The loss of the bread was less grave than the loss of the outraged customers, which Hiest estimated at twelve hundred livres, a sum that did not of course reflect damage to his reputation.

To serve the Paris market Hiest had to buy a horse, one that had the bad taste to die en route to the capital—but only after the baker had paid a smith to treat the animal (loss of 212 livres). Hiest's new oven broke down and he had to spend 250 livres to build a new one. The baker perceived himself as an alert, opportunistic businessman, so when he noticed that, as a result of the king's presence, prices at Fontainebleau were at twenty-six sous for the same loaf that sold for twenty-two sous at Corbeil, he tried to move into the new market. For reasons that are not clear, he ended up in jail—as a consequence of "calumnies," he insisted. Hard upon this fiasco, which cost him four hundred livres, he lost 820 livres on flour owing to "the bad faith" of a miller who first delayed grinding and then gave him someone else's inferior merchandise. "To crown his bad fortune," as he put it, when he visited Paris to "woo [faire la cour]" his creditors, one of them, unmoved by his charm, had him thrown in jail. Almost predictably, his servants profited from his incarceration by stealing Orléanais red evaluated at two hundred livres.

Nor does this story end with Hiest's first failure, which he turned miraculously into a moratorium enabling him to return to his business. Five years later he failed again, this time more deeply in debt with accounts payable swelling to 32,422 livres, less than a third of which seems to have to do with grain or flour. Actually, Hiest had worked hard to reduce his initial debt and had succeeded in paying off more than two-fifths in five years—years of difficult times in the Paris area. In the interim, however, he had contracted a maze of new debts and, in fact, filed his failure petition from Fort Levêque, the prison he had last visited just before his first failure.[30]

The failures involving very large sums of money are often more difficult to decode than the "ordinary" cases. In some instances, they appear to depend less on baking than on other professional activities, more or less obfuscated. Baker Charles-Joseph Hebert owed 65,332 livres, but it is impossible to pinpoint the sources of his debt. He had considerable resources, including a house with a mill and a laundry worth eighteen thousand livres. He appears to have had some commercial relations in Flanders. Charles Houlet was a superforain who exercised three other professions: wood merchant, coal merchant, and hay-straw and oats merchant for the royal court. He owed 53,597 livres, but only one-tenth of his accounts payable had to do with the bakery. His assets were equally impressive: fifty-one thousand livres, consisting of a house at Saint-Maur and sizable parcels of land.[31]

It is hard to tell why master Pierre Tollard owed over seventy-one thousand livres. Only 23 percent of it appears to be for merchandise (16,525 livres due to twelve flour merchants from Persan, Beaumont, Beauvais, Pontoise, Etampes, Melun, Chartres, Paris, among other places). It is possible that other accounts payable in the names of various merchants or bakers may also have concerned grain or flour. Certainly nothing on the assets side hints at an explanation for such mammoth liabilities. Tollard claimed six thousand livres in household furnishings, clothing, and linen and 14,001 livres in accounts receivable, only a part of which reflected bread sales. A fire destroyed Tollard's bakeroom, allegedly costing him the Homeric sum of ten thousand livres to reconstruct. The guildsman cited another 24,700 livres in losses attributed to commercial reversals, flour spoilage, illness, and "the rigidity" of his creditors.[32]

The Baker's Fall: Other Signs

"Failure," officially registered, vividly marks the route that led many bakers to ruin. There were, however, other signs, sometimes more subtle, that heralded a baker's fall. Indebtedness was of course the chief immediate cause of many of the wounded baker's woes. Instead of filing for failure (for fear, perhaps, of not obtaining palatable terms), or before such recourse, bakers frequently turned first to their family and then to fellow bakers for assistance. Baker Vincent Uttrope tried valiantly to rescue his brother's faltering bakery with a series of loans in cash and merchandise. Guy Leguay, one of four brothers in the bakery, lent his father, a master baker with a troubled business, five hundred livres in the hope of bailing him out. ("As a mark of filial love," he agreed to defer collection until the demise of both his father and his mother).[33]

Given the intense rivalry among bakers, it is surprising to see how often they come to one another's aid outside the framework of kinship. Toussaint Lapareillé, perhaps because his responsibilities as head juré elicited the statesman in him, lent four hundred livres to confrere Pierre Paul Leroux "to employ in his business dealings." The loan was heavily sheltered in mortgage, to be sure. But it is interesting to note that Leroux operated his shop on the very same street as Lapareillé and had had numerous clashes with him over the years. A baker on the rue Saint-Antoine advanced funds for buying flour to a colleague up the street. Master Jacques Fortin lent neighboring master Charles Bassille, whose reputation was rather unsavory, four hundred livres to pay his rent. Master J.-B. Mairet borrowed several thousand livres from three fellow masters. Pierre Deshayes owed money for advances in flour, bread, and cash to at least nine bakers, almost all of whom exercised near him in the faubourg Saint-Antoine.[34]

Without the consoling presence of family or professional friends, a needy baker could find himself obliged to turn to unscrupulous private lenders-for-profit. Desperate for eight hundred livres to keep his business afloat, master baker Antoine de La Vallée fell into the snare of a usurious *prêteur sur gages* named Regnier de Raffetange. While the lender promised to charge merely forty livres for a year's use, he demanded consignment of a large number of valuables, including jewelry, silverware, furnishings, and clothing, worth a total of 2,159 livres. And he took the interest off the top, according the baker 760 livres in cash. Four months later de la Vallée remitted two hundred livres toward reimbursement, and after another five months he was ready to acquit the balance and reclaim his possessions. Raffetange's demand for a further premium of two hundred livres in interest stunned the baker, who had been most pleased with himself for liquidating the loan in nine months and naively expected to receive ten livres in restitution of interest for the early settlement of the loan. After hard bargaining, the lender reduced his demand to a 140-livre supplement but warned the borrower that he was prepared to damage his collateral, sequester it, or deny ever having received it if he was not rapidly satisfied. It required the vigorous intervention of Commissaire Regnaudet to oblige Raffetange to return all de la Vallée's effects in return for the six hundred livres that he really owed him.[35]

By drawing on the experience of numerous bakers in trouble, one can reconstruct a composite itinerary that leads downhill to desolation. For Police Inspector Poussot, one of the first markers was the inability of the baker to obtain further credit at the Halles or the ports. Baker Sulpice Garin virtually announced that he was desperate by pawning many prized possessions. Marie-Madeleine Boudier, the widow of a baker, sadly took account of "the bad state of affairs" of her son's bakery: to insulate her estate, she changed her will, substituting her grandchildren for her son as the proprietary heirs. Hugues La Tour could not pay his rent. He negotiated what he thought was an arrangement with his landlord in order to be able to sustain his business long enough to sell it (with its customers) to a baker seeking establishment. The impatience of his landlord dashed his plans: he found himself locked out of shop and home, denied tools and clothes, and unable to bake bread. The abandonment of a fonds was a wrenching experience, especially if a baker had been rooted in the community. Driven by debts to abdication and to drink, (ex-)master Lefevre could not stop himself from returning to his old shop to reclaim, in a sorrowful stupor, his primal rights.[36]

Expulsion from one's shop usually meant expulsion from one's hearth as well. After dispatching his wife and daughters to their *pays* in Bourgogne, Pierre Duprat wandered the streets of the city in a state of homelessness. Baker Dupont was hardly better off, squatting in a battered old house on the verge of collapse that served as a refuge for "extremely suspicious persons." After having lived in a six-room house for many years, what melancholy master Jean-François Fourcy must have known sharing a single room with his no-nonsense sister, who did not shrink from suing him for payment of his share of the last term of rent.[37]

No longer moored to and supported by his own bakery and shop, the baker often lapsed into dependence upon fellow bakers who enabled him (more rarely her) to sustain a semblance of professional life. Leger Millerat hoped to keep many of his customers at the Palais-Royal market by baking in the cellar of neighbor baker Ragot, who charged him thirty sous an oven. Millerat's modest future was suddenly jeopardized when Ragot threw him out and confiscated his flour, on the grounds that he owed him money. A more amicable relationship enabled Foret to bake twice a week in master Bricard's shop in order to supply his stall at the Maubert market. It is probable that their bond derived from Foret's service in Bricard's shop as a journeyman. Elizabeth Joly, the wife of a master baker whose troubles had led him to abandon Paris provisionally, found herself in an extremely precarious situation. She lacked the wherewithal to keep the business going and so dismissed the journeymen. To keep alive the hope of resuscitation, however, it was imperative that she keep supplying her core of faithful customers with bread every day. She contracted with master Pileur, who lived nearby, to sell under her signature the bread that he would supply her. Her accord with Pileur foundered when the police fined her the hefty sum of fifty livres for selling bread of light weight, a fine that she earnestly and quixotically sought to have Pileur pay.[38]

An alternative survival strategy for some bakers was to "lend" or rent out their names. Psychologically not very satisfactory, for this solution estranged the baker from activity in his familiar professional milieu, this ploy had the merit of earning the baker something resembling a small annuity. Aging and without funds, cut off by his ambitious and vain daughter, who invested completely in her baker husband's rising star, Pierre François had no recourse but covering another with his name and title, a practice he clearly considered inglorious at best. Some master bakers lacked the audacity or the connections to exploit their own marginality. For them, if they wanted to continue to work in the bakery milieu, there was no choice but unqualified *déchéance*. The adult master regressed to symbolic and professional boyhood. One became, following Michel Guillaume, "a master baker without a shop serving the masters in the capacity of garçon" or, as in the case of Jacques Grégoire, "a former master baker and at present garçon baker." Indigence drove master Joseph Mechin to declass himself, at least temporarily, but he expected that his cousin, the master for whom he went to work, would not treat him superciliously. The guild statues consoled the fallen masters with the reminder that they could reestablish themselves as masters at any time (just as "derogated" nobles could "awaken from sleep"), yet the debts that plunged them into journeymanship were generally too staggering to overcome.[39]

The end of the road for some beleaguered bakers led to a total break with the profession. Etienne Thiery found some solace in drink, and in conveying his identity as "manual laborer [manoeuvre]" rather than "former master baker." His "state of poverty" reduced a forain named Dubrueil "no longer to follow the trade of baker but to undertake that of dayworker." While dropping out of the baking ranks entirely was for most the ultimate marker of ruin, for others it was the beginning of some sort of personal and professional renewal. Bideux lucidly decided that baking offered redemption no purchase. He owed almost a thousand livres to one broker and hundreds of livres to other suppliers, and his reputation was severely tarnished. Economically and morally, he found a way out. He went to work for a *tabletier*, who made games and decorative pieces, and offered to pay his creditors six livres a month, a substantial part of his modest wage. Girault had bolder and riskier plans. "No longer exercising his profession of baker," he freelanced as a grain- and flour-buying agent for a number of his former colleagues. Another dropout dabbled in the potentially lucrative business of buying and selling flour sacks. Ex-master Jean Desroches landed on his feet as kitchen aide in the house of the duc de Choiseul, a former minister and still an influential high courtier.[40]

There could be no symbolic exit more pathetic, in the grand baroque as well as the common sense of the term, than a fallen baker's entry into "l'hôpital," at once a mythical and very real place, a motley poorhouse where no one ever wanted to end up living out his final days (the specter continued to haunt workers of all sorts throughout much of the nineteenth century). A grave malady sunk the already foundering enterprise of baker Millerat. He saw no alternative but to demand "a document authorizing him to take refuge in the hospital," in this instance the Bicêtre house that admitted only indigent widowers and younger men who were either incapacitated or unable to work.

Within ten years after his marriage, master Pierre Augustin Gillet lost both his children and his shop. He died less than a year after entering Bicêtre.

Despair led a handful of bakers to a more tragic exit. Forced to abandon his old quarter in the Palais-Royal for business reasons and abandoned by his wife, a master baker cut his throat and bled to death in his cellar. Another baker with grave commercial difficulties attempted vainly to take his life with a dough knife.[41]

Failure reveals precious information about the resources of bakers and how they managed them, and about how state and society dealt with this manifestation of innocent deviancy, as they defined it. Many faillis had enjoyed successful careers with more or less robust asset histories. They remind us how easy it was to hurtle into the abyss. Others lapsed much more gradually into a kind of cumulative commercial incontinence. In the script of the bakers, the consumer public is frequently the chief malefactor (a further guarantee of the baker's innocence!), since the biggest chunk of congealed assets consisted of accounts receivable for bread sold on credit. Had the bakers been better at debt collection—had the public been more forthcoming—they would, so they argued, have spared themselves this trauma. Like the entrepreneurial millers, mealmen, and grain dealers, and certain brokers, a growing number of bakers wondered whether the authorities could not devise safeguards and buffers that would strengthen the always vulnerable credit configuration. The government remained torn between a regulatory propensity to resolve this problem from above, perhaps in emulation of the model that served the butchers, and a physiocratic disposition to do nothing at all.

Among the many reasons bakers evoked to account for their difficulties, very rarely did they point to police pressure through constrictive price controls. There was no particular reason to spare the police authorities in the independent consular jurisdiction where the regnant ethos was overwhelmingly inimical to (over)regulation. Bakers and others spoke candidly on most issues without fear of self-incrimination or official recrimination. There is no doubt that price ceilings, more or less rigorously imposed in crisis periods, circumscribed the adaptive latitude of the bakers in dealing with the upward surge of grain and flour prices, and with the more diffuse and invidious disorganizing impact of the crisis on market behavior, relations with suppliers, availability and cost of credit, quality of merchandise, accessibility of transport, and so on. Yet the Paris police remained extremely lucid, despite the febrile climate of stress, in their appreciation of realities. As we shall see shortly, they never lost sight of the peril of submerging the marginal bakers and handicapping the more rugged ones. They had a clear sense of the limits of authoritarian pricing in the capital, and of the ultimately symbiotic relation between consumer welfare and baker well-being. In Paris, price policy penalized the bakers but rarely maimed them. Other factors had more to do with the damage they sustained during these crises. Moreover, it is well to keep in mind both the short-term subsidies and other sorts of relief that the authorities made available and the profuse opportunities for compensation that the police accorded the bakers in the long phases of regulatory relaxation that characteristically followed dearths.[42]

In any event, baker failures cannot be explained by a more or less mechanical schema linked to crises. Frequently, bakers who failed to weather the storm had started out with serious or even fatal flaws. (The "inventory" figures suggest that bakers who were reasonably good managers—a majority of operators—got through the difficult times without accumulating heavy debts.) Subsistence disorders affected commerce and shaped individual comportment in highly variable ways. Some bakers experienced paralysis, and paid the price for fearful inaction. Others got in trouble through speculative excesses, seeking ways to profit from the disarray. Some got caught in a credit crunch, while others escaped the squeeze thanks precisely to the confused conditions marked, inter alia, by the influx of myriad new faces into the provisioning trade and the availability of supplies subsidized by the government. Nor was failure tantamount to dropping out: for numerous bakers, it was a strategy of recovery that was *easier* to pursue in the wake of crisis than during other periods because of the sympathy and credibility that the vivid memory of the crisis evinced. One must be cautious in deciding exactly what failure measures and denotes. It is not a simple narrative of the cumulative unraveling of the bakery trade.

Still, the heightened incidence of failure beginning in the early seventies is striking. Bakers turned to failure in part because they were better informed than earlier in the century about how to navigate the shoals of hard times and at the same time perhaps because a substantial number of them were less able businessmen than their predecessors (certainly the guild contended that after the reforms of 1776 favoring establishment, it was far easier for less competent players to enter the game). They failed in larger numbers because together they faced certain common problems in keeping afloat, problems doubtless associated with the rising costs of virtually all goods and services, the greater magnitude and fragmentation of accounts receivable for bread, more exigent creditors, and more aggressive competition from fellow bakers.

Bakers, like other businesspeople, had an ambivalent attitude toward failure. On the one hand, it afforded them protection and some breathing room. They could recharge their batteries by winning time and perhaps certain concessions on amounts due (interest if not principal). On the other hand, failure was embarrassing, for some even mortifying. It seemed hard to separate from failure in the vernacular sense of the word: inadequacy if not incompetence. To be sure, accidents of various genres and bad luck engendered losses leading to faillites. Still, official failure marked the baker psychologically and often damaged his social capital as well as his business.

Despite the wounds, bakers could recover from failure, and even regain their reputation. Other bakers fell into ruin without bothering to go through formal failure. For them, honor no longer mattered as a business asset. They merely sought to hang on.

Chapter 15

Reputation

Like many businessmen, bakers considered their reputation to be their most cherished capital. Some of them defined it in terms that conveyed both the moral and the material stake. The baker's reputation assured his "credit": the esteem in which he was held and the confidence he inspired in his product and his manner of dealing on the one hand, and the financial indulgence upon which he could count in conducting his business on the other. Certain bakers viewed sound reputation as a synonym of honor; as master baker Louis Tairet put it, "one has nothing more prized than one's honor." Another baker felt that he had lost his reputation as a result of the cascade of "the most dishonoring horrors" spread by his enemies. Few bakers, masters or journeymen, would have contested writer-reformer Philipon de la Madeleine's assertion that it was honor that transformed the worker from a "mercenary" into a "citizen."

Honor connoted an acknowledgment on the part of others that one's behavior conformed to the norms—norms of conduct in the entangled domains of public and private life, in particular probity and decency. To preserve their honor, bakers rarely fought duels but frequently went to court (or to the commissaire, who was the local and immediate judge); had wayward offspring incarcerated, just like their social superiors; practiced strategies of community implantation and public relations. They were highly susceptible to insults, rumors, or scandals that raised questions about their good name or their "good faith." When a baker (and/or spouse) irretrievably lost his (her, their) local reputation, he could not hope to survive professionally. Such was the fate of baker André Devaux and his wife, "forced to leave the faubourg Saint-Marcel for roguish behavior."[1]

Negative Representations

Forging and sustaining a good reputation was no easy task for the baker, for he began his professional life with a generically negative image throughout France (and other countries). The *Guide pratique de la meunerie* doubtless exaggerated when it evoked in the mid-nineteenth century the existence of a permanent war between innocent consumers and grasping bakers utterly indifferent to the public interest. Yet the burlesque song entitled "Le Rabais du pain," which reappeared with updated verses from the sixteenth through the eighteenth centuries, depicted bakers in harshly negative terms, as social criminals:

> qui, survendans trop leur Pain,
> Font mourir le Monde de faim
> [who, selling their bread so high,
> caused people of hunger to die]

The astonishing thing, according to this stereotype, was that the bakers believed they could act with impunity, disdainfully disregarding moral and positive law, and short-term police instructions. A mid-seventeenth-century version of the "Rabais" ditty imagines a trial whose verdict instructs bakers to "stop selling their bread at such a high price" and orders the banishment of six of the richest and most arrogant bakers (from Mont Parnasse) to serve as an example and warning to the others.[2]

The baker was reproached with the same failings of character and comportment throughout the Old Regime. He was quintessentially "greedy." He was prepared for any odious maneuver in order to gratify his lust for "exorbitant profits." A "usurer" according to a mid-sixteenth-century rendering of the "Rabais" and a "fiscal exactor" in the words of an eighteenth-century litany, the baker welcomed calamity and grew "angry with the advent of abundance."[3] An early-eighteenth-century reprise of the "Rabais" portrayed bakers as "truly inhuman" and "wicked," representations echoed in other texts that rehearsed examples of cruelty. Neither the authorities nor the bookseller Hardy, who reported the story, appeared surprised that a bakeress from Place Maubert pressed charges against a desperately hungry man who stole a loaf into which he dug his teeth even before he left the shop. His wife had just hung herself in despair and his two children were dying of hunger. The commissaire decried the bakeress's "paucity of humanity."[4]

The other salient trait of the baker, implicit in numerous denunciations, was his pervasive dishonesty. He was inveterately "corrupt," he cheated on small and great things (from ounces of bread to muids of grain), he could not be trusted on any account. In the eyes of many consumers, wealth was the irrefragable external sign of baker dishonesty. Most were disinclined to question the logic of Parisian rioters in the Flour War, who justified their pillage of certain shops by pointing to the alleged opulence of the baker (the widow Houdouart on the rue Mouffetard, for instance).[5] Nor was this image of the deceitful baker merely the hyperbolic view of a terror-stricken and naive public. A large part of the police shared the poor opinion in which bakers were held. These authorities censured in particular the "unscrupulousness" of the bakers, their unconscionable disdain for the public interest, and their stealthy efforts to "impose their absolute will."[6]

The mitron de Vaugirard, who was intimately familiar with the rampant dishonesty on the Parisian baking scene, agreed with the dark ambient diagnosis. "If Messrs. the rich Bakers of Paris want to cover themselves in glory and in particular to reap a great deal of money, I will reveal to them the most certain and rapid path," he advised, in implausibly mechanistic but reassuringly moralistic terms: *"Let them make good bread at low prices"* and then "wealth and honor will besiege their door" and "the Public will gradually become accustomed to praising them, to blessing them." Besides the bakers themselves, who spoke most eloquently through the voice of the chevalier Rutledge in 1789 ("we are the nourishers of the city"), Parmentier was the only prominent commentator who spoke out vehemently in their defense. Whereas Rutledge defended the bakers as unfortunate scapegoats in the specific pre-Revolutionary context caught between the Scylla of a company of official monopolists and the Charybdis of "the blind and famished multitude," Parmentier noted that they were in a structural sense doomed to suffer public obloquy—"injustice and ingratitude"—every time grain or flour was in short supply. "Let us not pay any more attention to these popular remarks, which have no plausibility," urged the scientist. Bakers deserved "consideration" precisely because they were courageous, willing to take risks in order to feed the people, a thankless and exhausting task conducted "in silence and in darkness, trapped in the stifling heat of the bakeroom, enveloped in smoke and flour dust, at a time of night when all of Nature is at rest."[7]

Insults and Innuendo

Most Paris bakers worried less about intermittent dearth-generated spleen and diffuse mistrust than about managing day to day in their neighborhood, residential and/or commercial. They tried to cultivate a positive local reputation that could protect them against the stereotypes that haunted them collectively. They toiled hard to prevent their names from being tainted on the local "wire" that carried news and rumor. They tried to imitate Jean-Baptiste Fontaine, "reputed to be an honest man, attached to his business, calm, peaceable," rather than Thomain, "known to be a violent and evil man." Bakers were hypersensitive to any public remarks that questioned their integrity (creditworthiness, reliability, honesty), accusations that could compromise their relations with their suppliers and/or their customers. Given the frequency of violent verbal confrontations, the generally accepted way of purging anger and suspicion, many bakers (among other socioeconomic actors in the community) had to engage in ongoing damage control. They could not afford to allow insinuations, imprecations, or aspersions to congeal into opinions that prudential citizens might feel obliged to take seriously.[8]

The "supply side" insult terrified the baker even as it enraged him. Jean Galland, a forain who had begun his career without any fortune as a struggling journeyman and had managed to build a going enterprise, found his future jeopardized "by talk as calumnious as it was damaging." The word spread that he was on the verge of bankruptcy, a rumor magnified by his sudden departure, a visit home to Burgundy to bury his father that was perceived as culpable flight. "Upon his return, Galland was struck by the impact of this maliciously propagated libel. . . . all the merchants with whom he had previously dealt henceforth refused to deliver any merchandise that he did not pay for in cash up front." Master baker Pierre Renault claimed he was the object of a similar "evil calumny," spread "throughout the market and the ports." The agent was a grain merchant who misconstrued his temporary absence as irrefragable proof of "terribly fraudulent bankruptcy," sounded the alarm, and invited all of the baker's habitual provisioners to cut him off. Faubourien Jean-Baptiste Guillaume feared a similar disaster as a result of flour merchant Meunier's public denunciation of his repayment delinquencies ("he owed money to all the merchants of the port. . . . he had no credit whatsoever"). Master Jacques Bebitte worried that he would not have enough flour to bake after a broker denounced him throughout the Halles as a "bankrupt, a thief, a damned rascal." All these bakers filed complaints with the police, demanding "redress" to clear their names, and sought on their own to convince the key opinion makers in the neighborhood of their innocence.[9]

Even more perilous to the bakers were aspersions that alienated bread-buying customers. Few accusations could have hurt master Adrien Loisel more deeply than the charge by a rancorous neighbor, vexed over a personal matter, that his bread was "filthy," that he put "garbage" into it. After she had been convicted of selling at short weight, it was hard for widow mistress Guillaume Lapareillé to escape denunciations of "thief" and "cheater," which resulted in an erosion in her clientele. More frustrating was the problem of master Charles Modinet, assailed "openly . . . in the street" by a vindictive neighbor and former friend as "a thief and rogue." Modinet had done nothing to detract from his professional standing, yet the affronts hurt his business, and his suit for reparation did not likely undo the damage. His honor blemished in the same way by a neighborhood wine merchant, master Pierre Leguay turned to the commissaire to help him stem a hemorrhaging of customers. A fellow master faced a situation familiar to everyone in the profession, the vague imputation from the particular to the general that put all bakers in the same corrupt sack. "It's this sonovabitch of a damn beggar of a baker," ranted a woman neighbor; "he's a thief, they're all thieves who work together." Questions regarding his wife's moral character without any direct link to the bakery nevertheless caused problems for baker Jacques Duhamel. If his wife was a "whore" and a "bitch," as several people in the quarter contended, then others apparently would not do business with the husband, who was deemed responsible for good order in his bedroom as well as his bakeroom.[10]

Masters Louis Billotte and Jacques Cousin fell victim to campaigns of character assassination launched by resentful ex-tenants who had been expelled from houses owned by the bakers. A widowed baker sought the eviction of a noble who ran a brothel in part of a house she owned because she feared the news of this immoral establishment would reach the public, threatening her honor and thus her business. Appearances mattered inordinately in the community. Master Charles Menuel had a wrathful argument with his own syndics, who attempted to extort money from him to register a fallen master working in his shop as a journeyman, an affair that had nothing to do with the baker's baking reputation. But the guild chiefs called the guard and had Menuel arrested and ritually humiliated, "so that he was compelled to cross the whole city to the border of the faubourg Saint-Antoine on foot like a criminal, while the said syndics and their deputies rode in a carriage." Menuel riposted by demanding redress from his guild because "he suffered the most searing affront for a man of honor without any remote justification. This affront was publicly administered and thus threatened to cause him the greatest damage—to his reputation, his status, and his credit."[11]

Certain bakers quite consciously sought to adorn their reputations while others seemed to have a reputation death wish. Those in the first category followed the itinerary adumbrated by the mitron de Vaugirard, who imagined well-behaved bakers attaining the social consecration of local and even supralocal "dignities," posts of honor and sometimes responsibility in the parish, in the commercial courts, and in municipal government. Faubourien Gabriel Mouchy was a warden in the Church of Saint-Etienne and Saint-Paul. Master

Gilles Pasquier served as marguillier and financial officer of the charitable works of the parish of Madeleine-de-la-Ville-Levesque and as director of the bureau des pauvres. Jean Guenon, a master, and Claude Pampelune of the faubourg Saint-Antoine each acquired a special bench in his church. Along with an elite of notaries and surgeons, baker Tribouillet was elected by a community "assembly of bourgeois" to be in charge of the illumination of the public lanterns. Many bakers took the idea of being "bourgeois" very seriously. For master Pierre Mandon, it meant a double contract: the bourgeois had to follow a code of honorable behavior, and others had to reciprocate by treating him respectfully ("one did not treat a bourgeois in this fashion," the baker complained to the police in reference to rough treatment he received).[12]

On the other side of the spectrum were the bakers whose apparent or alleged debauchery cast them into a netherworld of disrepute. In most cases the disgrace seemed to be the fruit of gradually accumulated sins that hardened into a crippling stigma. Numerous bakers were known in their quarters as inveterate drunkards. They caused trouble, attracted trouble, and were not worthy of trust. (They were carefully distinguished from the mass of bakers who regularly drank in the taverns while relaxing or doing business and who occasionally drank too much). A notch worse were those who gambled (cards, "Siam," "gallet," billiards, etc.) in their cups, for they frittered away their possessions (and those of their wives and children) as well as their time and health. Still more dangerous, to others and to himself, was the chronically drunk baker who became violent (forty-year-old baker Henriette was notorious for beating up tavern keepers who either refused to serve him more drink or demanded to see his money before they did so).[13]

In some instances, the baker fell brutally, practically without warning. In these situations, the stain was frequently sexual. Master Pierre Courreur, whose name may a priori have encouraged suspicion, insisted that he was a serious businessman who had always kept his reputation in mind, that "he had always conducted himself in such a fashion as to avoid any suspicion of loose behavior." One day he found himself accused of passing "the venereal vice" to the ten-year-old daughter of a nearby clothes dealer. In vain he swore that "he had never touched this child and had no disease like the one he was accused of transmitting." Despite his perfervid denials, "this unfounded chatter spread throughout the neighborhood, . . . [proving] more than sufficient to blacken his reputation. . . . he is a ruined man." Baker Jacques Desmeurs encountered similar difficulties when his own father-in-law launched the rumor that he had given his wife a venereal disease in return for her dowry.[14]

Quarrels among Bakers

Probably no one did more to discredit bakers than other bakers, who often had a vested interest in gaining an advantage in what they believed was a zero-sum game. Interbaker relations oscillated between poles of solidarity and rivalry. Precisely because they had a great deal in common, they were apt both to succor and to scourge one another. The need to defend shared interests (not all of their interests, to be sure, were shared) vis-à-vis outsiders (suppliers, other creditors, the public, the police) powerfully drew the bakers together. Similar personal and professional trajectories and lifestyles, articulated around the same rhythms of work, leisure, and responsibility, forged multiple links among them. At the same time these very bonds and intimacies generated tensions, jealousies, and misunderstandings. Professional and community codes of behavior, elaborated informally to complement the narrower public and corporate strictures that regulated daily life, did not always succeed in forestalling confrontations. By and large, bakers quarreled over the same types of issues that agitated other socioprofessional groups: competition for moral sway and reputation, for clients and markets, for labor, for raw materials, for provisioning priority, for space, for control of family resources; sexual rivalries; contention over prestatory obligations and other debts, and over matters of honesty and honor; disputes over the interpretation and application of the unwritten rules/norms of conduct.

A little more than twenty years ago, on the eve of the postmodern epoch, a fifty-year-old baker in a village close to the border between the Creuse and the Indre hovered on the brink of professional catastrophe. Inexplicably, his dough would not rise. Every other batch of bread came out badly, and more and more of his clients gravitated to the village's other bakery, run by an old woman. After vainly changing flour suppliers, introducing new leavens, and having his water analyzed, his jealousy of his rival led him to the conclusion that she had cast a spell on him in order to capture his customers and ruin his business. Counseled by some local *leveurs de sorts* (not, alas, *leveurs de pâtes*), he decided to respond in kind. He knew his suspicions had been well founded when the bakeress brusquely fell ill and had to leave her shop almost immediately after he sacrificed a black rooster according to a ritual that aimed the fowl's diabolical power against her.[15]

Sorcery *appears* to have had little purchase among the bakers of Enlightenment Paris, but "professional jealousy," characteristically turning on client fidelity, was probably the paramount cause of quarrels. Customers were considered to be a sort of property. In the division of the estate of a master baker, a brother and three sisters decided to reward their brother, the fifth heir, for having cared for their parents for many years by passing on to him "all the customers attached to the shop" of the deceased father. The most significant component in the sale of a *fonds de boutique* was the clientele. Thus the sense of outrage of master Louis Tairet, whose commerce was "entirely ruined" when the baker who sold him his business reestablished himself in a new shop near enough to the old quarter to draw his ex-customers back to the fold. Continuity in the service of the customers so as to rivet their fidelity served as the grand imperative (and the treasure trove of alibis) for the behavior of bakers. When a certain widowed mistress baker died, both her daughters, also widowed mistresses, rushed to sustain her trade, "for fear of losing the customers." Severely injured by a Gonessien, a faubourg baker worried not about his wounds, but about the grave risk to his business of "an interruption of service." Insisting that his purchases were legitimately made, master baker Martin Conty asked the lieutenant of police to override a seizure of grain conducted by the corps of officer-porters. His main argument turned on the pressing need to satisfy potentially volatile customers. In a similar vein, master Blondelle scoffed at the demand of his landlord to halt baking for a fire inspection: he had to "bake for his [faithful and expectant] customers."[16]

Bakers claimed that they acquired customers organically, over the course of time, and that customers, like other trothed actors, naturally tended to remain faithful. (Bravo for the lawyer who continued to use a faubourien bakeshop even after he relocated to the Cité and for the royal accountant who had his servant travel all the way to La Courtille to buy bread in the shop of the market baker from whom he purchased regularly in the nearby Marais market on Wednesdays and Saturdays.) Customers honored a commitment of reciprocity that guaranteed them credit facilities as well as good quality and honest measure, and moral support in the community. Most bakers made a conscientious effort to uphold their side of the bargain. Who could question the sincerity of master Charen when he beat his journeyman and his apprentice for producing a bread with "too much color" that alarmed numerous consumers? Bakers had no interest in assuming personal responsibility for poorly effected work: another master hastened to spread the word that it was the lack of adroitness of his journeymen in kneading and fashioning loaves that "drove his customers away from the shop."[17]

Occasionally a baker abjured an old relationship that had become too cumbersome, freeing the customer to seek new ties, or rejected an embryonic connection that seemed to reveal precocious fault lines. Problems of payment characteristically triggered such ruptures. Bakeress Barois preferred to believe her daughter in a dispute with the widow of a bourgeois de Paris over the amount of money the latter had remitted to the former. Overnight the client became "a thief," "a bitch." Umbrage sometimes overcame bakers when customers tried to return breads purchased in excess or when they squabbled over which bread on the shelf they wanted to take home. (Was it a banal and regrettable baker's tort when persnickety Marie-Madeleine Toutat rejected a hard, presumably stale mollet loaf that a bakeress handed to her and then fell through the trapdoor to the cellar while reaching for a more alluring bread on a shelf behind the counter—or was it poetic justice?)[18]

New customers habitually arrived from the outside as a result of word-of-mouth recommendations when they settled in the quarter. In any event, in principle one did not solicit them: such coarse courtship was unworthy of artisans (and not just incorporated ones) and could only issue in costly and permanent disorder. Nor did one suffer aspersions on one's honor, as a result of open verbal assault, insidious whispering campaigns of calumny, judicial actions, or other symbolic attacks, which could undermine confidence and (seemingly) justify customer abandonment.

Neighbors

Clashes often erupted between bakers residing on the same street. His coat and wig besmirched and his left eye painfully bruised by mud balls hurled at him "publicly" in his shop on the grande rue du faubourg Montmartre by André Legrand, a fellow baker from across the street, who punctuated his assault with a litany of insults (calling his colleague a "rogue" and a "screwup" "who had received charity from the parish," a sign of his parasitical fecklessness), master baker Antoine Soret filed a complaint to save "his honor and his reputation." Legrand had been harassing him for some time for having attempted to win over one of his best customers by offering "the gift of a sweetcake." Mud served, too, as the vector of the stigma conveyed by the Thuillier baker family, represented by two nieces and a nephew, who tossed handfuls of the slimy substance at neighbor Guillaume Lapareillé and his wife in their shop. Later they roughed up the wife and publicly questioned the quality of the Lapareillé loaves. Faubourien Lamare could not abide the perfidy/defection of a customer named Régis and the transgression/usurpation of Betemont, a colleague located on the other side of the often turbulent grande rue du faubourg Saint-Antoine. "Despite the deal that she struck with him to pay biweekly installments of thirty sous" on a bread debt of twenty-two livres, Régis paid nothing. Moreover, she began to buy from Lamare's rival Betemont. One day Lamare's wife confiscated two loaves that Régis's son purchased in Betemont's shop. When the latter's wife visited him to protest, Lamare "flew into a rage" and beat her unconscious.[19]

A three-year vendetta between two bakers' wives further marred the slippery tranquility of the grande rue du faubourg Saint-Antoine. Most recently Madame Royer expressed her "hatred" for Madame Le Prin and her jealousy for her commerce by striking her in the breast and telling people up and down the street that they should not do business with a woman who was "a knave, a thief, a robber of flour." Apparently because his rival Nicolas Pelletier had once worked for him, baker François Maillot could not forgive him for setting up on the same street (since it was the rue de Reuilly in the faubourg Saint-Antoine, the corporate stricture against a former journeyman establishing himself in the proximity of an ex-master did not apply). Maillot's frustration erupted brutally when he seized Pelletier "by his noble parts, which would have cost him his very life if [bystanders] had not rushed immediately to his assistance." A young baker from the faubourg Saint-Antoine delivered bread to city customers in a building occupied by a master baker. "For reasons of commercial jealousy," the latter's wife screamed at the interloper to move his cart from in front of their house. Upon his testy refusal, her husband surged out and attacked the faubourien from behind, assaulting him so badly that he died.[20]

The rue Mouffetard was a battleground between the two big baker clans, the Petits and the Lecoqs, whose members struck and cursed one another day in and day out. At the grande pinte de Bercy in the faubourg Sainte-Antoine the continued campaign of baker Coestien against his neighbor baker Gueret had begun to bear fruit in the community: certain colleagues and clients began to echo the rumor that he was in fact the "crook" denounced by his rival. Master Pierre-Nicolas Boutillier complained bitterly that for the past three months baker Caramel "maliciously makes the rounds of the neighborhood almost daily, visiting people he knows in order to destroy the reputation of the plaintiff and his wife by spreading the most damaging, libelous, and atrocious insults and then appearing several times a day in front of their shop to repeat these horrors openly and publicly in the street." Caramel's repertoire of outrages included the accusations that Boutillier had killed a fellow baker and purchased his freedom from the hangman and that he was an incorrigible thief.[21]

Beyond the shop and the street, a favorite forum for discrediting one's competitors was the local tavern, the locus of a great deal of extrajudicial litigation/wrangling and community decision making and the crucible of what one baker called "the spirit of the public." In a tavern situated near both their shops, before scores of people, time and again, master Louis Mandon of the rue de Loursine denounced baker Brichard, who resided on the same street as he, as "a brigand" (and the husband of a "fucking bitch"). For the forains, the bread markets were a natural site of contention. A Gonessien named Leray openly described a rival as a thief, causing a tumult around his stall at the Halles. At the same market, one bakeress characterized another as "a bitch" and "a whore" who "had been to Bicêtre to sweat out the pox."[22]

From 1759 at least through 1765, Jean Parain, a baker at Le Roule, harassed and mistreated André Devaux, a baker from the same place. Nor did repeated court orders to make restitution moderate the ardor of his charge. In the bread markets, at the grain market of the Halles, in the streets, wherever he accosted Devaux, Parain inveighed against him as "the biggest blackguard and knave that I have ever known," as a "bankrupter" and "rogue." Devaux feared that his suppliers might refuse him credit if Parain continued the refrain about his allergy to paying debts, and that his clients might stray to other stalls whose operators inspired more confidence. It is likely that the feud resulted from the fact that Devaux had taken over a shop at Le Roulle that had failed in the less steady hands of Parain. Like most people, Parain knew how to dish it out better than he could take it. He himself implored the police to save his reputation when widow Vassou, whose stall was quite close to his at the Halles, launched a campaign of discredit against him. Among other things, she announced that "he had taken the wife he currently has on a trial basis, that he had poisoned his first wife"; and she urged him, as he had Devaux, "to go pay his debts." The latter exhortation framed a paradox of neighborhood life: no one liked to pay off debts, certainly not in full (a redistributive fantasy lurked in this reckoning) and never prematurely (that is, too quickly), yet it was generally agreed that paying debts was a necessary act of honor and community.[23]

Himself the object of frequent complaints, master Julien Charles Blanvilain denounced repeated assaults made by the wife of a confrere who lived next door—the latest involving the usual public insults accompanied by a shower of rocks—which he attributed to "trade jealousy" in light of their extreme vicinity. Master Languelet tormented and threatened his colleague Riom by publicly upholding that the latter did not make nearly as good a bread as he, and pressing him to accept a wager to test this hypothesis. Emboldened by drink, Languelet escalated the attack by visiting Riom's shop and defaming his wife, for it seemed to follow that a baker who baked a sullied bread would marry a sullied woman ("a bitch, a whore").[24]

Sexual Aspersion

Because sexual dishonor was universally regarded as so grave, it served widely as metaphor and marker of disarray, dishonesty, and disrepute. Though aimed in the first instance at women, given their vulnerability (socially aggravated but biologically and psychologically grounded), sexual insult/innuendo struck at the men who were supposed to be their custodians, guarantors, and sole beneficiaries. Master Denis Leroy worried that his position in the community would become untenable when he repeatedly found horns placed above or painted on his shop door, notoriously the work of two other master families, bitter rivals of his wife's father, another master. Wherever they encountered Madame Leroy—even on the way out of church on Sunday—they shouted that she was "a bitch and a whore" who had slept with several of their journeymen.

Widows were attractive targets for aspersion, despite the absence of a male figure against whom the dissolute woman's evil would redound. Widow baker Marie-Jeanne Duvivier of the rue de Reuilly felt desperately undermined by the onslaught of baker Jacques Betemont of the same street, who seemed determined to shut her down. He insulted her wherever he encountered her. His standard affront had a double leitmotif: Duvivier was a "dirty bitch" and a "rotten fucking ass" who had no sexual honor; ergo, she was a "devourer of all possessions," incapable of managing a business honestly and successfully. More rarely the male himself was the agent of the sexual and moral trespass, the active transgressor of the taboo. Blanvilain felt that he could overcome his drunkard's reputation but that the rumor that "I screw my niece" would do his business far more damage.[25]

Disputes over Supplies

While the bulk of baker quarrels issued from competition for customers, other disputes turned on the allocation of supplies. Bakers fought quite often at the Halles and Grève markets over priority for purchasing and quotas on the amount that anyone could claim in a single day. Market officials called measurers spent much time refereeing these encounters and assuring all bakers an opportunity to acquire some merchandise. Bakers Graizel (one of two stormy brothers) and Fraineret frequently tried to impede merchants from selling to their rivals and/or enemies. Master Antoine Cerf struck a confrere in the face with a glass, endangering his eye, apparently over a question of access to flour, in terribly short supply at the end of 1740. Fifteen years earlier, during another acute dearth, a baker named Boucault had gone to the police to denounce the hoarding and speculation of two faubouriens, Haire and Bourgeois.[26]

Bakeress Marie-Angélique Chopineau complained that two other bakeresses tried to sap her capacity to buy on credit from the brokers and dealers at the Halles by spreading rumors that the sheets on which she slept were not her own and that she lacked the resources to honor her obligations. The wife of master Antoine Coquenpot had a similar experience. The spouse of a faubourien named Prix Paris deployed the classic tactic of linking sexual and commercial dissipation to defile her (since she was a "woman of immoral conduct" who offered her favors to the soldiers of the guard on the steps of the Jesuit church, it followed that she was "a devourer of all property"). Thanks to this campaign of rhetorical subversion, when Coquenpot bought some flour from mealman Nicolas Barbet, he refused to accord her credit. Money and space were two other permanent sources of discord. Bakers fought over the delimitation of market stalls, the use of shared cellars or granaries, and the repayment of loans made in cash or kind (grain, flour, or even fully baked bread).[27]

Tragedy

Some of the most acrimonious confrontations arose in the wake of tragedy. Matrimonial bonds frequently did not survive the eruption of tensions between the two families making claims on an estate. Speaking in the name of the children of his deceased daughter, the wife of baker Pierre Delamarre, master baker Jacques Lapareillé accused his (ex-)son-in-law of refusing to have an inventory made of his late wife's estate and of "dissipating" merchandise and other effects. Master Jean Cousin physically assaulted his son-in-law, master François Pillé, trustee of the estate of Cousin's deceased daughter, whom he denounced for dilapidating the legacy that he was supposed to preserve for the children.

The second wife of deceased master Nicolas Coutteux rejected the persistent demands of her stepson, master Jacques Coutteux, to have an inventory conducted. Vincent Constant, a master baker, contested the honesty of the widow of his brother, another master, some of whose property had mysteriously disappeared. It was relatively common practice for a surviving member of the broken couple to seek to shelter a portion of the estate from registration and division. Many surviving spouses must have followed the example of master baker Claude Thevenet's widow, who secreted two trunks with a neighbor just before the arrival of the commissaire, summoned to place seals on all the property associated with the deceased. The wife of master Claude Charretier instigated what could be called a presuccession quarrel. Implicitly, she taxed her mother, the widow of a master baker, with not dying, or at least retiring, promptly enough. Both daughter and son-in-law reminded the mother repeatedly that she was "already quite aged," "an old, rascally wench," a useless encumbrance. Another squabble involved a question of moral succession. Still grieving the death of her sister, the wife of a forain baker named Jean Parain, who apparently had not treated his spouse well, Agnès Genard, herself the wife of a master baker, publicly assailed Parain for "having killed his first wife, who was the sister of the said Genard woman," in order to marry Marie-Jeanne Barthoult, "a whore of a bitch" who "gave birth seven months after her marriage."[28]

Honor partook fully of two species: the moral and the material. It denoted traits that spoke to the spirit, and it referred to qualities that were readily marketable. Honor served at once as a repository of local capital and as a treasury of alibis. In defense of their honor, bakers would go to considerable extremes, transformed abruptly from night workers into community knights jousting with sundry infidels. Conversely, to assault another's honor in the service of their interests, they would go to commensurate extremes. So the everyday world of many bakers was a field of battle, more often with words than with blows, though it was hard to prevent the violent discourse from triggering bodily riposte (the discourse often pertaining to bodily function, the distance was not great).

Bakers fought over very concrete matters concerning immediate commercial advantage and right, and over more remote questions dealing with past events, future affronts, unsettled scores, and vague notions of justice. These encounters had important symbolic and practical stakes. Paradoxically, even as they fomented disorder, not infrequently they policed and purged the community, and recalibrated its norms and expectations. They cleared the air even as they polluted it (in Mary Douglas's sense). They tested neighborhood bonds. They invited the consumer public to decide whether the classical negative representations of baker character were well founded or not.

III

Police of

Bread and

Bakers

Chapter 16

Primer to Policing: Figuring Supply and Consumption

The uncontested premise of public policy was that nothing was more important than "the nourishing of the people."[1] The authorities construed the enormous task of provisioning Paris as a quintessentially political as well as a complex technical matter: the stability of the government and the well-being of the population were equally at stake. Inter alia, what distinguished eighteenth-century authorities from their predecessors was the growing conviction that a mastery of the technical issues could dramatically enhance their chances of mastering the political ones.

The technical perspective, which implied an intimate familiarity with the structures of the supply and demand systems, flattered their deepening sense of professionalism and resonated smartly with the Enlightenment vision of Perfected Public Administration. *Savoir* is *Pouvoir*, the exponents of the new(to-nian) science of government tirelessly reaffirmed. Officials at various levels believed that they had to constitute their science on the basis of an infinite number of observations, if possible corroborated and controlled. Inherited wisdom had proclaimed that nothing ever changes under the sun. While administrators did not wholly abjure the old frameworks, some of which continued (more or less unfortunately) to furnish the prisms through which the data was viewed, many came to believe that the decision-making process was ever shifting rather than fixed, problematic rather than clear-cut, intensely concrete and specific rather than abstract and general. Knowledge was the critical variable, social reality (like happiness, a relatively new idea in the European experience) the sole legitimate referent. So controllers general counted hearths and heads, registered transactions, measured distances, surveyed the production and flow of goods, and sought ways to universalize and standardize the search for information. Pondering the data (and the missing data), they drew certain lessons and ventured certain hypotheses and predictions. In their rudimentary fashion, they built models to govern their action.

In the provisioning domain, to conduct coherently "the police" of bakers and bread, the authorities had to know as much as possible about the total needs of the city in good times and bad and about the various sources and avenues of supply, and their dependability. The information they required was extremely difficult to collect, given the relatively primitive means of monitoring, unmasking, and verifying that they possessed. It was no less knotty to interpret, for it often came in forms that themselves demanded decoding, conversion, and/or disaggregation, operations that often proved to be equivocal and erratic. The conscientious administrator did not have the luxury of confronting clear and distinct evidence. Even if he managed to learn, to his satisfaction, more or less how many setiers of wheat the city could dispose of on a given day (or week or month), he could never be sure how much was in a (given) setier, how much (and what types) of flour the setier would (or could) produce, how much bread (of what sorts?) would result from it (and how many ovens—a measure of both time and space—and thus how many bakers would be required to produce it), what price would fairly cover commercial and manufacturing costs plus a reasonable profit and still be consonant with the public capacity to spend, and how much bread (of what sorts) the various categories of consumers (class, gender, age, state of health, state of mind) demanded. Provisioning was a defiantly intractable puzzle that challenged and frustrated the Enlightenment police, who navigated between excesses and deficits of intellectual skepticism and administrative modernism in their efforts to adjust their assumptions to the perceived realities of everyday life.

"I often heard the lieutenants general of Police and other persons in office say," wrote a Parisian interested in the subsistence question, "that it was impossible to know precisely what the consumption of flour in Paris was and that in a given year one could fall into an error of calculation of between twenty and thirty thousand sacks too many or too few."[2] If the police could not speak with confidence about the question, it is no wonder that, two centuries later, we find ourselves befuddled. Few of the critical elements have been illuminated by scholarly inquiry, and we remain dependent for the little firsthand data that survive on information generated by the police themselves for their own edification.

Why was, and is, the problem of bread consumption so difficult to unravel? First of all, the police were not able to collect all the information they needed concerning grain, flour and bread entries, and bread production and distribution, and they were not always sure of the accuracy of the data they gathered. A certain amount of grain and flour escaped detection and registration, either because the officers charged with the task (the measurers and porters primarily and in some instances the clerks of the general farm at the "barriers" or gates of the capital) were negligent or overtaxed at a given moment, or, more likely, because the merchandise entered the city without their knowledge: legally, in the case (say) of merchandise brought in by well-to-do Parisians for their private use from their own land or by religious communities, hospitals, or other institutional consumers; or illegally, in the case of bakers who failed to declare grain or flour that they purchased in the countryside and shipped clandestinely and directly to their shops without passing through a Paris market (*en droiture*).

Some "foreign," or forain, bread brought in from outside Paris may have been sold—illicitly—off the markets or delivered directly to customers' houses. In making their calculations, the police usually overlooked several informal bread marketplaces that developed almost spontaneously in the course of the century. It is highly unlikely that the police figures made allowances for the bread distributed by "illicit" bakers who operated in a sort of subsistence underground or the bread sold to Parisians by bakers located on the immediate outskirts of the city (La Courtille, La Chapelle, Bercy, Charenton, etc.). In determining the amount furnished by the bakers in the official bread markets, the police apparently extrapolated from an "average" week instead of drawing on entries day by day or week by week in order to establish the annual totals. Like the parlous protostatistical conception of "the common year," used for gauging harvests, the average weekly supply was a normative notion that may not necessarily have faithfully reflected commercial and human realities.

The lieutenants of police, according to our informants, braced themselves for an error of estimation in consumption of twenty to thirty thousand sacks a year. But how much flour was in each sack? Here is one question that we can answer with some confidence: beginning around 1730, the weight of the sack remained an invariable 325 pounds (though it is not always clear whether this includes the weight of the sack or not) till the early nineteenth century. Theoretically, sacks of 217 pounds were permitted because they were more manageable. Three of these sacks were considered equal to two 325-pound sacks. I have not found, however, a single instance in which sellers or buyers actually deployed the smaller sack.

How much wheat did one need to mill in order to produce a sack's worth of flour? Now we enter a much less certain terrain. Some specialists presumed that it required substantially more than two setiers of wheat to render 325 pounds, while the majority of the experts claimed that two setiers sufficed.[3] As late as 1769, a schedule of price-fixing prepared for the Paris police based its reckoning on the two-setiers-to-a-sack weight. Yet by this time, according to one well-informed observer, improvement in technique frequently permitted millers to draw as much as 370 pounds of quality flour from two setiers. Already in 1760, Poussot, an inspector of police assigned to provisioning questions, expressed his "astonishment" that "ministers and magistrates" continued to fall into the "extremely egregious mistake" that a setier and a half of wheat were necessary to produce a setier of flour. Despite the "*arrêts* or declarations of the [royal] council, of the parlement, and the regulations of the police," anyone familiar with the provisioning trade knew that "a setier of good wheat yields more than a setier of flour." "The error," concluded Poussot, "precludes a just calculation of the quantity of flour needed to feed Paris."[4]

Still, one might object, not every setier of wheat was "good wheat." Reckoned officially in terms of volume, a setier of good wheat, according to the Paris measure, could weigh as much as 250 or even 260 pounds, whereas a setier of mediocre wheat could weigh as little as 220 or 225 pounds. Rarely is the weight of the setier taken into consideration in eighteenth-century provisioning censuses. Usually contemporaries meant 240 pounds ("the average") when they spoke of a setier and two setiers when they spoke of a sack. But these were conventional assumptions rather than actual observations.

The Yield of Wheat or Flour in Bread

Nothing better illustrates the perils of estimating consumption than the question of the yield of bread from a given quantity of flour or wheat. Commissaire Delamare supervised a series of baking "trials" in 1700 meant to establish a standard for the elaboration of price and production criteria. Using three different grades of wheat, he obtained, on average, 181 pounds in loaves of white, mid-white, and dark bread. Parisian authorities in 1721 used 181 pounds as the standard yield of a setier in calculating the amount of bread furnished by bakers in the markets. Sixty-eight years after Delamare's tests, master baker Pierre Toupiolle, described by Commissaire Grimperel as "one of the best bakers in my quarter, whom I believe to be an honest fellow of good faith and an expert in his profession," conducted a trial at the request of Lieutenant General of Police Sartine. The results startled Grimperel, for when he compared them to Delamare's calculations, he discovered that his illustrious predecessor had allowed the bakers a much higher claim for expenses (in shop rent, in wood, in baking costs generally, in wages, etc.) than Toupiolle was declaring almost three-quarters of a century later, despite the fact that prices were considerably higher in 1768 than they had been in 1700. With some trepidation, for he knew the towering reputation of Delamare, Grimperel concluded that "the persons in charge of supervising the operations in 1700 allowed themselves to be fooled." Yet might it not be argued that Grimperel was himself a dupe to the extent that he accepted as normal Toupiolle's yield of 164 pounds? Even when one notes that Toupiolle baked only white and dark loaves and sold twenty-seven pounds of middlings in addition to seventy-four pounds of bran, it still seems surprising that his 235-pound setier produced a total of only 130 pounds of usable flour.[5]

Shortly after Grimperel's trial a hypothetical price schedule drafted for the Paris police assumed a yield of 216 pounds, 32 percent higher than Toupiolle's. Yet thirty years earlier, Courcy, the police commissaire then specializing in subsistence problems, had set the standard yield in three types of loaves at 232 pounds. An official test conducted in 1740 yielded 230 pounds of bread to the setier, but the bread consisted only of mid-white and dark loaves. It is extremely difficult to reconcile these rather hefty claims with the lesser estimate that one encounters elsewhere. In 1764, the deputies of commerce maintained that a setier yielded 182 pounds of bread—that is, about as much as in 1700.

Dr. Malouin, one of the first savants to become passionately interested in baking and milling, took a more generous and, I think, more realistic stand. While the yield was merely 180 to 186 pounds in 1700, in 1767, at the time Malouin wrote, it had swelled on occasion to 240, that is, one pound of bread for each pound of wheat. But this progress was the fruit of recent advances in bolting and milling technology, and it was confined to the most entrepreneurial and efficient bakers—those with the requisite capital in knowledge, in networks, and in matériel. Malouin would never have imagined a yield much above two hundred pounds before 1760.[6]

Pierre-Simon Malisset, an audacious and highly skilled baker-businessman, who was the first to practice the gradual-reduction milling technology (which he called "economic milling") in a systematic way, aroused considerable admiration when he managed to extract 223 pounds of three sorts of bread from a setier in a police-supervised trial in 1767. In other trials Malisset squeezed as much as 246 pounds out of a setier. At the Scipion bakery of the Hôpital général, also in 1767, lead baker Bricoteau obtained 245 pounds in white bread, and over 260 pounds of a darker loaf. At virtually the same time, Michel Louis Thevenon, a master baker on the rue de la Juiverie, contracted to furnish the Chapter of Notre-Dame with 130 pounds of fine bread per setier and 150 pounds of a coarser loaf for the servants—a dramatically lower range of yield. Trials conducted in 1793–94 at the Corbeil facilities first established by Malisset generated considerably less bread than the innovative baker had produced a quarter-century earlier, but conditions now were far less serene.[7]

The other promoters of economic milling—the pharmacist and chemist

Parmentier, the lawyer and subsistence writer Béguillet, the miller Bucquet, the physiocrat Abbé Baudeau—boasted of yields ranging from 230 pounds for a mélange of white and dark (Parmentier's prudential stand) to 265 pounds for a single sort of household bread (Baudeau's exuberance for popular political economy). But they all set the standard yield for ordinary milling significantly lower even as late as the 1760s and 1770s: 200 pounds, 201 pounds, 214 pounds, 225 pounds. Commissioned by the Paris Parlement to examine the price-yield ratio in the town of Rochefort in the early 1780s, the academician Tillet found that a setier (Paris measure) produced 210 pounds of bread. About the same time the historian-demographer Abbé J.-J. Expilly posited a yield per setier of between 270 pounds and a staggering 324 pounds.[8]

It may be that in the seventies and eighties experts (themselves newly emerged as a sort of school of subsistence science) were beginning to agree on the standards of productivity and production, but for most of the eighteenth century there was simply no consensus on yield. The most serious estimates of yields were based on "trials" conducted usually either with a view to constituting a reasonable price schedule or to test a new method in the production nexus. The specialists, however, were themselves rather skeptical of the reliability of these trials. "Just as the weighing of the dough never gives precisely the weight of the baked bread," noted Malouin, "so the trials never give surely and invariably the yield of grain into flour and then into bread, or the price of the bread."

It was impossible to establish "a just and stable" relationship because the trial gave different results at different times and in different conditions. The source of this inconstancy was not human error or bad faith, though the tests were easily botched or sabotaged. Rather, Malouin contended it was the nature of wheat itself: the yield varied with the quality of the soil from which it came, with the weather preceding or during the harvest, with its age, with the conditioning it received in storage, and so forth. Even if one could control and standardize milling and baking operations—an achievement Malouin considered extremely remote—one could not eliminate heterogeneity in wheat quality. Thus, concluded Malouin, a trial could never be permanently or universally applicable. Pressing this relativism to its logical extreme, Malouin hinted that a trial was merely relevant for the day on which it was made.[9]

Provided one repeated the tests at least once a year and made allowances for variations in the données, Parmentier was much less pessimistic about the "utility" of the trial than Malouin. But he viewed it as a kind of metaphor rather than a rigorously positivistic standard. He warned against confusing the metaphor (or the model) and the reality: "there is a great difference between what happens when one conducts a trial and the everyday work of the baker in his shop." The trial was a hothouse experience, a sort of mise-en-scène arranged with an exaggerated amount of vigilance and exactitude in every step in the process. Since no miller or baker, however honorable and meticulous, could be expected to conform to the normative criteria of the trial, the latter had to be deflated or relativized in some way in order to make it useful.[10]

Another authority deplored the confusion surrounding the question of bread yield and prices. It was urgent "once and for all" to "determine exactly the precise quantity of bread that one can make with a setier of wheat." Yet he was not sanguine about the prospect of accomplishing this imperative goal, for past experience suggested that no two trials ever generated the same results. Indeed, they produced "opposing" results, each of which claimed to be "indubitable."[11]

Faced with this "incertitude," what did one do? After undertaking a series of carefully programmed "new trials" in 1768, none of which allayed his doubts or inspired his confidence, the lieutenant general of police of Etampes, a market town in the fertile Beauce to the south of the capital, felt obliged to take refuge in the comforting old standards established in the trials of 1686. No wonder his bakers were "deeply satisfied." The anachronistic price schedule fixed by the police drastically underestimated the productive capacity of the bakers and thus assured them a huge profit margin. Even the most feckless of them was probably able to obtain at least 30 percent more bread than the authorities expected or demanded.[12]

Estimates of Per Capita Annual Wheat Consumption

Explicit or, more often, implicit or unexamined assumptions about yield shaped efforts to estimate Paris consumption. One common form of reckoning cast consumption in terms of setiers per capita per year. The considerable range of opinion underlines the lack of accord on yield and in certain instances the lack of understanding about the factors governing yield. There was general agreement that there had been progress over time, although the commentators differed on the periodization of the Dark Ages. Four setiers per person was the measure of barbarity. Several observers associated this onerous rate with the time of Guillaume Budé, but Bucquet claimed that it took four setiers per capita right up to the end of the seventeenth century. Indeed, Bucquet, seconded by Claude-Jacques Herbert, the author of an influential book on the grain trade written at midcentury, and by several other writers who were concerned with subsistence problems, contended that it took *more than* three setiers to feed Parisians on average throughout the better part of the eighteenth century.[13]

In order to feed 10,710 persons who lived under their stewardship, 207 religious communities on the Seine, Oise, and Aisne rivers alleged that they required 3,017 muids. Ninety-five houses on the Marne with a population of 4,530 indicated an annual consumption of 1,343 muids. In per capita terms, these figures worked out respectively to 3.38 and 3.56 setiers. Now, it is possible that these religious houses exaggerated their needs, that they tolerated extremely inefficient milling practices, or that they sanctioned waste by insisting on the highest quality of white loaf, extracted at low rates.[14]

Taking their cue from Vauban, who supposed annual Parisian consumption to be three setiers per person, a large number of authorities, including Parmentier, Béguillet, the economic commentator Dupré de Saint-Maur, the physiocrat trio of Quesnay, Dupont, and Baudeau, and the revolutionary Lebon considered that this was the rate at least until the late 1760s.[15] Nor were these exclusively estimates of the salon or the *cabinet scientifique*. The grain and flour measurers were among the best-informed police officials on matters concerning quality and productivity, for they supervised sales to bakers. Two of them, Alexandre and Cholet, fixed annual per capita consumption at around three setiers in 1757.[16] Another group of observers, including the foreign minister Choiseul (who headed up a ministry deeply interested in grain problems), the deputies of commerce in the mid-sixties, writers in the *Journal économique* and the *Journal de l'agriculture, du commerce et des finances,* and the scientist-economist Paucton, judged annual per capita consumption to be 2.5 setiers at least from midcentury on (too early to bespeak the impact of innovations in mill technology).[17]

Finally, a few observers contended that per capita consumption was only about two setiers. Two of them, the economist Forbonnais and an English traveler named William Mildmay, wrote in the 1750s before the influence of changes in the style of milling could be felt. The estimates of two others, Daniel Doumerc, an experienced victualer with a vast knowledge of subsistence affairs, and Desaubiez, a philosophe without compelling credentials in this domain, may very well have reflected the dramatic improvements in bread yield realized through the process called economic milling.[18] But banker-philosophe-de facto prime minister Jacques Necker and writer-administrator Moheau clearly suggested that two setiers was the average not merely around Paris, where economic milling may have been more or less widely implanted, but all over the kingdom.[19]

The exponents of economic milling believed that per capita consumption fell to two setiers a year after the late sixties. But they regarded this as a very recent triumph, a technological breakthrough followed by a campaign to convert mills to the new process and alert bakers to the advantages of bolting with greater discrimination and of using reground middling flour even when they did not have access to full-fledged economic mills. They continued to insist that ordinary consumption before the advent of economic milling varied between a minimum of two and one-half setiers and a maximum of four. They took enormous satisfaction in noting "the savings" or increased yields that economic milling could procure. The writer and economist Pierre Legrand d'Aussy put the figure at one-eighth, the Parlement of Provence at one-seventh, Malisset at one-sixth, and Parmentier, Béguillet, and Bucquet up to one-fourth. These savings, it must be remembered, were in large part hypothetical, for no one claimed that economic milling was universally adopted—even in the Parisian area. This meant that it still commonly required more than two setiers per capita to feed many Parisians and virtually all other Frenchmen.[20]

Estimates of Daily Per Capita Bread Consumption

Setiers per capita per year was a somewhat abstract concept. A number of eighteenth-century observers, followed by certain historians, preferred to calculate consumption in terms of daily per capita bread needs, because they seemed to be more concrete and immediate and thus easier to verify. But here, too, the spectrum of estimation is vast. The chemist–public affairs commentator Antoine-Laurent Lavoisier's claim of fifteen ounces is the lowest and is today the most prestigious, though it has never been subjected to serious critical scrutiny. Nicolas Dupré de Saint-Maur suggested eighteen ounces, while warning that such an average could not faithfully reflect the complexity of Parisian social reality. Dr. Malouin put consumption at eighteen and one-half ounces, while one of the great Revolutionary committees set up to study popular life and poverty determined that it was nineteen ounces. Necker, thinking in national rather than Parisian terms, placed it between twenty and twenty-four ounces. The Englishman Mildmay and the deputies of commerce agreed on twenty-one ounces. Béguillet calculated a pound and a half for Paris, while the future lieutenant of police Joseph d'Albert envisioned the same figure as a national or provincial standard. According to Abbé Jean-Baptiste Briatte, a student of popular life, the mass of people consumed on average one and a half pounds a day.[21] The mathematician Jean Louis Lagrange proposed twenty-eight ounces, probably as a national criterion, while a Revolutionary writer imagined that Parisians ate two pounds a day, more than twice Lavoisier's estimate.[22]

Most of these writers understood that consumption was both socially and demographically differential. The father with three children under ten, maintained Lagrange, will eat more than all of his children combined. "Average eaters," A.-J.-P. Paucton reported, needed eighteen ounces, but "gens de lettres," because they were sedentary, could get by with a pound. While Paucton's "big eaters" demanded only twenty-four ounces, Dupré de Saint-Maur and Lebon, followed by two of the best modern economic historians, affirmed that heavy laborers commonly required three pounds a day. Similarly, Du Vaucelles, a local Parisian Revolutionary leader, took umbrage at the commonly cited average daily ration of twelve ounces: "Go ask the Limousins, the Auvergnats, all the other mountain men who do the roughest work in Paris. They will reply that the dose is infinitely [too] small." Nor was anyone sure how much less women, especially working women, ate than men, or whether sick and elderly people should be classified as "small" or "ordinary" eaters. There was no consensus on per capita daily bread consumption, not only because contemporaries could not agree how much to assign to each sociodemographic category but also because they did not know how many people there were in each rubric.[23]

Consumption and Population

It goes without saying that the problems of consumption and population were intimately linked. Without an idea of the size of the population, per capita figures of yearly grain consumption or daily bread consumption cannot reveal the magnitude of aggregate annual consumption. Or, as the mitron de Vaugirard put it in a book about the subsistence question written in the 1770s: "Can one evaluate the quantity of wheat necessary for a year's consumption in the kingdom?" "No," replied his friend, "because they are still talking nonsense about the current population—some say it is eighteen million, others twenty-four million." Dupré de Saint-Maur looked at it the other way around. Rather than using population to get consumption, we can infer population from aggregate consumption: "I do not believe that there is a surer way to make a judgment of the number of inhabitants in the city than through its consumption."[24]

Dupré's method is full of perils, as I suggested already, because of all the uncertainties about yield and about per capita needs, as well as the difficulty of establishing a global consumption figure. If we were confident about our estimates of per capita consumption, the mitron de Vaugirard's procedure would prove more fruitful. But there was no less "nonsense" uttered about the population of Paris than about the population of France. Contemporary estimates vary inordinately, and they are often difficult to compare either because they were made at different times or because it is not clear to what precise epoch they apply. Messance, one of the first eighteenth-century writers to earn the title of demographer, was quoted in 1756 as calculating the population at slightly under 600,000. Another demographer, Expilly, had thought the capital contained 720,000 people—until a new method of calculation constrained him to reduce his estimate to 680,000 or less in the *Dictionnaire géographique* published in 1768. More than two decades earlier, Dupré de Saint-Maur had reached the same conclusion.

In the fifties, virtually at the same time that William Mildmay proposed a population of between 492,000 and 580,000, the Paris municipality concluded it was over a million. Commissaire Lemaire also said a million, but not until the 1770s, at which time the municipality, in fresh deliberation, revised its estimate downward to between 700,000 and 800,000, close to the figure that had been proposed in the *Encyclopédie* years before. Béguillet and two other acute observers of Parisian life, Brion and Mercier, all endorsed the municipality's revised figure, while Jean-Jacques Rutledge, a journalist with a special interest in the bakery, and Nicolas Des Essarts, the author of a police encyclopedia, resuscitated the municipality's earlier idea of one million. Upon this background, Paucton's figure of 589,000, published in 1780, seems niggardly. There was disagreement among all the commentators on the size of the fixed population, on the magnitude and the seasonal variations in the floating population, on the number of houses, on the number of families and their dimensions, and on the other demographic characteristics of the capital.[25]

It should be emphasized that these estimates were not merely academic exercises. For example, the population calculations of Isaac Thellusson, the man charged with conducting a large part of the mammoth provisioning campaign for Paris during the subsistence crisis of 1738–41, were of great significance in shaping his plans and the general strategy of the government. Thellusson, a Genevan diplomat and Paris-based banker with financial correspondents all over the world, supposed that Paris normally contained 800,000 persons, including the inhabitants of the immediate suburbs, but that it assumed the burden of feeding everyone within a twenty-league radius—1,500,000 people in all—during a serious dearth. Similarly, in 1731 the Assembly of Police calculated weekly consumption at eighteen hundred muids, an estimate that also had direct policy implications, for the Assembly was in the midst of organizing an emergency granary system, built into the religious communities, colleges, and hospitals of Paris. Indeed, most of the estimates we encounter exercised influence directly or indirectly on decision making.[26]

Estimates of Aggregate Consumption

Given the diversity of opinion on the amplitude of the population and per capita needs, it is not surprising to find an enormous range of appreciation of aggregate consumption. The highest estimate, 200,000 muids, was made at the beginning rather than at the end of the century by a correspondent of the procurator general. The lowest, 60,000 muids, was made in 1740 by a police agent who reported regularly on the state of public opinion. Five other appraisals placed consumption at over 150,000 muids. Two of them can perhaps be discounted on the grounds that their authors, Mercier (166,666 in 1780) and the Polish traveler Mniszech (150,000 muids in 1767), were not experts in the matter. But the three others were made by the bakers' guild in 1690 (182,000 muids), by Foucaud, an agent dispatched to observe harvest conditions and the provisioning process in the greater Paris region (150,000 muids in 1755, revised to around 180,000 muids in 1756), and by Béguillet, a specialist in subsistence technology (150,000 muids). Malouin assessed consumption at 133,333 muids, despite his conviction that the average Parisian needed only two setiers a year, because he calculated the population at 800,000 in 1767.[27]

Eight surveys placed consumption between 100,000 and 110,000 muids. They were made by a controller general, the minister specialized in economic and social affairs (102,200); a prévôt des marchands, the equivalent of mayor of Paris (100,000 in 1739); an intendant of Paris (104,000 in 1763); the Parisian municipality (over 104,000 in 1764); a civil and military victualer and government adviser on subsistence affairs (100,000 in 1771); an anonymous midcentury memorialist (more than 100,000); and two Revolutionary bodies (103,417 in 1791 and 109,500 in 1794–95).[28] In six other reports, again emanating for the most part from administrative sources, consumption is evaluated at between 90,000 and 100,000 muids: the Assembly of Police (93,600 in 1731); the Parisian municipality (90,000 to 100,000 in 1759); Trennin, a police official at Versailles (91,250 in 1768); the *Encyclopédie méthodique*, citing a Parisian police source, probably the grain measurers (about 100,000 in 1785); Du Vaucelles (98,246 in 1789); and Jean Bailly, a scientist and mayor of Paris during the early Revolution (between 91,000 and 97,000).[29] Seven estimates fell between 80,000 and 90,000 muids: an investigation ordered by Cardinal Richelieu

(84,000 in 1637); a report on the canal of Picardy (83,950 in 1728); Dupré de Saint-Maur, followed by Paucton (82,000 muids in 1746 and 1780); the forain bakers (at least 80,000 in 1738); Mildmay, who claimed, like Dupré, to have seen the entry records (82,000 at most in 1754); Lavoisier (80,640 in 1791).[30]

What can one make of this welter of contradictory data and conflicting opinion on the consumption question? A certain number of premises and evaluations can be discarded immediately as a result of scholarly research. No one today believes that the population of Paris could have been 800,000 in 1700 or a million at midcentury. It seems highly unlikely that per capita annual consumption was above three setiers at midcentury (or below two by the Revolution). On the basis of internal evidence we can surely eliminate aggregate consumption estimates below 75,000 and above 150,000 for the whole century. Yet it is worth emphasizing that we are still hostage to the same sociodemographic and technological uncertainties that fettered the statistical pioneers of two hundred years ago, who were as keenly aware then as we are now of the policy (police) implications of statistical analysis.

The Delalande Registers

Nevertheless, we have some data collected by police officials and their auxiliaries that may enable us, not to resolve the issue unequivocally, but to limit significantly the spectrum of credible options. In the eighteenth century the market officers called measurers registered the entry and sale of all the grain and flour arriving at the Halles and at the ports. The measurers submitted the itemized reports (*relevées*) to the senior commissaire of the Halles quarter, who was the field officer in charge of Parisian provisioning affairs. In addition to data on the influx of merchandise to the Halles and the ports, he collected information on the sale of bread in the Wednesday and Saturday bread markets from his colleagues who functioned in the quarters where the markets were located. He also obtained data on direct entries (grain and flour arriving en droiture) delivered to institutional consumers (religious communities, hospitals, colleges), shop bakers, and individuals (generally those who owned grain-producing lands) from the measurers and the porters, from the bureaus of the tax farms, and from other sources. On the basis of this information, which the Halles commissaire submitted to the lieutenant general of police, the Paris police were able to calculate what they believed were total supply and aggregate consumption.

To my knowledge, only nine of the manuscript registers containing this data have survived. They cover the years 1725 through 1733, and they were verified and approved by Delalande, a police official who worked directly for the lieu-tenant general. Delalande was acutely interested in procuring accurate consumption information, for he was in charge of the effort made by Paris police to establish an emergency granary system rooted in the religious houses and hospitals.[31]

Folio in format, each register contained several hundred pages of detailed information on incoming shipments (*arrivages* listing origin and commercial circuit as well as quantity), transactions, and prices. The rubrics were standardized and the pages printed with columns to be filled in by hand in each register. The police meant for the data to be comparable from year to year and obviously planned to continue compiling them indefinitely. It is likely that the authorities made several copies of every register each year. Delalande meticulously assessed the raw data, eliminating hidden double entries and making other adjustments. Though there are scores of long and tedious computations, the registers suffer from remarkably few arithmetic errors. They constitute the most graphic picture we have of the belly of Paris in the Old Regime.

We shall examine consumption calculations in depth for two years, 1727 (a "normal" year after two years of dearth) and 1733 (a year preceded by six years of relative subsistence tranquility). In 1727 the following sources accounted for a global wheat provision of 78,124 muids:[32]

Halles: grain and flour	29,078 muids
Port Gréve: grain	8,951 muids
Port Ecole: grain	5,102 muids
Religious communities and hospitals	5,124 muids
Bakers and individuals (droiture)	11,533 muids
Forain (second-class) bakers	18,336 muids

We are concerned exclusively with wheat (including flour expressed in terms of wheat measures) because Parisian bakers, save in crises, baked only wheaten loaves, and because the shipments of rye, barley, and other secondary grains (oats excepted) were generally negligible. In order to reduce the strain on the Paris market, institutional buyers were encouraged to buy directly in the countryside or to draw on their rents and revenue in kind. Bakers and individuals with lands and "rights" also imported grain (or flour) en droiture, directly to their houses from the countryside without passing through the Halles or the ports. I believe that the portion siphoned off from this rubric by individuals was relatively insignificant. Following Delalande, we will assume that all the wheat in the Halles and the ports was sold to bakers and converted to bread, along with the institutional wheat (though much of this bread did not pass through commercial circuits of distribution) and the droiture wheat brought in by Paris bakers.[33]

The administratively defined "first class" of bakers consisted of city (guild) and faubourg Saint-Antoine bakers for the most part, as well as some forain bakers established in the immediate environs who looked to the capital rather than the countryside for their raw materials. They used wheat or flour obtained either at the Halles or ports or from direct delivery (droiture) and registered by police authorities as part of the provisioning trade. The bakers of the "second class" were all forains who did not rely on Paris as the source of their grain or flour. These bakers made their bread from wheat that did not enter the city before conversion into bread and therefore was not registered in the Halles, port, or droiture entries.

Parisians purchased most of their bread from two sources: the markets and the shops. The police register informs us that 58,340 muids of wheat were converted into market bread. If we subtract this figure from aggregate consumption, we get 19,784 muids, the amount of wheat sold as shop bread. The register reveals that master bakers accounted for a little less than 35 percent of the market bread supply, or 20,228 muids. What portion did this represent of the total guild-baker contribution to Paris provisioning? According to a listing of all master bakers prepared for the police in 1724, average weekly guild consumption per capita amounted to 18.12 setiers, or a little more than a muid and a half.[34] On this base we could estimate total annual master-baker consumption in 1727 to be 39,036 muids. If one deducts their market supply, there remain 18,808 muids for baker shop usage, an amount quite close to the quantity of shop bread indicated by the police register (19,784 muids).

It is very likely, however, that the latter figure is too low, for it does not allow for *nonmaster* shop production. Most of the Saint-Antoine bakers, and a substantial number of close-by suburban bakers, maintained some sort of shop commerce. Let us imagine, conservatively, that 275 Saint-Antoine bakers used six setiers a week, or 7,150 muids a year for their shops (and let us presume that on average this figure covers the suburban shop bakers as well). This reasoning suggests that the 1727 register underestimates shop supply and therefore total consumption by 6,402 muids—surely a minimal figure.

Moreover, the register ignores the underground bakery. Though I would contend that the total number was larger and their production coefficient higher, let us estimate that there were eighty illicit bakers, doing business under various covers, and that they used a setier a day or 2,426 muids a year. If we add the supplementary shop supply and the illicit baker contribution to the official register total, we obtain 86,952 muids for 1727.

The "extra" grain and flour needed to account for this additional bread provision certainly came from unrecorded and/or surreptitious droiture shipments. In order to avoid controversies with the measurers or porters—the grassroots Paris market police—over fees that the latter claimed or to conceal purchases made illegally outside the official market places of the hinterland, the bakers did their best to keep their droiture traffic inconspicuous or secret. Frequent complaints by the commissaires and inspectors as well as repeated confrontations with the measurers and porters suggest that the bakers were in part successful in defying detection. Therefore it seems reasonable to suppose that enough "secret" grain and flour could have entered to account for the increase in consumption that I propose.

I must emphasize that 86,952 muids seems to me to be an absolutely minimal supply estimate for bread consumption. We have not allowed for shops at La Chapelle, Bercy, Le Roule, La Courtille, and other outposts on the border of the faubourgs and the suburbs that the Parisians regularly patronized. It is impossible to guess how much forain bread entered clandestinely, either for sale in the streets and squatter markets, for later regrating, or for home delivery (an intramural retail analogue to grain and flour droiture). We may have underestimated not only illicit and faubourg shop provisioning but also guild production, for at least one probing contemporary survey estimates average master-baker usage at two muids a week (over against our reckoning of 18.12 setiers).[35] We have neglected entirely home baking, though it is not likely to weigh heavily in our calculations. As a modest prudential buffer, I would be inclined to add to the 86,952 muids another five thousand muids, roughly the margin of error that our informant tells us the lieutenants of police learned to live with. In 1727, then, I would propose a global consumption of 91,952 muids, a figure almost 18 percent higher than the official registered calculations.[36]

According to the Delalande register, the total wheat provision for 1733 amounted to 78,236 muids, distributed as follows:[37]

Halles: grain and flour	22,808 muids
Port Grève: grain	9,505 muids
Port Ecole: grain	822 muids
Religious communities and hospitals	5,416 muids
Bakers and individuals (droiture)	16,620 muids
Forain (second-class) bakers	17,065 muids

This calculation is liable to the same criticism as the one in the 1727 register. Let us suppose that the total master-baker production is 39,264 muids, though this figure is probably too low, not only because average weekly usage was higher than we have allowed but also because there were probably more *active* masters in 1733 than in 1724, the only year for which we have a full corporate enumeration. Since the masters supplied 16,640 muids to the bread markets, their shop production came to 22,624 muids, a figure close to the 22,935 muids of shop bread indicated in the register for 1733.

But, as we saw in the case of the 1727 calculation, the register tally is too low, for it allows neither for nonguild shop production in the faubourg Saint-Antoine and the suburbs adjacent to the faubourgs nor for the bread made by illicit bakers. If we estimate 6,422 muids for the Saint-Antoine shops and 2,426 muids for the eighty illegitimate bakeries, then total consumption rises to 86,766 muids. But for all the reasons invoked in reference to the 1727 register, it is very probable that this figure is still too low. Annexing to it the same margin we permitted in 1727, we obtain an aggregate consumption of 91,766 muids for the year 1733. Consumption appears to have remained stable from 1727 to 1733. Neither estimate is far from the figure of 93,600 muids that the Assembly of Police used as a basis calculation in 1731.[38]

To my knowledge, for the period after 1733 we have neither serial data nor any fragments substantial enough to serve as points of extrapolation.[39] It is certain that consumption increased steadily across the century, for we know that population increased. We do not know, however, at precisely what pace because we do not know with any certainty the rate of Parisian demographic growth. If the population of the capital reached 600,000 on the eve of the Revolution, it is likely that aggregate consumption rose to between 115,000 and 120,000 muids, less if economic milling had a significant impact, as I am inclined to suspect.[40] In fact, population probably reached 700,000 or more.

During the six decades between 1730 and the Revolution, the patterns of provisioning changed substantially, though here too it is difficult to document the changes with exact figures. Flour eclipsed grain, the Halles overshadowed the ports. The brokers at the Halles became more influential than the grain barons of the rue de la Mortellerie. The sturdy and entrepreneurial miller emerged as a merchant who speculated, stocked, and financed sales. The Halles enjoyed a kind of renaissance, partly engineered by the police and capped by the construction of a new building that facilitated trading and surveillance (and that still stands today in the shadow of the Saint-Eustache church). One of the means envisioned by the police to promote the development of the Halles was the suppression of all droiture traffic. By concentrating all transactions at the Halles, where they would be visible and measurable, the police hoped to achieve greater control over the provisioning trade, especially baker grain and flour commerce. But the bakers stubbornly refused to capitulate, either to panopticon monitoring or to the constriction of their commercial freedom. They continued to expand the range of their buying operations, to scour the countryside and the collector markets of the supply zones, to eliminate middlemen, and to receive a substantial part of their stock in droiture.

In the end, the police never fully mastered the quantitative questions. Renowned for allegedly knowing what everyone who mattered was doing at any given moment, in Vienna or Versailles, Antoine Raymond de Sartine may have been more comfortable with the exotic than with the banal. There is no better metaphor of his relative powerlessness and of the burden of technological inertia than the bread production trials: for all of his finesse, Sartine could not obtain reliable figures on converting wheat to bread or on the cost of the operation.

Yet during the course of the eighteenth century, the provisioning authorities became reasonably conversant in the idiom of subsistence calculations. Sartine was far better armed to assay supply requirements and calibrate policy to productive and commercial capacities and consumption needs than had been Marc-René d'Argenson, the energetic police chief at the beginning of the century. Nothing was more corrosive of Delamare's timeless recipes than the experience of grappling with such apparently simple questions as how much bread people eat each day and how much bread will come out of a given measure of wheat and how much wheat (and flour) constitutes the city's annual treasury of necessity. Regulation evolved from a reflex and a profession of faith into a function of myriad contingencies, the treatment of which called for concrete knowledge of practices and outcomes. The authorities learned to deconstruct the categories of supply and demand, which turned out to be exceedingly complex. Though they did not free themselves entirely from normative and moralizing assumptions, they had a much more lucid conception of the sources and the significance of their data. And improved monitoring, both less obtrusive and more effective, and independent corroboration, yielded more and higher quality information.

The police made better sense of the data because they had a better understanding of how the system worked all along the way, from the time of the harvest to the moment of removing the loaves from the oven. More empirically inclined, the stewards of provisioning practiced a critical gaze, based less on extrapolations from a putatively immutable human nature than on an attentive examination of what they came to know as the facts. They forged a veritable concept of carrying capacity. An appreciation of weekly and yearly supply—a realistic order of magnitude if not a rigorous evaluation—enabled them to engage in far more rational short-term decision making and longer-term planning. They could intervene with more subtlety at lower cost with less market disruption. This knowledge gave them a keener sense of the political margin with which they could play in times of both stress and relative ease. Targeting their objectives more precisely, they applied controls with more flexibility and discrimination. The availability of these data helped to focus more attention on prevention and organization. At any given moment, the police knew how large a baker corps was necessary, how much incentive had to be offered to the forains, how much latitude could be granted to the bakers in their grain and flour buying.

Sartine was the first official who dared imagine that he could sustain more people on less grain. This was largely a matter of stunning technological innovation in the milling industry, but it also had something to do with overcoming inherited wisdom whose credibility had faded.

Chapter 17

The Police of

Bakers

"The subsistence of the people is the most essential object that must occupy the administration," wrote Jacques Necker, intimately familiar with the grain trade and the government. The part of the administration charged with this task of monitoring and regulating provisioning, and the very task itself, was called police. As noun, verb, and adjective, the term was used more broadly to describe the way in which social and civil life should be organized. Schooled in Delamare yet sensitive to the fashions of his own times, fellow commissaire Lemaire, writing in 1770, defined police as "the *science* of governing men." Beyond order and rationality, it was also the vocation of good police to assure, in the words of a jurist, that "harmony and concord" prevailed among citizens. Policing bakers and bread was a part of the larger process of public administration, a part located near the end of the administrative chain. It framed the most intimate and sustained point of contact between the police and the people.[1]

It could be said of police as it is said of politics, all police is local police. Certainly the police of bread and of bakers was a primordially local affair. Yet, given the way the subsistence question dominated public life at the center and the summit as well as at the periphery and the grass roots, and given the deep anxiety that the central government evinced for the tranquility of the kingdom's largest, most socially complex, and thus most vulnerable city, the entire police apparatus from the top down paid attention to Paris and in some way or another intervened in its life.[2]

The king exercised the supreme police power in the realm. Through royal legislation and the action of the royal councils, the monarch defined and refined rules and responsibilities, investing officials at the regional and local levels with authority to enforce existing codes and more broadly to assure the provisioning of their areas. In periods of dearth, symbolically and personally the king became more actively involved in the police of provisioning. His chief deputy for the police of provisioning, and indeed for virtually all matters bearing on the economy and public order, was the controller general. The latter's very omnicompetence limited his capacity to stay closely informed and intervene, but (as sometime controller general Necker intimated) he gave subsistence affairs the highest priority. Aided by the intendants of commerce and finance and by the "presumptuous clerks" of the grain department and the other "little kings of France" in the embryonic bureaucracy derided by the physiocrats, the controller general tried to anticipate deficits and surpluses and facilitate the regional distribution of supplies. In emergencies he coordinated regional and local efforts to cope with dearth and organized relief operations. He was interested in every dimension of the grain and flour trade and kept track of provisioning through the enormous correspondence he maintained not only with the intendants in the field who were directly responsible to him but with myriad other officials, including the lieutenants general of police.

The other figures at the center who exercised police were the controller general's ministerial colleagues, the secretaries of state, whose division of labor was both territorial and functional. Of special interest to us is the secretary of state for the royal household, in whose jurisdiction Paris loomed large. Although he did not deal directly with the provisioning trade, he kept a close eye on the subsistence situation, for he was personally accountable to the king for the tranquility of the city.

The intendants of the *généralités*, the administrative divisions that cut across the old provinces, constituted the first line of royal police in the field. The same type of constraints that inhibited the controller general at the center hampered the intendant: the overwhelming multiplicity of his tasks and an extremely modest staff with which to attend to them. In fact, he was dependent on a staggering array of local officials, each of whom retained substantial autonomy, for the conduct of the provisioning police. The intendant exercised his greatest leverage by regulating the circulation of grain and flour across time and space, impeding or facilitating movement through his jurisdiction. A king's man par excellence, the intendant was nevertheless capable of defying royal orders that he felt imperiled the well-being of *his* généralité. The intendant of Paris was the provincial head least competent and least inclined to obstruct royal will, largely because of his propinquity to Versailles. He also enjoyed little influence in the provisioning domain because his jurisdiction did not include the city of Paris—the capital was subject to no intendant, though the lieutenant general of police performed many of his functions—and because the Parisian police exercised extensive authority in the hinterland.

The police powers discussed so far emanated directly from the monarch and fit into a pyramidal hierarchy that extended from the pinnacle to the more than thirty généralités. There existed, however, yet another major police authority, exercised on a regional scale by the thirteen sovereign courts called parlements. Though royal in origin, the parlements possessed a generous measure of institutional and constitutional independence and practiced a sort of parallel police vis-à-vis royal administration, sometimes complementing it, sometimes supplanting or contesting it. The most familiar levers of parlementary power were the captious tactics by means of which the courts subverted royal legislation, be it a question of overt opposition (refusal to "register" laws) or of more subtle devices. Less well known was the parlements' license to promulgate quasi-legislative decrees called *arrêts de règlement,* regulatory and interpretive measures taken by the courts on their own initiative, often in the name of what they called "la grande police." The measures through which the Paris Parlement limited bakers in a time of scarcity to making only two types of bread and through which the Dijon court placed an embargo on the shipment of grain outside its jurisdiction were arrêts de règlement. Whereas the rights of registration and remonstrance endowed the parlements with enormous political influence on the national stage, this quasi-legislative power to enact arrêts permitted the parlements to function as genuine regional governing institutions. Until the sixties, the king and parlements fought bitterly over many issues (constitutional, fiscal, religious), but never about subsistence policy, a sacrosanct matter of public interest, nonpartisan by nature. The radical reforms of 1763–64 freeing the grain trade from the Bastille of regulation politicized the subsistence question, divided the parlements, and set the majority of them bitterly against the drastic royal departure from tradition.

Another powerful vehicle of parlementary police (and politics) was the juridico-administrative apparatus directed by one of the court's leading magistrates, the procurator general. It was particularly important in the Parlement of Paris, whose procurator general commanded exceptional prestige and whose jurisdiction stretched over a third of the kingdom. The procurator had *substituts,* or deputies, in scores of towns and hamlets—far more numerous and closer to the ground than the intendants' subdelegates—who functioned as local officers of justice and police as well as representatives of the parlement. They regarded the procurator as their protector and as the authority to whom they were ultimately accountable. It was they who enforced the law of the land as the procurator instructed them—another of the subterranean ways in which the parlements decisively influenced concrete political and social outcomes far from the glare of publicity given to the more spectacular clashes between king and courts. In many instances the procurator general was better informed than the intendants, whose jurisdiction he often challenged, or even the controller general, and better placed than they to see his orders executed.

The procurator general also took charge of organizing the informal Assembly of Police, composed of the first president of the Parlement of Paris, the lieutenant general of police, and the prévôt des marchands as well as the procurator. It met monthly or weekly or even daily, depending on the circumstances, to discuss all matters pertaining to the administration of the capital and evinced a special interest in the subsistence question, the core of the grand police. The lieutenant general reported to the Assembly on the state of public opinion, the situation in the Halles, or central markets, the conduct of the bakers, and the conditions in the markets and farms of the nearby hinterland; the prévôt passed on to his colleagues news about the port markets and the river-borne grain and flour trade; and the procurator communicated data from his substituts on harvests, stocks, and market performance throughout the jurisdiction. The officials evaluated the situation and attempted to fix upon a common course of action (or inaction). Consensus did not come easily, for the lieutenant and the prévôt were bitter rivals, and both resented the "tutelage" that the first president presumed to exercise over them in the name of the parlement. For the most part, the gravity of the stakes militated in favor of good coordination, despite the temptation of the parlement to swoop in like a deus ex machina to contest the monarch's vocation as father of the people, a title he brandished less and less serenely in the last third of the century.

Primary responsibility for the police of the capital's subsistence resided in the Lieutenance générale de police. Created in the 1660s to handle matters concerning the security of Paris, the lieutenance swiftly took charge of a sweeping range of functions, the crux of which turned on provisioning. The holder of this crucial office decided on day-to-day regulations, shared crisis management with the controller general and the procurator general, and occasionally ventured long-term planning. He reviewed the subsistence situation in depth four times a week, just before and after the formal market days, and kept his eye on events in between. One of the bureaus of his central office concentrated exclusively on provisioning affairs; one commissaire and an inspector worked virtually full-time on grain and flour marketing, while others carried out specific assignments concerning bakers, millers, and brokers; and a welter of minor officials (including market measurers and porters, police agents called exempts, and members of the watch, the guard, and the rural constabulary) gathered intelligence and implemented decisions in the capital and in the supply zones. Orchestrating the entire provisioning trade, the lieutenant kept track of entries into Paris and used his *droit de suite* to pursue suppliers into the barns, granaries, farms, and inns of the countryside. His authority superseded that of local officials; much to their indignation, he was not reluctant to use it to favor the needs of the capital. As supervisor of the entire corporate system, he paid special attention to the bakers' guild and monitored the faubouriens and forains as well. At once magistrate and administrator, he judged the cases of the bakers and others in his chamber of police, one of the branches of the Châtelet court.

The prévôt des marchands, who presided over the municipality, resisted—more or less vainly—the hegemonic claims of the lieutenance, specifically in the arena of subsistence and more generally in affairs involving economic, social, and cultural life. The prévôté drew its greatest strength from its traditional stewardship of waterborne commerce. Albeit in no position to set provisioning policy, the prévôt was left to manage relations between the city and the merchants, Paris-moored and itinerant, who shipped by water and sold at the ports. Like the lieutenant general, the prévôt issued quasi-legislative ordinances that had the force of law throughout the river system, and he adjudicated quarrels in his own court. The prévôt and the lieutenant learned to collaborate on matters of urgent public concern, often under the prod of the procurator general.[3]

Occasionally, the controller general intervened directly in the policing of bread and bakers. By and large the task remained in the hands of the lieutenant general, accountable both to the Assembly of Police (and through it to the parlement) and to the secretary of state for the royal household (when he was of vigorous bent). If the lieutenant had frequent contact with representatives of the bakery, his commissaires (and to a lesser extent his inspectors) dealt with the bakers on a daily basis. There were forty-eight commissaires assigned to twenty quarters. Each district had at least two of these officials, who were required to reside there, to keep their door open to the public twenty-four hours a day, and to develop an intimate familiarity with the life and business of the community, a mastery of local knowledge that their remarkable stability in the quarter (an average of twelve years per assignment) put within their reach. "The choice of individuals for these posts is of the greatest importance," wrote Mercier, "a greedy or perverse commissaire could occasion a host of small, hidden evils; judges of the first instance, counsels of first resort, they have the most immediate and direct influence on a multitude of affairs that find resolution or swell in significance according to their character; the public good is in their hands every single day."[4]

Charged with multiple responsibilities in the civil and criminal as well as the strictly "police" domain, the commissaires were supposed to give absolute priority, in their own words, to the mission "of containing, in their mores and their actions, a multitude impatient with the most legitimate checks and always ready to transgress them."[5] This quasi-apocalyptic description could have applied, depending on the circumstances, to consumers as well as bakers. Each commissaire was expected to know every baker in his territory: to have an appreciation both of his character and of his business. If a market existed within his jurisdiction, the commissaire had to acquaint himself with all of its regular denizens and their practices, and to visit it at least several times during each session of operation. It was obviously more difficult for him to learn about the thousands of consumers in his orbit. An adroit commissaire observed, asked questions, conducted rounds, visited taverns, and made himself visible and available. In a sense, the best way for him to get to know consumers was for them to get to know and invest confidence in him. The most successful commissaires were the ones who attracted the largest number of spontaneous collaborators from the public.

Bakers as Public Servants

It would not be fair to say that the bakers aroused the same deep mistrust in the eyes of the police as did the grain (and subsequently the flour) merchants. The authorities knew the bakers better, observed them more regularly, and had more immediate leverage over them. Nor could individual bakers, even the most commercially robust and viciously inclined among them, wreak as much social havoc, in the estimation of the police, as could many of the grain suppliers upon whom the capital depended. Yet the same underlying assumption about their psychology shaped the police attitude toward the bakers as conditioned the police view of merchants: that they were motivated exclusively by greed and self-regard, and that they would not perform the social service that the public had the right to expect of them, especially in adverse circumstances, unless they were offered incentives and/or subjected to surveillance, pressure, and (ultimately) harsh sanctions. "Temper as much as possible their greed for profit" served as Commissaire Delamare's watchword in his dealings with the bakers. His chief La Reynie never tired of reminding the bakers that they would be "severely punished" if they failed to fulfill their provisioning responsibilities. For d'Argenson, La Reynie's successor as lieutenant of police, the bakers often acted as "the cruelest enemies of the people." D'Argenson warned repeatedly that "it would be dangerous were we not in a position to repress the insolence of the bakers." The lieutenant general beginning in late 1725, Hérault saw an urgent need to apply "exemplary punishment" to the "bad faith" and "cupidity" of which "the public is victim."[6]

The same reasoning applied to bakers as to grain and flour merchants: on grounds of both humane consideration and political calculation, the subsistence of the people had to be protected. In their containment strategy, the police regarded bakers, like grain and flour dealers, as public servants in the first instance rather than as simple commercial actors. If they were allowed to fail in their task, the people would erupt more or less legitimately "in fury." The public order would be doubly imperiled, by the popular insurgency and by the dearth that occasioned it.[7]

As the police construed them, the duties of the bakers were to keep their shops and market stalls supplied regardless of difficulties and hardships. Nothing—certainly not an ordinary paroxysm of penury—dispensed the bakers from this obligation. If they had to cut back, they were expected first to slight their richest clients, by shifting all their resources away from luxury breads despite the briskness of demand for it, and subsequently by adjusting again away from common white and toward dark. Exhorted to use their ingenuity to ferret out supplies, they were provided with grain or flour if the authorities certified their incapacity to marshal merchandise on their own (for lack of credit or for lack of opportunity). In times of abundance, the police allowed them to enjoy a nice profit. Reciprocally, in times of crisis, "they must content themselves with a modest profit" or, if necessary, take a loss that would be compensated in some form or another. Until well into the nineteenth century, bakers had no right to quit the profession, especially under duress, without permission. Normally the authorities required a minimum of three to six months' notice, though in practice two weeks sufficed on occasion.[8]

Delamare elegantly reduced the "police of bakers," as the regulatory system was called, to the triple objective of "having good bread, having enough of it, and having it at a just price." Much of this police pertained to very prosaic matters that did not brusquely and directly imperil the commonweal. For example, bakers had to mark their loaves with their initials, with the initials of their market if they had an assigned stall, and with a simple code indicating weight. Police sentences periodically reminded bakers that they were obliged to file an "imprint" of their marks with the commissaire of their market. The real weight of the loaf had to correspond to the mark, and bakers had to provide scales, conspicuously placed in shop or stall, to enable resolute consumers to verify the claim. Market bakers had to post the number of their assigned place prominently on their stalls to facilitate consumer complaints. The bread had to be properly baked ("the correct degree of cooking") and of good quality ("their bread should not be made from rotten flour that would harm those who ate it"), a test that consumers were well placed to administer themselves. Bakers could not exercise certain professions judged "incompatible" with their function, in particular speculative grain and flour trading. All of these specific injunctions, however, were in the service of the larger, all-encompassing goal of controlling the price.[9]

The police applied the carrot and the stick to the bakers in varying mixes and doses. La Reynie made a point of praising them effusively and publicly when they showed sustained "obedience," especially in regard to price guidelines. Sartine tried to reward them for this docility with slight immediate increases or deferred but concrete pledges of adjustment. Like every lieutenant general of police who had to manage a crisis, d'Argenson subsidized zeal and abnegation. He distributed provisions of flour and monetary payments to both the weakest bakers, especially the fragile operatives in the bread markets, and submissive bakers who set the right example (he once issued a "supplement for the bakers who reduced their *bis* loaves from two sous three deniers to two sous for the poor"). Sartine followed the advice of an insider by playing off the "little bakers," whom he supported financially, against a powerful albeit small coterie of "grands," for whom he adduced a veiled contempt.[10]

Recalcitrant or rebellious bakers faced a range of collective and individual sanctions. D'Argenson, Marville, and Sartine threatened to set up public ovens and compete with the bakers directly for clients if they failed to increase or sustain their supply. In 1726 bakers worried aloud that Hérault's plan to organize a vast storehouse for the capital would lead to a rigorous taxation program. ("The government wants to hold us by the teeth," they complained.) By giving wide publicity to baker violations/ignominies—for example, posting two hundred notices throughout the city exposing a short-weight forain or 150 notices recounting the adulterations of a master—the police subjected the malefactors to public stigma. More rarely a baker could be humiliated by being locked in the pillory or being obliged to march through the market with a sandwich-board sign avowing his perfidy ("This Baker unwilling to serve justice, selling his bread above the price set by the police, and at false weight"). The idea was to damage baker reputations, though it involved the risk of imperiling their well-being if the public was sufficiently angered. Women who rioted in the bread markets in 1709 went first after those bakers whom d'Argenson had singled out.

Another way to torment wayward bakers was to hint that they would not be sheltered against the expected "rage of the populace" when they next raised prices—even if that augmentation was authorized. (As late as the 1850s, certain Parisian bakers pleaded in favor of an official price schedule because it protected them from public wrath.) Individual bakers faced hefty fines, characteristically enforced "de corps" in times of stress, which meant that bakers who failed to pay immediately went to jail and/or had their property seized and sold. Bakers could also lose coveted market stalls—coveted in good times and hard to obtain. Home-delivery tolerances could be forfeited. Shops could be walled for various periods, extending to two years, and/or stoves could be dismantled. Grave offenses could result in expulsion from the guild; provocatively mutinous bakers went directly to prison on royal orders, without the option of buying their freedom through the payment of a fine.[11]

The police of bakers suffered much less savage criticism than the police of grain and flour during the rise of the liberal tide in the second half of the eighteenth century. For the physiocrats, the bread question per se was a secondary matter. Still, the privileged situation of the guild bakers, according to the *économistes,* significantly exacerbated the deleterious impact of rising prices on the lives of consumers. The proverb "three dear years will raise a baker's daughter to a portion" could have been the shibboleth of these reformers. Their general hostility to the baker establishment enabled the physiocrats to assert that they were not, after all, indifferent to the (short-term) fate of urban consumers.

The most pointed attack on the police of bakers emanated from Dr. Malouin, author of the first comprehensive treatise on the baker's art. Eschewing theoretical considerations in favor of purely empirical ones, this savant of everyday life pointed to the historical record across hundreds of years as proof that draconian regulation was futile and counterproductive. Without attempting to explain the necessity that he posited as a dialectical law, he claimed that stringency always elicited rampant noncompliance. "One finds, in perusing the experience of France across the centuries concerning the police of bread, that the more one crafted laws against the bakers, the more one spurred transgressions of the law by them," he wrote; "one notes that the more severe were the Regulations on the bakery, the less well it worked."

As an object of police, Malouin argued, bread should be treated in the same way as money, "which the government touches as rarely as possible, paying attention only to verify that neither its quality nor its weight is altered." Bakers should be subject to sanctions only for false weight claims and defective quality, never for putative infractions of price, which should always remain subject to the give-and-take between buyer and seller. Caveat-emptor expectations would arm everyone for a fair combat, Malouin maintained, and spare society the terrible "wars" that often broke out between consumers and bakers. Self-defeating when promulgated with the intention of execution, regulations could be usefully issued "from time to time," allowed Malouin, not with the intention of rigorous implementation but "to serve as exhortation and as prod" to the bakers, who were somehow expected to tremble before a tiger they knew to be toothless.[12]

Parmentier, whose study of the bakery complemented yet superseded Malouin's, sympathized with the physician's attitude toward regulation, but viewed it, in the last analysis, as unrealistic. To be sure, it was shortsighted to "crush the bread maker" in the vain hope that such brutality would "relieve the situation of the distressed." The police should rather reinforce the process of "reconciliation of interests" that market exchange sought to achieve by "natural" means. Nor was it in the interest of social harmony for the police to subject bakers to ritual humiliation as a result of alleged deceits such as short-weight bread that could very well have issued from "good faith undermined by circumstances" rather than "premeditated fraud." Such administrative flamboyance, according to Parmentier, merely socialized the people into the dangerous belief that bakers were inherently inclined to cheat and harm them. Prudent authorities would confine themselves to punishing errant individuals, reproved as such, rather than casting "animadversion" so that it would "spill over onto the entire baking corps," stereotyping, stigmatizing, and discrediting it in the susceptible eyes of the public. Such irresponsible measures created a climate in which the bakery became "a labyrinth of trouble, fear, and humiliation." This state prevented it from perfecting its art and producing more, healthier, and cheaper bread.

Nevertheless, concluded the pharmacist, it was naive to imagine that the bakery could or should be spared incessant administrative monitoring, "because the matter which is the object directly concerns the life of the Citizens and the public peace." The fact of the matter, Parmentier acknowledged, was that the slightest fluctuation of price in a stressful period could "occasion alarms and disorders" that the police understandably sought to prevent. For regulation to work in everyone's favor, he suggested, the bakers would have to accept the fact that "this commerce will always remain under the safeguard of the laws," and the police had to have the good sense to apply those laws with discernment and discretion.[13]

Quality: Right and Obligation

In the eyes of the consumer public, it was no less imperative that the police assure that bread be of good quality than that it be available in ample quantity. The quality issue haunted the authorities through all the permutations of the famine plot persuasion. Just as it was widely believed that elements in government at various levels plotted to create and/or sustain shortage, so it was charged that authorities tolerated or even sponsored the production of inferior loaves for crassly venal reasons. The stubborn inclination of the consumers of 1789/93/95 to hold their commissaires and deputies and ministers responsible for myriad forms of (alleged) adulteration and corruption of the basic food supply was merely the culmination of more than a century of chronic suspicion and denunciation. Without in the least acknowledging the responsibility of officials, Controller General Terray readily conceded that "the bad quality of bread" was one of the principal causes of subsistence revolts in the civil (and not just in the prison) population. Even when the fault redounded directly to the bakers, no one questioned the legitimacy of the public outcry against the custodial authorities: thus the plaintive and angry voice of "a woman of the people" raised "against the magistrates, accusing them of negligence" in the market at Aix-en-Provence in 1773.[14]

Preoccupation with adulteration was by no means peculiar to France, though it did seem to be primarily an urban phenomenon. In the 1750s a London physician published a vitriolic attack against "the confederacy of bakers, corn-dealers, farmers and millers" entitled *Poison Detected*. In it he affirmed that "our bread is mixed with the most noxious & mobiferous matters," including chalk, lime, alum, and bone ash. Shortly afterward another physician, Joseph Manning, denounced by bakers as "the mad doctor," reiterated many of these charges in *The Nature of Bread, Honestly & Dishonestly Made*. Despite the publication of potent refutations, these accusations were echoed in the specialized French press, where they resonated with many deep-seated beliefs about the perfidies of meal manipulation and the hecatombs caused by "adulterated bread." The *Journal économique* conceded that the incorporation of bean flour did no harm (though London chemist Henry Jackson, a merciless critic of Manning, warned that "those who eat such bread are very apt to f——t [*sic*] soon after, and the more such admixture the more violent the explosion"), but noted that the other additives "tend manifestly to alter the temperament of the persons who eat it [the bread so altered], and often cause sudden death."[15]

Paris experienced no great bread-adulteration scandal, though on a number of occasions individual bakers and mealmen were accused of mixing into their dough potato flour, bone ash, beans and peas, wood chips, and "dust," perhaps including sawdust. (It was much more common for Paris subsistence dealers to cheat with substandard or debased cereals of various sorts than with frankly extraneous agents; and the relatively rare apocalyptic muckraking literature denouncing "poisoned bread" tended to hold rotten grain responsible rather than adulteration per se.) Rumors in September 1789 were rife that certain bakers were selling bread "mixed with lime." Not surprisingly, "the bad quality of bread" was one of the vehement leitmotifs of the women's march to Versailles the following month ("We'll make the Autrichienne swallow it [the rotten bread], and we'll cut her off at the neck," an inauspicious augur for the "good" early Revolution). The anti-Revolutionary writer Rivarol denounced the generalized, and presumably false, "accredited rumor" of bad quality grain and bread. Yet, "alum, soap, and flour in array, / such is the bread that you eat every day" remained a standard satirical refrain well into the twentieth century, intoned, among others, by bakers. In the middle of this century the adulteration story line seemed to come full circle. Scientific experts in France urged the addition of "a certain quantity of lime" to bread flour in order to counterbalance the decalcifying effects of a substance contained in bran that could not be eliminated in milling.[16]

"Persons who eat a good bread," wrote one of France's first political economists in 1600, "are always healthier, stronger, and more lively." From the perspective of public administration, he might have added "more submissive." The urgent need to protect the health and longevity of the consumer-people, and thus of the government that ministered to its needs, is one of the leitmotifs of Delamare's *Traité de la police*. The keystone in the regulatory arch remained the sixteenth-century ordinance requiring that "bread be without [foreign] mixture, well composed, fermented, and worked, as is suitable." The major provisioning role subsequently assumed by the faubouriens and forains incited Delamare to add the "special" injunction that prohibited them from using "bad or rotten flour, rewashed wheat, or reground bran," the latter sanction seriously hindering the rationalization of milling techniques. Not content to confine himself to judicial theory, Delamare cited specific recent cases to demonstrate how severely refractory bakers would be punished (e.g., for employing good foreign grain to mask the defects of native wheat, a baker had his oven demolished and his shop walled shut for half a year).[17]

The police commissaire articulated three criteria for ascertaining the quality of bread. First, the flour kneaded by the baker had to be of "good grain," neither "bitter" nor "overheated, fouled with odor, or spoiled." Second, the dough had to be "well risen." Third, the loaf had to be "cooked appropriately." Delamare maintained that the alert consumer could easily test for the first and third properties: "taste and sight readily reveal the one and the other of these defects." Such transparency did not apply, however, to the fermentation process, "for the worst [bread] is sometimes the one that projects more of an air of excellence." It was here that the police, assisted by the savant counsels of the medical and scientific academicians, could intervene to protect the public by establishing the standard procedure for salubrious and successful bread making. (For Delamare, this standard should have preempted the ravages that "molletization" would later perpetrate.)[18]

Though they may have been tempted to exploit opportunities for quick profits, or occasions to cut losses, most bakers did not underestimate the cost of a damaged reputation. It was more likely careful calculation than hot blood that induced master Laurent Lapareillé to challenge to a duel a neighborhood bourgeois de Paris who openly questioned the quality of his bread. Confrere Nicolas Eloy blamed dishonest flour suppliers for causing him to end up producing "very bad bread that discredited him and cost him the custom of his best houses," a loss that served as prelude to his business failure. "Destroyed by his own bad bread," Eloy reminded other bakers that price alone could not dictate their purchasing strategy. Forain Louis Hiest suffered a similar experience. During his absence, a supplier delivered "tainted flour." Either indifferent or insufficiently skilled, his journeyman did not realize the peril, and baked five hundred eight-pound loaves, primarily for workers on a large construction project. The bread proved unfit to eat, Hiest lost this major contract, and the rumor emerging from this incident took care of the rest. Another baker endured catastrophic defections because of a rumor that his well water was contaminated by the sewers, thus infecting his bread. The police warned bakers in the faubourg Saint-Jacques to avoid kneading during the hours of the removal of fecal waste because the potent, mephitic odors could ruin their loaves. (Mercier despaired for all the bakers in the faubourg Saint-Marcel: "bad air, thus bad bread.") In the fall of 1793, not only did Parisians rail furiously against bakers of "very bad bread," but they boycotted them despite the penury: "the throng of buyers is enormous at the shops with a reputation for making better bread than the others."[19]

Short Weight

Put a baker on the scales after he dies, runs the adage, and he will be found to be of short weight. The sale of bread light in weight was one of the most common offenses committed by bakers. It was "an almost general fraud," observed the author of a late-eighteenth-century Parisian guidebook. "To the shame of our bakers," remarked a Parisian in 1791, "one rarely finds a loaf [in Paris] that weighs its [announced] weight." To epitomize a baker's dishonesty and dishonor, consumers called him "a seller of underweight bread." Nor was the problem limited to the Old Regime. "Is it possible in a level of civilization such as ours," asked a member of the Parisian city council in 1859, "that when one wishes to buy a kilo loaf, . . . one has great difficulty in obtaining the right weight?"[20]

Twenty years later an editorialist in the *Petit Parisien* evoked the children's riddle, which weighs more, a pound of feathers or a pound of lead? It was not hard to fall into the trap of responding that the lead was heavier.

But if we are asked, what is the difference between a pound of household bread [*pain de ménage*] and a pound of white bread, wouldn't we be correct in replying: they have this in common, that neither one weighs a pound?[21]

In Paris the loaf was not weighed for sale. It was identified by a mark as weighing a certain amount and sold at a price reflecting the announced weight. The police regarded fraud on weight as a very serious infraction. Intolerable in itself, dishonesty was considered dangerous as well as reprehensible in matters such as provisioning that involved the "public good." It undermined the confidence that consumers had in the authorities as well as in the bakers. No wonder the police did not hesitate to characterize short weight as "a theft against the public" motivated by greed and ill will.[22]

The "theft" struck poor consumers harder than rich ones and was therefore an even graver offense that risked upsetting the delicate social balance of everyday life. Commissaire Dupuy denounced short weight as a sort of extortion practiced on "the downtrodden dayworker who had five or six children to feed." Bakers would quickly deny him a bread if he lacked a liard but would not hesitate to cheat him of two liards by selling him a short-weight loaf. The Paris Parlement called for weekly inspections of the shops and markets, and successive lieutenants general of police pressed sometimes complacent commissaires to show no indulgence to wrongdoers. There was no excuse for short weight, especially if it was "chronic," "exorbitant," and more likely the result of artifice than of accident. Even the *économiste* Mirabeau, a harsh critic of police regulation, called for tough repression of this abuse.[23]

Diderot told Empress Catherine the story of a Turkish cadi whose dinner was interrupted by the news that officials had apprehended a baker for selling a short-weight loaf. Recounted the cadi:

I went to his bakery. I had his bread weighed and found it light. His oven was still red hot. I had him thrown in, and my business was finished.

The cadi went on to explain to his shocked dining companions the rationale for his ostensibly cruel gesture:

His theft was a public theft that fell upon the most miserable portion of the people, those who buy their bread by the pound. You approve of the fate of the thief who robs the safe of a financier and is broken on the wheel, but you don't want me to burn the criminal who robs the bread of the poor.[24]

Parisian justice was far less brutal, though one could not infer this from the outraged protest of convicted bakers. The most severe statutory punishment for short-weight violation was the loss of one's mastership—a penalty applicable to only one category of bakers. The lieutenants general of police frequently threatened recidivists and brazen offenders with this mortification, but I have found not a single instance of execution. Closing and "walling" the bakeshop was the next harshest sentence. Those bakers who did not keep shops, forains and some Saint-Antoiners, obviously were unaffected by this sanction. The worst-case scenario prescribed permanent walling. In practice, it was usually for a fixed duration, long enough to dislocate the baker's business and compromise him financially.

Baker Goutard, a repeat offender who must have thought that his commission as a supplier to the royal court would exempt him from the common obligations of the trade, repeatedly sold unmarked and short-weight loaves in his shop and at his marketplace booth. His shop was ordered walled for a period of three months after he verbally abused Commissaire Sautel and struck his assistant with a sword as they sought to examine his bread. The receiver of fines was to advance the funds to pay for the walling so that it could be done immediately. Goutard would have to reimburse the receiver, and acquit a fine of two hundred livres, before his shop could be reopened at the expiration of the penalty period. The lieutenant general ordered baker Paulmier's shop walled for six months and fined him five hundred livres for "false weight" on twenty-five round and long mid-white loaves marked two to four pounds (light by one-half to four and a half ounces) and one four-pound dark loaf (light an ounce). In addition to a fine of one thousand livres, master baker Leroux suffered a six-month walling for baking forty underweight loaves, his second conviction.[25]

The "ordinary" penalty pronounced against all categories of bakers was the fine. The range of magnitude was staggering: from three livres to three thousand livres. For ninety cases scattered across the century, among all sorts of bakers, the mean fine amounted to 175 livres and the median to fifty livres. If one eliminates the five stiffest fines (ranging from 1,500 to 3,000 livres), the average sinks to seventy-three livres and the median remains unchanged. The "categorical" ranking of penalties is about what one would expect. Masters were fined most heavily, because they were believed to be the wealthiest and because they were supposed to set an example. They were followed by the Saint-Antoine bakers, who were assimilated to the corps of masters or to the itinerant contingent depending on the amplitude of their commerce. Simple bakers and forains shared the same degree of punishment.[26]

During the first half of the century, for thirty fines against masters ranging from five to two thousand livres, the mean was 325 livres and the median one hundred livres. If one eliminates the four biggest, the average drops to 125 livres while the median stays the same, within a range from five to five hundred livres. For seven cases after 1750, the mean and median were respectively forty-nine and fifty livres within a range of three to one hundred livres. Twenty simple bakers paid fines averaging twenty-six livres (median = 20 livres) in the first half of the century. Sentences pronounced against twenty-three Saint-Antoine bakers ranged from ten to three thousand livres, yielding an average of 211 livres (median = 50 livres). The mean falls to eighty-four livres (median stable) if one eliminates the extreme entry, setting the range from ten to five hundred livres. Seven forains suffered the least imposing fines, averaging twenty-seven livres (median = 20 livres).

The sentences inflicting fines and other penalties were pronounced by the lieutenant general in his police chamber, usually a few days after the commission of the infraction, often in the absence of the defendant ("defaulting"). But one punishment that befell virtually every baker caught with short weight was the immediate confiscation of his bread. For most other purposes, seizures of merchandise or property effected by the police were placed under seal awaiting final disposition by the magistrate. Short-weight bread was a different matter: here justice was expeditious and without appeal. The commissaire who issued the summons usually cut up the bread in a shaming ritual and distributed it immediately to "passing poor" or to a hospital, a convent, or a prison. This "military execution of the bread," as one observer styled it, was meant to send a reassuring signal to the public and a message of dissuasion to the bakers. This was the closest that eighteenth-century practice came to the old rite of public humiliation and expiation in which convicted short-weight bakers, bare of head, had to carry lighted two-pound candles (certified *poids de marc!*) to Notre-Dame "to beseech the pardon of God, the king, and the judicial system."[27]

One cannot always discern a correlation between the enormity of the offense and the level of punishment. In June 1726 a guild widow was fined thirty livres for twenty-four loaves of mixed sizes and shapes found to be short in weight. Thirteen years later a commissaire visiting the shop of baker Petiteau uncovered thirty-one loaves—all of it *gros pain* above four pounds—that weighed too little, including several twelve-pounders short almost a pound each. Petiteau sustained a fine of fifty livres. Yet, for very similar violations, five bakers in April and June 1742 suffered much heavier penalties. For thirty mixed breads averaging a little over two ounces in short weight per loaf, master Royer was fined two thousand livres. Master Charon was condemned to pay fifteen hundred livres for twenty-seven loaves of about the same degree of short weight. The fact that they were hidden in the bottom of a counter in his oven room did not strengthen his plea of innocence.[28]

Baker Lepage of the faubourg Saint-Antoine reeled under the blow of a fine of three thousand livres. In addition to short weight, he was guilty of having no scales or weights and of selling his luxury bread at prices above the established level. The magistrate levied fines of fifteen hundred livres against master bakers Boulanger and Leroux, the first for eleven mixed breads averaging less than two ounces in short weight each and the second for thirty-five breads, mostly long loaves of four pounds and round loaves of eight pounds, averaging almost four ounces in short weight. I have found no evidence to substantiate the claim of contemporary commentators that the police generally allowed a "tolerance" of one ounce per pound in baking loss. On numerous occasions authorities confiscated loaves lacking an ounce or even a half-ounce.[29]

These apparent inequities might perhaps be explained by factors to which we are not privy—recidivism, infamous reputation, the aggressiveness of police officials, and so on. Surprisingly, sanctions imposed in crisis periods do not seem to have been characteristically more unsparing than those levied in relatively easy times despite the fact that an offense committed in a dearth was always considered to be much more threatening and hence worthier of stern punishment than the same offense committed in "normal" circumstances. All fines over fifteen hundred livres were imposed in noncrisis moments. Of all the fines over one hundred livres in our sample, nine were levied in crisis periods and fourteen in noncrisis times. The authorities generally expected that bakers would seek to "indemnify" themselves on the weight for what they lost in profit as a consequence of stringent price controls during dearths. It may be that the police recoiled from rigor during crises for fear of driving bakers out of business when they were most keenly needed. It was easier to be draconian when the stakes were less elevated, though one wonders whether the baker victims of this severity actually acquitted their fines. Forains were generally "treated gently," as the officials put it, to avoid dissuading them from returning.[30]

It is safe to say that we have touched only the very tip of the iceberg. Defenders of the bakers admitted that short weight was rampant, though they insisted in the same breath that it was accidental and inevitable. In ninety-one cases I have examined, the police confiscated 2,337 loaves, the vast majority four-pounders and above. They were short a total of 9,633 ounces, representing over 150 four-pound loaves. It would not be unfair to estimate the total "daily loss" to Parisians from short weight at perhaps between twenty-five hundred and three thousand four-pound loaves on an ordinary day, substantially more on a market day—enough to feed several thousand families (allowing a liberal four-ounce "tolerance" for the baking of a four-pound loaf). Or a loss to Parisian consumers (and a gain to bakers) of something on the order of 27,500 sous (or 1,375 livres). Given the standard of living of most Parisians and the prevailing conceptions of economic regulation and social control, these figures were understandably significant in the eyes of the police. The four largest lots of bread in our sample were confiscated from forains at the markets: 402, 350, 302, and 130 loaves. (The fifth biggest batch belonged to a master who was also a juré.)[31]

Bread was supposed to be marked with its weight and with the initials of the baker to assure his accountability. Many of the bakers guilty of short weight also failed to mark their loaves. While they could blame the first of these two infractions on forces beyond their control, it was more difficult to explain away the second. When queried by a commissaire as to why her master did not seal his loaves, a *porteuse* (deliverywoman) casually replied "that all the bakers do not mark their bread." One reason for anonymity was of course to avoid detection in case of a problem. A less obvious but increasingly important reason for the absence of a mark was the fact that city bakers made more and more of their bread from soft (*molle*) or mixed (*bâtarde*) dough in which the mark did not fix permanently as it did in the bread of firm (*ferme*) dough. The penalty for unmarked bread was usually encompassed in the sentence for short weight rather than explicitly stipulated. For unmarked bread of proper weight, the usual fine was five livres; the statutory maximum was two hundred livres. Like short-weight loaves, unmarked ones were confiscated "to the benefit of the poor."[32]

More often than not, the police uncovered short weight in making their regular rounds of the shops (frequently conducted on Sundays) and bread markets. The jurés were also empowered to visit and denounce bakers for "false weight." The authorities preferred to see them accompanied in their rounds by a commissaire whose presence blunted the mission's corporate visage in favor of its juridical mandate and reduced the likelihood of flare-ups of jealousy or animosity between the baker leaders and the rank and file. The police also wanted to temper the zeal with which the jurés usually pursued the forains, whose loaves they almost never found to be heavy enough.[33]

Many bakers were denounced on the initiative of ordinary citizens—consumers—rather than professional monitors. They came from both ends of the social hierarchy. Upon returning from the Halles market, Marie-Jeanne Gadoue, a servant girl, weighed her six-pound dark loaf and found it three-quarters of a pound short. She rushed to Commissaire Courcy who confirmed her charge, inspected the baker's stand, and issued a summons. Another servant buying at the Halles stopped to weigh her four-pound loaves at a grocer's shop and found them each four to five ounces light. She got Commissaire Guyot to return with her to the market stall, where he discovered fifteen of the sixteen remaining loaves to be short.[34]

From the other side of the stairs the wife of a royal official noted that her household was consuming an implausibly large amount of bread. Her suspicions led her to have weighed the bread that master baker Moudeau delivered daily. She told Commissaire Machurin that the four-pounders weighed only three pounds. He convoked the baker, who arrived "in a fury," denied the truth of the allegation ("I bake three-pounders only for the taverns, not for ordinary customers"), cursed the woman ("whore," "slut"), and struck her even after she warned him that she was pregnant.[35]

That sort of violent reaction and resistance was not a rare occurrence. For reasons that we shall shortly examine, the bakers regarded the short-weight indictment as wholly unfair. And they deeply resented the "military execution" style of rendering immediate justice. Like baker Varlay, who begged Commissaire Roland "to leave the said eleven [short-weight four-pound] loaves in his keeping, for [a public distribution] would cause too much scandal and cause [the Varlays] to lose their credit in the community," many bakers worried about their reputations. When Commissaire Saulet showed up at his shop, master Goutard "raged like a madman," refused to allow his bread to be weighed, and obliged the commissaire, whom he threatened, to call the watch. Another baker first tried violence, invectives, and blows—and then offered a bribe to avoid all sanctions. A journeyman baker attempted to run away with a basket of bread on his back. Once captured, he refused "insolently" to give his name or that of his master. Nearly all of his loaves were short; the police assumed that he was privy to his master's "malfeasance."[36]

The least convincing argument deployed by the bakers in their defense was that short weight resulted from the perfidy or incompetence of their journeymen. Commissaire Roland found a considerable number of short-weight loaves in the shop of baker Félix on the rue Saint-Jacques. "Shocked that her bread did not weigh what it should, for it had never happened before," Félix's wife angrily concluded that it must be "the fault of her workers." Master baker

Pierre Hamart told Commissaire Regnault the same thing. Such alibis, true or false, made no difference, for the authorities always held the employers responsible for the work of the employees. The police model could have been master Jean-Claude Desmeures, who personally supervised his three journeymen in the bakeroom till midnight "in the task of weighing the dough and forming the loaves in order to conform to the ordinances and regulations of the police."[37]

Bakers who frankly avowed their technical helplessness to guarantee the right weight presented a much more compelling case. Master baker Estienne Mongueret protested that he should not be held responsible for short weight that he had not at all intended. It resulted from "an overbaking plus the heat that dries out [the bread] rapidly." In similar terms, his colleague Claude Ediere affirmed his innocence: "that [short weight] is due to the fact that the bread was overbaked, and then staled quickly." If one is to believe the bakers, the constraints were ever so slow to change. A century later baker Jacob made precisely the same argument about his inability to foresee and control the weight.[38]

The bakers vented their frustration in a collective petition addressed to the lieutenant general of police in 1743. They insisted on the extreme variability of the baking process. Since they were not alchemists, it was absurd to ask them to master "the four sovereign elements" of nature. The first was the *earth,* which produced grain (and then flour) of drastically different grades that behaved differently in every batch. Second was the *air,* which mediated the fermenting and proofing stages and ensured that these transformations were never exactly the same as the previous time. The third element was *water,* which combined with the flour in a marriage that on each occasion was truly unique. The fourth was *fire,* the heat of the oven, which constantly varied in intensity and was always unevenly distributed within the oven.

Since these elements combined in wholly unpredictable ways to produce different results at each baking, the bakers avowed that they had only one recourse to "correct the inequities": sale by exact weight. They wanted the right to protect themselves from the "tyranny" of the commissaires (who, they charged, sometimes acted out of "vengeance" and with the aim of "parading their authority") and from "the ill humor" of certain consumers motivated by "bad intentions" (who dried out their freshly purchased loaves over the fire in order to fabricate pretexts for complaint). The only way to prevent such "harassment" was to offer bread for sale on the scale: loaves lighter than marked would be reduced in price or supplemented in weight proportionately. The bakers could not understand why they were "the only trade excluded from the faculty of using scales and weights for the marketing of their merchandise."

This exclusion exposed them to "calumnies" and to "economic ruin." Scales would enable them to "recover their honor." A number of years later, in the first "scientific" study of the bakery, Dr. Malouin vigorously supported the baker argument. Reform of the police of bread had to begin with the conversion to sale "on the scales."[39]

A generation later, in 1778, the bakers renewed their demand in a strikingly similar petition to the same magistrate. They emphasized how deeply distressing it was for them to be held responsible, under the menace of humiliating and sometimes oppressive punishments, "for a precision that did not at all depend on their will." At this very moment the chemist Parmentier took up the bakers' cause.[40] It was up to Science "to bring to a definitive end this war of sorts that rages everywhere between the Police, the People, and the Bakers." Parmentier hoped that an "impartial" and "objective" assessment issued by an expert like himself could persuade the authorities to change their policy. While he criticized the bakers on many other issues, he espoused their position on short weight without reservation. A "shortage of weight" by itself "absolutely did not mean fraud." All the scrupulous care in the world could not prevent "the multitude of accidents that caused weight to vary infinitely from place to place and moment to moment," accidents contingent on air temperature and humidity, grain quality, milling procedure, water source and temperature, nature of yeast, size and arrangement of bakeroom, kind of oven, nature of wood, and so on.

Not even chemists and physicists could prescribe a foolproof method, affirmed Parmentier. Prudent bakers added a dough "bonus" to each loaf—ten ounces of *bon de poids* for a four-pounder, twenty ounces for a twelve-pounder, and so forth. Yet they, too, often found themselves victimized by unforeseeable "losses" in the oven and afterward. Nor would it be reasonable to ask bakers to further "load" the bonus because they operated on such narrow profit margins.[41]

Parmentier reproached the police for treating bakers unfairly. Chances are that the commissaire did not understand the enormous uncertainty of the baking operation. The loaves he chose to weigh at random could have been the first put in the oven and thus the last removed—probably, then, among the breads that suffered the greatest diminution. The draconian posture of the police could compromise the quality of the bread (and thus jeopardize the well-being of consumers) by inducing bakers to undercook in order to conform to weight standards.[42] Nor did the authorities ever give bakers "compensatory" credit for all the loaves that turned out to be a few ounces above the mark. Parmentier charged that by cutting up seized loaves and distributing them immediately, the police deprived the bakers of due process. The commissaire choreographed a scene that could only "stain the baker with dishonor," "discourage him," and invite public "insult."

Parmentier's solution was to call on "the most powerful mediator"—the scales. He ironized over "the bizarre and contradictory custom whereby bakers were forced to keep scales in their shops and forbidden to use them to sell their bread."[43] Let us put these scales to use, Parmentier proposed. Sale by weight would render special service to the laboring poor, whose "bread purchases drained the bulk of their budgets" and for whom "every ounce mattered." It would simplify police monitoring and restore a climate of confidence between the public and the bakers.

The "only possible objection" to the system envisioned by Parmentier was that bakers might add larger amounts of water to the dough and underbake, thereby increasing the weight of their bread at the expense of its nutritional value. Technically, Parmentier suggested that this fraud would be hard to realize because there was a threshold beyond which the dough simply would drink no more water—and ordinarily bakers were close to it. Moreover, caveat emptor! An alert customer would reject an underbaked loaf. And the police would continue to inspect for proper levels of baking. The *economiste* Baudeau and the physician-chemist Malouin, each a specialist in his own way on subsistence affairs, agreed wholeheartedly with Parmentier's analysis.[44]

In the fall of 1781 the bakers once again petitioned the lieutenant general of police to reconsider. This time Lenoir appointed a panel of experts, led by Tillet of the Academy of Sciences, to study the question.[45] Beginning in early October, Tillet arranged for a series of test bakings in the ovens of the Ecole de Boulangerie. "Despite all the precautions that we took to obtain four-pound breads of equal weight," reported Tillet, "we succeeded only with a few loaves." Tillet could not blame his failure on inept or indifferent journeymen. Nor did it result from *baisure,* the colorfully named phenomenon of sticking together that resulted from packing breads too tightly in the oven. Long loaves with more surface exposed lost more weight than more compact round shapes.

But Tillet could only speculate as to the reasons for the diminution of forty-four of his fifty four-pounders: perhaps the escape of "a gaseous air" or "a watery vapor." He concluded that the most adept "art" did not permit the baker to triumph over the intrinsic difficulties. The law had to be changed, Tillet esteemed, not merely because it was "impracticable in its rigor" and unjust to the baker but above all because "it did not accord with the facts of experience." Tillet endorsed sales by weight on the scale as the only workable solution. He acknowledged the risk of a caveat-emptor model: "used to being looked after," Parisians might not protect their interests as well as the police had. Yet he intimated that it was time for them to learn to take care of themselves. Meanwhile, he and other scientists would work on procedures that might enable bakers to regulate the heat of their ovens.[46]

Apparently Lenoir was not ready to devolve responsibility for policing the bakers upon the consumers. He called for further discussion even as he continued to implement the rules that Tillet had declared scientifically discredited. The bakers again demanded sales by weight with scales in a memorandum published in their behalf in 1789. They cited repeated tests showing that sample batches prepared together gave "disparate and uneven results in weight." In that same year the assembly of the district of Filles-Saint-Thomas received observations from a citizen arguing that the best way to end daily disputes over short weight was to authorize the bakers to sell by the pound and on the scale, and a leading newspaper announced that the bakers of Versailles were thenceforth authorized to sell by the ounce and the pound. At virtually the same time a Paris crowd threatened to hang a baker denounced for selling short.[47]

In the aftermath of the Revolution, Grimod de la Reynière, an enthusiastic patron of gifted makers and purveyors of food, complained angrily that despite "enormous profits," Paris bakers "still steal from the public with effrontery." In his famous food promenades, "we frequently found loaves of three pounds lacking nine to ten ounces and breads of two pounds that weighed only one pound nine ounces." Grimod deplored the "indifference" of the police to these "daily infidelities." In November 1840 a police ordinance, in response to reiterated pleas on the part of the bakers, ordered sales to be conducted by weight. To judge from the outcry of bakers registered in the *Enquête* of 1859, this new policy did not last long or was not faithfully implemented. On 14 November 1867 the prefect of police promulgated an ordinance instituting the sale of bread "at the weight conjointly verified by seller and buyer."[48]

Policing the Forains: Fugitive Bread

Though the poorest consumers, as we shall see, often waited until the end of the day to buy, sales were very brisk and haggling keen in the morning. Till noon, bakers were free to fix their own price in relation to demand (save in crisis time when the police imposed a ceiling). After 12:00 P.M., the market bakers could no longer raise the price, regardless of demand, above the "current" or "common" price that the morning transactions had established. Then after 4:00 P.M., the regulations required all market bakers to offer their bread at a discount, below the lowest price of the day. Bakers were not allowed either to store unsold loaves near the market or to take them back home. Nor could they delegate responsibility for market sales to anyone other than a family member or servant, a stricture that was contrived to encourage them to "let go a bit on the price" if they wanted to get home at a decent hour. Many bakers, facing a long trip home and anxious to leave promptly, offered "rebates" earlier in the afternoon. Thus, in principle, the buying day was socially stratified, the poorest waiting until last. But waiting could be very risky, for the markets sometimes sold out more or less completely by noon. Psychologically impatient, the poor were on occasion too hungry to wait until the end of the day. Many preferred to take their chances haggling while bread was still visibly abundant first thing in the morning.[49]

La Reynie, the first lieutenant of police, told Delamare that the surest way to guarantee that cheap bread would be available at the markets was to prevent the bakers from taking home unsold bread or storing it for the next market. This was the Old Regime analogue to the Revolutionary conviction that to allow large amounts of bread to leave the capital was "to organize famine in Paris." The market was the point of no return. Once it made its way there, bread was no longer the unqualified and inviolable property of the baker. Consumers had the right to claim *all* the bread displayed. La Reynie counted on the obligation to dispose of all the bread to force the baker to moderate his demands and, in the afternoon, to accept sacrifices. Theoretically, the baker with bread at the end of the day was in an utterly untenable situation. He had to dispose of it somehow or he would be in violation of the law. Better sell it cheaply than have it confiscated: that was the message of the police ordinances. The commissaires were empowered to force the sale of remaining bread at prices low enough to attract buyers, though they rarely resorted to this authoritarian measure.[50]

To assist them in their police of the markets, the commissaires depended heavily on the jurés of the guild. The corporate structure was an extension of the police apparatus. And this particular surveillance was only one of many police functions entrusted to the jurés. The prohibition against taking home or storing market bread was not contained only in the ordinances and sentences of the Châtelet. It was also written into the guild statutes. Bakers who failed to conform thus violated corporate as well as public law, though the masters regarded the measure as pertaining primarily to the forains, since the city bakers could reasonably move bread back and forth between shop and market without arousing suspicion. Not uncommonly masters stocked their market stalls with modest amounts of bread and sent to the shop for further supplies if demand proved to be brisk.

The jurés were quite zealous in fulfilling their responsibilities. In part this was a matter of self-image and sense of duty: as auxiliaries of the police, they took the matter of public order seriously. In fact, it was probably in the interest of the guild bakers to allow the outside bakers to take back their bread, for unsatisfied customers might be obliged to turn to the shops before the next market day. Yet the motives of the jurés were hardly altruistic. They deployed their police powers to control and subordinate noncorporate bakers. They intimidated and tormented them whenever the opportunity arose—usually in the name of the law. On Wednesday and Saturday afternoons, the jurés staked out the markets and prowled the streets leading to the city gates (especially the porte Saint-Antoine) searching for bakers or their agents who were taking bread home. Usually accompanied by a huissier from the Châtelet, the jurés stopped and challenged persons who appeared suspicious. If they discovered contraband bread, they took the malefactor to the nearest commissaire for arraignment. The commissaire confiscated the bread and issued a summons requiring the accused to answer the charges in the chamber of police.[51]

The persons most often caught in the jurés' ambushes were *the porteuses,* the women who delivered bread for the bakers and helped out with shop and market chores. Some submitted passively to the search and seizure orchestrated by the jurés. Others, more hard-nosed and devoted to their bakers, resisted in one way or another. Margueritte and Marie-Françoise Simon, probably sisters, caused a near riot by hurling their bread baskets to the ground and refusing to accompany the jurés to the commissaire's headquarters. A large crowd gathered quickly and the jurés had to call on the guard to escort them. When challenged, a porteuse from the Marché-Neuf with nineteen pounds of bread and two porteuses from the Cimetière Saint-Jean with thirty pounds between them claimed that the loaves were for their families' consumption. The jurés questioned their "sincerity" and suggested that they were merely trying to cover for their employers.[52]

Armed with twenty-two four-pound loaves, Marie-Anne Chevriot could not pretend that she bought them to eat. Instead, she feigned not remembering her baker's name. When pressed, however, she told the truth. Marie-Anne Tortillon, the thirty-year-old wife of a gagne-denier, was a more resourceful liar. She stated that she had purchased the twenty-six pounds of bread she carried from her employer, Dame Chauvin, a faubourg Saint-Antoine baker serving the Saint-Paul market. When the jurés tried to confirm her story—they took the trouble to hire a coach and drive with the porteuse through many streets in the faubourg Saint-Antoine—they discovered that Chauvin did not exist. Then Tortillon changed her story slightly: the bread was still her own but she did not know the baker for whom she worked and from whom she had purchased it. To the jurés' charge that she had far more bread than she needed for her family, she countered that she was expecting a visit from her parents for Saint Martin's. The jurés called her "a cheating rascal" and demanded that she be imprisoned.[53]

The jurés did not believe that another porteuse, Marie Deveau, bought for her own use from her employer the five loaves she transported because they were all *pain mollet* and a woman of her marginal socioeconomic status would not be nourishing her family on this sort of bread. But what seemed to trouble the jurés even more than the alleged violation was Deveau's contention that her employer was a master baker named Boucher. They suggested that the bread really belonged to Boucher's father-in-law, a baker from the faubourg Saint-Antoine who had no ties to the guild. The fidelity and discretion evinced by the porteuses was not always reciprocated by their baker employers. Convoked by the jurés to confront his porteuse, Martin of the faubourg Saint-Antoine selling at the Saint-Paul market declared "that he did not give the order to withdraw the five loaves in question; on the contrary he expected her to sell them at the market at discount until the end." Martin received a summons nevertheless on the grounds that he was "civilly responsible for the person he employed."[54]

Why did the bakers stubbornly continue to ignore the rules and try to sneak bread out of the markets? In case after case they complained that they could not find customers for the remaining loaves. One porteuse stayed till after 5:00 P.M. vainly trying to dispose of thirteen loaves. She ended up taking the bread home because she had no choice, she insisted. Dame Perrot, the wife of a forain from Bourg-la-Reine, protested sharply that "she did everything possible to sell what remained at a discount, even at a price two sous below the current, without success." The wife of another forain explained that "she found herself obliged" to take home sixty pounds of bread because there was nothing else she could do with them. She could not stay at the market after 5:00 P.M. because she was afraid of "the risks of returning home too late at night." Moreover, she could use the bread to feed her eight children. Yet another forain wife made a similarly pragmatic argument: unable to find buyers, she decided that it would be "useful" to take the bread home for sale in her shop, "which was not furnished sufficiently." Dozens of other bakers expressed outrage when their homebound bread was seized. Their goodwill effort to sell it should have spared them recrimination, they argued.[55]

Failing to find customers, what were they to do with their leftover bread? they asked. Some of them, however, were impatient. One Vincennes baker left the market at two-thirty and another at three-thirty, far too early to justify their contention that sales were impossible. What most bakers meant when they said they could not sell was that they could not find takers at a discount price they judged reasonable. The regulations placed the bakers in a cruel dilemma, for there was absolutely no provision made for unsold bread. The bakers were expected to make sacrifices; it was part of the Delamarist "contract." Unsold bread was prima facie proof of excessive obstinacy on the part of the baker. At six or seven at night, consumers took their revenge on the conscientious bakers. The others were willing to run the risk of summons rather than absorb the loss on price. The penalties perhaps were insufficiently dissuasive. The fines varied across the centuries from fifteen livres to thirty livres and recidivism does not seem to have been more severely punished.[56]

The commissaires were less concerned about petty or impromptu cheaters than about the bakers who seemed to premeditate and organize their transgressions. One woman baker at the Place Maubert arranged fictional purchases by specially recruited buyers who deposited the bread they bought in a grocer's shop that served as entrepôt. Widow Guichot from the Saint-Germain market was part of an illicit baker cooperative that rented a room for bread storage in a fruit vendor's apartment. Another group of bakers maintained a joint storage room near the Pillory. Joachim Roze hid his leftover bread until the next market day in a tailor's shop near the Saint-Paul market. A shrewd Gonessien had an arrangement whereby he consigned his unsold bread to a master who remained longer at the market and who was unlikely to be hampered if he took it back to his shop. In July 1750, the jurés discovered 1,431 pounds of bread that a group of thirteen forain bakers at the Halles had stored with a woman who regularly brought the leftover bread to their stalls for sale the next market day.[57]

The police were interested in making examples to deter other bakers from withdrawing or storing bread, but they had a longer-range view of supply needs, and that view often led them to moderate juré ardor. For example, instead of confiscating those 1,431 pounds of bread uncovered near the Halles and fining the clique of wrongdoers as the guild demanded, the lieutenant general of police let them go with a warning not to commit this infraction again. He feared that "too severe a punishment might induce the forains to quit serving the market altogether and thus compromise the provisioning of the capital." There is some scattered evidence that Sartine inaugurated a regime of tolerance for the removal of unsold bread beginning around 1760. There is no record of consumer complaints against bakers for absconding from the market with unsold bread, save in time of crisis. During the dearth of 1725, buyers in one market compelled a baker to stop for a search at the Richelieu barrier. Instead of bread in her cart, the vigilantes found only turnips.[58]

Illicit Sales: Regrate, Street, Home, Tavern

The aim of the police was to keep the bakers on the market and to keep their bread supply visible at all times. One of the practices the police feared if the bakers were freely allowed to remove bread from the market at will was regrating. Regrating meant primarily selling large amounts of bread to intermediaries who retailed it to the public at a price above the current price. But it also came to mean any illicit sale outside authorized time and space even when made directly by the baker-manufacturer. The first two lieutenants general of police regarded regrating as one of the major causes of recurrent periods of high prices (*cherté*) and repressed it energetically. Given the fact that the vast majority of bakers operated out of shops, and had no need for ancillary outlets, the authorities tended to regard regrating, like withdrawal of unsold bread from the market for storage or return home, as largely forain issues. The jurés, seconded by the commissaires, maintained their vigilance throughout the eighteenth century. A number of forains, usually in collaboration with persons who had a room to offer, established entrepôts that served as shops on market days and usually the day after a market as well. Nicolas Baron always dropped off a supply of bread on the way to the Carouzel market at the shop of Picot the watchmaker, who sold to clients in his neighborhood.[59]

Another forain named Ferry from the Saint-Paul market set up a bread outlet on the rue de la Perle. Ferry was shocked when the jurés seized his bread, for he had been doing this for twenty years. In return for a free supply of charcoal, a candlemaker allowed a faubourg Saint-Antoine baker named Thion to use his shop as a sales base on Wednesdays and Saturdays. Similarly, a Passy forain set up his alternative shop in a fruit vendor's lodging. A baker convicted of this sort of offense suffered the confiscation of all his bread, which could represent a large sum, and a fine usually ranging from ten to fifty livres for a first offense.[60]

There was another form of regrating that was retail in character. The wife of a used-clothes dealer bought substantial amounts of bread from various bakers and set herself up at empty spaces in the Halles market for resale. Viollet was a baker who had a stall at the Halles, but instead of marketing bread of his own baking, he bought bread from various colleagues and resold it to the public. Uncovered, Viollet lost his market spot, received a three-hundred-livre fine, and was threatened with jail if he continued selling others' bread. In this sentence against regrating the lieutenant of police warned the bakers that each one was expected to have his own oven in which to cook his own bread. Mahon brought his own bread to market, but since he baked only *gros* he purchased *petits pains* from a guildsman for resale at his stall. It was hard not to sympathize with his explanation (unless one was a juré): he needed to keep his customers happy.[61]

Another forain violation that authorities also viewed as a form of regrating was the sale of bread in the streets. Like entrepôt sale, this practice diverted supply from the market, thereby placing upward pressure on prices and perhaps causing anxiety among consumers. Jurés called this abuse "the moving shop" and they tracked down offenders vigorously. Given both the police injunction to sell everything during the day and their own desire to sell as large a volume of bread as possible, the more aggressive forains were not content to sell just in the marketplace. While the baker or his wife ran the market operation, an employee went off in search of business in the streets. Forains Picot from Vincennes, La Chambre from Chaillot, Deslions from Gonesse, and Goive from the faubourg Saint-Antoine arranged for their drivers to sell directly from their wagons in the streets and in the courtyards of some convents and hotels of grands seigneurs. The clients objected strenuously when the jurés interrupted the sale and confiscated the bread, which would suggest that they appreciated the convenience and that the price was probably at or very near the going rate. The porteuse of one Saint-Antoine baker sold from a horse that she paraded through the streets, while the porteuse of another carried the loaves on her back, which perhaps gave her greater incentive to sell them quickly.[62]

In 1737, the jurés repeatedly heard reports that the Versailles forains were selling en route to the market and arriving at their stalls "practically empty-handed." They set a trap and caught two bakers selling at dawn from their wagons near the ramparts. Even before they faced the lieutenant general in police audience, these bakers suffered a tremendous loss, for all their bread (2,404 pounds for one and 1,008 pounds for the other) was confiscated and an entire market day's income lost. What distinguished the case of Dame Landry from Gonesse is not that she offered her bread to passersby on the Pont-Neuf but that unlike all the other bakers who sold on the street, she had no space in any of the markets. Her spot was the pavement at the foot of the bridge; she is one of those phantom bakers whose contribution never appeared in the registers of supply and consumption.[63]

Another forain practice that siphoned bread off the market was home delivery. On the surface it was indistinguishable from street selling; indeed, the jurés contended that the former was a pretext to permit the forain to engage in the latter with impunity. In the view of the jurés, tolerating off-market sales was tantamount to authorizing fraud. As a group of masters noted during an intracorporate dispute in the sixties, home delivery enabled bakers to cheat on quality and weight because their bread escaped any risk of inspection.[64]

The regulation pertaining to forain home delivery was ambiguous: bakers could not send bread to their clients' homes before having offered it for sale on the market. The jurés gave this rule the strictest construction. When a porteuse, such as Marie Fleury, the employee of a faubourg Saint-Antoine baker, could not name specifically or supply a list of the addresses of the clients to whom she was delivering, the jurés presumed that she was merely selling illicitly to all comers; they seized the bread and had a summons issued to the baker. When the delivery persons—as in the case of Dame Jacob and Marie-Louise Anselet, both porteuses for faubourg Saint-Antoine bakers, and Jean Levadoux, a journeyman working for another Saint-Antoine baker—were able to provide names and addresses, the jurés seized the bread anyway on the grounds that it had not been previously placed on sale at the market or had not been left there long enough. Bakers found guilty of this alleged violation were generally fined between twenty and thirty livres plus court costs, and they lost the value of the confiscated bread. The forains resented this restriction deeply because they claimed that it cost them clients who offered their fidelity on condition that the baker spared them the need to go to the market.[65]

There was another kind of regrating that the jurés considered particularly vicious, less because of the fraud it theoretically invited than because it represented blatant encroachment on guild prerogatives. The offense was insinuating forain bread into the city for surreptitious sale on nonmarket days. The jurés regarded this black-market competition as a threat to their economic well-being as well as to their jurisdictional hegemony. Nothing deterred the jurés from combating this grave trespass. On a Tuesday morning in September 1762, two jurés provoked a small riot by seizing the still-warm bread that a forain named Tuniot was sneaking into the city. Baffled by the refusal of the jurés to allow a baker to enrich the city's bread supply, a crowd of local citizens rallied to his defense, necessitating the deployment of the guard. To underline its horror of this offense, the guild usually demanded a thousand livres in damages as well as a fine and confiscation. Habitually the police refused to fine forains more than ten livres and denied damages. The lieutenant general saw no need to gratify the corporate ego and purse at the risk of ruining a forain and disenchanting his colleagues.[66]

There was yet another forain transgression that aroused corporate passions because it threatened a piece of the guild monopoly. The forains were not allowed to sell bread to the hundreds of tavern keepers, cafe owners, wine merchants, innkeepers, and restaurateurs who served food in their establishments. If the forains obdurately continued to cultivate these tabooed clients, it was because the clients clamored for their bread—especially white Gonesse bread, no longer coveted in the market as it had been in the time of Retz, but still champion in the taverns. Explained the proprietor of a tavern on the rue Tirechappe: "everyone who comes to my place asks for Gonesse bread, especially on market days." A tavern keeper on the rue Betizy echoed this contention: "my customers want no other kind of bread."[67]

The forains, led by the Gonessiens, tried fruitlessly in the seventeenth century to get the tavern ban repealed, and they renewed their efforts in the eighteenth century.[68] Meanwhile they and their clients suffered an incessant cascade of convictions. The jurés visited tavern after tavern, and they seized all bread found to be from noncorporate sources. The tavern and innkeepers protested lamely that they were unaware that the regulations prohibited them from buying Gonesse or other forain bread. Yet intense public demand put pressure on the tavern keepers. The owner of Le Tambour on the rue Dauphine swore to Commissaire Courcy that "if he had known that it was not permitted [to sell forain bread], he would not have done it." Yet the very next week he was caught for the same infraction. The four-pound forain long loaf was so commonly found in public houses that it became generally known as "pub bread" (*pain de cabaret*).[69]

To protect themselves, neither the tavern keepers nor the bakers hesitated to lie. The tavern keepers pretended that they had bought the bread from a master baker (who would then be convoked and would deny it) or that they could not remember the name of the baker who sold it to them (the jurés could trace the provenance of the bread if the bread was marked with initials as it was supposed to be). Or the baker whom the tavern keeper named as his supplier would deny having baked the bread in question. Of course the jurés did not believe him, and since his bread was unmarked (otherwise he could not have rejected paternity), he received summonses for two violations rather than just one.[70]

Fire!

On the night of 23–24 August 1733 a fire emanating from the oven of a bakery quickly engulfed the entire building. Twelve people died in the blaze, which left several others seriously wounded. The magnitude of the tragedy was extraordinary, but not the source of the conflagration. Fires erupted relatively often in bakeshops, most commonly as a result of imprudence, according to the authorities. A bakeress died in the flames after returning to her burning shop to retrieve a sack containing twelve hundred livres. Nine years earlier the same house had burned to the ground, apparently as a result of a bakeroom accident.

Even when the fires spared life and limb, the cost to bakers was considerable. A fire that ruined most of his stock as well as his equipment in the 1780s prepared the way for master baker Pierre Tollard's failure, for reconstruction alone cost him around ten thousand livres. Fire struck the bakery of Pierre Dufour on the rue de Buci, exacting twenty-four hundred livres in damage. A fire forced master baker Bigre to change shops, at a cost of thirty-two hundred livres. The destruction of his *tailles* accounted for a substantial part of the thirty-five hundred livres in losses sustained by Jean Lanier as a result of a fire. An oven seemed to explode in the middle of the night in the Cloître Saint-Jacques de la Boucherie, devouring two floors and damaging another, with damage evaluated at three thousand livres. Other bakers enjoyed better fortune. Lardet lost only his oven and his kneading trough, along with a small quantity of wood and flour.[71]

The police intervened after the fact when negligence could easily be demonstrated and when neighbors vehemently complained. Baker Blanvilain's neighbors saved his shop by bringing water from nearby houses to extinguish an oven fire, not the first to occur in his shop. They called the commissaire's attention to the fact that the baker was of dubious competence and that his well did not work. For want of a well or another source of water in his cellar, master Chantard's personnel and his neighbors had difficulty extinguishing a fire incubated in the wood pile, which they were finally able to evacuate to the street. Lieutenant of Police Sartine pledged to inflict upon Chantard an exemplary punishment.[72]

Preoccupied with the ravages that fires often occasioned in the capital, the authorities preferred to take preventive rather than repressive measures. The guild statutes of 1783 required the syndics to visit each shop twice a year to certify that there was no risk of fire. Accompanied by a mason, an architect, or other experts, a commissaire regularly inspected ovens, new or refitted, that were under construction as well as ovens denounced for alleged inadequacies. Seeking to uncover "the danger," Commissaire Premontval examined the ovens themselves, the context in which they were situated in cellar or shop (or sometimes in courtyard), the place and manner in which the baker stored his wood, the nature and condition of the metal receptacle in which he stored his charcoal and embers after each recharging of the oven, and the means available to extinguish an incipient blaze. Master baker Maquart of the rue des Ecrivains had a well next to his shop and adequate storage facilities for wood and charcoal, but the commissaire summoned him to replace a wooden casing attached to the oven's chimney with an iron device. There was "no fire at all to be feared" in widow Colombet's oven room (located several doors away from her shop) because she had an oven mounted at a safe distance from the supporting beams of the cellar, an indoor well, and good storage practices.

The commissaire convoked a master locksmith to install on the spot an iron support and barrier to protect several joists next to master Delahogue's oven. To minimize the significance of the lack of a well, the baker introduced two neighbors who testified "that there was nothing to fear, the said Delahogue and his ancestors have been baking here for more than eighty years, and nothing has ever happened." Angry and anxious neighbors brought Commissaire Labbé to master Blondelle's shop in the Saint-Paul quarter because smoke continued to surge from the chimney, windows, and door. The guard and the fire patrol had ordered him to stop baking until the police inspected his oven and his manner of heating it. Inviting his neighbors to "fuck themselves," Blondelle continued to bake, and to emit billowing waves of smoke, in order to "serve his customers."[73]

As a rule the police acted expeditiously, but not without according a certain due process. Commissaire Pellerin visited the oven under construction in the cellar of François Leger, a privileged baker in the carrefour de la Pitié. He detected a peril of fire in the heat that would be generated next to walls containing combustible materials, and separating the bakeroom from shops storing large stocks of spirits and alcohol. In addition, the sustained heat threatened to dry out the eroded vaults of the cellar, whose ceiling could collapse. Pellerin recommended to Lieutenant of Police Marville "that it is extremely essential to prevent the utilization of this oven." Prior to rendering his decision, Marville enjoined Leger to obtain a counterevaluation that very day, or one would be conducted by the police architect.[74]

Punishing Wayward Bakers: Fines

The fine was one of the principal sanctions that the police used to punish and educate bakers. It was imposed for virtually every type of transgression. Scores of bakers suffered fines every year, ranging from petty sums to very considerable amounts. But were these fines actually collected? Did the police aggressively pursue the convicted bakers, or were they more interested in the symbolic dimension that stigmatized the baker in the public sphere and broadcast reassurance to the consumer-people that their interests were being protected?

The evidence suggests that the police used both approaches. More often than not they were ready to reduce the implausibly large fines they sometimes imposed—once they made their point and once the malefactor made his "submission." The lieutenant general of police may have levied a huge fine for political-didactic reasons, or merely to enhance his bargaining power in the negotiation for a mutually acceptable settlement that he knew would ensue. He was not indifferent to pleas for leniency, especially if they expressed contrition. The Assembly of Police was even willing to offer "grace" on fines to bakers who expiated by vigorously supplying their stalls at the bread markets. Speaking in general terms, a correspondent of the procurator general noted that "one overdoes it in the ordinances of police, and that it is not unusual to impose a fine of three thousand livres in order to obtain two hundred."[75]

Though sentences generally did not demand payment within an expressly stipulated time period, the collection arm of the police usually went into operation rapidly.[76] One day after master baker Nicolas Joly's conviction, Joly received an order from the receiver of fines to pay immediately or face imprisonment. Nor was this an idle threat. Master Jean Fontaine spent three days in jail until he paid a fine of thirty livres for surpassing the current bread price. Bakers Boulanger and Tribouillet spent only a day in jail. The first had his fine reduced from fifty to twenty livres and the second had his moderated from three hundred to one hundred livres. Master Etienne Felize remained in prison a full month until he obtained a parlementary arrêt lowering his fine, and an accommodation that enabled him to make an immediate down payment of fifty livres in exchange for his freedom. (Before the year's end he was back in jail for debt.) Unable to raise one hundred livres for a fine for short weight reduced from three hundred livres, another master suffered a seven-month sojourn in the Grand Châtelet. He was finally liberated because the police, obliged to pay for his food, determined that he had absolutely no resources and no prospects for help.[77]

The receiver of fines was a discreet but powerful figure. Nicolas Croizette, who occupied the post for a long period during the first half of the century, had a reputation for relentlessness. The receiver seems to have had the authority to transact compromises on his own. He also had the right to exact interest on overdue fines and "expenses" for his services. Although the police did not systematically cling to the letter of the sentence, it is fair to say that they considered fines to be serious business. Bakers understood that they could not take the matter lightly.[78]

The police of bread was the cornerstone of public policy in the Old Regime. It was construed as the guarantee of order, without which government could not endure and society could not hold together. Along with the more extensive commitment to the consumer-people embodied in provisioning regulation, the police of bread implemented the social contract of subsistence that bound governors and governed.

In a narrow sense, the police of bread was an elaborate exercise in social control. In broader purview, it was a rational and flexible effort to reconcile social and political contradictions as part of an integrated conception of urban life. On the one hand, this police was often perceived, quite plausibly, as parochial, repressive, and myopic. On the other, it took account of a multiplicity of competing interests instead of demonizing the bakers and other supply agents in order to win easy points with the public; it worked much harder at prevention and prophylaxis than at prosecution; and it strained, not always with success, to look beyond the short term.

The police of bread touched matters both of controversial political economy and of incontrovertible public security. Officials had to worry about the right mix of incentives and injunctions in order to sustain the flow of merchandise to the capital, and they had to reduce the risks of fire in the bakerooms. The police recognized and defended the rights of consumers yet at the same time discouraged them from taking those rights too literally. They insisted on the obligations of the bakers, who, in their eyes, exercised a public stewardship. Yet they had a realistic sense of just what bakers could do and could not do (and which bakers could do this or that at a given moment).

Concretely, three objectives focused their attention: assuring a sufficient supply ("abundance") of good quality bread at a fair price. The police deployed a panoply of means geared to attain these ends, sometimes making spectacular and didactic gestures (such as arresting a juré and parading him in the streets), more often acting with discretion (reprimanding bakers in police court or advancing money or merchandise). Everyday police had more to do with the prose of weighing loaves in shop after shop and stall after stall than with the poetry of making grand decisions that affected the destiny of the commonweal. Yet no commissaire ever forgot that if he failed to mobilize and discipline the bakers of his district, the consumers could eventually turn to their baker of last resort, a royal forain from Versailles.

Chapter 18

Setting the Price of Bread

As a way of dealing with the bread question, authorities practiced price-fixing in wide portions of the kingdom during the Old Regime. It was a time-honored strategy, heralded in the Bible, implemented in classical antiquity, and perfected in early modern times by the English and the Italians, among others.[1] Price-fixing, or *taxation* as it was called in France, could be systematically applied in a highly articulated tariff to cover all possible levels, or it could be used more simply to impose a ceiling, or maximum price, in certain or in all circumstances. Officials could tax bread on its own as a function of grain- and/or flour-price fluctuations or—more rarely—as part of an integrated and complex project of control based on the scheduling of grain, flour, and bread prices together.

Never the beneficiary of sociopolitical consensus, during the eighteenth century, *taxation* was usually justified by the police as a necessary evil. They shared the conviction of many moralists that merchants of goods of "first necessity" could not be allowed to hold society hostage to their unquenchable greed. But they did not uniformly endorse the view that "the avenue of authority" was the best way to attain "a just price," defined by Commissaire Lemaire as "a reasonable price . . . within the reach of the people." In response to the enthusiastic call by Narbonne, a commissaire at Versailles, for a permanent price ceiling for Paris ("it would be a benefit for the whole country"), subsistence specialists such as Commissaire Delamare and Inspector Poussot repeatedly warned that such a remedy could be worse than the disease, driving away supplies instead of drawing them in. Curiously, Turgot, a militant liberal, seemed more frankly resigned to the need for a bread tax—as long as guild bakers retained their supposed monopoly on production and distribution. Although intrinsically unfair and unnatural, in these circumstances the philosophe-administrator believed that the tax protected consumers against further "vexations."[2]

The Perils of Fixing Grain and Flour Prices

Although virtually all observers premised their reckonings on the intimate connection between grain and bread prices, there was much broader agreement on the peril and, ultimately, the futility of fixing grain prices than on the risks of bread taxation. While they construed bread price-fixing implicitly as an instrument of local sovereignty that could not be alienated, as a defensive measure to which one could resort plausibly in critical moments, grain taxation appeared to be both theoretically unwise (counterproductive in our modern jargon) and technically impossible to institute, given problems of scale, surveillance, diversity of measure, variety of type, and commercial practice. The leading figure in the Assembly of Police preferred competition to price-fixing as the foundation of dearth strategy in 1662. The banker and international trader Samuel Bernard persuaded the government during a similarly grave crisis in 1693 that taxation would drive away the foreign grain necessary to parry the shortage. In the midst of the near famine of 1709, the royal council conducted a survey among many of the kingdom's most knowledgeable public officials, the majority of whom rejected grain taxation on the grounds of its immense cumbersomeness and difficulty of execution, and the inevitable "disgust" it would inspire in merchants whose "interest" had to be mobilized for abundance to return.[3]

One of those queried in 1709 was Delamare, who adeptly unmasked the allure of grain taxing, which loomed as "a prompt remedy to pressing ills" but proved quickly to be a "specious remedy [that] halts the circulation of grain and destroys the foremost trade of them all, which is the commerce of grain." Not only would a "forced trade" fail to deliver the goods, but the elaborate organization it required would broadcast disaster cues likely to cause panic— "announce the dearth" and "annihilate public confidence," as a Revolutionary subsistence commentator observed some eighty years later. Delamare countenanced the occasional use of grain taxation applied tactically and limited rigorously in time and place, but even in this austere scenario Paris had to enjoy a special regime, with a higher price ceiling in order to attract a thicker and faster flow of grain.[4]

Many of the leading critics of deregulation in the eighteenth century nevertheless rejected grain taxation as a fruitless and wasteful policy. "Never fix the price of grain or bread even in the midst of the cruelest famine," Abbé Galiani, a merciless adversary of physiocratic liberal scholasticism, counseled Police Chief Sartine around 1770. The problem with grain price-fixing for Necker was not that it amounted to "theft," as the physiocrats clamored, but that it was not in the public interest because it was "unworkable." Untrammeled by any scruples about the claims of society on individual grain owners, Abbé Terray, the controller general during the terrible dearth years of the early seventies, argued that the way to influence grain prices was by shaping bread prices, not vice versa: "an increase in the price of bread triggers an increase in the price of grain: it is only by hampering the Baker that the latter himself contains the grain merchant; otherwise, if he is assured of always having the price of his bread in proportion to the price of grain, it will make no difference to him whether or not it rises, in fact often he will find it in his interest to see it go up so that he can profit from the provisions of grain that he has stored."[5] Terray's argument seems theoretically flawed from the outset, yet in practical terms it made sense, and many of the most effective lieutenants of police shared his viewpoint. Certainly it is worthwhile to think of price making as a complex process of exchange, tension, and feedback rather than as a linear, one-way outcome of fixed relationships. Terray believed that regulation could take many different forms. Instead of price-fixing by authority, he favored price making by convention—a more or less coercive convention to which most of the players subscribed. If it was widely and deeply believed that the bread price would regulate the grain or flour price, then that was what would happen, Terray suggested.

For Daniel Doumerc, an international grain trader whose dealings in behalf of Terray earned him incarceration and disgrace at the hands of his successor, Turgot, a bread tax implied a grain tax. Rehabilitated as a government subsistence counselor, Doumerc told the procurator general in 1789 that the "grand question" remained: "Should one seek to lower these [grain] prices by holding down the price of bread?" If somehow Paris could be wholly isolated from the rest of the realm, he reasoned, it might be possible to use a bread tax as a master gyroscope. But the capital could not be fed without "the help of its neighbors," near and far. "Let us not aggravate our ills for the sake of fleeting palliatives," concluded the man implicated in the notorious famine plot.[6]

Social Contract and Social Control

Officials remained conflicted on the critical issue of bread price taxation throughout the eighteenth century and during a large part of the nineteenth. On the one hand, the Assembly of Police reminded itself that draconian measures, commensurate with the demands of dire circumstances, would hardly be appropriate in relatively unstressful times. On the other hand, it reflected that since it would be "dangerous to hinder the price" during a crisis, then "it was necessary when there was no shortage at all to do everything to prevent an excessive augmentation."[7]

Nor did the authorities construe the taxation question as a narrowly objectifiable marshaling of supplies. They were acutely aware that the bread question was as much a matter of public opinion as of price (and supply), and that the latter shaped the former even as the former conditioned the latter. It was difficult to explain to the consumer-people that ostentatious government inaction would be more hospitable to their interests than aggressive intervention—price-fixing in this instance. Intimately familiar with the subsistence mentality, the working police were infinitely less sanguine than the reforming intellectuals about the prospects for engaging the people, as a Revolutionary commentator put it, "to look for natural causes behind the variations in prices without always wishing to believe in means for which we cannot account."

Louis Thiroux de Crosne, the lieutenant general of police at the end of the Old Regime, noted the tonic effect of his bread ceilings on "the workers and the gagne-deniers, who have barely emerged from an extremely harsh winter that deprived them of their capacity to subsist." His commissaires reported on their gratitude and on their confidence in the public administration—at least on the local level. De Crosne envisaged the tax not merely as a mechanical device that blunted the impact of rising prices according to a predetermined formula but as a tool of social policy that authorities could deploy flexibly to address changing conditions, political as well as economic.[8]

La Reynie, the first lieutenant general of police, argued that a tax system was required for the protection of the bakers as well as the public. If the price surpassed the critical psychological threshold of the moment—four and one-half sous per pound for the white loaf in 1694—the police chief confided to Delamare his fear that "the security of the bakers" would be jeopardized. Implicitly he hinted that the police would not be able to contain the avenging outrage of the people, whose rancor, he suggested, was well founded. (In 1790, the Section des Enfants-Rouges argued in a similar vein that without a tax, the baker was left exposed to spasms of popular vindictiveness and that a war would inevitably break out if one left the people "in uncertainty and in distrust of the bakers.")[9]

In a memorandum to the Academy of Sciences submitted in 1781, Mathieu Tillet developed the most complete argument for a bread-taxing policy. He predicated his analysis implicitly on the assumption that bread was too important a social good to be left to market allocation on its own and its price too sensitive a social sign and arbiter to be relegated "exclusively to competition within the bakery trade." Tillet alluded to a sort of social contract rooted in the night of time that afforded the people a sense of "security" regarding an adequate and accessible bread supply, a contract that the government could afford not to heed only at the gravest risk to its own stability.

Tillet offered science in the service of a more integrated vision of social control than the police ordinarily entertained. The object was to discover an equilibrium point at which the invincible "mistrust" of the consumer-people would not degenerate "into murmurs" of a potentially explosive sort and at which the avidity of bakers was reasonably mollified. Allying the intellectual power of science to the moral power of custom, Tillet maintained that his ultrarational solution would put the issue beyond dispute by producing "an exact equity."[10]

Tillet stressed the political stakes of bread price control. If taxation protected the public against uncertainty and anxiety, and protected the bakers against the vindictive fury of the public ("sheltered under the authority of the Magistrates," bakers living in a taxed environment "had less to fear from uprisings of the people than they would in an environment in which they themselves regulated the value assigned to bread"), it also protected the state—public authority from the bottom up—against the alienation of public opinion. For failing to attempt to contain the price of bread through a tax system, the magistrates risked far more than the ritual "insults of the low orders of the people." Tillet discerned in the taxation question the seeds of doubt and suspicion that burgeoned rapidly into the full-fledged famine plot persuasion that so deeply undermined the confidence of the public in the government and, ultimately, in the king. Without the sheltering structure of the tax, it was easy to suppose first "that the dearth grew as a result of their negligence" and then "that there is on the part of these same Magistrates some interest in maintaining it."[11]

In Tillet's view, a carefully constructed tax schedule would generate pedagogical and perhaps even therapeutic advantages for everyone. The processes of flour and bread production would become wholly transparent. Everyone would learn just how much it cost to make the goods and what constituted "a moderate profit" for the miller and the baker. (Certain anachronistic assumptions governed Tillet's analysis, in particular the expectation that the custom miller waiting passively to grind the client's grain would continue to be the baker's central partner rather than ceding place to the commercial miller committed to speculative grain buying and flour making.) In Tillet's sanguine design, the chief costs of production, once identified, "do not vary at all, whatever value grain may acquire as a result of the general accidents that the harvests sometimes suffer." It will thus be far easier to isolate the causes of upward price pressure, or at least to eliminate the responsibility of the bakers and (certain) millers.

Nourished by the old system, which encouraged dissimulation and manipulation, "the spirit of interest" that bound bakers together would wither, Tillet maintained, and the salutary "jealousy" that induced them to compete for business would reaffirm its vigor. Thus a taxing system would result in the need for less rather than more day-to-day police intervention by rendering the market far more predictable. Tillet contended that even where a tax system was not actually or regularly applied, it would be heuristically invaluable to publish a detailed tariff in order to educate the public and chasten the bakers, who would find it difficult to diverge from the unofficial benchmark.[12]

Hundreds, perhaps even thousands of copies of Tillet's elaborate tax tariff were printed and distributed throughout the realm. The document was discussed in several parlements and by scores of municipalities and police authorities. Rochefort, Chartres, and Bar-le-Duc were among the cities that adopted it; others adapted it to their specific needs. Though inclined toward subsistence deregulation, Controller General Calonne encouraged local officials to consider the Tillet tax schedule. Still, he admonished, it was only one possible policy model, since "an infinity of motives and circumstances can make inadmissible the arithmetic calculations that served as a base."[13]

The Critique of Taxation

Critics of price setting attacked on a wide front. Tillet himself vouched for one of their major strictures, the technical fragility of the whole taxation construction, a defect that he pledged to remedy with his brand of modern science. Abbé Baudeau, the accomplished physiocratic propagandist, developed this critique with his usual mordant flair, pointing to the inability to measure precisely any of the critical variables in the elaboration of a price schedule, including the productivity of grain and flour, the alternate technologies available, seasonable variations, and the multiplicity of costs (not merely the obvious items but the more subtle ones such as interest due on advances for the purchase of grain or flour, the cost of wood, the amount the baker owed his family for their unremunerated services, and so on).[14]

To extricate themselves from recurrent and intractable difficulties of governance, the authorities at Rochefort sought a scientific, quasi-mechanical solution to a political problem—that is to say, they groped for a utopian settlement.

Lack of certainty and confidence had become intolerable to all parties. With Tillet's framework and a series of controlled trials, they sought "the most precise [price] fixation" in order to "reestablish on solid ground evaluations that have been vacillating and uncertain for a long time." If they were successful, in a single stroke they would be able to purge lingering suspicions of monopoly and manipulation that plagued the bakers and persistent outcries of consumer complaint. While the plan hinged on "a wisely balanced equilibrium," the authorities did not conceal that in the end their principal devotion had to be to the consumers, "too frequently the victims of the cruelest scourges," in particular chronic dearths and recurrent price spirals. In anticipation of our discussion of trials, let us note that the Rochefort bakers contested the procedure from beginning to end. Criticizing the bakers' preoccupation with their narrow self-interest, the police reproached their unwillingness to expose themselves to the telling eye of independent monitoring and comparison.[15]

But the case was even more convincingly made by a public official, Boullemer, the lieutenant general of Alençon who had no scruples about the intelligent and efficient exercise of authority, all dogma to the contrary notwithstanding. (He hesitated to write because he felt ill at ease in "attacking Regulations made or adopted for the police by magistrates whom I respect.") Instead of defusing the confrontation between irate consumers and harassed bakers, in many places the tax exacerbated it because it was based on erroneous information, in particular "the lack of knowledge of the quantity and of the variation of the yield of the grain." The trials (*essais*) that generated the data for the tax failed to take into account the impact of year, season, and weather (including harvest conditions), factors that resulted in substantial disparities in production. Shaped by normative assumptions, the tax regulations tended to be rigid and unresponsive to shifting realities. (Boullemer deplored notably their ineluctable "for always" pretensions.) A setier could just as well yield 268 pounds as two hundred pounds of bread at a given moment. If the authorities wanted to rely on taxation strategies, they had to work harder at establishing a rigorous and reliable foundation for their prescriptions. "The administration of the police concerning the tax of bread," observed Boullemer, "can be compared to a veritable labyrinth from which one could not try to extricate oneself without the risk of entering into another one even more complex."[16]

The physiocrats riveted the claim that the tax paradoxically would never favor the (real) interests of the consumer-people. "The tax is always necessarily against the consumer," Dupont asseverated. Programmed to guarantee "a considerable profit" to the various producers along the line, it presented the consumer with a higher bill than would "unlimited liberty." The consumer would also suffer from the shoddy work that the tax might elicit, reasoned Parmentier, one of the vocal band of subsistence scientists. Constrained by the tax to buy mediocre wheat, to grind low and hot (and thus at a very high rate of extraction and at peril to the goodness of the meal), to forsake the recovery of flour-rich middlings, and generally "to neglect precisely [those expenditures in money and time] that can influence quality," the demoralized baker would produce mediocre bread at best. "Nothing resembles more closely the production of a poor worker," concluded the pharmacist–food chemist, "than that of a schooled man who is not adequately paid." (Other observers of the bakery were inclined to see bakers less as victims than as beneficiaries of the tax, for it furnished a context and a cover for them to engage quite deliberately in potentially lucrative fraud.) The bakers of Paris themselves did not miss the occasion to argue that the police would have to assume responsibility for popular disorders resulting from an unfairly constrictive tax because such a policy would force bakers "to make up [their losses] at the expense of loaf quality, and there would [then] be reason to fear that the complaints of the Public would be very loud and very well founded."

During the acute dearth at Chartres in 1770, bakers produced the sort of "tax bread" that Parmentier feared: surcharged with branny meal (the offal usually sold to starch makers), bloated with water, incompletely baked, unnourishing. Another way that the bakers "indemnify themselves for the diminution" imposed by the tax was to purchase inferior grain and flour outside the market, even as they agitated to drive up the market standard—the *mercuriale*—on which the tax was based. Such was the practice of the bakers at Caen during the crisis of 1770.[17]

Abbé Baudeau caustically exposed the slick urban regulators—"ces habiles taxateurs"—wholly unfamiliar with agriculture and commerce, who destroyed the very sources of consumer security with their "blind and ruinous taxes." The tax deterred growers from growing; it was an unnatural as well as a lethal policy, one that the *économiste* compared with trying to "prescribe to the moon and the sun the route that they must follow." Inefficient, the tax was also profoundly unjust, defying nature in yet another fashion according to physiocratic logic. Rationalized as a necessary social "ransom," the tax in fact violated the imprescriptible and natural laws of property and liberty and thus loomed in the eyes of Baudeau as nothing short of theft.

Bread price-fixing opened the door to despotism. "If you permit . . . the municipalities to tax at their will," warned the Revolutionary leader Rewbell in July 1791, "you open the door to arbitrary actions." At about the same moment the *Journal de Paris* celebrated the (ephemeral) rejection on the part of the maturing revolutionaries of "the old police," the regime of "the tax of goods," of a pervasive socioeconomic tyranny, in favor of "liberty," nostrum for all ills, guarantor of sustained abundance through competition in the sort of market that later would be called neoclassical. Even Abbé Galiani, a stringent adversary of grain liberalism, reproved the tax as "an atrocious injustice" to the bakers and a tactic likely to backfire on its exponents. Blending the idioms of physiocracy and Jansenism, his archcritic, Abbé Morellet, declared that the price of bread "can and should be fixed only by the liberty of conscience of commerce."[18]

Inverting Tillet's argument that the lack of a tax induced the people to hold the government responsible for shortages, another physiocratic sympathizer, the mathematician and political philosopher Condorcet, argued that it was the pernicious usage of the tax that accustomed the people "to regard its Magistrates as responsible for the high price of subsistence." Far from sheltering the leaders from the ravages of public opinion, price-fixing rendered them hostage to its caprices. Far from sparing the people from misery, the tax was capable of generating "a real dearth" (as opposed to a *disette d'opinion*) by frightening suppliers into hiding. More ominously, it afforded "factious troublemakers the only way of stirring up the populace" and thus undermining the stability of the state.

Despite his serious reservations about taxing policy, Parmentier rejected Condorcet's sweeping claim that it could never serve the *true* psychological and material interests of the mass of the people. The police should use the tax, however, only in a socially differential manner. They should strictly limit its application to the "[gros] loaf destined for the people." For their "hard and exacting work," bakers deserved a "legitimate profit." It was most appropriate, argued Parmentier, for the bulk of that profit to derive from the sale of "mollet and fantasy loaves." Indeed he suggested that the rich be required to subsidize the poor, "so that the production costs of the gros loaves were borne by the petit [or luxury] breads."[19]

Throughout the kingdom bakers decried taxing constraints and resisted them in various ways.[20] Some of them operated within the logic of the taxing system in which they were trapped, more or less successfully converting it into a kind of golden cage. These bakers controlled the terms of the tax by clinging to customary standards that they knew to be wholly anachronistic or by impugning the validity and relevance of the currently used basis for the tax or by seeking to manage and control the official trials that were to serve as a basis for the price schedule. The institutional memory at Châlons-sur-Marne recorded trials as early as 1436 to determine a pricing yardstick. In 1785 the town bakers angrily complained that they were still governed by a trial performed in 1623 despite the fact that all prices had risen by "at least two-thirds." Why was it, they asked, that "the Baker, alone enchained by the fixation of his wage, was not allowed to follow the cost of living?" Mobilized by a keener sense of far more recent, shorter-term changes, their colleagues at Troyes lamented at about the same time that they were "reduced to beggary." They inveighed against a trial of 1758, when wood, house-and-shop rents, and dayworkers' wages "cost half of what they are worth today."[21]

Since the tax became particularly constrictive during moments of penury, the clash between the bakers and the police characteristically broke out into the open when the crisis deepened, as occurred in the late sixties at Saumur and Senlis. Almost invariably the bakers inaugurated the confrontation by denouncing the "injustice" of the tax, both abstractly (as an infringement on their property and freedom of commerce) and concretely/specifically (for instance, the complaint of the bakers of Château-Thierry that the tax was based on artificially low grain prices, below the price at which bakers could obtain flour that could be made into acceptable bread, and that as a result the bakers faced "ruin"). Far from improving relations between bakers and their clients, the tax compromised and envenomed the latter, according to the bakers of Châlons-sur-Marne. It crystallized "a throng of prejudices," including the notion that the bakers should be held "responsible both for the inclemency of the season and for all the accidents that could result in a price increase."[22]

Nor did the indignation of the bakers subside as France entered the nineteenth century. In 1806 Jean-Louis Longprez, the grandson of one of the capital's best-known bakers in the previous century, blamed his business failure largely on "the manner in which bread is taxed." Speaking in behalf of the whole profession sixty years later, Victor Borie compared the fate of the bakers to that of the Jews ("admitted to benefit from the common law") and blacks ("emancipated"). The tax denied the bakers the "equality of rights" accorded all other citizens, reducing them to the status of "slaves" and making them "the pariahs of modern society."[23]

Bakers took their fight against the tax to the very centers of power. Two bakers from a small town in Champagne petitioned the procurator general of the Parlement of Paris for protection from price controls. From the other side of the kingdom, the bakers of La Rochelle appealed to the same court for relief. Failing to obtain satisfaction from the intendant of Languedoc in their ten-year struggle against the municipality, the bakers of Carcassonne turned to Necker, the de facto controller general. Necker disappointed them by ordering new trials and suggesting the need for a generally tighter price schedule. The authorities did not, however, behave in a predictably monolithic, hard-line manner. In 1725, the procurator general of the Paris Parlement concurred with the remonstrating bakers of Dreux that the local tax "is not sufficient to cover the price of wood, journeymen's wages, and other expenses."[24]

Frustrated, angry, and/or emboldened by crisis circumstances that seemed to enhance their short-term leverage, bakers occasionally deployed their ultimate weapon: the threat to stop baking. In 1725 the bakers of Rennes conducted a sort of strike, refusing to bake until the police withdrew price controls. When the judges of Guingamp in Brittany imposed a price ceiling in 1756 to thwart excessive profit taking, the bakers closed their shops. In 1770 the local authorities furiously denounced "the reiterated threats" of the bakers of Arpajon in the Parisian hinterland to quit their profession if a tax were definitively imposed. Sixteen years later at Luçon in the Poitou the bakers, according to the outraged report of a canon of the local cathedral, "adamantly and stubbornly refuse to sell bread according to the tax." Two centuries later bread mattered little in a strictly dietary sense, yet Paris was in ferment when the bakers struck for several days against the government-imposed price ceiling and other bureaucratic strictures. Gathering at the Mutualité, two thousand bakers, some of them brandishing empty flour sacks attached to sticks, demanded, in the standard Thomistic idiom tailored to their needs, "bread at its just price, the bakery wants to live."[25]

Under Surveillance: Paris and the "Current"

During the eighteenth century, as in recent times, the price of ordinary bread in Paris was officially "monitored" rather than "taxed." Commissaire Delamare declared that it was simply "impossible to fix its price in Paris." He adduced four overlapping categories of reasons, technical, logistical, commercial, and moral: purchased in myriad places throughout the sprawling supply zone at different prices and freighted with varying costs for transportation, storage, and so on, Paris-bound grain boasted no uniform price on which to base a tax, and it was impossible to gather comprehensive information rapidly enough to serve this end; bread itself, in very substantial quantities, came from many different places, near and far, burdened with the same diversity of costs; even if one managed to set a price, news of its weekly readjustments could not reach the entire corps of itinerant bakers in time, and the prospect of being surprised by a penalizing price upon arrival in the capital would deter many of the latter from contributing to the "abundance" necessary to meet Parisian needs; though the *gros pain* consumed by the bulk of Parisians was divided into three statutory classes—white (*blanc*), middling or bourgeois (*bis-blanc*), and dark (*bis*)—within each rubric there was considerable variation, in relative whiteness and goodness (*bonté*), the fruit of disparities in wheat quality or bakery style and skill, and as a result "it would be difficult to spell out all these differences and to proportion the price as it would be just to do."[26]

"Although the price of the gros loaf is not fixed, and the bakers have the freedom to put it on sale at the price they esteem appropriate, and the buyers to haggle for it," continued Delamare, "it is nevertheless important that the magistrates and the police officials who handle market inspection, to the extent possible, have knowledge of the just and true price that could be charged for bread according to the price of wheat." This knowledge would significantly strengthen the hand of the authorities in their ongoing dialogue with the bakers, enabling them confidently "to exhort the bakers, to press them, and to engage them, by way of an exact implementation of the police regulations to which they are subject, to slacken on the price as much as the good faith of commerce, reason, and justice demand of them, and to reconcile, in the end, their just and legitimate interest with that of the Public." To make the Paris system work the commissaire earnestly believed that the bakers required a large dose of commercial liberty. That freedom would be hedged, however, by the imperatives of the public interest, as construed by the police, whose regulations authorized all manner of coercion, and by the flow of market information, the knowledge coveted by Delamare. To structure and interpret that information and frame a strategy of (in)action, the police required the sort of yardstick that a master tax plan furnished, and that could only be obtained through carefully controlled trials. In a sense, then, the purpose of a (theoretical/speculative) tariff would be to obviate recourse to a (coercive, real-life) tariff.[27]

Thus the Paris bakers operated under the auspices of an unspoken or latent tax, a more or less shifting normative standard that was supposed to sustain a certain social and commercial equilibrium. In ordinary times, the police counted on brisk competition among hundreds of bakers of very different character, working out of shops, markets, and wagons, to keep the price of the big loaves ranging from three to twelve pounds within reasonable limits. Avid to extend their clientele or desperate to defend it, presumably these bakers would be inclined to "relax as much as they could on the price" and to compensate with gains on volume. Bakers were expected to conform to the *prix courant*, sometimes dubbed the *cours ordinaire*, the Parisian surrogate for a tax, defined as "the price at which the largest amount of bread was sold [at the previous market or during the early hours of the ongoing market]." In easy times the "current" was a self-adjusting price that expressed, in mutually satisfactory terms, the relation between bakers and consumers. Bakers who failed to respect it suffered the sanction of the market, imposed by discriminating consumers with deep memories and a vindictive itch.

The current or ordinary price represented a central tendency to which the bakery gravitated, with outliers on both sides, more of them above than below. News passed rapidly on the market wire, given the concentration of sellers and buyers, their physical propinquity, and the intensity of exchanges. It circulated somewhat more slowly through the shops, but still with surprising celerity, transmitted by consumers scouting for the best prices, by deliverywomen and journeymen, and by bakers (and their personnel) who worked both market stalls and shops. Even in relatively easy times, everyone talked about the bread price, a bit the way everyone still talks about the weather but with more urgency. One should not underestimate the swiftness with which word-of-mouth information can spread, even through a big city. In ordinary times, it is unlikely that either the police or the guild served as vectors of dissemination. Their mediating and communicating role became important only when the entire system suffered a shock.

In periods of growing stress, however, the current became a presumptively inviolable ceiling, a quasi-tax that was no longer necessarily the more or less direct reflection of the shifting prices of grain and flour. Pledging "justice" to the bakers, the police strained to infuse the ceiling with a consensual character by discussing it with the heads of the baker corporation and soliciting their endorsement and cooperation. By and large the jurés (later syndics) seemed more intimidated by the lieutenant general than by the prospect of a disavowal by their constituents. In ceding to police pressure, they tried to extract a promise of some form of future compensation. They used it to blandish their confreres, along with the reminder that the ceiling protected them from the savage ambition of many down-and-out consumers "to force them to sell the bread at a [still] lower price." Bakers who transgressed the current faced stiff fines or immediate imprisonment by royal order. At the pinnacle of a crisis, if the parlement ordered the baking of a constricted palette of bread, the current took on even more frankly the aspect and function of a tax.[28]

The strategy employed during the serious dearth of 1693–94 by the first lieutenant general of police, La Reynie, did not significantly change in the course of the following century. He relied on a ceiling (or a maximum—the word is too jolting because of its Jacobinic coloration) rather than an articulated tax. At crucial junctures, such as midsummer 1693, La Reynie fixed the price without discussion as a function of his reading of the gravity of the situation. He instructed Commissaire Delamare not to brook any resistance but to "oblige" the bakers "by all methods," including confiscation and imprisonment, to fall into line—in this instance, to lower their price despite the climate of rising grain and flour prices. Two weeks later, on 21 July, when he accorded the bakers a slight increase, he warned the commissaire to be alert to the possibility that "the common people might cause some disorder or commit violence against the bakers." While price information from the supply zone influenced his decisions, it did not determine them, for it was clear to him that bakers could manipulate the price of raw materials. Neither their "malevolence" nor the "monopoly" of the suppliers would prevent him from making bread available "at a just price," a price that conformed only rarely, following the Thomistic ideal, to the market price produced by the friendly encounter of supply and demand.

The word "juste" (or "iuste") recurred as a leitmotif in La Reynie's discourse as the rationale upon which all police action was founded. When the lieutenant general was convinced that the price of grain rose naturally (or legitimately) rather than artificially, he backed off from an authoritarian approach. Thus in April 1694, the bakers could not be forced to ignore the upward price surge. They could be subjected only to suasion cast in "general terms" rather than to more or less peremptory injunction. Instead of imposing a tax, the commissaires toiled to "reconcile the poverty of the people with [the requirements of] the bakers, in order to prevent the latter as much as possible from taking excessive advantage of the current conjuncture, because at the present time there is no sure and certain basis on which one could establish the price." (On the eve of the Revolution, the capital's last lieutenant general of police behaved in a strikingly similar manner: "The successive increases, Messieurs, of the price of wheat and flour," he told his commissaires in late January 1789, "have not allowed me to refuse bakers the permission that they requested to raise the price of the four-pounder to fourteen sous six deniers.")

Like Sartine, his successor seventy-five years later, La Reynie conducted a socially differential policy, focusing pressure on the Wednesday and Saturday bread markets, where the laboring poor paid cash, rather than on the shops, where bakers tried, with highly variable zeal and limited success, to restrict credit in crisis times to the better off, who could afford to pay a slight premium. While the police agonized over the decisive impact of the bread-market situation on consumer opinion and behavior, they also worried about the flight of the forain bakers, whose contribution to the supply pool was considerable. La Reynie feared the sort of boycott that later faced one of his eighteenth-century successors, Hérault, when the Versailles contingent stayed away because the commissaires "compel them to give up their bread at a loss as a result of the tax that they put on it." Although La Reynie posted his agents "at a very early hour" at the bread markets to announce and enforce the ceiling during the perilous moments, he conceived of these interventions as "extraordinary acts" rather than the normal course for the elaboration of bread prices. Ordinarily the police confined their role to "general inspection," to assure the reasonable empire of the current price.[29]

A lack of understanding about the precise character of the police bread price policy wrought a considerable amount of confusion during the eighteenth century (as it still does among historians at the end of the twentieth century). On the one hand, eminent public servants (Turgot) and writers (Macquer) referred repeatedly to "the tax utilized in Paris" or "the taxed price in Paris." (The former complained in fact that the schedule in the capital was excessively inflated as a result of the undue influence of the baker guild, a bête noire according to his liberal imagination.) Consumers, such as the twenty-three-year-old paver's aide Jacques Lemarchand, spoke instinctively of the price at which bread was "taxed," but much more commonly in crisis than in normal periods. In the 1780s the police in the suburb of Saint-Cloud emulated the Paris "tax," and the Parlement of Paris published an arrêt in which it prohibited the bakers of the capital "to sell bread above the tax that the Lieutenant general of police imposed." Authorities at various levels referred to a veritable Paris pricing model, if not precisely a tax, in which the bread price was calculated according to the price of flour rather than grain: two Paris setiers yielded a flour sack of 325 pounds, that sack produced either 400 or 408 pounds of bread, and an allowance of either 100 or 140 sous was added to the price of the flour to account for baker costs and profit.[30]

At the same time, doubtless bespeaking the inquietude and disappointment of many ordinary observers, including simple consumers, a police informant lamented the lack of police commitment to a strong control policy. Noting the effrontery with which bakers raised prices in October 1737, he commented somewhat artlessly on public reaction: "It is said that it is only in Paris that one suffers these liberties, and that the prévôt des marchands of Lyons habitually fixes the price of a loaf before each market day, as do everywhere else other judges and magistrates." Forty years later Lieutenant General of Police Lenoir confirmed this reading while at the same time injecting a note of temporal uncertainty regarding the evolution of policy and practice: bread "was no longer formally taxed by any police ordinance; the functions of the police in this domain are limited to an attentive surveillance to make sure that the market was always amply supplied in flour and to prevent any unjust swelling of the price of the common food of the people of the capital."[31]

To arm themselves with the intimate knowledge of bread making that the police required to make judgments about price, they organized more or less elaborate laboratory trials. They attempted to replicate real-life conditions, or to make allowance for deviations from them, though they were not always able to overcome the technical and moral obstacles.[32]

Lacking the resources of the Paris police, the lieutenant general of the supply zone market town of Provins complained of the failed trial he conducted in 1765—failed despite the infinite personal attention that he had devoted to it: "either in the milling process or during the preparation and baking of the bread, I was misled." He asked Sartine to share the results of the *essai* the latter was planning: "the one that is to be done under your orders will be much more reliable and will serve me as STANDARD, allowing of course for the fact that the costs of mastership, rent, wood, and wages constitute a much more considerable overhead for the bakers of Paris than for those of Provins." After profiting from a schedule that had favored them for over thirty years, the bakers of Caen quarreled with the authorities over the fairness of a new trial and then contested a second one conducted under the auspices of the Norman parlement. Though the intendant supported their appeal for a less draconian revision, much to the embarrassment of the government, the controller general rejected it out of hand in 1776.[33]

The earliest official Paris trials appear to date from the fourteenth century. At least four were held in the fifteenth century and one in the sixteenth. Delamare may have been involved in the testing conducted under the prod of the parlement during the crisis of 1693, though those experiments seem to have been geared less to charting a tax schedule than to discovering the formula for a sturdy household loaf that would be neither "too white nor too dear for the use of the Poor and the Artisans." At the end of the century the commissaire personally supervised one of the major trials that served as a reference through a large part of the realm.

From a *mine* (half a Paris setier) of good albeit not top quality wheat, Delamare obtained eighty-five pounds of bread of three sorts, a yield later criticized as excessively modest. He calculated that milling fees and wood would each cost about 5 percent of the purchase price of elite quality wheat, that house-and-shop rent claimed roughly 3.75 percent, and that the combined category of wages, food, family upkeep, and maintenance of bolters represented another 25 percent. In the unsettled terms of money of account in 1700, the commissaire estimated the costs of producing bread from a setier of grain to be six livres seven sous eight deniers. On his schedule, the price of a white loaf varied from two sous per pound when wheat commanded fourteen livres to five sous when it sold at forty livres (at the same grain price levels, bourgeois bread ranged from one sou eight deniers to four sous six deniers and the dark loaf from one sou one denier to three sous six deniers).[34]

Though Hérault, the lieutenant general of police who took over during the dearth of 1725, ordered testing in the flour-entrepôt towns of Pontoise and Beaumont in order to verify the pretensions of mealmen and bakers, Delamare's guidelines did not undergo a major revision until the dearth of 1739. The police subsistence expert, Commissaire Courcy, undertook several small trials, not for the purpose of elaborating a full-fledged tariff but in order to test baker claims on productivity. From an elite *mine*, he extracted eighty-five pounds of flour that in turn generated 116 pounds of bread of three types, over a third more than the 1700 yield. A lower quality *mine* that weighed slightly less rendered only sixty-four pounds of flour and eighty-four pounds of bread.[35]

At about the same moment, tests conducted at the Scipion bakery of the Hôpital général, a detention and relief center for the so-called deviant, the ill, the poor, and other pariahs, on foreign grain imported in the king's name produced the equivalent of 104 and eighty-seven pounds of bread derived from two different batches. A test the following year, on very dear wheat costing thirty-seven livres nine sous per setier, yielded 115 pounds of bread for a *mine*. The authorities calculated that five livres six sous (a little more than 14 percent of the price of the grain) would have to be allocated to cover wood, baking costs, and the baker's fair profit. Extrapolating from data apparently collected in the field, another report predicated pricing calculations on a *mine* yield of 118 pounds, but in the form of middling (60 percent) and dark (40 percent) bread only.[36]

"Since I have nothing else to do, Monsieur," the usually harried and overextended procurator general Joly de Fleury coyly wrote the lieutenant general of police at the height of the subsistence crisis of 1740, "I had fun putting together a calculation perhaps not as good as yours but in any event different." No trace of the police scenario seems to have survived, but it is interesting to note that they were attempting to work out coherent taxation schedules, as the implicit standard and justification for their normative expectations, even as they battled day by day to make supplies accessible. Based on the production of two types of bread (no white) rather than three, Joly de Fleury's tariff rose more slowly than had Delamare's of 1700. The commissaire's bourgeois loaf climbed characteristically by increments of one denier a pound (once by two, twice by one and a half) every time the price of grain increased by a livre per setier. The procurator clung to the one-denier cadence until the twenty-livres-per-setier mark and then granted a two-denier increase only once in the course of an augmentation of ten livres on the grain, despite his awareness that "by following this system above twenty livres the baker would always lose more and more, which is not fair." He reckoned on a yield of 160 pounds of bourgeois and seventy pounds of dark bread, or 115 pounds total per *mine.* The procurator allotted bakers four livres for all costs and profit (plus, the value of the sale of bran remains estimated at one livre per setier, an excellent inflation hedge for the baker), surely less than Delamare granted, allowing for differences in the value of money.[37]

The second half of the century was a time of great intellectual and political ferment in the subsistence arena. Reflection on grain-linked issues arrogated a cardinal place in the preoccupations of the philosophical and scientific communities (and, according to a semiserious Voltaire, the world of the salons and even of the opera). The central government placed the grain question at the core of its radical reformist agenda. Commercial and technological innovation began to transform the very nature of the provisioning trade. In the mid-sixties a grave and thickening dearth inaugurated the most turbulent decade of subsistence difficulties of the century. Police officials at all levels had probably never been so acutely self-conscious about the complexities and uncertainties of their charge. As part of a searching reassessment, not of the theoretical validity or coherence but of the efficacy of regulatory strategies, Sartine commissioned a series of *essais,* not all of which were directly concerned with price control.[38]

The advent of economic milling, a technique of gradual reduction of the wheat berry and the recovery of flour-laden middlings, raised the tantalizing prospect of producing more, better, and cheaper bread from a set amount of grain than the old methods yielded. In 1761 the lieutenant general ordered Machurin, the senior commissaire of the Halles quarter, and Poussot, his reform-minded inspector and subsistence specialist, to supervise, with the most elaborate precautions to assure "reliability and honesty," a comparative trial of flour produced "economically" by the miller-baker-entrepreneur Pierre-Simon Malisset matched against the flour processed at the Hôpital général, the results of which could incidentally be useful for future taxation calculations. The specter of "frauds" on the part of the hospital's bakers and millers, directly threatened by Malisset's innovations, worried Sartine, and his legendary intuition proved once again to be on the mark.[39]

Cheating brazenly on both the milling and bolting, the hospital amassed for the test far more flour than it normally extracted from two setiers. The police decided to pursue the trial anyway, Poussot consoling himself with the reflection that even if the hospital produced as much bread as Malisset, it would not attain his "superior quality." Commencing at 5:00 P.M. on 23 February 1761, Machurin and Poussot spent nineteen consecutive hours in the "unbearable heat" of the oven room of the Scipion annex, subjecting every step to scrutiny and taking the trouble to measure precisely every adjunction of water and to seal the three kneading troughs after each renewal of the starter leaven. Poussot's lyric enthusiasm matched his fatigue: "Since Paris has been Paris, never has there been a test on bread making . . . as there was done this night."[40]

The test required Malisset and the hospital to bake seventeen five-pound white loaves for the officers of the institution and to convert the remainder of the two setiers into five-pound dark loaves for the inmates. Using the amount of flour equivalent to its normal milling results (353 pounds), the hospital produced 475 pounds of bread (17 five-pound white loaves plus 78 dark), which was forty-seven pounds beyond its everyday average, once again provoking suspicions about the integrity of the procedure and perhaps causing some embarrassment to the institution's bakers, casting into relief their everyday lack of zealousness. Despite still another act of petty subversion—the hospital staff overheated Malisset's oven, subjecting his loaves to excessive weight loss during cooking—the economic flour (454 pounds) made 505 pounds of bread (17 plus 84). The experts judged Malisset's white "more handsome" than that of the hospital, and its dark "infinitely superior" to the institutional dark and little different from the white it served its officers. In an earlier test yielding three kinds of bread, Malisset drew at least 20 percent more flour from a setier than Delamare had in 1700, and whereas 41 percent of the commissaire's bread was white, 40 percent middling, and 19 percent dark, more than 90 percent of Malisset's loaves were white.[41]

Despite their pitfalls, Sartine continued to sponsor trials throughout the sixties, and the controller general invoked them to impugn the clamoring of bakers throughout the realm for an upward revision of the taxing structure. In April 1767, after conducting a comparative test of bread making from economic and conventional flour, Bricoteau, the chief baker at Scipion, concluded: "Economic milling yields a bonus of twenty-four to twenty-five pounds of bread per setier over the old way of milling." In November of the same year Machurin supervised an "experiment" at Scipion on economically ground grain that he himself had purchased at two different markets. Two setiers mixed together (weighing 235 and 228 pounds respectively) generated 445 pounds of bread. Substantially higher than bakers admitted to producing, the yield also contained a much larger proportion of white loaves than usually obtained (240 pounds of white, 101 mid, 104 dark). Milling and freight amounted to 2.5 livres per setier. Without breaking down the rubric or indicating whether it included a profit margin, Machurin reported the "cost of production" to be four livres per setier.[42]

In the first weeks of 1769, at the apex of one of the worst subsistence crises of the century, Sartine ordered a trial with a specifically taxationist goal: "to succeed as far as possible in fixing the relation between the price of bread and the price of wheat so that the public can obtain the relief it expects and at the same time the baker can find in his trade the means to subsist and continue practicing it." The lieutenant general himself attended parts of the test in order to emphasize the importance he attached to it, and he required the bakers' guild to assign senior observers to each phase of the operation in order to deny them the facility of rejecting a trial conducted, from their point of view, sub rosa. A real sense of urgency infused the task ("there is not a second to lose," Sartine repeatedly exhorted the actors): the idea seems to have been to apply the knowledge obtained from the "experiment" immediately to the badly dislocated provisioning arena. Le Ray de Chaumont, the head of the syndicate established to operate a royal emergency grain and flour reserve, and Duperron, an administrator of the Hôpital général with a special interest in subsistence questions, participated in the planning and analysis of the trial.

Elaborate precautions marked each stage: the police lieutenants in each of the five market towns where Machurin's agents purchased wheat testified to its range of quality, weight, and price; witnesses observed and certified the mixing process of the different wheats; seals were placed on the sacks and on the doors of storerooms; the baker delegates (jurés and anciens) were not permitted to approach the piles of meal. Milling took place at Malisset's economic factory at Corbeil, and for reasons that had to do both with the quality of the wheat and with the grinding operations, it did not go entirely smoothly. Bricoteau converted three sacks of flour (325 pounds each) into bread at the Scipion ovens: 404.5 pounds of white, 415 pounds of mid, and 440.5 pounds of dark. To demonstrate immediately, under real-life conditions, the viability of these loaves, Sartine ordered that they be sold the next day at the Saturday bread market at the Halles. Consumers purchased them at prices that appear to have been slightly below the going rate.

For reasons that are not clear, the bakers did not submit their formal evaluation of the trial until months later, perhaps because it was manifest to them that the authorities could not utilize this trial as a new standard. They denounced it as flawed in execution and misleading in results. They complained that the millstones had not been raised and properly cleaned before the grinding; that regrinding took place in a different mill from that in which the first passage under the stones took place, contrary to general practice; that the amount of grain/flour lost through ordinary manipulation was artificially reduced; that the jamming of the mill called "Economy" resulted from the miller's ill-considered lust to extract more than a reasonable amount of flour from the berry, which impelled him to place the stones too close together (grinding "tightly"); that this method issued in an increment of quantity at the expense of quality, for the resultant meal smelled of the stone and suffered a tainted whiteness; that the desire to maximize extraction slowed the operation to a caricatural pace and estranged it from the real conditions under which millers (and bakers) function: "But, one asks, who is the miller that can grind and convert forty-four setiers of wheat at three mills, as happened in this test, and spend two days and sixteen hours of time on just this task, that is to say, sixty-four hours at the price of thirty sous per setier delivered to the bakeroom? Where would such a miller find the wherewithal to pay the rent on his mills, the [major royal direct tax called the] taille, the standard overhead for the upkeep of the mills and their operation, the wages and food of his employees and servants, the sustenance of his horses, the cost of his wagons and equipment, not to mention the food and maintenance of his family?"

The bakers subjected the bread-making phase of the trial to equally serious strictures. Bricoteau inflated his yield by making an excessively "soft" dough in a manner never practiced by Parisian professionals. The results were particularly distorted in the case of the dark bread: "this quantity is exorbitant and could never be approached even remotely by ordinary bakers." The Scipion baker allegedly flooded the dough with water, formed the loaves in a mold instead of shaping them on the board, and used no dusting flour in fashioning them. Bricoteau's dark loaves had the color of gingerbread, "which the worker detests." Even worse, noted these self-appointed guardians of the people's well-being, this bread did not meet the physical requirements of manual workers, who needed energy and a persistent sense of satiety: "such a bread passes through the body without serving to nourish it." The consumer would end up buying an extra (third) pound in order to achieve the effect that two pounds of well-made bread normally provided in the course of a day. Finally, Bricoteau swelled the product by baking six- and eight-pound round loaves that suffered relatively less evaporation in the oven than did the long four-pounders. On these grounds, the baker guild rejected the trial as useless.

Sartine discarded the trial, ostensibly for very different reasons. He had no quarrel with the way it was conducted. On the contrary, he endorsed the methods used in both milling and baking, though he did not enter into specific details in emulation of the bakers. Accidental factors had compromised the test. Wheat from the '68 crop bore the deep-set traces of a wet harvest, and the problem of humidity was renewed and aggravated by the wet weather that accompanied the transport and milling of the grain. The initial infiltration of moisture diminished the productivity of the grain, while the recent humidity crippled the economic regrinding process. As a consequence the trial produced not more but less bread than it should have. "It would be appropriate to leave this trial in oblivion," the police reasoned, "since it could only mislead in the just evaluation of the conversion of wheat into flour."[43]

With a view toward escaping the hazards of hothouse trials in artificial circumstances, Sartine subsequently called on each of his senior commissaires to convoke "one of the best bakers in his district and one from the market located there and ask them to draft a statement containing the price of the wheat or flour that they have been buying, the costs of having the wheat converted into flour and delivered to the bakeroom, the costs of baking, the yield in bread, the price at which they sell the bread, and their weekly expenses for the wages, room and board, and other necessities of their journeymen and servants." Stressing the need for "exactitude" in the calculation of the bakers' profits, the lieutenant general reminded his agents not to overlook money made on the sale of charcoal, bran, and other residuals. It is not clear whether each commissaire was expected to participate personally in a trial undertaken by his baker in order to be in a position to verify the baker's claims step by step or if he was rather to identify bakers who kept reasonably complete records of their performance and in whose word and methods he could have confidence.

Commissaire Grimperel submitted "the account of the cost and yield of two setiers of wheat, according to the trial undertaken by Pierre Toupiolle, a master baker . . . and former juré of his guild, one of the best bakers of my district whom I believe to be an honest man of good faith and expert in his profession." Since Toupiolle conducted the trial several months before Sartine summoned his commissaires to collect this data, there is no compelling reason to presume that Grimperel was present. Despite its title, the *état* accounted for one setier weighing 235 pounds before milling. The baker wrought 124 pounds of white bread (from 96 pounds of flour) and forty pounds of dark bread (from 34 pounds), all in loaves of four pounds. In addition to remainders sold to starch makers and cowherds, he extracted twenty-four pounds of middlings that were probably to be reground into flour and thirty pounds of bran. Toupiolle reported total weekly expenses of one hundred livres.

In the surviving fragment of his commentary, Grimperel related that he was "frightened by the disparity" between Toupiolle's yield and costs and the results reported by Delamare from his trial in 1700, "which is all the more surprising since in 1700 wood, rents, and other expenses were significantly less substantial than in 1767." He speculated that "the persons charged with monitoring that operation in 1700 allowed themselves to be duped considerably." Though Grimperel did not work out disparities in the value of money, it was implausible to him that Delamare's baker spent 1,872 livres a year on wood vis-à-vis the sum of 1,589 that Toupiolle expended; that the baker's lodgings cost 1,404 livres a year in 1700 and 1,000 in 1767; that Delamare's baker disbursed almost three times more for miscellaneous costs, including wages, food, upkeep of equipment, and so on.[44]

An article in 1761 calling for a truly scientific grounding for the bread tax had already exposed the flaws in the Delamare trial, despite the fact that it was "carried out with much ceremony and many precautions." First, there were conceptual errors. For example, the commissaire did not distinguish what part of the meal he obtained was fit for bread making and what part fell into the category of unusable bran. Second, given the "quite feeble" yield in bread, it seemed likely that the ordinarily wily police official "was led astray by the juré bakers who presided with him over this enterprise." In order to get at the facts, "one must not ask questions of professional bakers [serving the public]." Instead this author recommended consulting institutional bakers, who presumably would not have the same investment in underestimating productivity. Yet bakers such as these labored to sabotage the Malisset test undertaken at the very moment this article appeared. Given the multiplicity of contradictory results and the disconcerting uncertainty they spawned, this author asked in urgent tones: "how could the magistrates of police ever fix the price of bread in an exact fashion during a calamity?"[45]

A tax proposition circulated in the Assembly of Police at the end of the sixties, predicated on a yield of 216 pounds of bread per setier, a comparatively robust result that the author of the piece in the *Journal économique,* and most contemporary subsistence scientists, would nevertheless have criticized as excessively modest. It assumed that a sack of flour, drawn from two setiers of wheat, would always be worth twice the price of a setier of wheat plus one livre to cover weather-linked interruption of grinding, freight costs, and market fees. Every time the sack price rose five livres, bread would sell for an extra sou per pound, more than twice the standard increment recommended by many authorities. For this same augmentation, the baker received an adjustment in compensation of eight sous (to cover the confection of 108 four-pound loaves). The baker's quite substantial reward "is necessary to encourage him in his trade, which becomes more difficult as the merchants raise prices." The authorities would be free, in the price intervals between the increments of 2.5 livres a setier, to favor either the consumers or the bakers as they saw fit. This schedule comforted the complaint of bakers articulated in the following years that they could not produce "excellent bread at two sous per pound"—the normative-fetishistic price considered through much of the century as the guarantor of social peace—"while buying wheat at twenty-two to twenty-four livres a setier, weighing 240 pounds."[46]

The "Rule" That Never Ruled

Throughout the century there was a temptation to think about the acceptable trajectory of the bread price in terms of a fixed increment (or reduction) in relation to grain, without regard for the critical factors that informed a properly drafted tariff, in particular productivity from grain to bread, milling methods and costs, and expenses for combustible and for lodging/shop. "Bread should be worth as many deniers per pound as the 240-pound setier of wheat is worth livres tournois," wrote a police agent, probably around 1750. Unlike many others who uttered this "rule," he offered "the proof." If a setier cost twenty-four livres, according to this logic a pound of bread (quality unspecified: it is fair to presume a good bourgeois loaf) would sell for twenty-four deniers, or two sous. A pound of wheat of sixteen ounces yielded twelve ounces in flour, three and a half ounces in bran and offal and half an ounce in grinding waste. But the twelve ounces of meal swelled into eighteen or nineteen ounces of dough, reduced by evaporation to loaves of between sixteen and seventeen ounces. Estimated at one-eighth the value of the wheat, and thus the bread, the bran would bring the baker three deniers "of profit," or 12.5 percent of the price of the bread. In the idiom of this analysis, profit ostensibly covered all the baker's costs, including milling fees, none of which was discussed.[47]

Delamare came reasonably close to this method in his schedule for bourgeois bread, not in setting the baseline but in charting the trajectory of augmentation. He followed it rigorously through the twenty-livres-per-setier mark, allowing the baker two extra deniers for every increment of ten livres per setier through thirty-six livres, and then slightly more for the next increment of ten livres. Joly de Fleury emulated Delamare but proved more faithful to the rule by granting the bakers slightly less of a bonus.[48]

Two ministers, at antipodes ideologically, propounded the same theorem. In the early seventies, Controller General Terray proposed to intendants a tax schedule based on the deniers-per-pound/livres-per-setier relation, with four deniers added for the costs of production. After sponsoring trials at Roissy, Controller General Turgot endorsed the bread-wheat price ratio, apparently without any bonus for the baker's costs. A religious historian and theologian whose language was permeated with grain metaphors, Abbé Para du Phanjas echoed the formula, stipulating that the grain must be "the handsomest" available.[49]

Foucaud, an itinerant grain specialist who toiled for the Paris police in the middle years of the century, envisaged a fixed norm that would reassure the producers, consumers, and administrators and against which one would take the measure of crises. In the best of all possible worlds, more or less everywhere in France, but particularly in the orbit of the capital, white bread should go for two sous six deniers a pound, middling at two sous, and the dark loaf at one sou six deniers. According to another controller general, Necker, the system actually practiced in Paris allocated the baker eight deniers per pound of white "for his profit and indemnification of his costs of production and sale above what the pound of wheat had cost him," four deniers for middling, and nothing for dark bread. "It is to be noted," explained the minister, "that, in making the first quality of bread bear the burden of the four deniers from which the bottom quality is spared, the intention was to ease the distress of needy persons; and that, in according in general four deniers per pound to the baker for each of the three sorts of bread, he has a certain and invariable profit, and thus he cannot increase the price on any pretext."[50]

Writing at the end of the century, in his celebrated *Leçons* on baking, milling, and rural piety, Louis Cotte treated four deniers a pound as the conventional amount accorded to bakers for all types of bread. Almost forty years earlier a writer in the *Journal économique* reflected a widespread conviction that the price of the bran and offal amply covered the baker's expenses; at higher prices, this compensation in kind constituted a further bonus. For his "honest profit" properly speaking, however, this writer granted the baker 13 percent of the price of the setier. "We believe sincerely that we must hold for constant," affirmed a memorandum dealing with subsistence issues at Troyes in the eighties, "that the milling by-products amount to and must compensate, at all times, in all circumstances, as in all countries, the costs, advances, and annual and daily expenses [of bakers] because they increase in proportion to these very elements of production." It was not clear, however, whether the bakers had a claim on a profit beyond this compensation.

Dr. Malouin distinguished the baker's "profit" (or "bénéfice" or "droit de négoce"), awarded to compensate "advances, misfortunes, losses, and waste," which he estimated at three deniers per pound of bread, from the baker's expenses, constituted by costs of management, labor, and production (including milling, bolting, kneading, cooking), which were always covered by the meal remains. Abbé Baudeau oscillated between the notion that three deniers was more than ample, for even in Paris real costs rarely surpassed one and a half deniers, and the more generous assessment that the Paris baker required four deniers to assure his costs and profit, one denier more than his provincial counterparts. Tillet raised the ante to four and a half deniers in his calculations for a universal tax model.[51]

None of these coefficients became the guiding standard for Parisian authorities, who preferred a wholly empirical and relativistic approach. Like fully ramified taxation schedules, the rule was both an intellectual exercise and a political instrument. It was a fixed theoretical model that entered into a permanent dialogue with the shifting terms of reality. It served as a convenient shorthand for measuring where things stood in absolute and in comparative purview. The police turned to the rule for normative inspiration and for argument for social rationalization. Yet, the Parisian authorities rejected the coercive invisible hand that operated in a mechanistic manner. This semihidden hand promised a kind of cover for all the actors, but it was predicated in the short run on a dubious collective willingness to embrace it as just, in the middle term on an equally tenuous practice of (escalating) reciprocal abnegation, and in the long run on a logic of repression. Locked into an itinerary whose meanders they could not influence, the police felt unbearably helpless. Even at the risk of appearing merely to extemporize, they preferred to retain their freedom to act and react according to changing circumstances and changing perceptions of circumstances.

Physiocratic animadversion to the contrary notwithstanding, the authorities responsible for governing the capital agonized over the question of how best to deal with price fluctuations. They did not belong to a single race of tropismatic regulators for whom *taxation* was both a philosophical credo and an instrumental panacea. They puzzled over the causes and the meaning of price formation. They groped toward an understanding of price as a complex sign system and a sensitive psychological construct as well as an expression of the physical encounter of supply and demand.

These officials knew that they could not be indifferent to prices, but most of them realized that prices could not be tamed by authoritarian means without exacting heavy costs, save perhaps in the very short run. Habitually, one's enthusiasm for fixing (which was not the same thing as regulating) Parisian prices varied inversely with one's proximity to the city's daily life. Controllers and procurators general were far more optimistic about imposing a single rule or model and treating Paris as if it were not sui generis, a sprawling agglomeration with a vast permanent and large floating population, a highly stratified social structure with a huge population of poor whose potential volatility worried everyone, and a very complex provisioning system built on a combination of ordinary and extraordinary supply zones and of decommercialized sectors and rival collector markets. The police never regarded the principles of political economy as mechanically antithetical to the imperatives of police. Preserving public order and fulfilling the social contract of subsistence were their cardinal objectives. If that required certain forms of intervention and regulation, it also implied nourishing and sustaining a vigorous provisioning trade, whose chief engine was acknowledged to be liberty.

The police realized that prices had reciprocal rather than linear relations and that the bread price shaped the grain (and flour) price even as the latter helped determine the former. For technical and political as well as intellectual reasons, the authorities rarely envisioned the direct taxation of grain and flour, though censuses, requisitions, and other regulatory measures had proxy effects. In practical terms, if one hoped to get at grain or flour, it made more sense to attack bread, a more accessible target. Beyond the logistical complexities of imposing a bread price schedule, however, if a viable one could be forged, the Paris police considered that a degree of freedom could more efficiently mobilize bakers of all stripes than an apparatus of permanent compulsion.

Contrary to general belief, bread prices in Paris were habitually *not* set by the police. In times of relative subsistence ease, they were largely the unmanaged (but never unmediated) fruit of what we call market factors. On the basis of the combined (and weighted) *mercuriales* of the capital and the hinterland, and the more or less relentless demand of Parisians, the bakers put a price on their loaves. Ordinarily, this price required neither police approbation nor guild vetting. A little lower in the market than the shop, a little lower in the late afternoon than in the morning, sometimes a little lower for the water carrier than for the banker or the marquis, the bread price also varied somewhat between the big Wednesday and Saturday market days as a function of news, rumors, leftover merchandise, new deliveries, and weather. There was no "normal" maximum bearing the lieutenant general's imprimatur. Provided the bakers respected standard operating procedures, they policed themselves (under the vigilant surveillance of their customers). Issuing from imperfect markets that indirectly reflected the regulatory presence of the police (as *idée force* as well as system of controls), the current or common price was not in some pristine sense unregulated. It was certainly not, however, either imposed by authority or controlled by design. The current or common price shouldered the destiny of Paris so long as calm prevailed in the provisioning arena. The commissaires and their staffs monitored the bakers and issued summonses to those who flagrantly defied the current or lapsed in their expected levels of bread supply.

The authorities prescribed ceilings reluctantly, under duress. They studied different ways to determine the fair price. They thought concretely in terms of cost-benefit calculations and abstractly in terms of justice. They worked out a host of detailed (sometimes dialectical) scenarios seeking various levels of social, economic, and political reconciliation and synthesis. The police imagined scientific frameworks that would place the question beyond contentious discussion. They conducted tests to try to arrive at credible conventions regarding cost and yield. They consulted with the bakers on a regular basis. The authorities improved their information collection and their tools of analysis as the century unfolded. Well before 1789, they had rehearsed in detail most of the subsistence issues that the revolutionaries would attack with such telescoped intensity. Ultimately, the police opted for an opportunistic and improvisational approach to prices that took account of as many of the operative variables as they could apprehend.

Chapter 19

Policing the Price of
Bread, 1725–1780

In a succinct and dramatic fashion, the price of bread represented many of the deepest apprehensions and fondest hopes of Old Regime Parisians. The product of many different forces, material and moral, often freighted with ambiguity and contradiction despite its ostensible precision and clarity, the price critically shaped the way in which the governors and the governed thought and acted. It introduced consumers precociously to inherently complex political arithmetic as well as to intuitive home economics. Even as it invited the police incessantly to intervene, it reminded them remorselessly of the limits of their efficacious power. The price changed, but in certain ways it always remained the same.

The Dearth in 1725

In 1725 Paris suffered from a grave subsistence crisis that produced one of the few major riots to jolt the capital during the eighteenth century prior to 1789. Beginning with an attack on bakeshops in the faubourg Saint-Antoine, the June uprising cast into relief the problem of mediating between the bakers and the consumers.[1] In its aftermath police tactics remained very much the same as they had been before its outburst. The major difference was a welcome increase in the flow of emergency grain and flour supplies, and a more active effort on the part of the commissaires to reassure consumers, especially at the Wednesday and Saturday bread markets, in the hope of reducing panic buying and of reestablishing the "confidence" that had been lost. The bakers "feared" René Hérault, the new lieutenant general of police, because he made it instantly clear, by arresting several "mutinous" bakers, that he would not tolerate brazen disobedience. But he also enjoyed an unearned increment of deference in the wake of the riot, which deeply alarmed bakers of all categories, and reminded them viscerally of their vulnerability.

Hérault regularly convoked not merely the guild jurés but leading bakers from each of the quarters, including the faubourg Saint-Antoine, to meetings, often in the context of the weekly sessions of the Assembly of Police, in order to inform them of the provisioning situation and to "announce his orders." Several times in September he "demanded" a lowering of the price; the vast majority of bakers complied. Systematically, he treated the city and faubourg bakers, who were "under his heel," more harshly than the forains, who were widely dispersed and thus difficult to control, and whose critical role required "that they be treated with great tact" on the part of the police, who did not want to deter them from continuing their supply. This solicitude explains the astonishingly light penalty of a thirty-livre fine levied in late October on a Gonessien who sold his white loaf "six deniers above the current price of the market" and on a confrere from Montmagny who asked a full sou per pound more than the others for his *bis-blanc*.

The forains understood that they enjoyed a certain leverage, and tried to take advantage of their position. Ordered by Commissaire Parent at the Saint-Germain market to roll back her six-denier-per-pound increase, Chamlire, a bakeress from Versailles, insulted him, swore to have him dismissed from office, and removed her loaves from display in order to take them where she could sell them at a price "that pleased her." On Parent's orders, Inspector Laurent followed her and watched her arrange to store her bread with a master barrel maker. The police seized the merchandise, arrested the artisan, and issued a summons to Chamlire, still too important an actor to send to jail with her accomplice despite her transgressions.

One of Hérault's commissaires, Delavergée, warned the lieutenant general of the perilous disorganizing implications of this sort of indulgence. Following Hérault's instructions, he held the line at the Carouzel market, constraining the bakers to maintain the price of the previous market. To underline the persistent gravity of the situation, Delavergée dispatched three bakers, who had vehemently demanded an increase, to suffer the personal reprimand of Hérault. Instead, they returned triumphantly to the marketplace announcing that the police chief had accorded them a three-denier augmentation. Infuriated at this dissonance, the other bakers brashly informed Delavergée that "they will go directly to [Hérault]" when they did not like the commissaire's instructions on pricing. By early December there were mounting complaints from consumers, corroborated by the grassroots police, that the bakers had fully recovered their "insolence." A plea for more vigorous action reached the lieutenant general in his capacity as "the father and protector of so many poor wretches who have been crying plaintively for so long."[2]

For the next several months we can track the evolution of the situation in the weekly (or twice-weekly) reports of Commissaires Labbé at the Saint-Paul bread market and Duplessis at the Halles bread market, supplemented by occasional contributions from their colleagues Menyer in the faubourg Saint-Germain and Divot at the Cimetière Saint-Jean. They lived each market day with overwhelming intensity, as if the fate of the kingdom would be played out within the week. There was no room for a middle- or semi-long-term perspective. What happened Saturday seemed to have a terrible finality about it, yet at the same time what happened the following Wednesday could radically alter mood and prospects. Often there was no articulated or objectified transition from gloom to joy or vice versa. The rhythm of their observations and anxieties underlines the extremely limited relevance of average monthly bread prices for an understanding of the lived experience of the different actors.

During the first half of January 1726, Labbé and Duplessis agreed that the bakers behaved "reasonably," furnishing ample quantities and asking a price commensurate with supply. Each boasted that his market "has never been more peaceful." Around the sixteenth the effects of brusquely inclement weather began to be felt. Prices rose. Labbé avowed that he failed to obtain the cooperation of the bakers, and Duplessis noted "much chagrin" among consumers, one of the most portentous manifestations of which was the "murmuring" of women against the price rise. Ever loath to bully the bakers, Duplessis uncharacteristically drew the line on the twenty-third: the bakers must not allow the price to mount any further, for the result would be the explosion of popular anger "with compelling reason."

The bakers and the weather suddenly turned more temperate, and by the last days of the month, prices had begun to erode. On the same day, 6 February, that Labbé complained that bread lagged behind the falling price of grain, provoking "the anger of the poor," Duplessis depicted burgeoning "tranquility." A week later, however, the commissaire of the Halles could not contain his frustration at the behavior of consumers who did not cleave to his rule of reason: "The populace is a ferocious beast who is very difficult to satisfy. Although the stalls of the bakers in the markets are abundantly furnished with handsome loaves of good quality that are being sold at three deniers per pound below the price of the previous market, the populace here is still murmuring and complaining more than it had on the preceding market days."

A month later, both commissaires reported "abundance," a downward price movement, and signs of collective composure. "Here we are saved!" Labbé proclaimed, "everything transpires as in the days of calm." By the end of the month he saw no need to maintain soldiers in the markets, posted there on highly visible patrol since the rioting of the previous summer. Duplessis registered successive price decreases, in part it seems, as a result of pressure placed on him by the lieutenant general to tighten the screws on the bakers, who habitually profited from a large margin of freedom in this commissaire's jurisdiction. While Labbé, whose bakers were decidedly less recalcitrant, spent less and less time in the marketplace, Duplessis somewhat timidly, and remorsefully, imposed a ceiling of three sous six deniers for the standard white, compelled bakers to reimburse overcharges, and dispatched the most rebellious ones to account for their deportment to Hérault.

In early April, bread gently followed the downward slope of grain and flour, though not without more pressure on the bakers than Duplessis liked having to apply. Still, he could not obtain their accord to drop below the psychologically critical threshold of three sous per pound. The climate in the market abruptly changed when illness obliged Duplessis to cede his surveillance for more than two weeks to his confrere Aubert of Saint-Denis, who took a much harder line with the bakers. "I disposed all the bakers of Gonesse, who have been selling their bread at three sous three deniers," the newly arrived commissaire wrote serenely, "to lower it to three sous the next market day. That is what they promised me." All the bakers complied, Aubert confirmed at the next market, except a woman from Gonesse, whose loaves he angrily threatened to cut into little pieces and distribute if she continued to resist.

At the Saint-Germain market, Commissaire Menyer repeatedly expressed deep skepticism about the civic potential of the bakers. They indulged their greed recklessly, ignoring orders and setting the price "as they pleased." With Hérault's support, he initiated a crackdown at about this time in order to "reduce the bakers to reason." Like Menyer, Divot at the Cimetière Saint-Jean detected signs of "cabal," of concerted action to reduce abundance and increase prices. To demonstrate their good faith, he required each of his bakers to agree to, and symbolically perform, an act of "submission" that obliged him or her not to surpass the "current price" or the closing price of the previous market. Bakers who defied him were either arrested immediately, in cases of special gravity, or subjected to daily police harassment that focused in particular on verifying the appropriate marks and weighing each loaf and confiscating every one that betrayed a mere fraction of an ounce of delinquency. (One day later in the year Divot confiscated over a thousand pounds in a few hours.) In April Commissaire Delavergée at Place Maubert also arrested bakers who failed to defer to the current price standard.

The bakers celebrated Duplessis's return at the beginning of May by again traversing the three-sous frontier. Despite governmental monetary manipulations that confused and hampered economic exchange, the commissaire believed that the crisis was virtually over. Although he remained deeply committed to a soft-sell approach in dealing with the bakers, he found himself again compelled to resort to "the avenue of authority" at the end of the month in order to obtain cooperation. He required bakers who charged over three sous to reimburse consumers, and he sent the culprits to Hérault "so that he could impose whatever penalty he deemed apposite." The Assembly of Police had begun to wonder whether Duplessis was capable of bringing his large market into line with the other bread markets, where more unsparing hands presided. He had promised the procurator general and the lieutenant general that he would not allow the bakers to maneuver around him: "I have kept my word," he announced, with a sigh of relief.

If the "perfect tranquility" of early June (abundance, brisk bargaining, "no overzealousness") gave way to weather- and harvest-linked inquietude at the end of the month, by mid-July abundance had returned, and bread stood steadily in the one-sou to three-sou range. When Hérault detected signs of another "plot" in late August, he ordered each commissaire to be present in the market no later than sunrise in order to "prevent them from attempting the least increase on any pretext whatsoever." Bakers who attempted to sell above three sous he branded as "rebels," who would have to answer directly to him. The willful determination of the bakers to raise prices, led by the faithless Gonessiens, who broke their morning pledge to stick to the current level, shocked Duplessis. He acted swiftly by arresting four of them: "the [continued] detention of the prisoners who are now behind bars will contain the others in their duty," he predicted. He turned out to be right: the others grumbled that they might have to give up supplying the market, but they all showed up and remained under the three-sou beacon.[3]

Grassroots Laissez-Faire

It is worth underlining the assumptions that governed the thinking, and acting, of Duplessis, who collaborated with Delamare and succeeded him as senior commissaire for subsistence questions. This police official did not have to await the instructions of the celebrated coterie of eighteenth-century *économistes* in order to practice a market-level brand of liberalism whose rationale he claimed to draw, like a good Newtonian, from the world that he experienced every day.

Though he agreed that the bakers had certain incontrovertible responsibilities, Duplessis saw them first of all as "businesspeople" who required a wide latitude in order to operate. Only the liberty to seek gain, and to seek relentlessly to enhance that gain, motivated them to act. To fail to take this psychology into account, or merely to revile it as antisocial and susceptible to sanction, was to turn away from reality and mortgage police strategy to obsolescent fears and self-defeating authoritarian impulses, the commissaire believed.

Thus Duplessis's watchword was laissez-faire. "All that I urge upon the bakers is a commitment to the abundance and the quality of bread," he informed Hérault; "as for the price, I leave it up to the [consumer-]people." When the people paid too much, it was frequently their own fault, for they forgot to play the role that market rationality assigned them. "It is the people themselves who are to blame for the high price at which the bakers sell their bread," he explained to the procurator general; "they wrench the loaves from the hands of the bakers and pay whatever price the bakers ask." Pressured by Hérault in early March 1726 to adopt a sterner approach, even as he acquiesced, Duplessis renewed his liberal profession of faith: "nothing should be forced, the reduction will come little by little of its own accord." He intoned this refrain time and time again. In mid-April he delicately cautioned the lieutenant general: "But I believe (if you will graciously permit me) that nothing must be forced and that it is a good thing for this diminution to take place gradually." A month later he wrote: "You will agree that a voluntary price diminution is much better than when [the reduction] is forced—it holds up far better." The following week, with deep conviction, he counseled Hérault "that nothing should be forced."

Unlike most of his colleagues, Duplessis did not envisage a permanent and obtrusive mediating role for the police between bakers (and other sellers) and consumers. The exponent of a low-profile administration, he hinted that the best sign of efficacy was the absence of any perceived need, and perhaps even of any inclination, to intervene. Within a broad framework of mutually acceptable rules, the bakers had to work problems out for themselves, save perhaps in truly catastrophic supply conditions. Even on matters of quality, where Duplessis believed that it was perfectly legitimate for authorities to act, he preferred to see the marketplace allocate the inputs and outputs. There was no reason to strike an exemplary blow against the bakers who offered a dark loaf with a "bad taste" during the late summer of 1726. Despite their need, consumers would avoid it because there were ample quantities of bread that was "well conditioned." They would remember which bakers proffered the inferior loaves, and they would punish them severely by denying them business and discrediting them in the community.[4]

Commissaire Duplessis did not share the skeptical and indeed negative view of the bakers that many of his colleagues entertained. He argued for the priority of an economic over a political logic in order to exploit the market forces that drove exchange and circulation. The only rational policy was to induce the bakers to collaborate on the basis of their own self-interest. Duplessis's colleagues retorted that the bakers were naturally unwilling to moderate their avidity and make compromises in the public interest. They pointed out, not without justification, that each gesture of appeasement during the crisis of 1725–26 had merely emboldened the bakers to take further advantage. At the grass roots as at the summit, the provisioning police was not a monolith.

The Decade After

During the largely tranquil decade following the crisis of 1725–26, the Parisian authorities allowed the bakers much more latitude in pricing. Still, during interludes of seasonal uncertainty, speculative tension, and/or psychological stress, they intervened to hedge baker proclivities deemed subversive of good order. For planning a "monopoly" in June 1727, faubourien Claude-Philibert Cousin went to jail until he had discharged a relatively light fine of one hundred livres. The very next market day Commissaire Divot arrested two widows from Gonesse for pursuing Cousin's path toward a price increase. Hérault berated market bakers in August for testing the will of the police and unsettling the mood of the public by announcing a forthcoming augmentation, as if to suggest that such an outcome were beyond their control and thus not their fault. In the fall bakers registered their continuing discontent, and effrontery, by posting a notice over a police sentence finding one of them guilty of price maneuvers that "insulted and threatened the government."[5]

In 1728 the police attacked another postharvest "monopoly" by imposing fines of three hundred livres on three shop bakers selling above three sous a pound. "Although all foods and goods have decreased considerably in price, and the harvests in wheat and all other grains this year and last were abundant," read the sentence, "nevertheless the majority of the bakers of this city, far from proportioning the price of bread to that of grain, have the temerity to sell at an excessive price." At the end of the year "their abject greed" erupted again. This time Hérault opted for a more measured response, taking account of realities to which the consumer-people had to learn to adjust, particularly when there was a little social margin with which to work. The leading baker rebel, a Gonessien, would be summoned for a personal reprimand, instructed the lieutenant general, "but without issuing a summons that would lead to a condemnation and to a fine because the public must learn to get used to price variations that we cannot avoid, especially when they are without [grave] consequence."[6]

Hérault judged it safe to allow the price to rise in the winter of 1729. Informants warned that the public was agitated by rumors of certain market bakers saying "that they will be lucky if the price does not climb to five sous a pound before the end of the winter." Some consumers blamed the increase on the timidity of the commissaires, whose vigilance wavered. Divot was one of the commissaires who felt uneasy with Hérault's sporadic liberalism, and who reacted to it by cutting an uncharacteristically low profile. After two successive market days in September 1730 during which the bakers raised prices, he wrote to Hérault announcing his inertness, soliciting orders, and implicitly distancing himself from a policy of indulgence: "Since I do not know if they have good reasons to justify their price increase, I stayed away from the marketplace for fear that my presence would only enhance the alarm of the public." Five years later a reinvigorated Divot denounced two masters and five forains for having increased prices on three consecutive market days and simultaneously reduced supply, "although there was no change at all in the price of wheat." There is some evidence that Hérault's balancing strategy commanded public confidence. According to a police agent, it was widely felt on the eve of the *soudure* of 1737—the hiatus between the exhaustion of the previous year's crop and the forthcoming harvest—that the "criminal" and "reprehensible" pricing practices of the shop and market bakers would imminently be checked by the lieutenant general, "who is going to tax the finest bread at eighteen deniers per pound."[7]

Keeping the Shops and Markets Stocked

One of the rules on which Commissaire Duplessis himself insisted was that bakers supply their stalls and shops in bad times as in ordinary times—unless they could demonstrate convincingly an incapacity to do so. The police viewed the interruption of supply not as a mark of strain but as the sign of a hoarding impulse, a desire to drive prices upward, and an effort to destabilize the market. When several bakers stopped furnishing, or cut back sharply, in near unison, the authorities immediately suspected a "plot." Punishment for abandoning supply—an invidious and furtive way of shaping police behavior—was far more severe than for overt attempts in the shop or market to extract a higher price from buyers. Upon instructions from the Assembly of Police in September 1725, Hérault issued an ordinance denouncing bakers who "by a criminal pretension did not adequately supply their stalls and shops" or who absented themselves completely "despite the recent return of abundant wheat stocks to the ports and the Halles." Such bakers would be liable to a fine of three thousand livres, loss of their mastership, and corporal punishment.

Several days afterward a baker on the rue de la Cordonnerie suffered the full fine, the immediate closing of his shop, and expulsion from the guild for failing to supply his customers ("he did not have a single baked loaf in his shop"). A month later the same fate befell Jean Lefevre, a baker from Les Porcherons who had left his stall at the Halles empty for a day. The tactics of two forains and two masters at the Carouzel were more subtle. They reduced their supply by about half, after having "menaced" the commissaire with a boycott. Reluctant forains from Versailles found some support from their hometown police chief, habitually a cooperative ally of the Paris lieutenant general, who reported that they could not absorb further losses imposed by the rigorous price ceiling: "I could not compel them to go to Paris when they tell me that the reason is that they no longer have the [financial] means to undertake their trade in light of the losses that they have suffered."

Toward the end of November the Assembly of Police let it be known that "if the bakers supplied the markets, it might be possible to pardon them," but it was not clear how retroactive this proposition was meant to be. Just before Christmas another ordinance reaffirmed the obligation to sustain an adequate supply and rehearsed the same penalties flourished in the September measure. In January 1726 two faubouriens who had boycotted the Cimetière Saint-Jean faced fines of three thousand livres and a loss of stalls. Several months later another baker at that market endured the same penalty. Wasting no time, Commissaire Divot immediately assigned her place to an apparently docile faubourien, who supplied the stall, but with unmarked loaves of a "defective quality."[8]

While the police tended to relax their rigor in executing the regulations in calmer times, the issue of regularity of supply was too delicate to accommodate laxity. The sanctions against reducing or abandoning bread furnishing served as reminders of the semipublic status that their provisioning responsibilities conferred upon the bakers. In November 1727 the police struck Michel Ferry, a faubourien, with a fine of three thousand livres and loss of market prerogatives for having boycotted the market for six consecutive exchange days. Commissaire Divot refreshed the minds of the bakers at Cimetière Saint-Jean that they had no right to sell or rent out their market stalls, which were not private property but conditional concessions "that belong to them only to the extent that they keep the said stalls copiously furnished with bread." Two Gonessiens suffered the same punishment as Ferry in July 1729 when they cut off their bread flow in order to express their anger at police price constraints. Two years later another Gonessien lost his space and received a fine of one thousand livres for abandonment. He had tried to sustain his legitimacy by sending a small child with four or five loaves of bread to occupy his stall every market day. In 1734 and 1735, a half-dozen bakers in Divot's precinct lost their spaces and faced fines of three thousand livres for failure to fulfill their mission.[9]

Bakers who removed bread at the end of the market day instead of selling it at a more or less substantial discount committed a less grave trespass in the eyes of the police but were nevertheless guilty of a kind of hoarding that had deleterious effects on prices and on consumer well-being. The police patrolled the streets leading away from the markets and inspected the carts of bakers, the baskets of servants, and other suspicious containers that passed on the backs of horses or persons. Alerted at 7:00 P.M. that large quantities of bread had remained unsold in the Halles and other markets at the beginning of May 1725 ("because of the excessive price they had wished to put on it"), Commissaire Laurent intercepted several persons, including a regrater bearing bread marked with the initials N. R. (for Nicolas Rafron, a baker at Saint-Germain-en-Laye) and a second woman who was apparently transporting the remaining loaves of another market baker to his shop in Les Porcherons. The bread was seized, and the bakers suffered fines of ten and twenty-five livres.

Militarizing the Markets

In the aftermath of the riot of June 1725, several detachments of royal troops were placed at the disposition of the lieutenant general in order to help "contain the people." They were assigned to the bread markets, where it was hoped that they would also assist in containing the bakers. "Sergeants of confidence" (five at the Halles, three at the Augustins, four at Maubert, three at Cimetière Saint-Jean, etc.) were asked to observe the behavior of the bakers, to testify to the practice of bargaining, and to track the flow of bread. Large numbers of troops were held in reserve for rapid deployment in one sector or another in case of trouble. At least one commissaire wondered whether the presence of soldiers did not spur bakers, who once again felt relatively secure, to hold out for higher prices.

It is certain that recourse to the soldiery was never an unequivocally positive step. From the vantage point of consumers, militarization of the markets seemed to confirm the disaster cues they read elsewhere. According to one police informant who followed consumer opinion, the people "say that as long as there are soldiers stationed in the markets, the outlook is very bad." The Assembly of Police had vivid memories of how easily the forces of order metamorphosed into the forces of disorder, as during the dearths of 1693, when groups of up to ten soldiers pillaged baker stalls in the markets, and of 1709, when soldiers watched their wives do the plundering while they themselves engaged in a profitable regrate trade in bread. To forestall serious dissatisfaction, the government significantly increased the wages of the gardes françaises in late August 1725, when a squad openly complained that "famine" had arrived.[10]

In at least one case, Dangoumois of the Chaumont company, one of the sergeants assigned to the Palais-Royal market, appears to have acted to increase the price rather than confine it. He instructed bakers to sell at almost one sou per pound above the going price. Several consumers "protested very vehemently against the sergeant and some even said that the rest of the bread should be pillaged." The commissaires received angry denunciations. Rumors spread linking Dangoumois with various schemes, including an elaborate bribery operation mounted by a cabal of bakers.[11]

Sullied Commissaires?

It was bad enough when outsiders such as soldiers became implicated in price-gouging designs. In terms of public and police morale, it was devastating when suspicions tainted the commissaires themselves, the principal protectors of the people in the markets. One recurrent bruit concerned Commissaire Labbé, who had been one of the major police figures in the riot in the faubourg Saint-Antoine. "The worthy Commre L'Abbé, who has never done anything but evil, arranged for an increase in the price of bread on several market days, either to get his share of the bakers' take or to hurt the public and push them to rebellion against the government," wrote an informant in September 1725. Two days later Labbé found himself again cited among "the bakers and commissaires who are crooks" and who collaborated so that "we will not see the price of bread go down in the near future."[12]

In November Hérault received a signed denunciation that impelled him to convoke the author and open an investigation, the results of which remain unknown. "All the efforts that you make to hold things in line in order to prevent bread from becoming even more expensive will be for naught," the witness told the lieutenant general, "if all the commissaires imitate the actions of Commissaire Tourton, who told the bakers at the Palais-Royal market this morning to augment their breads by two sous. The proof is that several persons who saw and heard what I have just reported ran off immediately to protest to the exempt stationed at that market." Excuse me for the liberty I have taken in sharing this information with you, added the correspondent, "but my conscience obliged me to do it."[13]

Another observer, this one anonymous, related an anecdote that further testifies to the diffusion of the representation of police corruption. In a tavern near the Halles, he sat near two bakers who were counting their money at the end of the market day. Each asked the other

combien gagne tu aujourdhuy. L'une disoit deux cent livres l'autre cinq cent. L'une disoit tu a deux boutiques et moy je nen ay qu'une. Elle se demandoit aussy ce quel avoit donné au commissaire. L'une dit un Louis lautre dis je nay donné qu'une ecus [*sic*].

[how much did you earn today. One said two hundred livres, the other five hundred. The first said, "you have two shops and I have only one." She wondered also how much the other had given to the commissaire. One said a louis, while the other said, "I only gave a sole écu."]

That the story contains implausible elements is less significant than that it was articulated and recounted, and linked to a larger pattern of suspect behavior, a sort of grassroots, modest version of the famine plot accusations that led to the fall of the duc de Bourbon, the chief minister, his mistress and adviser, Madame de Prie, and d'Ombreval, the lieutenant of police, at the end of the summer of 1725.[14]

It is significant that the charges of corruption touched only a small number of commissaires. There were more testimonials to the devotion of the commissaires than assaults on their honesty. By and large the commissaire remained an accessible and credible recourse for consumer complaints. The police encouraged the public to bring alleged infractions to their attention. Without a genuine complicity between the public and the commissaires, it would have been infinitely more difficult for the provisioning police to do their work. At the local level, consumers regarded the commissaire as more or less sovereign, capable of exercising summary justice on the spot. Take, for example, the case of a woman seeking to buy bread in a shop in the still tense postriot faubourg Saint-Antoine. Furious that the baker demanded a price she thought unfair, she hailed a passing man "wearing a robe" whom she took to be a commissaire and whom she implored to pressure the baker "to give it more cheaply." It turned out that the man was not a commissaire at all but a municipal huissier on the way to serve papers. Nevertheless, he intervened in favor of the woman and was supported by a large crowd in a vindictive humor that suddenly congealed outside the shop, much to the alarm of Commissaire Labbé, who arrived to investigate.[15]

The Bakers Seek Shelter

Some bakers tried to deflect suspicion from themselves by suggesting that the police—not at the relatively modest level of the commissaire but at the highest reaches of the lieutenance—participated in a heinous governmental plot to create and sustain shortage. The bakers hoped to convince the public that the abscess was not fully drained, that the famine plot had not ended abruptly with the disgrace of the duc de Bourbon and his accomplices, that Hérault, who cultivated an image of incorruptibility from his first moments on the job, had picked up where his predecessor d'Ombreval left off. Frustrated by his inability to find flour in sufficient quantity at a reasonable price, a master baker aptly named Bled intimated that Hérault protected "all the big grain merchants who enriched themselves as never before" by hoarding and manipulating huge stocks. Shop bakers in the Saint-Denis district denounced the severity of police sentences against their confreres as a smokescreen to conceal official perfidies: "the government . . . hurls powder in the eyes of the public in order to cover up the profitable trade that it continues to ply in grain and flour." Rumor had it that "Hérault cedes to the views of the government, in order to delude the coarse and uneducated public that the bakers deserve to be punished."[16]

Other bakers called on the police to aid them. Prices rose in part, they contended, because of their incapacity to bake, the result either of their inability to find supplies or their state of financial/commercial ruin, or a combination of both. The crisis subjected the bakers to a double credit crunch, one active and one passive. Given the paucity of goods, the speculative climate, and the pervasive uncertainty (compounded by acute monetary instability), grain and flour dealers, and millers, called in the paper they held on the bakers and refused to extend further credit, save to the minority of notoriously robust

Popular retribution for a baker's betrayal: baker François is summarily executed for hoarding fancy loaves reserved for wealthy customers—bread aristocrats—in October 1789. *Source: © cliché Bibliothèque nationale de France, Paris.*

practitioners. The bakers, in turn, pressed their clients to make good on their often substantial debts, at precisely the moment when many of them were least equipped to pay, or wily enough to realize how little leverage for recovery the bakers actually possessed. The jurés urged the lieutenant general to press suppliers, and their brokers, to renew lines of credit; there was even a plan for the government to guarantee repayment through the brokers. Individual masters petitioned Hérault to protect them against judicial seizures of their tools of work and their personal property initiated by impatient creditors. A priest at Gonesse called Hérault's attention to "the sad situation in which the corps of our Bakers find themselves today." Weakened by the disorganizing impact of the dearth ("debts for merchandise overwhelm them, and the cost of milling does them in completely"), and in some cases "reduced to the last misery," as well as panicked by rumors that they would be forced to sell at a loss, many of them would not be able to hold on much longer.

The prospect of significant erosion in the baker corps seriously worried the police. (The very large turnover in the forain corps, palpable by the end of 1726, proved that their fears were well founded.) The authorities extended various forms of assistance and pledges of future compensation (combined with dire threats of punishment for unjustified abandonment) in the hope of keeping bakers in the game. The government became a direct supplier of cheaper grain and flour. It distributed cash subventions to certifiably desperate bakers. It intervened to dissuade creditors from forcing the bakers to file for business failure. It helped bakers recover huge debts owed by flagrantly delinquent clients, and bakers continued to demand this help after the crisis ended in recognition of their civic zeal (thus the request formulated in 1727 for aid in recovering 808 livres from a weaselly wine merchant in light of the baker's efforts "to assist the poor to the best of his ability during the dear year of 1725"). Solicitude for baker survival led one police adviser to recommend a strategy utterly repugnant to most bakers. In order to prevent a hemorrhage of bread suppliers and "avoid the ruin of the bakers" that was increasingly plausible, he recommended the imposition of a strict tax ceiling on grain and flour (the elite no higher than twenty-five livres, which would mean top white loaves at no more than three sous a pound) in order to permit bakers to acquire merchandise.[17]

The stronger bakers needed protection against the police rather than assistance from them. To extricate themselves from difficulty or to fortify their positions, they called on influential patrons whose appeal to privilege the provisioning authorities resented. A close friend of a commissaire wrote him in July 1699, "Me Lepage, your bakeress and mine, was condemned at the police audience yesterday to a fine of 150 livres. She asks only a chance to explain herself," and thus begged for an audience with Lieutenant General d'Argenson. In 1709 the latter complained disgustedly that baker patrons in the parlement and in the court undercut his authority. "The bakers of Gonesse," he protested to the controller general, "continue with impunity to sell their white bread at an excessive price, . . . but they are *too powerful* and *too well protected* to be concerned about what the commissaires or I might have to tell them."

In November 1725 the bishop of Rennes interceded in behalf of two bakers and obtained a gigantic reduction in fines of one thousand livres, relying casuistically on the bakers' assurance "that they went astray as a result of ignorance." A little more than a year later the procurator general himself thanked Hérault for arranging the remission of a fine levied on baker Picard: "it is true that he married the governess of my children, but the more reasons I have to protect him, the more he must conform to the need to pay strict attention to observing the regulations. He is very fortunate, thanks to your good grace, to have escaped the rigorous punishment he merited."

In 1740 a landowner named Des Borus obtained an exceptional "indulgence" for his tenant farmer Joly, a forain who was heavily fined for selling a loaf six deniers above the tax. Two months earlier a Mlle Forcet failed in her effort to win grace for her baker, "who had always been reputed just, both for the excellence of her bread and for its price"—at least until she was caught violating the tax that morning. Apparently this patroness did not carry enough clout, for Marville, Hérault's successor, rejected her plea: "I believe that she is far too good a citizen to oppose an example that is necessary in order to suppress the avidity of the bakers." In 1742 a prominent client of a faubourien had no more success in sheltering his baker despite a thinly veiled hint that the baker might be compelled to stop furnishing his market stall.[18]

Restricting the Range of Breads

Bakers claimed to have been heavily damaged by the restrictions imposed on the kinds of bread they could make. To conserve grain and to impose a higher rate of flour extraction and thus an increase in productivity, in crisis periods the authorities prohibited the making of all types of white bread, ordinary and luxury loaves. Typically a parlementary arrêt enjoined bakers to produce only two sorts of bread, *bis-blanc* and *bis.* Quite rightly, the bakers viewed this measure as a displaced form of *taxation,* the purpose of which was to facilitate the control of prices and enhance available supply.

D'Argenson encountered sharp resistance from the bakers after the promulgation of the two-sort restriction in June 1709, a restriction apt, in his words, "to bring the bakers to their senses, to diminish the consumption of wheat, and to disconcert in advance those who hoard it." Forains from Corbeil deserted Maubert on the pretext that they lacked grain. Counting on the *fines bouches* of their grand seigneurial and parlementary clients, the Gonessiens defied the new rule by baking white and delivering directly to houses instead of furnishing the market. Barely concealing the message of blackmail, they petitioned the procurator general to exempt them from the two-sort requirement if he wanted to see them stay in business. With mixed results, d'Argenson fought hard to subjugate them. Bakers who brought no *bis* yet offered white loaves suffered fines of a thousand livres; unless they paid immediately—a difficult burden for most bakers to bear—they went to jail pending full settlement.[19]

During the crisis of 1725, after months of tergiversation, in August the police bustled through two stages of conservation/restriction policy. On 17 August a police ordinance required bakers to furnish one-third of their loaves in each of the three major classes of bread, white, middling, and dark. There had been an excessive amount of soft-dough luxury bread, now banned, and a radically insufficient amount of *bis-blanc* and *bis* on the market. "Given the fact that numerous individuals do not have enough money on market days to purchase an entire loaf at one of the stalls," the bakers were required to sell all loaves on request by the pound. Four days later the parlement reduced the range to two sorts, stipulating the precise recipe of flours that were to compose the middle and dark loaves. To prevent individuals with resources from contravening these strictures, the court also forbade millers to grind anything but dark (e.g., high extraction) flour for nonprofessional customers.[20]

It did not prove easy, in the words of the Assembly of Police, "to bring the bakers down to the *bis*." Many bakers fiercely resisted the prohibition on white. A common evasive tactic was to bake a small dose of household bread to serve as a cover for the continued sale of soft-dough loaves. Bardel of Gonesse was one of numerous bakers who defied the regulations across the board. Not only did he sell mostly white, but on the pretext that all his bread was "whiter" than that of his fellow suppliers, he demanded "a price so excessive that there is no proportion at all between the price of wheat at the ports and the Halles and the price of the bread that he sells." In addition to violating the two-sort injunction and the price code, Bardel raised his price in the course of the market in response to "the crush of people and the rush to buy," yet another serious infraction. Summoned to the police chamber to answer for helping to "maintain the shortage with its elevated prices," the baker did not deign to appear and suffered a fine of one thousand livres in absentia, a fine imposed surprisingly without the corporal constraint that would have subjected his property to immediate confiscation and his person to incarceration in case of failure to pay upon notification.[21]

In part because the authorities could not rigorously enforce this measure, and perhaps also in view of the increasing arrival of king's grain and flour, the parlement abrogated the two-sort injunction on 26 October. An informal understanding with the guild seems to have been reached: the bakers were expected in exchange for this recovered freedom to provide sufficient darker loaves to meet demand. The market commissaires threatened to harass bakers who did not offer a full range of breads at their stalls. To induce bakers to purchase the "grains from Barbary" imported by banker Samuel Bernard in the king's name, the police set up a model bakery, run by four bakers specially recruited in Marseille, where this sort of grain was commonly used. One of their chief objectives was to "disabuse the bakers of Paris of their widespread prejudice that these wheats were good only for making a dark loaf. Commonly they were used to make all three sorts, white, mid, and dark, all the loaves of which were pleasing to the eye and very good." Informants continued to warn Hérault in November and December that the public bridled at the swagger of bakers who featured fantasy breads in the midst of widespread misery, and offered an insufficient quantity of household loaves in a disdainful manner. During the next ten months the commissaires continued to find numerous bakers who baked no *bis*.[22]

Consumers complained not only of the paucity of household loaves but also of their dubious and sometimes even dangerous quality. Hard times did not attenuate the expectation of quality. On the contrary, precisely because consumers feared that unscrupulous merchants and bakers—perhaps in concert with faithless public officials—would try to profit from subsistence difficulties to pass along adulterated goods, the crisis sharpened their sensitivity to quality. Duplessis admitted that there were too many loaves in the market emitting "an odor [du nez]" and communicating "a bad taste." Despite the heralded reputation of the bread of the capital, Intendant d'Angervilliers of the generality of Paris had no trouble finding loaves from the provinces, habitually known for crude baking, that "would perhaps put our bakers to shame." Some bakers seem quite consciously to have practiced a *politique du pire*, even at the risk of confounding their own customers: they self-consciously produced mediocre bread in order to blame the lack of quality on the constraints that the police imposed on them. It is not clear how seriously the police took the infraction, for in October 1725 a Porcherons baker endured a fine of merely twenty livres "for having sold badly made bread that should not enter a human body."[23]

Consumer complaints persisted through 1726. "The difference in the taste of the loaves" worried Parisians. "Fortunate the buyer," noted an observer, "who meets a baker who sells honest bread." It was widely believed that bakers practiced suspect "mixtures" as well as "bad preparation." At about the same time that Duplessis boasted that he uncovered no loaf during his market inspection that was not "of very good quality," his colleague Prémontval at Maubert tasted bread presented to him by disgruntled consumers and pronounced it "very bad" and inedible. The two bakers responsible suffered hefty fines of three hundred livres each, payable within twenty-four hours.[24]

Other bakers, "who wished to preserve their clientele," toiled hard to assure a high standard. They stubbornly tracked down the remaining pockets of *vieux bled*, wheat from the crop of 1724, on the morrow of the harvest of 1725, whose product was despairingly "spongy." A setier of this new wheat yielded between one-sixth and one-third less flour than the old wheat. Moreover, its flour was of a significantly poorer quality: it stuck to the millstones, it bore the taste of grinding, it drank less water than usual, and it proofed unevenly. Along with domestic new grain, these bakers rejected the bulk of the foreign grain arriving at the port de l'Ecole, disparaged as "very badly treated and beginning to smell." The police feared that these bakers would use the quality argument as a justification for maintaining high prices. The lead ministry worried that the king's grain would rot, imposing major financial losses, instead of exercising downward pressure on prices by increasing abundance and allaying fears.[25]

The Crisis of 1738–42

Bad harvests once again plunged a large part of the kingdom, especially north of the Loire, into protracted dearth and serious economic and social dislocation beginning in 1738 and lasting in many places until the beginning of 1742.[26] In response to the usual mix of exhortation and intimidation, the bakers, including the forains, showed considerable restraint through much of 1738. In the wake of the poor harvest, bakers began preparing opinion for hard times, spreading the word in the markets and shops that the shortfall would necessarily drive prices up. Hérault acknowledged the reality of the penury by allowing the bakers to make small incremental adjustments. Only "excessive increases" galvanized the police into action. For having taken her top loaves over three sous, for example, a forain bakeress named Catherine Huron of the Quinze-Vingt market was brought before the lieutenant general in late November.[27]

A rigorous price police did not begin till late in the winter of 1739. Seconded by inspectors and exempts and occasionally by the watch, commissaires visited shops every day and markets on Wednesdays and Saturdays. (The replacement of Hérault by Claude Feydeau de Marville probably caused some disruption.) For fear of demoralizing the baker corps, and in recognition of the genuine duress under which they operated, the repression was muted: numerous convictions, but relatively light punishments, save in cases of unusual provocation or recidivism. On patrol on 4 March, Commissaire Cadot found three masters selling shop bread at between two and four deniers above the ceiling of two sous six deniers. He issued summonses for appearances at the police court and confiscated a part of the stock in each shop for immediate distribution to the poor of the quarter, directly as well as through the parish charity organization. Two days later at his audience, the lieutenant general sentenced these three bakers, and ten others remanded for similar offenses by Commissaires Regnault Senior, Parent, and Mouricault, to dwarfish fines of thirty livres.[28]

Among the culprits was Amant, one of the current jurés of the baker guild, whose conduct was supposed to set an example for all the masters, and whose functions were supposed to include the policing of the price maxima set by the lieutenant general. When he was caught barely three months later again transgressing the ceiling by three deniers a pound, he was summarily stripped of his juréship and fined three hundred livres. Ordered to convoke an electoral assembly to replace Amant within three days, the other jurés were enjoined "to make sure that none of the master bakers of their community sell above the current and common price of the market."[29]

Denunciations by Consumers

Consumers played an important role in the police of price. In many cases, they took the initiative and denounced extortionate bakers to the authorities. In others, the commissaires and their agents interrogated the consumers whom they encountered carrying loaves of bread, conducting a sort of exit poll outside the marketplaces. An episode in late March 1739 casts an interesting light both on buying and selling practices and on the alibi discourse of bakers. Denise Framboise had the status of a forain baker at the Petit Marché du Marais, yet she also maintained a shop at La Courtille. Margueritte Laguette, the servant of an important royal fiscal official, took the trouble to frequent Framboise both in the market, not far from her lodgings, and at the shop, which required a considerable displacement. On a Wednesday at the market she paid two sous nine deniers a pound, complaining to Framboise that the bread was "too expensive" and demanding to know when the price would decline. "When the raw materials go down in price," retorted the bakeress.

The next day the servant purchased another four-pound loaf at the shop for the same price, asking the same accusatory question and obtaining the same nebulous answer. This time, however, a passing exempt from the watch entered the shop and asked her how much she had paid. When Laguette revealed the price, the exempt immediately sent for Commissaire Defacq. Framboise, backed by her mother, insisted that she did not charge this price, which she acknowledged was above "common price." It was "the fault of the servant," the shopgirl who minded the store. She offered to return the litigious sous. The lieutenant general fined her one hundred livres and promised to shut down her business in the event of further violation.[30]

The offer of the wife of Gouffe, a Gonessien, to make the same reimbursement at his stall at the Tonnellerie did not spare him from the same fine. That same Saturday morning at the Saint-Paul market, a Gonessien named Gouffe received a summons, and ultimately a fine of one hundred livres, for selling above the current. Either this was the same forain, controlling two different stalls, one operated by his wife and the other by himself, or it was a hometown relative. In any event the Saint-Paul Gouffe admitted the facts but protested his innocence, "adding that he could not offer [his bread] for any less, since he was from Gonesse, where the local grain price was higher and whose distance from the capital swelled his costs." (Theoretically, the remove of the forain from the capital *reduced* his costs.) A fellow forain from Dugny, denounced by a woman buyer for having oversold a four-pounder by six deniers at the Halles, suffered confiscation of sixty-six four-pound loaves and a fine of one hundred livres.

Inquiring of a passing woman the price of the four-pounders she carried, Commissaire Courcy learned that she had paid two sous nine deniers a pound, substantially beyond the ceiling. Confronted by the commissaire, accompanied by the customer, the servant of a prominent lawyer, the baker's wife—he was a forain from Bonneuil—readily admitted that she had received this price but insisted that "it was the master of the said servant who voluntarily had her pay this price." The woman corroborated her assertion. Like many bakers, in the shops as well as the markets, she and her husband practiced a socially differential pricing policy. Where the traffic could bear it, and they were unlikely to suffer commercial reprisal, they obtained a higher price. "All the other loaves," she swore to Courcy, she sold "on the basis of two sous six deniers," the going level.[31]

Tactics and Sanctions, 1740

As the crisis deepened the following year, the sanctions became more rigorous, but the tactics remained largely the same. Marville ordered his agents to hold the line at two sous six deniers. Commonly, he imposed fines of three hundred livres on violators, who were more often forains than masters, in part because the police focused on the markets where the forains were concentrated and in part because the forains knew that they were better placed to take risks. Marville projected an image of severity that "so potently intimidated the bakers as a whole, that they no longer dare appear before this magistrate," reported an informant, who may very well have wanted to flatter his boss.

If Marville dealt sternly with the guild, whose leaders he frequently convoked, he was inclined, like his predecessors, to treat the forains more tactfully, given the relative fragility of their commitment to furnishing the capital. Still, forains such as Marie Magdelaine Fournier of Tillet ("near Gonesse") failed to talk their way out of a conviction. Fournier admitted that she had surpassed the ceiling, "but that she had only followed the current of the market and the price at which the other bakers had similarly sold [their bread]." Then she claimed that she had sold the bulk of her loaves at the two-and-a-half-sou maximum. Finally, she called attention to an increase in the wheat price at her home market of Dammartin, where "she had taken the top quality in order to satisfy her customers," another reminder that the forains did not aim exclusively at a popular clientele (and/or that the popular clientele was stubbornly exigent).[32]

Though less acute than in 1725, the problem of baker abandonment or boycott complicated Marville's task of price management. Like Hérault, Marville imposed dire sanctions: for discontinuing their supply, for instance, a forain and a faubourienne were faced with fines of three thousand livres apiece in September 1740. A palpable reduction in the quantity of bread in the shops and stalls caused alarm. If the bakers "bake as little as they did for the market day of the fourteenth [September], Paris will be starved," consumers gathering at the Halles were said to declare. An anonymous correspondent wrote Marville: "It seems even that the bakers want to increase our difficulties, for they hide their bread, and we can't get any even for cash." The lieutenant general intervened prophylactically as well as repressively. To deny forains a pretext to stay home during the terribly inclement weather toward the end of 1740, he dispatched horses and wagons to transport them and their bread.[33]

The bakers hid their bread by not bringing it to market, by taking it home with them, or by storing it rather than selling it at a price that they did not like at the end of the day. Marville tried to enforce the mandatory price reduction in the late afternoon by assigning soldiers and exempts to make sure that bakers did not remove bread from their stalls. The commissaires encouraged consumers, especially the poorest, to wait till the afternoon when conditions would be "more advantageous." Consumers themselves became an effective agency of police as substantial crowds gathered at all the marketplaces to observe the bakers as their day came to an end and to participate in the discount sale. Certain bakers attempted to sabotage the discount market by orchestrating an intense demand in the late afternoon. In order to accomplish this goal, they concealed the bulk of their bread during the course of the day and put it back on display in their stalls in the afternoon before anxious customers willing to bid the price up to the current if not higher, despite the rule against superseding the top morning price. A number of commissaires were accused of assisting bakers in hiding their bread, the only odor of market-level police corruption that wafted during this crisis.[34]

Beginning in the summer, Marville militarized the markets, but much more discreetly than had Hérault in 1725, calling on the Parisian guard and watch rather than soldiers from royal regiments. Their presence was meant to deter both bakers and consumers—the former from price and supply manipulations, the latter from spoliating the former. "The soldiers prevent the bakers from delivering bread to home customers before unloading everything at their market stalls, gestures that the public warmly applauds," related a police observer, "but the public was unhappy with the rough way these same soldiers treated them when they went to buy their bread." Individual soldiers got into trouble, as they were historically inclined to do. Vexed by a master baker's wife who held out for a higher price for a half-pound loaf than he felt reasonable, Mulaton, a soldier in the guard, severely beat her ("Give it to him, give it to him so he'll go away," she cried out) and bloodied her journeyman's nose. If the posting of soldiers helped to keep the marketplaces tranquil, it did not address the problem of what might be called subsistence security in the streets. According to the lawyer Barbier, "cooks had themselves escorted by a lackey and some [other] man to go shopping for bread."[35]

Marville stood firm on the two-and-a-half-sou mark through mid-July, when he felt obliged to cede ground in light of continuing increases in grain and flour. A month later the ceiling climbed to three sous, the fatidic threshold. A faubourienne who demanded three sous three deniers—despite the remonstrances of an exempt stationed at the Cimetière Saint-Jean market who exhorted her to obey the rule—drew a fine of four hundred livres and a caustic reprimand. Three other women bakers (two faubouriennes and a Gonessienne) received fines of two hundred livres the same day, including one who committed precisely the same offense. Consumers hastened to denounce bakers who violated the maximum.

By the first days of September, the level of contravention had risen to three sous six deniers. Fined three hundred livres for selling at this price, bakeress Bethmont the younger of Gonesse invoked the implacable rise in the price of wheat. Ironically, she had recently leveled a complaint against Fressé, a hometown laboureur, for offering to sell her grain above the current price. On 9 September three Gonessiens at the Halles reeled under the blow of fines of four hundred livres for failing to honor the ceiling.[36]

By the end of September, Marville succeeded in forcing (and enforcing) a current of three sous. Police informants reported, however, that discontent and anxiety were widespread. People in the markets called for a sterner taxation schedule. Though the price had stabilized, it still desperately strained the resources of "the artisans who claim that they do not earn enough to purchase sufficient bread for themselves and their children." There was nostalgia for Hérault, "on the grounds that the current lieutenant general of police, despite his goodwill, cannot prevent what his predecessor managed to prevent when he was in office." Marville himself reported to the procurator general that a woman at the Halles market was arrested for shouting "that it was necessary to organize a revolt against the bakers."[37]

The Two-Loaf Limit—Again

It was an unmistakable mark of the gravity of the situation that the parlement, citing the "advantageous fruits" procured in 1709 and 1725, issued an arrêt on 22 September limiting bakers to *bis-blanc* and *bis* loaves in order to conserve grain, increase supply, and depress prices. The procurator general estimated that it could save as much as one muid in twelve. The measure also imposed a one-year moratorium on the use of any bread-making grains, including barley, by starch makers. In order to economize at least a hundred muids during Epiphany, another measure forbade the baking of *gâteaux du roi*.[38]

The two-sort restriction seems to have been strictly enforced. According to Narbonne, a police official at Versailles keenly interested in subsistence affairs, in the first few months following its promulgation, Marville levied twenty-five thousand livres in fines. One of his first victims was a governing juré of the guild, Gillet, whose shop was full of white bread two weeks after the publication of the arrêt. Gillet's explanation was at once feeble and ironic: "having been obliged by order of M. L. L. G. [the lieutenant general] to conduct the visit of the bakeshops last night in his capacity as juré charged with confiscating mollet breads, he was not able to supervise the operations of his journeymen, who baked during his absence." He was lucky to get away with a fine of six hundred livres. At about the same moment, two fellow standing jurés, Noirot and Lenepveu, were caught with loaves composed of the whitest flour and were sentenced to fines of five hundred livres. During the week before these convictions, all three guild officers had accompanied various commissaires in shop inspections, conducted in the early morning hours, that resulted in the conviction of at least a dozen masters for kneading "a bastard dough [paste bastarde]."

Mortified and discredited, the three jurés appealed their verdicts to the procurator general in the hope of a rapid rehabilitation after the lieutenant general rejected their plea for mercy. They affirmed that the allegedly mollet one- and two-pound loaves and other breads esteemed "too white" seized by the police in their shops were in fact "made from firm doughs, unalloyed by milk or salt, composed of the different flour types prescribed by the arrêt of the Court." "The lack of knowledge of these commissaires concerning the production of bread" misled them into a hasty and unfounded appreciation. Presumably the commissaires should have asked one of the guild jurés to accompany them and provide technical advice. If the fines were executed, the appellants pleaded, they would be "totally ruined," materially as well as morally, for, like many of their confreres, they had suffered "considerable losses" in the crisis and in addition had responsibility for large families.

Most of the other early fines were for three hundred livres, more lenient than they had been in 1725. Master Jean Chocarne, in whose shop "there was found neither dark bread nor the flour to make any of it," suffered a fine of three hundred livres. Visiting shops on 13 October, Commissaire François Leblanc found loaves of "a mollet dough" in five bakerooms, resulting in fines of three hundred livres each for the masters. Baker Alexandre had ample *bis-blanc* but no *bis*, an infraction rated at only fifty livres. If master Denizet had to pay four hundred livres, it was because of a double violation: selling white and selling above the ceiling. Widow Aubert faced the same bill but for a different pair of offenses: mollet loaves and false weight. Yet master Guillaume Neveu repined (briefly) in jail until he could raise over a thousand livres to pay two fines for baking mollet loaves and bread of soft dough.

Commissaire Cadot was careful to correct an error that could have been costly to master Audry. "Tender" as they came out of the oven, his one-pounders appeared to be kneaded with milk. Upon cooling and drying out, however, the bread was found to be within the rules: "we believe that it is of firm dough and that there is no violation in this regard." He was not the only baker exonerated upon further investigation. Denounced by an informant who sent a small white loaf to the procurator general that allegedly came from his shop, Lambert risked serious punishment until three days of surveillance by "reliable persons" convinced Marville of his innocence. While master Picard delayed Cadot at the door, the baker's wife concealed his white and soft loaves. Unable to prove his case, Cadot settled for a judgment of one hundred livres against Picard for refusing to let him in. Surrounded by an ovenful of one-pound mollet loaves ("made with milk"), master Megret had no grounds for denial or for disputing a fine of three hundred livres. Commissaire Defacq uncovered seven bakers with "soft dough" or "beaten dough" and "white flour."[39]

These conservation measures did not arrest the upward movement of bread prices. To the growing alarm of consumers, some bakers were asking five sous and more for *bis-blanc* loaves by the end of October. Informants warned Marville that there was increasing public doubt about his mastery of the situation. "Either the bakers see no cause to worry about paying the [substantial] fines levied upon them by the magistrate every day," reported one observer, "or the judgments that are pronounced contain merely threats of [future] punishment." Nevertheless, the commissaires continued to track down violators of the ceiling and Marville continued to fine them, yet not nearly as heavily as Hérault had done in 1725. The lieutenant general's goal was to keep *bis-blanc* below four sous six deniers a pound and *bis* under four sous. Not only did the more aggressive commissaires issue immediate summonses, but they forced the wayward bakers to make cash refunds and they confiscated the bread and distributed it to the ever-growing number of poor. "The people have had enough of this dearth and the high price of bread," noted a police intermediary, "and the same people say out loud that they fear they will not have bread to feed their families."[40]

The situation remained quite strained during the last two months of 1740. Bakers continued asking five sous for *bis-blanc* when the official norm was four sous six deniers, and three and a half or even four sous for *bis*, whose legal ceiling was three sous three deniers. Nor did all the bakers renounce offering white and soft-dough loaves. Certain bakers, such as widow Balbien at the Cimetière Saint-Jean market, defied both the price schedule and the injunction limiting bread types. Inn- and tavern keepers pressed their bakers to continue furnishing the little luxury breads so ardently desired by their habitués. Many bakers responded similarly to the persistent demand for white evinced by their individual clients, who were ready to pay whatever was necessary in order to indulge their appetite. Marville's Christmas message to the bakers announced increasing stringency: he imposed fines of five hundred, six hundred, and a thousand livres on those who revolted against the austerity program.[41]

The Central Government's Role

The new year witnessed growing tensions between Controller General Orry and Lieutenant General Marville. The organizer of a prodigious operation to import foreign grain and marshal available domestic surpluses, Orry wanted to restore abundance gradually, without further dislocating the regular grain and flour trade, and in so doing to lose as little royal money as possible on the operation. That meant a very gradual reduction in prices. The controller general was largely insulated from the daily pressures and the contact with widespread misery and fear that rendered Marville less patient. The lieutenant general believed that Paris had suffered enough, and that it was increasingly risky in political terms as well as unconscionable in human terms to allow the crisis to linger. The economic crisis engendered by the dearth had begun to affect even "the most affluent bourgeois." In the markets the mood of women consumers was increasingly desperate and angry. Marville warned the ministry that rumors of a famine plot masterminded by Cardinal de Fleury and Orry, with the help of the company of the Indies, abounded. To protect his own position, the lieutenant general spread the word that he was urging Fleury to override Orry's extreme circumspection.[42]

In the field, Marville exercised a heightened severity, which was widely advertised and which enlisted a positive reaction from consumers in the marketplaces and taverns. Bakers risked increasingly grave penalties for spurning price directives. A sweep of the shops in the Saint-Denis quarter in early March "concerning the bread tax" found every baker in order except Nozery, who was arrested by lettre de cachet on the spot for selling between one and two liards above the current. The next day, another patrol in the Saint-Honoré quarter unearthed only one violator, Mandon, "who tried to rebel along with his journeymen," and who was immediately incarcerated. Two others went to jail a few weeks later. "M. le lieutenant general is inexorable with all the bakers who sell their bread above four sous," a report on the state of opinion noted; "he had one put in jail on the eighth for having sold at four and a half sous." Optimists predicted that the price would soon drop to three sous six deniers. Perfectly informed of the official price standard, consumers contested bakers who refused to cede bread at that price and turned to the commissaires for instant justice.[43]

In mid-March Orry suggested that the restriction on bread making could be lifted, since "we are no longer beset by the prospect of lacking grain—we are on the contrary now abundantly supplied." Though he would have preferred to see a more significant decline in the mercuriale, Marville concurred. Citing, inter alia, a promising decline in grain prices and signs that the forthcoming crop would be plentiful, on 23 March the parlement published an arrêt permitting bakers to make all sorts of bread. According to a chronicler of opinion, "the Public learned this news with joy." The news convinced many skeptics that the days of reasonable prices were not far off. The city seemed to cross a psychological Rubicon, as grain and flour dealers rushed to place their hoards on the market before the price declined further. The downward price mood provoked frictions among the bakers. Old-timers at the Cimetière Saint-Jean excoriated a relative newcomer who broke ranks "by offering his bread at a cheaper price than the other bakers."[44]

Marville tolerated no resistance on the part of the bakers. Those who were caught superseding the ceiling—three sous nine deniers for white at the beginning of April—suffered a fine of five hundred livres, and in some cases went right to jail. A month later the ceiling fell to three sous, the symbolic divide that led toward relaxation, and Marville let it be known that he would be inclined to set it at two and a half sous, if Orry cooperated by accelerating the normalization of the grain price.[45]

The easing of conditions did not authorize bakers to take advantage. To sustain public confidence, Marville deemed it essential to keep the bakers in line. In 1742 he was harsher with individuals than he had been earlier, partly because there were fewer violations and he had less to fear from concerted baker action. He condemned widow Aubert, a guild bakeress, to a fine of one thousand livres for exacting "an exorbitant price." For not respecting the current, failing to provide scales, and selling lightweight loaves, a faubourien named Lepage was subjected to a fine of three thousand livres. Consequent upon Orry's agreement to reduce the price of the king's grain in February, Marville lowered the ceiling of the finest loaves to two sous six deniers. In response to a rumor diffused by the bakers in mid-April 1742 that the lieutenant general would authorize a price increase if the drought continued, Marville promised "to give no quarter" to bakers who asked a higher price. In July the bakers again threatened to increase prices, but a reasonably good harvest thwarted their plans.[46]

The Inalienable Right to Quality

Dearth meant poor quality bread: the crisis of 1739–41 was no exception. Consumers complained intermittently of a serious degradation of the goodness of their bread. The bakers blamed it on the defective or substandard grain that had flooded the markets, in particular the king's grain, or on the obligation to give up baking white and soft-dough loaves. The police exposed as spurious the notion that dark bread could not be exquisitely prepared, and suspected that the bakers deliberately furnished flawed loaves as a sort of blackmail (or *bis*-mail) to force the police to give up the baking restrictions. Rumors also spread of adulterated merchandise—dough bloated with water and/or sawdust, and flour tainted with peas or beans or chestnuts, or whitened with various concoctions. The marquis d'Argenson, a writer and sometime minister, reported a near riot in September 1740 as a result of the detection of rotten flour used by bakers.[47]

Poor quality bread, and an abruptly reduced ration, provoked a major riot at the Bicêtre prison in September 1740. Bicêtre had an unsavory reputation as a "house of correction or rather of torments." Merely to pronounce the word, wrote Mercier, induces repugnance and horror, for the institution contained "everything that society has that is most disgusting, most vile." Yet it also housed many "worthy poor" and honorable older infirm people (such as Antoine Rayter, a veteran of twenty-two years of labor in the Scipion granaries who was rewarded with a retirement in Bicêtre, which included a bed to himself and a double portion of white bread) alongside the vagabonds, mad persons, child abusers, thieves, and so-called libertines. It was probably the prison's not wholly justified image of promiscuity rather than the living conditions themselves that shocked humane observers such as Mercier. About four thousand prisoners, all categories combined, usually occupied the prison.[48]

The bread of the prisons, and to a large extent the hospitals as well, was characteristically portrayed as highly flawed, though the supercilious subsistence reformers who voiced these charges most vehemently probably overstated their case. "The detestable quality of prison bread" led physicist Cadet de Vaux, Parmentier's collaborator in the creation of an Ecole de Boulangerie in the 1780s, to call upon the authorities to entrust the baking of jail rations to this new institution ("Long live the king!" the prisoners at the Châtelet were said to have shouted when they first tasted the scientific loaves). The *Journal de Paris,* close to Cadet, attributed the string of uprisings in the prisons to bad bread. No friend of Cadet, innovative miller César Bucquet, contended that "the frequent revolts that used to occur at Bicêtre and at the Hôpital général never had any other pretexts if not causes: 'there's the bread,' that was the rallying cry." A year and a half before the Bicêtre riot of 1740, baker Felize, who held the concession for providing bread to four other Parisian prisons, was sentenced to a fine of twenty-one hundred livres for producing "defective bread, composed of bad wheat, appearing in addition to be of light weight."[49]

Without consulting the Assembly of Police, in the fall of 1740 the procurator general, "the temporal father of the hospitals," cut back the bread ration in the prisons, including the Hôpital général, by either a third or a half—from one and a half pounds to one according to some sources, from a pound to a half-pound according to others. Apparently, he construed this as an analogue to the two-sort restraint on bakers: like the general population, the prisoners had to share the burden of a forced reduction of consumption. The dearth had resulted in a significant decline in quality, surely beyond what the baker-concessionaire could legitimately have claimed as ransom conceded to the crisis. Already angry about the unpleasant taste and uneven texture of their ration, the prisoners exploded in violent rage when they learned about the reduction in quantity. The uprising lasted for a day and a half. According to d'Argenson, fifty of these "poor wretches" fell to bullets or bayonets; other observers put the toll at fifteen or twenty during the riot, and up to seven in the exemplary punishment that quickly followed. ("It is sad to annihilate men who cry out for bread," conceded the lawyer Barbier, but the harsh truth was that "one man hung contains ten thousand.")

Marville, who had jurisdiction over the prisoners albeit not over the institutions, protested acerbically that "all of Paris knew about this revolt before I received any news of it." There is reason to believe that he shared the widely held view that Joly de Fleury, the procurator general, had "lost his head," that the decision was "inhuman," that it would have been more prudent and (somehow) less cruel to release a large number of inmates ("useless mouths") who did not represent a danger to the public. The real peril, noted a police informant, was that all the city could become "a prison" and suffer the fate of Bicêtre if the government did not parry the dearth rapidly.[50]

Relative Tranquility, 1742–64

For most of the next quarter-century, the subsistence landscape was remarkably placid. By and large the authorities allowed bakers and consumers to set the price, in collaboration with the grain and flour dealers. As always, bakers were expected to respect the current market price, but in relatively unstressful times that price had a certain lability and the police did not insist on a ruthless fidelity. Save in brief moments of resurgent anxiety, the lieutenance itself did not proclaim a ceiling. As long as the players remained reasonable, they were permitted to play with little direct interference.

A shortfall in 1756 led to difficulties that persisted for about a year. The police imposed a maximum in the fall. Machurin, the commissaire in charge of the subsistence department, met frequently with the jurés to discuss the situation—frequently enough to give them the flattering impression that they were involved in veritable negotiations. He agreed to monitor carefully the grain and flour prices in the supply zone in order to have a full understanding of the real strictures that framed the behavior of the bakers. His associate, Inspector Poussot, visited each bread market personally and announced to the concession holders in unequivocal terms that they would be punished for failing to stock their stalls as usual and to defer to the boundary price.

Poussot was inclined to give first-time offenders the benefit of the doubt and to take into consideration the economic stoutness of delinquent bakers before recommending punishment. He protected one of four contravening bakers because he was "the least rich" and thus the least able to sustain the blow of a hefty fine. The inspector's brother, also a police agent, recommended "grace" for a bakeress from Bonneuil who sold at a liard over the two-sou-six-denier ceiling in late November. She argued that the wheat price at Dammartin left her no latitude and that the news of the "tax" had not reached her. Bakers cheated at the margins by baking two grades of white bread, "the fine [le beau]," which they reserved for their regulars who were willing to pay slightly above the maximum, and "the other [l'autre]," which they offered at the official price to casual buyers who "bargain."

Poussot reported a very high level of compliance, which he attributed to the rigor and fairness of the policy of containment. "Almost all the bakers sell the [four-pound] loaf in their shops at no more than ten sous," he informed the lieutenant general in mid-December, and the forains followed suit. This restraint exercised "a marvelous effect" on the surrounding grain markets, where the price began to drop. Around Christmas, bad weather slowed the circulation of merchandise and the production of flour, resulting in an upward surge. The police hesitated, and bakers began selling substantially above the ceiling, pushing the price in some shops and stalls to three sous. The price remained elevated well beyond the next harvest, whose yield was uneven. With the help of modest government purchases of grain and flour, the police managed to keep the price from transcending the three-sou frontier. Still, Damiens could not have failed to hear the din of "murmurs" arising from discontented consumers as he undertook to assassinate the (culpably distracted) father-king in early 1757.[51]

The Crisis of the Late Sixties

The state remained steadfastly committed to the consumer interest, with two spectacular exceptions, one in 1763–70 and another in 1774–76. (A third exception, in the 1780s, turned out to be more timid in inspiration and in resonance.) On these occasions, the royal government broke drastically with provisioning tradition and recanted its covenant with the consumer-people. Grain and flour were liberated from virtually all the controls that had trammeled their movement, spied on their whereabouts, and governed the conditions of their exchange. Most of the practices that had heretofore been decried as antisocial monopoly were now rehabilitated and legitimized in the name of natural law, laissez-faire efficiency, agricultural renewal, and commercial revitalization.

Even as grain and flour were freed, the hands of the police were manacled and the consumers were invited to fend for themselves. On both occasions the old supply system disintegrated. In part because both the police and the people passionately resisted this radical new departure, neither of these liberal regimes endured. Each experience seemed to confirm the basic premise of the police of provisioning: that unless the suppliers were "contained," the people could not be "contained."

In order to permit the radical measures to take root in the rest of the kingdom and to shelter the government against the gravest possible backlash, the liberalizing laws of 1763–64 provisionally exempted Paris. In practice, it proved impossible to maintain an island-cocoon of regulation in a stormy sea of liberty. The capacity of the lieutenant general to direct the provisioning trade throughout the supply zones and assure the primacy of Parisian interests was severely diminished. Under the combined disorganizing impact of several feeble harvests and the dramatic policy change, much of the kingdom suffered an acute and protracted subsistence crisis in the late sixties; other parts were not touched until the beginning of the seventies. Despite certain steps taken to insulate Paris from the expected turbulence engendered by the dismantling of the regulatory apparatus, including an emergency grain and flour reserve and the maintenance of some controls, the capital reeled under the blows of penury, compounded by a generalized economic crisis. In the absence of the normal capacity to mobilize grain and flour merchants, the lieutenant general focused more keenly than usual on the bakers.

During 1766 the price remained accessible to the bulk of the Paris population—eight sous six deniers between Easter and the end of the summer. After reaching ten sous at Christmas, the price gently declined during the first half of 1767 and then rose steadily after the poor (but not catastrophic) harvest. In September the procurator general warned Laverdy, the controller general deeply committed to the policy of liberalization, that the increase to three sous per pound was driving the people to the brink of revolt.[52]

Sartine operated on many fronts at once. He lobbied the ministry, vainly to restore his authority to mobilize the resources of the grain and flour trade, and successfully to obtain larger doses of royal grain from the strategic reserve. He used myriad "little ruses of war" to create the impression of abundance and allay anxieties, including the tactical requisition of stocks stored in monasteries and hospitals and the ostentatious deployment of piles of grain and flour sacks filled with rags and gravel. Fascinated with the new subsistence science, he promoted experiments in milling and baking to enhance productivity and lower cost. And he opened a long and difficult dialogue with the bakers.[53]

Sartine had enjoyed good relations with the guild and forain practitioners since his ascension to the lieutenance in 1759 largely because he allowed them to go about their business with very little interference. His men had continued to monitor weight and quality, and to control grain- and flour-buying practices while allowing great range for country buying, but they had refrained from interfering with the price, which had remained within limits deemed reasonable. The first serious signs of tension appeared in September 1767 when the lieutenant general regretfully but firmly rebuffed a guild request for permission to raise bread prices in response to increases in the price of flour at the Halles and grain in the hinterland. Though the master bakers avoided a solemn pledge, in light of his optimistic assessment of the prospects for a leveling off of grain and flour, they "made us hope that they would not take the four-pound loaf to twelve sous." Like his predecessors, Sartine did not feel that he could expect "the same sacrifices on the part of the forains. . . . for the most part these are people living quite precariously, unable to abide the least loss, who would be forced to quit the baking business if we compelled them to sell their bread at a price below what it cost them to make it."[54]

Sartine appears to have been less authoritarian in his relations with the bakers than most of his predecessors—liberalization *oblige?* Until well into 1768, he continued to give them a great deal of latitude. He expected them to play the game fairly, according to a sort of reciprocity of reason. When they did not, he raged. "From the end of December and throughout January, Monsieur, the bakers should have moderated the price of their loaves," he wrote to each of his commissaires at the end of February 1768. The bakers used the slight upward push caused by the March plantings as a pretext to raise the best four-pounder to thirteen sous six deniers and to announce an imminent augmentation to fourteen sous. "This is harassment on their part," erupted Sartine, "that I will certainly not tolerate." He instructed the commissaires to convoke the bakers in their districts and convey his anger. "I am all the more displeased with their guild," he wrote, "since *they owe me a debt of gratitude for not yet having recourse to a taxing of the price,* and for having up till now shown them indulgence." He ordered the commissaires to fix the ceiling at thirteen sous, and to issue a summons to any baker who violated the maximum or threatened to do so.[55]

Sartine held the line at thirteen sous through the early summer. Bakers who transgressed were either jailed on lettres de cachet or fined up to a thousand livres, or they endured both sanctions. Stiff penalties faced bakers, including forains, who cut back on their provision without compelling justification. Exempts patrolled the bread markets and their peripheries to thicken the police presence after another harvest shortfall propelled grain and flour upward. While Sartine acknowledged the legitimacy of baker requests for an increase— "I would like to render them justice and even service"—he insisted that he could not allow bread go any higher at the present moment. "Ceaselessly divided between them and the Public," Sartine opted for the latter, as more and more Parisians grew desperate for work and for food. The very short run preoccupied the lieutenant general: he needed to hold the line for a month. If the price of grain and flour did not fall by then, he promised either material compensation from the king or, if he failed to obtain royal largesse, "a reasonable increase."[56]

Like his colleagues, Commissaire Grimperel convoked the bakers in his district (fourteen showed up) and transmitted Sartine's sentiments. "They appeared to me to be full of gratitude for the favors with which you have graciously honored them," reported the commissaire to his chief, with perhaps a wisp of irony. While they pledged to conform to his instructions, they stressed their increasingly precarious situation. Losing money on their sales, they were very low on cash and running out of credit with which to buy merchandise.[57]

A day later, however, after receiving the visit of "a very large number of bakers," the lieutenant general ceded to their importunate request for a new ceiling, according them not the sou they sought but an augmentation of six deniers per pound. Ever a realist, he capitulated not merely because "their current situation is truly deplorable" but because numerous bakers already demanded the highest level (thirteen sous six deniers per four-pounder) and significant numbers in the shops and markets had reduced their production. Two days later he informed the procurator general that "the people murmur and complain vociferously of the high cost of bread," hardly a surprise. Increasingly inured to this rumbling discontent, two weeks later, with some relief, he reported that "the markets were calm, and there were only the *ordinary* complaints concerning the high price of bread."[58]

Bread continued its inexorable ascension, reaching fifteen sous for the best white in the first fortnight of October. Still, the markets were generally well supplied, and there was considerable variation, even for the same quality loaf, from one market to another, a phenomenon that probably accounts for the internal migration of consumers within the capital, appearing as "strangers" in the eyes of veteran police agents of the quarter where they ended up purchasing their bread. The authorities seemed increasingly helpless to arrest the upward course. In November the parlement announced the convocation of "all the Estates of Paris" in a grand assembly of police, a congress of calamity last convened in the tragic waning years of the reign of Louis XIV.

Sartine recast the ceiling at sixteen sous, and fumed at the news that seventeen sous was a price commonly found in the shops and stalls by the middle of the month. He enjoined his commissaires to inform each baker personally that the ceiling was absolutely infrangible, and to be pitiless in striking at bakers of all sorts who violated the boundaries, but he insisted that such action be taken with discretion. Encouraged by the vigorous onslaught of the parlement of Paris against the liberal regime and the significant dip in the grain and flour prices in the very first days of December, the lieutenant general called upon the bakers immediately to pass on the benefit to consumers by lowering the ceiling to fifteen sous or fifteen sous six deniers. Bakers had to resist the enormous and justified temptation to indemnify themselves for their losses: "bread is now too dear. . . . they must await a less difficult moment."

By 10 December the maximum stood at fourteen to fourteen sous six deniers, climbing two liards higher the following week. The commissaires reported widespread compliance; only about a dozen bakers received summonses to answer to Sartine's court. Perhaps to avoid any suspicion of baker-police collusion, as well as to sustain the image of baker cooperation, the lieutenant general exhorted his men to act "so that the public is as little aware of it as possible" and to "avoid allowing the public to perceive what is happening." Yet they distributed printed *billets* admonishing bakers not to surpass the maximum.[59]

Bakers as Victims

Even at the zenith of the crisis, the bakers did not become the object of collective rancor, partly because of the dispositions taken by the police, but perhaps also because the public viewed the bakers, like other citizens, as the victims rather than the authors of the dearth. Anger did not crystallize around the manipulative cupidity of the bakers as in the faubourg Saint-Antoine in 1725. It focused on real and imaginary grain and flour speculators, and on the government as coordinator of the heinous famine plot. The bakers did not always show, as Sartine urged, "gentleness toward the public." The incidents of friction, however, were dispersed in time and place, and never acquired any momentum. Almost invariably they turned on the question of price. Refusing to cede any ground, widow baker Tessaux at the Saint-Paul market was insulted by one buyer-bargainer, who refused to pay fifteen sous six deniers, and then physically assaulted shortly thereafter by a second "who exploded over the price of bread." A jobless postilion entered into a bilious quarrel with a journeyman baker, the son of a widow mistress baker, who demanded two sous six deniers for a half-pound of bread that generally sold for two sous. First the baker refused to return the client's money when he rejected the price. Then the baker insulted his sense of honor by disdainfully offering "to make him a handout of this liard." Indignant, the postilion retorted that he was not one to accept charity. The journeyman demonstrated his philanthropic inclination by plunging his bread knife into the client's hand. A woman named Catherine Picot, seeking to purchase a top quality four-pound loaf, expressed her outrage with baker dishonesty more sedately. Aware that the price was fixed at fifteen sous, she visited four successive shops in which the baker was asking fifteen sous six deniers. Picot denounced the bakers to Commissaire Grimperel, who immediately convoked the bakers for a reckoning.[60]

Another reason for public restraint vis-à-vis bakers may have been the manifestly crippled state to which several score were reduced by the crisis. Numerous bakers filed for business failure. Particularly hard hit was René Morin, jailed for debts, who "declares to have been without any furniture or furnishings for the past year and a half, who has been sleeping atop his oven during all this time, having been forced to sell his possessions because of the harshness of conditions." Described by witnesses as "a very decent man," Charles Desauge, a faubourien mired in debt and facing a suit filed by his wife for separation of property, blamed his "ruin" on "the current time of misfortunes," on "the high cost of goods," and on "the bad faith" of grain and flour dealers who sold him poor quality merchandise. Another baker, Jean Simon, had no operating cash "because of the high cost of merchandise, having earned barely enough to keep his household going, with difficulty."[61]

Even before the guild sent a blue-ribbon panel of thirty-six anciens to call Sartine's attention to the deepening misery of "a great number among them [who] would find themselves in the necessity of closing down their shops," the lieutenant general had begun to provide assistance in money and flour to stricken individuals, forains as well as masters. A commissaire first investigated their claims; if approved, the bakers had to agree, in most cases, to consider the advance a repayable loan, and in all cases to pledge to sustain their regular level of supply. Virtually destitute bakers received "flour gratis," without obligation to restitute. "My intention is not to give twice to the same person," decreed the lieutenant general, but he found himself compelled to contravene his own policy on numerous occasions in order to salvage a marginal baker's contribution to the capital's supply. Though Sartine favored modernization and rationalization of the bread industry, and did not recoil from the politically risky prospect of concentration and even vertical integration, he still identified security for the city with the multiplication of sources—even very small sources—of provision.[62]

Commissaire Grimperel provided Sartine with a pithy sketch of each of the candidates in his quarter seeking certification as "one of the poor and distressed bakers." Pierre Amar, a master in the rue Montorgueil, "is extremely poor and, lacking in means, is reduced to baking merely a half sack of flour a day." Seconded by their curé, his wife testified that misery had forced them to place a sickly daughter with a good samaritan and that "she and her husband were reduced to sleeping above the oven, having been forced to rent out their bedroom in order to pay the rent." It was in the public interest to assist widow mistress Provendier, "who is known as an honest woman, but in straitened circumstances," because "a little help would enable her to continue baking." Grimperel endorsed the petitions of three bakers burdened with seven, eight, and nine children respectively, and those of others who had been victim of a credit cutoff by their brokers as well as their flour dealers. (The inability or unwillingness of brokers to sustain and/or extend credit during the long crisis that persisted till the early seventies impelled Lenoir, Sartine's successor, to plan a major reform of the brokerage system.)

Limited to baking two small ovenfuls a day, master Etienne Guillaumé of the rue aux Ours, the father of six young children, hovered on the brink of disaster because his merchant miller at Persan refused to deliver a new load of flour, purchased on credit, until he paid an old debt for nine hundred livres, and his flour merchant at Versailles declined to furnish twenty sacks without both a down payment *and* the liquidation of an older obligation of 940 livres. Pierre Tauvin, who entered the guild in 1767, claimed to be "in a state of very great need, having been ruined by his entry into the guild, which he claims cost him fifteen hundred livres, and by the misery of the time." His broker had denied him further credit.

Grimperel did not simply wait for supplicants to seek him out. "Despite their misery, these bakers still have a great deal of pride," he noted, and thus many truly needy ones did not come forward seeking help. The commissaire sought them out, on the basis of his own observations and knowledge of the quarter, and on the recommendations of a senior guildsman. Nor did Sartine grant assistance to every baker who applied. He rejected bakers known to be fundamentally "well-off" and others who had committed serious infractions, in particular regarding the price ceiling. The lieutenant general turned down Gaspard Thevenot, a fallen master who borrowed a confrere's bakeroom twice a week to make bread for his stall in the market at the Halles. A baker without his own oven was not worth investing in, and in any event, according to the regulations of the guild, had no right to exercise the trade.[63]

Sartine claimed to have aided about five hundred bakers during 1768. Abbé Roubaud, a physiocrat and close observer of the subsistence scene, evoked the "large sums of money" distributed to "a bevy of bakers" during the crisis. A police document referred to over a hundred needy bakers who had received aid and had contractually engaged to reimburse on installments on a weekly basis as soon as they were on their feet. Apparently a large number of these bakers later joined in a "cabal" to refuse repayment, presumably on the grounds that they had suffered substantially at the hands of the regulators and deserved to take the putative advances as compensatory damage. Others were willing to reimburse but contested the conversion rate of flour into bread (105 four-pound loaves per sack) by which the authorities determined their debt for advances in merchandise. The lieutenance refused to tolerate their reasoning and their recalcitrance. In order to persuade the others to pay, the police arrested four of "the most mutinous." They languished in prison for almost two weeks until they agreed to "fulfill the commitments they had contracted." Reproaching Sartine his failure to honor his promise to render them "justice," many bakers vowed never again to make unrequited sacrifices.[64]

The next massive request for aid from the bakers did not occur until 1789. The context was tense and controversial, not only because of the growing momentum of the Revolution and the pain of an acute subsistence crisis but also because of a political confrontation pitting the bakers against the Leleu victualing company charged by the government with the task of providing Paris with emergency supplies. Accused of hoarding and other activities geared to raise prices, the baker's guild attempted to turn the charge against the Leleu "monopolists," whom it denounced for organizing a sort of famine plot. An adroit polemist and journalist called the chevalier Rutledge spoke in behalf of the bakers. Through a contact at court who had easy access to the chief minister, Necker, it was also he who urged needy bakers to sign a register requesting government assistance. Two hundred thirty-eight bakers signed up for differing levels of aid in proportion to the amount of bread they normally produced. Before Rutledge obtained a response from Versailles, he found himself arrested, denounced in the streets "to the people" on his way to jail as "a grain hoarder and aristocrat who went around at night distributing money to bakers in order to get them not to bake." The man whom the guild hired to extricate the masters from various difficulties had momentarily plunged them into even deeper trouble.[65]

Though prices gradually decreased, the situation remained strained through the early seventies. The police did not discourage bakers from diversifying their production: a differential schedule compensated those who used "elite" flour to compose a "superior" bread. Sartine evinced no tolerance, however, for bakers who refused to play according to his rules. After learning that many bakers failed to implement his instructions to reduce prices by six deniers a pound as a function of falling grain and flour prices, he ordered the commissaires in January 1770 to tell them "that *I absolutely demand this diminution.*" Disobedient bakers faced "dire consequences." If mill-interrupting inclemency constituted the "physical cause" of a price spurt in May, "the moral cause lay in the greediness of the bakers," a senior government official observed. But in June Sartine conceded an increase of six deniers in light of "the progressive increase that has made itself felt on wheat and flour." He predicated this increase on the relationship of reciprocity that he expected the bakers to honor: "I am happy to be able to accord them this new easement—but on the condition that they reduce the price on their own initiative when the situation becomes more favorable."[66]

Given the continued upward slant of the mercuriale, Sartine found it "impossible to prevent" the bakers from effecting two successive increases of six deniers each in late August and mid-September. Hoping to profit from an improbable discretion in this highly attuned world of bread buyers and sellers, the lieutenant general counseled his commissaires: "If the bakers of your market have not asked for an increase, you are to tell them nothing." By mid-November the top price had fallen from fifteen and a half to fourteen sous for the four-pounder, and Sartine was determined to offer Parisians a Christmas present of the market loaf at thirteen sous. A number of bakers at the Palais-Royal market resisted, including a woman named Ranteliere "who said some very wicked things, such as that she would not sell her bread at less than thirteen sous six deniers. She found a cache to store forty four-pounders and she incited her colleagues to disobey." Ordered to sell at the new marker, she suffered "a very severe reprimand" and a threat of "stiff punishment" if she emitted the slightest peep.[67]

Sartine continued to deploy the same tactics for the next several years. During the cold wave of January 1771, he froze prices. In March and April he practiced the same rigor, administering formal reprimands to market rebels and threatening arrest in case of persistent indocility. In May, while a thirty-year-old woman killed herself because her baker refused to extend further credit, Sartine called upon the bakers to be patient. In June, however, he continued to deny them the adjustment they inferred would be forthcoming. In July "rebellion" erupted among forains at several marketplaces. They pushed the price to thirteen sous, six deniers above the ceiling. The fact that Sartine merely fined them one hundred livres each suggests that he did not dismiss their impassioned protests that even at thirteen they were losing money "in light of the prices of the wheat and flour that they buy." At least two similar outbreaks occurred in October and November, with bakers professing that they merely wanted to offer the loaf at "the price that it was worth." The police upbraided, fined (up to two hundred livres), and in at least one case incarcerated.[68]

Between Christmas of 1771 and May 1772 Sartine gently brought the four-pounder down another 17 percent. Bakers such as Marie-Thereze Labbé who openly defied the schedule went to jail "in order to make an example." The commissaires convoked the most intractable bakers at each downward arbitrage to warn of the consequences of refractoriness. The bakers held at ten sous until the end of August, when rising grain prices placed them under considerable pressure. A half-sous concession by Sartine did not allay their anxiety and anger. A delegation of twelve bakers at the Palais-Royal informed Commissaire Trudon that they could not continue to supply under conditions that had become draconian as flour soared toward sixty livres a sack. They were perfectly happy to settle for a decrease in the price of raw materials if the government could not accord them an increase in the price of bread. Cumulative losses, climbing prices, and the inability to obtain credit for merchandise placed many of them on the brink of shutting down.

Trudon's vague assurance that prices could not remain so high for a very long time "did not touch them in the least." Confident that he would not provoke a flood tide of defections, Sartine held his ground: the weak ones could leave the markets provided they gave the standard two-week notice. As the new grain reached the markets in the fall of 1772, the mercuriale began to register a decline, and the threatened shakeout of the baking corps never took place: the police had won another round.[69]

Problems erupted again two and a half years later at the time of the Flour War. In Turgot's liberal (macro)vision, the bakers represented perfidious anachronisms and self-regarding parasites. They had no moral or political-economic claims to exercise the liberty that he bestowed on many other socioeconomic actors, including grain and flour dealers. His lieutenant general of police, Joseph d'Albert, had no scruples about applying more brutally authoritarian measures than his putatively antiliberal predecessors had undertaken. In the aftermath of the riots that buffeted the Paris region and beyond in May

1775, grain and flour prices remained quite high. Not only did Albert impose a ceiling on the bakers without discussion or explanation, but he peremptorily abrogated the traditional six-denier differential between market and shop, meant to subsidize the higher fixed costs of the shop bakers. He also made adjustments by increments double the size of those applied by Sartine.

The bakers tried to resist following canonical patterns of insubordination: the forains by threatening to stay away and the city bakers by speaking in the corporative voice that Turgot regarded as profoundly illegitimate and baleful. The chronicler Bachaumont reported that the guild defied Albert by refusing to reduce the price from fourteen to thirteen sous in mid-July until the flour price fell commensurately, and in addition by refusing to make use of cheaper "bad flour," probably from stocks held by the government at Corbeil ("for fear of losing their customers"). The lieutenant of police, "albeit a very great partisan of liberty," responded by threatening to hang the first baker who stopped baking.[70]

Jean Lenoir was Albert's successor; he had also been his predecessor. He had first become lieutenant general when Sartine, his protector, joined Turgot in the first ministry convened by the new monarch, Louis XVI, in 1774. Tensions had flared from the beginning between the controller general and the police chief when the former announced a reversion to the ultraliberal grain politics of the sixties without consulting the latter on the implications for the capital. Their relations reached a breaking point in the spring of 1775 during the episode of proliferating subsistence disorder called the Flour War that cascaded across the kingdom's core, from Burgundy to Normandy. As always, the precise causal dosage is hard to recalibrate. As in the sixties, however, it seems clear that a blend of growing penury and radical policy fanned the fear and drove the disorder. Lenoir fell into disgrace when he failed to contain the Paris riots. What appeared to Turgot as a quasi-treasonous form of inertia bespoke for Lenoir the impossible double bind into which he had been trapped by the profoundly contradictory injunction to "tend to the security of the capital" yet "not to meddle with [the question of] bread."[71]

Recalled to the lieutenance after Turgot's fall in 1776, Lenoir returned to a more supple and less strident style of price control. He met frequently with delegations of both the forain and guild bakers, and he repeatedly emphasized that he understood the problems of the bakery. As long as the self-adjusting market price was just, he suggested in Thomistic terms, he would allow it to reign. He would intervene, however, whenever he discerned an incongruity between the price demanded by the bakers and "the interest of the people."

Lenoir reasoned in terms of a broad equilibrium: bakers should be allowed to extract a little more profit in easy times and should be prepared to suffer some pain in hard times. Refusing a demand for an increase formulated by the guild in July 1777, the police chief reminded the bakers that they had just enjoyed nine to ten months of profit taking. Now he would not accord an increase until "it became just and legitimate." Regularly, he sent inspectors to buy bread in various shops and stalls to check on pricing. Lenoir also encouraged consumers to denounce price extortion or exaggeration, and he tried to make it easier for poor people to file complaints by streamlining the process so that they would not lose time from work.[72]

As a measure of public security and a technique of price stabilization, Lenoir, along with Necker, explored the idea of imposing a storage requirement on the bakers. Theoretically the authorities could call upon two emergency reserve systems in times of grave need, one lodged in the religious communities, educational institutions, and hospitals, the other maintained more or less discreetly by the royal government. The former was always an uncertain source, contingent on the often precarious economic well-being of the participating houses and not rigorously monitored, whereas the latter's strength was marred by scandal and sapped by liberal doubts within the ministry. Anticipating a scheme that would not be realized until the first empire, a number of writers suggested that the bakers themselves were the best placed agents to undertake an efficient and permanent revolving storage mission. A modest stock of two to four months' worth of flour in each bakery would significantly insulate the capital from crisis shock.

Lenoir consulted bakers, millers, and subsistence specialists on the feasibility of the project. While few questioned the (abstract) merit of the idea, virtually everyone agreed that it posed quasi-insuperable problems of application. At best a fifth or a sixth of the baking corps could undertake and sustain such a stocking obligation. The others would have to borrow funds whose cost would inevitably be passed on to consumers (since the government lacked the wherewithal to subsidize this huge enterprise). Logistical problems compounded the financial ones, for most bakers lacked the facilities to store. Even if they could locate space near their shops, they would lack the funds necessary to rent them. Moreover, maintaining a stock of grain or flour implied costly care to prevent spoilage caused by insects, rats, heat, and humidity. Nor was it evident that the bakers could keep their granaries full without disrupting the provisioning markets, and thus placing upward pressure on the prices. In light of these difficulties, Lenoir merely encouraged bakers to act on their own, promising to protect them from the taint of hoarding and hinting at symbolic if not material rewards in recognition of their civic efforts.[73]

Prices bore a heavier, more complex connotative charge in the eighteenth century than they do today. They were not mere indicators of cost or of the sums for which a thing could be bought or sold. They were not just signifiers but also signifieds; they addressed ends as much as means. Prices bespoke moral and political issues as much as they addressed economic ones. While prices in day-to-day usage had more prosaic utility functions, arguments about prices could be quite far-reaching. They dealt with questions of right and obligation, of legitimacy and transgression, of custom and innovation, of social organization and the proper relation of state to society. To police the price of a good as genuinely precious and as fetishized as bread was to undertake a solemn and arduous responsibility that had a resonance transcending market and shop.

With the metamorphosis of the philosophical insect into the economic insect, as philosophe and antiphysiocrat Simon Linguet put it, price became a contentious intellectual issue in the second half of the eighteenth century. Liberals wanted to purge the notion of price of its thick and ostensibly distorting cultural overlay, to restore it to its natural vocation. For them, the price of a good was a quasi-scientific measure, unburdened with emotional freight, when it was allowed to form on its own, unimpeded and untainted by any sort of social engineering (here we hear the voice of a modernist-primitivist, laid-back Enlightenment against that of an intrusively manipulative Absolutism). This debate cannot be ventilated here. Suffice it to note that the issue of price would never again shrink into consensual eclipse. With the physiocrats came to an end the once largely unchallenged apothegm that high prices for goods of first necessity should be avoided at all costs.

There were, however, certain dissonances or nuances (if not contradictions) in liberal high-price boosterism. The *économistes* openly espoused high prices for raw materials but not for manufactured goods, for grain but not for bread. The liberals were not entirely devoid of political sense, for they maintained, thanks to a somewhat circuitous logic, that higher grain prices would (eventually) result in lower bread prices as well as other consumer advantages, such as the evening out of prices and the regularizing of supply. The physiocratic discourse reminds us that no one—not even economic illuminati—was flat out in favor of a more expensive loaf, and that no one needed to like bakers in order to sustain grand reforming schemes. Paradoxically, then, bread stood at the center of the great debates of the sixties and seventies and at the same time remained to some considerable extent estranged (or sheltered) from them.

In terms of their economic, political, social, and psychological implications, the measures of 1763–64 liberating the provisioning trade were the most radical reforms of the eighteenth century before 1789.[74] They changed fundamentally the relations between the governors and the governed, and the rules structuring community life. In deference to the capital's gigantic size and social combustibility, these reform laws temporarily exempted Paris from entering the new regime, though it was virtually impossible to prevent contamination by the virus of liberty. In any event, the new regime did not survive: after years of surprisingly resolute engagement in the liberal camp, the government abjured the radical turn in 1770 and reverted to a policy of apprehensive paternalism. To summarize a complicated argument, I would maintain that on the ground things returned largely to the status quo ante, but that in the air things would never again be the same.

In terms of our immediate concerns, that meant that there was no serious discontinuity in the practice of police, in strategical approaches and tactical actions, between the long "first" eighteenth century (1700–60) and the short "second" eighteenth century that ended with Napoleon's coronation. That does not signify that the police had not changed since Delamare. On the contrary, imbued with Enlightenment preoccupations and skepticism, better informed, more tolerant of ambiguity and more open to internal debate, more sophisticated, and more patient, the police by the time of Sartine and Lenoir had matured considerably. Beyond their capacity to make more subtle connections, to distinguish more sharply between different kinds of reality, the police still served the same ends.

What changed definitively and dramatically was the political and psychological context. Subsistence was no longer a nonpartisan issue. Suddenly and profoundly, it had become politicized. The consumer interest no longer reigned uncontested. Police legitimacy could no longer be taken for granted. In addition to the chronic uncertainty of the harvest, the police authorities would now confront a permanent political uncertainty: how much longer would there be a political will capable of justifying and sustaining regulation, intervention, dirigisme? Utterly discredited and in disarray in 1770 after a half-decade of chaos in the land, the liberal camp nevertheless regained power only a few years later, a testimony to the vitality (and corrosiveness) of its ideology and to the persistent (and deep) doubts about the old way.

Out of phase with the lean, spectator king and his ministry in the sixties, and again with the new, still unknown king and his reformers in the mid-seventies, the police could console themselves with the plausible claim that they remained fully synchronized with the beat of the consumer-people. Bread served as the signal bond of communication between the police and the people. In terms of its magnitude and its grueling, relentless periodicity, there was nothing comparable to the massive daily mobilization over which the police presided more or less discreetly. Literally tens of thousands of people marched on the shops every day and on the markets twice a week, a usually peaceful uprising that left no street—probably no single building—of the entire city untouched. They were as much the clients of the police as the customers of the bakers. Their demands were simple, immutable, and only marginally negotiable: bread of good quality in sufficient quantity at a reasonable price.

Mediators between the buyers and the sellers, from the apex the police orchestrated a triangular relationship of reciprocity and suspicion, of expectation and anxiety, of submission and rebellion. In ordinary times the role of the police was to reduce frictions, to keep material and psychological transaction costs as low as possible. Symbolically and practically, the commissaire and his agents had to know what was allowed and what was forbidden, what were the thresholds of tolerance of each of the parties, in what manner must each of the parties be talked to. The operation was extremely delicate, because everything generally turned on nuances and deniers in a highly ritualized dance. There was little room for maneuver between bakers for whom it was business (as usual) and consumers for whom it was partly business (for they had their feet on the ground) but partly rights (for they had a strong sense of survival entitlement). As guarantors of equity but also of peace, the police had to pay attention to the transient and concrete realities of markets and mercuriales but also to the deeper realities of custom and custodianship. At once tutor, partner, and hostage to each of the parties, the police used a strained, syllogistic reasoning to try to convince both of them that at bottom their interests converged.

The police had more leverage with the bakers than with the consumer-people, who remained hard to engage as an interlocutor, despite (or perhaps because of) their massiveness and ubiquity. So it was with the bakers that they tended to treat, and to deal, in the business sense of the term. For much of the work of policing the price was cutting deals—deals for restraint, for subsidies or advances in cash or in kind, for immediate gratification, for future compensation, for remission of penalties, for extension of grace periods, for (re)distribution of market stalls, and for other forms of collective protection or individual privilege. In ordinary periods the police left the bakers considerable latitude, as a function of the grand deal, the structuring entente that governed their relation. The flip side of that deal gave police the license in times of stress to impose ceilings, more or less in reference to grain and flour availabilities, but also to impose sacrifices as required by the imperatives of social control. On balance, shifting between the roles of private merchant and public servant, the baker was expected to "trouver son compte"—to make an honest living—as

Sartine repeatedly said. Sartine and Lenoir were by and large less authoritarian than La Reynie and d'Argenson, and far less so than the liberal Albert, and they consulted more frequently with the bakers in a dialogical mode, but at bottom they did not do things much differently.

The overriding goal of the police was to keep the shops and stalls regularly furnished. The single most contentious issue, price, was ultimately open to negotiation, but not the principle of provisioning fidelity. Unless they obtained authorization after notice or had some other compelling excuse, the police expected bakers to keep their shops and stalls furnished with the regular dose of bread regardless of the circumstances. Hard times might oblige them to curtail the range of breads they produced, the lower end displacing the fine whites as the core of everyone's production, but no troubles dispensed the baker from baking bread of a quality sufficient not to shock or provoke consumers. Mutinous bakers suffered a range of penalties, whose inflection depended both on personal history and general circumstances. Fines, often aggravated by the automatic seizure of goods and/or incarceration for failure to pay immediately, most commonly sanctioned indocility. Occasionally, bakers found themselves jailed for infractions on supply and price. More rarely, the police closed down and walled shops, dismantled ovens, or drove bakers from the guild and even from the profession. There are very few episodes of serious baker resistance through boycott, "cabal," or other forms of collective defiance; given their greater degree of freedom, forains were more likely to act out their resentment than Parisian bakers physically under the heel of the lieutenance. The key to dealing with us, master baker Toupiolle told Commissaire Grimperel, is to offer the carrot more than the stick. "We are not evil persons," he reassured the commissaire.

Conclusion

During the Old Regime, people did not live by bread alone. It went a long way, however, toward keeping them alive. For most people, it was the daily ration of survival. It was the center of pressing material preoccupations, but at the same time it bore an immense symbolic charge. It was not just (urgently needed) fuel. It carried a host of meanings, sacred and profane, solemn and casual, that further distinguished it from all other goods, including those of "first necessity."

Even as bread was the body of Christ, so it was also the social body.[1] It was associated with sacrifice in the secular as well as the spiritual domain. Bread stood for what had to be given up in order for society to remain a whole, and in order for the state, as trustee of social order and cohesion, to sustain itself. This was the sacrifice through which the state obtained/retained its legitimacy. Offering the guarantee of bread to the people on the altar of social peace, the state gave up the prerogative to act neutrally or indifferently, following either the logic of a myopic absolutism or a scientific liberalism. The state assumed obligations and constraints, and forced a part of the citizenry to give up (partially) rights of property and liberty labeled fundamental or inviolable by certain theorists. The sacrifice was an act of atonement for the ordinary abuses to which the state subjected the ordinary subject; and it was an act of conciliation through which the socially consecrated bread affirmed a sort of moral equality.

In the Parisian case, the equality represented by regulatory leveling, and acted out by communion ritual, had a basis in social practice. For Parisians ate wheaten and (mostly) white loaves, within a much narrower range of differentiation than should have obtained if culture always ceded to economics. Consumers confined themselves within a sort of Eucharistic cage: nothing else was fit to eat, not in the purely formal hygienic sense but in terms of sociopsychological variables of self-representation. Identity as well as morality were bound up with breadways. The rules and boundaries seemed most rigid precisely where one might have expected them to be most supple, given the empire of necessity, the structural menace of recurrent penury, and the uncertainty of day-to-day provisioning. Provincials and moralists had it figured out from the beginning: in foodways as in all ways, Paris embodied egocentricity, corruption, self-indulgence, extravagance.[2] From the government's perspective, the Parisian predilections were costly if not perverse: ersatz was unacceptable, even in dearth, and a hearty household bread was a nefarious provocation.

The philosophes were right to rail against popular "prejudices": there is no doubt that they obstructed change. (They were wrong to imagine, however, that all such prejudices were inherently unfounded and that the way to overcome them was through bludgeoning.) Few prejudices, or habits, are more deeply rooted than those concerning food. After nascent social science (in the form of physiocracy) failed to vanquish them, hard science took over. The chemists displaced the economists; mean Cadet followed lean Turgot. The former was even more confident in experimental reason than the latter had been in infallible nature. The scientists deployed a diabolical strategy: they assailed the bakers in their self-esteem and cast into irredeemable discredit the popular craft culture in which they were socialized. In a word, the bakers learned that they made bad bread because they did not understand the principles of making good bread. But the bakers resisted this discourse of progress with the same ferocity with which consumers repulsed the new loaves of progress. In none of its many phases or inflections and on none of the major issues of the time did the Enlightenment successfully come to grips with its People Problem.

The scientists did not play favorites: bakers of all types used flawed methods to attain outdated objectives. On the surface, however, all the bakers were not the same. The only bakers who sported official guarantees of certified training were the masters, grouped together in a guild. Portrayed in clichéd terms that often seriously deformed them, the guilds were among the leading targets of the liberal onslaught of the eighteenth century. The incorporated bakers constituted an especially attractive adversary because they were a guild operating in the subsistence arena, the privileged sphere of liberal criticism. The liberals accused the bakers of holding the people of Paris hostage through the exercise of a monopoly that enabled them to impose a high price and an uneven standard of quality. It turns out that this monopoly was a fiction. The periphery surpassed the core: the masters contributed less bread to the provisioning of the capital than the forains and the faubouriens. The shops furnished substantially fewer loaves than the markets. Against the city's settled ways, the faubourg Saint-Antoine played a dynamic counterpoint: it was a frontier territory that favored freedom, a refuge for innovation and insurgency well before Balzac pronounced it the Revolutionary quarter par excellence. Its role reminds us of the profound ambivalence of the Colbertian development strategy, wagering on a fecund tension between incorporated and laissez-faire poles of production and distribution in the Parisian economy. The spirit of Saint-Antoine infiltrated the city as the underground bakery thrived in corporate space through various illicit arrangements: systematized regrating, squatter markets, makeshift shops, rented or clandestine ovens, silent partnerships, and so on.[3]

Competition among bakers was brisk. "Professional jealousy," as they called it, induced them to fight over baking and delivery prerogatives, customer fidelity, access to supplies, and turf, physical and moral. The guild tried to impose its control over the rival contingents. Vis-à-vis the faubouriens, it followed the outraged line of the entire corporate world: the *faux* privileges of the faubourg did not exempt the latter from the inspection and jurisdiction of the guild wardens. More remote and dispersed, the forains faced even greater restrictions on their commercial freedom. The guild sought to exclude them from producing certain types of bread, supplying inns, storing unsold loaves, delivering to customers' homes, and so on. To prevent the guild from dissuading the forains to enrich the city's provision with their bread, the police seriously tempered its hectoring zeal. Hybrid creatures born of the loam but familiar with the city street, the forains evinced a quasi-corporate capacity to resist collectively, one reinforced by their strongly dynastic-familial character and their multisecular tradition of confecting a highly prized item. The faubourg Saint-Antoine, to which many Gonessiens and other forains gravitated over time, also betrayed a capacity for collective action that hints at informal organization.

Over the course of the century, the total baker universe contracted; in particular, markedly fewer forains supplied the capital. Baker productivity, however, increased, as the population of Paris continued to grow and demand for bread intensified. Rationalization, in part spurred by the striking development of the flour trade, rather than innovation, accounted for the increased baking capacity. Bakers used their resources more efficiently, but they did not change their techniques. The mollet mutation proved irreversible: bakers kneaded less and less hard dough and profited more and more from the time-saving and lightness-enhancing properties of brewer's yeast. Loaves became less weighty but more voluminous, longer and narrower in form, more inclined to stale rapidly and disagreeably, and probably somewhat whiter. Total annual consumption was probably closer to a hundred thousand muids than to the ninety-plus figure I offered above: I am more and more impressed by the contribution of illicit and undeclared sources. The amount continued to rise in proportion to population, though after the 1760s increased (and better quality) yield from wheat to flour allowed bakers to feed more people with the same quantity of wheat (but this would not have dampened the total entries, since the bulk of supplies arrived in the form of flour).

The physiocrats taunted the subsistence police that a single sheaf of wheat had never grown in a city. Yet in place of an annual crop, the city boasted a daily harvest in bread. The circadian rhythms of bread production and distribution further distinguished the city from the countryside. Hundreds of bakerooms hummed with activity through much of the night. With an odd mix of violence and gentleness, the baker or his geindre prepared to give life to the sleeping city by infusing life into the inert flour. As the flour stubbornly took its time to rise in the city shops, enveloped in the vinegary odor of fermentation, the forains were already on the road. By 4:00 A.M., as the city ovens disgorged their first batches, baker carts clogged the arteries leading to the markets. Tension built in the stalls as the first customers arrived. In the city bakeries the workers rushed to get the bread moving—to market, to homes, to the shop shelves. The pace at the markets was hectic; unable to raise prices above the morning level after the noon bell and obliged to offer the loaves at a reduction later in the day, the bakers sold as hard as they could in the early hours. With any luck a master could be at the tavern by 4:00 P.M.; having already imbibed and unwound by then, the journeymen napped near the oven.

In terms of frequentation and fidelity, no other urban institution could compete with the bakers. Bread set the city in teeming motion every day, with especially massive mobilizations on Wednesdays and Saturdays. Given its scale and its intensity of purpose, it evokes the image of a sort of daily transhumance, or perhaps a city in permanent procession. From interactions between bakers and patrons grew the bonds of clientage, clientalism, and community, and the sense of mutual claims that buyers and sellers had upon each other. The exchanges were not confined to goods and money, and they were frequently ritualized and coded. They were more complex at the twelve official marketplaces than at the hundreds of shops because they were less strictly bilateral and less narrowly focused. At the market, for slightly less money, after more or less earnest haggling, in addition to his or her loaf a consumer generally obtained more information of all genres and a richer dose of sociability than the shops offered. But the two sites were complementary: the shop remained open every day, usually made credit more readily available, and anchored relations in a fixed neighborhood setting rather than a transient if recurrent bivouac. Bread structured space as well as time. The local police did not conceal their suspicion of unfamiliar buyers at the market or in the urban village located around the shop.

During the Old Regime, people did not live by cash alone. The *taille* that mattered to urban folk was not the onerous royal direct tax on property and persons but the tally stick that inscribed bread acquisition on credit. Buyers at all levels, but especially grassroots consumers, regarded credit as the first of their natural rights. Credit was the coinage of everyday social and economic relations from the bottom to the top. Without the framing functions of the Great Chain of Credit, it is hard to imagine how things could have worked. For the people, credit was the hidden lubricant of the moral economy; along with the just price, it was the chief gyroscope of socioeconomic stability. It was no less imperative for economic actors, big and small, who needed it to finance their operations across time and space. Here was one market where the government intervened rarely and at best on the margin. Supply and demand tended to allocate the input and output of credit. Credit bespoke reciprocities that transcended the marketplace; it was never a purely commercial affair. It reduced transaction costs and forged precious links, but it also generated tensions, asymmetrical expectations, and terrible double (or triple) binds. The danger of credit was that it imperceptibly turned into its insidious other: debt. Whereas the one dilated the vital channels, the other occluded. Still, like bread itself, credit was a great leveler in this highly stratified world.

It was less altruism, as a perverse effect of reciprocity, than bad management that led the baker into business failure. Bakers went under because they could not pay their suppliers. Debts to merchants swelled beyond measure, sometimes as a result of speculative errors, subsistence crises, or family difficulties, but most often as a consequence of the inability of bakers to collect from delinquent consumers of bread. (Was this poetic justice? The ironic payback in the name of those consumers who had suffered at the hands of greedy bakers? It was a far more efficient settling of scores than, say, a riot.) More often than not, it was cheaper for the parties, and the public, to find an accommodation than to wipe out the adversary definitively. Institutionally, the consular court, composed of businessmen, favored this sort of politic solution, while, curiously, the jurisdictions closest to the police, composed of professional magistrates, favored a much more mechanical, draconian, and myopic denouement. Merchants had more leverage over bakers than the latter had over consumers. But the bakers did not lack in strategies of legal contestation and delay and in tactics of guerrilla resistance. Despite the communitarian ethos that theoretically infused the guild, it made no effort to develop collective remedies. Nor did the authorities frankly address this permanent source of destabilization until late in the century, when a series of grain and flour brokerage failures shook the subsistence arena to its foundations. Even then, they lacked the audacity, the capital, and the conceptual tools to envision a global reform.

If the guild failed the bakers on this critical issue of debt and credit, can it convincingly claim to have defended the interests of its members? There is no simple answer to the question, either for the bakers or for other corporations. Although the history of work has undergone a profound renewal in recent years, the guilds remain on many counts an enigma. In some ways their influence was pervasive and profound in forging the identities of their members. In other ways they could not contain or keep pace with the vitality and the diversity of their adherents. The baker case cautions us not to conflate trade and corporation. Not only did more bakers operate outside the guild than within it, but numerous baker-masters conducted their businesses without the least regard to the sacrosanct strictures of the community statutes. Nor is it clear, in so doing, that they had a cognitive sense of defying the rules: in their view, their quest to succeed economically did not diminish their social and political status as masters. Objectively, the lines of cleavage now passed through the guild itself, not between the masterly insiders and the tainted outsiders. Since masters had palpably different economic interests—far more so in many other corporations, where a much greater range and scale of activity obtained— the guild could not speak simultaneously for all of their concerns.

Many of the guilds understood this new fact of life, and adjusted to it, averting their eyes from areas of divergence and focusing instead on matters of general accord, such as the need to repulse outside (i.e., royal) intervention in internal affairs, to fend off outside challenges to guild monopoly, or to subordinate chronically indocile journeymen in a highly contentious and disruptive labor market. The oligarchy that ran the bakers' corporation tended to take this modernist/prudential stance, though it was not loath to pursue mutinous masters when it felt it could marshal a coalition against them (on such grounds as collaboration with forains or faubouriens or complicity with millers or conspiracy with workers or connivance with illicit bakers). Masters who resented juré policies could either ignore them or contest them. A substantial number of masters challenged the leadership of the jurés on issues ranging from constitutional construction to financial administration to the conduct of elections. Something resembling a democratic movement took shape within this guild and many others, acting out locally the sort of thing that R. R. Palmer described on grander levels.[4] Here was continuing education for future sansculottes, a practical experience in political socialization in which masters learned much about parliamentary rules and manners, deploying rhetoric, interpreting law, lobbying, electioneering, and intimidation.

We still tend to see the guilds in terms of stigmatizing stereotypes emitted in physiocratic press releases and echoed by their nineteenth- and twentieth-century liberal epigones. We should have no illusion about the rigor and reliability of the Enlightenment gaze: it was often as deeply ideological, as tendentious, and/or as intemperate as that of its adversaries/targets. There was a great deal wrong with the guild system; but the systematic nature of the militant physiocratic critique has badly misled historians. Like many other corporations, the bakers were less corrupt, prodigal, and tyrannical than the liberal line would have us believe. Partly as a result of royal purging, the bakers' guild was not in disarray, financially or administratively, on the eve of its (first) abolition. It had reimbursed its most onerous debts and its budget was in balance. After the death of Louis XIV, it was no longer a permanent fiscal milch cow. (And, if we can digest the absurdities provoked by Louis's wars, it is time to rethink our understanding of Colbert's conceptions of liberty, enterprise, regulation, and incorporation.) The guild of bakers was miserably unsuccessful in practicing monopoly. It was not a closed community strictly reserved for the sons of masters: the majority of its recruits, even before the reforms of 1776, came from outside. Nor did it reduce apprentices to protracted torment and journeymen to slavery.

One of the rare eighteenth-century figures about whose apprenticeship and journeymanship we know something is the Parisian master glazier Jacques-Louis Ménétra. His autobiography, for all its Rabelaisian hyperbole, gives us an intriguing clue to what the study of life histories and socioprofessional life cycles could reveal.[5] How much better could we understand the baker if we could not only learn the details of his often Sisyphean trajectory on the way up but also assess how the memory of that experience shaped his attitude and behavior toward his own apprentices and journeymen? Or if we could interrogate the journeyman—man in translation, eternal boy in the common, lived French idiom—who sensed from the beginning that he would remain a lifelong worker, about the evolution of his attitude toward his (various) bourgeois, as he referred to his boss, and on the way in which he negotiated entry and exit and forged solidarities, horizontal as a rule but perhaps occasionally vertical?

Regardless of the disposition of the master, apprenticeship would ordinarily have been a hard time had it taken place at the canonical moment: the inherently jolting transition from childhood to adolescence. Aged twenty, the average baker apprentices were remarkably old, far too mature to train and tame as children, indeed almost indistinguishable from the garçons. By contemporary standards, their sojourn was brief: three years in practice as well as theory. As to their relations with their masters, one can do little more than lamely observe that they varied enormously, from experiences of genuine learning and growth marked by asymmetrical mutual respect to situations of chronic abuse, physical, moral, and professional.

To the extent that a licensed apprentice now dubbed compagnon or garçon could no longer (but precisely since when?) define himself automatically as someone "awaiting mastership," according to the venerable formula, I suppose that one could in rancor denounce journeymanship as slavery. Certain masters assimilated their journeymen to the ranks of domesticity, a symbolic step toward a sort of servitude that the workers profoundly resented yet took very seriously as a strategy of demoralization and subjection. Their working conditions probably disposed garçon bakers to a certain alienation. Relentless, physically draining, likely to compromise their health if not shorten their lives, their intense labor took place in exiguous, poorly ventilated cellars, at night as well as through much of the day. With little leisure time, they had to live these brief interludes rapidly, intensely, and artificially, with little possibility of sustaining durable relations of a familial character. In addition, relative to other skilled workers, they were paid a low wage, which was often withheld to incite them to prolong their service.

The "social question," as it was later called, resonated with particular virulence throughout the bakery during the eighteenth century. The police viewed work as the primary mechanism of social control; in their utopian fantasy, work would be universal and ceaseless, subject to panopticon surveillance from which no one could escape for an instant. They warmly seconded the efforts of the guilds to seize control of the labor market, a major site of the burgeoning worker insurgency that so many commentators denounced as the century wore on. Few guilds knew more wrangling tumult in the labor market than the bakers. Linking placement and departure in a tightly knit control system epitomized by the livret, with its blackballing or exclusionary devices, the guild sought to determine where, when, and under what conditions every journeyman would (or would not be allowed to) work.

To defend their rights, primarily the freedom to dispose of their labor as they saw fit, the workers elaborated various forms of resistance, violent and serene, ranging from individual gestures of bakeroom subversion and neighborhood vindictiveness to collective inn- and tavern-based cabals to place workers or to extract higher wages, sometimes issuing in veritable organizations to undertake strikes and pressure or punish especially overbearing or refractory masters (and traitorous journeymen), to patient litigation in the courts in the name of both abstract principles and statutory and positive law precedents. For reasons of interest and ideology, a substantial number of masters defied corporate rules and dealt directly with journeymen, often on their terms. Most masters, however, including the independent-minded ones, recoiled in horror at the apocalyptic moment in 1776 when Turgot suppressed the corporate system and the workers took their insubordination to its ultimate stage, symbolically reducing the masters to dependency, shutting down their shops by boycotting work, and setting up rival baking and distributing operations, fragile but functional. The social question, along with the bread question, contributed to the outbreak of the Revolution, gained great political and psychological momentum from it even while exacerbating its difficulties, and emerged from it ready to envenom the next century, which alternately and vainly tried to resolve both questions with similar market-based and authoritarian/paternalist policies.

Most of the baker boys who declared themselves masters overnight had little chance of surviving. Emancipatory exuberance could not compensate for lack of capital and connections. Establishment was complicated and costly, even if one could skip the onerous and expensive (sometimes extortionate) stages of mastership candidacy, including the preparation of the masterpiece, scrutiny by the elders, payment of corporate fees and diverse informal exactions, and formal initiation. The path to establishment through mastership was the most prestigious: one became part of the great corporate chain that led through successive rings, in the standard image of the day, directly to the king, who thus became one's interlocutor as well as one's protector.[6] According to corporate theory, this road was supposed to be the natural, organic itinerary of the licensed apprentice-become-journeyman. In fact it was treacherous, largely exclusionary, and highly discriminatory. Outsiders had to work harder and pay more than sons of masters, although they constituted the bulk of the corps of renewal. Mastership was supposed to be tantamount to the end of hierarchy among brothers, but in practice it inaugurated a new ordeal, a kind of second journeymanship, with all its humiliations, to which the new masters who wished to enjoy full civic rights in the guild had to submit.

Establishment, save for the forains and some faubouriens, implied the need to acquire a shop. That meant not only a lease on a site but a clientele as well. The average cost, 1,662 livres, made it a dear-bought investment. Many bakers had no chance of attaining it on their own: they looked toward inheritance or association (both of which were often part of family regulation strategies for old-age care) or marriage, or they envisaged an illicit solution either inside or outside corporate frontiers. Marriage was a primordially family affair, and the choice of a spouse the most important single business decision a baker would ever make. Although it characterized not quite half of the unions, endogamy produced a dense honeycomb of interlocking alliances and dynasties that exercised enormous influence in the trade. Apart from invaluable human and commercial skills, frequently nurtured in a family bakery, that they would deploy in their new shop, wives brought dowries of variable size rich in cash and trousseau. Their husbands contributed slightly larger and complementary portions containing business investment and annuities as well as some money. The most lucrative marriage, uniting masters who were sons of masters to daughters of masters, represented ten thousand livres, a very hefty capital. The average marriage lucre, however, a little over five thousand, was in most cases an ample launching treasure.

It appears, in any case, that marriage wealth is not a good predictor of the fortune that the couple will attain once they have been in business for some time. Average net worth gravitated around 12,500 livres, a comfortable sum placing the bakers in the lower-middle range of artisans and merchants. One-third of the bakers possessed mediocre holdings worth under two thousand, while another third stood in a category of some affluence, having amassed more than ten thousand. Fortunes were larger in the second half of the century than the first, though they did not grow continuously from the beginning to the end of the century. The economic and subsistence crises that ushered in the Revolution hurt numerous bakers, but on the whole the three contingents resisted rather well. Throughout the century, the masters vaunted quite substantial wealth, but they did not command the greatest fortunes. The most robust bakers of the faubourg Saint-Antoine, brandishing the unofficial title of merchant baker to distinguish themselves, surpassed them, symbolically marking the domination that the faubourg came to exercise over the city in a number of economic domains. Baker assets were generally well balanced: a fifth in hard investments (real estate, annuities, and offices), another 20 percent in paper (most accounts receivable for bread furnished), less than 10 percent in equipment, 13 percent in stocks (mostly flour), and most of the remainder in household and personal items, including cash. Under eighteen hundred livres, liabilities were quite small, about half of them representing accounts payable for commercial services and merchandise.

Highly inelastic demand did not make the baker trade "the best of them all," as Marat suggested, though as a group its members were moderately well off. It did, however, make them subject to almost ceaseless public scrutiny. Since baker success depended on the opinion consumers had of bakers, they regarded their reputations as a precious form of capital. A good reputation assured their credit in a double sense, morally and materially, with suppliers as well as with customers. Bakers had to struggle against an ambient negative stereotype that portrayed them as grasping, manipulative, and dishonest. Awareness of this representation heightened their sensitivity to insults, a widely practiced idiom of exchange and retribution, whose commercial implications the bakers feared, especially when they were uttered in public. Sexual innuendo was one of the most feared themes of aspersion because it drew connections between private and public behavior, suggesting a pervasive dissoluteness and untrustworthiness.

Bakers often turned to the police for redress, but this was not the major reason that the lives of the two groups were intertwined. "As the most essential concern that must occupy the administration," the subsistence issue preoccupied the entire policing structure, from the king on down. The police of bakers, and of bread, represented in some sense the bottom line, the final act in the complex drama of provisioning. Like the bakers and the people, the bread question mobilized the police virtually every day.

"Bread . . . belongs to police and not to commerce," observed the Neapolitan philosophe Abbé Ferdinando Galiani.[7] Yet the police could no more do the baking than they could do the grain and flour trading. Galiani of course understood the limits of intervention. What he meant was that those who dealt in grain or in bread were accountable to the police, and through them to the people, not merely to themselves. The police viewed the bakers, like the grain and flour dealers, as amphibious beings who worked to gratify their self-interest but at the same time were public servants. The bakers had a straightforward duty: to furnish an adequate quantity of good quality bread at a reasonable price. In order to make sure that the bakers fulfilled their responsibilities, the authorities had at once to assure them a profit, in one way or another, that was, on average, capable of motivating them and to threaten them with an array of sanctions sufficiently unpleasant to deter them from lapsing into disobedience. Punishments, usually immediate, and rewards, sometimes deferred, vied for primacy in the context of a vast and permanent negotiation that took place both individually (say, between commissaire and forain at the market) and collectively (say, between lieutenant general of police and corporate delegates).

On the one hand, the police pledged support and shelter to the bakers. They offered permanent assistance in procuring supplies in the open trade in the countryside as well as in the towns and in the Paris markets, including refuge from local authorities, seigneurs, and venal officeholders; freedom to store large stocks of grain, incidentally an incentive to speculation but above all a precious component of an unofficial emergency reserve for a city that lacked granaries; immunity in certain circumstances against seizures for debt; occasional subsidies in kind or in cash; and the opportunity to compensate themselves in interludes of subsistence calm for periods of self-denial during hard times. On the other hand, the police warned of (and carried out) heavy fines, confiscations, shop closings, market-stall cancellations, expulsions from the profession and/or the guild, incarceration, and public shaming rituals. In its most elaborate form, this ongoing negotiation involved the people, clients of the police as

The women of Paris returning triumphantly from their expedition to Versailles on 6 October 1789, with their royal trophy, the "baker," Louis XVI, whose link to subsistence is figured by the carts full of sacks of flour that precede and follow him. *Source:* © *cliché Bibliothèque nationale de France, Paris.*

well as of the bakers. The police found themselves conveying contradictory cues. To the bakers: we are a bulwark against the wrathful vengeance of the consumer-people. To the people: we are a bulwark against the chronic perfidy of the bakers. To give voice to the people was to court dire risks, for the "rage of the populace" against the bakers always proved extremely difficult to contain. Yet the authorities counted heavily on consumer cooperation in order to unmask bakers of all stripes who cheated on quality (adulteration or flaws in cooking) and quantity (short-weight loaves) or maneuvered to keep bread off the market in order to dispose of it illicitly.

One of the most durable arguments in favor of a price-control (*taxation*) system was the cover that it would afford bakers vis-à-vis highly susceptible consumers. For all the parties, price was the decisive marker, even when it was recognized to be a symptom or an artifact rather than a cause. Yet there emerged considerable resistance to the unalloyed authoritarian approach, not only among participants in the provisioning business and many theorists but within the ranks of the police. For the police were not tropismatic regulators wrought of a single strain and invested in a monolithic credo. Increasingly well informed across the century and attuned to the complexities of price making as a psychological as well as an economic process, the Paris police carefully weighed the costs and benefits of coercion and of freedom. Technically, the evidence suggested that it was probably impossible to establish a fair and viable price schedule given the diversity and intractability of the variables concerning provenance, quality and price of wheat, cost of transportation and storage, methods of milling and bolting, recipes and protocols of baking, and so on. The many trials undertaken by the authorities suggested both the fragility of any effort at systematization and the estrangement of any system from shifting everyday realities. Commercially, the tax could backfire by driving supplies into hiding, as the bread tax put downward pressure on the mercuriale. Practically, bakers would often be in a position to hijack the tax, and use it to enhance their profits, or indemnify themselves by diluting quality. Politically, the tax engendered false expectations that led them to impute responsibility for every fluctuation to the government.

In light of these factors, and in acknowledgment of the power and necessity of a certain dose of liberty in the operation of the provisioning trades, the Paris police did not regularly tax the price of bread. In ordinary times, bakers were free to post the price that demand would tolerate. The price of reference was known as the "current," a self-adjusting level that expressed, in mutually satisfactory terms, the relation between sellers and buyers. The bakers who traduced this standard suffered market sanctions. The market wire, relayed by the shop network of word-of-mouth signaling, spread news rapidly. Only in moments of mounting stress did the current become a presumptively inviolate ceiling. In a crisis the police manipulated the maxima on a day-to-day basis as conditions changed, without fixed deference to any model or tariff. Sartine modernized La Reynie's notion of the just price, and took seriously into account the needs and prospects of the bakers, yet in the end the old sociopolitical criteria for interventionist legitimacy prevailed. This empirical, pragmatic, and relativistic strategy gave the police optimal flexibility but at the same time required incessant surveillance and a capacity for swift readjustment.

During the short century surveyed here, many fundamentals besides the repertoire of the police did not change. The people of Paris were more numerous and probably more fragile. Yet Commissaire Duplessis's "ferocious beast" does not seem more volatile or more dangerous in the aftermath of the disorders of 1775 than it had been in the wake of the riots of 1725. At once banal and aberrant, each uprising took shape in a specific and circumscribed context of stimuli and responses; neither bespoke seething and/or cumulative alienation. At the end of the period the people still saw themselves primarily as consumers rather than as producers, and specifically as bread eaters, eaters of quality loaves, the mark of their conceit and their identity as Parisians. And they regarded the price of bread in normative terms as more or less immutable. Like the police and the master bakers, both of whom ascribed much of what went wrong in their respective (and overlapping) worlds to cabals, the people continued to explain much of their experience, and the loftier happenings of which they caught only glimpses, in terms of plots of all sorts.[8] This disposition did not connote a deficit of rationality. Rather it signaled a number of distinct and complementary ways of managing their interests and dealing with different sorts of reality. Finally, the consumer-people still saw themselves as parties to the social contract of subsistence and children of a nourishing prince, despite one or two baffling and painful lapses in paternal coverage, the proverbial exceptions that confirmed the rule of rules.

Those solecisms seemed equally inconceivable to the police (and perhaps to the bakers as well, at least the bulk of them, who were heavily invested in the protectionist system despite its chronic frictions and anomalies). Superficially expiated and repaired, they were nevertheless harbingers of profound change. The political and ideological basis for the old social contract was unraveling. The impasse in which the state had been mired for so long seemed to result more or less directly from the bloated representation of its vocation that the contract implied. The stagnation in which the economy languished seemed to result more or less directly from the sorts of entitlements that the contract guaranteed. The archaism of the social structure seemed to draw sustenance from the contract's deeply conservative ethos. The newer version of natural law, individualistic and libertarian, most exquisitely defined by the physiocrats, displaced the older collectivist and organicist rendition that framed the contract. The sheer force of capitalism, which first exhilarated and then terrified Diderot, proved increasingly irresistible, intellectually if not yet institutionally. In the desacralizing mood that was gaining momentum, the salus populi no longer could afford to obey a Eucharistic logic.

Appendix A *List of Inventaires Après Décès and Complementary Materials (Deposited in Minutier Central of the Archives Nationales unless Otherwise Indicated)*

XXVIII-265, 11 August 1740

LXXXIX-484, 4 April 1743

XXXV-626, 21 June 1741

LXXXIX-592, 7 July 1760

VII-291, 27 June 1754

VII-288, 1 August 1753

LXXXV-581, 9 July 1764

VII-287, 17 May 1753

LXI-630, 14 September 1739

XXVIII-34, 24 April 1699

XIII-252, 30 August 1734

XXVIII-290, 11 February 1745

CXI-96, 30 September 1719

CXI-95, 26 August 1719

VII-417, 21 July 1775

XVII-553, 16 April 1714

LXXXV-551, 15 December 1757

XCIV-297, 5 May 1760

XXVIII-114, 24 October 1711

VII-329, 4 November 1760

CXI-321, 13 July 1755

LXVI-550, 25 October 1765

CI-279, 19 January 1730

VII-307, 1 March 1757

VII-398, 21 June 1771

XXVIII-255, 4 September 1738

VII-354, 28 March 1765

XXXV-552, 30 October 1714

VII-402, 21 February 1772

LXXV-523, 25 April 1724

XXVII-174, 26 October 1726

XIX-612, 27 July 1715

XXVIII-253, 10 May 1738

XXVIII-274, 12 April 1742

XIII-252, 25 September 1734

XLIX-539, 19 March 1731

VII-291, 28 May 1754

XX-558, 6 February 1736

XII-411, 19 August 1730

LXI-492, 3 September 1761

CXXI-300, 25 October 1734

XXVIII-290, 15 February 1745

VII-309, 21 July 1757

XXXIV-388, 31 August 1716

LXV-227, 2 March 1730

XVII-561, 12 July 1714

LXXXIX-431, 18 August 1735

V-524, 22 April 1762

CI-470, 13 October 1755

XXVIII-216, 16 March 1730

XXVIII-252, 15 April 1738

XIII-164, 16 June 1710

XXVIII-43, 26 January 1699

XXVIII-233, 26 May 1734

LXXXIX-453, 9 September 1738

LXXXIX-212, 15 May 1709

XXVIII-266, 22 December 1740

XXX-437, 24 April 1773

XXVIII-131, 10 October 1714

XIX-589, 17 January 1709

XIX-469, 26 January 1741

XXVIII-109, 24 November 1710

XIX-610, 16 February 1715

XLVII-109, 26 April 1745

XCV-54, 29 August 1709

CXXI-292, 4 June 1731

VII-309, 29 August 1757

CI-602, 4 February 1775

VII-417, 12 September 1775

XXVIII-128, 3 April 1714

VII-347, 8 November 1763

XCIX-470, 18 May 1741

XCIX-477, 28 January 1743

CXI-92, 24 March 1719

XXVIII-224, 7 January 1732

VII-288, 15 July 1771

CI-540, 7 October 1765

XLIII-232, 4 July 1797

CXVIII-357, 21 August 1728

XXVIII-128, 22 March 1714

CXVIII-351, 6 November 1728

XXVIII-260, 22 October 1739

CIX-412, 14 November 1712

CXV-785, 7 July 1767

VII-386, 23 August 1769

XXVIII-251, 23 January 1738

XXVIII-131, 17 September 1714

LXVIII-374, 27 April 1729

LXXXIX-655, 12 October 1767

LXIX-317, 30 January 1734

XXXIX-441, 30 June 1757

VII-288, 6 July 1771

VII-400, 12 October 1771

VII-404, 21 August 1772

CIX-490, 3 August 1735

XXX-438, 13 July 1773

V-291, 19 July 1754

XXXVIII-98, 21 June 1711

LI-799, 18 May 1736

XLIX-689, 22 April 1751

CVIII-488, 3 March 1750

CXII-704B, 5 April 1751

XX-605, 23 September 1751

CVIII-494, 19 July 1751

XXVII-257, 29 January 1751

CXI-234, 12 July 1751

CXXII-682, 5 November 1751

LXI-603, 14 September 1784

XII-411, 19 August 1730

LXXXV-583, 22 November 1764

LXXXIX-592, 7 July 1760

XCIV-301, 28 January 1761

Archives Seine-Paris, DE[1] 3, 4 February 1755

BHVP, NA 156, fol. 384, October 1776

AN, Y 15621, 18 December 1750

AN, Y 14940, 21 December 1723

AN, Y 11233, 1 August 1746

AN, Y 15616, 6 August 1748

AN, Y 11587, 31 March 1769

AN, Y 13524, 29 December 1762

AN, Y 15624, 22 July 1752

Appendix B *List of Scellés (Preinventories) from the Y Series of the Archives Nationales*

13926, 31 May 1741

12729, 1 March 1735

11232, 27 August 1745

13100, 21 March 1747

11222, 22 September 1735

13099, 16 January 1745

14964, 5 June 1748

13089, 24 August 1738

11388, 29 May 1772

12597, 15 December 1752

11043, 15 March 1728

14948, 18 January 1733

11045, 9 March 1730

12602, 1 May 1755

14536, 9 June 1740

14536, 26 June 1740

14535, 28 January 1739

15615, 25 January 1748

15603, 5 May 1742

15252, 2 January 1743

15247, 3 May 1740

15063, 13 March 1762

15241, 10 June 1735

11449, 4 November 1727

12574, 29 January 1729

12575, 19 March 1731

14536, 9 June 1740

14777, 7 September 1727

11171, 5 June 1754

14084, 22 December 1757

12625, 6 June 1775

11074, 1759 (Jean Gouillard)

13376, 15 January 1771

12729, 1 March 1735

13366, 22 August 1742

14964, 5 June 1748

13080, 12 April 1733

11309, 14 September 1748

13741, 11 March 1741

13741, 22 March 1741

11673, 30 September 1749

14948, 8 January 1733

15236, 1 July 1729

15240, 18 July 1733

15241, 6 April 1734

11240, 24 December 1753

Appendix C *List of Marriage Contracts (from the Minutier Central of the Archives Nationales unless Otherwise Indicated)*

CI-211, 14 October 1719
CXI-74, 29 July 1715
CXI-96, 30 September 1719
XXIV-646, 2 June 1735
VII-404, 30 July 1772
CXXII-653, 4 April 1742
L-175, 21 February 1683
XLI-355, 16 September 1714
XLIV-377, 27 May 1746
XVI-695, 12 January 1738
CXVII-290, 15 January 1718
XXVIII-47, 26 September 1699
VII-297, 1 June 1755
XIII-180, 24 June 1714
CXVI-270, 1 October 1730
XXXVIII-300, 12 October 1738
VII-284, 21 September 1756
LXXXII-363, 13 March 1757
VII-283, 18 June 1752
LII-278, 1 February 1738
CI-602, 23 February 1775
XXXVIII-208, 8 December 1720
(double marriage Félix-Delamarre)
CI-402, 17 January 1744
VII-400, 26 September 1771
LXXXI-416, 17 January 1771
XVII-846, 16 January 1759
VII-386, 17 August 1769
IV-311, 9 July 1702
XXIV-553, 2 June 1712
VII-297, 12 June 1755

VII-376, 27 April 1768
LXXVII-202, 1 March 1731
VII-282, 10 February 1752
CIX-412, 17 November 1712
XLIX-538, 19 November 1730
CXXI-294, 16 March 1732
XXVIII-166, 11 February 1720
XXVIII-178, 24 April 1721
VII-301, 28 March 1756
XXXVIII-209, 19 January 1721
LXXXI-416, 24 January 1771
LXV-149, 6 July 1700
LXV-193, 3 July 1717
CIX-490, 12 December 1730
LXV-193, 22 July 1717
XXXVIII-282, 23 June 1735 (also AN, Y 11587)
LXXXIX-83, 18 May 1686
XXXVIII-279, 19 December 1734
XXXVIII-362, 9 November 1747
CXXI-294, 6 April 1732
VII-402, 5 February 1772
VII-309, 23 July 1757
LXV-197, 21 July 1718
XLIX-539, 5 April 1731
LXXVII-118, 28 December 1710
LIII-261, 8 April 1732
LXXXV-582, 20 September 1764
VII-293, 10 October 1754
XLV-492, 27 October 1753
XIII-241, 1 March 1731

XCIV-302, 26 February 1761
VII-329, 17 December 1760
VII-388, 8 August 1771
CI-133, 20 January 1710
C-536, 28 January 1731
VII-293, 12 September 1754
XIX-594, 27 October 1711
XXI-559, 12 August 1736
XXVIII-224, 17 January 1732
XIX-609bis, 23 April 1714
XXIV-572, 3 June 1715
LI-800, 8 May 1716
XXXIV-501, 17 June 1725
XII-179, 15 April 1714
XXXIV-439, 20 October 1726
XXVIII-44, 26 April 1699
XXVIII-45, 12 May 1699
XXVIII-217, 20 March 1730
XXVIII-254, 14 August 1738
XXVIII-97, 7 May 1711
XXXIV-391, 3 September 1717
XXIV-566, 12 July 1714
XLIII-291, 5 August 1714
XXVIII-258, 11 June 1739
XXX-147, 6 April 1698
CI-537, 15 June 1765
XX-559, 11 November 1736
XLI-356, 16 February 1715
XXVIII-104, 18 February 1710
XXVIII-246, 27 February 1737
XXVIII-282, 16 August 1743
XXXVIII-98, 6 June 1711
VII-335, 24 September 1761
XXVIII-282, 18 August 1743
XXVIII-255, 9 October 1738
VII-309, 17 August 1757
XXXVIII-97 (Ph. Lambert-Riberette)
L-175, 21 February 1683
XXVIII-43, 30 January 1699
XXVIII-216, 18 February 1730
LXXXIX-655, 1 October 1767
XIII-134, 25 January 1699

XV-469, 7 July 1715
XLIII-289, 22 April 1714
LXXVII-118, 31 December 1710
XXVIII-210, 26 September 1728
XXVIII-210, 7 July 1728
XXVIII-208, 28 January 1728
LXXXIX-132, 19 August 1694
XXXVIII-209, 5 January 1721
LXXXIX-131, 12 May 1694
XIX-589, 19 January 1709
XXVIII-217, 15 May 1730
XXVIII-43, 17 January 1699
XXVIII-174, 7 October 1720
XXVIII-185, 17 October 1722
LXXXIX-469, 17 July 1740
LXXXI-294, 23 June 1742
XIX-609ter, 23 November 1714
XXVIII-250, 21 November 1737
XXVIII-178, 8 May 1721
V-438, 17 March 1755
XXVIII-282, 1 September 1743
XXVIII-131, 25 October 1714
LIX-162, 2 September 1714
LXXXIX-422, 9 May 1734
LXXXIX-666, 30 March 1769
XV-466, 23 February 1715
XLVIII-65, 5 October 1732
VII-178, 23 October 1706
LXXXV-581, 28 June 1764
CX-260, 25 January 1707
LXXXIX-175, 25 May 1702
XXVIII-248, 25 August 1737
XCII-361, 1 September 1712
LXXXVI-649, 7 July 1751
XV-690, 10 January 1751
LXXIX-72, 10 May 1751
CXI-234, 12 July 1751
XC-369, 10 August 1751
CIX-585, 13 November 1751
LIV-849, 11 June 1751
XII-516, 17 February 1751
LXVI-490, 30 December 1751

CIX-580, 17 February 1751
CXXII-682, 5 November 1751
XIII-293, 16 August 1751
AN, Y 13084, 25 January 1680
AN, Y 15621, 28 April 1731
AN, Y 14940, 5 March 1699

AN, Y 15616, 1 February 1720
AN, Y 14775, 1 October 1692
AN, Y 11571, 18 August 1755
Archives Seine-Paris, 6 AZ 1462, 10 January 1765

Marriage-contract data has also been drawn from a large number of the inventaires après décès cited in appendix A.

Appendix D *List of Faillites (Drawn from the D4B⁶ series of the Archives de la Ville de Paris et de la Seine)*

3-170, 17 January 1742
4-183, 16 April 1742
4-235, 23 July 1743
6-275, 30 July 1745
11-526, 3 May 1765
11-526, 28 June 1770
11-526, 28 November 1752 and 3 May 1765
12-576, 12 July 1753
13-587, 28 August 1753
13-589, 1 September 1753
13-630, 26 June 1754
14-653, 5 November 1754
16-788, 17 December 1756
20-985, 8 July 1759
20-997, 31 September 1759
20-1004, 26 September 1759
21-1075, 29 July 1760
21-1029, 5 January 1760
24-1218, 11 August 1762
24-1249, 1 September 1763
24-1262, 7 March 1763
25-1306, 9 September 1763
26-1378, 23 June 1764
27-1424, 28 February 1765
28-1476, 14 September 1765
29-1530, 6 May 1766
29-1559, 8 August 1766
29-1585, 11 November 1766
34-1860, 8 May 1769
36-1958, 7 December 1769
38-2106, 21 August 1770

39-2138, 1 October 1770 and 30 March 1772
41-2260, 15 June 1771 and 25 June 1784
41-2283, 22 July 1771
42-2307, 16 August 1771 and 20 March 1776
42-2386, 26 November 1771
43-2403, 11 December 1771
43-2410, 19 December 1771, 13 August 1779, and 22 July 1782
43-2420, 8 January 1772
43-2435, 24 January 1772
43-2468, 17 February 1772
44-2512, 30 March 1772
44-2524, 6 April 1772
44-2555, 7 May 1772 and 9 September 1774
44-2557, 7 May 1772
45-2610, 14 July 1772
46-2706, 23 November 1772
47-2825, 24 April 1773
48-2850, 5 June 1773
48-2863, 25 June 1773
48-2869, 5 July 1773
48-2915, 20 August 1773 and 22 January 1776
48-2926, 31 August 1773
49-2955, 2 October 1773
50-3046, 19 January 1774
50-3061, 4 February 1774
50-3076, 17 February 1774
50-3079, 20 February 1774

51-3117, 28 March 1774
51-3137, 22 April 1774
51-3156, 3 May 1774
53-3251, 26 September 1776
53-3308, 29 November 1774
54-3390, 18 March 1775
56-3530, 10 October 1775
56-3555, 14 December 1775
57-3613, 9 March 1776
57-3645, 13 April 1776
58-3699, 12 June 1776
58-3700, 12 June 1776
60-3870, 2 December 1776
60-3872, 4 December 1776
61-3883, 13 December 1776
61-3937, 7 February 1777
62-3961, 4 March 1777
62-3973, 20 February 1777
63-4036, 17 May 1777
63-4040, 22 May 1777
63-4095, 2 July 1777
63-4102, 7 July 1777
65-4192, 23 September 1777
65-4234, 22 November 1777
66-4278, 8 January 1778
66-4294, 27 January 1778
67-4379, 29 April 1778
71-4710, 5 May 1779
75-4997, 25 April 1780
76-5023, 18 February 1780
79-5251, 13 October 1780
79-5271, 16 November 1780
79-5281, 29 November 1780
79-5287, 6 December 1780
79-5294, 26 January 1781
80-5305, 24 January 1781 and 6
February 1782
80-5345, 6 March 1781
82-5476, 7 August 1781
83-5537, 5 October 1781
83-5562, 11 December 1781

83-5592, 14 January 1782 and 23
September 1783
83-5593, 18 January 1782
84-5659, 26 April 1782
85-5759, 26 September 1782
85-5766, 3 October 1782
87-5884, 10 April 1783
88-6016, 22 October 1783
88-6048, 24 November 1783
89-6074, 31 December 1783
89-6093, 5 February 1784
89-6111, 27 February 1784
90-6140, 20 March 1784
90-6146, 23 March 1784
90-6176, 3 May 1784
90-6179, 4 May 1784
90-6211, 16 June 1784 and 4 September
1787
91-6237, 20 July 1784 and 16 January
1786
92-6324, 11 November 1784
93-6406, 2 March 1785
95-6623, 24 February 1786
98-6883, 20 April 1787
98-6896, 4 May 1787 and 5 May 1788
101-7102, 14 March 1788
101-7128, 22 April 1788
101-7168, 17 June 1788 and 18 April
1789
102-7232, 9 September 1788
102-7241, 16 September 1788
103-7253, 24 September 1788
103-7304, 13 November 1788
104-7356, 31 December 1788
104-7371, 19 January 1789
104-7395, 4 February 1789
105-7431, 10 March 1789
106-7518, 4 June 1789 and 1 July 1791
106-7557, 22 July 1789
113-8056, 28 November 1791
113-8084, 7 February 1792

Appendix E

Bakers with Short-Weight Loaves

I. Archives Nationales

Y 9538	29, 30 May	1725 (and *notes* for whole month)
Y 9538	9, 24, 25, 28, 30 June	1725
Y 9538	28 June	1726
Y 9434	25 June	1733
Y 9435	26 February	1734
Y 15242	23 July	1735
Y 15244	6, 12 March	1737
Y 15244	7 December	1737
Y 9440	19 April	1739
Y 9440	2, 4, 16, 29 May	1739
Y 9619	29 May	1739
Y 9440	27 June	1739
Y 9440	31 October	1739
Y 9619	11 December	1739
Y 12142	31 January	1740
Y 9441	12 February	1740
Y 9441	27 April	1740
Y 9441	15 July	1740
Y 9441	11, 14, 24 October	1740
Y 9441	3 March	1741
Y 9499	3 March	1741
Y 9442	4, 28 April	1741
Y 9619	28 April	1741
Y 9442	9 June	1741
Y 9442	4 August	1741
Y 9621	18 August	1741
Y 9443	12 January	1742
Y 9621	13 April	1742
Y 9443	13, 26 April	1742

Y 9443	27 June	1742
Y 9499	3 August	1742
Y 9443	(?) August	1742
Y 9499	16 November	1742
Y 9499	28 May	1743
Y 9499	13 December	1743
Y 9621	2 December	1746
Y 9621	13 December	1748
Y 9533	16 January	1750
Y 9621	24 November	1752
Y 15359	25 March	1758
Y 9539	16 December	1759
AD 1 23[A]	1 July	1767
Y 12826	11 April	1772
Y 12826	30 June	1772
Y 12826	11 April	1777
Y 12826	30 June	1777
Y 12826	13 February	1778

II. Other Sources

Arsenal, ms. Bastille 10027, pièce 389 (anon. to Marville, 23 September 1740).

BN, ms. fr. 21640, fol. 145 (27 June 1724).

Nicolas-Toussaint L. Des Essarts, *Dictionnaire universel de police* (Paris, 1786–90), 2:255 (3 August 1742); 2:256 (16 November 1742); 2:257 (5 May 1739).

Lettres de M. de Marville, lieutenant général de police, au ministre Maurepas (1742–1747), ed. A.-M. de Boislisle (Paris, 1896–1905), 1:226 (13 December 1743) and 2:222 (30 June 1745).

Notes

Introduction

1. On these questions, see Steven L. Kaplan, *Bread, Politics, and Political Economy in the Reign of Louis XV*, 2 vols. (The Hague, 1976), and *Provisioning Paris: Merchants and Millers in the Grain and Flour Trade* (Ithaca, N.Y., 1984).

2. Archives nationales (hereafter AN), Y 10558, 5 May 1775; Edgar Faure, *La Disgrâce de Turgot* (Paris, 1961), pp. 304–6.

3. Vladimir S. Ljublinski, *La Guerre des farines*, trans. Françoise Adiba and Jacques Radiguet (Grenoble, 1979), p. 107; AN, Y 10558, 5, 27 May 1775.

4. Jacques Necker, "Mémoire instructif" (August 1788), in *Les Elections et les cahiers de Paris en 1789*, ed. Charles L. Chassin (Paris, 1888–89), vol. 2, p. 550; Richard Du Pin, *Le Boulanger et la boulangère* (Paris, 1789), p. 4; Rubigny de Berteval, *Avis au peuple sur les premiers besoins* (Paris, ca. 1793), p. 4. Cf. the latter on p. 6: "Le pain de Paris, autrefois si vanté, est aujourd'huy, lourd, noir et mal-sain."

5. Blaise Pascal, *Pensées* (Paris, 1961), pensées 304, 311, pp. 154–55.

6. See the themes developed by Claude Macherel and others for the Belgian-based exposition and album called "Une Vie de pain" (sponsored by the Institut de sociologie of the Université Libre de Bruxelles and the Crédit Communal de Belgique and planned for 1995).

7. Interview with R. Calvel in *Michel Montignac Magazine*, no. 1 (May–June 1993), p. 35. For a fuller expression of Calvel's viewpoint, see his *Le Goût du pain: Comment le préserver, comment le retrouver* (Paris, 1990). See also the older but vigorously argued "L'Autre bout de la chaîne: L'évolution de la qualité du pain," *Actualités agricoles*, 30 April and 7 May 1976. Significantly, one of the first to denounce the decline was an American journalist in Paris, Meg Bortin. "One thing is certain," she wrote, "the quality of the bread here is going downhill. . . . le pain n'a plus de goût." Bortin, "Give Us Our Daily Bread," *Paris Métro*, 26 October 1977.

8. Jean-Pierre Coffe, *Au secours le goût* (Paris, 1992), pp. 58–59, 61–66. The stakes were drastically different in the 1760s when Abbé Armand-Pierre Jacquin deplored "the lack of police concerning the quality of bread" that allowed "the ignorance and the dishonesty" of the bakers to imperil the health of the inhabitants by exposing them to "a loaf that was badly fermented, poorly cooked, ponderous, and filled with particles of stone and earth." When Drs. Galippe and Barré denounced the "imperfection" of bread a century later, they had in mind the hygienic consequences of its excessive whiteness rather than any technical or aesthetic defect. Abbé Armand-Pierre Jacquin, *De la Santé, ouvrage utile à tout le monde* (Paris, 1771), p. 116; Victor Galippe and G. Barré, *Le Pain:*

Alimentation minéralisateur—physiologie—composition—hygiène—thérapeutique (Paris, 1895), pp. 109, 112, 116.

9. M. Montignac writing in the inaugural issue of his own organ, *Michel Montignac Magazine,* no. 1 (May–June 1993), p. 21. Today the baking profession paid the price, concluded this author, for "40 ans de sabotage, puisque la consommation a considérablement diminué d'année en année." Ibid., p. 23.

10. Ibid., pp. 42, 44–45.

11. Michel Perrin from Mussidan, cited in ibid., p. 23.

12. Interviews with bakers, June and July 1995.

13. Alain Schifres, "Laissez-nous notre pain quotidien," *Nouvel Observateur,* 25–31 December 1987.

14. Baker testimony and miller/baker fliers and other publicity; J.-P. Coffe cited in *Michel Montignac Magazine,* no. 1 (May–June 1993), p. 37. Cf. Coffe's not wholly convincing claim for a diffuse public nostalgia for the folkloric baker, a sort of flour-covered Paul Bunyan of the kneading trough, powerful, shrewd, forthright, and congenial. Coffe, *Au secours le goût,* p. 74. A point in Coffe's favor: the effusively positive response of millions of televiewers to France 2's series *Les Maîtres du pain* (though Jérôme the baker-hero was hardly a giant of a man).

15. *Michel Montignac Magazine,* no. 1 (May–June 1993), p. 23; Schifres, "Laissez-nous notre pain quotidien." Cf. Bernadette Gutel, "Les Français veulent manger du bon pain," *L'Hôtellerie,* no. 2197 (1991), p. 43. Refusing to underwrite the rhetorical posturing, Jean-Pierre Coffe suspects (intuitively?) that in all France and Navarre there are no more than a thousand artisans "besogneux et respectueux des traditions." *Au secours le goût,* p. 52.

16. Bortin, "Give Us Our Daily Bread"; Schifres, "Laissez-nous notre pain quotidien."

17. See Pierre Mayol's interesting remarks on the negative charge of bread today, very close to one pole in the ambivalent representation that I am inclined to impute to many Old Regime consumers. A symbol of the harshness of life in general and the hard times of life and work in particular, bread evoked "la mémoire d'un mieux être durement acquis au cours des générations antérieures." By "sa présence royale sur la table où il trône," bread reassures "qu'il n'y a rien à craindre, *pour l'instant,* des privations d'autrefois." But it was impossible, Mayol maintained, presumably as a result of interrogating his contemporaries, to dissociate the staff of life from visions of "la souffrance et la faim." Mayol, "Habiter," in Luce Girard and Pierre Mayol, *Habiter,* vol. 2 of *L'Invention du quotidien,* ed. Michel de Certeau (Paris, 1980), p. 109.

18. See on this point Marcel Lachiver, *Les Années de misère: La Famine au temps du grand roi, 1680–1720* (Paris, 1991), pp. 22, 37–41.

19. Interrogation of Jean Thomas, domestique, nineteen years old, AN, Y 10558, 9 May 1775 (cf. the testimony of a master butcher's wife on baker treachery, ibid., 13 May 1775); René François Lebois, *Lettre des boulangers de Paris au peuple* (Paris, 1789); [J.-J. Rutledge], "Mémoire pour la communauté des maîtres boulangers de Paris présenté au roi, 19 February 1789," in Chassin, *Les Elections et les cahiers de Paris en 1789,* vol. 2, p. 556; Garin and sixteen other bakers, petition to the National Assembly, n.d. (ca. November–December 1789), AN, Y 10506 (deploring the way in which the great scarcity has "exposed the bakers to the greatest misfortunes"). Without, unfortunately, citing any documentation or cases, Jean Delumeau evokes rampant anger toward bak-

ers: "everywhere one threatened to grill them and bake them in their ovens." Delumeau, *La Peur en occident (14e–18e siècles)* (Paris, 1978), p. 166. Jean Meuvret showed brilliantly how fear as well as calculation shaped the decisions of grain growers. His peasant was "a businessman, [but also] a man of fear [un homme d'affaires . . . un homme de peur]." He desired to keep a grain reserve for seed and food, and he remained reluctant to sell even when grain became scarce and prices rose substantially. Meuvret, *Le Problème des subsistances à l'époque Louis XIV: Le Commerce des grains et la conjoncture* (Paris, 1988), vol. 1, p. 147.

20. Pitirim Sorokin, *Hunger as a Factor in Human Affairs*, trans. Elena P. Sorokin and ed. T. Lynn Smith (Gainesville, Fla., 1975), pp. 88ff. With clear-cut objectives in mind but at the risk of mobilizing a crowd to inflict on me a charivari for theoretical posturing, in this same sentence I allude to basic conceptual idioms of two other eminent, and fashionable, sociologists, Pierre Bourdieu and Norbert Elias, the notoriety of whose formulae obviate precise documentation. For recent theoretical perspectives that I find congenial and stimulating, see Alf Lüdtke, ed. *The History of Everyday Life: Reconstructing Historical Experiences and Ways of Life*, trans. William Templer (Princeton, 1995).

21. With a lack of proper deference to the author's sense of the term, I have appropriated *everyday order* from Pierre Bourdieu, *Outline of a Theory of Practice*, trans. Richard Nice (Cambridge and New York, 1977), p. 170. I am equally tempted to borrow Michel de Certeau's *ordinary culture* in order to envisage the discourse of the everyday order. See de Certeau, "Pratiques quotidiennes," in *Les Cultures populaires: Permanence et émergences des cultures minoritaires locales, ethniques, sociales et religieuses*, ed. G. Poujol and R. Labourie (Toulouse, 1978), pp. 23–29.

22. See Kaplan, *Bread, Politics, and Political Economy.*

23. AN, Y 10589, 4 and 10 May 1775; Hardy's journal, Bibliotheque Nationale (hereafter BN), ms. fr. 6682, p. 64 (6 May 1775); George F. Rudé, "The Bread Riot of May 1775 in Paris and the Paris Region," in *New Perspectives on the French Revolution*, ed. Jeffry Kaplow (New York, 1965), p. 198.

24. François Emmanuel Guignard, comte de Saint-Priest, *Mémoires, la Révolution et l'émigration*, ed. Baron de Barante (Paris, 1929), vol. 2, p. 15; Francois-Marie Arouet de Voltaire, *Diatribe à l'auteur des "Ephémérides"* (May 1775), in *Oeuvres complètes*, ed. L. Moland (Paris, 1877–85), vol. 29, p. 368.

25. "Manoeuvre faite par les Srs Leleu et Cie—hauteur de l'augmentation du blé," ca. fall 1789, AN, Y 10506. Glossing the celebrated "Give us this day our daily bread," Luther argued that the coat of arms of a pious sovereign should be decorated with bread rather than the canonical lion or crown. The social contract would not be the appanage of Catholics. See Delumeau, *La Peur en occident*, p. 162.

26. In addition to the two big books cited in note 1 above, I published two more modest monographs: *The Famine Plot Persuasion in Eighteenth-Century France* in *Transactions of the American Philosophical Society* 72, part 3 (1982), and *La Bagarre: Galiani's "Lost" Parody* (The Hague, 1979).

27. Antoine-Alexis Cadet de Vaux, *Discours prononcés à l'ouverture de l'Ecole gratuite de Boulangerie (8 juin 1780)* (Paris, 1780), p. 61.

28. See S. L. Kaplan, *Farewell, Revolution: The Historians' Feud, France, 1789/1989* (Ithaca, N.Y., 1995).

29. See, for example, Pierre J.-B. Legrand d'Aussy, *Histoire de la vie privée des*

François (Paris, 1783; and the edition edited by J.-B.-B. de Roquefort in 1815), and Armand Husson, *Les Consommations de Paris,* 2d ed. (Paris, 1875).

30. Among many useful works, see Pierre Bourdieu, *La Distinction: Critique sociale du jugement* (Paris, 1979) and *Le Sens pratique* (Paris, 1980); Claude Grignon and Christiane Grignon, "Styles d'alimentation et goûts populaires," *Revue française de sociologie* 21 (1980), pp. 531–69; Claude Grignon and Jean-Claude Passeron, *Le Savant et le populaire: Misérabilisme et populisme en sociologie et en littérature* (Paris, 1989); and Claude Grignon, Maurice Aymard, and Françoise Sabban, "A la recherche du temps social," in *Le Temps de manger: Alimentation, emploi de temps et rythmes sociaux,* ed. Aymard, Grignon, and Sabban (Paris, 1993), pp. 1–37.

31. Meuvret, *Le Problème des subsistances à l'époque Louis XIV;* Jean-Marc Moriceau, "Au rendez-vous de la 'révolution agricole' dans la France du XVIIIe siècle: À propos des régions de grande culture," *Annales: HSS,* January–February 1994, pp. 27–63; Moriceau and Gilles Postel-Vinay, *Ferme, entreprise, famille: Grande Exploitation et changements agricoles. Les Chartier, XVIIe–XIXe siècles* (Paris, 1992); Moriceau, *Les Fermiers de l'Ile-de-France* (Paris, 1994); Philip T. Hoffman, "Land Rents and Agricultural Productivity: The Paris Basin, 1450–1789," *Journal of Economic History* 51 (December 1991), p. 803; George Grantham, "Agricultural Supply during the Industrial Revolution: French Evidence and European Implications," *Journal of Economic History* 49 (1989), pp. 43–72, and "The Growth of Labour Productivity in the 'Cinq Grosses Fermes' of France, 1750–1933," in *Productivity, Change, and Agricultural Development,* ed. Bruce Campbell and Mark Overton (Manchester, 1991); Jean-Michel Chevet, "Le Marquisat d'Ormesson: Essai d'analyse économique," thèse de troisième cycle, Ecole des hautes études en sciences sociales, Paris, 1983, "Les Crises démographiques en France à la fin du XVIIe siècle et au XVIIIe siècle: Un Essai de mesure," *Histoire et mesure* 8 (1993), pp. 136–44, and "Production et productivité: Un Modèle de développement économique des campagnes de la région parisienne aux XVIIIe et XIXe siècles," *Histoire et mesure* 9 (1994), pp. 101–45; David Weir, "Les Crises économiques et les origines de la Révolution française," *Annales: E.S.C.* 46th year (July–August 1991), pp. 917–47; Gérard Béaur: "L'Histoire et l'économie rurale à l'époque moderne, ou les Désarrois du quantitativisme: Bilan critique," *Histoire et sociétés rurales* 1 (1994), pp. 67–97; Jean-Yves Grenier, "Croissance et déstabilisation," in *Histoire de la population française,* ed. Jacques Dupâquier (Paris, 1988), vol. 2, pp. 460–62; Julian L. Simon, "Demographic Causes and Consequences of the Industrial Revolution," *Journal of European Economic History* 23 (spring 1994), p. 145.

32. Chevet, "Production et productivité," p. 116.

Chapter 1 Breadways

1. Ferdinand Brunot, *Histoire de la langue française des origines à 1900* (Paris, 1905), vol. 6, p. 261.

2. *Encyclopédie méthodique* (Paris, 1782–1832), Jurisprudence, police et municipalités, vol. 10, p. 168. Though circumstances and practices had radically changed, in the Paris of 1970, in the midst of a bakers' strike, there was a note of anguish redolent of the bread *taxis*—or preoccupation—of the eighteenth century. See *Le Figaro,* 20 April 1970.

3. Louis Caraccioli, *Lettres récréatives et morales sur les moeurs du temps* (Paris, 1767), vol. 1, p. 76; Antoine-Augustin Parmentier, *Recherches sur les végétaux nourrissans, qui, dans les temps de disette, peuvent remplacer les aliments ordinaires, avec des nouvelles observations sur la culture des pommes de terre* (Paris, 1781), pp. 387, 389, 397; Le Camus, "Mémoire sur le bled," *Journal économique,* November 1753, p. 117; Toupet, "Mémoire concernant la culture d'un grain dit seigle de Russie," ca. 1793, AN, F^{10} 226. Cf. Mac-kenzie's "Règles pour conserver la santé," *Journal économique,* November 1764, p. 517. On the way in which the bread image is riveted into French consciousness, see Bache-lard's *cogito pétrisseur.* "Tout m'est pâte, je suis pâte à moi-même, mon devenir est ma propre matière, ma propre matière est action et passion, je suis vraiment une pâte première." Gaston Bachelard, *La Terre et les rêveries de la volonté* (Paris, 1948), p. 80.

4. Antoine-Augustin Parmentier, *Examen chymique des pommes de terre dans lequel on traite des parties constituantes du Bled* (Paris, 1773), p. 182; *Journal de l'agriculture, du commerce, des arts et des finances,* July 1773, p. 130; Parmentier, *Recherches,* p. v; "Avis économique sur la boulangerie" [attributed to M. de B***], *Journal économique,* Sep-tember 1757, p. 43. On French bread dependence/myopia, cf. A. Maurizio, *Histoire de l'alimentation végétale depuis la préhistoire jusqu'à nos jours,* trans. F. Gidon (Paris, 1932), p. 168.

5. Pierre Durand, *Le Livre du pain* (Paris, 1973), pp. 30–31; George Galavaris, *Bread and the Liturgy: The Symbolism of Early Christian and Byzantine Bread Stamps* (Madi-son, Wis., 1970), pp. 3–4; William Addis and Thomas Arnold, *A Catholic Dictionary* (London, 1951), pp. 323–25; A. Kennedy and B. J. Roberts, "Bread," in *Dictionary of the Bible,* ed. James Hastings, rev. ed. Frederick C. Grant and H. H. Rowley (New York, 1963). On the medieval preoccupation with "food miracles," see Jacques Le Goff, *La Civilisation de l'occident médiéval* (Paris, 1967), pp. 290–93. The Paris bakers developed the link between bread and resurrection by taking Lazarus as one of their patron saints. On at least one occasion, bakers' wives were accused of manipulating bread totems in order to achieve some sort of magic. Bibliothèque de l'Arsenal (hereafter Arsenal), archives de la Bastille (hereafter ms. Bast.) 10111, 11 October 1771. For thoughtful and moving remarks on the sacral and liturgical dimensions of bread, see Maurice Lelong, *Célébration du pain* (Revest-Saint-Martin, 1963). On saintly connections with bread, see W. Ziehr and E. M. Buhrer, *Le Pain à travers les âges* (Paris, 1985), pp. 160–61, 179. See also the suggestive remarks of Barbara B. Diefendorf on the corporate dimension of the Eucharist, which figures French society as an organic whole ("one bread, and one body"). Diefendorf, *Beneath the Cross: Catholics and Huguenots in Sixteenth-Century Paris* (New York, 1991), pp. 32, 38.

6. G.-H. Rivière, "Le Folklore des objets domestiques: Le Pain," course in the Ecole du Louvre (January–May 1944), p. 127, Musée national des arts et traditions populaires, ms. 74, pp. 136–38; Paul Sébillot, *Traditions et superstitions de la boulangerie* (Paris, 1891), pp. 37–38, 41–57; Sébillot, *Légendes et curiosités des métiers* (Paris, 1894–95), pp. 33–34; Robert Mandrou, *Introduction à la France moderne, 1500–1640: Essai de psychologie historique* (Paris, 1974), pp. 17–18; Edwin Radford, ed., *Encyclopedia of Superstitions* (New York, 1949), pp. 65–66; A. Varagnac, in Lucien Febvre and Gaston Berger, eds., *Encyclopédie française* (Paris, 1935–66), vol. 14, 14.38.14; Albert Maugarny, *La Banlieue sud de Paris: Histoire, onomastique, langage, folklore* (Le-Puy-en-Velay, 1936), p. 58;

Alfred Franklin, *Les Repas*, in *La Vie privée d'autrefois* (Paris, 1887–1901), p. 208; Voltaire, *Oeuvres complètes*, vol. 17, pp. 349–50, and vol. 18, p. 13; H. G. Ibels, "Le Costume et le meuble," in Henri Bergmann, ed., *La Vie parisienne au XVIIIe siècle* (Paris, 1914), p. 104; *Gazette de l'agriculture, du commerce, des arts et des finances*, 5 January 1768; Edme Béguillet, *Traité des subsistances et des grains qui servent à la nourriture de l'homme* (Paris, 1780), pp. 410–12; and the passionate attack on "l'exécrable mode de la poudre" that directly caused the misery of the poor ("les boutiques des perruquiers sont plus enfarinées que des Moulins") by a senior priest in an anonymous letter to Hérault, 20 April 1731, Arsenal, ms. Bast. 10027, no. 229. Cf. the eighteenth-century philosophes—blasphemers *sans le savoir?*—who used bread as an eraser on their manuscripts. David Pottinger, *The French Book Trade in the Ancien Régime* (Cambridge, Mass., 1958), p. 46. See also H. E. Jacob, *Six Thousand Years of Bread: Its Holy and Unholy History*, trans. R. Winston and C. Winston (Garden City, N.J., 1944), p. 145, and Ronald Sheppard and Edward Newton, *The Story of Bread* (London, 1957), p. 88.

7. Pierre Richelet, *Dictionnaire de la langue françoise ancienne et moderne* (Lyon, 1758), vol. 3, p. 7; Joseph Bricourt, *Dictionnaire pratique des connaissances religieuses* (Paris, 1925–28), vol. 5, pp. 198–99; E. Mangenot and A. Vacant, eds., *Dictionnaire de théologie catholique contenant l'exposé des doctrines catholiques, leurs preuves et leur histoire* (Paris, 1908–50), vol. 11, p. 1731; Louis-Sébastien Mercier, *Tableau de Paris* (Amsterdam, 1782–88), vol. 2 (1783), pp. 87–89; Denis Diderot et al., *Encyclopédie, ou Dictionnaire raisonné des sciences, des arts et des métiers, par une société de gens de lettres*, ed. Diderot and Jean d'Alembert (Paris, 1751–65), vol. 11, p. 751; *Mercure de France*, November 1747, p. 133; "Mémoire pour M. Philippe Boidot, prêtre . . . contre les Sieurs curés et marguilliers de la paroisse de Sainte Marine en la Cité," Archives de la Seine et de la Ville de Paris (hereafter ASP), 3AZ 109, pièce 1F; BN, Collection Joly de Fleury (hereafter Joly) 2428, fol. 2; police sentence, 29 May 1756, in Edmé de la Poix de Fréminville, *Dictionnaire ou Traité de la police générale des villes, bourgs, paroisses et seigneuries de la campagne* (Paris, 1758), pp. 394–96; deliberations, Bureau de l'Hôtel de Ville, 16 August 1732, AN, H* 1854, p. 433; Hérault convocation, 28 December 1725, Arsenal, ms. Bast. 10149. Though he was not Catholic, when in Paris Benjamin Franklin insisted on offering blessed bread: thirteen brioches, one for each colony. The priest objected to his desire to wrap each one in a ribbon inscribed with the word "liberté." Louis Petit de Bachaumont, *Mémoires secrets pour servir à l'histoire de la République des lettres en France depuis 1762 jusqu'à nos jours, ou Journal d'un observateur* (London, 1780–86), vol. 11, p. 59 (10 January 1778).

8. *Dictionnaire universel françois et latin, vulgairement appelé Dictionnaire de Trévoux, contenant la signification et la définition des mots de l'une et de l'autre langue* (Paris, 1771), vol. 6, p. 454 (cf. "c'est pain bénit que d'escroquer un avare" in Richelet, *Dictionnaire de la langue françoise*, vol. 3, p. 7); Paul-Jacques Malouin, *Description des arts du meunier, du vermicelier, et du boulenger, avec une histoire abrégée de la boulengerie et un dictionnaire de ces arts* (Paris, 1761), p. 1; AN, Y 15869, 4 May 1775, and Y 9619, 29 May 1739; Voltaire to Charles Bordes, October 1766, no. 12811, in *Voltaire's Correspondence*, ed. Theodore Besterman (Geneva, 1953–65), vol. 63, p. 135; *Dictionnaire de Trévoux*, vol. 6, pp. 453–54; *Le Babillard*, 25 June 1778, p. 167; Mercier, *Tableau*, vol. 3 (1783), p. 230; Sébillot, *Traditions*, pp. 55, 57; Olivier Biffaud in *Le Monde*, 8 October 1994. On the metaphors of vital life processes, see Claude Macherel, "Le Corps du pain et la maison du père," *L'Uomo* 3

(1990), p. 125, and Piero Camporesi, *The Magic Harvest: Food, Folklore, and Society* (Cambridge, 1993), p. 15. (Although Camporesi rightly insists on the profound "classness" of bread, he tends to see it in monolithic terms. Ibid., p. 25.)

9. Richelet, *Dictionnaire de la langue françoise*, vol. 3, pp. 6–7; *Dictionnaire de Trévoux*, vol. 6, p. 454; *Le Babillard*, 30 April 1778, p. 393; Georges d'Avenel, "Le Mécanisme de la vie moderne," *Revue des deux mondes* 147 (15 June 1895), p. 829; Maria Leach, ed., *Funck and Wagnall's Dictionary of Folklore* (New York, 1949–50), vol. 1, p. 162; Bernard Dupaigne, *Le Pain* (Paris, 1979), pp. 191–95.

10. Simon-Nicolas-Henri Linguet, *Annales politiques, civiles et littéraires du XVIIIe siècle* (Brussels, 1777–92), vol. 5, pp. 431–36, and *Dissertation sur le bled et le pain* (Neufchâtel, 1779), pp. 8–15; Voltaire, *Oeuvres complètes*, vol. 17, pp. 348–49, and vol. 18, p. 7. Cf. the commentary of Charles Gide, in the spirit of Linguet and Charles Fourier, who affirmed that an extraplanetary visitor arriving on earth "s'étonnerait surtout de voir les hommes peiner depuis des siècles pour faire pousser sur cette terre une humble graminée dont *le hasard a fait une plante sacrée*, et qui n'a peut-être pas rendu des services proportionnés aux labeurs et aux sueurs qu'elle a infligés aux humains par la charrue, par la meule, par le pétrin." "Note additionnelle" to Weichs-Glon, "La Municipalisation de la boulangerie," *Revue d'économie politique* 10 (1897), p. 983. For a wonderfully stimulating reading of Linguet's life and thought, see Darline G. S. Levy, *The Ideas and Careers of Simon-Nicolas-Henri Linguet: A Study in Eighteenth-Century French Politics* (Urbana, 1980).

11. Linguet, *Annales politiques*, vol. 5, pp. 443–49, *Réponse aux docteurs modernes, ou Apologie pour l'auteur de la "Théorie des loix," et des "Lettres sur cette théorie," avec la réfutation du système des philosophes économistes* (London and Paris, 1771), pp. 158–90, and *Du pain et du bled* (London, 1774), pp. 9–41. Throughout his writings, Linguet played on the Rousseauian theme that practices conventionally associated with the civilizing process in fact undermined it. (He also developed the "prison de la longue durée" image that Braudel made famous two centuries later.) Mercier depicts the harshness of grain making in terms redolent of Linguet, to whom he makes veiled allusion, and welcomes the prospect of finding "des moyens de nourriture moins dispendieux, moins fatigans pour l'espèce humaine." *Tableau*, vol. 12 (1788), pp. 150–51. Parmentier's description of the onus of grain cultivation also suggests that Linguet's picture is not purely polemical caricature. *Le Parfait Boulanger, ou Traité complet sur la fabrication et le commerce du pain* (Paris, 1778), pp. 516–22.

12. Linguet, *Du pain et du bled*, p. 28, *Annales politiques*, vol. 5, pp. 431–32, 437–38, letter to M. Tissot, 2 November 1779, *Annales politiques*, vol. 7, pp. 169–70. See in the "catalogue des livres imaginaires" figuring in Turgot's library a volume entitled *Dangers du pain* by S.-N.-H. Linguet. Anne-Robert-Jacques Turgot, *Oeuvres de Turgot et documents le concernant, avec biographie et notes*, ed. Gustave Schelle (Paris, 1913–23), vol. 3, pp. 683–85.

13. Linguet, *Annales politiques*, vol. 7, pp. 165–69, *Réponse aux docteurs modernes*, p. 177, *Du pain et du bled*, pp. 30–34.

14. Linguet, *Annales politiques*, vol. 7, p. 165; Paul-Jacques Malouin, *Description des arts du meunier, du vermicellier, et du boulanger*, ed. Jean-Elie Bertrand (Neufchâtel, 1771), p. 3; Le Camus, "Mémoire sur le bled," p. 117; *Dictionnaire de Trévoux*, vol. 6, p. 453. Dr. Malouin admitted that bread, "plus qu'on ne le croit," causes illnesses,

especially those of the stomach. He cited an Arab philosopher who warned that "de toutes les indigestions, la plus mauvaise est celle du pain." Paul-Jacques Malouin, *Description des arts du meunier, du vermicellier, et du boulanger, avec une histoire abrégée de la boulangerie et un dictionnaire de ces arts* (Paris, 1767), p. 2. Cf. the story of the young man who died in large measure as a result of his horror of bread. *Journal historique et politique des principaux événements*, no. 14 (10 May 1774), pp. 232–33.

15. Antoine-Augustin Parmentier, *Mémoire sur les avantages que le province de Languedoc peut retirer de ses grains, avec le mémoire sur la nouvelle manière de construire les moulins à farine par M. Dransy* (Paris, 1787), p. 390; Bachaumont, *Mémoires secrets*, vol. 15, pp. 109, 114 (7 April 1780); Abbé André Morellet, *Théorie du paradoxe* (Amsterdam, 1775), pp. 57–71, and *Mémoires inédits sur le XVIIIe siècle et sur la Révolution* (Paris, 1822), pp. 227–30. César Bucquet could not figure out how to react to Linguet, the writer "qui vouloit nous faire passer le goût du pain en le décriant comme un poison, tandis que c'est ce même pain qui donne tant d'esprit à M. Linguet & à tous les Docteurs." *Observations intéressantes et amusantes du Sieur César Bucquet, ancien meunier de l'Hôpital général, à MM. Parmentier et Cadet* (Paris, 1783), pp. 11, 24n.

16. Jean-Baptiste Briatte, *Offrande à l'humanité* (Amsterdam and Paris, 1780), pp. 166–68, 200. Given his trenchant demonstration of the "inconvéniens du bled," Abbé Ferdinando Galiani might have been tempted to move in Linguet's direction. After listening to the chevalier's persuasive analysis, the marquis remarks: "Je commence à me dégouter tellement du bled, que je crois que j'en reviendrai aux glands, illustre et fort amère nourriture de nos premiers pères." Yet the chevalier bespoke Galiani's impatience with Rousseau-like atavisms: "En attendant de vous voir réinstallé dans l'âge d'or . . ." At bottom Galiani remained a civilized man, with all the constraints and costs that this membership placed upon him, including those imposed by bread, emblem of that perpetually troubled but resourceful and rich civilization. *Dialogues sur le commerce des bleds, 1770*, ed. Fausto Nicolini (Milan, n.d.), p. 169; but see also pp. 162–72. In the early nineteenth century, Sylvester Graham suggested that refined wheaten white bread was the source of most of the "obstructions and disturbances in the stomach and bowels, and other organs of the abdomen," of the sort that may have afflicted Linguet. *Treatise on Bread and Bread-Making* (Boston, 1837). Or did Linguet have celiac disease?

17. On the provincial consumption of wheaten bread, see Henri-Louis Duhamel du Monceau, *Traité de la conservation des grains et en particulier du froment* (Paris, 1753), p. 2 ("Les habitans des villes ne connoissent presque que le pain de froment"); Abbé Noël Chomel, *Dictionnaire oeconomique* (Paris, 1740), vol. 2, p. 141 (the text is identical: "Les habitans des villes ne connoissent presque que le pain de froment"); Missonnet to Joly de Fleury, 16 April 1753, BN, Joly 1113, fol. 4 ("Le petit peuple des villes et des campagnes . . . n'usant presque de ces grains [wheaten]"); questionnaire, 1772, Archives départementales (hereafter AD) Indre-et-Loire, C 95 (but the wheaten bread eaten by town dwellers often contained "une pointe de seigle"); Jean-François de Barandiéry-Montmayeur, comte d'Essuile, *Traité des communes, ou Observations sur leur origine et état actuel* (Paris, 1779), p. 151; Louis Stouff, *Ravitaillement et alimentation en Provence aux XIVe et XVe siècles* (Paris and The Hague, 1970), pp. 48–50; Joseph Benzacar, *Le Pain à Bordeaux* (Bordeaux, 1905), p. 13; Georges Jorré, "Le Commerce des grains et la minoterie à Toulouse," *Revue géographique des Pyrénées* 4 (1933), p. 42; J. Meuvret, "Les

Mouvements des prix de 1661 à 1715 et leurs répercussions," *Journal de la Société de statistique* 85 (May–June 1944), p. 119. Voltaire claims that the English peasant ate white bread. *Lettres philosophiques*, in *Oeuvres complètes*, vol. 22, p. 109. Cf. the article "Fraude" in Voltaire's *Dictionnaire philosophique* (Paris, 1964), p. 5: "Il faut du pain blanc pour les maîtres et du pain bis pour les domestiques." On the English consumption of wheat, see also Arthur Young, *Travels in France during the Years 1787, 1788, and 1789*, ed. Constantia Maxwell (Cambridge, 1929), p. 314; Phyllis Deane and W. A. Cole, *British Economic Growth, 1688–1959: Trends and Structure* (Cambridge, 1967), pp. 62–63; John Burnett, *Plenty and Want: A Social History of Diet in England* (London, 1966), p. 3; William J. Ashley, *The Bread of Our Forefathers: An Inquiry in Economic History* (Oxford, 1928), pp. 1–2, 42–44, 47; Sylvia Thrupp, *A Short History of the Worshipful Company of Bakers of London* (London, 1933), pp. 32–33. For a more prudent view of wheat as an exceptional and elite breadstuff, "the food of a privileged minority," see Alexander Gerschenkron, *Bread and Democracy in Germany* (Berkeley, 1943), p. 8; Georges d'Avenel, "Paysans et ouvriers depuis sept siècles: Les Frais de nourriture aux temps modernes," *Revue des deux mondes* 168 (15 July 1898), p. 433; Pierre Binet, *La Réglementation du marché du blé en France au XVIIIe siècle et à l'époque contemporaine* (Paris, 1939), p. 17.

18. Wilfred J. Fance and B. H. Wragg, *Up to Date Breadmaking* (London, 1968), pp. 12–13; Y. Pomeranz, ed., *Wheat Chemistry and Technology* (Saint Paul, Minn., 1978), pp. 9–10; H. B. Lockart and R. O. Nesheim, "Nutritional Quality of Cereal Grains," in *Cereals '78: Better Nutrition for the Millions* (Saint Paul, Minn., 1978), p. 217; M. Pontas, *Le Dictionnaire des cas de conscience,* cited by Jules Gritti, "Deux Arts du vraisemblable: La Casuistique, le courrier du coeur," *Communications*, no. 11 (1968), pp. 99–121; E. J. T. Collins, "Why Wheat? Choice of Food Grains in Europe in the Nineteenth and Twentieth Centuries," *Journal of European Economic History* 22 (spring 1993), pp. 26–27; Polycarpe Poncelet, *Histoire naturelle du froment* (Paris, 1779), pp. 232–47.

19. Parmentier, *Parfait Boulanger*, pp. 13–14; Poncelet, *Histoire naturelle*, p. xxii; *Journal économique*, April 1751, pp. 3–24. According to Abbé Tessier, wheat was "presque le seul grain dont on fait usage à Paris." *Encyclopédie méthodique*, Agriculture, vol. 3, p. 472. Also on the goodness of white and its reliability as a marker, see Joyeuse l'aîné to Duhamel du Monceau, 30 August 1756, AN, 127 AP 6. On the symbolic power of wheat, particularly its Christian connections, see Jacob, *Bread*, pp. 163–64; Addis and Arnold, *Catholic Dictionary*, p. 328. On the foundering of the wheaten El Dorado on human greed, see J. W. Brinton, *Wheat and Politics* (Minneapolis, 1931), p. 20. There still rages a battle in scientific and metascientific quarters over the goodness of white. During the past decade, white once again fell under suspicion as a vehicle of corporate greed and processor spoliation. On white bread as a symbol of high-tech, unnatural manipulation, see James Whorton, *Crusaders for Fitness: A History of American Health Reformers* (Princeton, N.J., 1982), p. 331. Grahamites and other natural-food exponents have long exasperated food scientists, especially those close to the food industry. "No article of food has been so maligned as white flour," contended Harry Snyder, an eminent food chemist and industrial consultant active in the twenties and thirties. In his view, bran and feed middlings were of low digestibility and nutritional value for humans. White bread supplied a larger amount of available energy and more available nutrients than coarse flour containing fibrous elements. Snyder, *Bread: A Collection*

of Popular Papers on Wheat, Flour, and Bread (New York, 1930), pp. 63–87. English scholars D. W. Kent-Jones and John Price echoed Snyder's attack on "ill-informed" and "incorrect" criticism of "much-maligned" white bread in *The Practice and Science of Bread-Making* (Liverpool, 1951). Another scholar has insisted that white is not a choice of snobbery or social mimicry/vengeance but a rational selection "absolutely preferable" in terms of baking technology (especially before the twentieth century) and in terms of quality and benefit for the consumer. Hans D. Renner, *The Origin of Food Habits* (London, 1944), pp. 174–78.

20. Cf. Bordeaux, where the bulk of the population apparently ate wheaten loaves but where white bread was nevertheless described as "le pain des riches." Mathieu François Pidansat de Mairobert, *Journal historique de la révolution opérée dans la constitution de la monarchie française, par M. de Maupeou, chancelier de France* (London, 1774–76), vol. 4, p. 172 (18 May 1773).

21. "Typical wheaten breads before the nineteenth century were very different from those of today. Many would now be regarded as unacceptable, being much darker, coarser, and heavier than those to which we have become accustomed. This raises the interesting question of the extent to which, in the past, the benefits of wheaten bread may have been overstated." Collins, "Why Wheat?" pp. 28–30.

22. Adrian Bailey, *Blessings of Bread* (New York, 1975), p. 26; Renner, *Food Habits*, pp. 175, 196; Robert A. McCance and Elsie Widdowson, *Breads, White and Brown: Their Place in Thought and Social History* (Philadelphia, 1956), pp. 5–7; Maurizio, *Histoire de l'alimentation*, p. 653; *Magasin pittoresque* 43d year (1875), p. 110. On the psychological and moral connections between darkness in food and defilement, see Renner, *Food Habits*, pp. 114–15, and McCance and Widdowson, *Breads, White and Brown*, p. 4.

23. Abbé Nicolas Baudeau, *Avis au peuple sur son premier besoin, ou Petits Traités économiques, par l'auteur des "Ephémérides du citoyen"* (Amsterdam and Paris, 1768), 3e traité, pp. 134–36; Bucquet, "Mémoire," 2 March 1769, AD Isère, C 47; Bucquet, *Observations intéressantes et amusantes*, p. 64; "Réponse au mémoire du Sr Bucquet," ca. 1768, Arsenal, ms. 7458; Daure, "Mémoire," 1771, AN, F[11] 264; Edme Béguillet, *Traité de la connoissance générale des grains et de la mouture par économie* (Dijon, 1778), p. 71; Parmentier, *Mémoire sur les advantages*, p. 400.

24. AN, X[2B] 1307, 7 December 1751.

25. *Révolutions de Paris*, no. 4 (2–8 août 1789), 10 vols. (Paris, 1789–93), pp. 7–8; *Enquête sur la boulangerie du Département de la Seine* (Paris, 1859), pp. 14, 56, 73, 117, 126, 213, 642, 768. This mid-nineteenth-century inquiry also underlined the fact that white bread "trempait mieux la soupe," the core of the worker's diet. Ibid., pp. 357, 426. On the early-nineteenth-century passion for white bread, see Maître Chauveau-Lagarde, *Mémoire pour les boulangers de Paris* (Paris, 1817), Cornell University Library, p. 5. On the "désolante ténacité du préjugé du pain blanc," which the author regarded as an "obstacle au progrès" in the same class as alcoholism and "carnivorisme," see Dr. Albert Monteuuis, *Le Vrai Pain de France, ou la Question du pain sur le terrain pratique* (Paris, 1917), pp. 109–11, 134.

26. Husson, *Consommations*, p. 146; Léon de Lanzac de Laborie, *Paris sous Napoléon* (Paris, 1905–13), vol. 5, p. 159; "Mémoire sur le prix du bled en la généralité d'Orléans," ca. 1715–20, AN, G[7] 1704; Du Vaucelles, "Offrande civique," ca. late 1789, AN, F[10] 215–

16; AN, BB³ 76, 21 nivose an II; Monchanin, "Quel est le moyen de pourvoir à l'approvisionnement . . . ?" 20 October 1791, AN, F¹⁰ 215–16.

27. Nicolas Delamare, *Traité de la police* (Paris, 1705–38), vol. 2 (1729), p. 760; *La Misère des garçons boulangers de la ville et faubourgs de Paris* (Troyes, 1786), p. 5; Thellusson to Joly de Fleury, 5 December 1740, BN, Joly 1109, fol. 70; Antoine-Louis Séguier, in *Recueil des principales lois relatives au commerce des grains avec les arrêts, arrêtés, et remontrances du Parlement sur cet objet et le procès-verbal de l'assemblée générale de la police* (Paris, 1769), p. 122 (Assembly of Police, November 1768); mémoire de Monchanin, 20 October 1791, AN, F¹⁰ 215–16. See also *Encyclopédie méthodique*, Agriculture, vol. 3, p. 377. On the generic attack on white bread consumption as the appanage of the powerful and rich, and the pillar of the traditional social structure, see Louis-Sébastien Mercier, *L'An deux mille quatre-cent quarante, rêve s'il en fût jamais*, new ed., 2 vols. (London, 1785), p. 150; Mercier, *Tableau*, vol. 10 (1788), p. 147; *Révolutions de Paris*, no. 7 (22 August 1789), p. 3; circulaire du 28 pluviôse an II (15 February 1794) aux administrateurs de district, in Pierre Caron, *Commission des subsistances de l'an II: Procès-verbaux et actes* (Paris, 1925), p. 366 ("Il est temps que tous les frères de la même famille mangent le même pain, et que ces diverses distinctions dans la préparation de la première nourriture et la plus importante, inventées par le luxe et la mollesse, soient bannies, pour faire place à une juste uniformité"); Ambroise Morel, *Histoire abrégée de la boulangerie en France* (Paris, 1899), p. 274.

28. Comte d'Essuile, *Traité des communes*, p. 151; Abbé Nicolas Baudeau, *Résultats de la liberté et de l'immunité du commerce des grains, de la farine et du pain* (Amsterdam and Paris, 1768), p. 19; Baudeau, *Avis au premier besoin*, 3e traité, p. 107; "Réponse au mémoire du Sr Bucquet," ca. 1768, Arsenal, ms. 7458; Parmentier, *Recherches*, p. 413; Béguillet, *Traité des subsistances*, p. 651. In London, the fascination for white led to a potentially perilous complicity between producers and consumers: "Yet as the People in general prefer the whitest [bread] (which they ignorantly conceive to be the only mark of purity), . . . the bakers are compelled to use a small admixture of Alum to please the fancy of their customers rather than lose them." In this scenario, "the Buyer is equally culpable, as the whiteness is his pride as well as the Baker's." H. Jackson (a chemist), *An Essay on Bread* (London, 1758), p. 13.

29. Delamare, *Traité*, vol. 2 (1729), p. 760. Cf. the verses on "l'homme sensuel" who coveted milk and salt in his bread in *La Misère des garçons boulangers*, p. 5.

30. Nicolas-Toussaint L. Des Essarts, *Dictionnaire universel de police* (Paris, 1786–90), vol. 1, pp. 254–70; *Année littéraire* 7 (1770), p. 332; Henri d'Almers, "Le Pain de Paris," *Magasin pittoresque*, 65th year, 2d ser., 15 (1897), p. 315. Olivier de Serres reported on a *pain mollet* in his time. Parmentier, *Mémoire sur les avantages*, p. 333.

31. Edouard Fournier, *Le Roman de Molière* (Paris, 1863), pp. 191–227; Mercier, *Tableau*, vol. 8 (1783), p. 299.

32. La Condamine's poem cited in *Almanach des muses* (Paris, 1770), pp. 85–88. For a slightly different version of these verses, see Emile Raunié, ed., *Clairambault-Maurepas: Chansonnier historique du XVIIIe siècle* (Paris, 1879–84), vol. 8, pp. 37–39 (e.g., "Et de janvier jusqu'en décembre / Licenciés et bacheliers / Et présidents et conseillers / Des Enquêtes, de la Grand-Chambre / En prenant du café au lait / Rendent hommage au pain mollet"). Cf. Friedrich Grimm et al., *Correspondance littéraire, philosophique et*

critique par Grimm, Diderot, Raynal, Meister, etc., ed. Maurice Tourneux (Paris, 1877–82), vol. 6, pp. 249–52 (April 1765); Pierre-Jean Grosley, ed., *Oeuvres inédites* (Paris, 1812–13), vol. 3, pp. 164–73; Alan Charles Kors, "The Coterie Holbachique: An Enlightenment in Paris," Ph.D. diss., Harvard University, 1968, p. 456.

The poem is delightful and revealing. Here is a full translation of the Raunié version cited above:

> They knew about pain mollet
> A century before Nicolet;
> "Fit for the queen," they did determine:
> Médicis, who was our sovereign,
> Liked it and decided to
> Use some in her first ragout.
> So did the city folks, those at court;
> Each one believed in the new sort.
> It all took place quite tranquilly
> Till 1668, you see,
> Gonesse bakers to each other say
> They're enemies of pain mollet.
> They saw its taste pleased everyone,
> And so for malice, not for fun,
> Denounced the bread to Parlement,
> As harmful and bad aliment.
> Then the leaders of the land,
> Health protectors of each man,
> Asked the famous faculty,
> "Announce to us, no flattery,
> What must we think about this bread?"
> Some folks have chewed it sixty years.
> It could well be completely sour!
> Gui Patin, sage of his hour,
> And leader of those who opposed,
> Spoke out thus and did expose,
> Haranguing in his brotherhood:
> "I predict that there's no good
> Can come from this bad industry,
> Which does appeal to glutton's taste;
> Yes, even those who think clearly
> Will find their mind has gone to waste:
> This poison does its work slowly,
> To undermine life without haste."
> He concluded that death's way

Was quickly found with pain mollet.
Then Perrault, his antagonist,
Said loud and clear: I'm a *"Pain molletist."*
"Gentlemen, I do attest,
This bread is easy to digest."
Patin then answered: "But the yeast,
And that from Flanders, to say the least,
Made from an impure beer, that's certain,
Breeds a germ that is quite rotten
And against man's staying well.
What kind of devil did instill
This modern, evil, bad invention?"
"Modern!" Perrault's exclamation.
"Your memory has made a wrong turn,
It came from the Swiss canton of Berne;
They made it in Holopherne's time.
But even closer to our day,
Into Pliny, it finds its way.
I quickly see that Master Patin
Does much better in Greek than Latin."
Patin stepped back, to say the least,
And all because of brewer's yeast.
Each of the men was ready there
To take the other by the hair.
Then the court joined in the fight
And with a law ended the plight:
We forbid you to buy or sell
Yeast from Flanders, hear ye well;
And we condemn those who insist
To a fine of five hundred francs at least.
So consequently, from that day,
For a hundred years, that is to say,
in France's capital quite hidden
There entered yeast that was forbidden.
Each year fined twenty thousand ecus, remember;
And from January to December
Those with "licences," bachelors,
Presidents, and counsellors
of the Grand-Chambre of Parlement,
While drinking their café au lait,
Render hommage to *pain mollet.*

33. Jean Gauthé was one of many Parisian bakers who offered classical mollet loaves larded with milk and who softened his regular doughs with brewer's yeast in order to produce lighter and more attractive standard white and mid-white breads. ASP, D5B⁶ 3542. On the proliferation of "demie molé [*sic*]" loaves, see the evidence in the transfer of a shop, AN, Minutier central (hereafter MC), XCIV-310, 22 May 1762. Though strictly banned from the marketplaces, mollet loaves occasionally turned up in the stalls of both city and forain bakers. BN, Joly IIII, 172 (10 July 1760). Mollet luxury breads, and presumably soft-dough derivatives, were commonly demanded throughout the provinces in the second half of the eighteenth century. Colin des Murs to Sartine, 8 January 1769, AN, Y 12618; règlement for Paray, 26 October 1767, AD Côte d'Or, B II 39/6; subdelegate Masson to intendant of Champagne, 17 July 1770, AD Aube, C 299. Cf. Mercier's evocation of the insupportable arrogance of the mollet loaf: "Le pain mollet, parce qu'on le paye un peu plus cher avec sa croûte ferme & dorée, semble insulter à la miche du Limousin. . . . le beau pain mollet a l'air d'un noble parmi les roturiers; il va descendre dans les estomacs de qualité: la présidente, la duchesse & la marquise ne veulent tâter que de celui-là; elles regardent le pain de pâte ferme comme si c'étoit du foin." *Tableau*, vol. 12 (1788), p. 147. Husson considered inflated the (putative) administrative statistic announcing that one-third of all the bread produced in Paris in 1811 fell under the "fantasy" rubric. By midcentury, however, he felt confident that the luxury market share had attained this level. *Consommations*, pp. 144–45.

34. AN, Y 15062, 4 February 1761; Charles Estienne and Jean Liébault, *L'Agriculture et la maison rustique* (Lyon, 1698), p. 497; J.-L. Muret, *Mémoire sur la mouture des grains, et sur divers objets relatifs* (Berne, 1776), pp. 7, 11, 25; *Journal économique*, May 1769, p. 196. Cf. Des Essarts, *Dictionnaire universel*, vol. 2, p. 195.

35. Malouin, *Description* (1761), p. 214; Des Essarts, *Dictionnaire universel*, vol. 2, pp. 232, 242; Diderot et al., *Encyclopédie*, vol. 11, pp. 749, 752; *Magasin pittoresque*, 65th year, 2d ser., 15, p. 315; "Statuts des maistres boulangers du faubourg Saint-Germain des Prez," 23 May 1658, AN, K 1030–31; Richelet, *Dictionnaire de la langue françoise*, vol. 3, pp. 6–7; *Dictionnaire de Trévoux*, vol. 6, p. 452; Alfred Franklin, *Dictionnaire historique des arts, métiers et professions exercés dans Paris depuis le XIIIe siècle* (Paris, 1906), pp. 96–97. It seems likely that what passed for a "pain bourgeois" in Rabelais's time probably did not meet the eighteenth-century middling standard in the capital. Lazar Saineanu, *L'Histoire naturelle et les branches connexes dans l'oeuvre de Rabelais* (Paris, 1921), p. 404. On the range of medieval bread types throughout France, see Françoise Desportes, *Le Pain au moyen âge* (Paris, 1987), pp. 92–94.

36. Delamare, *Traité*, vol. 2 (1729), p. 936; *Gazette de l'agriculture, du commerce, des arts et des finances*, 3 January 1769, p. 4; AN, Y 13114, 21 September 1762; Diderot et al., *Encyclopédie*, vol. 11, pp. 749–50; after-death inventory of wife of baker G. C. Rousseau, 15 July 1771, AN, MC, VII-288; ASP, D5B⁶ 1968 and 3338 (baker Rousseau) and 4288 (baker Modinet); *Journal économique*, January 1754, p. 101; Malouin, *Description* (1761), pp. 215–19; Mathieu Tillet, "Expériences et observations sur le poids du pain au sortir du four," *Journal de physique*, February 1782, p. 101; Parmentier, *Mémoire sur les avantages*, p. 331; and *Parfait Boulanger*, pp. 543–44; *Le Patissier en colère sur les boulangers et les taverniers en vers burlesques* (Paris, 1649); AN, Y 9434, 7 August 1733; AN, Y 9387, 6 February 1761; petition by jurés en charge de la communauté des maistres paticiers, ca.

September–November 1732, AN, Y 13743; Roze de Chantoiseau, *Tablettes royales de rénommée, ou Almanach général d'indication* (Paris, 1791). Parmentier deplored the desire of every Parisian "jusqu'aux gens du peuple" to have his own bread ("chacun a son pain"), a bread of small weight, a *petit pain*, rather than a chunk of a larger bread, a phenomenon that increased consumption of flour and wastage in baking. *Parfait Boulanger*, p. 436. Cf. "Rapport sur un déficit de 100 sacs de farine dans les magasins de Scipion," BN, ms. n.a.f. 1434; Busuel, "Plan d'un règlement pour les approvisionnements des grains et des farines," December 1791, AN, T 644[1-2]. Labeling uncertainties caused a certain confusion at the highest levels of public administration, inhibiting viable comparisons and evaluations, for it was not always clear whether the *blanc, bis-blanc,* and *bis* triumvirate referred to wheaten, maslin, or rye loaves or to composites. See, for instance, controller general to intendant, January 1768, AD Haute-Garonne, C 2063. Despite the multiplicity of names, Parisians in practice did not seem to have much trouble in distinguishing bread types and in making themselves understood. See the interrogation of a suspected pillager, an unemployed journeyman mason, AN, Y 10558, 3 May 1775.

37. François Cretté de Palluel, "Mémoire sur les moyens d'économiser la mouture des grains and de diminuer le prix du pain," in *Mémoires d'agriculture, d'économie rurale et domestique publiés par la Société royale d'agriculture* (Paris), autumn 1788, pp. 20–27 ("Le pain blanc est celui dont on fait la plus grande consommation à Paris"); Citoyen Du Vaucelles, "Opinion sur les subsistances," AN, F[10] 215–16 ("Depuis longtems, on n'use plus à Paris que du pain blanc et du pain mollet"); Boullemer, lieutenant general of Alençon, "Mémoire," July 1765, BN, ms. fr. 11347 ("le pain des places de Paris vaut au moins le pain blanc qu'on mange en la pluspart des villes de province"); ASP, D5B[6] 6230; BN, ms. n.a.f. 6765 (June 1768); Poussot's journal, 11 December 1756, Arsenal, ms. Bast. 10141; Parmentier, *Parfait Boulanger,* p. 625; Malouin, *Description* (1761), p. 264. Cf. the continued preference in Paris for petit pain through the Revolution. ASP, F[1C] III-27 (18 June 1793). Cf. Abbé Jacquin's plea for a middling loaf, far healthier than the flimsy mollet or the ponderous Gonesse loaf. *De la Santé,* p. 118. See the eighteenth-century call for a return to *bis,* echoed a hundred years later in a more rigorously scientific/medical voice (arguing that the *bis* contained a precious phosphoric acid, essential for growth and cell replacement, that was absent in the white). François Lacombe d'Avignon, *Le Mitron de Vaugirard, dialogues sur le bled, la farine et le pain, avec un petit traité de la boulangerie* (Amsterdam, 1777), pp. 43, 45; Victor Galippe and G. Barré, *Le Pain: Technologie, pains divers, altérations* (Paris, 1895).

38. Nicolas Edmé Restif de la Bretonne, *La Vie de mon père,* ed. Gilbert Rouger (Paris, 1970), p. 130; *Courrier de France,* no. 88 (15 June 1790), p. 165; La Reynie to Delamare, 19 March 1694, BN, ms. fr. 21643, fol. 65; "Réponse au mémoire du Sr Bucquet," ca. 1768, Arsenal, ms. 7458; Voltaire, "Petit Écrit sur l'arrêt du Conseil du 13 septembre 1774" (January 1775), in *Oeuvres complètes,* vol. 30, p. 541; César Bucquet, *Manuel du meunier et du constructeur de moulins à eau et à grains, par M. Bucquet,* new ed. (Paris, 1790), p. 109; Malouin, *Description* (1761), pp. 262, 311, 324; Lacombe d'Avignon, *Le Mitron de Vaugirard,* pp. 35, 43; Busuel, "Plan d'un règlement pour les approvisionnements des grains et des farines," December 1791, AN, T 644[1-2]; Auguste-Philippe Herlaut, "La Disette de pain à Paris en 1709," *Mémoires de la Société de l'histoire de Paris*

et de l'Ile-de-France 115 (1918), pp. 19, 42; Jacques Saint-Germain, *La Vie quotidienne en France à la fin du grand siecle d'après les archives . . . du lieutenant général de police, Marc-René d'Argenson* (Paris, 1965), p. 211. It is virtually certain that the *bis* of Voltaire and Maintenon was a largely rye loaf with some wheat to help give it volume. See Briatte, *Offrande,* p. 133. Briatte was preoccupied with the very poorest people of France, who would be content to eat a dark loaf of the 1709 sort "non pas dans les années de *disette,* mais même dans les années d'abondance." Saint-Germain affirms that Maintenon ate an oat bread, *Vie quotidienne,* p. 193. Even in the provinces, reported a subdelegate in the Alençon intendancy, "la troisieme de pain Bis quoyque fort bon n'est plus d'usage. Le peuple la dedaigne." Legendre to intendant, 28 October 1773, AD Orne, C 89. Journey-men on the Tour de France referred to the dark loaf as "le brutal." Lucien Carny, ed., *Les Compagnons en France et en Europe* (Eyrein, 1973), vol. 2, p. 152. An angry consumer in the mid-nineteenth century rehearsed the Old Regime accusation that it was the bakers who "indisposed the public" to eat lesser loaves. Entering a shop that posted prices of a range of breads, he wrote: "J'ai demandé un kilo de pain de seconde qualité. Le boulanger m'a répondu, en se moquant de moi et d'un air de dédain, qu'il ne faisait pas de pain de seconde qualité." The consumer noted that the shop boasted an opulent interior decoration, including mirrors, painted and gilded panels, and an ornamented ceiling, an ensemble lovelier than the Louvre. He vigorously contested "le droit" of this baker and numerous confreres not to make a second quality loaf available. 23 November 1867, probably to the editor of *Le Courrier français,* in the Collection E. Thomas, c.p. 4804, Bibliothèque historique de la Ville de Paris (hereafter BHVP), series 114.

39. Mémoire de Monchanin, 20 October 1791, AN, F10 215–16; J. Bertrand in Ma-louin, *Description* (ed. Bertrand, 1771), pp. 41, 232, 324; Jackson, *Essay on Bread,* pp. 14–15; Polycarpe Poncelet, *Mémoire sur la farine* (Paris, 1776), pp. 69–70, and *Histoire naturelle,* pp. 172–73, 188–91; Abbé Antoine Pluche, *Le Spectacle de la nature, ou Entretiens sur les particularités de l'histoire naturelle* (Paris, 1732–50), vol. 6, p. 423; BN, Joly 122, fols. 60–61, November–December 1740 (on the stale *bis* that had "meilleure mine" and was healthier). Cf. John Kirkland, *The Modern Baker, Confectioner, and Caterer* (London, 1927), vol. 1, p. 124, on the perfectionist representation of the proper (holistic) bread. See also the argument made for a middling loaf whose "*bis*ish" components made a bread that "nourrit davantage que le pain blanc, rend les corps plus robustes . . . et fait aussi que le retour de la faim est moins prompt." *Journal économique,* February 1754, pp. 109–10. On the eighteenth-century French passion for daily-fresh bread, see Galiani's remarks in *Dialogues entre M. Marquis de Roquemaure et M. le Chevalier Zanobi,* ed. Philip Koch (Frankfurt, 1968), p. 43. On the mid-twentieth-century American suspicion that fresh bread was unwholesome, see Snyder, *Bread,* p. 234.

40. Parmentier, *Eloge de M. Model* (Paris, 1775); Parmentier in *Journal de physique,* February 1773, pp. 158–59; Parmentier in Abbé François Rozier, *Cours complet d'agriculture théorique, pratique, économique et de médecine rurale et vétérinaire* (Paris, 1781–96), vol. 7, pp. 374, 378, 385; Béguillet, *Traité des subsistances,* pp. 81, 324–27n, 555–71, 628–32; Kaplan, *Provisioning Paris,* pp. 440–49. Parmentier also chastened those who believed that a *pâte ferme* provided more nutrition than softer doughs: the difference between most mollets and firm doughs was not a large quantity of flour but four pounds of water per hundred pounds of flour! *Parfait Boulanger,* p. 390. On the "illusion" of the superi-

ority of brown over white, see: John Storck and Walter D. Teague, *Flour for Man's Bread—A History of Milling* (Minneapolis, 1952), pp. 303–5 (insisting in particular on the relative digestibility of "white" and "dark" proteins); Fance and Wragg, *Breadmaking*, pp. 147–48; Kirkland, *Modern Baker*, vol. 1, pp. 36ff. (emphasizing the undesirable influence of the aleurone layer, alleged seat of proteins and nutrients, on the dough in fermentation); Thomas J. Horder et al., *The Chemistry and Nutrition of Flour and Bread* (London, 1954), pp. 144–45; Philip Wood, *The Story of a Loaf of Bread* (Cambridge, 1913), p. 127; Snyder, *Bread*, pp. 63–70, 82–83 (rebutting the "dire calamities [that] are attributed by the uninformed to the use of white flour and bread," the true staff of life); Alfred Gottschalk, *Le Blé, la farine, et le pain* (Paris, 1935); *Lancet*, 27 April 1940, p. 791 ("the dietetic gain from a thorough extraction is by no means as great as is often maintained"). For a fleeting exposition of the case that refinement fleeces wheat of its subtle yet crucial nutritional properties, see Peter Farb and George Armelagos, *Consuming Passions: The Anthropology of Eating* (New York, 1980), p. 218. Maurice Aymard makes the historical case for the serious deficiencies of a diet based largely on white bread, deprived of a good part of its richest proteins and minerals and most niacin and other B vitamins. "Pour l'histoire de l'alimentation: Quelques remarques de méthode," *Annales: E.S.C.* 30th year (March–June 1975), p. 437.

41. Legrand d'Aussy, *Histoire de la vie privée*, vol. 1, pp. 133, 150; Philippe Macquer, *Dictionnaire portatif des arts et métiers, contenant l'histoire, la description, la police des fabriques et manufactures de France et des pays étrangers* (Paris, 1766), vol. 1, p. 148; Le Camus, "Mémoire sur le bled," pp. 147–48; Abraham du Pradel, *Le Livre commode des adresses de Paris pour 1692*, ed. Edouard Fournier (Paris, 1878), vol. 1, p. 306 (this reference kindly furnished by P. Hyman); *Dictionnaire de Trévoux*, vol. 6, p. 452; Delamare, "Propositions," BN, ms. 21635, fol. 2; Cleret to Hérault, 1 October 1725, Arsenal, ms. Bast. 10270; Duplessis to Joly de Fleury, 17 September 1725, BN, Joly 1117, fols. 205–6; Joly de Fleury to Hérault, 15 September 1740, ms. Bast. 10277; Orry to Joly, 25 September 1740, Joly 1121, fol. 102; Joly draft to Orry, Joly 1121, fols. 71–72; Thellusson to Joly, 26 October 1740, Joly 1109, fols. 49–51; *Journal économique*, March 1762, p. 111; note to controller general, ca. December 1767, AN, F[11] 1194; Joly 1138, fol. 82; Hardy's journal, BN, ms. fr. 6680, fol. 136 (4 November 1775); Turgot to Dupleix, 30 December 1774, AD Somme, C 87; Pidansat de Mairobert, *Journal historique*, vol. 6, p. 229 (21 October 1774); Turgot to intendants, circular, 17 September 1775, in *Oeuvres de Turgot*, vol. 4, p. 495; note to controller general, 6 December 1767, AN, F[11] 1194; Arthur Michel de Boislisle, "Le Grand Hiver et la disette de 1709," *Revue des questions historiques* 74 (1903), p. 516; Bertrand's note in his edition of Malouin, *Description* (1771), 18–19n; Béguillet, *Traité des subsistances*, pp. 100–101. To underline the grave distress of the people, a delegate from the Besançon Parlement dispatched to Versailles in 1782 brandished a piece of a nasty oat loaf. Félix Rocquain, *L'Esprit révolutionnaire avant la Révolution, 1715–1789* (Paris, 1878), p. 403. Despite their state of hunger and misery in 1770–71, the poor inhabitants of Vitry-le-François protested the heavy use of barley in their dearth bread. *Gazette de l'agriculture, du commerce, des arts et des finances*, 5 February 1771, pp. 84–85. See also the proposition in 1790 to put rye into loaves collectively prepared for large numbers of public works laborers, nonworking poor, and prisoners. Sigismond Lacroix, ed., *Actes de la Commune de Paris pendant la Révolution* (Paris,

1894–98), vol. 4, p. 234 (27 February 1790). Cf. the great unpopularity of a "trifling admixture" of maize flour (10 percent) to the standard wheaten loaf in England during the First World War. Ashley, *Bread,* pp. 1, 20.

42. Malouin, *Description* (1761), pp. 214, 260, 263–64.

43. Pluche, *Spectacle de la nature,* vol. 6, pp. 423–24; Abbé Baudeau, *Avis aux honnêtes gens qui veulent bien faire* (Amsterdam and Paris, 1768), p. 91, and *Avis au premier besoin,* 3e traité, pp. 134–36; Malouin, *Description* (1761), p. 261; "Réponse au second mémoire," Arsenal, ms. 7458. Cf. the strictures of the subdelegate of Avesnes against the fascination for "pretty bread" that undermined social and personal health. Ca. 1773, AD Nord, C 6690.

44. Malouin, *Description* (1761), p. 261, and *Description* (1767), p. 12; "Réponse au premier mémoire," Arsenal, ms. 7458; Lacombe d'Avignon, *Le Mitron de Vaugirard,* pp. 44–45; Cretté de Palluel, "Mémoire sur les moyens," pp. 23–24.

45. Bucquet, "Mémoire," 2 March 1769, AD Isère, II C 47, and *Observations intéres- santes et amusantes;* Antoine-Auguste Parmentier, *Avis aux bonnes ménagères des villes et des campagnes sur la meilleure manière de faire leur pain* (Paris, 1772), pp. 103–4; Edme Béguillet, *Mémoire sur les avantages de la mouture économique et du commerce des farines* (Dijon, 1769), Arsenal, ms. 2891, fol. 211, and *Traité des subsistances,* p. 71; Louis Cotte, *Leçons élémentaires sur le choix et la conservation des grains, sur les opérations de la meunerie et de la boulangerie et sur la taxe du pain* (Paris, 1795), p. 67; "Cahier d'observations" addressed to Sartine, August 1769, AN, Y 12618. Cf. a late-eighteenth-century text that may have been composed by Bucquet: "J'appelle *pain du peuple* celui qu'on devroit faire de toutes farines bien assorties sans tirer la première fleur et en n'ôtant que le son et les recoupettes." "Réponse au second mémoire," Arsenal, ms. 7458.

46. Pradel, *Le Livre commode,* vol. 1, p. 307; Pluche, *Spectacle de la nature,* vol. 6, pp. 314–15; *Gazette de l'agriculture, du commerce, des arts et des finances,* 22 June 1771, pp. 398–99; "Cahier d'observations" to Sartine, August 1769, AN, Y 12618; Mercier, *Tableau,* vol. 4 (1782), pp. 129–30; Parmentier, *Parfait Boulanger,* pp. 544–47. Cf. the seventeenth-century English recipe for a multigrain household bread and for a whole meal flour made without removing any bran. John Evelyn, *An Account of Bread: Pan- ificium,* in the Houghton Collection for Improved Husbandry and Trade (16 January 1683), pp. 133–37.

47. Boislisle, "Grand Hiver," p. 515; Auguste-Philippe Herlaut, "Disette," pp. 86– 87. Adrien-Henri Thery, *Gonesse dans l'histoire: Une Vieille Bourgade et son passé à travers les siècles* (Persan, 1960), p. 91; Joly de Fleury to royal procurator at Moulins, 20 May 1750, BN, Joly 2432, fol. 214; T. de Montigny to Miromesnil, 22 March 1768, Armand Thomas Hue de Miromesnil, *Correspondance politique et administrative de Miromesnil, premier président du Parlement de Normandie,* ed. P. Le Verdier (Paris and Rouen, 1899–1903), vol. 5, p. 131. Dr. Malouin explained very clearly that "en général on n'a travaillé utilement pour ce qui concerne le pain, que lorsque les grains ont été effectivement chers, parce qu'on n'y a fait assez d'attention, que dans ces temps-là." Paul-Jacques Malouin, *Description et détails des arts du meunier, du vermicellier, et du boulanger* (Paris, 1779), pp. 38–39, 85. Cf. the subtle glissando from experimentation with permanent, everyday application to experimentation geared to "faire du pain plus à

la portée du peuple dans les temps de disette." Marcel Arpin, *Historique de la meunerie et de la boulangerie, depuis les temps préhistoriques jusqu'à l'année 1914* (Paris, 1948), vol. 1, p. 248.

48. Saint-Florentin to B. de Sauvigny, 1 December 1768, AN, O¹ 410; de Montvallier to Sartine, 28 June 1768 and accompanying dossier, AN, Y 12617; Arsenal, ms. Bast. 12353 (1768). Cf. the efforts in the Rouen area to promote a "pain économique" and a "pain de tous les produits du bled." *Journal économique,* July 1768, p. 311, and Laverdy to Miromesnil, 29 March 1768, in Miromesnil, *Correspondance,* vol. 5, p. 138. While the authorities encouraged innovation, they also worried about its perverse effects. Fearful of exciting popular anger—and lending credibility to deeply seated conspiratorial beliefs—the lieutenant general of police of Rouen ordered a baker to renounce his production of a "charity" bread made largely of bran. Initially destined for a poor woman with eight children and sold at about a quarter of the going price for white bread, the loaf "n'etoit pas sain" in the judgment of the police. Conseil secret, Rouen Parlement, 24 March 1768, AD Seine-Maritime. For crisis-driven innovation in the Toulouse area, in particular the elaboration of a "pain économique," see Raynal to intendant Saint-Priest, 24 April 1773, and Vandour to same, 25 April 1773, AD Hérault, C 2914.

49. Boulangerie de Vaugirard, ASP; Hardy's journal, 6, 15 October 1774, BN, ms. fr. 6681, fols. 426, 430; Georges Weulersse, *La Physiocratie à la fin du règne de Louix XV, 1770–74* (Paris, 1959), p. 190. Cf. Grimm et al., *Correspondance,* vol. 8, p. 421. In 1993 bakers introducing new (in fact "old") sorts of bread have again turned to pedagogy through the classical vehicle of the small brochure, invariably prepared by the mill supplying the baker and directing the marketing strategy. See, for example, the campaign for the *banette* and for other loaves said to re-create the taste of the 1930s.

50. Essais Moulinot–de la Jutais, 1770–71, AN, F¹¹ 265, and Y 12621, 27 July and 3 August 1771; *Année littéraire* 8 (1764), p. 40. Cf. a Parisian official, apparently in the Châtelet, who claimed in 1725 to possess "un secret pour faire multiplier le bled qu'on semoit, si considerablement . . . que ce secret consistoit uniquement à faire tremper le grain qu'on vouloit semer dans une certaine liqueur." Report from the "caffé de Baptiste," Arsenal, ms. Bast. 10277. See also the petition of a Paris merchant, Lebercher, to Hérault advancing claims for a similar capacity to enhance bread production. Ibid. It would be worth studying the whole range of "magic potions" presented to public authorities for different purposes during the Old Regime.

51. Edme Béguillet, *Discours sur la mouture économique* (Paris, 1775), p. 167; *Enquête sur la boulangerie,* pp. 56–58; Malouin, *Description* (1761), pp. 260, 262–63; "Le Pain du peuple," ca. 1768, Arsenal, ms. Bast. 12353. Most of the Revolutionary projects for "le pain républicain" or "le pain égalitaire" were predicated on a strange alloy of authoritarianism and civism. See, for example, the project from the district of Crépy in Year II, AN, F²⁰ 292. The Swiss J.-L. Muret was one of the rare subsistence commentators who promoted home baking: to get a household bread one must return to the *ménage. Mémoire sur la mouture,* pp. 52–58. Like Parmentier, Baudeau rejected home baking as a regressive and uneconomical project. Parmentier, *Parfait Boulanger,* pp. xxxviii–xxxix, and Baudeau, *Avis au premier besoin,* 3e traité, p. 101. Béguillet encouraged "le bas

peuple, les artisans et les journaliers" to eschew bakeries, but he was thinking in terms of the provinces rather than the capital. *Traité des subsistances*, p. xxxix. The intendant of Caen was one of many public officials who saw home baking as a way to check baker "friponnerie." "Observations sur un projet de loi," October 1770, AD Calvados, C 2623. On the widespread practice of home baking in the provinces, see A. Subtil, "Réglementation municipale de la distribution des grains et de la boulangerie," *Actes du 81e Congrès national des sociétés savantes* (congress held in Rouen-Caen, 1956) (Paris, 1956), p. 281.

52. Baudeau, *Résultats de la liberté*, p. 8; Marie Jean Antoine Nicolas Caritat, marquis de Condorcet, eulogy to Malouin, in *Histoire de l'Académie des sciences, 1778* (Paris, 1781), p. 63; Parmentier, *Parfait Boulanger*, pp. xxii–xxiii. In the bakery, unlike the other arts, noted Bachaumont, "l'amélioration, au lieu d'avoir augmenté la dépense, a donné des bénéfices considérables, parce qu'en boulangerie l'économie marche de front avec la perfection." *Mémoires secrets*, vol. 21, p. 9 (24 July 1782).

53. Evelyn, *Panificium*, p. 132; Le Camus, "Suite du mémoire sur le pain," *Journal économique*, December 1753, p. 115; Abbé Gabriel-François Coyer, *Voyages d'Italie et de Hollande* (Paris, 1775), vol. 1, p. 56; Lacombe d'Avignon, *Le Mitron de Vaugirard*, p. 26; Jean-Charles-Pierre Lenoir, *Détail sur quelques établissemens de la ville de Paris demandé par sa majesté impériale la reine de Hongroi à M. Lenoir* (Paris, 1780), p. 51; Malouin, *Description* (1761), pp. 115–16; mémoire of forains to procurator general, ca. 1738, BN, Joly 1314, fol. 50. Parmentier, *Parfait Boulanger*, pp. 211, 441; Bucquet, *Observations intéressantes et amusantes*, p. 96n; Mercier, *Tableau*, vol. 8 (1783), pp. 135–37. Cf. the modern insistence, in the spirit of Mercier-following-Parmentier, that bread making is an extremely complicated scientific process involving general chemistry, rheology, and biochemistry and that "each and every day the baker makes bread he undertakes to perform a complex biochemical experiment." Kent-Jones and Price, *Bread-Making*, preface.

The parochial self-regard of Parisian bread scientists and commentators exasperated the Swiss writer Bertrand, editor of an edition of Malouin's *Description*. He called on them to overcome their ethnocentric purview and adopt a more relativizing address. He cited a moderate and reasonable colleague who wrote as follows: "Il faut laisser chacun dans son opinion sur la meilleure sorte de pain. . . . Quelqu'un n'a pas craint de dire que c'est à Paris qu'on mange le meilleur pain de tout le monde. J'en suis charmé pour cet auteur, mais je crois que s'il avait mangé en Hongrie des *Zipoliten*, à Vienne des *semmeln*, . . . il aurait oublié le Pain de Paris." Malouin, *Description* (ed. Bertrand, 1771), pp. 230–31. Mercier's French bread chauvinism, especially inhospitable to the Swiss, would surely have provoked Bertrand's ire: "Les prisoniers de Paris mangent un pain beaucoup meilleur que celui qu'on mange dans les cantons Helvétiques." *Tableau*, vol. 12 (1788), p. 146. It is rarely clear which specific bread(s) the commentator/traveler has in mind when he lauds Parisian (or French) virtuosity. The common loaf? The middle or the top-of-the-line? Specialty breads? Nor do we usually know the predominant criteria for excellence (taste? color? hygiene/nutrition?). Moreover, one knows nothing about the provenance of the bread that is celebrated, in what conditions it was confected, etc. Thus the historian must use this impressionistic "survey" data with extreme care.

54. Pluche, *Spectacle de la nature*, vol. 6, pp. 310–13; Condorcet, eulogy to Malouin, in *Histoire de l'Académie des sciences*, p. 63; Lenoir, *Détail*, pp. 50, 51; Parmentier, *Parfait*

Boulanger, pp. xvi–xx, xlii–xliii, 209–11, 422, 536–37; Cadet de Vaux cited by Des Essarts, *Dictionnaire universel,* vol. 7, p. 473; Bucquet, *Observations intéressantes et amusantes,* pp. 7, 15, 18–21, 68, 121, and *Traité pratique de la conservation des grains, des farines et des étuves domestiques* (Paris, 1783), pp. 42, 52; Kaplan, *Provisioning Paris,* pp. 457–63. On the need for a union between the bakery and experimental physics/chemistry, see Le Camus, "Suite," p. 89, and Malouin, *Description* (1767), p. 12.

55. *Journal de Paris,* 4 July, 6 August 1782; Bucquet, *Observations intéressantes et amusantes,* pp. 84–85; Cadet de Vaux, "Mémoire sur l'Ecole gratuite de Boulangerie," ca. 1783, Archives des affaires étrangères, France 1395, fols. 311–14, and *Traités divers d'économie rurale, alimentaire et domestique* (Paris, 1821), pp. 39–41; "Projet d'Ecole de Boulangerie," AD Loiret, HH15, in V. du Murand, ed., *Inventaire sommaire des Archives communales antérieures à 1790* (Orléans, 1907), p. 7. For instances of the traditional system for awarding prison baking concessions, see AN, Conseil secret, 3 September 1726, X^{1B} 8907; July 1746, X^{1B} 8923; 8 January 1759, X^{1B} 8937; 20 June 1767, X^{1B} 8954. For an example of the sort of chronic problem that the administration encountered in dealing with prison suppliers, see Secretary of State to D'Ombreval, 2 August 1725, and to Hérault, 9 September 1725, AN, O^1 372. For an example of the sort of chronic problem that suppliers encountered in their dealings with the administration, see Parlement criminel, 17 July 1773, X^{2B} 1049. In behalf of the new Ecole, Cadet followed his attack on prison provisioning by seeking to take over the concession for the Conciergerie jail. BN, Joly 1426, fol. 61.

56. Bachaumont, *Mémoires secrets,* vol. 21, p. 8 (4 July 1782); Lenoir, *Détail,* pp. 521–54; *Journal de Paris,* 4 July 1782; Brocq to Ministry of the Interior, 21 fructidor an VII, AN, F^{11} 1230; deliberations, Bureau of the Hôpital général, 15 January 1787, new ser., no. 47, Archives de l'Assistance publique (hereafter AAP); Parmentier, *Manière de faire le pain de pommes de terre sans mélange de farine* (Paris, 1779), p. 23; "gratification" for Bricoteau, AN, F^{11} 1195; Baudeau, *Avis au premier besoin,* 3e traité, p. 57; Antoine-A. Cadet de Vaux, *Avis sur les blés germés par le comité de l'Ecole gratuite de Boulangerie* (Paris, 1782); Maxime de Sars, *Lenoir, lieutenant de police, 1732–1807* (Paris, 1948), p. 98; André J. Bourde, *Agronomie et agronomes en France au XVIIIe siècle* (Paris, 1967), vol. 3, p. 1331; Arthur Birembaut, "L'Ecole gratuite de boulangerie," in René Taton, ed., *Enseignement et diffusion des sciences en France au XVIIIe siècle* (Paris, 1964), pp. 498–99, 501–2. Cf. the baking school established by the emperor Trajan in 100 A.D., *Encyclopedia Britannica* (1966), article "Bread," vol. 4, p. 133.

57. Scipion, the bakery for several of the hospitals, merits a study on its own. Beginning with Malisset and Commissaire Poussot in the early sixties, it became something of a preschool of milling and baking and a testing laboratory. Bricoteau continued the process of organizational rationalization, shifting from a system of "boulange par four" to a more efficient division of labor in a "boulange par brigades." Wages were increased and linked directly to an incentive system to increase productivity through the application of new methods, largely the repercussions of economic milling. Scipion increased its productivity significantly, decreased its wastage, and (despite Cadet's sweeping critique) generally baked a much finer quality bread at lower cost than it had prior to the innovations introduced by Malisset. Scipion also served as a storage facility where new methods of conservation and rehabilitation were tried, including the uses of various

étuves, or drying ovens. AAP, Hôpital général–Scipion, no. 105, liasses 4, 16; new ser., no. 4, carton 1; deliberations, Bureau of the Hôpital général, 24 February 1783, 15 January 1787, no. 14. See also BN, Joly 1229, fol. 231, and 1247, fol. 21; Poussot's journal, Arsenal, ms. Bast. 10141 (27 February 1761); Lefèbvre, "Rapport sur un déficit de 100 sacs de farine dans les magasins de Scipion," ca. 1800, BN, n.a.f. 1434.

58. Bucquet, *Observations intéressantes et amusantes,* pp. 100–105. On Cadet's arrogance ("nous vous éclairerons du flambeau de la théorie. . . . nous obtiendrons des succès; vous les partagerez. . . . Votre état est fait pour être honoré, & il ne l'est pas; il est déchu de ses droits, & on veut lui rendre cette considération à laquelle il a lieu de prétendre"), see his inaugural address in *Discours prononcés à l'ouverture de l'Ecole gratuite de Boulangerie,* pp. 98–99.

59. Cadet de Vaux, "Mémoire," fol. 313; *Journal de Paris,* 4 July, 6 August, 22 October, and 5 November 1782; Bachaumont, *Mémoires secrets,* vol. 21, pp. 147–49 (21 October 1782); Archives communales de Beauvais, HH 6, in R. Rose, ed., *Inventaire sommaire des archives communales antérieures à 1790* (Beauvais, 1887), p. 219; Arpin, *Historique,* vol. 2, pp. 238, 242, 265; Morel, *Histoire illustrée de la boulangerie en France* (Paris, 1924), p. 249; Birembaut, "L'Ecole gratuite," pp. 496–99; Sars, *Lenoir,* p. 99; Mercier, *Tableau,* vol. 8 (1783), pp. 138–39. Cf. in reference to the question of language and the relation of the savant and the popular artisanal cultures Galiani's mordant assessment of Abbé Baudeau's semiscientific "charlatanry" on the bakery, in which he "affectoit un stile populaire et bas." *Dialogues sur le commerce,* p. 19.

60. Lenoir papers, Bibliothèque municipale Orléans, ms. 1421; Cadet de Vaux, *Avis sur les blés germés; Journal de physique,* December 1782; Birembaut, "L'Ecole gratuite," p. 502; Arpin, *Historique,* vol. 1, p. 247; Alfred O. Aldridge, *Franklin and His French Contemporaries* (New York, 1957), pp. 176–77.

61. Lenoir papers, Bibliothèque municipale Orléans, ms. 1421; Cadet de Vaux, "Mémoire," fol. 312; Parmentier, *Mémoire sur les avantages,* p. 416; Bourde, *Agronomie et agronomes,* vol. 3, p. 1331. Cadet obliquely acknowledged baker resistance in his inaugural speech at the Ecole. See *Discours,* pp. 98–99.

Some of Galiani's trenchant critique of Baudeau's bakery recommendations, based on "un monde idéal" nurtured "dans sa tête," could probably apply to the Parmentier-Cadet approach as well. These two observers paid too little attention to the bakery as a complete business enterprise, to the enormous and unpredictable influence of what the Neapolitan called "les hasards," which could range from the death of a mule or a horse that had to be replaced immediately to the costs of recovering accounts receivable to the burden of various repairs. In quasi-Labroussean terms, Galiani argued that these "cas fortuits," taken en masse, constitute "une quantité constante, réglée, périodique, toujours égale" in the course of a year or a cycle of a few years. Together they probably increased the "dépenses ordinaires" by a third. Without taking these costs into account, none of the subsistence reformers could calculate a convincing coefficient of "savings" that the application of rationality-plus-science was bound to realize. Galiani, *Dialogues sur le commerce,* pp. 180–84.

62. Henri Pigeonneau and Alfred de Foville, eds., *L'Administration de l'agriculture au contrôle général des finances: Procès-verbaux et rapports* (Paris, 1882), pp. xvi and 70 (29 September 1785); remarks by Fougeroux de Bondaroy, Duhamel du Monceau's

nephew, and Duhamel's marginalia in his copy of Parmentier, *Parfait Boulanger,* in AN, 127 AP 6; Du Vaucelles, "Offrande civique," ca. late 1789, AN, F^{10} 215–16; Commission des Revenus nationaux to Commission de l'Instruction publique, 29 brumaire an IV, report of the Bureau d'agriculture to minister of the interior, 19 pluviôse an IV, and accompanying correspondence of Parmentier and Cadet and Lecesne, head of the granary, AN, F^{10} 257; accounts of the Ecole, verified in part by Brocq, AN, F^{11} 1230; Arpin, *Historique,* vol. 1, pp. 248–49; Birembaut, "L'Ecole gratuite," pp. 503–6; Rapport, Ministry of the Interior, 9 germinal an V, AN, F^{10} 257; Parmentier and Brocq to Bureau d'agriculture, 5 ventôse an V, AN, F^{10} 250; chef de la 4e division to Cadet, 16 vendémiaire an V, AN, F^{10} 257.

63. A. B. L. Grimod de la Reynière, *Almanach des gourmands, servant de guide dans les moyens de faire excellente chère; par un vieil amateur* (Paris, 1803–12), vol. 3 (1806). (I am grateful to P. Hyman for this reference.) Added Grimod: "la destruction de l'Ecole de Boulangerie, fondée d'après les principes et les soins du célèbre et savant M. Parmentier, . . . en est probablement la cause." Echoing Grimod de la Reynière, but with a much more sanguine opinion of the true taste of most consumers, the popular writer Jean-Pierre Coffe yearns for the days when Parisians ate the world's finest bread: "Masturbation makes one deaf, so it seems. One might say that the bakery has the same effect. For ages, the majority of bakers have decided to pay no more attention to the complaint, the laments, and the anger of their customers. The consumers get angry, rage, and vituperate. Nothing works. The quality of bread declines dangerously." Coffe, *Au secours le goût,* p. 51.

Chapter 2   Bread Making

1. Parmentier, *Parfait Boulanger,* pp. 273–75; AN, Y 11388, 7 August 1772; Arsenal, ms. Bast. 10080, 7 August 1772; Y 10311, 31 August 1770; Y 15238, 12 September 1731; partage, AN, MC, VII-288, 6 July 1771; Y 11456, 22 July 1737. On the uterine image of the oven as sacred, life-giving mother, see Erich Neumann, *The Great Mother: An Analysis of the Archetype* (Princeton, N.J., 1963), p. 286.

2. Arsenal, ms. Bast. 10099, 4 September 1757; AN, MC, VII-288, 15 July 1771; AN, Y 11309, 14 September 1748.

3. Malouin, *Description* (1779), pp. 33, 71–73; AN, MC, CXXX-292, 4 June 1731; Arsenal, ms. Bast. 10141, fols. 274–75 (16 November 1759); MC, LXXXIX-592, 7 July 1760. Cf. *Journal économique,* 15 March 1774, pp. 528–33, and Rivière, "Folklore des objets domestiques," p. 20.

4. Malouin, *Description* (1761), pp. 116–24; Macquer, *Dictionnaire raisonné universel des arts et métiers, contenant l'histoire, la description la police des fabriques et manufactures de France et des pays étrangers,* new ed., rev., ed. Pierre Jaubert (Paris, 1773), vol. 1, p. 298; AN, Y 11225, 5 August 1738; ASP, D5B^6 (June 1785); D4B^6 11-526 (June 1770); D4B^6 13-587 (August 1753); D4B^6 43-2423 (January 1772); D4B^6 39-2156 (October 1770), and D51E^3 carton 1 (1792); BN, Joly 1743, fols. 137-44; AN, MC, XCIV-304, 2 May 1761.

5. Lacombe d'Avignon, *Le Mitron de Vaugirard,* p. 69; Le Camus, "Suite," p. 98–99 (echoed verbatim by Jacques Savary des Bruslons, *Dictionnaire portatif de commerce*

[Copenhagen, 1761–62], vol. 5, p. 119); Malouin, *Description* (1761), pp. 125–30; Evelyn, *Panification*, IV, p. 133.

6. Parmentier, *Parfait Boulanger,* pp. xxxii, 254–58, and *Dissertation sur la nature des eaux de la Seine* (Paris, 1787). Parmentier was followed by Cadet de Vaux and Louis Bosc. Bucquet, *Observations intéressantes et amusantes* and the *Encyclopédie méthodique*, Agriculture, vol. 5, p. 521.

7. Le Camus, "Suite," p. 100; Lacombe d'Avignon, *Le Mitron de Vaugirard*, p. 75; *Journal économique*, January 1766, p. 7; *Mercure de France*, August 1765, pp. 156–57; Lenoir to procurator general, 11 January 1781 and notes of procurator general's office, BN, Joly 1426, fols., 64–66 ("que les puits ayant souvent communication avec des fossés d'aisance voisines"); Malouin, *Description* (1761), pp. 125–26; Mercier, *Tableau*, vol. 1 (1782), p. 129, and vol. 3 (1783), p. 135 ("Ces fossés, souvent mal construites, laissent échapper la matière dans les puits voisins, . . . et l'aliment le plus ordinaire est néces-sairement imprégné de ces parties méphitiques et mal-faisantes"); AN, MC, VII-386, 23 August 1769; AN, Y 12159, 14 August 1756. Cf. G. Thuillier, "Water Supplies in Nineteenth-Century Nivernais," in *Food and Drink in History: Selections from the Annales,* ed. Robert Forster and Orest Ranum (Baltimore, 1979), p. 120. On the defense of the Seine, see Parmentier, *Dissertation sur la nature des eaux,* p. 4, and "Dissertation physique, chymique et économique sur la nature et la salubrité de l'eau de la Seine," *Journal de physique,* February 1775, pp. 161–86; and "Qualité de l'eau qu'on peut boire à Paris," Arsenal, ms. 3945.

8. Malouin, *Description* (1761), pp. 125–30; Baudeau, *Avis au premier besoin,* 3e traité, pp. 54–56.

9. Parmentier, *Parfait Boulanger,* pp. xxx, 278–80, 293; AN, Y 11237, 18 June 1750; Baudeau, *Avis au premier besoin,* 3e traité, p. 34. For an evocative depiction of the power and the mystery of the life-giving, transmogrifying leaven (as well as the "religious dance of kneading"), see Karel Capek, "The Needle," in *Tales from Two Pockets,* cited by Neumann, *The Great Mother,* p. 286. For a technical discussion of the options for fermentation from a mid-twentieth-century perspective, see Centre national de coordi-nation des études et recherches sur la nutrition et l'alimentation, *La Qualité du pain, novembre 1954–août 1960* (Paris, 1962), vol. 2, pp. 609–705.

10. Malouin, *Description* (1761), pp. 135–48.

11. Parmentier, *Parfait Boulanger,* pp. 281–82; Baudeau, *Avis au premier besoin,* 3e traité, p. 47; Baltasar-George Sage, *Analyse des bleds* (Paris, 1776), p. 47; *Journal éco-nomique,* March 1771, p. 143; Malouin, *Description* (1761), p. 152. Cf. Parmentier, *Parfait Boulanger,* pp. 329–30.

12. Malouin, *Description* (1761), pp. 155–56. Malouin anticipated some of the so-called poolish methods of kneading widely used in Paris today. The technician, chemist, and healer was also a moralist. We live in "a century of flabbiness [mollesse]," wrote Ma-louin, playing on the *pâte molle* so coveted by Parisians. No one is any longer willing to put out the work necessary to accomplish the duties of his profession. The baker's distaste for kneading on sour dough is merely one sympton of a general trend of corner cutting and indolence. *Description* (ed. Bertrand, 1771), pp. 406–42.

13. Arsenal, ms. Bast. 10155, fol. 75, 26 September 1725; Macquer, *Dictionnaire portatif,* vol. 1, p. 148; Cotte, *Leçons élémentaires,* p. 45; Jacques Saint-Germain, *La Reynie et la*

police au grand siècle, d'après de nombreux documents inédits (Paris, 1962), p. 280; Diderot et al., *Encyclopédie*, vol. 12, p. 450; Parmentier, *Parfait Boulanger*, p. 318; letters patent of 25 August 1789, Collection Marcel Arpin. On the lingering mistrust of barm, see Graham, *Treatise on Bread*, pp. 75–76 ("the impure and poisonous substances" endowed with "disagreable properties"); Dauglish's carbonic gas project and other nineteenth-century debate in Sigfried Giedion, *Mechanization Takes Command: A Contribution to Anonymous History* (New York, 1948), pp. 184–87, and Maurizio, *Histoire de l'alimentation*, p. 169.

14. Parmentier, *Parfait Boulanger*, pp. xxx–xxxi, 317–29, 407–8.

15. AN, MC, CIII-21, 16 January 1751; AN, Y 15232, 10 August and Y 12592, 14 July 1748; AN, MC, XXX-437, 24 April 1773; Y 12595, 3 March 1751; Parmentier, *Parfait Boulanger*, p. 317; Malouin, *Description* (1761), p. 152n.

16. Malouin, *Description* (1761), pp. 159–62; Macquer, *Dictionnaire portatif*, vol. 1, p. 300 (Baudeau shared this pro-salt point of view. *Avis au premier besoin*, 3e traité, pp. 63–64); Parmentier, *Parfait Boulanger*, pp. xxxiv–xxxv, 348–61; Cotte, *Leçons élémentaires*, p. 46; *Journal économique*, February 1754, pp. 100–101; AN Z^{1H} 417, 19 December 1744 (Charon).

17. Parmentier, *Parfait Boulanger*, pp. 368–90; Malouin, *Description* (1761), pp. 163–70. The description of kneading (and of the preparation of leavens) in the early twentieth century is not very much different from that of the processes used two centuries earlier. See the military intendant Eugène H. L. Sérand, *Le Pain: Fabrication rationnelle: Historique* (Paris, 1910), pp. 17–18. Compare a mid-twentieth-century treatment of kneading, Centre national de coordination, *Qualité du pain*, vol. 2, pp. 565–608.

18. Baudeau, *Avis au premier besoin*, 3e traité, p. 52; Macquer, *Dictionnaire portatif*, vol. 1, p. 301; Malouin, *Description* (1761), p. 137, and Malouin, *Description* (ed. Bertrand, 1771), pp. 439–41; Legrand d'Aussy, *Histoire de la vie privée*, vol. 1, pp. 89–90; Alfred Carlier, *Histoire du pain* (Cannes, 1938), p. 1; Joseph Barbaret, *Le Travail en France: Monographies professionnelles* (Paris, 1886–90), vol. 1, p. 481; Raymond Calvel, *Le Pain et la panification* (Paris, 1964), pp. 28–29. See the expression of resistance to mechanical kneading (unnecessary, even subversive) in early-twentieth-century Normandy in Sylvie Anne's novel *Victorine, ou le Pain d'une vie* (Paris, 1985), p. 236. Parmentier observed that economic milling also eased the burden of kneading by regrinding and purifying the middlings that had previously resisted incorporation into the dough. *Parfait Boulanger*, p. 363. On the lyrical character of bread making, as exalting as it was exhausting, see the evocative remarks of baker-tycoon Lionel Poilâne, *Guide de l'amateur de pain* (Paris, 1981).

19. Parmentier, *Parfait Boulanger*, pp. 391–413; Malouin, *Description* (1761), pp. 197–200; Baudeau, *Avis au premier besoin*, 3e traité, pp. 15–16.

20. Durand, *Livre du pain*, p. 102; *La Vie illustrée*, January 1906, p. 22. Cf. A. Gottschalk, who worried that the germs introduced by the kneaders would not be killed in the oven because it did not get hot enough. *Le Blé, la farine, et le pain*, p. 37. In the television series, *Les Maîtres du pain*, produced for France 2 that regaled millions of French viewers during the Christmas season of 1993, the heavily perspiring master baker Jérôme assures his apprentices, only half facetiously, that his sweat imparts the distinctive taste to his loaves. A test conducted in 1908 by the Syndicat de la boulangerie

de Paris measured the number of grams lost by the geindre kneading between 172 and 340 kilos. Much of the lost weight passed into the dough as verifiable perspiration. Paul Nottin, *Le Blé, la farine, le pain, les semoules, les pâtes alimentaires* (Paris, 1940), p. 65.

21. Lacombe d'Avignon, *Le Mitron de Vaugirard*, p. 67; Malouin, *Description* (1779), p. 93; Mercier, *Tableau*, vol. 3 (1783), p. 179; cf. Béguillet's remarks on the baleful vapor of excrement. *Traité des subsistances*, p. 582.

22. AN, Y 11449, 4 November 1727, and Y 11673, 30 September 1749. The long four-pounder measured between twenty and twenty-two inches. *Journal de physique*, February 1782, p. 90.

23. Malouin, *Description* (1761), pp. 208–9, 266; Malouin, *Description* (ed. Bertrand, 1771), pp. 16–17; Baudeau, *Avis au premier besoin*, 3e traité, pp. 70–73; "Mémoire pour les boulangers de Chaalons-sur-Marne," 30 July 1785, BN, Joly 1742, fol. 43; "Mémoire pour les boulangers de Troyes," 1789, BN, Joly 1743, fol. 134; Morel, *Histoire abrégée*, p. 212; Parmentier, *Parfait Boulanger*, pp. 426–36, 621–22; Macquer, *Dictionnaire portaitif*, vol. 1, p. 301; Tillet, report for Bailly, 1790, cited by A. Birembaut, "Le Problème de la panification à Paris en 1790," *Annales historiques de la Révolution française* 37 (July–September 1965), p. 366; Edmond Villey, "La Taxe du pain et les boulangers de la ville de Caen en 1776: Documents," *Revue d'économie politique* 2 (1888), p. 187. Husson's estimates of loss strike me as too low. *Consommations*, p. 143.

24. Malouin, *Description* (1761), pp. 206–12; Parmentier, *Parfait Boulanger*, pp. 434–43; Poncelet, *Histoire naturelle*, pp. 200–206.

25. Parmentier, *Parfait Boulanger*, pp. xxxviii, 484–90.

26. AN, Y 11673, 30 September 1749, 2 October 1733; Y 15114, 26 October 1780; Y 11385, 9 August 1771; Y 14961, 8 April 1745 (Bossu); Y 13494, 1729 (Mongueret); Y 15355, 12 November 1755. Cf. the price of ovens in Year IX. ASP, D5B⁶ 4728. On the shop-oven separation, see the cases of bakers Leprince and Colombet, AN, Y 14948, 2 October 1733. On the hostility to ovens evinced by other tenants, see Y 15354, 4, 6 February 1755.

27. Parmentier, *Parfait Boulanger*, pp. 454–78; Baudeau, *Avis au premier besoin*, 3e traité, pp. 74–75; Diderot et al., *Encyclopédie*, vol. 7, p. 222; Malouin, *Description* (1761), pp. 242–46; Rivière, "Folklore des objets domestiques," pp. 3–4, 99; BN, Joly 1743, fol. 134 (1786).

28. ASP, D5B⁶ 1829; Parmentier, *Parfait Boulanger*, pp. 475–80; AN, Y 13500, 17 May 1734; deliberations, Bureau de l'Hôtel-Dieu, 11 July 1727, AAP, no. 96; *Journal économique*, November 1757, pp. 47–53; Arpin, *Historique*, vol. 2, pp. 202–5; *Journal de physique*, December 1783, pp. 433–47. On the tools used to treat the oven floor, see H. Tremand, "Les Objets en rapport avec le pain," Musée national des arts et traditions populaires, Archives ms. no. 49, December 1945, pp. 608–48.

29. Deliberations of the Paris municipality, AN, H* 1866, fol. 175 (August 1755); Des Essarts, *Dictionnaire universel*, vol. 1, p. 617; cf. Diderot et al., *Encyclopédie*, vol. 2, pp. 305–6, and vol. 7, p. 129; William Mildmay, *The Police of France, or An Account of the laws and regulations established in the kingdom for the preservation of peace and the preventing of robberies, to which is added a particular description of the police and government of the city of Paris* (London, 1763), pp. 92–93; and François Gosset, *Inventaire des arrêts du conseil du roi* (janvier-février et mars-avril 1760) (Paris, 1938), pp. xii, xxiii; secretary of state for the royal household to prévôt des marchands, 14 July 1726, AN, O¹

373; Cherrière, *La Lutte contre l'incendie dans les halles, les marchés, et les foires de Paris sous l'ancien régime,* in *Mémoires et documents pour servir à l'histoire du commerce et de l'industrie en France,* ed. Julien Hayem, 3d ser. (Paris, 1913), pp. 170–81; AN, H 2192, fol. 50; *taxation* schedules, 8 March 1760 and 28 January 1768, ASP, 6AZ 249/7 and 9; BN, Joly 2432, fol. 210; ordinance of prévôt des marchands, 2 February 1704, BHVP; BN, ms. n.a.f. 1032, fol. 93; Assembly of Police, 2 July 1739, BN, ms. fr. 11356, fols. 400–401; Jean-Pierre Gay, "L'Administration de la capitale entre 1770 et 1789, la tutelle de la royauté et ses limites," *Mémoires de la Fédération des Sociétés historiques et archéologiques de Paris et de l'Ile-de-France* 10 (1959), pp. 229–30; Joly 1742, fol. 42 (July 1785); sentence of prévôt des marchands, 4 October 1720, AN, F^{11} 264, 4 October 1720; dispatches to prévôt des marchands, 10, 20, 26 October 1731, O^1 378, pp. 299, 307–8, 311; Assembly of Police, 6 December 1731, ms. fr. 11356, fol. 174; Mercier, *Tableau,* vol. 12 (1788), p. 148; *Encyclopédie méthodique,* Jurisprudence, police et municipalités, vol. 9, p. 440; Albert Babeau, *Paris en 1789* (Paris, 1892), p. 332.

30. *Mémoire pour les Srs. Buvry, Bouchot [et al.], maîtres boulangers* (Paris, 1786), BN, 4° Z Le Senne 2263; Lefèbvre, "Rapport sur un déficit de 100 sacs de farine dans les magasins de Scipion," ca. 1800, BN, n.a.f. 1434, fols. 10–12; Parmentier, *Mémoire sur les avantages,* pp. 411–12; Cotte, *Leçons élémentaires,* p. 76; "Mémoire pour les boulangers de Troyes," 1786, BN, Joly 1743, fol. 132; "Avis économique sur la boulangerie," pp. 66–67; "Explications," ca. September 1725, Joly 1116, fol. 240. Cf. E. Labrousse in C.-Ernest Labrousse and Fernand Braudel, *Histoire économique et sociale de la France* (Paris, 1970–82), vol. 2, p. 399.

31. ASP, D_5B^6 1547 and 5253, and D_4B^6 21-1075; *Mémoire pour les Srs. Buvry.*

32. Parmentier, *Parfait Boulanger,* pp. 482–503; Malouin, *Description* (1761), pp. 246–54; Parmentier, *Mémoire sur les avantages,* p. 327; Baudeau, *Avis au premier besoin,* 3e traité, pp. 79–80.

33. Kent-Jones and Price, *Bread-Making,* pp. 118–27; Kirkland, *Modern Baker,* vol. 1, pp. 77–78; Percy A. Amos, *Processes of Flour Manufacture* (London, 1912), pp. 211–12; Malouin, *Description* (1761), pp. 256–59; Malouin, *Description* (ed. Bertrand, 1771), p. 392; Parmentier, *Parfait Boulanger,* pp. 539–40; Abbé Noël Chomel, *Dictionnaire Oeconomique,* ed. M. de la Marre (Paris, 1767), vol. 2, p. 799; Giedion, *Mechanization,* pp. 204–5. On the organoleptic traits of a fine loaf, see also "Receuil des usages concernant les pains en France," in *Le Pain,* Actes du Colloque du CNERMA (a conference held in Paris, November 1977), ed. Jean Buré (Paris, 1979), pp. 280–82.

Chapter 3 Baker Shops and Bread Markets

1. Delamare, *Traité,* vol. 2 (1729), p. 755; Diderot et al., *Encyclopédie,* vol. 2, pp. 360–61. In a petition written in 1686, the Paris master bakers indicated that there were 250 bakers of *petit* in the city, fifteen hundred bakers of *gros* in the faubourgs. BN, ms. fr. 21640, fol. 73. These figures correspond roughly to the ones Delamare elaborated on some twenty to thirty years later. The Paris bakers may merely have lumped together the forains with the faubourg bakers under the latter's rubric. Remember, too, that the Paris guild had not yet absorbed the faubourg masters. In a petition drafted four years later, the Paris bakers estimated that the capital was supplied by a corps of two thousand

bakers. Ms. fr. 21640, fol. 50. Cf. an undocumented claim of 1,150 bakers in 1686. Eugen Schwiedland, "Les Industries de l'alimentation à Paris," *Revue d'économie politique* 9 (1895), p. 300. In a memoir written in 1737 or 1738, the guild claimed that in the 1690s there had been 250 bakers of *petit* and nine hundred bakers of *gros*, for a total of 1,150, a figure that may or may not have included the forains. BN, Joly 1314, fol. 70.

2. Delamare, *Traité*, vol. 2 (1713), p. 898; Henri Sauval, *Histoire et recherches des antiquités de la ville de Paris* (Paris, 1724), book 6, p. 656; *Journal économique*, January 1754, p. 106; Des Essarts, *Dictionnaire universel*, vol. 2, p. 250; BN, ms. fr. 21640, fol. 35. Market for market the numbers that comprise the subtotal of 1,534 are Delamare's from the *Traité*. Cf. an undocumented estimate for 1709 of fifteen hundred bakers, a thousand from Paris and five hundred from the country. Saint-Germain, *Vie quotidienne*, p. 210.

3. BN, Joly 1428, fols. 9–10. The faubourg masters were apportioned as follows: Saint-Laurent: 69; Saint-Germain: 54; Saint-Denis: 36; Saint-Honoré: 27; Saint-Marceau: 20; Temple: 18; Saint-Jacques: 15; Saint-Antoine: 4; Saint-Victor: 1. Tessier appears to have used this same document for his calculations. *Encyclopédie méthodique*, Agriculture, vol. 3, p. 483.

4. BN, Joly 1428, fol. 7. It is possible that this report was written later than I believe, in which case its estimates could be reconciled with the 1,607 estimate, for the number of forains did indeed decline later on and it is probable that the number of nonmaster faubourg bakers increased. The faubourgs probably absorbed a number of forain families. Levasseur apparently drew on the first estimate for his calculation of 757 bakers in 1721, one for every 792 inhabitants given an estimated population of six hundred thousand. He contrasted this ratio with that of one baker to 1,608 people in 1896 (1,522 bakers for 2,488,000 persons). Emile Levasseur, *Histoire des classes ouvrières et de l'industrie en France avant 1789* (Paris, 1901), vol. 2, p. 762. Bouteloup estimated that in 1863, before the "freeing" of the Parisian bakery, the ratio was 1:1,800; in 1890, 1:1,600; and in 1908, 1:1,290. Maurice Bouteloup, *Le Travail de nuit dans la boulangerie* (Paris, 1909), p. 12n.

5. BN, Joly 116, fols. 234–369 (the distribution by quarters is the same as in the 1720 mémoire); état des noms et demeures des mâitres Boulangers, 15 October 1724. The faubourg apportionment is as follows: Saint-Germain: 64; Saint-Laurent: 47; Saint-Lazare: 24; Saint-Marceau: 22; Temple: 20; Saint-Jacques: 18; Les Porcherons: 15; Saint-Martin: 15; Saint-Denis: 13; Saint-Honoré: 11; Saint-Victor: 7; Saint-Antoine: 6. The distribution by faubourg did not change significantly from 1720 to 1724, despite appearances. The 1720 *état* included the faubourg Saint-Lazare masters (about twenty-four) in the Saint-Denis category and the faubourg Saint-Martin masters (about fifteen) in that of Saint-Laurent.

6. AN, Y 11306. One can judge the relative size of the baker guild by noting that at this time there were 86 *apoticaires*, 140 *chaircuitiers*, 250 *pâtissiers*, 320 *libraires-imprimeurs*, 500 *chirurgiens*, 1,600 *marchands de vin*, 1,700 *couturiers*, 1,800 *tailleurs*, and 2,500 *merciers*.

7. The Bureau of the Hôtel de Ville estimates 597 masters in 1759. Deliberations, AN, H* 1868, fols. 311–16, 20 September 1759. According to one edition of Macquer's *Dictionnaire*, there were 585 guildsmen. Philippe Macquer *Dictionnaire raisonné*, vol. 1, p. 149. Abbé J.-J. Expilly reported the number at approximately 580 in his *Dictionnaire*

géographique, historique et politique des Gaulles et de la France (Paris, 1762–70), vol. 5, p. 420. Of course it is possible that Macquer and Expilly simply copied dated sources in the manner of Des Essarts. In a work published in 1779, Béguillet counted six hundred, but his estimate must have pertained to a period before 1776. Edme Béguillet, *Description historique de Paris et de ses plus beaux monuments* (Paris, 1779–81), vol. 1, p. 160. In 1763 the intendant of Paris, who should have known better, claimed that "the baker guild . . . is composed of 1,000 or 1,200 persons." "Observations sur le commerce des grains," 1763, BN, ms. fr. 11347, fol. 227. The anonymous author of a mémoire on "le prix du pain" who contended that there were 450 masters in 1774 was also mistaken, though less egregiously than the intendant. "Le Prix du pain," AN, F^{11} 265, September 1774.

8. Lalanne's estimate of 183 masters on the eve of the Revolution is patently absurd. Ludovic Lalanne, *Dictionnaire historique de la France* (Paris, 1872), p. 347. A. Feyeux's figure of six hundred in 1790 is slightly more credible but wholly undocumented, as is his claim that the numbers swell overnight to two thousand with the abolition of the guilds in 1791. "Un Nouveau Livre des métiers," part 1, *La Science sociale*, 2d year, 4 (October 1887), p. 336. Cf. the similar figures in Lanzac de Laborie, *Paris sous Napoléon*, vol. 5, p. 160. According to the *Tableau des boulangers de Paris pour l'an 1811* (Paris, 1811), there were 653 bakers officially recognized at that time. Cf. the estimate of eight hundred bakers in the capital and faubourgs in 1786 in *Mémoire pour les Srs. Bandy, et al.* (Paris, n.d.), BN, 4° F^3 35341, and also Morel, *Histoire abrégée*, p. 287.

9. Bibliothèque de l'Institut de France (hereafter Institut), mss. 515, 521. Léon Cahen used the Delalande registers, but too rapidly. He appears to have forgotten that there were masters as well as faubourg bakers who did not furnish the market. See "A propos du livre d'Afanassiev: L'Approvisionnement de Paris au debut du XVIIIe siecle," *Bulletin de la Société d'histoire moderne* 22d year (5 March 1922), pp. 165–66. He also underestimates the baker population in "La Population parisienne au milieu de XVIIIe siècle," *Revue de Paris* 5 (September–October 1919), p. 161.

10. *Statistique de l'industrie à Paris résultante de l'enquête faite par la chambre de commerce pour les années 1747–48* (Paris, 1851), p. 43.

11. Turgot, Baudeau, and others contended that with the abolition of the corporations, many more bakers would set up independently and their competition would result in lower prices. Yet it has been argued that in fact the unregulated competition inaugurated in 1791 led to a serious atomization of the trade and to an increase in production costs per shop that were passed on to consumers. Feyeux, "Un Nouveau Livre," part 1, pp. 336–39.

12. Since there were only a handful of Saint-Antoine masters, when I say faubourg Saint-Antoine bakers without further specification, I mean nonmasters.

13. We lack the data to determine whether the upward movement of the shop supply constituted the beginning of the trend. I doubt that the shop component grew much larger in the course of the century.

14. For some indication of the existence of many Saint-Antoine nonguild shops, see the listing in the "Limites de la Ville et faubourgs de Paris" drafted in response to the royal declaration of July 1724. AN, Q^{1*} 1099$^{158-75}$. There were at least twenty-one bakers on the rue de Reuilly and twenty on the rue du faubourg Saint-Martin in 1724. AN, Q^{1*} 1099$^{159-64}$. I am grateful to my friend Isabelle Dérens for this reference.

15. One even finds three-pound loaves offered at the markets. See, for example, the procès-verbal of the baker-jurés, AN, Y 13098, 1 June 1746.

16. The police were reassured when the people "bargained for their bread." That meant that there was no press, that the supply was more than adequate, and that there would be pressure on the bakers to relax the price. See Duplessis to procurator general, 22 May 1726, BN, Joly 1118, fol. 146. For an interesting discussion of the significance of haggling in markets of preindustrial cities, see Gideon Sjoberg, *The Pre-Industrial City: Past and Present* (New York, 1960), pp. 204–8. On the new sense of *marchander* in the eighteenth century, see Ferdinand Brunot, *Histoire de la langue française*, vol. 6, p. 368; plainte to Maillot, October 1771, AN, Y 9474; Y 10116, 31 January 1748. For other testimony to the practice of market haggling see Arsenal, ms. Bast. 10141, 11 December 1756, and Y 15929, 10 September 1727.

17. AN, Y 9442, 10 March 1741. This was a year of serious dearth and exorbitant prices.

18. AN, Y 12158, 21 June 1756.

19. Sartine to commissaires, 22 January 1771, AN, Y 15114; Butay to lieutenant general of police, 25 September 1725, Arsenal, ms. Bast. 10270. A nineteen-year-old servant explained that if his baker at the Palais-Royal market charged as much as his shop baker, he would buy from the latter. Y 10558, 9 May 1775. In addition to bearing greater costs, the shop bakers, so the police claimed, "fashioned their bread with greater care." A. Gazier, ed., "La Police de Paris en 1770, mémoire inédit, composé par ordre de Marie-Thérèse," *Mémoires de la société de l'histoire de Paris et de l'Ile-de-France* 5 (1878), p. 126. The forains rebutted the claims of the shop bakers for higher indemnities, arguing that unlike the shop bakers, they had to pay the taille, maintain horses and wagons, and absorb other costs unknown to Parisians. Anonymous mémoire, BN, Joly 1116, fol. 255. During the dearth of 1740, the police prohibited the shop bakers from selling "above the common price of the market." Joly 1111, fol. 172.

20. AN, Y 15241, 20 November 1731; Y 13926, 31 May 1741; ASP, D5B⁶ 87 (January–April 1753 entries have a clear-cut market rhythm); D5B⁶ 5290; AN, MC, V-539, 28 July 1764. It goes without saying that if one seeks credit at the market, one is not in a strong position to haggle on the price.

21. ASP, D5B⁶ 3338; D5B⁶ 4119 (e.g., May, July 1766); D5B⁶ 4288 (e.g., January 1752). The bakers may have favored friends or steady customers or bulk buyers such as inn-keepers.

22. ASP, D5B⁶ 4419, 3338, 1061, and 1003.

23. AN, Y 10215, November–March 1760; Y 15929, 10 September 1727; Y 13525, 5 February 1761.

24. See, for example, the entries in the registers of bakers Modinet and Rousseau, ASP, D5B⁶ 4288 and 3338; deliberations, Bureau of the Hôtel de Ville, AN, H* 1867, fol. 445, 22 May 1758; D5B⁶ 1968; D5B⁶ 2532, 19 July 1770; D2B⁶ 737, 29 May 1741; AN, Y 15067, 22 September 1734; Malouin, *Description* (ed. Bertrand, 1771), p. 381 and plate viii; Y 12625, 6 June 1775.

25. Lefèbvre, "Rapport sur un déficit de 100 sacs de farine dans les magasins de Scipion," ca. 1800, BN, n.a.f. 1434, p. 11; BN, Joly 1743, fols. 136–37; ASP, D5B⁶ 322, 1829, 2208, 3338, 4119; DE¹ carton 7, dossier 71; AN, MC, VII-329, 4 November 1760. An

expert testified in 1859 that the retail sale of charcoal remained crucial for bakers, in some cases practically covering their fuel costs. *Enquête sur la boulangerie*, p. 705.

26. See Joseph Barbaret, *Le Travail en France*, vol. 1, p. 468. On the "ruinous luxury" of the late-nineteenth-century shops, see Feyeux, "Un Nouveau Livre," part 1, p. 341. Renovation comes in waves. In the last ten to fifteen years scores of Parisian shops have undergone modernization featuring an opulent cosmetic front. The shops to avoid today are the ones that not only look like pharmacies on the outside but smell like them on the inside. See also *Enquête sur la boulangerie*, p. 65.

27. For an idea of the range of bread typically displayed, see AN, Y 14524, 31 July 1727. For an example of the grill, see Y 11242, 16 June 1755, in which master Côte filed a complaint for damage to his shop. Sometimes the bread was displayed on a shelf hanging on the grill overlooking the street. See the trouble this provoked for the wife of master Georges Fremin, rue Saint-Honoré, Y 12159, 7 February 1756.

28. I defy anyone to demand—as one is entitled to by law—that the baker's wife weigh the bread, however. Weight is still a highly sensitive issue. See the discussion of short weight below.

29. For the sad tale of a customer who fell down the trap (she was the third such victim in recent times), see AN, Y 12141, 22 January 1739.

30. AN, Y 15611, 31 May 1746; Y 19605, 23 March 1743; Y 12602, 17 June 1755.

31. See, for example, AN, Y 12599, 29 August 1753; Y 11457, 16 April 1738; Y 15431, 6 September 1726; Y 11242, 7 August 1755; Y 12741, 10 May 1755. These cases concern a butcher, a fruit seller, and a milk dealer who blocked bakers' doors or obscured their grills and a horse trader who trimmed his horses' tails in front of a shop, casting ugly blotches of blood and hair on the displayed bread.

32. See, for instance, AN, Y 9441, 19 May 1740; Y 9443, 18 January 1742; Y 9621, 20 October, 20 November, 2 December 1746. The fines varied from three to five livres, perhaps depending on the degree of grimness.

33. Delamare, *Traité*, vol. 2 (1715), p. 898. Here is Delamare's enumeration of markets with the number of bakers attached to each:

aux grandes halles: 342

aux halles de la Tonnellerie: 104

Place Maubert: 159

Cimetière Saint-Jean: 158

Marché-Neuf de la Cité: 89

Devant l'église des Jésuites, rue Saint-Antoine: 148

Quay des Augustins: 92

Petit marché du faubourg Saint-Germain: 147

devant l'église des Quinze-Vingt, rue Saint-Honoré: 95

Place du Palais-Royal: 40

devant l'hostellerie des Bastons royaux, rue Saint-Honoré: 30

Marché du Marais du Temple: 46

devant le Temple: 22

Place de la Porte Saint-Michel: 36

la halle du faubourg Saint-Antoine: 16

Vincent claims sixteen markets for the beginning of the eighteenth century, but he provides no documentation. François Vincent, *Histoire des famines à Paris* (Paris, 1946), p. 17.

34. Delamare to procurator general, 11 November 1721, BN, Joly 1428, fols. 12, 19, 26. He added that no bread was sold on the rue Sainte-Nicaise, where an informal market may have occurred sporadically. Delamare insisted, by contrast, that there was still an active bread market at Quinze-Vingt, a place that would disappear from police lists in a few years. Delamare's original table of markets, drawn from the *Traité*, was republished time and again later in the century to determine whether it still described reality. See, for example, *Tableau de Paris pour l'année 1759* (Paris, 1759), p. 38, and *Tableau universel et raisonné de la Ville de Paris* (Paris, n.d. [ca. 1757]), p. 22.

35. Arrêt of the Paris Parlement, BHVP, 30 August 1723, and deliberations, Bureau of the Hôtel de Ville, AN, H* 1861, fol. 499, 20 December 1745; Pierre Narbonne, *Journal des règnes de Louis XIV et Louis XV de l'année 1701 à l'année 1744 par Pierre Narbonne, premier commissaire de Police de la ville de Versailles* (Versailles, 1866), pp. 347–51. AN, Y 11169, 4 October 1752; Arsenal, ms. Bast. 10055, fol. 250, 12 April 1760. Apparently the Daguesseau was moved to a new location in 1745. Jacques-Antoine Dulaure, *Histoire physique, civile et morale de Paris* (Paris, 1823–25), vol. 6, p. 2. The fact that a patrol of baker-jurés visited the Daguesseau market in 1754 is quasi-definitive proof that it functioned as a bread market. Y 11241, 5 January 1754.

36. Narbonne, *Journal*, pp. 347–50; Arsenal, ms. Bast. 10141, fols. 404–5, 12 November 1760; Paul Leblond, *Le Problème de l'approvisionnement des centres urbains en denrées alimentaires en France* (Paris, 1926), p. 258. Narbonne also mentioned the market of the faubourg Saint-Antoine as a bread entrepôt, though in fact it had not offered bread for sale—at least officially—for some fifteen years.

37. Arsenal, ms. Bast. 10041, fol. 409, 20 November 1760; ms. Bast. 10141, fol. 497, 9 January 1761; ms. Bast. 10141, fols. 403–7, 6, 15 November 1760. The Quinze-Vingt may already have been reborn—more than once. See the ephemeral reference to it in the *scellé* of baker Claude Thevenet, AN, Y 15603, 3 May 1742.

38. Arsenal, ms. Bast. 10141, fols. 404–5 (8 October, 12 November 1760).

39. Sars, *Lenoir*, p. 100; Georges Touchard-Lafosse, *Histoire de Paris et de ses environs* (Paris, 1851), vol. 3, p. 539; Arsenal, ms. Bast. 10141, fol. 378 (25 September 1760); A. Gazier, "La Police de Paris en 1770, mémoire inedit, composé par ordre de Marie-Thérèse." *Mémoires de la Société de l'histoire de Paris et de l'Ile-de-France* (1878), p. 126; Archives de la Préfecture de police (hereafter APP), A/a 84, no. 202. Theoretically marketplaces had to be numbered beginning early in the century, but the requirement was not rigorously enforced. See, however, the police sentence of 12 February 1734, in Des Essarts, *Dictionnaire universel*, vol. 2, p. 259. Divot, the commissaire in charge of the Cimetière Saint-Jean market in the thirties, anticipated many of Poussot's reforms. He insisted that all stalls be numbered and he exhorted the public to denounce bakers who sold short weight or badly made bread. He also asked the lieutenant general to require the market bakers to mark their loaves not only with the weight and their initials but also with the initials of the market they served. He ordered all bakers to file their

stamps or "prints" with him. See Divot's reports in AN, Y 9435, 26 February 1734, and Y 9431, 16 November 1731.

40. Delamare, *Traité*, vol. 2 (1729), p. 763. Virtually word for word, later police texts and commentaries on police regulations rehearsed Delamare's dictum. Duchesne, *Code de la police, ou Analyse des règlements de police*, 4th ed., rev. (Paris, 1767), p. 117; *Journal économique*, January 1754, p. 102; Des Essarts, *Dictionnaire universel*, vol. 2, p. 248; *Encyclopédie méthodique*, Jurisprudence, vol. 2, p. 94, and vol. 10 (1789), p. 227. The application of the notion of public service to grain and flour merchants was far more controversial, because it seemed to impinge on an economically more important form of private property. Liberals who were willing to see bakers "reduced" to public service refused to tolerate the same treatment for merchants. See Kaplan, *Provisioning Paris*.

41. Macquer is wrong to suggest that there was a rigid quota. Macquer, *Dictionnaire portatif*, vol. 1, p. 302.

42. BN, Joly 1117, fol. 54, and ms. fr. 21651, fol. 336; sentence against Jean Lefevre, AN, Y 9538, 25 October 1725. Jeanne-Nicole Poiry, a faubourg Saint-Antoine baker, committed a double infraction. She not only stopped supplying the market but also transferred her stall illicitly to another baker, who used it to sell bread "of defective quality." Poiry was fined three thousand livres and lost the stall irrevocably. Police sentence, 23 August 1726, BHVP.

43. BN, Joly 1117, fol. 63, 20 November 1725; La Reynie's exhortation to Delamare, 12 July 1693, BN, ms. fr. 21638.

44. Couet de Montbayeux to procurator general, 11 November 1725, BN, Joly 1117, fol. 248; Labbé to procurator general, 13 April 1726, Joly 1118, fol. 128. Bakers were careful to cover themselves. Pierre Brion of the faubourg Saint-Antoine filed a preemptive complaint with the police against a transporter who had reneged on a promise to bring a muid of Brion's grain for milling and was thus responsible for the baker's inability to furnish the Saturday market. AN, Y 12571, 25 September 1725.

45. Police sentence, 28 November 1727, BN, ms. fr. 21640, fols. 57–58; police sentence, 22 July 1729, BHVP; police sentence, 16 November 1731, ms. fr. 21640, fols. 59–60; AN, Y 9435, 26 February 1734; Y 9618, 26 February 1734; ms. fr. 21640, fol. 62, 2 September 1735; Fréminville, *Traité de la police*, p. 97; police sentence, 14 September 1740, BHVP.

46. See the police sentences, 2 December 1729, BN, ms. fr. 21640, fols. 170–71, and 22 July 1729, BHVP; AN, Y 13634, 7 January 1730.

47. AN, Y 11218; Y 11221; AN, MC, XCIX-470, 18 May 1741.

48. AN, MC, LXXIX-453, 21 September 1738; ASP, D4B⁶ 48-2869, 5 July 1773; MC, XXVIII-210, 25 September 1728.

49. See Courcy's denunciation of the dower practice, AN, Y 11219, 14 August 1732; AN, MC, XXXVIII-208, 8 December 1720; MC, XXXVIII-246, 27 February 1737; MC, XXXVIII-282, 16 August 1743. Dowry evaluations frequently reflect familial concerns and tensions rather than real market value. Market stall prices varied because no one dared publicize the amount of sales. On the selling of market positions and their use as dowry by Gonessiens, see J.-P. Blazy, *Gonesse, la terre et les hommes: Des origines à la Révolution* (Meaux, 1982), p. 209. There are interesting similarities between the practice of allocating market slots in the Parisian bread markets and that of distributing selling space in the twentieth-century market of Bagio City in the Philippines. See

William G. Davis, *Social Relations in a Philippine Market: Self-Interest and Subjectivity* (Berkeley, 1973), pp. 106, 123–24, 128.

50. AN, MC, XXXVIII-282, 23 June 1735; AN, Y 11587, 31 March 1769; MC, XXXVIII-259, 3 July 1751; Y 11218, 14 May 1720. Petition paid dearly: 360 livres. The contract read like a *fonds* or business transfer. Jacquet ceded his clientele as well as his physical space and pledged not to supply this or nearby markets in the future.

51. Police sentence, 2 December 1729, BN, ms. fr. 21640, fols. 170–71.

52. See, for example, "Ordonnance servant de règlement aux boulangers," BN, ms. fr. 21651, fol. 295, 4 May 1725.

53. BN, ms. fr. 21633, fol. 134; AN, Y 11229, 3 October 1742; Trudon to Sartine, 12 September 1772, Y 15114. One of her two markets was Daguesseau, another proof of the revival of this entrepôt.

54. See, for example, AN, Y 11225, 31 December 1738; Y 11226, 7 March 1739; Y 11232, 7 August 1725; Y 11233, 12 March 1746.

55. Police sentence, 9 June 1724, BHVP; AN, Conseil secret, X^{1B} 8907, July 1726; AN, H* 1861, fol. 499, 20 December 1745; BN, ms. fr. 21633, fols. 23–27, 6 February 1723; Arsenal, ms. Bast. 10055, p. 250, 12 April 1760.

56. Interrogation Louis Bernac, December 1766, AN, Y 18662; Arsenal, ms. Bast. 10041, fol. 102, 24 October 1753; Y 13634, 2 April 1729. Y 13751, 12 January 1746. Félix's father was robbed the very same day, also on his way home from market.

57. See, for example, the cases of Etienne Faron of Passy and Pierre Delamare, a master baker in the faubourg Saint-Laurent. AN, Y 12619, 17 February 1770; MC, XXXIX-441, 30 June 1757. Also: Y 11224, 19 January 1737; Y 18674, 9 October 1771; Y 13634, 9 April 1729; Y 11224, 19 January 1737.

58. AN, Y 12157, 20 May 1754, and Y 11236, 15 November 1749; Y 11225, 22 November 1738; Y 11240, 19 December 1753; Y 11225, February 1738; Y 15607, 13 January 1744.

59. AN, Y 11242, 14 June 1755; Y 11226, 15 April 1739; Arsenal, ms. Bast. 10141, 21 May 1758. Cf. a similar feud between brothers that seems to have been crystallized by their wives, who ran their stalls at the Halles. Y 12614, 20 June 1765.

60. AN, Y 15242, 23 July 1735.

61. AN, Y 15244, 7 December 1737.

62. AN, Y 15244, 6 March 1737. According to a modern authority, average baker production at the beginning of the twentieth century was 300 to 350 kilograms of bread per day from 1.5 to 1.75 sacks of flour containing 157 kilos each. Bouteloup, *Travail de nuit*, p. 12.

63. Pierre Samuel Dupont de Nemours, *De l'exportation et de l'importance des grains* (Paris, 1764), in *Collection des économistes et des réformateurs sociaux de la France*, ed. Edgard Depitre (Paris, 1911), p. 33n; Moncouteau to Delamarche, 12 January 1792, AN, F^{30} 157.

64. Malouin, *Description* (1767), pp. 252, 284; Malouin, *Description* (ed. Bertrand, 1771), p. 201; AN, Y 13926, 31 May 1741; Y 10558, 10 May 1775. The average fournée, Malouin suggests, was composed of ten round loaves of six pounds, ten round loaves of twelve pounds, and sixteen long loaves of four pounds. Nine bakers in the section of the Arsenal in 1793, probably baking a *ménage* bread of high-extraction flour, averaged about 273 pounds per fournée. AN, AF^{iv} 1470, 16, 17 April 1793.

65. Malouin, *Description* (1767), p. 284 (his baker, then, would convert between 2.4

and 2.98 muids a week, quite a high figure); Rutledge papers, AN, Y 10506, 19 December 1789.

66. *Révolutions de Paris,* 17–24 October 1789, p. 27; Guérin to Delamarche, 28 January 1792, AN, F^{30} 157; AN, Y 15595, 22 September 1728; AN, AF^{iv} 1470, 16–17 April 1793; Parmentier, *Mémoire sur les avantages,* p. 410, and *Parfait Boulanger,* pp. xxxix, 537; Y 14095, 28 October 1768. In fact Gibert's daily average was lower, for he surely cooked much more bread on Tuesdays and Fridays in preparation for the market day.

67. AN, Y 15060, 7 November 1759; Malouin, *Description* (1767), p. 180n; Y 15063, 5 May 1742; Y 15261, July 1753. A fournée every two hours was a much more common cadence. AN, F^7 3688^3, 29 September 1793.

68. AN, Y 12140, 5 July 1738; Arsenal, ms. Bast. 10060, 25 September 1761. On the "pénible métier" of the porteuse, see also Feyeux, "Un Nouveau Livre," part 1, p. 345.

69. For examples of journeyman *porteurs,* see AN, Y 12594, 22 April 1750, and Y 15628, 12 December 1754. Toward the middle of the nineteenth century, when as much as one-half of the total bread production was home-delivered, there were 1,643 porteuses and 712 porteurs working for Paris bakers. Barbaret, *Le Travail en France,* vol. 1, p. 440; *Enquête sur la boulangerie,* p. 263.

70. AN, Y 13123, 25 April 1771 (Prot); Y 14963, 8 February 1747 (Hornet); Y 12590, 24 August 1746 (Fouard); Y 12140, 5 July 1738 (Chartier); Y 15613, 27 June 1747 (Barois); *Enquête sur la boulangerie,* pp. 630–31; Y 15235, 20 September 1728; Y 11225, 22 February 1738, and Arsenal, ms. Bast. 10060, fols. 356–57, 28 September 1761; Y 11221, 15 September 1734; ASP, D5B^6 5290 (1759–60); Y 9539, 12 September 1739. For nearby deliveries of a few loaves, baker Saint-Denis used his eight-year-old niece. Y 15238, 18 June 1731.

71. AN, Y 11233, 1 October 1746; Arsenal, ms. Bast. 10060, fols. 356–57, 25 September 1761; Y 11221, 10 April 1734; ASP, D5B^6 5290, ca. 1759–60, and D5B^6 82; Y 12597, 16 September 1752; Y 11225, 22 February 1738; Y 15941, 30 January 1742.

72. AN, Y 14963, 8 February 1747. Baker Prot's porteuse was less fortunate. The male servant of one of her customers got her pregnant and then abandoned her. Y 13123, 25 April 1771.

73. AN, Y 15063, 22, 25 May 1762; Y 12597, 16 September 1752; ASP, DC^6 242, fol. 227, 29 April 1758 and DC^6 240, fol. 237; Y 15061, 11 August 1760; Arsenal, ms. Bast. 10138, 10 September 1748.

74. AN, Y series, 4 September 1747. Cf. the case of the daughter of a porteuse, a sixteen-year-old, who was coerced into "carnal relations" by the seventeen-year-old son of a master baker for whom she may have delivered bread. Y 15068, 7 April 1765.

75. Police sentence, 22 June 1740, AN, AD XI 10; François Olivier-Martin, *L'Organisation corporative de la France de l'ancien régime* (Paris, 1938), p. 225; AN, Y 15267, 23 November 1759; Y 15235, 23 November 1728; Y 15238, 13 October 1731.

76. BN, Joly 1111, fol. 171; AN, MC, VII-297, 31 July 1755; AN, Y 13746, 17 January 1739; Institut, ms. 521, provisioning registers for 1733.

77. *Encyclopédie méthodique,* Jurisprudence, police et municipalités, vol. 9, p. 438; Charles Desmaze, *Les Métiers de Paris d'après les ordonnances du Châtelet* (Paris, 1874), p. 159; AN, Y 13396, September 1767; Y 15235, 5 September 1728.

78. AN, Y 15240, 23 July 1733; Y 9434, 18 December 1733; Y 9441, 19 February 1740; Y 15244, 11 April 1737; Y 15236, 1 July 1729.

79. AN, Y 15233, 12, 16 July 1726; Y 12605, 21 December 1757.

80. AN, Y 11238, 2 August 1751.

81. AN, Y 9474, 1771; Y 12164, 27 May 1762; Arsenal, ms. Bast. 10112, 22 April 1772.

82. AN, Y 18665, 29 November 1767, and 11 May 1968; ASP, DC⁶ 24, fol. 86, 27 July 1781; DC⁶ 227, fol. 156, 1739; Ganneron to procurator general, 13 June 1769, BN, Joly 1146, fols. 102–3; ASP, D4B⁶ 287, 1 February 1746.

83. AN, Y 12617, 9 September 1765; ASP, D2B⁶ 1079, 22 November 1769; D2B⁶ 842, 4 February 1750; Y 12608, 15 April 1760.

84. AN, Y 12605, 21 December 1757; Hardy's journal, BN, ms. fr. 6681, p. 344 (6 April 1777); ASP, D5B⁶ 2642 and D4B⁶ 92-6324, 11 November 1784; Y 18667, 18 November 1768, and Y 18668, 29 October 1768.

85. "Explication des statuts faits en la confrairie des marchands de bleds," BN, ms. fr. 21635, fol. 118; AN, MC, XX-605, 23 September 1751; MC, XXXVIII-362, 31 December 1747; AN, Y 15353, 30 April 1755; MC, LXV-193, 12 July 1717; MC, VII-291, 1 June 1754; ASP, 6 AZ 1462, 10 January 1765; MC, VII-376, 27 April 1768; Y 15621, 18 December 1730; MC, XXXIV-551, 5 December 1744; MC, XXXIV-501, 16 May 1725; MC, VII-283, 29 June 1752; MC, VII-301, 28 March 1756; MC, XXXV-570, 30 January 1731; MC, CIX-581, 19 March 1751; MC, LXI-449, 14 January 1751; ASP, DC⁶ 234, fols. 211–12, 30 May 1743 and 13 April 1747; MC, CXV-594, 21 November 1751; MC, CIX-490, 12 December 1730; MC, LXXXV-582, 20 September 1764.

Chapter 4   The Forain World

1. Delamare, *Traité,* vol. 2 (1713), p. 888. In order to protect the in-city supply from excessive pressure and to draw upon untapped country supplies, the police regulations forbade forain bakers to buy grain at the Paris Halles and ports. See, for example, the police sentence of 30 March 1635, BN, Joly 1111, fol. 174. Yet it is clear that the police made exceptions. See the purchases of the bakers from Pantin and Vaugirard. AN, Y 13648, 14 August 1752, and ASP, D2B⁶ 800, 31 August 1746. The authorities were especially indulgent in periods of dearth, when grain was extremely scarce even for bakers. "L'Approvisionnement de Paris," Joly 1116, fol. 4, August 1725; "Mémoire," Arsenal, ms. Bast. 10270, p. 313, 1725; Y 11227, 1 June 1740; Y 12619, 17, 21 February 1770.

2. Faubourg Saint-Antoine bakers often referred to themselves as *boulangers forains.* See, for instance, Jacques Terre, AN, Y 15060, 26 December 1759.

3. As I noted earlier, some of the "country" forains, especially close to the Paris "border," also maintained shops. See, for example, Michaux from la Haute Borne, AN, Y 9539, 10 August, 12 September 1760.

4. See, for example, police sentence, 19 January 1769, and arrêt du parlement, 11 April 1769, BHVP fichier.

5. Delamare, *Traité,* vol. 2 (1713), p. 888, and Des Essarts, *Dictionnaire universel,* vol. 2, p. 238. Cf. an echo of this line in the article "Boulanger" in Diderot et al., *Encyclopédie,* vol. 2, pp. 360–61. Yet see an old case in which the police did support the forains against the city bakers. Delamare, *Traité,* vol. 2 (1729), p. 761.

6. See Herlaut, "Disette," pp. 46–47. Cf. Saint-Germain, *Vie quotidienne,* p. 211, and Boislisle, "Grand Hiver," p. 514.

7. Saint-Germain, *Vie quotidienne,* p. 211; Conseil secret, AN, X¹ᴮ 8935, 16 May 1757;

Thery, *Gonesse dans l'histoire*, pp. 91, 93; Le Paige cited by A. Gazier, ed., "La Guerre des farines," *Mémoires de la Société de l'histoire de Paris et de l'Ile-de-France* 6 (1879), p. 11; AN, G⁷ 444 (on the route from Gonesse); Delavergée to Hérault, 31 October 1725, Arsenal, ms. Bast. 10271. In this instance, it appears that the bakers were "rogues," for the governor of Versailles claimed that he had never ordered them to renounce their Paris obligations. Blouin to Hérault, 4 November 1725. Arsenal, ms. Bast. 10271.

8. Arrêt of the Paris Parlement, 10 July 1760, AN, AD XI 14; AN, MC, VII-291, 8 April 1754; MC, VII-178, 23 October 1706. On a list of ninety-five Gonesse bakers drawn up by the Paris police in 1693, there were sixty-four different family names. See Claude Gindin, "Les Principaux Laboureurs de Gonesse à la fin du XVIIe siècle et dans les premières années du XVIIIe siècle." D.E.S. thesis, May 1967, Sorbonne.

9. Thery, *Gonesse dans l'histoire*, p. 83.

10. AN, Z¹ᴳ 291ᴮ, 298ᴮ, 300ᴮ, 310, 326ᴮ, 342ᴮ, 397ᴬ.

11. AN, MC, XLVI-799, 18 May 1716; MC, XXVIII-131, 10 October 1714; MC, LXXXIX-592, 7 July 1760; MC, VII-178, 23 October 1706; MC, XXVIII-217, 15 May 1730; MC, VII-417, 12 September 1775; ASP, D4B⁶ 7-128, 22 April 1788; ASP, 6 AZ 1490, 30 December 1714. Forain daughters also moved toward Paris. Faubourg Saint-Antoine bakers Goiffe, Lambert, and Pignon all married daughters from the Gonesse area. MC, XLI-56, 16 February 1715; MC, XXXVIII-97, 7 May 1711; MC, XXVIII-43, 17 January 1699.

12. I have combined into one market Delamare's grandes halles and his halles de la Tonnellerie.

13. The renewed guild offensive against what it regarded as forain encroachment on its territory may have deterred certain forains from continuing to supply Paris.

14. Yet, if one is to judge by the minutes of the Assembly of Police, the contraction of the baker corps did not cause anxiety among officials responsible for provisioning.

15. Institut, mss. 513–21. The figures used in table 4.5 reflect official tabulations. They are subject to the qualifications I developed above. There were doubtless other country forains whose supply was not officially reported because they were not regular market bakers.

16. See Jean Lebeuf, *Histoire de la ville et de tout le diocèse de Paris* (Paris, 1883), vol. 2, pp. 59–73. Julien Philippe de Gaulle, *Nouvelle Histoire de Paris et de ses environs* (Paris, 1839–42), vol. 5, pp. 153–55. The Hôtel-Dieu of Paris, *engagiste* of the seigneurial domain, controlled several mills and large farms. See, for example, Sommier de l'Hôtel-Dieu, fols. 169–76, AAP, n.s. liasse, 28, no. 1. Cf. also Richard Cobb, *Les Armées révolutionnaires: Instrument de la terreur dans les départements* (Paris, 1961–63), vol. 2, p. 377n.

17. Delamare, *Traité*, vol. 2 (1729), p. 823; Savary des Bruslons, *Dictionnaire portatif* (1761–62), vol. 6, p. 119. *Journal économique*, December 1753, p. 99; Thery, *Gonesse dans l'histoire*, p. 88; AN, Y 12512, 12 August 1752; ASP, D4B⁶ 83-5537, 5 October 1781; Blazy, *Gonesse*, pp. 129–30, 201, 203, 275.

18. For this paragraph see Gindin, "Principaux Laboureurs."

19. AN, Z¹ᴳ 291ᴮ, 298ᴮ, 300ᴮ, 310, 326ᴮ, 397ᴬ.

20. Blazy, *Gonesse*, pp. 208, 215–18, 246–49.

21. Ibid., p. 259.

22. Jean François Paul de Gondi, cardinal de Retz, *Mémoires du Cardinal de Retz*

adressés à Madame du Caumartin, ed. Aimé Louis Champollion-Figeac (Paris, 1912–13), vol. 1, p. 19 (24 February 1649); Mercier, *Tableau,* vol. 12 (1788), p. 83; *L'Ami du Roi,* 19 August 1791; Thery, *Gonesse dans l'histoire,* p. 80; Blazy, citing a late-sixteenth-century source, *Gonesse,* pp. 140–41. He also cites the burlesque comparing "l'absence des pains de Gonesse" to the estrangement of a mistress and other unbearable torments. Ibid., p. 148. According to John Kirkland, in the fifteenth and sixteenth centuries the bread supply of London depended almost entirely on merchandise brought in by the bakers of Stratford (in Essex) in long carts, *Modern Baker,* vol. 1, p. 13.

23. Evelyn, *Panificium,* p. 137. See the recipes here and also in Diderot et al., *Encyclopédie,* vol. 11, p. 749. See also ibid., vol. 7, p. 379, and Expilly, *Dictionnaire géographique,* vol. 3, p. 620. On the bread's international fame, see also Jacques Savary, *Le Parfait Négociant* (Paris, an VIII, 1721), vol. 1, p. 795. According to Alfred Franklin, Gonesse bread was "much prized" as early as the thirteenth century. Franklin, *Dictionnaire historique,* p. 96. Also: Thery, *Gonesse,* p. 86; Martin Lister, *A Journey to Paris in the Year 1698* (1699), ed. Raymond P. Stearns (Urbana, Ill., 1967), p. 148; Olivier de Serres, *Théâtre de l'agriculture,* cited by Expilly, *Dictionnaire geographique,* vol. 3, p. 620; *Dictionnaire de Trévoux,* vol. 6, p. 452; Vigneul de Marville, *Mémoires,* cited by Lebeuf, *Histoire,* vol. 2, p. 270; Collection E. Thomas, series 114, cote provisoire 4804, BHVP; Saverio Manetti, *Traité des Grains,* cited by Malouin, *Description* (ed. Bertrand, 1771), pp. 554, 569. Bertrand maintained that it was widely believed in Germany and Switzerland that "bad water yields bad bread." Ibid. On the centrality of water in bread making see also Estienne and Liébault, *Agriculture.* Cf. the dithyrambic verses of François Colletet, *Le Tracas de Paris, ou la Seconde Partie de la ville de Paris, en vers burlesques* (Paris, 1650), cited by Henri Bunel, "Rapport présenté par M. Bunel sur le carrefour Pirouette," *Commission municipale du Vieux Paris* (procès-verbaux, 1899), no. 10 (7 December 1899), pp. 346–47. A recent discussion of the forain world can be found in Didier Martin, "Les Boulangers de Gonesse et leur rôle dans l'approvisionnement de Paris," mémoire de maîtrise, Université de Paris VII, 1986, p. 25.

24. The argument for middling bread is made by Arpin on the basis of texts from 1788 and 1805. It is hard to believe that such specialists as Malouin, Béguillet, Parmentier, and Malisset would have ignored this practice if it was as salient as Arpin hints. I suspect that certain Gonessiens, converted to economic milling, began to use middlings systematically in the seventies and that this example led a few contemporary observers to assume mistakenly that this technique had always been used. Arpin attributes the decline of Gonesse bread to the introduction of economic milling in the 1770s, which allegedly made middling bread widely available. There is reason, however, to date the beginning of the Gonesse decline much earlier. Arpin, *Historique,* vol. 2, pp. 86–87. See also the inconclusive bread analysis of Geoffroy in the *Histoire de l'Academie royale des sciences: Mémoires, 1732* (Paris, 1735), pp. 28, 30, 47. According to François Cretté de Palluel, "on ne séparait pas les gruaux, ils y entraient en totalité & donnaient au pain cette couleur blanc et jaunatre, & cette saveur exquise de noisette." "Mémoire sur les moyens," pp. 24–25. Didier Martin follows Arpin. "Les Boulangers de Gonesse," p. 29. Without adducing any evidence, Arpin maintained that the Gonessiens kneaded with their legs and feet in order to obtain a firm dough and a tight crumb. Also at a loss for documentation, Martin suggests either kneading with a paddle called a *brie* (albeit he

found no trace of the tool in the Gonesse bakerooms) or by sabot-clad feet. "Les Boulangers de Gonesse," p. 27.

25. Arpin insists that the Parisian copy was totally unlike the Gonesse original, but his reasoning is circular and he adduces no hard evidence to support his conclusion. *Historique*, vol. 2, pp. 87–88. Also: René de Lespinasse, ed., *Les Métiers et corporations de la Ville de Paris*, vol. 1, *Ordonnances générales, métiers d'alimentation* (Paris, 1886), p. 212 (article 15); F. Foiret, "Deux Marchés de 1644 pour la fourniture du pain de Gonesse," *Bulletin de la Société historique du VIe arrondissement de Paris* 24 (1923), p. 85; AN, Y 15241, 21 April 1734; Malouin, *Description* (1761), p. 212. An event in 1783 helped to reinforce the bumpkin image of the Gonessiens. A Montgolfier-type balloon fell at Gonesse about forty-five minutes after launching from the Champs de Mars. Allegedly "the terror-stricken inhabitants promptly destroyed it with pitchforks." Claude Anne Lopez, *Mon Cher Papa Franklin and the Ladies of Paris* (New Haven, Conn., 1966), p. 216.

26. Parmentier, *Parfait Boulanger*, p. 392 (Parmentier's claims are striking, but he did not document them and they remain to be proven); Legrand d'Aussy, *Histoire de la vie privée*, vol. 1, p. 107; Lebeuf, *Histoire*, vol. 2, p. 271. Lebeuf's contention was challenged by Gonesse's modern historian. See Thery, *Gonesse dans l'histoire*, p. 88, and Collection E. Thomas, series 114, cote provisoire 4804, BHVP. On the imitation of Gonesse bread by Parisians, see also Lebeuf, *Histoire*, vol. 4, p. 69.

27. Unable to present a coherent explanation, Blazy points to a number of factors that may have contributed to the decline of the Gonesse bakery at different points in the century: the delayed impact of the impoverishment of the last years of Louis XIV; profound shifts in the grain and flour trade, and in the organization of the Paris supply system (but there is no inherent reason why these changes should not have favored the Gonessiens); the displacement of small and middling laboureurs by the stronger ones; the land-rent "offensive" of the last quarter of the century; the cumulative burden of rising overhead in the last part of the century (especially housing and wood); the movement of bakeries toward the capital (cause or effect?), partly in response to Parisian demand for fresh bread every day. Blazy, *Gonesse*, pp. 309–15, 335.

28. Delamare, *Traité*, vol. 2 (1729), pp. 755–57; BN, Joly 1116, fol. 255; BN, ms. fr. 21636, fol. 355 and ms. fr. 21638, fols. 336–37; A. J. Sylvestre, *Histoire des professions alimentaires dans Paris et ses environs* (Paris, 1853), pp. 63–66. On the early hostility of city bakers to forains, see Desportes, *Le Pain au moyen âge*, pp. 122–23. On the nineteenth-century attitude toward the forains, see *Enquête sur la boulangerie*, p. 115.

29. Delamare, *Traité*, vol. 2 (1729), pp. 726–27; Sylvestre, *Histoire des professions alimentaires*, p. 74.

30. Mme de *** to M. de Mopinot, 24 Septembre 1757, in J. Lemoisie, ed., "Sous Louis le Bienaimé," *Revue de Paris* 4 (1 July 1905), p. 160.

31. AN, X¹ᴮ 3685, 20 September 1757; X¹ᴮ 3732, 10 July 1760, also in AN, AD XI 14 (for this and the following paragraphs); BN, Joly 1111, fol. 171. In a luminous article, Carlo Poni reminds us that forain-city quarrels characterized many other trades. See the efforts of the Bolognese master shoemakers to restrict the rights of production and the conditions of sale of their rural counterparts. "Norms and Disputes: The Shoemakers' Guild in Eighteenth-Century Bologna," typescript, pp. 5–17.

32. There is some evidence that faubouriens and nearby forains were selling *pain*

mollet both before and after the parlementary decree. Lepage of the rue de Reuilly was cited before the chamber of police in 1742 for price and weight infractions, but nothing was said about the fact that his bread was mollet. AN, Y 9443, 13 April 1742. In August of 1760, the jurés found only one- and two-pound soft-dough breads in the shop of Michaux, a forain at the Haute Borne. Y 9539, 12 September 1760.

33. In the seventeenth century the guild had apparently been willing to tolerate forain home deliveries, provided the bread was covered with a cloth and not "cried out" in the streets. Delamare, *Traité*, vol. 2 (1713), p. 842.

34. Delamare, *Traité*, vol. 2 (1729), p. 759. Eighteenth-century experts concurred with the Delamarist view. If the forains could bake fantasy loaves, Malouin observed, no one would bake a bread for the people. *Description* (1761), pp. 263–64. Malouin's Swiss editor noted that this problem was pan-European: city bakers as a rule scorned the task of perfecting a common bread. Malouin, *Description* (ed. Bertrand, 1771), p. 149n.

35. Albert to Sirebeau, 19 March 1776, AN, Y 15606; Y 15853[A]; AN, X[1B] 8983, 13 December 1786. In the nineteenth century, the police were reluctant to authorize forain commerce on the grounds that it was fundamentally a form of regrating. Prefect of police to sous-prefect of Sceaux, ASP, Dm[6] 2, 25 January and 13 March 1822.

Chapter 5 Bread on Credit

1. Abbé Ferdinando Galiani, *Dialogues sur le commerce*, ed. Nicolini, pp. 180–81. Cf. the testimony of a specialist on the bread question in 1859: "Il y a pour le boulanger un très grand écueil, c'est le crédit qu'il peut faire aux consommateurs nécessiteux. A Paris, les pertes sur les crédits doivent être énormes. Il y a tant de facilité pour les mauvais débiteurs de changer de quartier et d'échapper aux réclamations que les boulangers de Paris doivent faire beaucoup de pertes." *Enquête sur la boulangerie*, p. 434.

2. AN, Y 15961, 24, 26 October 1764. Cf. the insightful remarks of Crossick and Haupt on the ambiguity of the daily solidarities between shopkeepers and the working class in the nineteenth century: the shopkeepers served as "the bankers of the poor," yet they took advantage of the dependence of their clients in numerous ways. Geoffroy Crossick and Heinz-Gerhard Haupt, *Shopkeepers and Master Artisans in Nineteenth-Century Europe* (New York, 1984), pp. 18–19.

3. AN, MC, CXI-92, 24 March 1719. Le Pelletier served as controller general from 1726 to 1730. He owed Rousseau 5,069 livres in 1719.

4. AN, Y 15932, 29 April 1732; ASP, D4B[6] 7395, 4 February 1789. Cf. Mercier's warning to the bakers to offer credit only to consumers with fixed residences. Mercier, *Tableau*, vol. 10 (1788), p. 354. I have found no example in the eighteenth century of the shame evinced by the carpenter's wife who had no choice but to accept the baker's offer of credit when her husband suffered an incapacitating accident at work: "Elle ne comprenait plus rien. Elle se sentait lasse, et au lieu d'être satisfaite, elle sentait comme une atroce blessure qui s'ouvrait. 'Le crédit, c'est la perte de la dignité!' Elle avait menti en prétextant le besoin de sortir. Elle cacherait l'achat du pain à crédit, à son mari. Jusqu'où irait-elle de déchéance en déchéance?" Henry Poulaille's "populist" novel set around

1900, *Le Pain quotidien* (Paris, 1934), cited by Benigno Cacérès, *Si le pain m'était conté* (Paris, 1987), p. 143.

5. AN, MC, LXI-601, 26 June 1784; MC, XXVIII-216, 16 March 1730; Mercier, *Tableau*, vol. 4 (1783), p. 204.

6. Mercier, *Tableau*, vol. 7 (1784), p. 114. Down and out in Paris, the writer Marmontel never lost hope, because the baker never withdrew his credit. Jean-François Marmontel, *Mémoires de Marmontel*, ed. Maurice Tourneux (Paris, 1891), vol. 1, pp. 149–51. Cf. Maître Chauveau-Lagarde, *Mémoire ampliatif pour les boulangers* (Paris, 1818), p. 46. See the dialogue between two fathers charged with large families in Lebois, *Lettre des boulangers*, p. 5, in the Cornell University Library:

C. Hélas! Je viens de chercher du pain chez mon boulanger, & je n'ai pas pu en avoir.

G. Mais il y a d'autres boulangers dans Paris.

C. Sans doute, mais je n'ai pas crédit ailleurs; je ne sais comment faire.

7. *Enquête sur la boulangerie*, pp. 45, 62, 122, 385, 722–23. The former mayor of Rouen, by contrast, condemned baker credit as "a cause of demoralization." When workers "did not pay for their bread in cash, they took their money to the tavern, where they spent it on debauchery." Ibid., p. 293. See also the complaints of medieval bakers against the pressure to extend credit. A. Maurizio, *Histoire de l'alimentation*, p. 501. Yet see the desperation imputed to early-nineteenth-century consumers should their bread credit be cut off: "Or, il est impossible de calculer le degré de misère où la majeure partie de la population serait réduite, si demain les Boulangers supprimaient les tailles et toute espèce de crédit." Chauveau-Lagarde, *Mémoire ampliatif*, p. 46.

8. BN, ms. fr. 6680, 1 May 1771; Nicolas Restif de la Bretonne, *Les Nuits de Paris*, in *L'Oeuvre de Restif de la Bretonne*, ed. Henri Bachelin (Paris, 1930), vol. 1, p. 103. See the reflections on the *dettes actives* in AN, MC, XXVIII-128, 13 April 1714, and MC, VII-291, 9 May 1754. Of course, from a strictly business perspective, selling to insolvent customers would quite rightly be considered—as in the case of master baker Joachim Decaut—as proof "of the derangement of the master baker." AN, Y 13386.

9. AN, Y 13926, 31 May 1741; MC, V-539, 28 July 1764; ASP, D5B⁶ 5290; Y 15241, 20 November 1731; D5B⁶ 87, 1753. See Sartine to the commissaires, 24 June 1771, Y 15144.

10. ASP, D5B⁶ 4199; AN, Y 13644, 4 July 1749; D5B⁶ 3416; D5B⁶ 1968.

11. AN, MC, XIX-589, 17 January 1709. Master Jacob sued an innkeeper for 664 livres for bread supplied on the basis of a purely verbal record—he had no documentation. ASP, D2B⁶ 736, 26 April 1741.

12. ASP, D4B⁶ 90-6211, 14 September 1787; AN, MC, XCIV-310, 22 May 1762, and D5B⁶ 4289, 10 August 1725. The taille survived well into the nineteenth century in Paris and to the Second World War in the countryside. Chauveau-Lagarde, *Mémoire pour les boulangers*, p. 46; J.-M. Simon et al., "La Taille du boulanger," *Bulletin folklorique de l'Ile-de-France*, 4th ser., 31 (1969), pp. 115–16; sample tailles, Archives du Musée national des arts et traditions populaires. London bakers used a "tayle," but it is not clear how long this practice survived. Thrupp, *Worshipful Company*, p. 137.

13. AN, MC, LXI-492, 3 September 1761; ASP, D4B⁶ 50-3079, 20 February 1774; D4B⁶ 43-2435, 24 January 1772 (master baker Robert), and D4B⁶ 11-526 (master Nicolas Guenee). Baker Desfosses could not sign at all. D4B⁶ 21-1075, 3 July 1760; AN, Y 9441,

14 October 1740. D5B⁶ 3385, 3006, and 3106; MC, CVIII-494, 19 July 1751; Y 11045, 9 March 1730. Cf. Mercier, *Tableau*, vol. 12 (1788), p. 147.

14. AN, Y 12594, 25 November 1750; Y 12138, 29 February 1736; ASP, D5B⁶ 1003.

15. Models for effective bookkeeping abounded. See Pierre Bernard d'Henouville, *Le Guide des comptables, ou l'Art de rédiger soi-même toutes sortes de comptes* (Paris, 1709); Claude Perrache, *Nouvelle Méthode de tenir les livres marchands d'une manière très sûre et très facile* (Brussels, 1748); Savary, *Parfait Négociant*, vol. 1, p. 288.

16. ASP, D5B⁶ 4289, 87, 2532, 5347.

17. ASP, D5B⁶ 87; AN, MC, VII-291, 28 May 1754; Y 11233, 1 August 1746; MC, VII-291, 9 May 1754; Y 16396, 5 March 1779.

18. ASP, D5B⁶ 3338; AN, MC, XXX-437, 24 April 1773; D5B⁶ 2619 and 4133.

19. AN, MC, XCIV-310, 22 May 1762; ASP, D5B⁶ 87, 3009; AN, Y 15961, 24, 26 October 1764; D5B⁶ 5347, 20 April and 9 June 1754. A customer owing baker Lapareillé seventy-eight livres also agreed to repay at six livres a month. Ibid., D5B⁶ 2611, 1 August 1786.

20. ASP, D5B⁶ 4289, 10 August 1725; D5B⁶ 5437, 1753.

21. AN, Y 13155, 23 January 1763 (he owed 590 livres and waited two years); MC, XVIII-556, 7 July 1741 (master baker Fouquet).

22. AN, MC, XCIV-297, 5 May 1760; MC, LXV-227, 2 March 1730; ASP, D5B⁶ 1829; MC, CXXI-292, 4 June 1731; Y 9535, 25 November 1775.

23. ASP, D5B⁶ 4449, 10 November 1756; D5B⁶ 2618, 2619; AN, Y 11170, 29 December 1752–12 January 1753.

24. ASP, D5B⁶ 3010, 2712, 3760, 3416, 3329, 1061. It is true that these debts covered the late sixties and early seventies, years of economic disarray and acute subsistence difficulties.

25. ASP, D5B⁶ 4837; AN, Y 13397, 26 October 1769.

26. ASP, D5B⁶ 4288, 4527, 1829.

27. AN, Y 13905, 31 May 1726; Y 12141, 22 July 1739; Y 12741, 28, 30 December 1755; Y 12605, 27 August 1757; and Arsenal, ms. Bast. 10153, January 1728; curé to Hérault, 2 December 1728, ms. Bast. 10274 and Y 15254, 18 February 1746; ms. Bast. 10153, January 1728.

28. AN, Y 12141, 22 July 1739.

29. AN, Y 13905, 31 May 1726; Y 12605, 27 August 1757.

30. AN, Y 15622, 14 February 1751; Y 12595, 14 December 1751; MC, VII-288, 14, 15 September 1753.

31. Guerin's parish priest endorsed his petition. We do not know Hérault's response. Ms. Bast. 10151, May 1727; AN, MC, XXIII-506, 12 July 1739.

32. AN, Y 14094, 18 May 1767.

33. *Encyclopédie méthodique*, Jurisprudence, vol. 2, p. 94; Jean-Baptiste Denisart, *Collection de décisions nouvelles et de notions relatives à la jurisprudence actuelle*, 9th ed. (Paris, 1777), vol. 1, p. 271, Savary, *Parfait négociant*, vol. 1, p. 673; Dantan to Voltaire, 8 March 1768 in *Voltaire's Correspondence*, vol. 68, p. 211 (no. 13890); Morel, *Histoire illustrée*, p. 265; AN, Y 11311, 22 March 1734.

34. AN, MC, XXVIII-216, 16 March 1730; Y 15246, 9 February 1739; ASP, D4B⁶ 48-2915, 20 August 1773.

35. AN, Y 7488, 22 August 1759; procès-verbal, 26 November 1766, in ASP, D5B⁶ 3337; Y 7488, 18 July 1759 (Chaveau v. Dernic), and 22 August 1759 (Avez v. Bordelet).

36. ASP, D2B⁶ 1019, 12 November 1764; D2B⁶ 801, 26 September 1746; D2B⁶ 842, 4 February 1750; D2B⁶ 841, 16 January 1750.

37. ASP, D2B⁶ 809, 19 May 1747, and D2B⁶ 801, 7 September 1746. Many fines ranged between one hundred and two hundred livres. For instance, D2B⁶ 841, 16 January 1750 (Boucher v. Morland); D2B⁶ 950, 5 February 1759 (Lapareillé); D2B⁶ 1075, 2 December 1769 (Lecuit v. Bonhomme).

38. ASP, D2B⁶ 736, 26 April 1741 (Jacob v. Rolland) and 28 April 1741 (Forget v. Sardan); D2B⁶ 762, 14 June 1743 (Picard v. Masset); D2B⁶ 800, 19 August 1746 (Jamard v. Clement); D2B⁶ 801, 7 September 1746 (Moisson v. Derambe); D2B⁶ 802, 14 October 1746 (Cadoret v. Picard); D2B⁶ 809, 19 May 1747 (Gregoire v. Leclerc) and 31 May 1747 (Deline v. Crepy); D2B⁶ 842, 6 February 1750 (Guerbois v. Dolley) and 9 February 1750 (Guerbois v. Belleville); D3B⁶ 51, 17 February 1750 (Chevalier v. Grumeau); D2B⁶, 26 February 1762 (Cousin v. Lecomte); D2B⁶ 1080, 29 December 1769 (Crette v. Ray); D2B⁶ 1101, 20 September 1771 (Refroigniet v. Brioux); D2B⁶ 1124, 13 August 1773 (Joubert v. Nicolas); D2B⁶ 1126, 8 October 1773 (Lemerle v. Lebeau); D2B⁶ 1128, 31 December 1773 (Garnier v. Dufresne); and AN, Y 11369, 27 May 1768 (Lesimple).

39. BN, Joly 2324, fol. 283 (Giraud).

40. For an example of Grezel's tenacious style, see AN, MC, VII-402, 21 February 1772.

41. AN, Y 14951, 31 October 1736; Arsenal, ms. Bast. 10152, January 1728; Y 11369, 27 May 1768; Y 7488, 1 August 1759.

42. See below, chapter 11, and Kaplan, *Provisioning Paris.*

Chapter 6   The Guild

1. René de Lespinasse and François Bonnordot, eds., *Le Livre des métiers d'Etienne Boileau* (Paris, 1879), pp. xx–25; Lespinasse, *Métiers de Paris: Ordonnances générales*, pp. 195–228.

2. Abbé Annibal Antonini, *Mémorial de Paris et de ses environs*, ed. Abbé G. Raynal (Paris, 1749), vol. 2; Diderot et al., *Encyclopédie*, vol. 2, p. 360, and vol. 9, p. 817; Jean Dutillet, *Recueil des Roys de France* (Paris, 1602), pp. 286–88; Delamare, *Traité*, vol. 2 (1713), pp. 846–47, 851–55; BN, ms. fr. 21639, fols. 30–31; Delamare, *Traité*, vol. 2 (1729), pp. 1725–34. The prévôt des marchands also disputed the grand pannetier's jurisdictional claims.

3. Delamare, *Traité*, vol. 2 (1713), p. 847. Cf. the somewhat quixotic assessment in "Histoire des Boulangers," *Magasin pittoresque* 25th year (1857), p. 137, and the muddled treatment by Pierre Vinçard, *Les Ouvriers de Paris: Alimentation* (Paris, 1863), pp. 21–24.

4. Delamare, *Traité*, vol. 2 (1713), pp. 858–60. Apparently the faubourgs Saint-Denis and Saint-Martin also remained "free" through most of the seventeenth century, because the police wanted to encourage the implantation of new bakers and they found that the faubourg guilds no less than the city corporation tended to inhibit new establishments. Thus the largest faubourg, Saint-Germain, had only seventy bakers, while

there were over five hundred in the faubourgs Saint-Antoine, Saint-Martin and Saint-Denis in the early seventeenth century. Delamare, *Traité,* vol. 2 (1729), pp. 736–37.

5. See the *règlement* of 9 March 1681, AN, MC, XLIII-176, and the statutes and regulations of the baker guild of the faubourg Saint-Germain, remarkably similar to the statutes of the city masters, AN, K 1030–31, 23 May 1658.

6. Delamare, *Traité,* vol. 2 (1713), pp. 861–63.

7. Delamare, *Traité,* vol. 2 (1729), pp. 736–39.

8. Delamare, *Traité,* vol. 2 (1713), pp. 863–73; AN, K 1030–31, no. 19; BN, ms. fr. 21639, fols. 8–9. On the fiscal aspects of the edicts of 1673 and 1678, see "Histoire des Boulangers," p. 134.

9. These fiscal motives are explicitly stated in the "reunion" edict of August 1711, BHVP.

10. They were invited by the edict to set up shop anyplace they wished, but the guild elders vigorously discouraged attempts by new masters to establish themselves at locations judged too close to the old masters.

11. Arrêt du conseil, 28 August 1716, BHVP.

12. Arrêts du conseil, 20 January 1719 and 3 April 1719, BHVP.

13. The statutes are in AN, AD XI 14. The parlement did not register them until 4 April 1721. Apparently the examination by the court was not a merely formal and casual manner. Rather, the long period of examination between submission and approval seems to have opened an arena for lobbying by various special-interest groups, public and private (to the extent that one can sustain this distinction). The parlement made a number of modifications in the text. For instance, it suppressed article 47 enabling Paris-licensed apprentices to be received as masters in all other cities of the realm without paying any fees beyond those required by letters of reception. AN, X[1B] 8902, 4 April 1721. See also Lespinasse's emphasis on the similarity between these statutes and those of the thirteenth century. *Métiers de Paris: Ordonnances générales,* p. xxii. This line is reiterated by the Syndicat de la boulangerie de Paris, *Histoire* (Paris, 1900), p. 5. There are also remarkable similarities between these statutes and those of the old faubourg Saint-Germain guild. See 23 May 1658, AN, K 1030–31.

14. For a sample of elections see AN, Y 9326, 7 October 1745; Y 9327, 8 October 1750, 5 October 1752, 8 October 1754; Y 9329, 16 October 1755, 7 October 1756, 6 October 1757, 7 October 1758, 15 October 1759; Y 9330, 16 October 1760, 8 October 1761, 7 October 1762, 6 October 1763, 11 October 1764, 7 October 1765; Y 9389, 11 October 1766; Y 9390, 8 October 1767, 6 October 1768; Y 9391, 5 October 1769, 4 October 1770, 15 October 1772; Y 9333, 8 October 1777, 6 October 1778, 21 October 1782; Y 9334, 23 October 1784.

15. AN, MC, VII-301, 28 March 1756. On the oligarchic proclivities of a contemporary (Belgian) guild, see Anne Godfroid, "La Corporation des boulangers à Namur au 18e siècle," mémoire de licencié en histoire at the Université Libre de Bruxelles, 1992–93, pp. 43–44, kindly communicated by Claude Macherel.

16. See, for example, Deschard, named as *arbitre,* 21 November 1746, ASP, D6B[6] article 3.

17. AN, Y 15359, 25 March 1758; Y 15616, 7 September 1748.

18. On the exercise of police powers by the corporation, see Olivier-Martin, *Organisation corporative,* pp. 179, 322, 448–49. The juré visits were remarkably thorough. See, for example, the itinerary of Commissaire Courcy and jurés Claude Lapareillé and

Charles Grand on 5 January 1751: all the stalls in five marketplaces, scores of bakeshops, and all the inns and wineshops on the route. AN, Y 11238. On the mediating functions of the jurés, see, for example, ASP, D2B⁶ 7321, 3 February 1741 (plumitif).

19. On the police of the forains and illicit bakers, see chapters 4 and 19 herein. On similar tensions between city and country bakers in London, see Thrupp, *Worshipful Company,* pp. 6, 55–56, and Sheppard and Newton, *Story of Bread,* pp. 37–39. Brittany witnessed the same phenomenon. Jean Letaconnoux, *Les Subsistances et le commerce des grains en Bretagne au XVIIIe siècle* (Rennes, 1909), p. 105. On the starch makers, for example, see the inspection rounds by jurés Gibert and Hubert, accompanied by Commissaire Regnard le jeune, during which they demolished, on the spot, eight illegal ovens. AN, Y 15233, 20 July 1726.

20. See Gazier, "La Police," p. 125, as well as my treatment of bread quality and short weight below, chapters 17 and 19. The jurés also made sure that the bakers did not bake fancy breads when they were banned in periods of dearth. See, for example, AN, Y 9441, Y 9441, 14 October 1740. For examples of illegal traffic in grain and flour, see Y 15246, 12 December 1739; Y 9441, 19 February 1740; Y 11589, 17 March 1772.

21. AN, Y 15238, 19 November 1731; Y 15233, 27 May and 12 July 1726. Cf. BN, Joly de Fleury, 1111, fol. 173, and Kaplan, *Provisioning Paris.*

22. AN, Y 9440, 11 September 1739. See also Y 9441, 9 February 1740, and Y 15233, 19 July 1726. See the complaint of two flour merchants who claimed that jurés "ruined" them by seizing most of their stock. Y 11226, 13 August 1729.

23. See AN, Y 9385, 29 May 1753.

24. Inventaire et recollement des effets, 11 October 1749, AN, MC, XXXIV-576; MC, VII-285, 14 October 1752; MC, VII-328, 22 October 1760; MC, VII-335, 10 October 1761; MC, VII-378, 7 October 1769. Neither the physical arrangement of the offices nor the furnishings changed much when the guild moved in the early fifties to the quai de Conti. AN, Y 14995, 12 March 1776.

25. Delamare, *Traité,* vol. 2 (1729), pp. 734–45; letters patent, 8 October 1439, BN, ms. fr. 21639 fols. 67–69; *Journal économique,* January 1754, pp. 96–97, Vinçard, *Alimentation,* pp. 62–64; Henri d'Almers, "Le Pain de Paris," *Magasin pittoresque,* 65th year, 2d ser., 15 (1897), p. 315; Pierre Desnoyers, "St. Firmin, patron des boulangers de l'Orléanais," *Mémoires de la Société archéologique et historique de l'Orléanais* 27 (1898), p. 159; Charles Cahier, *Caractéristiques des saints dans l'art populaire* (Paris, 1867), vol. 1, p. 144, and vol. 2, pp. 597, 641, 681; Arthur Forgeais, *Collection des plombs historiés trouvés dans la Seine* (Paris, 1862). On the great range and vitality of confraternal life in the Middle Ages, consult Desportes *Le Pain au moyen âge,* pp. 189–95; Catherine Vincent, *Les Confréries mediévales dans le royaume de France: XIIIe–XVe siècle* (Paris, 1994). On the association of the bakers and Saint Honoré in the nineteenth century, see C. D. Félix, *Saint-Eustache et la Chaumière, pot-pourri en deux parties à l'occasion de la Saint-Honoré en 1823* (Paris, 1823).

26. AN, Y 9379, 20 May 1738; Delamare, *Traité,* vol. 2 (1729), pp. 734–35; Jurés versus Mongueret, Y 13494, 12 September 1721; quittance in inventaire, AN, MC, CI-540, 7 October 1765. In the seventeenth century there were more than two *confréries,* for each of the faubourg guilds had its own pious association. Jean-Baptiste Le Masson, *Le Calendrier des confréries de Paris,* ed. Valentin Dufour (Paris, 1875). The guild of the faubourgs Saint-Michel and Saint-Jacques funded their *confrérie* in part by attribution

to it of one-half of the revenue it collected in fines and fees levied on masters. Règlement, 9 March 1681, AN, MC, XLIII-176.

27. AN, V⁷ 423, 10 June 1753. The bakers appear to have named an "administrateur juré de la confrairie du Saint Honoré" in 1736 who was not concurrently the juré-accountant of the guild. AN, Y 13745, 26 June 1736.

28. See, for example, AN, Y 9379, 13, 20 May 1738; Y 9379, 15 May 1739; Y 9442, 17 November 1741; Y 9368, 14 November 1748; Y 9385, 29 May, 14 December 1753; AN, MC, VII-328, 22 October 1760, and MC, VII-335, 10 October 1761. See also the mention of the dues obligation in a *cession* of the business, MC, XXIV-501, 17 June 1725. There is no mention of *confrérie*(s) in the statutes and regulations of 1785, AN, AD XI 14. Cf. the similar connections and tensions between the guild of masons and its confraternity in J.-J. Letrait, "La Communauté des maîtres maçons de Paris au XVIIe et XVIIIe siècles," *Revue historique de droit français et étranger* 26 (1948), p. 125.

29. The family of master Pierre Noel Bossu borrowed the corporate silver and funeral pall, which enabled them to provide a proper burial for under a hundred livres. AN, MC, CXXII-682, 5 November 1751. See also MC, VII-325, 5 February 1760, in which a baker family rejected these trappings in order to bury their loved ones "in the greatest modesty." See the suit initiated by elder and doyen Claude Larticle to force the guild to pay for a mass in one of the confraternal churches in memory of his wife. AN, Y 9378, 7 November 1742.

30. AN, Y 9387, 17 December 1762. Cf. the more elaborate social services offered by the baker guild at Bordeaux. Benzacar, *Le Pain à Bordeaux,* p. 84.

31. On the capitation tax, see Marcel Marion, *Dictionnaire des institutions* (Paris, 1923), pp. 90–91, and Mercier, *Tableau,* vol. 11 (1783), p. 97, and vol. 3, pp. 84–98.

32. See, for example, the opposition to the removal of seals in AN, Y 12729, 1 March 1735 (Goujet), and Y 11673, October 1749 (Bellotte).

33. See, for instance, the case of Claude Millot, a master who refused to return an apprenticeship license despite repeated convocations. AN, Y 15356, 22 March 1756.

34. AN, Y 11228, 14 December 1741. See "the last and final warning" issued to master Petitot in 1776. His imposition was based on an assessment drawn up by Lieutenant General Berryer in 1750. BHVP, series 118, c.p. 1482. See also Marion, *Dictionnaire des institutions,* pp. 556–57.

35. Police sentence, 17 July 1744, AN, AD XI 14; arrêt, 5 January 1746, AD XI 14.

36. Baker petition, BN, Joly 400, fol. 967.

37. AN, MC, VII-285, 11 December 1752; MC, VII-292, 30 July, 15 November 1756; AN, X¹ᴮ 8935, 6 September 1757; MC, VII-303, 30 July 1756, and MC, VII-305, 15 November 1756.

38. Letters patent of 1 April 1783, registered in parlement on 13 December 1785. AN, AD XI 14 and AN, X¹ᴮ 8983, 13 December 1785.

39. See Lespinasse, *Métiers de Paris: Ordonnances générales,* p. 196n. On the general question of fiscality and the guild, see Etienne Martin Saint-Léon, *Histoire des corporations de métiers* (New York, 1975), p. 545. On the salience of financial issues in another baker's guild, see Godfroid, "La Corporation des boulangers à Namur," pp. 55ff.

40. See the *quittances* in the guild papers, AN, MC, VII-285, 14 October 1752, and MC, XXXIV-576, 11 October 1749; AN, Y 13494, 28 May 1729. This was probably the

droit de confirmation des privilèges that was traditionally levied on the guild at the ascension of a new monarch. See Lespinasse, *Métiers de Paris: Ordonnances générales,* pp. 148–50.

41. Lespinasse, *Métiers de Paris: Ordonnances générales,* p. 159; AN, MC, VII-285, 14 October 1752. See the suits against masters Garin, Planquet, Fontaine, and Leroux in the royal procurator's chamber. AN, Y 9385, 14 December 1752.

42. Edict of August 1758, BHVP. The tax took the form of an "augmentation de gages" on the previously purchased offices. The preamble struck a sharply modern chord: this money was not meant to prosecute an ugly war but to assure "a glorious and solid peace." See also the subscription of rentes, 23 August 1759. Three of the masters advanced eight thousand livres and two others four thousand livres apiece.

43. Conseil secret, 15, 17, 19 June 1767, AN, X^{1B} 8954.

44. AN, V^7 423, 28 January 1720.

45. Edict of 20 January 1719, BHVP.

46. Arrêts du conseil, 5 April and 3 May 1729, BHVP.

47. Arrêt du conseil, 28 January 1731, AN, AD XI 14 and AD XI 38.

48. Arrêt du conseil, 5 April 1731, AN, AD XI 14.

49. AN, V^7 423, 12 March 1732. Cf. the arrêt du conseil of 28 March 1730, which enjoined all corporate officers, not just baker-jurés, to submit all their financial records on payment of a fine of one thousand livres. AN, AD XI 10.

50. AN, V^7 423, 14 August 1737.

51. AN, AD XI 14.

52. Lespinasse, *Métiers de Paris: Ordonnances générales,* pp. 147–48.

53. See the "état des sommes que les corporations et communautés d'arts et métiers de Paris pourront employer dans leurs comptes pour la dépense des Tedeum" (22 June 1756), which fixes the baker quota at three hundred lives, considerably below the wine merchants' quota (1,000 livres) but far above that of the shoemakers (50 livres), the masons (60 livres), and the glove makers (150 livres). This listing represents an official view of the hierarchy of corporate wealth. APP, Collection Lamoignon (hereafter Lamoignon), XL, fols. 418–22.

54. Nigeon made this unwarranted assumption and naturally found that the financial status of most of the guilds was "quite good." Even if the former jurés ultimately paid their debts, those payments were generally not completed until ten or twenty years after the fiscal year in question. Thus it is very difficult to determine what real surplus the guild might have disposed of in any given year. Moreover, Nigeon made numerous errors in calculating the surplus, in some cases because he failed to verify the arithmetic of the commissioners. René Nigeon, *Etat financier des corporations parisiennes d'arts et métiers au XVIIIe siècle* (Paris, 1934), pp. 73–79.

55. Total projected surpluses were 46 percent higher in the period 1762–75 than in 1744–61.

56. Arrêt du conseil, 21 January 1749, APP, Lamoignon XXXVIII, fol. 574; AN, V^7 423. It is not clear what impact the comprehensive reorganization of 1775–76 had on these procedures. The matter was quantitatively far less serious than before, because corporate expenses and revenues were both drastically reduced. The new statutes emphasized fiscal responsibility and built many checks into the administrative process. Yet

on the eve of the Revolution the police and the parlement complained about the irresponsibility of the syndics. Police sentence, 5 April 1788, and arrêt du parlement, 5 May 1788, BHVP.

57. AN, V⁷ 423.

58. AN, Y 9380, 27 May 1742, 1 June and 6 July 1743; AN, MC, LXV-227, 2 March 1730.

59. AN, Y 11222, 27 January 1735; Y 11228, 9 December 1741; Y 11220, 2 September 1733; Y 15340, 22 March 1744.

60. AN, Y 9379, 5 July 1735; Y 9440, 18 December 1739, and Y 9443, 26 January 1742; Y 15603, 18 May 1742.

61. AN, Y 13640, 8, 11 March 1745; Y 12622, 22 April 1772. Insisting on the solemnity and dignity demanded at executive meetings, a dozen elders expressed their shock at this "unheard-of indecency and insolence," regardless of the motive.

62. See, for example, AN, Y 9441, 29 April 1740; Y 9443, 19 January, 7 April, 20 July 1742.

63. AN, Y 9440, 4 September 1739 (for a similar case see Y 9443, 17 August 1742); Arsenal, ms. Bast. 10150, August 1726.

64. AN, Y 15236, 9 June 1729; Y 13742, 3 July 1730; AN, MC, LXV-227, 2 March 1730.

65. See, for instance, BN, ms. fr. 11356, fol. 140 (14 December 1730). I have found no examples of controversies provoked by banquets for the elders held at the expense of the guild as commonly occurred in the seventeenth century. Cf. ms. fr. 21639, fols. 132–33. In 1789 certain masters accused the police of "maneuvering" to cause divisions in the guild in order to weaken it. Jean-Jacques Rutledge, *Second mémoire pour les maîtres boulangers, lu au bureau des subsistances de l'Assemblée nationale, par le chevalier Rutledge* (Paris, 1789), p. 41. On the general issue of dissension in the corporations, see Steven L. Kaplan, "The Character and Implications of Strife among Masters inside the Guilds of Eighteenth-Century Paris," *Journal of Social History* 19 (summer 1986), pp. 631–48. See also Levasseur, *Histoire des classes ouvrières*, vol. 2, p. 467.

66. AN, Y 15397, 19 September 1785.

67. On the attack on the guild see Francis Veron de Forbonnais, *Recherches et considé-rations sur les finances de France, depuis l'année 1595 jusqu'à l'année 1721* (Liege, 1758), vol. 3, pp. 110–16; Faiguet de Villeneuve, "Maîtrises," in Diderot et al., *Encyclopédie*, vol. 9, p. 915; *Journal de l'agriculture, du commerce, des arts et des finances* 7 (November 1766), pp. 25–26; G.-F. Coyer, *Chinki, Histoire cochinchinoise* (London, 1768); Simon Clicquot de Blervache *Considérations sur le commerce et en particulier sur les compagnies, sociétés et maîtrises* (Amsterdam, 1758); Bigot de Sainte-Croix, *Essai sur la liberté du commerce et de l'industrie* (Paris, 1775); Emile Coornaert, *Les Corporations en France avant 1789* (Paris, 1941), pp. 172–78.

68. Baudeau, *Avis au premier besoin*, 3e traité, pp. 91–106; Turgot to intendant, 24 July 1775, AD Orne, C 90; Turgot, "Septième Lettre sur le commerce de grains," in *Oeuvres de Turgot*, vol. 3, pp. 350–51 (2 December 1770); Lacombe d'Avignon, *Le Mitron de Vaugirard*, pp. 43–44; *Le Babillard* 36 (30 June 1778), p. 188; Marie Jean Antoine Nicolas Caritat, marquis de Condorcet, *Le Monopole et le monopoleur* (n.d.), in vol. 14 of *Collection des principaux économistes*, ed. Eugène Daire and G. de Molinari (1847; reprint

Osnabrück, 1966), p. 452n; Fontette, "Observations sur le projet de loi," October 1770, AD Calvados, C 25223; Georges Weulersse, *La Physiocratie sous les ministères de Turgot et de Necker (1774–1781)* (Paris, 1950), p. 19.

69. Assembly of Police, 14 December 1730, BN, ms. fr. 11356, fol. 140; Vantroud to Marville, 13 November 1742, Arsenal ms. Bast. 10012.

70. The guild itself, of course, had to be policed. The authorities had no illusions about its devotion to the public interest if left entirely to its own devices. See Delamare, *Traité,* vol. 2 (1713), pp. 846–47.

71. Among the writers commissioned or inspired to defend a particular guild or the corporate system in general, see Linguet, *Réflexions des six corps de la ville de Paris sur la suppression des jurandes* (Paris, 1776); André Lethinois, *Apologie du système de Colbert* (Amsterdam, 1771); *Ephémérides du Citoyen* 7 (1771), pp. 150–71; A.-L. Séguier, *Discours sur la nécessité du rétablissement des maîtrises et des corporations,* speech delivered on 12 March 1776 (Paris, 1815). Wrote Abbé Galiani: "Concerning the suppression of the *jurandes,* I say it straight to the faces of all the fashionable reasoners and of all the *économistes:* it is a stupidity, an error, an absurdity. They do not know men." Galiani to Madame d'Epinay, 13 April 1776, in Abbé Ferdinando Galiani, *Lettres de l'abbé Galiani à Madame d'Epinay,* ed. Eugène Asse (Paris, 1881–1903), vol. 2, p. 222.

72. "Le Prix du pain," AN, F[11] 265, September 1774; Parmentier, *Parfait Boulanger,* pp. 611–12.

73. "Le Prix du pain," AN, F[11] 265, September 1774.

Chapter 7 From Apprentice to Journeyman

1. On apprenticeship, see Steven L. Kaplan, "L'Apprentissage au XVIIIe siècle: Le Cas de Paris," *Revue d'histoire moderne et contemporaine* 90 (July–September 1993), pp. 436–79.

2. Nicolas Delamare, *Traité,* vol. 2 (1713), pp. 841, 845. On the quasi-familial vocation of apprenticeship, see its portrayal as a stage between childhood and adolescence in England. S. R. Smith, "The London Apprentices as Seventeenth-Century Adolescents," *Past and Present,* no. 61 (November 1973), pp. 157–61.

3. AN, MC, CIX-190, 29 November 1650; MC, VII-390, 31 May 1770.

4. AN, MC, VII-285, 23 November 1752; MC, XXXIV-551, 26 November 1744 (two separate contracts, Dubequet and Bonjean).

5. See Bigot de Sainte-Croix's stringent critique of the practice of requiring apprentices to pay "for the services they render." *Essai sur la liberté,* pp. 47–48. Hélène Davet reports that half of all apprenticeships in the north of Paris were *payants. La Population de l'axe nord de Paris, 1749–74,* thèse de troisième cycle under the direction of Pierre Goubert, Paris I, n.d., pp. 35ff. AN, MC, VII-283, 13 April 1752. Cf. the master baker Jacques Mire, who had to pay six hundred livres to place his son with a master epicier. Ibid., 16 March 1742, MC, XXXIX, 367. About 41 percent of the apprentices in Kaplan's study (drawing on a vast array of Parisian métiers) required payment to the master. "Apprentissage," pp. 448–50.

6. AN, Y 15061, 4 July 1760.

7. Martin Saint-Léon, *Histoire des corporations,* p. 534, and *Dictionnaire portatif du commerce* (Paris, 1777), p. 172; AN, MC, VII-316, 12 October 1758.

8. Bigot de Sainte-Croix, *Essai sur la liberté;* AN, MC, IV-344, 28 October 1708; MC, XVI-642, 6 June 1714; MC, XXIII-419, 25 June 1714; MC, CIX-184, 29 July 1647; MC, XIX-440, 5 August 1649; MC, XIX-438, 1 December 1648; MC, CIX-176, 16 April 1643; CIX-186, 13 September 1648; Charles Ouin-Lacroix, *Histoire des anciennes corporations d'arts et métiers et des confréries religieuses de la capitale de la Normandie* (Rouen, 1850), p. 578; Pierre Lefèvre, *Le Commerce des grains et la question du pain à Lille de 1713 à 1789* (Lille, 1925), p. 98; Martin Saint-Léon, *Histoire des corporations,* p. 534. According to Delamare, the guild had fixed the term at three years in the first part of the seventeenth century and then at three and a half years in 1675. Delamare, *Traité,* vol. 2 (1713), p. 844. Despite the overwhelming evidence for the three-year term in the eighteenth century, misinformation was rife. See, for instance, the announcement of a five-year baker apprenticeship in the widely used reference manual *Tableau universel et raisonné de la Ville de Paris,* p. 211. The *Dictionnaire portatif du commerce,* p. 172, also wrongly claimed that baker apprenticeship and journeymanship were each of five years duration. See also the arrêt du conseil that generalized a four-year apprenticeship apparently for all trades in the jurisdiction of the Paris Parlement (but presumably excluding the capital). AN, AD XI 11, 1 May 1782.

9. In the north of Paris, Davet found average ages of 16.5 for native Parisians and 18.5 for provincials. *La Population de l'axe nord, Hélène de Paris, 1749–74* (thèse de 3ème cycle, Universite de Paris-I, n.d.), pp. 35ff.

10. Kaplan, "Apprentissage," p. 453. Davet reported that 42 percent of apprentices came from areas near Paris: Champagne, Picardy, Normandy. *La Population de l'axe nord,* pp. 35ff.

11. AN, MC, XXIII-419, 25 June 1714; MC, XXXIV-528, 17 November 1740. Cf. the advertisement placed in the London press offering a reward for the return of a fugitive apprentice. Mary Dorothy George, *London Life in the Eighteenth Century* (New York, 1965), pp. 234–27.

12. AN, Y 9387, 23 April 1762; Y 9385, 2 December 1755; Y 9390, 4 December 1767, and Y 9385, 2 December 1755; Y 9379, 14 February 1738. Cf. a similar sentence in Y 9379, 23 July 1737.

13. AN, Y 9385, 5 December 1755, and Y 9390, 14 July 1767; AN, MC, VII-283, 29 May 1752.

14. AN, Y 9390, 20 February 1767; Y 9391, 5 December 1769.

15. AN, Y 9376, 6 July 1723; Y 9380, 14 April 1741; Y 15608, 19 August 1744; Y 15356, 22 March 1756.

16. AN, Y 10264, and AN, X^{2B} 1033, 22 February 1765; Y 9377, 14 March 1727.

17. George, *London Life,* p. 418, and Max Beloff, *Public Order and Popular Disturbances, 1600–1714* (London, 1938), pp. 27, 30–31; AN, Y 11221, 1743; Y 12154, 15 March 1750.

18. Jacques Savary des Bruslons, *Dictionnaire universel de commerce* (Geneva, 1742), vol. 1, pp. 82, 987. *Encyclopédie méthodique,* Jurisprudence, vol. 139, p. 287; Denisart, *Collection de décisions,* vol. 1, p. 74.

19. Arsenal, ms. Bast. 10038, 28 January 1754, and 10086, 3, 7 June 1774; AN, Y 11388,

5 June 1772; Y 9474, 18 January 1771; AN, X²ᴮ 1055, 23 December 1776. A journeyman baker who assaulted several men in a tavern was recognized by his dress. Y 15453, 15 September 1755. For another case of denunciation by appearance, see APP, A/a 137, 3 February 1791.

20. AN, X²ᴮ 1046, 4 April 1772; X²ᴮ 1061, 29 September 1778; X²ᴮ 1060, 18 May 1778; X²ᴮ 1022, 4 July 1759; AN, Y 12614, 15 November 1765; Y 11229, 8 June 1742; Y 15237, 27 August 1760; Y 15392, 20 September 1781; Y 15246, 13 January 1739.

21. AN, Y 13761, 21 August 1754; Y 12143, 24 December 1741; Y 15250, 18 June 1742. On the operation of the labor market as I construe it, see my "La Lutte pour le contrôle du marché du travail à Paris au XVIIIe siècle," *Revue d'histoire moderne et contemporaine* 36 (July–September 1989), pp. 361–412. My notion of an "institutional" as opposed to a "free" labor market is inspired by Clark Kerr, "Can Capitalism Dispense with Free Markets? Labor Markets: their Characteristics and Consequences," *American Economic Review,* papers and proceedings 40 (May 1950), pp. 279–83, and "The Balkanization of the Labor Markets," in *Labor Mobility and Economic Opportunity* ed. E. W. Bakke (New York, 1965).

22. According to the guild officers, they were acting only to protect "the liberty of the masters to make decisions without consulting their workers." See, for example, the police sentence of 29 January 1700, APP, Lamoignon, XX, fols. 665–68. On the attitudes of masters, see the ire of Charles Menuel of the Place Maubert who was willing enough to register the journeyman whom he had hired on his own but was outraged by the obligation to pay the guild a fee, since they had not assisted in finding the worker. The master denounced the jurés (then called syndics) for their "monopoly" and extortion, and they retorted by calling him "pauper" and "scoundrel" (the first epithet surely wounding Menuel more deeply than the second). AN, Y 11731, 4 August 1788. See also Steven L. Kaplan, "Reflexions sur la police du monde du travail, 1700–1815," *Revue historique* 261, no. 1 (January–March 1979), pp. 17–77.

23. See, for example, police sentence of 29 January 1700, APP, Lamoignon, XX, fols. 665–68; arrêt of the Paris Parlement, ibid., 28 February 1763, XLI, fols. 402–26; letters patent, 2 January 1749, François-André Isambert, ed., *Recueil général des anciennes lois françaises* (Paris, 1821–33), vol. 12, p. 221; letters patent, 12 September 1781, Isambert, *Recueil des lois,* vol. 28, p. 78–80; *Encyclopédie méthodique,* Jurisprudence, vol. 148, p. 576.

24. AN, Y 11229, 8 June 1742.

25. AN, Y 12604, 8 October 1756.

26. AN, Y 11572, 25 October 1756.

27. AN, Y 12604, 30 September 1756; Y 11571, 11 September 1755.

28. See AN, Y 11732, 25 February 1788, and Y 11590, 1 September 1746; *Encyclopédie méthodique,* Jurisprudence, vol. 140, p. 131; Diderot et al., *Encyclopédie,* vol. 2, p. 475; Kaplan, "La Police."

29. AN, Y 9457, 6 August 1756. Cf. the bitter protests of nineteenth-century garçon bakers against the brutal, extortionate, discriminatory practices of the *placeurs,* who virtually controlled the labor market. George Sand, *Questions politiques et sociales* (Paris, 1879), p. 28.

30. AN, Y 12612, 3 September 1763; Arsenal, ms. Bast. 10141, fol. 265, 29 October 1765; Y 12613, 23 June 1764.

31. Ordinance of 17 August 1781, BHVP.

32. Lacroix, *Actes,* vol. 5, p. 552 (29 July 1791). The story of the baker boys in the labor market still lacks two important chapters. The first deals with labor relations in the eighties, in the wake of the first abolition by Turgot in February 1776. The second turns on the everyday impact of the resolutions taken on the night of 4–5 August 1789: the two years between the promise of the end of the guilds made in August 1789 and their final abolition may have witnessed important changes in practices of entry and exit.

33. Baker Statutes of 1719 and 1757, BHVP.

34. Voltaire cited by Michel Foucault, *Histoire de la folie à l'âge classique* (Paris, 1972). See also the illuminating remarks of Foucault, ibid., pp. 83–84, and Jean-Pierre Gutton, *La Société et les pauvres: L'Exemple de la Généralité de Lyon, 1534–1789* (Paris, 1970), pp. 217, 244–47.

35. On the perils and implications of idleness, see Des Essarts, *Dictionnaire universel,* vol. 3, pp. 459ff., and M. Guillaute, *Mémoire sur la réformation de la police de France soumis au roi en 1749* (Paris, 1974), p. 87; Denisart, *Collection de décisions,* vol. 4, p. 564. For Delamare, the duty to work is imposed by God. *Traité,* vol. 3 (1719), p. 446. For Des Essarts, in the fashionable language of the High Enlightenment, it is imposed "by Nature." *Dictionnaire universel,* vol. 7, pp. 346–47.

36. See Des Essarts, *Dictionnaire universel,* vol. 8, p. 72. Idleness also led to physical as well as moral pollution and deterioration, as we shall note later. See the early-nineteenth-century police report published by G. Vauthier, "Les Ouvriers de Paris sous l'empire," *Revue des études napoléoniennes* (1913), p. 429. The link between moral and physical pathology, between individual and social pathology, and between these pathologies and crime is brilliantly developed in Louis Chevalier's pioneering *Classes laborieuses et classes dangereuses à Paris pendant la première moitié du XIXe siècle* (Paris, 1958).

37. Police ordinance of 10 September 1783, BHVP; police sentence of 20 July 1743, APP, Lamoignon, XXXV, fols. 203–7; Bertrand Gille, *Les Origines de la grande industrie métallurgique en France* (Paris, 1947), p. 155; arrêt du conseil, 8 March 1713, Arsenal, ms. Bast. 10321 and 11546, August 1744; police sentences of 15 February 1738 and 6 August 1756, BHVP; statutes of 1719 and 1757, BHVP; police ordinances of 8 November 1776 and 17 August 1781, BHVP.

38. Police sentence of 5 September 1720, APP, Lamoignon, XXVII, fols. 131–37; police ordinance of 10 September 1783, BN, fichier central.

39. See, for example, the account books of master bakers Tayret and Dufour, ASP, D5B⁶ 4119 and D5B⁶ 3328.

40. See, for example, garçon baker J. Broux and master baker Jacques Quillet, ASP, D2B⁶ 736, 5 April 1741; garçon baker Louis Chator and baker Adme Pollard, AN, Y 14084, 22 December 1757; garçon baker Joseph Thevenin and master baker widow Villemain, AN, MC, CXXII-753, 23 November 1770; scellé, Y 13741, 11 May 1741.

41. See, for example, police ordinance, 13 April 1785, AN, AD XI 25.

42. Savary des Bruslons, *Dictionnaire universel,* vol. 2, pp. 1017, 1019. Cf. Diderot et al., *Encyclopédie,* vol. 3, p. 186; police sentence of 14 October 1741, in Fréminville, *Traité de la police,* p. 239; Delamare, *Traité,* vol. 4 (1738), p. 91; Alfred Des Cilleuls, *Histoire et régime de la grande industrie en France au XVIIe et XVIIIe siècles* (Paris, 1898), p. 166; Edouard Dolléans, *Histoire du travail en France: mouvement ouvrier et législation sociale* (Paris, 1953–55), p. 97; Pierre Hurtaut and P. N. Magny, *Dictionnaire historique de la ville de Paris*

et de ses environs (Paris, 1779), vol. 1, p. 653; Edmond Soreau, *Ouvriers et paysans de 1789 à 1792* (Paris, 1936); Levasseur, *Histoire des classes ouvrières*, vol. 2, p. 384; "Edit du Roy pour le règlement des imprimeurs" (1686), article XXXV, p. 277, BHVP; Jean Jacques, *Vie et mort des corporations: Grèves et luttes sociales sous l'ancien régime* (Paris, 1970), p. 122; letters patent of 2 January 1749 and edict of August 1776, in Isambert, *Recueil des lois*, vol. 21, p. 221, and vol. 24, p. 85; *Encyclopédie méthodique*, Jurisprudence, vol. 141, p. 85. The "liberals" in the ministry in the second half of the eighteenth century vigorously criticized the restrictions that governed entering and leaving work. Trudaine de Montigny denounced the obligation to ask for congé: it "destroyed the equality that must exist between two free men." Turgot, Roland, and Trudaine père, among others, claimed that the entire regulatory apparatus concerning placement was irreconcilable with the principle "that workers are not slaves in France." Levasseur, *Histoire des classes ouvrières*, vol. 2, pp. 667, 803, and Des Cilleuls, *Histoire de la grande industrie*, pp. 108, 168.

43. Ordinance of 23 June 1671, BHVP.

44. Ordinance of 25 September 1728, BHVP. Police sentence of 15 February 1738, BHVP.

45. Police sentences of 18 June 1749 and 6 August 1756, BHVP.

46. Police sentence of 19 January 1769, homologated by a parlementary arrêt of 11 April 1769, BHVP, and BN, 4° Z Le Senne, 2263 (13).

47. AN, Y 15375, 3, 31 July 1772, 3–7 August, 19 September 1770; avis des députés de commerce, 5 December 1788, AN, F[12] 724; arrêt du parlement (Paris), 12 November 1778, AN, AD XI 11; AN, F[12] 1372, 7 September 1779.

48. AN, Y 10996, 30 June, 14 August 1755; Y 10994, 3 July 1752.

49. AN, Y 14424, 21 February 1776; Hardy's journal, 7, 24 March 1774, BN, ms. fr. 6682, pp. 191, 194; "Mémoire des Boulangers," 1776, BN, Joly 462, fols. 106ff.

50. Police ordinance of 8 November 1776, BHVP; prefect Dubois's ordinance in the *Tableau des boulangers, 1811; Enquête sur la boulangerie*, pp. 72, 800, 805. One wonders to precisely what epoch the chief of the bureau of military subsistence referred when he invoked "la rupture de l'ancienne solidarité qui existait entre eux [the journeymen] et leur patron."

51. Police ordinance of 17 August 1781, BHVP; letters patent of 12 September 1781, which uses the term "livre ou cahier" rather than *livret*, in Isambert, *Recueil des lois*, vol. 27, p. 79.

52. Guillaute, *Mémoire*, p. 87; Des Essarts, *Dictionnaire universel*, vol. 3, pp. 459–563.

53. AN, Y 10292, May–June 1767; AN, X[2B] 1039, 1 February 1768; Y 15040, 26 August 1782.

54. AN, Y 9474, 12 July 1771, and Y 9391, 14 August 1772; Y 9391, 14 August 1772; Y 9387, 28 July 1761; Y 9390, 14 August 1767; Y 9390, 20 March 1767; Y 9391, 21 July 1770; Y 9391, 5 December 1769 (Thierry); Y 9391, 18 June 1771 (Berton and Chatelin); Y 9391, 27 July 1770 (Deshaye); Y 9391, 27 July 1770 (Michaux); Y 9387, 9 January 1761 (Legrand); Y 9387, 27 November 1761 (Delessat); Y 9387, 3 September 1762 (Jouy); Y 9387, 23 April 1762 (Leduc); Y 9387, 17 April 1761 (Coquasspot); Y 9387, 24 July 1761 (Dubreuil); Y 9292, 13 May 1774 (Rousseau).

55. See, for example, the case of master baker Joly who accused master baker Bonnome of having "seduced" his garçon. Arsenal, ms. Bast. 10056, 29 August 1760. Cf. a similar case in the clockmakers' guild. Ms. Bast. 12081, 1729.

56. AN, Y 15376, 26 July and 3, 6 August 1770; Y 16022^1, 14 November 1774; Y 10233, 17 March 1767; Y 16022^1, 21 February 1774; Y 9377, 14 March 1727; Y 12617, 14 September 1768; Y 9441, 8 July 1740; Y 9377, 14 March 1727; Y 9391, 14 August 1772; Y 9387, 28 July 1761. On occasion the guild asked for much higher fines for infractions of guild rules than the procurator was willing to assess. E.g., on 3 September 1762, the jurés requested a fine of one hundred livres but the procurator imposed a fine of only ten livres. Y 9387.

57. AN, Y 9335, February 1780; Y 9484^A, 16 March 1781; Y 16022^A and Y 16022, 21 July 1777; Y 9390, 20 March 1767; Y 9387, 7, 24 July 1761.

58. AN, Y 9387, 12 June 1761. Cf. the conviction of master Alliance. Y 9392, 30 May 1774. Y 9390, 19 February 1768. For another example of a master's refusal to authorize departure, see Y 11440, 28 November 1778.

59. Arsenal, ms. Bast. 10142, 29 September 1769; AN, Y 13748, 21 June 1743. Cf. the case of journeyman Belvazet, who was accused of theft, denied the right to defend himself, dismissed, and refused a certificate on the grounds of his unworthiness. Y 11385, 22 August 1771.

60. AN, Y 11237, 18 June 1750. See also the case of the journeyman who could not find work after his master branded him as dishonest. Y 15616. One result was that a black market sprung up in the sale of authentic certificates from one baker boy to another and probably also in the sale of forged and stolen papers. See the arrest of journeyman Jean Diat, called Auvergnat. Y 10323, 29 February 1772.

61. Police sentence, 25 September 1728, BHVP; Y 15246, 13 January 1739; Y 9539, 15 January 1762; Y 9465^A, 4 February 1763; Y 15375, 4 August 1769 and 4–6 March, 4 August 1770.

62. BN, ms. fr. 21646, fol. 4, 23 June 1771; AN, Y 12606, 16 March 1758; Y 13167, 28 December 1760; Y 13167, 7 December 1760; Y 16022, 21 July 1777.

63. Police sentences: 9 August 1700, 22 August 1727, 9 January, 13 February, 23 April, 7 May 1728, 21 April, 7 November 1730, 21 June 1732, BN, ms. fr. 21710, fols. 54–55, 221, 273–78, 289–92, 387–90; royal ordinance, 1 August 1733, Isambert, *Recueil des lois*, vol. 31, p. 379; Fréminville, *Traité de la police*, pp. 177–85; Des Essarts, *Dictionnaire universel*, vol. 1, p. 471. Commissaire Divot specialized in tavern and inn affairs in the thirties. See, for example, AN, Y 9434 and 9394, 20 November 1733; BN, ms. fr., 21646, fol. 4.

64. Sartine to Dudoigt, 11 September 1765, AN, Y 14685.

65. Police ordinance, 17 June 1741, and police sentence, 19 February 1745, BN, ms. fr. 21710, fols. 32–41; Gazier, "La Police," pp. 50, 78; Des Essarts, *Dictionnaire universel*, vol. 1, p. 466; *Encyclopédie méthodique*, Jurisprudence, vol. 139, p. 569; Lenoir to Commissaire Gillet, 12 February 1777, AN, Y 13728; BN, Joly 2412, fols. 234–71; Fréminville, *Traité de la police*, p. 169; Benzacar, *Le Pain à Bordeaux*, p. 80; arrêt of the Paris Parlement, 12 November 1778, in Isambert, *Recueil des lois*, vol. 25, p. 452; arrêt of the Paris Parlement, 3 December 1781, Isambert, *Recueil des lois*, vol. 27, p. 125; *Encyclopédie méthodique*, Jurisprudence, vol. 147, p. 576. The *Encyclopédie méthodique* rightly characterized the prohibition against receiving more than four journeymen at a time as "an impracticable and vexatious thing" that was rarely implemented.

66. AN, X^{2B} 1063, 8 June 1779; Y 12741, 9 December 1755; Y 10258, 4 May 1764.

67. Arsenal, ms. Bast. 10012, 6 August 1760. On two other occasions, Wagon was unable to exculpate himself. Unemployed for five months and "without a fixed resi-

dence," he was arrested in 1774 and convicted of stealing meat from a butcher shop. AN, Y 10350, 13 October 1774. Though he was banned from the capital for nine years, Wagon returned in 1778 and was accused of stealing twenty-eight pieces of silver. He was sentenced to be hanged and strangled after submission to torture. AN, X²B 061, September 1778. Hardy reported that the sentence was carried out on the Grève. Hardy's journal, 1 October 1778, BN, ms. fr. 6683, p. 52. Another baker boy was hung not long afterward for theft against his master. Hardy's journal, ibid., p. 176.

68. AN, X²B 1032, 13 July 1764; AN, Y 10258, April–May 1764. Cf. another example of the inference of criminality from unemployment. Y 13172, 30 January 1789.

69. AN, Y 15246, 20 March 1739; Y 11384, 7 February 1771; Y 10171, May 1755 and AN, X²B 1015, 29 November 1755; X²B 1035, February–March 1766 (Swire's father was a relatively well-off master baker, but the son seems to have been estranged from his parents); Y 10059, August 1737; X²B 982, 27 February 1738. For a useful exploration of food theft, see Arlette Farge, *Délinquance et criminalité: Le Vol d'aliments à Paris au XVIIIe siècle* (Paris, 1974).

70. Arsenal, ms. Bast. 10054, 8 March 1760; ms. Bast. 10102, 3 August 1761; ms. Bast. 10038, 31 March 1754; and AN, X²B 1031, 8 June 1764, and AN, Y 10258; Y 1033, 26 February 1773; Y 10039, 29 November 1730; Y 13121, 27 December 1769; Y 12619, 29 May 1770; X²B 1048, 12 March 1773; Y 10324, 11 March 1772; X²B 985, 2 July 1739; Y 10333, 23 February 1773; Y 10079, 27 September 17; X²B 1046, 4 April 1772 and Y 10324; X²B 1060, 18, 29 April and 18 May 1778; X²B 1055, 23 December 1776; Y 10370, 2 September 1776; X²B 967, 27 September 1730; Y 10039, 5 September 1730; Y 15255, 8 July 1745; Y 11572, 12 August 1756; Y 10397, 29 April 1779; and X²B 1063, 8 June 1779.

71. AN, Y 13530, 30 July 1763; Y 14949, 24 January 1734; Y 13115, 26 June 1763; Y 12141, 16 March 1739; Y 15356, 13 March 1755; Y 12741, 9 December 1755.

Chapter 8   At Work

1. Parmentier, *Parfait Boulanger*, pp. 378, 634; Macquer, *Dictionnaire portatif*, vol. 1, 303; Bouteloup, *Travail de nuit*, p. 59, and Blois l'Ami-du-Travail, "Le Boulanger," in *Compagnonnage*, pp. 214–15; Delamare, *Traité*, vol. 2 (1729), p. 760; Malouin, *Description* (1761), p. 113; Sand, *Questions*, p. 34. On the burden of kneading, see also the *Bulletin de la Société d'encouragement de l'industrie nationale*, no. 26 (March 1806), p. 235. Parmentier overlooked neither the hurly-burly of the multiple bread-making operations nor the decisive part played by kneading. Remaining consistent with his emphasis on knowledge and finesse, however, he maintained that this work required "plus d'adresse et d'agilité que de force et de courage"—a perspective that seems excessively intellectual. *Parfait Boulanger*, p. xxxiii. Legrand d'Aussy stressed not the difficulty of the labor but its protracted length. "De la nourriture des français," BN, ms. n.a.f. 3328, fol. 10.

2. See the bitter complaint of the Paris baker boys, who not only compared themselves unfavorably with the journeymen in other crafts, whose tasks were so easy, but also were envious of the forain bakers, who bake only for the two weekly market days and have time to stroll in the village, stop at the tavern, and play *boules*. Dufrene, *La Misère des garçons boulangers de la ville et faubourgs de Paris* (Troyes, 1715). Aside from the ratio-

nalizing "tableau régulateur," the organization of work in the *fournil* of 1830, in the hands of a brigadier and a first aide, does not seem to have differed greatly from the practice of the eighteenth century. S. Vaury, *Le Guide du boulanger* (Paris, 1834), pp. 6–11.

3. See "Mémoire pour les Boulangers de Chaalons-sur-Marne," 30 July 1785, BN, Joly 1742, fol. 43; Georges Blond and Germaine Blond, *Histoire pittoresque de notre alimentation* (Paris, 1960), vol. 2, p. 258; Mercier, *Tableau*, vol. 12 (1788), p. 145; Barbaret, *Le Travail en France*, vol. 1, pp. 412, 414, 452; Dufrene, *Misère*; Barbaret, *Le Travail en France*, vol. 1, p. 451; Bernardino Ramazzini, *Diseases of Workers*, trans. W. C. Wright (New York, 1964), p. 225. Cf. Marx's reflections in *Le Capital*, chap. 10, in *Oeuvres*, ed. Maximilien Rubel (Paris, 1963), especially vol. 1, pp. 830–31. Apparently in order to expose night work as a relatively recent innovation, a departure from tradition, certain bakers in the late nineteenth century wrongly claimed that night work was only introduced during the reign of Louis XVI. Bouteloup, *Travail de nuit*, p. 2. Cf. Maurizio's description of "le metier où on ne dort pas." *Histoire de l'alimentation*, p. 501. On the obligation to be forever "en caleçon et en bonnet," see Delamare, *Traité*, vol. 2 (1713), p. 839. The evocation of night work as corrosive of the laws of nature was the only example I have found of an explicit journeyman appeal to the natural law that Michael Sonenscher contends became such a fundamental if not determinant part of their idiom. See Michael Sonenscher, *Work and Wages: Natural Law, Politics, and Eighteenth-Century French Trades* (Cambridge, 1989).

4. Prefect of police to minister of the interior of 30 May 1908, in Vauthier, "Ouvriers," p. 430; Bouteloup, *Travail de nuit*, pp. 1–11, 257–58; Léon Bonneff and Maurice Bonneff, *La Classe ouvrière, les boulangers . . .* (Paris, 1912), p. 28; AN, Y 15348, 1 May 1751; Y 18672, 30 July 1770, and Y 12596, 29 March 1752; Y 18670, 30 September 1769, and Y 18670, 10 July 1769; Arsenal, ms. Bast. 10061, 7 October 1761; Y 15093, 5 June 1785. Cf. a recent Reuters dispatch: "Workers at a London bakery threatened to strike because they said long shifts were leaving them too tired to make love to their wives." *New York Times*, 1 March 1973.

5. AN, Y 14953, 15 May 1738; Y 12729, 1 March 1735, and AN, MC, VII-329, 4 November 1760; MC, CIX-412, 14 November 1712, and Y 11171, 5 June 1754; MC, VII-287, 17 May 1753. Widow Louis Leseigle offered the same accommodations. Y 12575, 19 March 1731. In 1848 the journeyman bakers of Frankfurt unsuccessfully struck in order to obtain beds in the premises of the bakery where they could sleep or rest. They resented having to improvise uncomfortable and humiliating solutions on tables and in troughs. W. Ziehr and E. M. Buhrer, *Le Pain à travers les âges*, p. 170.

6. Prefect of police to minister of the interior, 30 May 1807, in Vauthier, "Ouvriers," p. 43; Sand, *Questions*, p. 30; Lacombe d'Avignon, *Le Mitron de Vaugirard*, pp. 66–68. See also Vinçard, *Alimentation*, pp. 52–53; Weichs-Glon, "Municipalisation," pp. 968–69. Cf. Bouteloup's "white miners." *Travail de nuit*, p. 1.

7. AN, Y 14779, 14 November 1729. Cf. AN, MC, XXVIII-114, 14 September 1711. Enriched with the journeyman's sweat and perhaps with the residue of his ablutions, did the flour contain vestiges of other (human) organic elements? A lubricious bakery song reminds us, inter alia, that baking bread was symbolically homologous to engendering a child . . . and that the arduous craft of baking did not leave ample time for canonical forms of sexuality:

C'était un garçon boulanger, ohé, ohé

Qui se masturbait dans la farine

Une belle dame vint à passer, ohé, ohé

Et lui dit vot' pain sent la pine

I am grateful to my friend Daniel Roche, whose streetwise erudition knows no bounds, for this fascinating reference.

8. Sage, *Analyse des bleds*, p. 46; Mercier, *Tableau*, vol. 12 (1788), p. 146.

9. Ramazzini, *Diseases of Workers*, p. 231; Philibert Patissier, *Traité des maladies des artisans et de celles qui résultent des diverses professions, d'après Ramazzini* (Paris, 1827), p. 195; Sheppard and Newton, *Story of Bread*, p. 45; Ramazzini, *Diseases of Workers*, pp. 225–29; Jacob, *Bread*, p. 138.

10. Collection E. Thomas, cote provisoire 4804, series 114, BHVP; Bouteloup, *Travail de nuit*, pp. 52, 54; Jacob, *Bread*, pp. 131, 138; Ramazzini, *Diseases of Workers*, p. 229; Delamare, *Traité*, vol. 2 (1729), p. 529 (livre 4, titre 12, chap. 1). Nor was their reaction purely mystical; they also provided material support to the leper hospices.

11. Bouteloup, *Travail de nuit*, pp. 52–53; Vinçard, *Alimentation*, p. 55; Patissier, *Traité des maladies*, p. 195. The baker boys were among the workers most frequently subject to venereal disease, according to the information provided by Dulaure, *Histoire civile*, vol. 9, p. 41. On the basis of a sampling of the documents concerning the treatment of male victims of venereal disease in 1765, the late Professor E.-M. Benabou indicated that the baker boys are very well represented, though they fell far behind the tailors and they also lagged behind the domestics and the shoemakers. Personal communication. Cf. the story of Joseph Leiber, a former army hospital worker, who agreed to supply a cure to two syphilitic journeymen bakers for twenty-four livres apiece and later was beaten and robbed by them and a third baker boy. Arsenal, ms. Bast. 10086, 7 June 1774.

12. Mercier, *Tableau*, vol. 12 (1788), p. 146; Macquer, *Dictionnaire portatif*, vol. 1, p. 303; Sand, *Questions*, pp. 32–33.

13. AAP, deliberations, Bureau de Hôtel-Dieu, no. 137, 24 February 1768; AN, Y 13117, 29 March 1765; Dufrene, *Misère*.

14. Patissier, *Traité des maladies*, p. 195; Barbaret, *Le Travail en France*, vol. 1, p. 411; Vauthier, "Ouvriers," p. 429; Bouteloup, *Travail de nuit*, pp. ii, 48; Charles-Louis Cadet de Gassicourt, "Considérations statistiques sur la santé des ouvriers," *Mémoires de la Société médicale d'émulation* 8 (1817), pp. 160–74, cited by Michael Sibalis, "The Workers of Napoleonic Paris, 1800–1815," Ph.D. diss., Concordia University, 1979, p. 104. A recent general work on the life of journeymen put modal mortality in the nineteenth century at thirty-eight to forty-two years, with very few bakers living beyond fifty. Carny, *Compagnons*, vol. 2, p. 150; AN, Y 15093, 5 June 1785.

15. See, for example, the bakery of master Etienne Huin, where François Sivry served as "gindre" (*sic*) or "first journeyman" and Denis Chapuzot as "second journeyman." AN, MC, VII-291, 28 May 1754. I have found only one instance where a garçon baker is referred to as a "mitron"—and the speaker was a carriage renter, not a baker. AN, Y 11229, 26 December 1742.

16. Delamare, *Traité*, vol. 2 (1729), p. 717; Malouin, *Description* (1761), p. 113: Parmentier, *Parfait Boulanger*, p. 378. For a modern, medicalizing, functionalist explanation of

geindre, see Paul Nottin, *Le Blé, la farine, le pain,* p. 66: "Le pétrissage soulève des poussières de farine qui pénètrent dans les poumons de l'ouvrier et y produisent à la longue des lésions; c'est pour se protéger contre ces poussières que le boulanger, au cours du pétrissage, fait entendre une sorte de sifflement ressemblant à un gémissement. A ce bruit certains attribuent l'origine du mot geindre; d'autres tirent cette expression de junior, le plus jeune des deux ouvriers de l'équipe, chargé du pétrissage."

I have found a number of cases in which journeymen were related to the master. Two of Jean-François Fourcy's three journeymen were his sons. AN, MC, VII-354, 28 March 1743. The widow Dugland's son was her journeyman. AN, Y 9623, 4 May 1743. In addition, in five cases baker boys were the nephews of masters, in three instances the cousins, and in another the brother-in-law. Y 12141, 15 August 1739; Y 12604, 4 April 1756; Y 12596, 21 March 1752; Y 15628, 12 December 1754; Y 12141, 15 September 1939; Y 13536, 25 September 1766; Y 12616, 13 December 1767; Y 13110, 12 April 1758; Y 15263, 3 May 1755. Needless to say, journeymen with close relatives as protectors and patrons had a much greater chance of rising to mastership.

17. AN, Y 11229, 23 December 1742; Parmentier, *Parfait Boulanger,* pp. 365–67; Malouin, *Description* (ed. Bertrand, 1771), pp. 386, 394–98. Malouin warned masters to be careful not to hire journeymen whose sweat was so foul and whose breath was so bad that the yeast would spoil under their touch. *Description* (1761), pp. 113–14.

18. AN, Y 15359, 8 May 1758; Y 15239, 15 February 1732; AN, MC, CXXII-753, 23 November 1770. For other cases in which a baker boy acted as *porteur,* see Y 13117, 29 May 1765; Y 12594, 22 April 1750; Y 13536, 25 September 1766.

19. Thrupp, *Worshipful Company,* pp. 17–18; Malouin, *Description* (1761), p. 113 (I suspect that Malouin counted the apprentice as a regular journeyman); Raymonde Monnier, *Le Faubourg Saint-Antoine, 1789–1815* (Paris, 1981), p. 73; District Saint-Etienne du Mont, 8 April 1790, AN, F[30] 156. (Did widow Brille exaggerate the magnitude of her work force in order to swell her claim for currency in small denominations?) In a smaller sample (n = 16) based on an analysis of *scellés,* or preinventories drawn up by police commissaires rather than notaries, the modal number of baker boys is two per master (mean = 1.7; median = 1.8). In both the inventory and scellé studies, master bakers have on average slightly more journeymen per shop than their faubourg counterparts (1.96 versus 1.86 and 1.8 versus 1.5 respectively).

20. ASP, D5B[6] 2532; DC[6] 244, fol. 240, and DC[6] 242, fol. 236. See also DC[6] 246, fol. 166, and AN, MC, XXXIV-722, 4 March 1780. That is not to say that in no cases did poor bakers have servants. See, for example, the cases of Claude Thevenet and Pierre Tranchard. AN, Y 11571, 5 July 1755; Y 15603, 5 May 1742.

21. AN, Y 14949, 24 January 1734; Y 11440, 28 November 1778; Y 14948, 29 June 1733; Y 13122, 10 January 1770; ASP, D4B[6] 11-138, 27 March 1782; Y 18669, 18 January 1769; AN, W 39, no. 2634 (prairial an II). From a traditional corporate perspective, Regnier was "debauched" from his old master by his new master. The guild labored hard to prevent this sort of competition for labor. On the rapid turnover of journeymen in Rouen and elsewhere, see the insightful discussion in Sonenscher, *Work and Wages.*

22. ASP, D5B[6] 3328, 1702, 4118.

23. AAP, deliberations, Bureau de l'Hôtel-Dieu, no. 95, 25 June 1726, and no. 107,

27 June 1738; Hôtel-Dieu, Panneterie, journals for 1743 and 1755–61, 33d liasse; "Second mémoire" and "Réponse au second mémoire," Arsenal, ms. 7458 (which suggests that the wage varied as a function of the delicacy of the bread made, a "mollet" garçon earning twice the sum of a maker of a "pain du peuple"); AN, Y 12416, 11 January 1751. The wage of thirty sous a day that journeyman François Sorlin of the faubourg Saint-Laurent allegedly received seems implausibly high. Y 12416, 21 January 1751. In Bordeaux the geindre earned three livres a week in 1748 and the other garçons two livres and a food allowance. Benzacar, *Le Pain à Bordeaux*, p. 90. At Troyes the garçons earned ten livres a month toward the end of the century, also not counting food. "Mémoire," ca. 1786, BN, Joly 1743, fols. 137–44.

24. AN, MC, XIX-610, 16 February 1716; MC, XIX-612, 22 July 1715; MC, XXX-437, 24 April 1723; MC, XXXIX-441, 30 June 1757; MC, LXVI-550, 25 October 1765; AN, Y 11237, 18 June 1750; Y 10318, 10 October 1770; Y 11440, 28 November 1778; Y 10396, 5 March 1779; Arsenal, ms. Bast. 10039, 19 August 1754; ASP, D2B⁶ 736, 7 April 1741; D2B⁶ 841, 23 January 1750; D4B⁶, carton 13, dossier 587, 21 August 1753; D5B⁶ 4119; D5B⁶ 4246; D6B⁶ 11, no. 138, 27 March 1782; D5B⁶ 2532; D6B⁶ 1702; D4B⁶ 113-8056, 28 November 1781; AAP, Incurables, no. 118; AAP, n.s. no. 47, Scipion, 31 March 1786 (after-death inventory); Marcel Marion, "Les Lois de maximum et la taxation des salaires sous la Révolution," *Revue internationale de sociologie* 25 (1917), p. 487. Despite their sallow look, baker journeymen were reputed to have a keen taste for meat, an "excessive" portion of which they attempted to extract from their bourgeois. See article 7 of the Règlement concernant les garçons boulangers à Paris (1 and 19 floréal year III) published by the Bureau central du canton de Paris ("Les garçons boulangers ne pourraient exiger de ceux chez lesquels ils sont employés, des rations de viande plus fortes que celles qui sont accordées aux ouvriers travaillant de leurs bras").

25. See, for instance, Arsenal, ms. Bast. 10039, 19 August 1754 (bimonthly); AN, MC, XXXIX-441, 30 June 1757 (monthly); and ASP, D5B⁶ 4240 (monthly and yearly); D5B⁶ 4118 (yearly); D4B⁶ 13-587, 2 August 1753 (yearly); D5B⁶ 4118; D5B⁶ 4119; D5B⁶ 3328; D5B⁶ 1702. For an eerie modern echo of traditional practices, see the recent suit of a fifty-nine-year-old bakery worker in Bayonne against his baker employer, who had not paid him a regular wage for almost a decade. The boss gave him spending money and paid his tax bill but retained his wages. The Conseil des prud'hommes characterized this incident as an affair "d'une autre époque." *Le Monde*, 11 March 1992.

26. AN, MC, VII-291, 28 May 1754; ASP, D5B⁶ 4119; MC, XVI-550, 25 October 1765; AN, Y 13741, 11 May 1741; D5B⁶ 2532; Y 14084, 22 December 1757; MC, CXXII-753, 23 November 1770.

27. AN, Y 11440, 28 November 1778; ASP, D2B⁶ 736, 5 April 1741; D2B⁶ 7336, 7 April 1741; D2B⁶ 841, 23 January 1750; D2B⁶ 644, 31 August 1733; D2B⁶ 1125, 1 September 1773. Another baker boy won a seizure of property against his master, Y 13741, 21 June 1727.

28. ASP, D6B⁶ 11, no. 138, 27 March 1782.

29. See, for example, the protest of the bakers of the Hôtel-Dieu. AAP, deliberations of the Bureau de l'Hôtel-Dieu, no. 110, 7 July 1741.

30. Police sentence of 24 April 1719, AN, AD XI 14; police ordinance of 8 November 1776, BHVP; police sentence of 12 December 1704, BN, ms. fr. 21640, fol. 143; police

sentence of 6 August 1756, AN, Y 9457. The police opened an investigation and forbade any assembly of baker boys on pain of corporal punishment. They also reaffirmed the rule prohibiting journeymen from leaving masters without notice.

31. AN, Y 11229, 26 December 1742; Arsenal, ms. Bast. 10011, fol. 28, 12 January 1742 (known as a "good subject," Nicolas came from a peasant family in a village located in a seigneurial jurisdiction belonging to Marville, the Paris police chief); BN, Joly 1101, fol. 163. Cf. the fidelity of mistress baker Petit to her journeyman, accused of theft. Y 13749, 15 September 1744. Also: Arsenal, ms. Bast. 10068, fols. 656, 701, 7 October 1764; Y 12589, 2 August 1745; AN, MC, XXVIII-166, 11 February 1720. On the collective convulsion of 1750, see the rich study of Arlette Farge and Jacques Revel, *The Vanishing Children of Paris: Rumor and Politics before the French Revolution* (Cambridge, Mass., 1991). See the night when Debure beat up a journeyman who had the bad taste to leave the table a winner before the game was declared over. Y 12605, 24, 27 April 1757.

32. Police ordinance, 8 November 1776, BN, fichier central. This refrain was not the monopoly of the bakers. Masters from many other professions echoed their anxiety, especially in the second half of the century. See Kaplan, "La Police." Hardy deplored the "spirit of insubordination" rampant among journeyman blacksmiths, masons, and bakers, inter alia. Hardy's journal, BN, ms. fr. 6685, p. 315 (23 March 1786). See also ibid., fols. 191, 194. Hardy reported that a baker boy slapped an exempt who tried to force him to leave a tavern in the Marais and return to work. The two most celebrated observers of the Paris scene, L.-S. Mercier and Restif de la Bretonne, detected the same "insolence," "insubordination," and "spirit of revolt." Mercier, *Tableau*, vol. 12 (1788), pp. 323–24; Restif de la Bretonne, *Nuits de Paris*, vol. 1, pp. 58–59.

33. Delamare, *Traité*, vol. 2 (1713), p. 121; Gazier, "La Police," p. 56; Des Essarts, *Dictionnaire universel*, vol. 3, pp. 459–61, and vol. 8, p. 401; AN, Y 14542, 26 August 1747. See the case of the butchers who called their journeymen "lackeys" and the police interrogation concerning a journeyman baker and "his bourgeois." Arsenal, ms. Bast. 11727, 10 August 1750; Y 10318, 22 June 1771. The guild referred to journeymen collectively as "domestiques," whose wayward condition jeopardized the industry. Mémoire (to lieutenant general?), September 1743, Arsenal, ms. Bast. 10024. Cf. Feyeux, "Un Nouveau Livre," part 1, pp. 343–44, part 2, pp. 518–23. The assimilation worker-domestic deeply wounded the self-esteem of the journeymen, the elite in the world of work. For one example in which a master joins his journeyman to protest vehemently against journeymen being treated by the police as "lackeys," see Poussot to Marville, Arsenal, ms. Bast. 11596, 11 October 1746.

34. ASP, 3 AZ 101, pièce 1.

35. See, for example, AN, Y 13744, 19 February 1755; Y 12138, 18 August 1736; Y 11237, 18 June 1750.

36. AN, Y 14778, 29 August 1728; Y 10171, 26 April 1755; Y 11220, 29 June 1733; Y 11212, 12 May 1724; Y 13748, 21 June 1743. Construed as a reward for fidelity, the certificate of congé was bound to cause problems. Since Dubreu fired Berri for stealing, he could hardly attest to his fidelity. Yet Berri needed a certificate to avoid falling into outlawry. Many misunderstandings and conflicts probably arose over the interpretation of the significance of the certificate. Some masters clearly were disingenuous in their assessment of their departing journeymen. Master Lemaitre noted that "he had always

known (his journeyman) to be a libertine and a bad subject." But then the question is, why did he keep him in his employ for over two years? Y 13112, 27 February 1760.

37. AN, Y 15250, 13, 19 June 1742. For another incident of a physical attack by journeymen, see Y 14084, 20 February 1757.

38. AN, Y 12592, 3 July 1748; Y 11226, 4 August 1739.

39. ASP, D4B⁶ 43-2410, 22 July 1782; D4B⁶ 83-5592, 23 September 1783; D4B⁶ 106-7557.

40. See, for example, Arsenal, ms. Bast. 10086, 3 June 1774.

41. AN, Y 11385, 22 August 1771; Y 12142, 6 October 1740; Y 13121, 21 August 1769; Y 10322, 29 February 1772; AN, X²ᴮ 1046, 10 April 1772.

42. AN, Y 10318, November 1770–January 1771; AN, X²ᴮ 1045, 11 July 1771; APP, A/a 74, 9 March 1790.

43. AN, Y 10337, 7 July 1773; AN, X²ᴮ 1049, 14 July 1773; APP, A/a 215, 18 January 1791.

44. AN, X²ᴮ 1048, 12 March 1773; AN, Y 11229, 23 December 1742.

45. AN, X²ᴮ 1063, 7 May 1742. Hardy reported Gamain's execution for stealing 3,401 livres from his master. BN, ms. fr. 6681, 27 July 1779; Arsenal, ms. Bast. 10038, 12 March 1754; AN, Y 12618, 18 August 1769.

46. Arsenal, ms. Bast. 10069, 3 December 1764.

47. See, for example, Arsenal, ms. Bast. 10056, 20 August 1760; ms. Bast. 10086, 3 June 1774; and AN, Y 10311, 17 December 1769; Y 13121, 27 December 1769; AN, X²ᴮ 1045, 14 May 1771; ms. Bast. 10187, 4 July 1750; Y 11338, 5 June 1772. Though in theft cases masters were habitually the accusers and journeymen the defendants, it ought to be pointed out that there were instances in which the roles were reversed. After only two weeks in the shop, baker boy Jules Quissier accused the master's wife along with the master of stealing his trousers (with a fortnight's wages in the pocket) while he was asleep atop the oven. Y 13761, 19 August 1754; ms. Bast. 10039, 19 August 1754.

48. Hardy's journal, BN, ms. fr. 6680, p. 56 (11 February 1767); AN, X²ᴮ 1033, 20 March 1765.

49. AN, Y 11229, 23 December 1742.

50. Arsenal, ms. Bast. 10124, 10 July 1769; AN, Y 15595, 22 September 1738. Worse, after taking a reprimand, Lebesse announced that he was quitting forthwith, without the required notice. His mistress was furious, for she knew she would not be able to find a replacement in time to make the six fournées needed for the next day.

51. AN, Y 12741, 9 December 1755; Y 13761, 21 August 1754; Arsenal, ms. Bast. 10080, 2 September 1722; Y 13536, 25 September 1766. In addition, Hebert took spirits from his master's wine closet without permission.

52. AN, Y 11237, 18 June 1750.

53. AN, Y 11377, 1 December 1769 (the fact that Madame Mathias had an alcoholic and negligent husband did not help business either); ASP, D4B⁶ 42-2307; AN, W 39, no. 2634. With revolutionary zeal, the evidence—burned four-pound loaves—was fished out of a mass of fecal matter. Although the bread was in a state of "putrefaction," the experts determined that the crumb had been good, despite the overdone crust. One of André's coworkers testified that he had been shocked by André's behavior, though he had been too afraid of the geindre to denounce him. "Coquin," he claimed to have said to André, "is that how you betray the fatherland?" Said André, "A bread was not a big thing worth fussing over."

54. AN, Y 14954, 23 September 1739; Y 15356, 4 July 1756; Marville to Maurepas, 6 March 1743, ms. 719, reserve 21, no. 77, BHVP.

55. Parmentier, *Parfait Boulanger*, pp. 380, 620.

56. Ibid., pp. xii–xiii, 633.

57. Brocq cited by Parmentier, *Parfait Boulanger*, p. 635. Cf. ibid., p. 380. On Brocq, see also Parmentier, *Manière de faire*, p. 23; AAP, deliberations of the Hôpital général, 15 January 1787, n.s. 47; AN, F^{11} 1230; Bureau d'agriculture to Brocq, 21 nivôse an IV, ASP, 4 AZ 794.

58. Parmentier, *Parfait Boulanger*, pp. 637–39.

Chapter 9   The Journeyman's World outside the Shop

1. This may also be because our sampling was too narrow. Remember that there are thousands of *liasses* in the Minutier central and that there were no subject indexes at all when this research was undertaken.

2. AN, MC, C-536, 14 September 1731; MC, CXXXIV-286, 8 June 1759; MC, XCIII-21, 29 March 1751; MC, XXXVIII-391, 6 September 1751.

3. AN, MC, VII-208, 2 March 1715; MC, XXVIII-247, 10 June 1737; MC, VII-309, 5 July 1757; MC, IV-483, 9 June 1734; MC, LXV-311, 18 May 175?; MC, IV-569, 24 July 1750; AN, Y 10243, September 1762; Arsenal, ms. Bast. 10066, 8 November 1763; Y 12594, 20 May 1750; Y 12155, 23 January 1751; Y 12148, 14 December 1745; Y 15068, 7 April 1765; AN, Y 9539, 12 September 1760; ms. Bast. 10117, 31 August 1774; ms. Bast. 10054, 7 January 1760; Y 10269, 28 November 1767. Other documents indicate journeyman baker marriages without, however, revealing anything about the wives or the households. E.g., Y 12611, 18 March 1762, and Y 12605, 24 April 1757.

4. Thereze was probably related to Genevieve Le Guay, who married seventeen years earlier.

5. AN, Y 12142, 14 July 1740.

6. AN, Y 10269, 22 November 1765. Mainguet claimed that the marriage bans had already been published. On the dependence of baker widows on their garçons and on the remarkable devotion of some of latter, see Y 12740, 13 May 1754; Y 15350, 25 October 1752 (widow Petit); Y 15340, 4 May 1744 (widow Sens).

7. AN, Y 10243, September–November 1762; Sentence of the Grande Panneterie, 17 July 1710, BHVP; Y 10369, 28 November 1767.

8. AN, Y 12148, 14 December 1745; Y 12155, 23 January 1751.

9. Perhaps it is unfair to exclude the possibility that some of these women were victims of their own promiscuity.

10. AN, Y 11242, 3 May 1766.

11. The designation "marchand de grains" covered an extraordinary range of activity, from the international dealers with huge fortunes to the petty *blatiers* who eked out a humble existence. Chances are that Gobelet's husband came from the latter side of the spectrum. Otherwise it is hardly likely that his widow would have succumbed to the charms of a single journeyman baker.

12. AN, Y 15946, 20 November 1751.

13. AN, Y 15069, 7 April 1765; Y 9539, September 1760; ASP, DC⁶ 227, fols. 115–16.

14. AN, Y 9649, 2, 8 January 1757; Y 11243, 13 July 1756; ASP, DC⁶ 227, fols. 115–16, 14 April 1733.

15. AN, Y 15272, 8 December 1764; AN, MC, XXVIII-241, 5 November 1735; MC, XXVIII-216, 29 January 1730.

16. Arsenal, ms. Bast. 10034, 5 September 1752; AN, Y 15629, 5 May 1755; AAP, n.s. 47, Scipion, 31 March 1786 (after-death inventory); AN, X²ᴮ 1039, 1 February 1768.

17. See AN, MC, XIII-252, 25 September 1734; MC, XXIII-622, 12 March 1759; AN, Y 12729, 1 March 1735; Y 12575, 11 March 1731; MC, VII-287, 17 May 1753.

18. See, e.g., Arsenal, ms. Bast. 10080, 15 August 1772. Journeyman Pierre Philipaux, by contrast, kept his locker in his master's bedroom, where it was broken into. Ms. Bast. 10063, 14 October 1762.

19. AN, Y 15063, 13 March 1762; AN, MC, XXVIII-216, 16 March 1730. On occasion, even married journeyman bakers managed to establish something resembling a household outside the shop. See Arsenal, ms. Bast. 10054, 7 January 1760, and Y 12611, 18 March 1762.

20. AN, Y 13761, 19 August 1754. Cf. Arsenal, ms. Bast. 10039, 19 August 1754; Y 11229, 26 December 1742; ms. Bast. 10073, 15 November 1765. For other examples see ASP, D6B⁶, carton 4, June 1764 (Noel); Y 11388, 3 August 1772 (Sivry).

21. AN, Y 15238, 9 October 1731; Arsenal, ms. Bast. 10061, 30 June 1762; ms. Bast. 10064, 29 October 1761, and ms. Bast. 10034, 5 September 1752. For examples of the journeymen who lodged in the inns see the discussion of journeymen without papers above and Y 12614, 15 November 1765; Y 15233, 19 December 1726; Y 10370, 2 September 1776; AN, X²ᴮ 967, 27 September 1730; X²ᴮ 1031, 8 June 1764; Y 13121, 27 December 1769; Y 15238, 15 October 1731; Y 13115, 26 June 1763; X²ᴮ 1032, 13 July 1764.

22. AN, Y 11229, 26 December 1742; Y 15233, 19 December 1726; Y 10397, 29 April 1779; Y 10079, 26 August 1740; Y 11224, 2 March 1737; Y 15376, 27 July 1770; AN, X²ᴮ 1022, 5 July 1759; Y 10209, 19 June 1759; Y 11229, 26 December 1742; Y 13110, 17 April 1758; Y 11224, 2 March 1737; Y 15453, 27 October 1755; Y 12416, 11 January 1751; Y 11384, 7 February 1771; Arsenal, ms. Bast. 10064, 7 January 1763. Cf. the list of lodgings for "ouvriers" in the *Tableau universel de Paris,* pp. 31ff.

23. AN, Y 10209, 19 March, 19 June 1759; Y 13550, 6 December 1772; Y 11229, 26 December 1742. On the population density of rooms, see AN, X²ᴮ 1022, 4 July 1759, and Y 13110, 17 April 1758. On the hôtesse-mère, see *L'Express,* 11–18 October 1977.

24. AN, Y 12609, 5 November 1760.

25. AN, Y 14778, 29 August 1728; Y 10397, 29 April 1775; AN, X²ᴮ 1063, June 1779; X²ᴮ 1022, 4 July 1759. See the police ordinance of 17 August 1781, BHVP, which empowered the guild to "break" the boycott by conscripting garçons to work in damned shops.

26. Arsenal, ms. Bast. 10078, 20 December 1772; AN, Y 10621, 1 January 1770; ms. Bast. 10024, 14 February and July 1769; ms. Bast. 10078, 12 February 1772; Y 11578, 29 October 1761, and Y 12614, 15 November 1765; ms. Bast. 10054, 13 February 1760; ms. Bast. 10111, 27 July 1771; ms. Bast. 10078, 12 January 1772; AN, X²ᴮ 1022, 4 July 1759; Y 10209, March 1759.

27. AN, Y 11364, 12 August 1767; Y 11385, 28 June 1771; Arsenal, ms. Bast. 10080, 7 August 1772; Y 11388, 7 August 1772; ms. Bast. 10127, 16, 17 January 1772; ms. Bast. 10050, 12 February 1769; ms. Bast. 10083, 7 April 1773.

28. Arsenal, ms. Bast. 10076, 24 September 1766; AN, Y 14953, 15 May 1738; Y 15233, 12 December 1726; Y 15357, 27 April 1757.

29. AN, Y 15955, 14 December 1758; Y 15070, 12 January 1767.

30. AN, Y 14531, 20 May 1737, and Y 14532, 10 February 1738; AN, X^{2B} 1023, 30 June 1760; X^{2B} 991, 6 October 1742; Y 12140, 8 February 1738; Y 11571, 11 September 1755; Y 15355, 23 June 1755; Arsenal, ms. Bast. 10166, 29 June and 6 July 1737; Y 11568, 7 December 1752; Y 15239, 23 February 1732; Y 12138, 8 June 1736; Y 12151, 26 September 1748; ms. Bast. 10141, 4 February 1761.

31. AN, X^{2B} 1025, 12 June 1761.

32. For example, see AN, X^{2B} 1038, 4 December 1767; X^{2B} 1024, 16 October 1760; X^{2B} 1024, 5 August 1760; AN, Y 9634, 29 February 1769; X^{2B} 987, 26 October 1740; X^{2B} 988, 22 April 1741; X^{2B} 1007, 4 September 1750; Y 9538, 19 March 1723; Arsenal, ms. Bast. 10055, 6 June 1760; ms. Bast. 10038, 29 January 1754; APP, AB/119. A journeyman's wife had the bad judgment to try to rob the duc de la Vrillière, the secretary of state responsible for the Parisian police. Ms. Bast. 10117, 31 August 1774.

33. AN, Y 14962, 4 November 1746.

34. Patissier, *Traité des maladies*, p. 195; prefect of police to minister of the interior, 30 May 1807, in Vauthier, "Ouvriers," pp. 426–51; Bouteloup, *Travail de nuit*, p. 63. For a thoughtful view of the socializing and desocializing impact of ambient violence, see Daniel Roche, "Pleasures and Games: Jokes, Violence, Sexuality," in Ménétra, *Journal of My Life*, pp. 270–74, and *Le Peuple de Paris* (Paris, 1981), pp. 242–75. See also the insights and powerful vignettes in Arlette Farge, *La Vie fragile: Violence, pouvoirs, et solidarités à Paris au XVIIIe siècle* (Paris, 1986).

35. Drink also provided a caloric supplement and helped the journeyman to replenish some of the strength that he lost as result of unremitting perspiration in the intense heat of the bakeroom. It is well known even today that chefs working over hot ovens are inclined to consume large amounts of alcohol in order to sustain themselves.

36. AN, X^{2B} 1037, 26 May 1767; AN, Y 10233, November 1766; Y 14948, 29 June 1733; Y 15337, 27 August 1730; Y 11213, 31 August 1726; Y 15248, 19 May 1750; Y 13550, 6 December 1772; Y 15233, 19 December 1726; Y 11241, 12 December 1754; Y 93094, 4 March 1743; Y 10620, 2 August 1768; X^{2B} 1046, 20 May 1772; Y 10325, December 1771; Y 15617, 16 January 1749; Y 13535, 22 August 1766; Y 14531, 24 November 1737; Y 11580, 31 March 1762; Arsenal, ms. Bast. 10104, 23 December 1763; Y 15269, 30 January 1761; Y 13125, 5 March 1773. Guillaume, incidentally, was a "master baker without shop serving the masters in the capacity of journeyman baker," a striking example of downward mobility and its attendant woes.

37. AN, Y 15070, 12 January 1767, and Y 14531, 24 December 1737; Y 13110, 31 August 1758; Y 11229, 26 December 1742, and Y 10370, 2 September 1776. The church-tavern couple was not peculiar to France. See Keith Thomas, *Religion and the Decline of Magic* (New York, 1971), pp. 161–62. Note also the remarks of Le Bras on the opposition *cabaret-église*, or *taverne-tabernacle*, equally valid in many ways for the city as for the village: "The tavern keeper is the antipriest of the village. He has the same clientele as the priest, the same hours of opening, and he tries to keep his clients there before,

during, and after services. There is more than one book to be written on the social importance of the tavern, on its role as counterchurch." Gabriel Le Bras, *Etudes de sociologie religieuse*, vol. 2, *Sociologie de la pratique religieuse dans les campagnes françaises* (Paris, 1955), p. 240.

38. Delamare, *Traité*, vol. 1 (1705), p. 580. Cf. Mercier, *Tableau*, vol. 2 (1782), p. 138, and vol. 4 (1782), p. 159.

39. Fréminville, *Traité de la police*, p. III; Des Essarts, *Dictionnaire universel*, vol. 1, p. 472; *Encyclopédie méthodique*, Jurisprudence, vol. 140, pp. 132–33, and vol. 147, pp. 463–66; AN, Y 10324, 11 March 1772; Y 13536, 22 April 1776.

40. Arsenal, ms. Bast. 12369 (1769) and 10846 (1724); AN, Y 12826, 18 October 1776; police sentence, 12 December 1704, APP, Lamoignon, XXII, 88–92; police sentence, 31 October 1739, Lamoignon, XXXII, 613–18; police sentence, 20 July 1742, Lamoignon, XXXI, 203–7; ms. Bast. 10846 (1724); ms. Bast. 12202, December 1764; Y 9523, 23 February 1724.

41. Cf. the role of association for London bakers. Thrupp, *Worshipful Company*, pp. 108ff.

42. Elsewhere I have speculated that the confrérie may have been more important than the classical *compagnonnage* in the capital. See my "La Police." David Garrioch and Michael Sonenscher, "Compagnonnages, Confraternities and Associations of Journeymen in Eighteenth-Century Paris," *European Historical Quarterly* 16 (1986), pp. 25–45. On nineteenth-century compagnonnage for bakers—the *chiens blancs*—see Carny, *Compagnons*, vol. 2, pp. 150–51. The most probing general work on compagnonnage is Cynthia M. Truant, *The Rites of Labor: The Brotherhoods of Compagnonnage in Old and New Regime France* (Ithaca, N.Y., 1994).

43. AN, AD XI 14, 24 April 1719.

Chapter 10   Establishment

1. I found no documentation to support the claim made in the *Tablettes royales de renommée, ou Almanach général d'indication des négocians, artistes célèbres et fabricans* (Paris, 1772) that bakers characteristically required a four-year probationary period for journeymen. One finds a surprisingly large number of Protestants in other trades, especially the highly skilled ones, but I have no evidence of an avowed Protestant baker boy.

2. AN, Y 15599, 4 August 1740.

3. Legrand d'Aussy, "De la nourriture des Français," BN, ms. n.a.f. 3328, fol. 10; Franklin, *La Vie privée d'autrefois*, p. 194. Cf. the same test prescribed by the master bakers of the faubourg Saint-Germain-des-Prés, 23 May 1758. AN, K 1030–31.

4. Diderot et al., *Encyclopédie*, vol. 3, p. 273.

5. Statutes and regulations of 1719 and of 1746/57, and police sentence, 16 May 1721, BHVP.

6. AN, Y 9387, 17 December 1752.

7. AN, Y 15242, 30 December 1735; police sentence of 29 July 1720, AN, AD XI 10. Cf. the complaint in the nineteenth century that 80 percent of the masters were nothing more than "bread merchants" who did not know how to make bread. Barbaret, *Le Travail en France*, vol. 1, p. 392.

8. AN, AD XI 14, 12 June 1740. Though the arrêt of 1740 was apparently not abrogated until May 1745, it is likely that the guild continued to insist on the old fee schedule during the entire period.

9. AN, AD XI 11. According to this same table, the butchers had paid fifteen hundred livres, the brewers eleven hundred livres, the shoemakers 550 livres, the masons seventeen hundred livres, and the coopers 1,250 livres. In the bakers' guild at Bordeaux, while sons paid only thirty livres, outsiders paid four hundred livres in 1773 and six hundred livres in 1786. Benzacar, *Le Pain à Bordeaux*, p. 76. On the London guild, see Thrupp, *Worshipful Company*, pp. 41 and passim. Hurtaut and Magny say nine hundred livres, but their figure may be drawn from the edict of 1776. *Dictionnaire historique*, vol. 1, p. 659. A leading contemporary reference guide that was published before the edict also indicates nine hundred livres, but it was wrong on a number of other facts concerning the bakers. *Tableau universel de Paris*, p. 211. See also the *Guide des marchands* (1766) cited by Lespinasse, *Métiers de Paris: Ordonnances générales*, p. 197.

10. AN, MC, XXIV-553, 2 June 1712, and MC, XCI-74, 29 July 1713. (Cf. the evaluation of four hundred fifty livres in MC, XIII-180, 24 June 1714); MC, XIII-241, 1 March 1731; BN, Joly 1733, fol. 17; AN, Y 15242, 30 December 1735; Y 11233, 1 August 1746; MC, XCIV-301, 28 January 1761, and MC, V-539, 28 July 1764; Y 9292, 27 October 1775.

11. ASP, D4B⁶ 41-2260, 15 June 1771; D4B⁶ 46-2706, 23 November 1772; D4B⁶ 39-2138, 30 March 1772; état des boulangers, September 1767, AN, Y 13396; D4B⁶ 63-4102, 7 July 1772.

12. See, Robert F. Gould, *The History of Freemasonry* (New York, 1884–87), vol. 1, p. 191, who perceives in this ceremony a striking likeness to the attestation of the Chinese, to wit, "may I be as broken as this pot if I break my oath."

13. See, for example, the edict of August 1776, BHVP.

14. Arrêt du conseil, 14 September 1728, BHVP; AN, V⁷ 423, 5 August 1733. Apparently Royon had successfully completed his masterpiece.

15. Diderot et al., *Encyclopédie*, vol. 9, pp. 911, 913; ibid., vol. 3, p. 273; Clicquot de Blervache, *Considérations sur le commerce*, p. 18; Clicquot de Blervache [M. Delisle, pseud.], *Mémoire sur les corps de métier* (The Hague, 1758), pp. 13–23 and passim; M.J. Auffair, "Le Commerce abandonné à lui-même, ou Suppression des compagnies, sociétés et maîtrises," *Journal de commerce*, March 1761, pp. 120–24.

16. Edict of February 1776, Isambert, *Recueil des lois*, vol. 23, pp. 372–73.

17. Coyer, *Chinki*, p. 30. Cf. the following remarks written in 1771 by Daure, former Paris baker, subsistence entrepreneur, and future head of the bakery of the French expeditionary force in America—a well-informed expert: "the ease with which an individual can set himself up as a baker results in the operation of many more than necessary and thereby brings about a quite satisfactory competition." "Mémoire," AN, F¹¹ 264.

18. It would be wrong to envision the arrêt du conseil exclusively as an instrument of redress used by the king to mitigate corporate arrogance and inbreeding. Ordinarily it was utilized to enable deserving journeymen whose advancement may have been obstructed to obtain mastership. But it was also called upon in some instances to circumvent ordinary corporate procedures with the consent and perhaps the encouragement of the guild, for the guild stood to profit financially from the exceptions. See, for example, the case of Louis Guerineu, who is dispensed from serving the three-year journeyman's training period in return for the payment of a supplementary impost of

five hundred livres over and above the mastership fee to be applied to the corporate debt amortization fund. Arrêt du conseil, 15 January 1760, in Gosset, *Inventaire des arrêts,* p. 40. Masters' sons born before their fathers' mastership and thus constrained to follow the outsiders' path to upward mobility were the ones most likely to have been tempted by this venal disposition.

19. For an example of the liberal argument, see Clicquot de Blervache, *Considérations sur le commerce,* p. 18. Martin Saint-Léon also takes the liberal position. *Histoire des corporations,* pp. 519–20. See edict of 25 June and arrêt du conseil of 9 July 1726, AN, AD XI 10.

20. See the edict of March 1767, BN, fichier central. Obviously the consequences of this legislation must be assessed on a guild-by-guild basis.

21. Edict of August 1776, AN, AD XI 11. A recent study of the bakers' guild of Namur invalidates "l'hypothèse xénophobe et héréditaire" of a corporation closed to all but insiders. See Godfroid, "La Corporation des boulangers à Namur au 18e siècle," pp. 24–26.

22. ASP, D4B⁶ 88-6016, 22 October 1783, and D4B⁶ 90-6211, 14 September 1787; AN, AD XI 11. The increase was to help the guilds pay for the loan they needed in order to raise 1,500,000 livres for a vessel of war that they offered the king. In principle, three-quarters of all mastership fees went into an amortization fund to pay debts. The remaining quarter was to be used for current administrative expenses. Of the latter, 20 percent was allotted for the honoraria of syndics and their helpers (adjoints).

23. Martin Saint-Léon claimed that the edict of August maintained the requirement to complete apprenticeship and journeyman's probation as in the past. *Histoire des corporations,* p. 586. I find nothing in the edict that perpetuates these requirements. The guild bylaws issued several years later seem to sustain my view.

24. Statutes and regulations of 1783, AN, AD XI 14. A royal declaration of 18 August 1777 allowed husbands to provide in advance for the automatic incorporation of their widows as masters by paying a special premium equivalent to one-quarter of the mastership fee they paid at the time of their admission. Reminiscent of certain legislation regarding the transmission of venal offices, this declaration served the fiscal interest of the guild and the appetite for security of the bakers. By an arrêt of 29 December 1783, widows were exempted from the obligation to pass the standard examination before being received as masters. AN, AD XI 77.

25. In fact, however, it is not at all certain that the abolition of the corporations would have offered the journeymen the opportunities that they vaguely glimpsed. Turgot had no intention of turning the socioprofessional structure on its head overnight. The journeymen may have become quickly disenchanted had Turgot's reforms survived.

26. AN, Y 15267, 23 November 1759.

27. AN, MC, VII-325, 25 January 1760; MC, VII-325, 25 January 1790; MC, VII-325, 29 September 1761; MC, XXXIII-622, 12 May 1759; MC, LXXXIX-453, 21 October 1738; MC, XXXIII-580, 11 October 1770; MC, VII-309, 22 July 1757; MC, VII-291, 29 June 1754; MC, LXXXII-606, 24 February 1786; MC, VII-329, 17 December 1760; MC, VII-329, 21 December 1760; MC, VII-291, 1 June 1754; MC, XXVIII-338, 10 August 1754; MC, VII-369, 18 February 1767; MC, VII-301, 28 March 1756; MC, VII-308, 8 May 1757; MC, XCIV-310, 22 May 1762; MC, XCIV-303, 28 March 1761; MC, LXXXIX-664, 5 January 1769; MC, VII-390, 23 May 1770; MC, VII-402, 5 February 1772; MC, VII-386, 19 October 1769; MC, XXX-438, 6 September 1773; MC, VII-404,

21 August 1772; MC, XIII-241, 1 March 1731; MC, XXIV-646, 2 June 1735; MC, VII-404, 30 July 1772; MC, XLV-492, 27 October 1753; MC, CIX-490, 3 August 1735; BHVP, n.a. 29, fol. 105, 26 August 1783; BHVP, ms. ser. 118, cote provisoire 1482; AN, Y 11232, 18 June 1744; Y 15257, 20 June 1749; Arsenal, ms. Bast. 10073, 31 December 1765; ASP, D5B⁶ 2707, 1778; D4B⁶ 83-5592, 3 September 1783; D4B⁶ 75-4997, 27 January 1780; D4B⁶ 91-6237, 20 July 1784; D4B⁶ 90-6140, 20 March 1784; D4B⁶ 90-6211, 14 September 1787; D4B⁶ 88-6048, 24 November 1738; MC, X-619, 31 May 1770 (attermoiement); MC, XXIV-727, 17 August 1751; MC, XXVIII-321, 6 May 1751; MC, IX-677, 11 November 1751. Beyond the thirty-five-hundred-livre "pot de vin," master Loiselle listed 950 livres in miscellaneous setting-up costs, including the interest on the loan required for the acquisition of the fonds. MC, XXXIII-586, 20 November 1771.

28. One is reminded here of Giovanni Levi's land market in his superb *Inheriting Power: The Story of an Exorcist*, trans. Lydia G. Cochrane (Chicago, 1988).

29. In two separate cases, drawn from documents that are not themselves fonds transfer contracts, the pot de vin alone is reported as thirty-five hundred livres. AN, MC, XXXIII-586, 20 November 1771, and ASP, D4B⁶ 42-2386.

30. Note that the evaluation in this case was not left merely to the parties but was undertaken by two master bakers, one of whom was probably the brother of the seller. AN, MC, XCIIV-310, 22 May 1762.

31. AN, MC, XXVIII-251, 26 February 1738.

32. Among the most instructive contracts are AN, MC, VII-390, 9, 23 May 1770; MC, LXXXIX-453, 21 September 1738; MC, VII-325, 25 January 1760.

33. A good example is AN, MC, LXXIX-664, 5 January 1769. In the nineteenth century the buyer still had no right to examine the books of his prospective shop, but he commonly did enjoy the "quinzaine de garantie" during which time he could take the measure of the business and decide whether he really wanted it. Obviously the seller was tempted to manipulate conditions to his advantage. One authority claimed that baker-sellers sometimes called on specialists, experts in promotion and puffery— "entrepreneurs de garantie"—to help them increase sales during the two-week trial period in order to dupe the buyer. Bouteloup, *Travail de nuit*, p. 11n.

34. AN, MC, VII-325, 25 January 1760, and MC, XXVII-338, 10 August 1754.

35. AN, MC, XXIII-622, 12 March 1759; MC, VII-309, 22 July 1757; MC, VII-390, 9, 23 May 1770.

36. AN, Y 12148, 30 September 1745; AN, MC, LXXXIX-453, 21 September 1738; MC, IX-677, 11 November 1751.

37. AN, MC, VII-291, 1 June 1754.

38. AN, MC, VII-291, 1, 19 June 1754.

39. ASP, D4B⁶ 87-5884, 10 April 1783; D4B⁶ 88-6048, 24 November 1783; D4B⁶ 83-5592, 14 January 1782; D4B⁶ 41-2260, 15 June 1771. In the 1880s the estimated cost of opening up a bakeshop in Paris, *without* counting the purchase of a clientele, was eight to ten thousand francs. Barbaret, *Le Travail en France*, vol. 1, p. 469.

40. ASP, D4B⁶ 88-6016, 22 October 1783; D4B⁶ 104-7356, 31 December 1788; D4B⁶ 102-7168, 18 April 1789; D4B⁶ 41-2260, 15 June 1771 and 25 June 1784; D4B⁶ 113-8056, 28 November 1791, and D4B⁶ 24-1249, 1 September 1763; D4B⁶ 63-4095, 2 July 1777.

41. I am alluding to the cost of grain and flour for starting off. A baker would need at

least twelve sacks of flour his very first week (say, eight white and four medium-white), representing an expense of about three to four hundred livres. If he had family recommendations and other patronage, the new baker could usually obtain credit from merchants and brokers. Remember that he also had to buy wood, yeast, candles, salt, and other basic goods and material.

42. See appendix A. I have found only one instance where an owner demanded key money, a "pot de vin," for the lease rather than the fonds. Master baker Jean-Baptiste Vosniez paid six hundred livres for access to quarters on the rue Saint-Jacques. ASP, D4B⁶ 90-6211, 14 September 1787.

43. ASP, 6 AZ 180; D4B⁶ 20-1004, 26 September 1759; AN, Y 14942, 30 July 1729. For comparative perspectives on middling masters' living space, see Davet, *La Population de l'axe nord,* pp. 247–48.

44. AN, Y 15609, 18 September 1742.

45. AN, Y 13634, 27 April 1729. (Is this Delu the Mortellerie grain baron?)

46. AN, MC, CXXI-294, 23 April 1732.

47. AN, MC, V-539, 28 July 1764.

48. The array of utensils enumerated in the Chartie contract along with the inclusion of a market as well as a shop clientele suggests that the price was seriously underestimated. Leguay estimated his equipment at three thousand livres, charged his son four hundred livres for the cost of his mastership, saddled him with onerous debt recoveries, and generally seems to have tried to inflate the price of the transfer in order not only to cover the share he owed his son from his wife's estate but also to assure himself a sweet annuity. A possible counterproof or hint of excessive disproportion in the fonds evaluations: Chartie rented his house and shop to his son for thirteen hundred livres, while Leguay rented his for eight hundred livres.

49. AN, MC, VII-325, 25 January 1760; MC, XCIV-310, 22 May 1762; MC, VII-308, 8 May 1757. Like Leguay, Constant evaluated her tools at three thousand livres. But she also included in the price a very large amount of flour.

50. AN, MC, V-539, 8 July 1764; MC, XCIV-310, 22 May 1762; MC, LXXXIX-581; MC, VII-325, 27 February 1760.

51. In some cases it appears that the son, living at home, engaged independently in the baking trade at the same time and in the same place as his father. See, for example, the case of Martin, father and son, in the grande rue du faubourg Saint-Antoine. ASP, D2B⁶ 736, 7 April 1741. Neither, of course, was a master, and I suspect that the bulk of their commerce was concentrated in the markets rather than in the shops. It was illegal for corporate bakers to share the same oven and shop. The jurés enforced this interdiction energetically, as we shall see. *Encyclopédie méthodique,* Jurisprudence, vol. 9, p. 438.

52. Marriage contract of 28 April 1739 in *compte,* AN, Y 15621, 18 December 1750; AN, MC, XXIV-639, 11 January 1732. The late father of Saunier's niece was a master baker.

53. AN, MC, VII-329, 17 December 1760; MC, XXIV-646, 2 June 1735; MC, VII-404, 30 July 1732; MC, CI-402, 17 January 1744; MC, LXV-193, 22 July 1717; MC, XLI-416, 17 January 1771; MC, VII-402, 5 February 1772; MC, VII-293, 10 October 1754; MC, VII-329, 17 December 1760; MC, LXXXV-581, 28 June 1764; MC, XLV-492, 27 October 1753; MC, XIX-317, 30 January 1734.

54. AN, Y 15615, 25 January 1748; Y 15608, 30 August 1744. Garin and his sister-in-law

quarreled violently over the management of the shop. He claimed that she incited the journeymen against him, that she disturbed the "peace of his home," and that she was "dirty" (she fed the cat on the table, for example).

55. AN, MC, VII-297, 31 July 1755.

56. There was also the possibility of an establishment in one of the so-called privileged places, such as Saint-Jean-de-Latran, the enclos du Temple, or the Quinze-Vingt. But a young baker would find it rather difficult to set up shop in one of these citadels without the encouragement of a protector. See Delamare, *Traité*, vol. 2 (1729), p. 754.

57. AN, MC, LXVIII-374, 12 May 1729; MC, VII-386, 12 August 1796. There are of course exceptions, especially in the hamlets in the Parisian sphere of influence, in which the bakers cultivated local clientele at the same time that they served the capital on market days. Thus Jacques Pisson purchased the business of Jean Cosme (who dared to call himself "master baker") at Vaugirard for fifteen hundred livres: nine hundred livres for the equipment (which included a bolter, two kneading troughs, a counter, and a host of tools) and six hundred livres in pot de vin. MC, LXXXII-606, 24 February 1786.

58. See, for example, Jean and François Bonnion, sons of master François Bonnion, of the faubourg Saint-Martin, who set themselves up as plain bakers in the faubourg Saint-Antoine. AN, Y 14535, 28 January 1739.

59. See the statutes and regulations of 1719 and 1746/56, BHVP. See also police sentence, 22 June 1740, AN, AD XI 10; BN, Joly 1111, fol. 171; *Encyclopédie méthodique*, Jurisprudence, vol. 9, p. 438; Olivier-Martin, *Organisation corporative*, p. 225; Desmaze, *Métiers*, p. 159. On the incursion of illicit work into other métiers, see Davet, *La Population de l'axe nord*, p. 368.

60. AN, MC, VII-301, 28 March 1756. Master Paul Chevillot openly sold his fonds (for eleven hundred livres) to a baker boy in a formal notarial transaction. MC, XXIV-727, 17 August 1751. So did merchant baker Claude Leprince of the faubourg Saint-Antoine—for the same price. MC, XXVIII-321, 6 May 1751. Baker boy François Alix bought the tools and equipment of a privileged baker on the rue Mouffetard. MC, XLVII-135, 26 August 1751. Journeyman baker Jacques Bourdon rented a shop in a notarial contract for ninety livres a year. MC, XXVIII-332, 26 June 1751.

61. AN, MC, VII-329, 21 December 1760. Since Leblanc paid only six hundred livres for the fonds—in fact four hundred fifty livres, for he imposed a penalty of one hundred fifty livres on Lepileur for breaking the lease and for damage to be repaired—one may infer that either the value of the business had deteriorated radically in Lepileur's hands or that Lepileur was in desperate financial straits (or both).

62. AN, Y 15355, 12 November 1755; Y 15242, 30 December 1735. Obviously this procedure was unusual. It may have had something to do with Fleury's influence. Or perhaps Aubert was a master's son.

63. AN, MC, XCIX-469, 26 June 1741.

64. AN, Y 15244, 20 December 1737; Y 15242, 30 December 1735.

65. AN, MC, XXXIV-501, 17 June 1725.

66. See Delamare, *Traité*, vol. 2 (1713), pp. 877–78. There were officially twelve privileged bakers following the court at any given moment, but the number of bakers claiming the title was usually greater. Bakers continued to claim the privilege even after

they retired or passed it on to a relative. Moreover, other royal jurisdictions attached to the royal court also claimed the right to grant privileged status to merchants and artisans.

67. AN, Y 15241, 6 April 1734; Y 15242, 6 April 1734; Y 11221, 6 April 1734; Y 15235, 12 February 1758.

68. AN, Y 15236, 1 July 1729; Y 13494, 11 December 1729.

69. AN, Y 11222, 25 November 1735; Y 15599, 4 August 1740; Y 9434, 20 November 1733. Frambois's operation was by no means marginal and thus could hardly have been discreet. He had seventy-one sacks of flour, a huge inventory, in his storeroom above the shop.

70. AN, Y 15240, 18 July 1733; Y 9434, 20 November 1733; Y 9377, 19 June 1725; Y 11222, 25 November 1735; AN, Y 15244, 14 March 1737; AN, Y 15233, 20 July 1726; AN, Y 9377, 19 June 1725. Cf. the cases of Hautier and Larticle in ASP, D2B⁶ 1080, 29 December 1969, and Arsenal, ms. Bast. 10149 (Josse to lieutenant general of police, 11 April 1726).

71. AN, Y 9434, 20 November 1833; Y 9377, 19 June 1725.

72. AN, Y 15236, 1 July 1729; Y 15235, 23 November 1728; Y 15237, 25 September 1730; Y 9379, 18 September 1740.

Chapter 11 Marriage Strategies and Family Life

1. For a theoretical discussion of endogamy see Robert King Merton, "Intermarriage and the Social Structure: Fact and Theory," in *The Family, Its Structure and Functions,* ed. R. L. Coser (New York, 1964), pp. 140–41. Among the extremely variegated group designated "masters and merchants," Daumard and Furet found that 60 percent married in the same category (only a relatively faint reflection of endogamy) while 20.2 percent married with the same corps de métier (a useful measure of endogamy according to our criteria). Adeline Daumard and Francois Furet, *Structures et relations sociales à Paris au milieu du XVIIIe siècle, Cahiers des Annales,* no. 18 (Paris, 1961), pp. 74–75, 78–79. In her study of northern Paris, Hélène Davet found 52 percent endogamy among the broad "maîtres de métiers" category (n = 61). *La Population de l'axe nord,* p. 190.

2. See the case of master baker Adrien Loisel, who refused the persistent entreaties of a mason for his daughter's hand, for he wanted her to marry a master baker like himself. Ultimately he realized his wish, for his daughter married master Jean Solle. AN, Y 13572, 30 April 1762.

3. AN, MC, XXVIII-208, 8 December 1702; MC, LXXXI-416, 24 January 1771; MC, VII-301, 28 March 1756.

4. AN, MC, VII-301, 28 March 1756.

5. AN, MC, XXXVIII-279, 19 December 1734; MC, XXXVIII-362, 9 November 1747; MC, LXVIII-374, 12 May 1729; MC, LXXXI-416, 17 January 1777; MC, CI-402, 17 January 1744; MC, VII-400, 26 September 1771.

6. ASP, DC⁶ 19, fol. 200, 17 May 1772; DC⁶ 19, fol. 274, 6 October 1775; DC⁶ 18, fol. 218, 12 June 1773; DC⁶ 234, fol. 28, 16 April 1749; DC⁶ 22, fol. 159, 4 March 1780; DC⁶ 150, fol. 55, 7 September 1723. AN, MC, LXXXVI-649, 7 July 1751; Gindin, "Principaux Labourers," p. 170.

7. AN, MC, XXVIII-338, 28 July 1754; AN, Y 15373, March 1767; ASP, DC^6 240, fol. 112; LXIV-341, 5 July 1751; MC, CIX-412, 1 December 1742 (it is possible that there was a third son who was a flour dealer); MC, XXVIII-251, 10, 13 February 1738; Y 15338, 30 April 1742. There are scores of other examples of the constitution of familial mafias in the bakery. See, inter alia, ASP, 5 AZ 1430, 15 May 1754; DC^6 20, fol. 95, 12 December 1776; DC^6 259, fol. 262, 4 September 1780; Y 14541, 16 January 1746; Y 12597, 15 December 1752; Y 14964, 5 June 1748; Y 11063, 18 July 1748; Y 15258, 10 December 1750.

8. AN, MC, XII-412, 10 October 1730; ASP, 6 AZ 843, 19 October, 28 November 1776 and 9 January, 28 April 1777; AN, Y 14084, 22 December 1757, and Y 12625, 6 June 1775. Remarriage also caused serious frictions on occasion, most commonly a kind of Oedipal-professional jealously of the widow's son for her second husband. See, for example, Y 12599, 23 July 1753.

9. AN, Y 15261, July 1753; AN, MC, LXVII-292, 1 April 1731.

10. Given the extremely incomplete data, the marriage contracts seriously underestimate the dynastic case. The sample consists of 175 contracts, but in over 80 percent of the cases we have no information concerning either siblings or their professions. There is every reason to believe that a great many of these brides and grooms had brothers and sisters in the bakery.

11. Though it should be noted that in most instances only a part of the contribution from each side legally "entered," or became part of, the community established between the husband and the wife.

12. In 1749, Daumard and Furet found that there were contracts for about 61 percent of all marriages celebrated. Allowing for lost documents and contracts signed by Parisians in the countryside, they estimate that about three-quarters of all Parisians signed contracts. I suspect that the proportion is higher for all bakers above the journeyman level and that it was probably 100 percent for masters. Daumard and Furet, *Structures sociales*, pp. 8, 9, 39.

13. Daumard and Furet argue the dissimulation thesis. *Structures sociales*, p. 21.

14. In general, times were easy for bakers between 1750 and 1765. But many of them suffered greatly during the dearth of the late sixties and early seventies. See Kaplan, *Bread, Politics, and Political Economy*, vol. 1, pp. 328–43.

15. One cannot rely on the size of the husband's marriage contribution to predict the size of the dowry, or vice versa. There was only a modest association between them: changes in one account for no more than 26 percent of the variation in the other ($r = .51$ at .001).

16. See Daumard and Furet, *Structures sociales*, pp. 30–32. The bakers' distribution is surprisingly close to that of the wage earners in industry and commerce, the category ranked below the master-merchants in the 1749 study. Three-quarters of the wage earners were located between five hundred and five thousand livres, while about 65 percent of the bakers were in this range. Ibid., p. 28. In the 1749 study, 61.7 percent of all fortunes were situated between five hundred and five thousand livres. Ibid., p. 23. Cf. the awkward category of "maîtres de métiers" and "petits marchands" in Davet's study of the north of Paris, whose average marriage contributions were three thousand livres (bride) and thirty-five hundred (groom), for a combined mean marriage fortune of 5,430 livres. Elsewhere the author says that about two-thirds of all the couples in this

category had fortunes at the level of five thousand livres or less. Davet, *La Population de l'axe nord,* pp. 197, 397.

17. AN, MC, VII-293, 10 October 1754; AN, Y 15621, 28 April 1739; MC, LXXXV-581, 28 June 1764; MC, VII-329, 17 December 1760; MC, XXXVIII-208, 8 December 1720; MC, XXVIII-246, 27 February 1737; MC, XXVII-282, 16 August 1743; VII-329, 17 December 1760. Other contracts made reference to the transfer of portions of arable land and mills. See, for instance, MC, XII-179, 15 April 1714; MC, XXVIII-217, 14 August 1738.

18. In principle the widow enjoyed the usufruct of the *douaire,* which was supposed to revert eventually to the children born of the marriage. *Dictionnaire de Trévoux,* vol. 3, p. 441. See also Diderot et al., *Encyclopédie,* vol. 5, pp. 68–71. There is a moderately strong correlation between the douaire and the male marriage portion: .63 at the .001 significance level. Although there is no juridical link between the wife's dowry and the douaire, the two were even more closely associated: .87 at the .001 level. It is quite likely that bakers calculated the douaire as a function of the bride's contribution as well as of their own wealth.

19. *Dictionnaire de Trévoux,* vol. 5, p. 956, and Diderot et al., *Encyclopédie,* vol. 13, pp. 272–73. The préciput was more strongly associated with the douaire (r = .86) than with the husband's contribution (r = .63) at the same confidence level (.001).

20. I discounted the five-thousand-livre fortune of a baker, son of a master, marrying the daughter of a baker, because there was only one case.

21. Jean-Baptiste Jobidon found women for successive marriages on the same street where he had lived since acquiring his mastership. AN, MC, CIX-412, 17 November 1712, and MC, XLIX-538, 19 November 1730. For a rich discussion of the characteristics of neighborhood, see David Garrioch, *Neighborhood and Community in Paris, 1740– 1790* (Cambridge, 1986).

22. Jean Dutarte and Marie Guenn lived in different quarters of Paris, but both were children of deceased Burgundian laboureurs, perhaps from the same village. AN, MC, XXIV-566, 12 July 1714. Charles Desfosses and Maris Hareux resided in different neighborhoods in Paris, but their fathers were both laboureurs in the same village in Picardy. AN, MC, LII-278, 1 February 1738.

23. Ninety-five percent of middle-level masters in the north of Paris signed as did 89 percent of their wives. Davet, *La Population de l'axe nord.* Harvey J. Graff argues that signatures underestimate the level of reading competence. *The Literacy Myth: Literacy and Social Structure in the Nineteenth Century* (New York, 1979).

24. AN, MC, XXVIII-282, 18 August 1743; AN, Y 13651, 6 December 1755.

25. AN, MC, LXV-149, 6 July 1700; MC, LXV-193, 3 July 1717; MC, CLX-490, 12 December 1703. The stepson of Denis Larcher from his second marriage, Vincent Second, abandoned his debutant mason's career to become a baker. Thanks to his stepfather, he rapidly became a master. MC, LXV-196, 31 July 1718.

26. Baker Claude Panplune (or Pampelune?) married a baker widow who had six minor children, and master Jacques Legros married the widow of a master baker who brought him four minor children. AN, MC, LXXXIX-131, 12 May 1694; MC, XXXVIII-209, 19 January 1721.

27. For over two-fifths of the bakers, the investment in goods and equipment repre-

sented more than 50 percent of the marriage fortune, whereas for 90 percent of the brides there was no commercial rubric in the dowries at all. There was a relatively strong association between the dowry and the trousseau: r = .6 at .001. It is possible that both the husband's commercial and the wife's clothing and furnishings categories were bloated in order to give the marriage contributions a more respectable appearance. But it should be remembered that each family served as a sort of control on the other. In most cases the equipment and the trousseau were examined by all parties and evaluated by common accord. Given the enumeration occasionally made, it seems to me that it is quite possible that these categories were sometimes underestimated—preemptively, as it were, by overprudent witnesses. In fact only 5 percent of the grooms and 9 percent of the brides had any real property at all. The average for rentes is misleading, for in fact over three-quarters of the bakers had none at all, while only 7 percent of their future wives owned any annuities.

28. In these instances, however, it is likely that the wife simply did not count her bakeshop as part of her dowry, either because her late husband's estate was not settled, or because she did not want any part of the business capital to enter into the matrimonial community.

29. Aside from the wife's father's profession, we rarely know very much about bakers' wives. In only three instances have we encountered prospective wives who specifically indicated professions. The Bassille sisters, daughters of a baker, were dressmakers, and another bride was a seamstress. AN, MC, VII-297, 1 June 1755; MC, VII-329, 17 December 1760; MC, VII-282, 10 February 1752. A fourth wife seems to have worked as a practical nurse after her marriage, but her husband may only have been a journeyman. Arsenal, ms. Bast. 10083, 28 May 1773. Since other brides indicated that parts of their dowries came from "savings and earnings," it stands to reason that they exercised some occupation. The bakers' daughters probably worked in their fathers' shops. Of the 175 brides in our study, we know the exact age of only sixteen: ranging from eighteen to twenty-five, they averaged twenty-two years old. Twenty-three others were minors, that is to say, under twenty-five years of age. Save for the forty widows, who were all over twenty-five, it is impossible to conjecture about the ages of the others. For the handful of men whose ages are known the average was 26.2 years. Another ten were known to be minors. On the wife's critical role in the market operation, see for example, AN, Y 9441, 26 August 1740.

30. Undoubtedly some of the wives took charge of the books because their husbands could neither read nor write. For example, Dupuis, master baker, and Prenat, faubourg Saint-Antoine baker. ASP, DB⁶, carton 5, Gordrin report, 21 May 1767; AN, Y 15941, 30 January 1742. But, as we have seen, in the great majority of cases, bakers were literate in fundamentals. The wife's role seems to be the fruit of a voluntary policy decision in the household.

31. The following is an example of a simple *billet:* "arate de conte avecque Madame Cary a la some de 79 livres a Paris ce 18 avril 1759." ASP, D5B⁶ 4449. See also D5B⁶ 3015, 2942 (especially), 2712, 2007, 3010, 3327, 3337, 3416, 3760, 4134, 5297.

32. See, for example, AN, Y 15961, 24, 26 October 1764; Y 13115, 28 January 1763; Y 15932, 29 April 1732.

33. ASP, D5B⁶ 87, 5347, 4308, 3015; D3B⁶ 73, 12 May 1772; AN, Y 12143, 2 October 1741;

Y 9632, 14 June 1765; BN, Joly 133, fol. 332, and ms. fr. 21635, fol. 118. Cf. the poignant and perhaps sexist frustration of a consular *arbitre* appointed to mediate between two wives representing their husbands' commerce: "but, Messieurs, you gave me two women to reconcile; that surpasses my power. They listened neither to me nor to each other, yet they both chattered a great deal and said nothing. That is why I have sent them back to you the court." ASP, D6B⁶ carton 5, 27 April 1769.

34. AN, Y 12608, 31 March 1760; Y 1127, 17 September 1740; Y 12615, 3 October 1766. For other examples, see Y 12608, 23 February 1760; Y 12619, 17, 21 February 1770; Y 11226, 21 July 1739; ASP, D5B⁶ 1963 (1759–60); D3B⁶ 64, 4 May 1763; D3B⁶ 74, 24 September 1773; D2B⁶ 1125; D3B⁶ 74, 31 March 1772; DC⁶ 21-158.

35. AN, Y 15961, 24 October 1764; ASP, D3B⁶, 11 June 1767. When bakers fell chronically ill, their wives commonly continued to run the business. See Grimperel's report, September 1767, Y 13396.

36. AN, MC, LXV-227, 2 March 1730. The counterproof is provided by Marie-Louise Noel, the fifty-two-year-old widow of a *gagne-denier* who married a master baker widower named Seurre. The police reported, after her husband fled his creditors, "that she has no idea of the names of the customers of her husband or of how much they might owe." One of the likely reasons for the difficulty in which Seurre found himself was his wife's indifference toward the business. Their marriage had been strained from the very beginning. Noel was obviously not the sort of a wife a baker needed. AN, Y 12616, 13 December 1767; Y 10264, 21 December 1764. See the characterization of the ideal baker couple in "Avis économique sur la boulangerie," p. 50.

37. Unfortunately the data rarely furnish the age of the (re)marriage partners. Thus we cannot estimate how old was too old for remarriage or with what celerity a widow remarried. While rapid remarriage made sense in most instances, see the case of infinitely patient Geneviève Nescot, who waited twenty-three years before taking a second husband, another master baker. AN, MC, LIV-849, 11 June 1751. Note also the case of master widower Nicolas Guerbois, age fifty-six, who married a twenty-five-year-old woman. AN, Y 10215, November 1759–March 1760.

38. AN, MC, XVI-748, 28 September 1761; MC, XCIX-470, 18 May 1741; MC, XXX-437, 12 January 1773; MC, XXX-438, 13 July 1773; AN, Y 14777, 7 September 1727.

39. AN, Y 11587, 31 March 1769.

40. For remarriages resulting in dynasty fusion, see AN, Y 13080, 12 April 1733, and Y 14084, 22 December 1757 (it is not clear in these cases whether the bakers moved to the "city" or whether the widows joined their husbands in the faubourg Saint-Antoine). For master widows who married below their station, see Y 13651, 6 December 1755 (widow Dubois waited one year before marrying her journeyman). See also AN, MC, LXI-449, 14 January 1751.

41. AN, Y 13741, 11 May 1741; *Almanach des gourmands* 7 (1810). Rumors concerning sexual relations between widows and journeymen may have been more common in the bakery than in other métiers because of the image of the bakeroom as particularly sensual space: a dimly lit, overheated, humid place where half-naked men labored.

42. AN, MC, LXV-311, 18 May 1751.

43. AN, Y 15242, 30 December 1735; Y 13119, 26 October 1767. See the attack in the

Encyclopédie on the guild for depriving widows of the right to exercise their former husbands' profession after remarrying nonmasters. Diderot et al., *Encyclopédie*, vol. 9, p. 912.

44. AN, Y 15621, 18 December 1750; ASP, D4B⁶ 43-2435, 24 January 1772.

45. AN, Y 12599, 23 July 1753.

46. AN, Y 12138, 29 February 1736; Y 12141, 20 November 1739; Y 12153, 6 November 1749. Two years later, still living with his mother, Langlois was involved in a nasty fight with a relative. Y 12155, 10 August 1751.

47. AN, MC, CXXII-753, 23 November 1770; MC, VII-297, 31 July 1755.

48. AN, Y 15242, 30 December 1735; Y 15603, 5 May 1742; ASP, D2B⁶ 801, 7 September 1746; AN, MC, XCIV-303, 28 March 1761; Y 15244, 14 March 1737 and 20 December 1737; MC, LXXXII-315, 13 October 1751.

49. AN, Y 12141, 22 January 1739; Y 13926, 31 May 1741.

50. AN, MC, XXVIII-241, 22 October 1735; AN, Y 11388, 25 April 1772; Y 15620, 14 December 1752; Y 11388, 3 April 1722; Y 12157, 25 October 1754.

51. Diderot et al., *Encyclopédie*, vol. 15, p. 209; Mercier, *Tableau*, vol. 12 (1788), p. 103; Edme Goncourt and Jules Goncourt, *La Femme au XVIIIe siècle* (Paris, 1862), p. 209.

52. Claude-Joseph de Ferrière, *Nouvelle Introduction à la pratique* (Brussels, 1739), vol. 2, pp. 699–700; Diderot et al., *Encyclopédie*, vol. 11, p. 60; *Dictionnaire de Trévoux*, vol. 7, p. 654.

53. For persons of "condition," atrocious insults and repeated threats were often sufficient to warrant separations, because "for these persons insults wound as deeply as brutalities would ordinary people." Diderot et al., *Encyclopédie*, vol. 15, p. 60; *Dictionnaire de Trévoux*, vol. 7, p. 655.

54. AN, Y 12594, 13 September 1750; Y 14289, 31 July 1726; Y 11571, 5 July 1755.

55. AN, O¹ 397, 22 August 1753; AN, Y 15608, 24 July, 30 August, 3, 19, 18 September 1744; Y 11556, 13 February 1726. Garin claimed that he loved his wife and that he treated her "with good manners," that she suffered the miscarriage as a result of being struck with a flour sack accidentally, and that the chief source of tension in the household was her sister, to whom he granted a one-third share in the business at the time of his marriage, apparently at the behest of his father-in-law.

56. AN, Y 11310, 9 August 1749; Y 15053, 23 August 1749; Y 15265, 4 December 1757; Y 13524, 18 August 1760; Y 11377, 1 December 1769; Y 13386, 17 April 1760; Y 12732, 20 May 1740; Y 11451, 1 April 1730; Y 13548, 18 December 1771.

57. AN, Y 15257, 12 June 1749; Y 11569, 11 February 1753.

58. AN, Y 12146, 30 March 1743.

59. Arsenal, ms. Bast. 10017, pièce 47; AN, Y 15239, 29 August 1732. Y 11219, 16 September 1732; Y 11219, 12 October 1732. The girl and the mother avenged themselves by assaulting Ursule and calling her "gueuse" and "garce de putain." Y 11220, 2 March 1733.

60. AN, Y 15241, 25 February, 21, 24 April and 7 August 1734; Millet and her family and friends continued to riposte with physical and verbal violence. See Y 11221, 13 August 1734.

61. AN, Y 11221, 31 August, 20 October and 22 November 1734.

62. AN, Y 11222, 22 January and 10 February 1735; Y 15242, 1 June 1735; Y 15243, 12, 14 January 1736.

63. AN, Y 11224, 12 July 1737. It was this time that Blanvilain was arrested for trying to run pedestrians over on horseback on the rue du faubourg Montmartre. Y 15245, 28 May 1738.

64. AN, Y 11232, 24 May 1745.

65. AN, Y 11231, 12 October 1744; Y 11232, 24 May 1745. The Blanvilains had children, though only vague allusion is ever made to the impact of the sempiternal battle on the family.

66. AN, Y 11236, 9, 19 April 1749.

67. AN, Y 11236, 17 July 1749.

68. Ibid.

69. AN, Y 11237, 12 June 1750; Y 15258, 30 June, 3 July 1750.

70. Arsenal, ms. Bast. 10141, 3, 16 July 1756.

71. AN, Y 13450, 16 December 1768.

72. AN, Y 12148, 30 November 1745; Y 11219, 1 December 1732; Y 12571, 27 January 1725; Y 13093, 5 November 1742; Y 12571, 1725.

Chapter 12   Fortune

1. In the cases of second or third marriages obviously one cannot speak of the very beginning of a career. As for the inventories, let us note that premature deaths are not infrequent, thus provoking a midcareer reckoning, and that in many instances widows take over the business, thus forestalling a career "end." In general the interval between the marriage contracts and the after-death inventory is long enough to constitute what we may call a "career." In the eighty instances for which we have information, the average period between marriage contract and after-death inventory is 18.5 years, the mode and the median 16 years.

2. Adeline Daumard and François Furet, "Méthodes de l'histoire sociale: Les Archives notariales et la mécanographie," *Annales: E.S.C.* 14th year (October–December 1959), p. 678; Maurice Garden, "Les Inventaires après décès: Source globale de l'histoire sociale lyonnaise ou juxtaposition de monographies familiales?" *Cahiers d'Histoire* 12 (1967), p. 157. See also M. Garden, *Lyon et les lyonnais au XVIIIe siècle* (Paris, 1970), pp. 223–26. By comparing the number of inventories drawn up with the total number of deaths in a given year, Daumard and Furet probably give an unduly pessimistic picture. Total deaths presumably include substantial numbers of children, of old and retired persons, and of members of the floating population, most of whom would have no reason to have inventories performed. It would probably make more sense to compare the number of inventories with what we might call the "active" deceased population. In the Chartres area, Michel Vovelle found that the inventories represented about 30 percent of adult mortality. Vovelle, review of A. Daumard and F. Furet, *Structures et relations sociales à Paris au milieu du XVIIIe siècle* (Paris, 1961), in *Annales historiques de la Révolution française* 34 (January–March 1962), p. 99. See also Vovelle, *Ville et campagne au 18e siècle (Chartres et la Beauce)* (Paris, 1980). For a thoughtful assessment of the inventories as a guide to the "popular" history of Paris, see Daniel Roche, *Le Peuple de Paris* (Paris, 1981), pp. 58–63 and passim. On the different ways historians have used/

read inventories, consult also Jean Sentou, *Fortunes et groupes sociaux à Toulouse sous la Révolution* (Toulouse, 1969); Micheline Baulant, "Niveaux de vie des paysans autour de Meaux en 1700 et en 1750," *Annales: E.S.C.* 30th year (March–June 1975), pp. 505–18; Richard Lick, "Les Intérieurs domestiques dans la seconde moitié du XVIIe siècle d'après les inventaires après décès à Coutance," *Annales de Normandie* 20, no. 4 (December 1970), pp. 293–302. On the limits and problems as well as the opportunities that inventories present, see also Anton J. Schuurman, "Probate Inventory Research: Opportunities and Drawbacks," in *Inventaires après-décès et ventes de meubles: Apports à une histoire de la vie économique et quotidienne (XIVe–XIXe siècle)*, ed. Micheline Baulant, A. J. Schuurman, and Paul Servan (Louvain-la-Neuve, 1988), pp. 19–28, and Paul Servan, "Inventaires et ventes de meubles: Apports à l'histoire économique," in ibid., pp. 33–34.

3. AN, MC, XXXIX-441, 30 June 1757; partage, MC, VII-288, 6 July 1771.

4. Denisart, *Collection de décisions*, vol. 3, p. 27. Cf. ibid., vol. 3, pp. 27–35, and vol. 4, p. 351. Counseled by experts and (in the best of cases) steeped in experience, the notaries had to know, among other things, how to apply the *crue*, the premium in value (generally 25 percent in Paris) accorded to certain items in the household to compensate for potential underevaluation. See Annik Pardailhé-Galabrun, *La Naissance de l'intime: 3,000 foyers parisiens, XVIIe–XVIIIe siècles* (Paris, 1988), p. 121.

5. The experts were named by the notaries upon consultation with the "interested parties." To preempt potential conflict, the notary allowed each side to have its expert when there was a clear-cut division. See the case of widow Bethmont and her brother-in-law, AN, MC, LXXIX-592, 7 July 1760. Nor were the parties always satisfied with the experts' recommendations. Louis, a journeyman working for widow baker De Sens, vehemently abused master Jacques Delahogue, who had just finished appraising his mistress's shop and bakeroom, on the grounds that he had "come to rob his mistress." AN, Y 15340, 4 May 1744.

6. Mercier, *Tableau*, vol. 3 (1783), pp. 109–10.

7. AN, Y 15252, 2 January 1743.

8. It is interesting to note that in no inventory is the value of the fonds, or business, included. In virtually all cases the surviving spouse or the children continue the business at least for a time. Presumably all the heirs had a claim on its ultimate disposition. Daniel Roche reminds us that, following the rules of devolution in the custom of Paris, it is quite possible that *biens propres* are not accounted for in the inventory. He also discusses other sources of underevaluation, in particular the fact that the notarial estimation *precedes* the sale of estate goods at auction (which would presumably affirm their (higher?) market value or their division among heirs. *Le Peuple de Paris*, pp. 60–61.

9. I examined fifteen other baker inventories after I completed these calculations. They conform to the pattern elaborated here. Cf. the baker "fortune" calculated by Garden, "Les Inventaires après décès," pp. 163–65. Garden does not specify the exact composition of these fortunes. On methodological grounds, the discussion of forty-seven inventories of various masters in northern Paris does not inspire confidence. Davet, *La Population de l'axe nord*, pp. 227–33.

10. In table 12.1 and in other tables, as a result of missing values and variations in the

number of cases for each category, there may appear to be discrepancies in the figures. In this table, for instance, net worth calculations do not precisely represent assets minus liabilities.

11. See Kaplan, *Bread, Politics, and Political Economy.*

12. On the long-term economic trends and the crisis at the end of the Old Regime, see C.-Ernest Labrousse, *La Crise de l'économie française à la fin de l'ancien régime et au début de la Révolution* (Paris, 1944) and *Esquisse du mouvement des prix et des revenus en France au XVIIIe siècle* (Paris, 1933), and Georges Lefèbvre, "The Movement of Prices and the Origins of the French Revolution," in *New Perspectives on the French Revolution: Readings in Historical Sociology,* ed. Jeffry Kaplow (New York, 1965), pp. 103–35.

13. Certain masters called themselves "master and merchant baker," presumably to further enhance their prestige, but they are located in the master rubric.

14. Cf. the more sophisticated categories for the analysis of fortune used by A. H. Jones, "Fortune privée dans trois colonies d'Amérique," *Annales: E.S.C.* 24th year (March–April 1969), pp. 246–47, and "Wealth Estimates of the American Middle Colonies," Ph.D. diss., University of Chicago, 1967. Given the method used by Annik Pardailhé-Galabrun and her team, I cannot fruitfully compare the results of my analyses of baker inventories with the conclusions they draw from hundreds of inventories of Parisians of all social categories. They calculate fortunes exclusively on the basis of *biens mobiliers,* thus excluding not only real estate and rentes but also all paper assets. Privileging entry into the household, this approach eviscerates the professional dimension and invites misleading extrapolation concerning net worth. Moreover, the extremely broad and overlapping socioprofessional categories used by Pardailhé-Galabrun further estrange the researcher from economic realities. In exchange one is regaled by an excellent discussion of furnishings, cultural goods, domestic architecture, etc. See Pardailhé-Galabrun, *La Naissance de l'intime,* especially pp. 120ff., 159–60, 211.

15. The average number of children before 1745 was 2.68; after, 3.3. There were also three times as many children in the baking business in the second half of the century as in the first. In my analysis of fifty scellés, a document similar to the inventory, I found 1.5 (average and median) children entering the baking profession.

16. In the scellés, seven bakers were owners, three tenants, forty unknown.

17. Yet the correlation coefficient does not reveal a great strength of association between rents and rooms: r = .55 at .001. The highest rent payers had the highest liabilities.

18. The relation between household and family fortune emerges clearly from cross-tabular analysis, though the correlation coefficient is only .58 at .001.

19. R = .75 at .001.

20. The forains average 3.33 beds per house, presumably because they had at least two more children than the others. For a thoughtful commentary on the centrality of the bed-as-investment and marker (site of birth and death, fortress against the cold and the indiscreet glance), see Pardailhé-Galabrun, *La Naissance de l'intime,* pp. 276–84.

21. For a wonderful discussion of clothing and dress, see Daniel Roche, *La Culture des apparences: Une Histoire du vêtement (XVIIe–XVIIIe siècles)* (Paris, 1989).

22. AN, Y 14963, 8 February 1747.

23. Among the masters who owned jewelry, the average value was 219 livres. Silver was much more strongly associated with total assets (r = .77 at .001) than was jewelry (r = .45 at .001).

24. A systematic study of baker wills would help to assess religiosity. See, for example, the will of relatively affluent Nicolas François, who endowed two hundred low masses in one church and fifty in another. AN, MC, XXXVIII-387, 14 January 1751.

25. Our study of the marriage contracts revealed that a large majority of the bakers could sign their name. In the inventories, an average of 2.7 bakers signed each contract and an average of 1.23 bakers avowed that they could not sign—a higher ratio of "illiterate" to "literate" than we would have expected. Curiously the merchant bakers had the highest ratio of nonsignatures to signatures. In another study, based on the Châtelet archives, I found that 92 percent of masters signed their names (98 of 107), 83 percent of faubourg bakers (39 of 47), 33 percent of forains (3 of 9) and 67 percent of journeymen (29 of 43). The sources are over two hundred procès-verbaux of various sorts.

26. Variation in bakery investment did not account for a large part of the variation in total assets, for r = .4 at .001. The correlation coefficients between assets on the one hand and the components of bakery investment on the other are even lower. Yet the total assets of families in the highest two quartiles of baker investment (assets of 14,817 and 23,311 livres on average respectively) were far more substantial than those in the lowest two quartiles (assets of 5,778 and 6,781 livres respectively).

27. For equipment and total stock, r = .67 at .001; for equipment and journeymen, r = .6 at .001.

28. Inventories: on average one finds stocks of middling flour worth seven livres in 1711–25, twenty-seven livres in 1726–40, twenty-three livres in 1741–50, and 327 livres in 1751–60. In the preinventories, which permitted coding by measure of grain, one finds .28 setiers in 1731–40, .5 in 1741–50, and 2.7 in 1751–60.

29. In the inventories, horses and rural property ownership had a modest association: r = .54 at .001.

30. The correlation coefficient between journeymen and baker tools is .64 at .001. For journeymen and total bakery capital investment, r = .51 at .002.

31. In the preinventories the correlation coefficient relating stock and journeymen is .63 at .031, while in the inventory pool for the same variables r = .67 at .001. The coefficient relating journeymen and bread placed under seal in the early morning is .63 at .008.

32. There was a moderately strong association between total grain and flour stocks and total assets: r = .6 at .001. There was no significant relationship between horse ownership and the ability to stock (r = minus .1), but the number of sacks was connected with stocking ability (r = .6 at .001), as was baker capital investment (r = .67 at .001).

33. See Steven L. Kaplan, "Lean Years, Fat Years: The 'Community' Granary System and the Search for Abundance in Eighteenth-Century Paris," *French Historical Studies* 10 (fall 1977), pp. 197–230.

34. To what extent do inventories give a convincing picture of *everyday* stocks? Inventories that were performed immediately probably offer us as "authentic" a glimpse as we could hope to get into the baker granaries. Yet even a delay of a week or two after death probably results in an erosion of normal stock levels, for it is very unlikely that the

survivor-caretakers of the business would stock beyond the current needs until they had a clear idea of the status of the estate. I suspect the inventories underrepresent actual average holdings. Of course there are many facts to be taken into account in speculating about stocks: the credit position of the baker, whether he died suddenly and unexpectedly or after a long agony, whether his wife was capable of running the business efficiently, etc.

35. One might be tempted to interpret the fall in grain and flour holding as a sequela of the deep subsistence crisis of 1738–41. Yet to account for the copious stocks of 1727–30, one would have to suppose a very rapid recovery from the jarring dearth of 1725–26. Until we have a larger number of socioprofessional/trade studies and a more intimate grasp of the respirations of the economy (not the epic ones, but the short- and middle-run movements), it will be extremely difficult to explain variations in the values that make up net worth.

36. But there is no significant correlation between stock stored and number of granaries in the inventory study (r = .03). In the preinventories, however, there was a stronger link between the two (r = .57 at .002). Not surprisingly, there is a moderately strong correlation between the number of flour rooms and the number of bolters (r = .66 at .001).

37. Rentes holding appears to be a good predictor of assets and net worth: for both r = .84 at .001. Paris and non-Paris real estate are much less closely associated with net worth: r = .58 at .001 and .44 at .001 respectively.

38. For this tax, see the excellent description in Marion, *Dictionnaire des institutions,* pp. 59–71.

39. The correlation coefficient between capitation and assets is .68 at .002. The capitation was not, however, a good predictor of net worth: r = .33 at .117. Bakers in the third capitation quartile had one-third greater net worth than those who paid the most capitation—a suggestive measure of the "social justice" of the tax schedule.

40. The discrepancy is much more marked for the bride's than for the groom's portion. All three baker types have substantially larger marriage contributions in the marriage study than in the inventories.

41. R = .36 at .002 for bridal dowry and assets; r = .35 at .020 for groom's contribution and assets; r = .32 at .006 for bridal dowry and net worth; and r = .32 at .030 for groom's contribution and net worth.

42. Statistically, "total marriage fortune" is a parlous and rather jerry-built category, because it is founded on unequal numbers of cases of male and female marriage contributions. Let us note, nevertheless, that it correlates with total assets and net worth as follows: r = .34 at .003 and r = .31 at .006.

43. The standard deviation of mean liabilities was 3,659. The wholly unrealistic figure for the mean percentage that liabilities were of assets, 76.2 percent, underlines the dispersion. The median percent liabilities of assets was 13.8 for the whole universe (15 percent for masters, 6.2 percent for fauboriens).

44. In twenty-eight of the fifty preinventories, there were claims made by the suppliers against the estate for money due for merchandise. In eight cases the creditors were grain merchants, in twelve instances flour merchants, in one occurrence a miller, on five occasions Paris factors, and in six events bakers.

45. R = .26 at .012. Similarly, r = .28 at .007 for flour liabilities and flour stocks.

46. For these variables and stock liabilities, r = minus .27 at .051; r = minus .05 at .326; r = .13 at .452; and r = .07 at .283, respectively.

47. R = minus .08 at .246.

48. R = .3 at .051.

49. R = .36 at .008.

50. R = .67 at .001. Yet there was a weak association between net worth and medical care costs: r = .19 at .250.

51. AN, MC, CI-279, 19 January 1730. Denise's family nevertheless spent sixty livres on his funeral.

52. AN, MC, CXI-234, 12 July 1751.

53. AN, MC, LXXXV-551, 15 December 1757. Chocarne's family paid fifty-nine livres for his funeral.

54. AN, MC, XXXV-626, 21 June 1741.

55. AN, MC, CXXII-682, 5 November 1751.

56. AN, MC, LXXXV-581, 9 July 1764.

57. Many of Delahogue's accounts receivable were owed by simple people who settled on *tailles* (though one *taille* must have reached gargantuan proportions: 774 livres). Other business was conducted with aristocrats, and Delahogue might have flattered himself by indulging them excessively. The princesse de Conty owed 3,711 livres and the comtesse de Polignac was 1,200 livres in arrears. But Delahogue had already taken legal steps to collect from the last-named.

58. AN, MC, LXVI-550, 25 October 1765.

59. AN, MC, VII-417, 21 July 1775.

60. AN, QI* 1099, 159–62.

61. One of Lepage's baker sons, Bonpierre, styled himself "marchand bourgeois de Paris."

62. In the inventory of Lepage's wife conducted eight years later there is a more complete list of paintings, most of which probably were part of her husband's household. Among the tableaux not mentioned specifically in the first inventory are a Nativity, a Sainte Cécile, an animal, and a King David. AN, MC, XXVIII-252, 15 April 1738.

63. Again the wife's inventory is more detailed. AN, MC, XXVIII-252, 15 April 1738.

64. At the time of her death Madame Reverard had clothing worth 412 livres. AN, MC, VII-417, 12 September 1775.

65. AN, MC, VII-347, 8 November 1763.

66. Jean-Paul Marat, in *L'Ami du peuple,* no. 402 (17 March 1791), p. 5.

67. For the middling level of most bakers in medieval towns, who were less well-off than butchers or tavern keepers, see Desportes, *Le Pain au moyen âge,* pp. 187–88.

Chapter 13   Bakers as Debtors

1. On the bakers and the grain and flour trade, see Kaplan, *Provisioning Paris,* especially chapter 12.

2. The butchers of Paris had a long history of credit assistance/intervention from the

central government. The eighteenth-century incarnation was the Caisse de Poissy, created by a royal edict of 1719 and then refined in arrêts du conseil of 1723 and 1724 (which, incidentally, evoked two other caisses, one devoted to beverages, primarily wine, and the other to fish and poultry). The grain-flour-bread world enjoyed no such bureaucratic-fiscal organizing of its credit needs, nor even was it targeted by the opportunistic scavenging of the independent "bankers" called "grimbelins" (apparently after the verb for "to sift," perhaps also "to cut a deal") who lent money for meat transactions on market days at allegedly "usurious" rates. For this information I am grateful to my student Sydney Watts, who is writing a thesis on the butchers of Paris and the meat question.

3. Clifford Geertz, *Peddlers and Princes: Social Change and Economic Modernization in Two Indonesian Towns* (Chicago, 1963), p. 39. Cf. John Waterbury, *North for the Trade: The Life and Times of a Berber Merchant* (Berkeley, 1972), pp. 179–81.

4. Cyril Belshaw, *Traditional Exchange and Modern Markets* (Englewood Cliffs, N.J., 1965), p. 56; Davis, *Social Relations,* pp. 216–44; Sidney W. Mintz, "The Employment of Capital by Market Women in Haiti," in *Capital, Savings, and Credit in Peasant Societies,* ed. Raymond Firth and B. S. Yang (Chicago, 1964), pp. 261–63, and "Pratik: Haitian Personal Economic Relations," in *Symposium: Patterns of Land Utilization and Other Papers,* proceedings of the 1961 Annual Spring Meeting of the American Ethnological Society, ed. Viola E. Garfield (Seattle, 1961), pp. 54–63; Maria C. B. Szanton, *A Right to Survive: Subsistence Marketing in a Lowland Philippines Town* (University Park, Penn., 1972), pp. 97–116; Clifford Geertz, Hildegard Geertz, and Lawrence Rosen, *Meaning and Order in Moroccan Society* (Cambridge, 1979), pp. 218–20.

5. Margaret Katzin, "The Business of Higglering in Jamaica," *Social Economic Studies* 9 (September 1960), p. 317; Shepard Foreman and Joyce Riegelhaupt, "Marketplace and Marketing System: Toward a Theory of Peasant Economic Integration," *Comparative Studies in Society and History* 12 (April 1970), p. 197; Ralph L. Beals, *The Peasant Marketing System of Oaxaca, Mexico* (Berkeley, 1975), p. 207; R. J. Bromley and Richard Symanski, "Marketplace Trade in Latin America," *Latin American Research Review* 9 (fall 1974), p. 13; Szanton, *A Right to Survive,* pp. 90–93; "Second Mémoire des boulangers," ca. 1789, Arsenal, ms. 7458.

6. Fuad I. Khuri, "The Etiquette of Bargaining in the Middle East," *American Anthropologist* 70 (August 1968), p. 702.

7. Geertz, Geertz, and Rosen, *Meaning and Order,* pp. 223, 261n; Geertz, *Peddlers and Princes,* p. 39; Davis, *Social Relations,* p. 215; Burton Benedict, "Capital Savings, and Credit among Mauritian Indians," in *Capital, Savings, and Credit in Peasant Societies,* ed. Raymond Firth and B. S. Yang (Chicago, 1964), pp. 342–43; Ju-K'ang T'ien, *The Chinese of Sarawak: A Study of Social Structure,* London School of Economics and Political Science, Monograph on Social Anthropology no. 12 (London, n.d.); Barbara Ward, "Cash or Credit Crops? An Examination of Some Implications of Peasant Commercial Production with Specific Reference to the Multiplicity of Traders and Middlemen," *Economic Development and Cultural Change* 8 (January 1960), p. 154.

8. ASP, D4B⁶ 49-2956, October 1772; AN, Y 12619, 14 March 1770; D5B⁶ 1902, 3118, 179; Y 13741, 24 December 1740; Y 12605, 18 November 1757; Y 11384, 2 March 1771; Y 12604, 12 August 1756; widow Leger's relations to Provendier and Chenart in D4B⁶ 25-1306,

9 September 1763, and D4B⁶ 24-1249, 1 September 1763; D2B⁶ 1103, 8 November 1771 (baker Deschard standing in for flour merchant Mabille); BN, Joly 1829, fol. 296.

9. AN, Y 12625, 30 March 1775; AN, H 1867, fol. 432 (22 May 1758); Y 12606, 24 April 1758; Y 11227, 15 December 1740; Y 12604, 12 August 1756; Y 12605, 18 November 1757.

10. ASP, D5B⁶ 733, 21 March 1788.

11. Delamare, *Traité,* vol. 2 (1729), p. 935. Delamare traced the custom of extending credit to the year 1652. Yet there is evidence that it existed earlier. See, for example, AN, MC, XIII-9, 16 January 1629. The usage in Brittany was to deny credit to bakers, thus deterring men without substantial resources from engaging in this craft. Letaconnoux, *Les Subsistances,* p. 96.

12. When a Mortellerie merchant suspected that a refractory baker was really solvent and was merely trying to slither out of a pratik commitment, he held him hostage by threatening to have him jailed for indebtedness if he connected up with another dealer. Thus his indebtedness constrained the baker to remain faithful to his major creditor. Duchesne to lieutenant general of police, 25 September 1698, BN, ms. fr. 21643, fol. 373.

13. AN, Y 12618, 8 September 1769; Y 13516, 13 July 1757; Y 11226, 21 July 1739.

14. If one counts only those bakers who actually owed money for grain rather than the whole group, the average debt was 957 livres (mean = 618).

15. If one counts only those bakers who actually owed money for flour rather than the entire group, the average debt was 1,604 livres (median = 794).

16. Another source, fifty-eight lawsuits for recovery, indicates that the average amount owed by a baker to a flour dealer was 508 livres (428 before 1750, 782 after). Remember that it was not uncommon for a miller or a mealman to face three or four delinquent bakers at a time, in addition to brokers representing other bakers. See Kaplan, *Provisioning Paris,* pp. 356–58.

17. See Kaplan, *Provisioning Paris,* pp. 518–23, 532–42. In the last decades of the Old Regime a number of projects surfaced dealing with ways to stimulate the provisioning trade by granting public credit to bakers and advances to merchants. AN, F¹¹ 1179. Brokers did not obtain hard guarantees against baker default till 1811. Morel, *Histoire abrégée,* p. 295. Cf. Mercier's general appreciation of the deleterious influence of credit on the character of the trade and of the tradesman. *Tableau,* vol. 11 (1788), p. 40.

18. Arrêt du conseil, 18 December 1642, BN, ms. fr. 21639, fol. 167; ms. fr. 21649, fols. 127–28; ms. fr. 21635, fol. 118. Since the seventeenth century, the Paris-based grain traders had been complaining about the generalized (quasi-contagious) insolvency of the bakers—allegedly up to half the guild in the 1690s. See Meuvret, *Le Problème des subsistances à l'époque Louis XIV,* p. 127.

19. Petition to procurator general, 1733, BN, Joly 28, fols. 29–31, and Joly 133, fols. 330–32.

20. Assembly of Police, 25 June 1733, BN, ms. fr. 11356, fols. 209–11.

21. See Paul Dupieux, "Les Attributions de la juridiction consulaire de Paris (1563–1792)," *Bibliothèque de l'Ecole des Chartes* 95 (1934), pp. 134–44; *Journal de commerce et d'agriculture,* September 1762, p. 124.

22. See AN, Z¹H 239, 3 September 1726 (Gilbert v. Tavernier for 350 livres) and 12 November 1726 (Bourgeot v. Boulanger for 1,196 livres). Cf. Arsenal, ms. Bast. 10150, August 1726 (Lory and Gaultier v. Fremin).

23. See, for example, ASP, D2B⁶ 762, 9 June 1743 (Hamony v. Thevenot for six hundred livres); D2B⁶ 768, 18 December 1743 (Calleau v. Gaudet and Aubry for 735 livres); D2B⁶ 801, 9 September 1746 (Vastel, known as Picard v. Gueret for 342 livres); D2B⁶ 802, 5 October 1746 (Hamony v. Veron for 216 livres); D2B⁶ 1009, 23 January 1764 (LeGouge v. Ducastel for 150 livres); D2B⁶ 1125 and D3B⁶ 64, 24 September 1773 (Fournier v. Lionnet for 650 livres).

24. ASP, D2B⁶ 809, 10 May 1747; D2B⁶ 798, 8 June 1746; D2B⁶ 762, 7 June 1743; D2B⁶ 1125, 24 September 1773; D2B⁶ 800, 12 August 1746; D2B⁶ 768, 20 December 1743 (Louis v. Lambon for 100 livres); D2B⁶ 842, 4 February 1750 (Bernardin v. Lepar for 244 livres); D2B⁶ 843, 6 March 1750 (Gaudumier v. Dugland for 236 livres).

25. For instance, ASP, D2B⁶ 768, 18 December 1743 (master bakers Gaudet and Aubry who bought "in the granaries") and D2B⁶ 809, 10 May 1747 (master baker Richer Senior, who bought grain through an agent).

26. ASP, D2B⁶ 1012, 13 April 1764 (Benard v. Tremblay for 324 livres); D2B⁶ 1102, 11 October 1771 (Collinet v. Garmont for 216 livres); D2B⁶ 843, 6 March 1750 (Robin v. Lecocq for 180 livres); D2B⁶ 986, 15 February 1762; D2B⁶ 800, 31 August 1746 (Pied v. Mouchy for 54 livres 10 sous); D2B⁶ 841, 23 January 1750 (Choconin v. Félix for 95 livres); D2B⁶ 809, 10 May 1747.

27. ASP, D2B⁶ 736, 10 April 1741; D2B⁶ 809, 19 April 1751. In the mid-nineteenth century, millers expected to be paid three to four weeks after delivery. *Enquête sur la boulangerie*, pp. 230, 331.

28. ASP, D2B⁶ 809, 10 May 1747; D2B⁶ 1128, 1 December 1773; AN, MC, XXXV-626, 18 August 1741.

29. ASP, D2B⁶ 800, 31 August 1746, and D2B⁶ 754, 22 October 1742. Versailles brokers such as Quevanne also used the consular court to pursue Parisian and forain bakers. D2B⁶ 1034, 3 February 1766 (v. Decant for 669 livres), and D2B⁶ 1128, 31 December 1773 (v. Langlois for 160 livres).

30. ASP, D2B⁶ 809, 3 May 1747; D2B⁶ 801, 28 September 1746; D2B⁶ 1104, 4 December 1771. There were of course more stringent terms: D2B⁶ 679, 23 July 1736 (three months, Roche v. Leduc for 320 livres); D2B⁶ 801, 19 September 1746 (two months, Dupin v. Legresle, 58 livres).

31. ASP, D2B⁶ 1102, 21 October 1771; D2B⁶ 801, 19 September 1746.

32. ASP, D2B⁶ 802, 10 October 1746; D2B⁶ 754, 5 October 1742; AN, MC, XVI-748, 28 September 1761.

33. ASP, D2B⁶ 841, 21 January 1750; D2B⁶ 801, 7 September 1746, and D2B⁶ 802, 7 October 1746.

34. The curé was the archetype of the sagacious nonexpert. He was called upon for "country" cases—suits rooted in the hinterland whence he came. See Dupieux, "Juridiction consulaire," p. 125. I have found one instance of priestly arbitration in a provisioning affair. A cleric from Dammartin entered a case pitting a grain merchant against a Gonesse baker. Sentenced to pay sixty-one of the seventy livres sought by the plaintiff, the forain claimed that he had never been "heard" by the arbitre. ASP, D2B⁶ 768, 18 December 1743; D6B⁶ carton 2, 28 May 1743.

35. ASP, D6B⁶ 11, 6 November 1783; D6B⁶ 2, 26 September, 14 October 1746.

36. ASP, D6B⁶ 3, 14 December 1757, 16 January 1758.

37. ASP, D6B⁶ 6, 7, 23, January 1771; D6B⁶ 3, 1 April, 27 May 1761; D6B⁶ 6, 5 March, 2 April 1773; AN, Y 15595, 22 September 1738. In a particularly difficult case between a miller and a baker, the baker-jurés arbitres resigned and asked the court to name the entire guild assembly to hear the evidence and issue a recommendation. D6B⁶ 3, 21 November 1746.

38. ASP, D6B⁶ 5 May, 21 June 1773; D6B⁶ 6, 13 September, 1 October 1773.

39. ASP, D6B⁶ 5, 11, 23 December 1767; D6B⁶ 7, 15 October, 18 November, 10 December 1773; D6B⁶ 11, 27 November 1783.

40. ASP, D6B⁶ 2, 29 May, 31 July 1741; D6B⁶ 6, 22, 29 March, 2 April 1773.

41. ASP, D6B⁶ 3, 14 October, 4 November 1759; D2B⁶ 802, 10 October 1746.

42. Cleret to Hérault, 30 October 1725, and Hérault's marginal notes, Arsenal, ms. Bast. 10271.

43. Arsenal, ms. Bast. 10150, July–August 1726; ms. Bast. 10027, fol. 68, ca. 1725; petition, January 1726, ms. Bast. 10149 and Gesvres to Hérault, 24 January 1726, ms. Bast. 10017.

44. ASP, D2B⁶ 768, 20 December 1743; AN, X¹ᴮ 3489, 10 January 1781; X¹ᴮ 3482, 7 July 1740.

45. AN, X¹ᴮ 3691, 2 March 1758; AN, Y 12616, 11 December 1743. The parlement granted permission to enter one's home, but not on a Sunday or holiday.

46. AN, Y 12588, 4 September 1744; Y 15241, 15 July 1734; Y 14942, 30 July 1729; Y 15944, 18 November 1746. Cf. widow baker Petit, who also had her bakery dismantled to pre-empt action by creditors. She was jailed. AN, Y 13761, 9, 10 April 1733. In several other cases, suspicious landlords called in authorities to catch fleeting tenants in *flagrant délit*, only to discover that they had powerful alibis: one baker was simply moving out a buffet and nothing else, and another was having his oven rebuilt. Y 12145, 26 October 1742, and Y 15616, 7 September 1748. Cf. the Compiègne baker who sought admission to a hospital in order to escape his creditors. AN, O¹* 390, fol. 52 (1 February 1745).

47. AN, MC, LXXXIX-666, 9 March 1769; ASP, 6 AZ unclassified, 6 February 1772; ASP, don 5822, uncatalogued, 24 April 1748, 23 January 1749; MC, XCVIII-444, 17 November 1731.

48. AN, MC, XXVIII-41, 27 October 1698; MC, CXVIII-464, 16 February 1751; MC, VII-282, 6 February 1752, and MC, VII-301, 18 March 1756.

49. AN, MC, X-619, 31 May 1770.

50. AN, MC, XXXIII-586, 20 November 1771; ASP, DC⁶ 18, fol. 4, 20 November 1772.

51. Among master bakers, all in ASP, series DC⁶: 18, fol. 200, 17 May 1772 (Chareret); 18, fol. 203, 15 May 1773 (Martin); 18, fol. 218, 12 June 1773 (Destor, known as le Borgne); 19, fol. 21 (Allyot); 19, fol. 178, March 1775; 19, fol. 290, 30 January 1776 (Lecoq); 20, fol. 41, 7 June 1776 (Gallard); 22, fol. 159, 4 March 1780 (J.-F. Destors); 27, fol. 55, 24 November 1783 (Meusnier); 27, fol. 131, 23 January 1784 (Deshayes). Among the other bakers in cession, same series: 18, fol. 22, 18 January 1772 (baker Robert); 18, fol. 249, 21 August 1773 (baker Dugland); 18, fol. 261, 2 October 1773 (baker Cholan); 19, fol. 172, 15 February 1775 (baker Jeulin); 22 fol. 4, 5 June 1779 (baker Auger); 20, fol. 81, 17 October 1776 (forain Godillion); 19, fol. 21, 12 February 1774 (merchant baker Flechel); 19, fol. 133, 5 November 1774 (privileged baker Bernier).

52. ASP, DC⁶ 17, fol. 252, 16 June 1770, and DC⁶ 28, fol. 14, 16 June 1784.

53. AN, Y 11390; AN, MC, VII-279, 20 July 1751; Y 14529, November 1734; Y 15603, 3 February 1742; Y 14290, 1727; Y 13522, 5 September 1759. For a very similar confiscation against baker Mongrade, see Y 13741, 12 September 1727.

54. Simon Linguet compared conditions in "civil" jails to the miserable Bastille. *Observations sur l'histoire de la Bastille* (London, 1783), p. 148.

55. Mercier, *Tableau,* vol. 3 (1783), p. 329; "Mémoire pour diminuer le nombre des procès," cited by Shelby McCloy, *The Humanitarian Movement in Eighteenth-Century France* (Lexington, Ky., 1957), p. 144; Dupont to Carl Ludwig von Baden, 15 January 1773, in Carl Knies, ed., *Carl Friedrich von Baden, Brieflicher Verkehr mit Mirabeau und Dupont* (Heidelberg, 1892), vol. 2, pp. 25–28; Simon Linguet, *Théorie des loix civiles, ou Principes fondamentaux de la société* (London, 1767), vol. 2, pp. 398–410. Cf. Brissot's argument against debt prison, in McCloy, *Humanitarian Movement,* p. 146.

56. APP, A B/191, 11 March 1736; A B/185, 24 January 1727; A B/188, 28 March 1730; AN, MC, LXXXV-465, 20 February 1738; A B/185, 24 December 1725.

57. APP, A B/321, 22 September 1788; A B/96, 7 June 1725; AN, MC, CVIII-409, 17 September 1732; AN, Y 11232, 24 May 1745.

58. APP, A B/191, 12 November 1735; A B/189, 10 August 1731; A B/321, 25 April 1788; A B/193, 10 May 1738; A B/196, 12 July, 11 August 1742, 7 July 1743; A B/321, 14 November 1791.

59. Denisart, *Collection de décisions,* vol. 3, p. 426; APP, A B/185, 15 February, 7 March 1726, 8 March 1721; A B/195, 10 July 1740, 3 February 1741; ASP, D4B⁶ 25-1269, 1 July 1773.

60. The purpose of the edict was to confide the task of execution to experienced and capable men instead of risking disorder when the processes were served by agents commissioned by the plaintiffs. Henceforth the huisser could enter the house of the convicted debtor to execute the sentence, save at night (unless accompanied by the commissaire) and on holidays and Sundays (unless specified exceptionally by the magistrates). For each capture, the garde was to receive sixty livres. Isambert, *Recueil des lois,* vol. 22, pp. 551–54.

61. AN, Y 13648, 19, 29 August, 7 September 1752; Y 13290^A, 17 March 1788.

62. BN, Joly 1416, fol. 126, 19 January 1781, and fol. 224, 18 March 1789.

63. Hardy's journal, BN, ms. fr. 6681, fol. 158 (18 February 1773); ASP, D4B⁶ 25-1269, 1 July 1773; D2B⁶ 1009, 9 January 1750; D2B⁶ 841, 15 January 1750. Desrues pursued Constant for over three years. See D7B⁶ 337, 16 January 1750.

64. Arsenal, ms. Bast. 10041, fols. 275–76, 18 September 1755.

Chapter 14   Failure

1. Savary, *Parfait négociant,* vol. 1, p. 19; Denisart, *Collection de décisions,* vol. 1, p. 214, and vol. 2, p. 304; Ferrière, *Nouvelle Introduction,* vol. 1, pp. 628–29; *Dictionnaire de Trévoux,* vol. 4, p. 10; *Révolutions de Paris,* no. 109 (6–13 August 1791), p. 216. For a pioneering use of failure records, see Jean Sentou, "Faillites et commerce à Toulouse en 1789," *Annales historiques de la Révolution française* 25 (July–September 1953), pp. 217–

56. On the practices of the consular jurisdiction in which the failure procedure occurred, see Dupieux, "Juridiction consulaire," pp. 116–48, and Jacqueline L. Lafon, "L'Arbitre près la Juridiction consulaire de Paris au XVIIIe siècle," *Revue historique de droit français* 51 (April–June 1973), pp. 217–70.

2. Savary, *Parfait négociant*, vol. 1, p. 657 and passim; Denisart, *Collection de décisions*, vol. 4, p. 134; Guillaume François Letrosne, *De l'administration provinciale et de la réforme de l'impôt* (Basel, 1779), vol. 2, pp. 413–14; Jean-Nicolas-Marcellin Guerineau de Saint-Péravi, *Principes du commerce opposé au trafic* (Paris, 1787), pp. 108–16; Mercier, *Tableau*, vol. 2 (1782), pp. 71–73.

3. Denisart, *Collection de décisions*, vol. 3, p. 32, and vol. 4, p. 344.

4. On *attermoiement*, see ibid., vol. 1, p. 142.

5. ASP, D5B⁶ 85-5766, 3 October 1782. But in certain cases because of the "public interest" nature of his profession, the baker could not be forced to *cession*. Savary, *Parfait négociant*, vol. 1, pp. 699–703; Denisart, *Collection de décisions*, vol. 1, pp. 267, 326–27; Ferrière, *Nouvelle Introduction*, vol. 1, pp. 242–45.

6. See, for example, ASP, D5B⁶ 66-4294, 27 January 1778; D5B⁶ 50-3079, 19 February 1776; D5B⁶ 56-3530, 10 October 1775. One baker now installed in Paris had been "ruined," he claimed, by the bread riots in Bordeaux in May 1773, during which his shop was pillaged and sacked, D5B⁶ 51-3156, 3 May 1774.

7. See appendix D for the faillite sources.

8. Yet the relationship between total assets and bread accounts receivable was not as strong as one might have imagined: $r = .58$ at .001.

9. Bakery investment was rather strongly associated with total assets: $r = .65$ at .001. Note, however, that the median value of masters' investment, 2,375 livres, is substantially higher than the nonmaster median, 1,157.

10. Stock holding was very strongly associated with total assets: $r = .8$ at .001. Yet it seemed to be only marginally influenced by bakery investment: $r = .3$ at .057.

11. ASP, D5B⁶ 17-5294, 26 January 1781. But he owed over twenty thousand livres for his flour and grain in addition to another eighteen thousand livres in other debts, and he needed to recover almost fifteen thousand livres in accounts receivable for bread.

12. It is very unlikely that the faillis had already disposed of all these assets. Was their discretion, on the contrary, a wishful (and surely vain) effort to keep the household entirely out of the discussion of business?

13. Property was strongly associated with total assets: $r = .7$ at .001.

14. $R = .2$ at .150.

15. $R = .51$ at .001.

16. There was a relatively strong association between total losses and total liabilities: $r = .63$ at .001.

17. There was a weak relationship between these bad bread debt losses and bread accounts receivable listed by the bakers: $r = .36$ at .027.

18. It is possible that this apparent decrease in liabilities reflects "arrangements" made with creditors, i.e., reduction in debt as an incentive to pay, remission of certain debts or interest and damage penalties, etc., rather than a serious effort at rebuilding.

19. ASP, D4B⁶ 39-2138, 1 October 1770 and 30 March 1772.

20. ASP, D4B⁶ 29-1530, 6 May 1766.

21. See Kaplan, *Provisioning Paris,* especially chapters 4 and 12.

22. ASP, D4B⁶ 24-1218, 11 August 1762.

23. ASP, D4B⁶ 43-2468, 17 February 1772.

24. ASP, D4B⁶ 90-6146, 23 March 1784.

25. ASP, D4B⁶ 90-6140, 20 March 1784.

26. ASP, D4B⁶ 103-7304, 13 November 1788.

27. ASP, D4B⁶ 51-3137, 22 April 1774.

28. ASP, D4B⁶ 66-4278, 8 January 1778. Master baker Jean Lemerle of the rue Saint-Victor blamed Flour War insurgents for stealing the huge amount of ten fournées' worth of bread. D4B⁶ 66-4294, 27 January 1778. Fellow guildsman Jean Macquard alleged that he suffered three thousand livres in losses attributable to bread thefts during the same disorders. D4B⁶ 79-5271, 16 November 1780. Masters Guiolot and Joquain each cited losses due to Flour War "pillage." D4B⁶ 56-3530, 10 October 1775, and D4B⁶ 50-3079, 19 February 1776.

29. See Kaplan, *The Famine Plot,* especially chapters 8 and 9.

30. ASP, D4B⁶ 42-2307, 16 August 1771 and 20 March 1776.

31. ASP, D4B⁶ 28-1476, 14 September 1765; D4B⁶ 83-5537, November 1781.

32. ASP, D4B⁶ 101-7102, 14 March 1788.

33. AN, Y 11673, 30 September 1749; AN, MC, XXVIII-210, 30 September 1728.

34. AN, MC, VII-322, 22 August 1759; MC, LXV-227, 2 March 1730; MC, VII-329, 17 December 1760; MC, LXV-227, 2 March 1730; MC, LXI-492, 3 September 1761; ASP, D5B⁶ 2707; D4B⁶ 75-4997, 25 April 1780; D4B⁶ 90-6140, 20 March 1784; D4B⁶ 25-306, 9 September 1763; D2B⁶ 1022, 13 February 1765. Bakers did not shrink from pressing their colleagues for reimbursement through the courts, either as the result of a falling out or because they found themselves in need. See, for example, Levesque v. Mallet in D2B⁶ 679, 13 July 1736. Cf. the case of the notorious baker François, decapitated in the first days of the French Revolution, who had been very generous with needy confreres. Morel, *Histoire illustrée,* p. 255. It seems quite logical that bakers frequently asked other bakers to serve as the executors of their wills. See widow Durans-Vassou, DC⁶ 229, fol. 229, 28 June 1740.

35. AN, Y 15349, 24 November 1751, and Y 15350, 18 January 1752.

36. Arsenal, ms. Bast. 10141, fol. 337 (10 April 1760); ASP, D2B⁶ 1109, 25 May 1772; DC⁶ 257, fol. 67 (13 April 1774); AN, Y 15349, 18 August 1751; Y 14942, 3 March 1729. Cf. Parmentier: "Qui se ressent le premier de la misère des boulangers? Le peuple." *Mémoire sur les avantages,* p. 412.

37. AN, Y 9441, 29 March 1740; Y 14538, 20 August 1742.

38. AN, Y 13762, 4 February 1755; Y 11229, 26 December 1742; Y 13521, 12 July 1759.

39. AN, Y 11222, 25 November 1735; Y 15617, 6 January 1749; ASP, D4B⁶ 48-2869, 5 July 1773; Y 11731, 4 August 1788.

40. AN, Y 13506; ASP, D6B⁶ carton 2, 7 September 1747; Y 9621, 16 July 1761; arbiter Langin to consular court, 2 October 1783, ASP; Arsenal, ms. Bast. 10105, 15 June 1764, and 10087, 17 January 1774.

41. Deliberations of the Bureau of the Hôpital des enfants-trouvés, 10 September 1751, AAP; AN, Y 15967, 22 June 1768; Hardy's journal, ms. fr. 6684, fol. 143 (14 April 1782), and 6681, fol. 158 (18 February 1773).

42. In a recent piece Judith Miller has posed some interesting questions concerning the bakers and their relations with both the police and the consuming public. I take issue with many of the answers she proposes, and with certain aspects of her methodology. In my view her article suffers from an insufficiently critical reading of contemporary testimonies, a tendency to impute causality through affirmation rather than demonstration (see, for instance, the post-hoc arguments on p. 253), and an impatience to connect virtually all the parts of her discussion with the Revolution (e.g., pp. 229–33, 260–62) even at the cost of forcing and dramatizing issues and voices (as if her fresh research on Old Regime Normandy could not stand powerfully on its own). In reference to faillites, I do not think that she makes her case. She does not show that "Parisian policies . . . *resulted in* a bankruptcy pattern similar to that of Rouen" (my emphasis). I'm not even certain how intimately the Parisian pattern mimicked the Rouen distribution, for we disagree radically on the sheer number of veritable failures that occurred. In my view it is not at all "clear that the bakers were finding it more and more difficult to remain in business." Nor is it at all clear that public policy more than anything else drove them to this apocalypse. Moreover, failures were by no means confined to guild bakers; the Paris balance sheets were not nearly as "mute" on the sources of distress as Miller claims; one needs to take into account the protocol of the failure filing, the "bargaining" conventions, and the way to measure total indebtedness while allowing for various tactical and ritual forms of puffery; the decision to offer a failure petitioner some sort of grace or terms depends not on the court but on the creditors (just "by filing" one was not sheltered from pursuits). Much of Miller's treatment is predicated on unexamined assumptions about the resourcefulness and resiliency of bakers—that is, about the economy of the bakery, the cost of manufacturing different sorts of bread, the thresholds of toleration, etc. The author also premises her case on the sustained application of laissez-faire policies from the sixties onward ("Thwarted in their efforts to provision their cities by overseeing the grain trade, local administrators concentrated their energies on controlling bread prices in the last decades of the Old Regime," p. 241). In my view this sweeping generalization cannot be documented, not merely in Paris, always in some way or another sui generis, but in many other towns and cities. It is crucial to distinguish between discourse and practice, and to subject the discourse to a rigorous deconstruction. In the same spirit, the "market realities" of the Old Regime were very often normative imperatives rather than the allocation of inputs and outputs by the free play of supply and demand. The treatment of administrative-juridical relations involving the ministry, parlements, intendants, and local police authorities is rich in anecdote, but often vague in explanation (with parties "pressuring" or "convincing" others to take this or that action, sometimes in extremely awkward terms, e.g., the controller general on p. 259 did not have to "convince" the police to issue an ordonnance but merely order them to do it). Concerning Paris, Miller makes claims that I believe are erroneous or misleading. Paris never had a price-fixing *tarif.* There is no such thing as "the guild bakers' legal price." Masters were not systematically undersold by nonmasters, in part because a large proportion of the guild bakers sold in the bread markets and many forains exercised out of shops as well as in the markets. I have found no evidence of a secret "taxe" that was "smoothly" communicated from the police to the guild and that enabled bakers to steal a march on suppliers: this is

neither how bread price making worked in Paris nor how the baker-driven grain and flour trade worked. The account of the crisis of 1789 and the clash between the bakers, represented in writing by the vulpine chevalier Rutledge, and the Leleu company (more or less backed by the ministry), seems to me to be oversimplified. Albert was not the lieutenant general of police in 1770. See Judith A. Miller, "Politics and Urban Provisioning Crises: Bakers, Police, and Parlements in France, 1750–93," *Journal of Modern History* 64 (June 1992), pp. 227–62.

Chapter 15   Reputation

1. AN, Y 12159, 14 August 1756; Y 13761, 21 August 1754; Y 12144, 2 June 1742; Arsenal, ms. Bast. 10101, 1 August 1760; Philipon de la Madeleine, *Vues patriotiques sur l'éducation du peuple* (Lyon, 1783), pp. 221–25. Madeleine added: "Communément on croit que le peuple veut toujours étre conduit par l'effroi du supplice ou par l'appât du gain: erreur. Il a son point d'honneur, tout comme les personnes d'un rang plus élevé." According to contemporary jurists, "c'est toujours une matière grave & importante que celle où il s'agit de la réputation." There were three categories of "injures" that damaged persons morally and/or physically: words, writings, and actions (in particular, blows). The commissaire usually tried to settle a matter of insult and honor (it was relatively easy, for instance, when the insults were reciprocal and thus "negated" each other). Formal "réparation d'honneur," in the simplest cases, required the affronting party to "reconnoitre en présence d'un certain nombre de témoins au choix de la personne injuriée, qu'elle n'est point tachée des injures qui ont été proférées contr'elle." Sometimes the offending party would be expected to ask "pardon," depending on the "qualités" of the persons involved. Denisart, *Collection de décisions*, vol. 2, p. 577, and vol. 4, pp. 188, 201. On the question of honor, see the suggestive insights in the essays collected by John G. Peristiany, ed., *Honour and Shame: The Values of Mediterranean Society* (London, 1965), especially pp. 11, 19, 21, 27–28, 30.

2. Pierre Marmay, *Guide pratique de la meunerie et de la boulangerie* (Paris, 1863), p. 86; *Arrest notable de la Cour des Muzes . . . pour le rabais du pain* (n.p., n.d.). Cf. the persistent suspicion of bakers in England: "Considérations sur la cherté en Angleterre," *Gazette de l'agriculture, du commerce, des arts et des finances*, September 1772, and McCance and Widdowson, *Breads White and Brown*, p. 14.

3. "La Cherté du pain," 21 January 1792, AN, T 644$^{1-2}$; Controller general to Cypierre, 10 May 1770, AN, F^{12*} 153, fols. 197–98; "Le Rabais du pain" (Paris, 1549), p. 7; ASP, 4AZ 794; Sébillot, *Traditions*, pp. 28–32; Christian Bouyer, *Folklore du boulanger* (Paris, 1984), pp. 94–96; Des Essarts, *Dictionnaire universel*, vol. 2, p. 267; Vincent, *Famines*, p. 71; Paul Bondois, "La Misère sous Louis XIV: La Disette de 1662," *Revue d'histoire économique et sociale* 12 (1924), p. 111. On the medieval version of baker miscreancy—dishonesty, fraud, manipulation of quality and quantity, etc.—see Desportes, *Le Pain au moyen âge*, pp. 171–72, 179.

4. "Chanson nouvelle sur le rabais du pain," *Magasin pittoresque* 43d year (1875), p. 110; Fernand Evrard, *Versailles, ville du roi (1770–1789): Etude d'économie urbaine* (Paris, 1935), p. 409; Des Essarts, *Dictionnaire universel*, vol. 2, pp. 264–67; AD Finistère, B

2370, 22 March 1755; BN, Joly 1123, fol. 209 (23 August 1740); Hardy's journal, BN, ms. fr. 6684, fol. 409 (31 January 1789).

5. Lacombe d'Avignon, *Le Mitron de Vaugirard*, p. 35; BN, Joly 1135, fol. 70 (September–October 1767); anon. to Marville, 23 September 1740, Arsenal, ms. Bast. 10027; Comité de salut public to Comité civil de la section des Lombards, 1 floréal an III, ASP, 4AZ 1130; Abbé Chomel, *Dictionnaire oeconomique*, vol. 2 (1767), p. 799; AN, Y 10558, 4 May 1775. For an example of analogous consumer complaints against bakers in a very different culture, see "The Song of the Baker" cited by Clifford Geertz, "Suq: The Bazaar Economy in Sefron," in Geertz, Geertz, and Rosen, *Meaning and Order*, p. 310.

6. Corchons (?) to Hérault, 1739, Arsenal, ms. Bast. 10009, fol. 687; Levesque to procurator general, 15 November 1767, BN, Joly 1135, fol. 54; Controller general to Turgot, 2 September 1770, AN, F^{12} 155, fol. 36.

7. Lacombe d'Avignon, *Le Mitron de Vaugirard*, p. iii; Jean-Jacques Rutledge, *Mémoire pour la communauté des maîtres boulangers de la ville et des faubourgs de Paris, présenté au roi le 19 février 1789* (Paris, 1789), pp. 6–7, 11–13; the phrase on nurturers, appropriated by Rutledge, belonged to baker leader François Garin, AN, Y 10506, 5 December 1789; Parmentier, *Parfait Boulanger*, pp. viii–xiii. Cf. the hatred for the bakers that exploded in the early days of the Revolution. *Révolutions nationales*, 16–19 September 1789, pp. 192–94.

8. AN, Y 15354, 4, 6 January 1755; Arsenal, ms. Bast. 10153, January 1728.

9. ASP, D4B^6 87-5884, 10 April 1783; AN, Y 11220, 27 January 1733; Y 11226, 21 July 1739; Y 11221, 2 August 1734.

10. AN, Y 13527, 30 March 1762; Y 15248, 1 April 1741; Y 13100, 20 January 1747; Y 12150, 2 February 1747; Y 15269, 18 July 1762; Y 12599, 16 September 1753; Y 14542, 25 August 1747.

11. AN, Y 13120, 17 December 1768; Y 14536, 29 September 1740; Y 12141, 16 September 1739; Y 11731, 4 August 1788.

12. Lacombe d'Avignon, *Le Mitron de Vaugirard*, p. iii; AN, MC, XXVIII-131, 10 October 1714; MC, LXI-601, 26 June 1784; MC, XIX-610, 16 February 1715; AN, Y 14940, 1724 (partage Guenon); Y 9442, 3 August 1741; Y 12138, 21 August 1736; Y 12135, 5 February 1734.

13. AN, Y 13744, 26 August 1735; Y 15263, 21 July 1755; Marville to Maurepas, 23 August 1742, Arthur-Michel de Boislisle, ed., *Lettres de M. de Marville, lieutenant général de police, au ministre Maurepas (1742–1747)* (Paris, 1896–1905), vol. 1, p. 67; Y 15235, 30 March 1728; Y 13530, 2 August 1763; Y 15061, 19 June 1760; Y 15248, 27 January 1741; Y 15239, 15 February 1732; Y 15947, 3 August 1752; Y 15268, 12 May 1760; Y 15350, 24 June 1752; Y 15627, 8 April 1754.

14. AN, Y 13392, 28 January 1766; Y 11240, 15 May 1753. The imputation of other diseases also threatened bakers. Customers fled after a neighbor announced that a widowed bakeress had a pox coming out of her eyes and ought to have her arms and nose amputated. Y 15248, 1 May 1741.

15. *Paris-Match*, 30 June 1973, pp. 72–75.

16. AN, MC, CXXI-294, 23 April 1732; AN, Y 12148, 30 September 1745; Y 13926, 31 May 1741; Y 11240, 19 December 1753; Y 15244, 17 September 1737; Y 12575, 13 June 1730.

17. AN, Y 11237, 18 June 1750; Y 11377, 1 December 1769; Y 14963, 8 February 1747; Y 9440, 20 March 1739.

18. AN, Y 15613, 27 June 1747; Y 12141, 22 January 1739. For another case of ruptured baker-client relations involving a *porteuse*, see Y 11224, 20 July 1737. Bakers rarely gave the benefit of the doubt to servants, whether they were buying for their own (very modest) needs or on behalf of their masters (who were expected to accept the word of the artisan). Y 14523, 1 June 1726; Y 12158, 30 June 1755; Y 15255, 10 October 1747. Cf. the case of the servant the validity of whose money a baker questioned. Y 13118, 12, 18 February 1766.

19. AN, Y 11218, 6 January 1731; Y 15941, 13 October 1742; Y 15961, 24–26 October 1764. In filing for damages and interest, Betemont testified to his dependence on his wife in the conduct of his business (he had been away on a buying mission the day she was beaten; she managed the shop on a day-to-day basis): "qu'en outre le commerce du plaignant exige la présence de sa femme, [actuellement] en souffrance, ce qui lui fait un tort considérable."

20. AN, Y 15589, 26 April 1735; Y 13082, 2 August 1734; Hardy's journal, BN, ms. fr. 6686, fol. 240 (30 November 1785). Cf. Morel, *Histoire illustrée*, p. 279.

21. AN, Y 12601, 13 December 1754; Y 12595, 29 November 1751.

22. AN, Y 12741, 24 April 1755; Y 13386, 9 September 1760; Y 13761, 18 January 1754; Y 11241, 16 December 1754; Y 11228, 25 October 1741.

23. AN, Y 12610, 29 July, 23 September 1761; Y 12611, 18 March 1762; Y 12614, 15 July 1765. Cf. another case of succession jealousy: master Delogny took over the shop of a failed baker named Le Fevre, who returned in a drunken wrath on several occasions to express his anger at having been usurped ("Je veux entrer chez toy, j'y ay entré avant toy et j'y entrerez toutes les fois qu'il me plaira"). For another example of an embittered baker, down and out, who wanted others to remember when he flourished, see the tale of Pierre Ferrière: Y 14942, 3 March 1729, Y 13746, 6 August, 2 November 1739, and Y 12141, 11, 12 October, 2 November 1739.

24. AN, Y 11562, 9 April 1740; Y 15344, 28 December 1750.

25. AN, Y 14778, 6 September, 20 October 1728; Y 15062, 3 November 1761; Y 15236, 28 August 1729.

26. AN, Y 12605, 21 December 1757; Y 12608, 23 February 1760.

27. AN, Y 9638, 10 November 1719; Y 9440, 3 August 1739; Y 13741, 24 December 1740; Arsenal, ms. Bast. 10271, 11 December 1725; Y 11305, 21 February 1742; Y 11219, 23 October 1732; Y 13819, 1 August 1771. Cf. the hypersensitivity of the bakers of Arras to territorial matters, as bespoke article 10 of their statutes: "Les boulangers paieront amende lorsqu'ils dérangeront les étaux de leurs confrères, ou qu'ils prendront leurs ustensiles sans leur permission, lorsqu'ils urineront à quatre pieds près de leurs étaux ou de ceux de leurs voisins, ou qu'ils cracheront avec violence." Ouin-Lacroix, *Histoire des anciennes corporations*, pp. 735–36.

28. AN, Y 15252, 2 January 1743; Y 14777, 7 September 1727; Y 13366, 22 August 1742; Y 11673, 30 September 1749; Y 15603, 5 May 1742; Y 11241, 26 September 1754; Y 12606, 11 March 1758. Cf. the even more complicated disputes between the children of a first bed and the surviving stepparent, or between children of several successive unions. E.g., Y 13080, 12 April 1733.

1. Kaplan, *Bread, Politics, and Political Economy,* introduction, and *Provisioning Paris,* pp. 23–29.

2. "La Cherté du pain," 21 January 1792, AN, T 644[1–2].

3. A Paris setier consisted of 1.56 hectoliters or 4.43 U.S. bushels; a muid, which contains twelve setiers, equals 18.72 hectoliters or 53.12 U.S. bushels. See Ronald E. Zupko, *French Weights and Measures before the Revolution: A Dictionary of Provincial and Local Units* (Bloomington, Ind., 1978).

4. BN, Joly 1143, fol. 107 (1769); Sartine to syndics des commissaires, 12 March 1768, AN, Y 13396; Edme Béguillet, *Traité des subsistances,* p. 675; Abbé Tessier cited in *Encyclopédie méthodique,* Agriculture, vol. 3, p. 473; Arsenal, ms. Bast. 10141, fol. 309 (20 January 1760).

5. Delamare, *Traité,* vol. 2, pp. 1071–78; BN, Joly 1428, fols. 11, 13; AN, Y 13396, December 1767–April 1768.

6. BN, Joly 1743, fol. 107; Courcy to Joly, 29 January 1739, BN, Joly 1119, fols. 41–42; Joly 1122, fol. 65; Joly to lieutenant of police, 29 September 1740, Arsenal, ms. Bast. 10277; "Mémoire," 1764, BN, ms. fr. 14295, fol. 6; Malouin, *Description* (1767), p. 295 (in the edition of 1761, p. 274). Another estimate brought to the attention of the procurator general fixed the yield at 235 pounds, without indication of the sorts of bread. BN, Joly 1121, fol. 18 (September 1740). An Englishman visiting Paris in the early 1750s also supposed that a setier rendered 235 pounds of bread. Mildmay, *Police.* Moheau, an acute observer of Old Regime society, maintained that a pound of wheat ordinarily rendered more than a pound of bread, a claim that seems extravagant to me, unless it is restricted to wheat milled in the most efficient "economic" mills and bread of a single sort. *Recherches et considérations sur la population de France* (Paris, 1778), in *Collection des économistes et des réformateurs sociaux de la France,* ed. René Gonnard (Paris, 1912), p. 34. Cf. the estimates reported by A.-J.-P. Paucton, *Métrologie, ou Traité des mesures, des poids et des monnaies des anciens peuples et des modernes* (Paris, 1780), pp. 489–90, 507; the calculations of the bakers of Troyes (1786), BN, Joly 1743, fol. 131; and the suggestions registered by Cotte, *Leçons élémentaires,* p. 75.

7. AN, Y 12616, November–December 1767; Arsenal, ms. Bast. 7458; AN, F[12] 1299[B] and L 530; Edme Béguillet [with César Bucquet?], *Manuel du meunier et du charpentier de moulins, ou Abrégé classique du traité de la mouture par économie* (Paris, 1775), extracted in *Nouvelles Ephémérides économiques* 10 (1775), pp. 156–57. In 1769 Malisset had already obtained as much as 252 pounds from a setier, Y 12610. In this trial at the Hôpital général, however, the yield of the setier milled by traditional means turned out to be 236 pounds in one run and 265 in another. Malisset and Machurin, the police commissaire who supervised the trial, both cried fraud and disallowed the findings for ordinary milling.

8. Parmentier, *Parfait Boulanger,* p. 595; Béguillet, *Discours sur la mouture économique,* pp. 165–70, and *Traité des subsistances,* pp. 658–64; Béguillet [and Bucquet?], in *Nouvelles Ephémérides* 10 (1775), pp. 156–57; Baudeau, *Avis au premier besoin,* 1er traité, pp. 17, 23–24; "Mémoire," BHVP, series 118, cote provisoire 1482; Des Essarts, *Dictionnaire universel,* vol. 2, pp. 273–77; Abbé J.-J. Expilly, *Tableau de la population de la France*

(Paris, 1780), p. 1973. Cf. the dramatically more modest results reported by the Roman Annona in Jacques Revel, "A Capital City's Privileges: Food Supplies in Early Modern Rome," in *Food and Drink in History: Selections from the Annales,* ed. Robert Forster and Orest Ranum (Baltimore, 1979), p. 40. For a household bread made of whole flour, Parmentier fixed the setier's yield at 249 pounds. *Parfait Boulanger,* pp. xlii–xlv.

9. Malouin, *Description* (1767), pp. 274–80, and *Description* (ed. Bertrand, 1771), pp. 49–50 and 278–79.

10. Parmentier, *Parfait Boulanger,* pp. 597–98, 601.

11. "Réflexions sur le commerce du pain," *Journal économique,* July 1761, pp. 319–20.

12. Picart to Sartine, 5 January 1769, AN, Y 12618. Cf. the request of the lieutenant general of Provins for copies of the Paris trials, since the tests conducted locally produced highly dubious results. Colin des Murs to Sartine, 8 January 1769, ibid.

13. "De la mouture," *Journal économique,* May 1771, pp. 216–17; Bucquet, *Manuel du meunier,* pp. 3–4; Claude-Jacques Herbert, *Essai sur la police générale des grains, sur leurs prix et sur les effets de l'agriculture* (Paris, 1775), in *Collection des économistes et des réformateurs sociaux de la France,* ed. Edgard Depitre (Paris, 1910), pp. 36–39; "Mémoire important sur les causes de la cherté," BN, Joly 1116, fol. 278; Etienne Chevalier, *Mémoire sur les moyens d'assurer la diminution du pain* (Paris, 1793), BHVP. In the mid-nineteenth century, an expert projected annual per capita consumption at three hectoliters of wheat—under two setiers (1.92). *Enquête sur la boulangerie,* p. 326.

14. BN, Joly 1428, fols. 103–57. See also Kaplan, "Lean Years, Fat Years." It seems just as likely that the houses understated their needs in order to reduce the burden that the government imposed on them of stocking three times their annual grain consumption. Moreover, religious communities were not commonly known for eating *pain de fantaisie* or other opulent forms of white.

15. Sébastien Le Prestre de Vauban, *De l'importance dont Paris est à la France et le soin que l'on doit prendre de sa conservation* (Paris, 1821), p. 29; Nicolas-François Dupré de Saint-Maur, *Essai sur les monnaies, ou Réflexions sur le rapport entre l'argent et les denrées* (Paris, 1746), pp. 21, 24, 53, 59, 101; François Quesnay, "Fermier," in Diderot et al., *Encyclopédie,* vol. 6, p. 533; Georges Weulersse, *Le Mouvement physiocratique en France de 1756 à 1770* (Paris, 1910), vol. 2, p. 544; Baudeau, *Avis au premier besoin,* 1er traité, pp. 23–24; Dupont cited by Pigeonneau and de Foville, *L'Administration de l'agriculture;* Parmentier, *Parfait Boulanger,* pp. xliii–xlv; Parmentier, *Avis aux bonnes ménagères* and *Recherches,* p. 442; Béguillet, *Traité des subsistances,* pp. 354–55, 459–60, and "Avantages de la mouture économique," 1769, Arsenal, ms. Bast. 2891; G. de Molinari, "Céréales," in Charles Coquelin and Gilbert-Urbain Guillaumin, eds., *Dictionnaire de l'économie politique* (Paris, 1852–53), vol. 1, p. 308; Arnoult cited by Abbé Henri-Alexandre Tessier, *Journal économique,* July 1759, p. 52; *Encyclopédie méthodique,* Agriculture, vol. 3, pp. 482–83; "Mémoire important sur les causes de la cherté," BN, Joly 1116, fol. 278. In a mémoire written in the early eighteenth century, the bakers of Paris calculated consumption on the basis of three setiers a head. But in another mémoire drafted shortly afterward, they reduced their coefficient to two setiers without explanation. Joly 1428, fols. 5–7.

16. AN, Y 9538, 6 May 1757.

17. Etienne-François, duc de Choiseul, *Mémoires de M. le duc de Choiseul . . . écrits par*

lui-même et imprimés sous ses yeux, dans son cabinet à Chanteloup, en 1778, ed. Jean-Louis Soulavie (Chanteloup and Paris, 1790), vol. 1, p. 45. (Choiseul allowed that "le peuple ouvrier" ate three setiers, but women and children compensated for their overconsumption by eating only two setiers.) Députés de commerce, "Premier supplément au premier mémoire," ca. 1764, BN, ms. fr. 14295, p. 6; "De la mouture," pp. 216–17; *Journal de l'agriculture, du commerce, des arts et des finances* 7, 2d pt. (November 1766), pp. 120–21. Concurring with the estimate of 2.5 setiers were L.-J. Bourdon-Desplanches, *Projet nouveau sur la manière de faire utilement en France le commerce des grains* (Brussels and Paris, 1785), p. 44; F. Aubert, "Observations sur la liberté du commerce des grains," *Journal du commerce,* September 1759, p. 79. A study written for Richelieu in 1637 fixed Paris consumption at the remarkably low figure of 2.52 setiers (based on a population of 400,000). "Mémoire du dénombrement fait par le commandement de Mr le Cardinal de Richelieu," BN, Joly 1428, fols. 1–2.

18. Francis Veron de Forbonnais, "Mémoire sur la police des grains," BN, ms. fr. 11347, fol. 181 (April 1758); Mildmay, *Police;* Doumerc to procurator general, 17 February 1789, BN, Joly 1111, fol. 204; Vatar Desaubiez, *Le Bonheur public* (London, 1780–82), p. 138. Regardless of milling technology, a nineteenth-century economist contended that three setiers could never have been a *national* per capita average, because such a figure implied an annual crop far beyond what French arable land was capable of producing (according to Lavoisier's figures). M. A. Moreau de Jonnès, "Statistique des céréales de la France: Le Blé, sa culture, sa production, sa consommation, son commerce," *Journal des économistes* 4 (1843), pp. 155–57. There is still no agreement among scholars on the exact extent of cultivated surface and on productivity and production (not to mention population). Eventually we should be able to create a much more finely tuned model than Moreau de Jonnès fashioned and subject it to his tests.

19. Jacques Necker, *Sur la législation et le commerce des grains* (Paris, 1775), p. 59; Moheau, *Recherches,* p. 34. Moheau took the trouble to specify that in his judgment economic milling was not an influential force. Both Moheau and Necker seem to have been heavily influenced by consumption models drawn from military provisioning experience and public-assistance institutions. Neither may have been appropriate for civilian extrapolation. In another work attributed to Moheau, the ration was set at three setiers because of the limited diffusion of economic milling technology; because rye, still widely consumed, yielded less bread than wheat; and because the quality of the wheat and/or rye utilized was often mediocre. "De l'étendue de la France et de son revenu territorial," 1789, AN, F[20] 403. Cf. also Paul Emile Girod, who claimed that the average consumption was set at two setiers by "the statisticians of the eighteenth century." "Les Subsistances en Bourgogne et particulièrement à Dijon à la fin du XVIIIe siècle, 1774–1789," *Revue bourguignonne de l'enseignement supérieur* 16 (1906), p. 24.

20. Legrand d'Aussy, *Histoire de la vie privée,* vol. 1, p. 75; Parlement of Provence to Louis XV, ca. 1768, AD Bouches du Rhône, C 2420; Arsenal, ms. Bast. 10141, fol. 474 (20 April 1761); *Journal économique,* August 1761, p. 363; Parmentier, *Parfait boulanger,* p. xxvii, and *Avis aux bonnes ménagères,* p. 105; Béguillet, Mémoire sur l'art," AN, F[10] 256, and *Traité des subsistances,* pp. 354–55, 399–402, 638–44; Bucquet, *Manuel du meunier,* p. 117. Cf. Mirabeau's sanguine estimate of a one-third increase in yield. Weulersse, *Mouvement,* vol. 2, pp. 526–27. Moheau scoffed at all these exuberant appre-

ciations. He maintained that economic milling increased bread yields by no more than one-twentieth. *Recherches,* p. 34. Malouin was the only expert in the economic-milling generation who emphasized the gradualness of the technological ameliorations and who hinted that they had begun well before the physiocratic propaganda in the sixties made the "economic" label fashionable. Various improvements in grinding, bolting, and baking were sufficiently diffused, in Malouin's view, to reduce real consumption to two setiers in Paris. *Description* (1767), p. 294. On economic milling, see Kaplan, *Provisioning Paris,* especially chapters 10 and 11.

Louis Stouff claimed that the per capita ration in Provence in the fourteenth and fifteenth centuries was three hectoliters, i.e., under two setiers. *Ravitaillement et alimentation,* pp. 78 and passim. If it is correct, this estimate testifies not to technological mastery but to an extremely varied and relatively rich popular diet. In 1909, one expert reported that the average annual consumption in France was still three hectoliters of wheat. Edmond Rabate, *Le Blé, la farine, le pain* (Paris, 1909). This estimate seems high to me. Cf. André Cochut, who puts per capita Parisian consumption at 1.33 setiers (a little over two hectoliters) toward the end of the nineteenth century. "Le Pain à Paris," *Revue des deux mondes,* 33d year, 2d period, 46 (15 August 1863), p. 978. Another writer contended that there were a number of departments in the nineteenth century where four setiers were still required to feed an individual as a result of the stagnation of milling techniques. J. E. Horn, *L'Economie politique avant les physiocrates* (Paris, 1867), p. 186. We have reason to expect a steady decline in per capita consumption because of both technological improvements and changes in socioeconomic status and diet. Yet we should contemplate Jean-Jacques Hémardinquer's reminder that Frenchmen still drew on average one-third of their calories from bread in the 1950s. "En France aujourd'hui: Données quantitatives sur les consommations alimentaires." *Annales: E.S.C.* 16th year (May–June 1961), p. 560.

21. Antoine-Laurent Lavoisier, *Résultats extraits d'un ouvrage intitulé: De la Richesse territoriale du royaume de France* (Paris, 1791), p. 38; Alfred de Foville, *La France économique: Statistique raisonnée et comparative* (Paris, 1890), p. 122; Dupré de Saint-Maur, *Essai,* pp. 56–57; Malouin, *Description* (1767), pp. 293–94; François Furet, "Pour une définition des classes inférieures à l'époque moderne," *Annales: E.S.C.* 18th year (May–June 1963), p. 463; Camille Bloch, *L'Assistance et l'etat en France à la veille de la Révolution* (Paris, 1908), pp. 4–5; Necker, *Législation,* p. 59; Jean-François de Tolozan, following Necker, *Mémoire sur le commerce de la France et de ses colonies* (Paris, 1789), p. 4; Mildmay, *Police,* p. 73; députés de commerce, "Premier Supplément au premier mémoire," ca. 1764, BN, ms. fr. 14295, p. 6; Philippe Benoist and Jean-Julia de Fontenelle (who estimated twenty-one ounces), *Nouveau manuel complet du boulanger, du négociant en grains, du meunier et du constructeur de moulins* (Paris, 1856); Béguillet, *Traité des subsistances,* pp. 658–64; Albert to Sartine, 10 May 1770, AN, F^{12} 153, whose estimate is repeated in "Mémoire," 1761, BN, ms. fr. 11343, fol. 217; Briatte, *Offrande,* p. 166. For an example of an insouciant (and flawed) approach to Lavoisier, see Robert Philippe, "Une Opération pilote: L'Étude du ravitaillement de Paris au temps de Lavoisier," *Annales: E.S.C.* 16th year (May–June 1961), pp. 564–68. On the uses of Lavoisier, see also Herbin de Halle, *Statistique générale et particulière de la France et de ses colonies,* ed. P. Etienne (Paris, 1803), vol. 7, pp. 273–74.

22. Jean Louis, comte de Lagrange, *Essai d'arithmétique politique*, in *Collection des principaux économistes*, ed. Eugène Daire and G. de Molinari (1845–47; reprint Osnabrück, 1966), vol. 14, pp. 609–10; R. Durin, *Le Boulanger et la boulangère, ou le Pain au mois de novembre au plus tard ne vaudra que 8 sols les quatre livres* (Paris, 1789). Modern historians have been generally content to draw on the estimates of contemporaries. Focusing on the *menu peuple*, George Rudé, following Auguste-Philippe Herlaut, calculated the ration at one pound. Herlaut, "Disette," p. 9; Rudé, "Prices, Wages, and Popular Movements in Paris during the French Revolution," *Economic History Review*, 2d ser., 6 (1954), p. 248. François Furet relies on the official Revolutionary estimate of nineteen ounces. "Pour une définition," p. 463. Fernand Braudel cites a number of different estimates, mostly in the low range. *Civilisation matérielle et capitalisme, XVe–XVIIIe siècles* (Paris, 1967), vol. 1, p. 99. T. J. Markovitch, "Histoire quantitative de l'économie française," *Cahiers de l'Institut de science économique appliquée*, no. 173 (May 1966), p. 192. Pierre Lefèvre proposed an average consumption of 1.5 pounds at Lille. *Le Commerce des grains*, p. 4. Written in the nineteenth century, Armand Husson's discussion is one of the most probing and intelligent reflections on the problem. He felt that consumption had been generally overestimated by contemporaries. *Consommations*, pp. 170 and passim. For estimates of daily consumption—norm and practice—in the Middle Ages, see Desportes, *Le Pain au moyen âge*, pp. 202–4.

23. Lagrange, *Essai*, pp. 57–59; Paucton, *Métrologie*, pp. 498–99; "Mémoire de la consommation de pain qui se fait tous les jours à Paris," BN, Joly 1428, fol. 13 (ca. 1720); Moheau, *Recherches*, p. 34; Dupré de Saint-Maur, *Essai*, pp. 56–57; Lebon, "Observations," AN, F[10] 226; Labrousse, *La Crise de l'économie*, p. xxiv; Cahen, "A propos du livre," p. 165; Du Vaucelles, "Offrande civique," ca. late 1789, AN, F[10] 215–16. See also Briatte's strikingly insightful remarks, taking account not only of age, sex, and work but also of lifestyle, metabolism, weather, etc. *Offrande*, pp. 141, 155. How much nourishment did a pound of bread provide? A wheaten pound loaf furnished approximately twelve hundred calories, though the exact amount varied with the rate of extraction and the type of bread. Around 88 percent represented starch, while protein made up 10 percent. Kent-Jones and Price, *Bread-Making*, p. 175; Harry C. Sherman and Constance S. Pearson, *Modern Bread from the Viewpoint of Nutrition* (New York, 1942), p. 13; Marian Apfelbaum and Raymond Lepoutre, *Mangeurs inégaux* (Paris, 1978), p. 25; Colin Clark and Margaret Haswell, *The Economics of Subsistence Agriculture*, 3d ed. (New York, 1967), p. 54. On the difficulties of calculating caloric content from wheat to bread, and on inferring nutritional information beyond calories, see Maurice Aymard, "Toward the History of Nutrition: Some Methodological Remarks," in *Food and Drink in History: Selections from the Annales*, ed. Robert Forster and Orest Ranum (Baltimore, 1979), p. 2.

24. Lacombe d'Avignon, *Le Mitron de Vaugirard*, p. 49; Dupré de Saint-Maur, *Essai*, p. 59. I regret that Eric Brian's brilliant new book did not reach me in time to help me better understand the discourse of counting. See *La Mesure de l'Etat: Administrateurs et géomètres au XVIIIe siècle* (Paris, 1994), especially pp. 153–78, 256–92.

25. For all these estimates, see Kaplan, *Bread, Politics, and Political Economy*, vol. 1, p. 34, and Roche *Le Peuple de Paris*, chapter 1. See also Expilly, *Dictionnaire géographique*, vol. 5, pp. 400–402, 808.

26. Thellusson to procurator general, 5 December 1740, BN, Joly 1109, fol. 69; Béguillet, *Traité des subsistances*, p. 664; Assembly of Police, 15 April 1731, BN, ms. fr. 11356, fol. 154; Kaplan, "Lean Years, Fat Years." The inelasticity of the demand impressed Thellusson. Hard times forced consumers to abandon certain noncereal foods that they ate regularly or occasionally. In this banker's view, they took refuge in their bread consumption as a sort of biosocial last stand. This position was challenged by Abbé Tessier, who argued, somewhat vaguely, that people decreased their bread consumption in a dearth by shifting to certain vegetables or by tightening their belts. *Encyclopédie méthodique, Agriculture*, vol. 3, pp. 472–76.

27. "Mémoire de la consommation de pain qui se fait tous les jours à Paris," BN, Joly 1428, fols. 9–10 (ca. 1720); Gazetins, 1–2 October 1740, Arsenal, ms. Bast. 10167, fol. 151; Mercier, *Tableau*, vol. 4 (1782), p. 195; *Revue rétrospective*, January–June 1887, p. 76; BN, ms. fr. 21640, fol. 50; Foucaud to procurator general, August 1755 and 12 November 1756, BN, Joly 1113, fols. 222, 236; Béguillet, *Traité des subsistances*, pp. 162–64; Malouin, *Description* (1767), p. 291. Without convincing documentation, Cochut claims to have found a late-seventeenth-century source that fixes consumption at 208,000 muids. "Le Pain à Paris," p. 973. Arthur-Michel de Boislisle cites another contemporary source for 150,000 muids. *Mémoires des intendants sur l'état des généralités dressés pour l'instruction du duc de Bourgogne* (Paris, 1881), vol. 2, p. xxii. Expilly also cited 150,000 muids, but it appears that he is presenting a generally held view rather than stating his own. The estimate is based on a per capita annual consumption of three setiers, whereas in 1780 Expilly set per capita consumption at two setiers. *Dictionnaire géographique*, vol. 5, p. 303, and *Tableau*, pp. 12 and 27 (where Expilly rejects Sauval's estimate of one hundred thousand muids for 1714). Presumably the *Almanach parisien en faveur des étrangers* (Paris: 1772, 1786) meant to endorse the figure of 150,000 muids. But its 1772 edition cited 15,000 muids for annual consumption (p. 20), for which the authors compensated in 1786 by indicating 1.5 million muids!

28. Controller general to M. de Luxembourg, 4 August 1725, AN, G[7] 34, "Observations sur le commerce des grains," 1763, BN, ms. fr. 11347, fol. 225; Charles Philippe d'Albert, duc de Luynes, *Mémoires du duc de Luynes sur la cour de Louis XV*, ed. L. Dussieux and E. Soulié (Paris, 1860–65), vol. 3, p. 78 (November 1739); "Mémoire et avis" (1764), BN, ms. fr. 14296; AN, H* 1870, fols. 374–84 (February 1764); Daure, "Mémoire," 1776, AN, F[11] 264; "Mémoire," Arsenal, ms. Bast. 5282, fol. 355; Conseil général de la commune de Paris, 27 September 1791, AN, T 644[1–2]; Commission du commerce et des approvisionnements, "Etat général des besoins de Paris," 1794, AN, F[10] 215–16. The Revolutionary estimates are in sacks of 325 pounds, which I calculated at two setiers on the grounds that the bulk of informed opinion still believed that this was the correct relation. In fact I think that two setiers rendered 10 to 15 percent more even in mills not converted to "economy."

29. Assembly of Police, 15 April 1731, BN, ms. fr. 11356, fol. 154; deliberations, Hôtel de Ville, 20 September 1759, AN, H* 1968, fols. 311–16; Trennin to procurator general, 25 October 1768, BN, Joly 1142, fol. 166; *Encyclopédie méthodique, Agriculture*, vol. 3, p. 476; Du Vaucelles, "Offrande civique," ca. late 1789, AN, F[10] 215–16. Du Vaucelles sharply criticized the commonly invoked early Revolutionary estimate of 91,250 muids (1,500 sacks a day). See, for example, Morel, *Histoire abrégée*, p. 275; Louis Bergeron,

"Approvisionnement et consommation à Paris sous le premier empire," *Mémoires de la Fédération des sociétés historiques et archéologiques de Paris et de l'Ile-de-France* 14 (1963), p. 208; Seymour E. Harris, *The Assignats* (Cambridge, Mass., 1930), p. 154; Jean-Sylvain Bailly, *Mémoires,* ed. Saint-Albin Berville and Jean-François Barrière (Paris, 1822), vol. 2, pp. 280–81. Cf. ibid., p. 96, where Bailly proposes 720,000 pounds as daily Parisian bread consumption. At a yield of 225 pounds of bread to the setier, this amount is consistent with Bailly's estimate of daily aggregate flour needs.

30. Mémoire du dénombrement fait par le commandement de Mr le Cardinal de Richelieu," BN, Joly 1428, fols. 102; "Instruction générale pour les intéressés au canal de Picardie," BN, ms. fr. 21690, fols. 9–10 (1728); Dupré de Saint-Maur, *Essai,* p. 59n (but on the basis of records dating from 1729–30, not 1750 as claimed by Tessier, *Encyclopédie méthodique,* Agriculture, vol. 3, pp. 472–76); Paucton, *Métrologie,* p. 508; "Mémoire des boulangers forains," BN, Joly 1314, fol. 50 (March 1738); Mildmay, *Police;* Lavoisier, *Richesse,* pp. 36–37.

A word must be said about Lavoisier's calculations because of the scientific cachet that his brilliant reputation seemed to give them and because later analysts often used them. He drew his figures from a retrospective inquiry ordered by Turgot on the amount of grain and flour that entered Paris in the period 1764–74. But six, perhaps seven of those years were years of crisis, of dearth, of acute disorganization of the ordinary provisioning channels, and of a disruption of consumption habits. Moreover, it is very unlikely that the official registers reflected all entries, for a great deal of grain and flour came from "extraordinary" sources and was not recorded either by conscious design or by accident. Lavoisier claimed that 66,289 muids of flour entered each year. In 1767 approximately 24,444 muids were sold at the Halles. BN, Joly 1139, fol. 129. If Lavoisier was right, that means that about forty-two thousand muids arrived *en droiture* for the institutional buyers and the bakers. Though flour had long eclipsed grain in the Paris provisioning pattern, Lavoisier's figure seems high. He assumed that each pound of wheat yielded a pound of bread and that each 325-pound sack of flour produced 416 pounds of bread. It was commonly believed by police officials, however, that a sack of flour was the fruit of two setiers of wheat. Why didn't the scientist–farmer general claim, then, 480 pounds of bread for each sack of flour? If he in fact thought that two setiers produced more than 325 pounds (as a number of specialists affirmed), then we cannot assimilate the grain and flour entries without making corrections. The flour muids must be deflated, as it were, or made lighter to reflect the higher yield of grain into flour and to enable us to combine flour and grain in the same measure. In this case, Lavoisier's estimate would have to be lowered to something like 59,660 muids for the flour entry. There is one further problem with Lavoisier's appreciation. He supposed that the amount of bread brought by the forains was "just about compensated for" by the amount of bread (and grain and flour) purchased by non-Parisians for consumption outside the capital. Now, while it is true that many outsiders came to Paris to buy bread, especially in dearth years, there is no doubt that the forain supply was of substantially greater significance. Cf. *Encyclopédie méthodique,* Agriculture, vol. 3, pp. 472–76, and Husson, *Consommations,* pp. 133–34.

31. The police assigned a weekly wheat consumption figure for each market baker that serves as the basis for annual aggregate calculations. It seems extremely likely that the

amount attributed to each baker was empirically determined. But it acquired normative value in the minds of the authorities. It became a quota in difficult times and ordinarily served as a measure for judging the fidelity of bakers from week to week. On average I believe that it faithfully reflected baker supply. We have no guarantee, however, that each baker filled his/her quota (rather than surpassing it or not attaining it) every week. Clearly there is room for error in our annual calculations of market bread supply. The Delalande registers are probably the same ones that Dupré de Saint-Maur consulted in the 1740s. See his *Essai,* p. 59.

32. There are two copies of the 1727 register, one in the BN where it is curiously considered an *imprimé,* and one at the Bibliothèque de l'Institut. I have corrected certain errors conveyed in Institut, ms. 515.

33. No allowance is made for wheat usage by pastry makers. It may have been assumed that they employed reground middlings (*gruaux*) and thus did not cut into the primary wheat supply. Apparently it is also assumed—probably wrongly—that dogs and cats, like chickens and pigs, ate bran and other issues or secondary grains. The police also appear to have supposed that neither starch makers nor brewers encroached upon primary, bread-destined provisions. I suspect that pets and pastry makers did divert some primary wheat supply. If I were interested in determining *total* wheat supply, I would have to suggest some compensation for this. Since my concern, however, is with bread consumption, I will not make any adjustments in the calculations given.

34. "Etat des noms et demeures des maîtres boulangers de la Ville et faubourgs de Paris et de ce qu'ils employent de bled et de farine par semaine," 15 October 1724, new ser. ms. 116, BHVP. In 1724, twenty-six of a total of 512 master bakers were listed as using no grain or flour. A number of these bakers may have retired; in some cases, however, it is possible that the police simply were unable to get the necessary information. I have assumed five hundred active masters for 1727, a figure that is perhaps a bit low.

35. See BN, Joly 1428, fol. 506. There is another indication that the master weekly consumption estimate may be low. In 1733, the average master usage *just at the markets* was 1.47 muids, only a little lower than the figure we have used for total weekly master consumption. Now it is true that fewer than half of the masters served the market. Nevertheless the 18.12 setier figure (1.51 muids) seems too low.

36. Population extrapolation, as I mentioned, is an extremely parlous enterprise, but let us note the possible range according to eighteenth-century criteria. If mean annual per capita consumption was three setiers (too high a coefficient in my opinion), then the population of the capital turns out to be 367,808 (if I am right about aggregate consumption). If per capita consumption was, as I suspect, closer to 2.5 setiers, then the population hovered around 441,370. Based on the admittedly uncertain information we have about the capital's population and its evolution, both extrapolated estimates are too low. That probably means that there were hidden supplies that escaped the Delalande net and my prudential supplement. (It seems to me unlikely that fewer than 2.5 setiers were required per capita.) At one hundred thousand muids, the population would stand at 480,000 (with a coefficient of 2.5 setiers), a quasi-plausible figure.

Shortly after I completed my calculations and analyses of the Delalande registers, I learned that two historians, soon to become my close friends, had begun working on the same materials. On a number of important points, including the figures they transcribed and the use they made of them, and the interpretive context into which they put them, I take issue with their work. See Jean-Michel Chevet and Alain Guéry, "Consommation et approvisionnement de Paris en céréales de 1725 à 1733," in *Les Techniques de conservation des grains à long terme: Leur Rôle dans la dynamique des systèmes de cultures et des sociétés*, ed. Marceau Gast, François Sigaut, and Corinne Beutler (Paris, 1985), vol. 3, fasc. 2, pp. 463–77.

37. The total of 78,236 muids is a corrected figure based on a recalculation of the register data. The register is Institut, ms. 521.

38. Assembly of Police, 15 April 1731, BN, ms. fr. 11356, fol. 154.

39. The bits and pieces that we have provide information that is both incomplete and unconvincing. For example, the *Gazette du commerce, de l'agriculture et des finances*, 6 January 1767, p. 12, indicates a combined port and Halles grain and flour provision of 22,836 muids for 1765. Though it is allegedly based on monthly *relevées*, this figure is far too low. It is true that the ports continued to decline in importance and that droiture trade continued to expand, but the Halles also flowered as a market. Already in the early thirties the Halles turned over more (wheaten) merchandise than the *Gazette* imputes to both Halles and ports a generation later, when the population was substantially larger. Juxtapose the *Gazette* tally to the total of 45,105 muids of grain and flour arrivals at the ports and the Halles for 1769 contained in a police report. BN, Joly 1143, fol. 166. This was a year of dearth and disorder, but it is inconceivable even in the crisis that transactions doubled in the space of a few years.

40. My estimate is high; it is based on a per capita consumption coefficient of 2.3 to 2.4, which allows for a gradually cumulative amelioration in milling technology (between, say, 1730 and 1780) rather than a sudden breakthrough in productivity.

Chapter 17   The Police of Bakers

1. Necker to Sartine, 14 February 1778, AN, F^{11*} 1, fol. 258; Kaplan, *Bread, Politics, and Political Economy*, vol. 1, pp. 11–14. "De tous les objets qui regardent essentiellement la police, il n'en est pas de plus intéressant qui méritent autant votre attention, la vigilance et le zele de vos commissaires que les alimens de première nécessité pour la nourriture des citoyens," wrote Jacques Dupuy, himself a commissaire during the closing years of the Old Regime. AN, Y 12826, 1 September 1780.

2. As Mercier put it, Paris owed its privileged position in the competition for subsistence to the conviction held by the government that "outcries of need there would be more dangerous than anywhere else and would set a fatal and contagious example." Mercier, *Tableau*, vol. 4 (1782), p. 203. On the preoccupation with Paris and its acute political and provisioning vulnerability, see Kaplan, *Bread, Politics, and Political Economy*, vol. 1, pp. 32–37.

3. Kaplan, *Bread, Politics, and Political Economy*, vol. 1, pp. 14–28; Kaplan, *Provisioning Paris*, pp. 33–38.

4. Mercier, *Tableau*, vol. 9 (1788), pp. 129–30. On the origin, recruitment, organization, and functions of the commissaires, see S. L. Kaplan, "Note sur les commissaires

de police de Paris au XVIIIe siècle," *Revue d'histoire moderne et contemporaine* 28 (October–December 1981), pp. 669–86.

5. "Mémoire pour le Doyen, les Syndics et la Compagnie des Commissaires contre les substituts du Substitut," ca. 1781, ASP, 2 AZ 13, pp. 62–63. Cf. the similar vein in which Commissaire Lemaire evoked the work of his colleagues: to "contenir" Parisians "sans qu'ils s'en aperçoivent, sous le joug de la subordination et de l'obéissance si nécessaires pour les gouverner." Gazier, "La Police," pp. 57–58.

6. Delamare, *Traité*, vol. 2 (1729), p. 868; La Reynie to Delamare, BN, ms. fr. 21639, fol. 171 (24 October 1693), and ms. fr. 21638, fol. 396 (12 July 1693); Boislisle, "Grand Hiver," p. 516; Saint-Germain, *Vie quotidienne*, pp. 218–19. Hérault to procurator general, BN, Joly 1117, fol. 193 (26 September 1725). Cf. the remarks of the lieutenant criminel in the Châtelet in 1666 on "la malice des boulangers" and "leur avidité à gagner." Ordonnance, 15 October 1666, ms. fr. 21639, fol. 134. The prévôt des marchands similarly denounced the "avidité" of bakers who bloated the price far beyond the warrant of the harvest at the end of 1737, the eve of a grave and protracted subsistence crisis. AN, H* 1857, fols. 232–33 (2 December 1737). The bakers were so cunning in intrigue, suggested an "indigent citizen" in 1791, that only bakers were adept enough to prevent and repress the crimes in the field of subsistence: "Qui peut mieux connoître la source des accapareurs que les Boulangers?" Address to Pethion (*sic*), December 1791, AN, T 644¹⁻². On police atti-tudes toward grain merchants, see Kaplan, *Provisioning Paris*. For similar attitudes toward bakers elsewhere in France, see Maurice Bernard, *La Municipalité de Brest de 1750 à 1790* (Paris, 1915), p. 322; "Sur les commerces de la viande et du pain," *Précis analytique des travaux de l'Académie de Rouen* (Rouen, 1777), vol. 4, p. 157; Villey, "Les Boulangers de Caen," p. 185; Girod, "Les Subsistances en Bourgogne," pp. 30, 32.

7. Des Essarts, *Dictionnaire universel*, vol. 2, p. 233. Napoléon simplified the issue of the Paris bread police in the extreme: "It is unjust that bread be kept at a low price in Paris when it is expensive everywhere else, but that is because the government is there, and because soldiers do not like to fire on women who come to wail at the bakers' doors with children in their arms." Cited by Sibalis, "The Workers of Napoleonic Paris," p. 346. Cf. the bitter complaint of baker champion Ambroise Morel that even after the Revolution bakers were treated as "fonctionnaires subordonnés au préfet de police" rather than as "commerçants." *Histoire abrégée*, pp. 296–97.

8. Colin des Murs to Sartine, 8 January 1769, AN, Y 12618; Maxime du Camp, Paris, vol. 2, p. 80. The prohibition against quitting the bakery without seeking approval from and giving notice to "la partie publique" was applied throughout the realm. See, for instance, the regulation of the bailliage of Paray, 26 October 1767, AD Côte d'Or, B II 39/6. The obligation to keep the bakeshop constantly stocked was also quasi-universal. See, for example, the ordinance of the police of Chartres, 25 July 1769, AD Eure-et-Loire, B 3935.

9. Delamare, *Traité*, vol. 2 (1729), p. 759; police sentence, 16 November 1731, BN, ms. fr. 21640, fol. 60; lieutenant general to Parisot, 15 September 1734, AN, Y 15934; Y 9435, 26 February 1734; Y 9431, 16 November 1731; Y 9441, 24 January 1740; Y 12826, 19 April 1776 and 1 September 1780; Malouin, *Description* (ed. Bertrand, 1771), p. 481; Diderot, et al., *Encyclopédie*, vol. 2, p. 360.

10. La Reynie to Delamare, 24 July 1693, BN, ms. fr. 21639, fols. 169–70; ms. fr. 21640, fols. 44–45; Mercier, *Tableau,* vol. 12 (1788), p. 148; [probably Daure, a subsistence specialist and dealer], "Mémoire," ca. 1711, AN, F¹¹ 264. On the strategy of mobilizing "les pauvres boulangers" (who constituted "le plus grand nombre") to force "les boulangers plus aisés à baisser aussi le prix de leur [pain]," see Narbonne, *Journal,* p. 128.

11. AN, Y 9632, 18 June 1773; Gazetins, 14 May 1726, Arsenal, ms. Bast. 10156, fol. 210; Charles Musart, *La Réglementation du commerce des grains en France au XVIIIe siècle: La Théorie de Delamars, étude économique* (Paris, 1922), p. 158; Y 15047, 25 October 1740; police sentence, 17 November 1730, BHVP file; Carlier, *Histoire du pain,* pp. 10, 14; La Reynie to Delamare, 24 July 1693, BN, ms. fr. 21639, fol. 169; mémoire of forains to procurator general, ca. March 1738, BN, Joly 1314, fol. 50; Saint-Germain, *Vie quotidienne,* p. 230; *Enquête sur la boulangerie,* p. 162.

12. Malouin, *Description* (1761), pp. 259–60, 272. *Encyclopédie*-like, after denouncing most of the tools still commonly employed by authorities in France, Malouin praised the sageness of the police of bakers—but not without ambiguity in tone and language: "On peut dire qu'il n'y a dans aucun pays, si ce n'est peut-être en Chine, autant de police qu'il y en a en France, & jamais elle n'a été aussi bien administrée qu'elle l'est aujourd'hui à Paris." Ibid., p. 312, and *Description* (ed. Bertrand, 1771), p. 392.

13. Parmentier, *Parfait Boulanger,* pp. xiv, 592, 604.

14. For the general context, see Kaplan, *The Famine Plot,* and for specific instances Bureau central de police de Paris, division du commerce, to the section de l'Unité, 13 pluviôse an IV, ASP, 4 AZ 794; Terray to jurats of Bordeaux, 16 May 1773, AD Gironde, C 1441; Bailly, *Mémoires,* vol. 2, pp. 406–7; and *Journal historique et politique des principaux événements* (Geneva), no. 14 (6 May 1773), p. 55. On the recurrence of quality-associated riots in the prisons, see Hardy's journal, BN, ms. fr. 6681, fol. 115 (23 November 1772), and Pidansat de Mairobert, *Journal historique,* vol. 6, p. 61 (18 June 1774).

15. *Poison Detected* cited by Jackson, *Essay on Bread;* "Réflexions sur la nature du pain," *Journal économique,* September 1761, pp. 429–31 (review of Manning); Frederick A. Filby, *A History of Food Adulteration and Analysis* (London, 1934), pp. 80–104; Sheppard and Newton, *Story of Bread,* pp. 42, 72–73; Thrupp, *Worshipful Company,* p. 34; *Journal économique,* August 1762, p. 382, and January 1766, pp. 35–36; Burnett, *Plenty and Want,* p. 72; Kirkland, *Modern Baker,* vol. 4, p. 19; E. Parmalee Prentice, *Hunger and History: The Influence of Hunger on Human History* (New York, 1939), pp. 102–3; Giedion, *Mechanization,* p. 181. Cf. Sylvester Graham's portentous evocation of bread "too frequently rendered the instrument of disease and death, rather than the means of life and health, to those that eat it." *Treatise on Bread,* pp. 42–47. Fava-bean meal is commonly—and legally—added to bread-making flour today. In the intensified kneading method it tends to whiten the flour and to favor the appearance of a lovelier "grigne" or swelling on the top of the loaf. The other acceptable additives today are ascorbic acid, malt, and lecithin. R. Guinet, "Evolution de la qualité du pain: Incidence de l'équipement et des méthodes de fabrication," in *Le Pain,* ed. Jean Buré, pp. 121–22. Cf. "Recueil des usages concernant les pains en France," in ibid., pp. 266–67 ("adjuvants de panification").

16. Béraud's report, 15 September 1793, AN, F⁷ 3688³; "Plainte d'un boulanger, rue

St.-Antoine," in Alexandre Tuetey, *Répertoire général des sources manuscrites de l'histoire de Paris pendant la Révolution française* (Paris, 1890–1914), vol. 2, p. 137; Chauveau-Lagarde, *Mémoire pour les boulangers*, pp. 31–32; F. Funck-Brentano, "'Gazette de police,' 24 September 1725," *Revue retrospective*, December 1885–July 1886, p. 157; M. J. D. L., *Le Pain à bon marché, ou le Monopole terrassé et le peuple vengé* (Paris, n.d. [ca. 1790]), p. 2; Bordeaux Parlement, lettre-représentations to king, ca. fall 1748, AD Gironde, C 1439; chevalier de Nicolay to Marville, 20 November 1742, Arsenal, ms. Bast. 10019, fol. 150; *Journal de l'agriculture, du commerce, des arts et des finances*, January 1774, pp. 84–101; Letaconnoux, *Les Subsistances*, pp. 104, 114; AN, Y 10001, 25 September 1789; *Journal de la Ville*, cited by Lacroix, *Actes*, vol. 2, p. 63; Bailly, *Mémoires*, vol. 3, pp. 80–81 (5 October 1789); Charles-Elie de Ferrières, *Mémoires du marquis de Ferrières*, ed. S.-A. Berville and J.-F. Barrière (Paris, 1822), vol. 1, p. 306; Antoine de Rivarol, *Mémoires de Rivarol*, ed. S.-A. Berville (Paris, 1824), p. 254; Paul Olivier, *Les Chansons des métiers* (Paris, 1910), p. 293 ("alun, savon et farin' mélangés, / voilà le pain que chaq' jour vous mangez"); Charles Richet and Antonin Mas, *La Famine* (Paris, 1965).

17. Antoine de Montchrétien, *Traicté de l'oeconomie politique*, ed. F. Funck-Bretano (Paris, 1889), p. 261; Delamare, *Traité*, vol. 2 (1729), pp. 497, 504, 760.

18. Delamare, *Traité*, vol. 2 (1729), pp. 498–503. Cf. ibid., pp. 763–64; Des Essarts, *Dictionnaire universel*, vol. 1, p. 253, and vol. 2, p. 249; BN, Joly 1428, fols. 64–67 (on the police preoccupation with water quality in bread making).

19. AN, Y 13772, 1763; ASP, D4B⁶ 83-5592, 14 January 1782; arbitre's report, 6 November 1783, D6B⁶ 11; D4B⁶ 42-2307, 20 March 1776; BN, Joly 1426, fols. 64–66 (8 January 1782); Malouin, *Description* (1779), p. 93; Mercier, *Tableau*, vol. 3 (1783), p. 179; rapports, 15, 19, 21, 28, 29 September 1793, AN, F⁷ 3688³.

20. Maurizio, *Histoire de l'alimentation*, p. 502; *Les Numéros parisiens par M. D**** (Paris, 1788), p. 7; Busuel, "Plan d'un règlement pour les approvisionnements des grains et des farines," December 1791, AN, T 644¹⁻² (cf. a similar accusation two years later: AN, AF^IV 1470, 18 March 1793); AN, Y 15248, 1 May 1741; *Enquête sur la boulangerie*, p. 340. Mercier was virtually alone among commentators in claiming that "our bakers do not at all sell at false weight." *Tableau*, vol. 12 (1788), p. 147. An English traveler allowed that French tradesmen were "in general downright cheats." Philip Thicknesse, *Useful Hints to Those Who Make the Tour of France* (London, 1767), letter XIX, p. 191. For the portrayal of the cheating baker in folklore, see Bouyer, *Folklore du boulanger*, pp. 93–94.

21. 30 October 1879, in Collection E. Thomas, cote provisoire 4804, série 114, BHVP. One should note that underweight bread was not peculiar to Paris or to France. The Parlement of Bordeaux banned *michots*, little loaves advertised as one-pounders that barely weighed three-quarters of a pound. 1 August 1764, arrêts parlementaires non cotés, AD Gironde. The Rouen Parlement pronounced against short weight. 13 October 1769, Conseil secret, AD Seine-Maritime. "False weight" practices enraged consumers at Dourdan and at Châlons-sur-Marne, especially in times of high prices. 1 August 1770, AD Seine-et-Oise, 4B 1161, and July 1768, BN, Joly 1140, fols. 80–90. The subdelegate at Bar-sur-Aube complained to the intendant of Champagne that local authorities did not repress the fraud with sufficient diligence. Gehier to intendant,

10 January 1775, AD Aube, C 299. London had problems with faithless bakers cheating on weight from the early modern period through the twentieth century. Sidney Young, *The Worshipful Company of Bakers of London* (London, 1912); Thrupp, *Worshipful Company*, p. 53; Kirkland, *Modern Baker*, vol. 4, p. 226. Like Mercier, John Houghton demurred: "I hear no complaint that the poor are cheated in weight of bread." *Husbandry and Trade Improv'd* (London, 1727), vol. 1, p. 111 (5 May 1693). Colonial Boston bakers committed the same infraction. Karen J. Friedmann, "Victualling Colonial Boston," *Agricultural History* 97 (1973), p. 194.

22. For instance, AN, Y 11395, July 1729 (Mongueret).

23. AN, Y 12816, September 1780; Des Essarts, *Dictionnaire universel*, vol. 3, p. 60, and Berryer to Mutel, 4 October 1752, Y 11169; Weulersse, *Mouvement*, vol. 2, pp. 529–30. Cf. Busuel's evocation of "a father and mother with family or a worker" who suffered from short weight more than others. "Plan d'un réglement," December 1791, AN, T 644[1–2].

24. Paul Ledieu, ed., "Observations sur l'instruction de l'Impératrice de Russie aux députés pour la confection des loix (1774)," *Revue d'histoire économique et sociale* 8 (1920), p. 392. For other examples of the image of the baker as purveyor of short-weight loaves, see Jean Bruyère, *Histoire littéraire des gens de métier en France* (Paris, 1932), p. 13.

25. Police sentence, 27 June 1724, BN, ms. fr. 21640, fol. 145; AN, Y 9441, 11 October 1740; Y 9499, 13 December 1743. See also the sentences pronounced by Marville in the forties. *Lettres de M. de Marville*, vol. 1, pp. 213, 226.

26. For the archival documentation for these fines, see appendix E. A fine for short weight was one of the commonest sentences pronounced by the medieval police. In Dijon the fine loomed as nothing less than "une taxe supplémentaire exigée de tous les boulangers." Desportes, *Le Pain au moyen âge*, pp. 181–83.

27. Arsenal, ms. Bast. 10155, 23 September 1725; police sentence confirmed by parlementary arrêt of 30 October 1521, cited by Des Essarts, *Dictionnaire universel*, vol. 2, p. 248.

28. AN, Y 9538, 28 June 1726; Y 9440, 4 May 1739; Y 9443, 13 April 1742.

29. AN, Y 9621, 13 April 1742; Y 9443, 26 April 1742; Dupré de Saint-Maur, *Essai*, p. 57. Cf. baker Jacob's report of a two- to four-ounce tolerance on a loaf of two kilos during the first empire. *Enquête sur la boulangerie*, pp. 605–6. On nineteenth-century practice, see also Cochut, "Le Pain à Paris," p. 414, and Rabate, *Le Blé*, p. 109.

30. Gazetins, 25–26 April 1741, ms. Bast. 10168, fol. 179. Cf. Chauveau-Lagarde, *Mémoire ampliatif*, p. 27, and *Mémoire pour les boulangers*, p. 31. Commissaires pressed for harsh sentences against repeat culprits like master baker Domaget, who "seemed to make a game of cheating the public." AN, Y 9479, 2 June 1776. One police document claims that there was a fine schedule: fifty livres for the first offense, three hundred livres for the second, six hundred livres for the third, followed by walling and deprivation of mastership. "Affaires générales de la police," 1753, p. 10, BHVP, ms. 29736. I have found no documentation to suggest that such a pattern was followed, either in the first or in the second half of the century.

31. "Les Boulangers gagnent beaucoup sur le poids . . . ," *Les Numéros parisiens*, p. 7; AN, Y 15242, 23 July 1735; Y 15244, 7 December 1737; Y 15244, 6 March 1737; Y 15359, 25 March 1758 (the juré was J.-G. Mouchy). Still, short weight was not an absolutely

universal occurrence. Upon completion of his rounds one day in September 1780, Commissaire Dupuy reported that twenty of twenty-three shops he inspected were "very much in order" in regard to weight. Another commissaire found most of forain Nicolas Royer's fifty-eight loaves to be "overweight"; only "a few" were light by an ounce. Y 12826, 1 September 1780; Y 15244, 6 March 1737. Cf. the situation in 1862: "Le pain réglementaire de deux kgs. pèse rarement son poids, et la petite différence dont le marchand profite constitue à la fin de l'année un bénéfice important." Cochut, "Le Pain à Paris," p. 424.

32. Police ordinance, 10 March 1649, BHVP; Desmaze, *Métiers,* p. 159; AN, Y 13516, 13 July 1757; Y 9538, 28 June 1725; Y 9534, 14 November 1764; Malouin, *Description* (1761), pp. 212, 265. A forain chose a path of discretion. He marked his bread in code with "a sort of fleur de lys." Y 11221, 27 November 1734. Cf. the London "seal," Thrupp, *Worshipful Company,* p. 45. The practice of the mark was at least as old as the bakers of Pompeii.

33. For examples of Sunday visits, see AN, Y 12142, 31 January 1740, and Y 9499, 2 May 1743. Why Sunday? Perhaps because the bakers were less likely to be on their guard on the quietest day of the business week. Or perhaps because there was a greater likelihood of finding short-weight bread in the shops of those bakers who did not bake on Sunday mornings, instead displaying Saturday's bread, still shrinking in weight as a result of the evaporation/staling process. Critics of the juré police charged that it was a form of pressure used to coerce the bakers to toe the corporate line. Baudeau, *Avis au premier besoin,* 3e traité, p. 120. For mixed juré police inspection, see Y 15233, 19 July 1726, and Y 9539, 12 September 1760. See the quarrel between a presiding juré who wished to treat the forains with flexibility and a former juré, who insisted on a rigorous application of the rules down to the half-ounce. Y 15340, 22 March 1744. The wardens of the London guild of bakers were also responsible for inspecting for short weight. Thrupp, *Worshipful Company,* p. 25.

34. AN, Y 12607, 22 August 1759, and Y 9539, 22 August 1759 (the baker claimed that the loaves found short belonged to a Gonesse baker who had returned home early); Y 13416, 13 July 1757. Cf. the case of a fruit seller whose twelve-pound loaf from the Halles was almost a pound short. Y 12604, 4 September 1756, and Y 9538, 10 September 1756.

35. AN, Y 12591, 7 October 1747.

36. AN, Y 12741, 20 March 1755; Y 9538, 23 June 1724 (for another example of violent resistance to weight verification, see Y 15582, 8 February 1715); Y 12143, 2 July 1741; Y 12143, 24 December 1741. Baudeau contended that some bakers bought protection against short-weight summonses by paying off police subalterns. *Avis au premier besoin,* 3e traité, p. 112. Cf. Jacob, a seasoned baker, who quit the trade in the mid-nineteenth century because he could not stand being "treated as a thief" over the short-weight issue. *Enquête sur la boulangerie,* p. 909.

37. AN, Y 12740, 6 June 1754; Y 15356, 4 July 1756; Y 11229, 23 December 1742.

38. AN, Y 13494, July 1729; Y 12741, 10 December 1755; *Enquête sur la boulangerie,* p. 609. Cf. Marmay, *Guide pratique,* p. 34: "No. He [the baker] acted in good faith, but the flour was not like him, it was faithless." Antoine Boland, author of another baking manual, concurred. *Traité pratique de la boulangerie* (Paris, 1860), p. 115.

39. Arsenal, ms. Bast. 10024, September 1743; Malouin, *Description* (1767), p. 260. The bakers did not fail to excoriate their journeymen, though this issue did not command pride of place. They did not desire to stand as "guarantors" of the behavior of their "domestics" whose "drunkenness" or "insouciance" or "ill will" resulted in burned ovenfuls or rotten proofs.

40. Baker mémoire cited by Boland, *Traité pratique de la boulangerie*, p. 115. The following discussion is drawn from Parmentier, *Parfait Boulanger*, pp. 614–32. There are sufficient similarities between the guild's petition and Parmentier's exposition to suggestion complicity (or plagiarism).

41. On this safety margin, see Cotte, *Leçons élémentaires*, p. 73. Slightly higher supplementary doses were prescribed in Diderot et al., *Encyclopédie*, vol. 2, p. 750. For two recipes to reduce loss in baking, one by kneading with water boiled with bran and the other by using a firmer dough, see Narbonne, *Journal,* and Busuel, "Plan d'un règlement," December 1791, AN, T 644$^{1-2}$. Unlike another defender of the bakers, the mitron de Vaugirard, Parmentier was unwilling to admit that there was a more or less reliable threshold—five or six ounces short weight for a four-pounder, said the mitron—over which fraud was a very likely explanation. Lacombe d'Avignon, *Le Mitron de Vaugirard,* p. 46. Critics of the bakers did not care how much bonus dough the bakers used. The only thing that mattered to them was how much the finished bread weighed at the moment of sale. Francois Aubert, "Réflexions," 1775, BN, ms. n.a.f. 4433, fols. 48–49.

42. On this tactic, see Lacombe d'Avignon, *Le Mitron de Vaugirard,* p. 45, and Marmay, *Guide pratique,* p. 90. For a case in which a commissaire seized bread that was "insufficiently cooked," see AN, Y 12816, 11, 19 April 1776.

43. All bakers were required to have scales permanently affixed in their shops along with appropriate weights. AN, Y 9443, 13 April 1742, and Duchesne, *Code de la police,* p. 116. The police conducted occasional inspections to ascertain whether the weights displayed by bakers were too heavy or too light. Y 13082, 20 September 1734.

44. Baudeau, *Avis au premier besoin,* 3e traité, pp. 119–21; Malouin, *Description* (1761), pp. 265–67. Béguillet endorsed this position, somewhat less emphatically. *Traité des subsistances,* pp. 161–76. A Swiss writer claimed that bakers should develop the "habit" of foreseeing short weight, yet he was impressed by the considerable variation in evaporation in a single ovenful. Muret, *Mémoire sur la mouture,* pp. 78, 119. Cf. Julius Wihlfahrt, *A Treatise on Baking* (New York, 1935), p. 166.

45. Bachaumont, *Mémoires secrets,* vol. 18, pp. 223–25 (31 December 1781).

46. "Expériences et observations sur le poids du pain au sortir du four," *Journal de physique,* February 1782, pp. 90–103.

47. Bachaumont, *Mémoires secrets,* vol. 18, p. 225 (31 December 1781); Rutledge, *Mémoire pour les maitres boulangers,* pp. 9–10; C.-L. Chassin, *Les Elections et les cahiers de Paris,* p. 497; *Versailles et Paris* 11 (11 August 1789), p. 8; Ferrières, *Mémoires du marquis de Ferrières,* vol. 1, p. 285.

48. Boland, *Traité pratique,* p. 124; Morel, *Histoire illustrée,* pp. 248–49; Grimod de la Reynière, *Almanach des gourmands,* vol. 7 (1810), p. 197; Cochut, "Le Pain à Paris," p. 415; *Enquête sur la boulangerie,* p. 609; ASP, DM^6 2. According to an American agricultural chemist, highly critical of U.S. state laws requiring loaves to conform to weight specifi-

cations, the debate over weight loss reached the U.S. Supreme Court. Snyder, *Bread*, pp. 154, 299.

49. See, for example, the Vincennes baker who wanted to leave the market at 2:30 P.M. AN, Y 15060, 7 November 1759. See also Gazier, "La Police," p. 127; *Gazette à la main*, 31 March 1743, BHVP, ms. 623, fol. 243; Delamare, *Traité*, vol. 2 (1729), p. 756.

50. La Reynie to Delamare, 9 July 1693, BN, ms. fr. 21642, fol. 193; Agence des subsistances et approvisionnements to commissaire de la police de la section des Arcis, 28 frimaire an III, ASP, 6 AZ 701; ordonnance servant de réglementation aux boulangers, AN, AD I 23ᴬ, 4 May 1725.

51. The procès-verbaux of the jurés on patrol often read: "[We were] on the alert for violations by the forain bakers." See, for instance, AN, Y 15062, 4 February 1761.

52. Widow Pairpart and Anne Jamain, who was only twelve years old. AN, Y 15060, 8 February, 14 July and 26 December 1759; Y 15061, 21 October 1760.

53. AN, Y 15060, 7 November 1759; Y 9434, 7 November 1764.

54. AN, Y 15062, 4 February 1761; Y 15061, 31 October 1761.

55. AN, Y 15060, 8 February 1759; Y 11590, 24 February 1773; Y 15063, 24 February 1762; Y 15061, 29 October 1760; Y 15101, 20 July 1748; Y 15060, 28 November and 1, 7, 26 December 1759; Y 15061, 29 October, 27 December 1760; Y 15069, 31 May 1766. About half the bread seized in the several dozen cases discussed above was found to be light in weight and almost the same proportion was unmarked with bakers' and markets' initials. Thus many of these bakers were liable for prosecution on two entirely different matters.

56. AN, Y 15060, 7, 28 November 1759; police sentence, Y 9538, 30 June 1725; police sentence, Y 9632, 18 June 1773; police sentence, 11 March 1727, BHVP; Gazetins, Arsenal, ms. Bast. 10155, nos. 72–73, 26 September 1725. Police sentence, 11 March 1727, BHVP (Guichot received a fifteen-livre fine); police sentence, 30 June 1725, Y 9538 (Nicolas Lapareillé suffered a fine of thirty livres for his participation); Y 15060, 7 November 1759; Y 12607, 22 August 1759 (Bruley and Playette); Y 11238, 23 July 1750.

57. AN, Y 11238, 23, 28, 29 July 1750.

58. See BN, Joly 1111, fol. 171, and the cases of bakers who openly report the theft of bread that they were bringing home from market without the fear of incrimination: Arsenal, ms. Bast. 10124 (8–10 July 1769), and AN, Y 13542, 9 July 1769. See also ms. Bast. 10270, p. 275, 3 October 1725; Gazetins de police, ms. Bast. 10155, 26 September 1725.

59. BN, ms. fr. 21638, fols. 341–48, and ms. fr. 21639, fols. 178–79; AN, Y 15224, 7 December 1737.

60. AN, Y 15594, 12 March 1738; Y 15242, 26 February 1735; Y 11241, 5 January 1754; Y 9440, 16 January 1739.

61. AN, Y 11229, 3 October 1742 (the woman regrater in question here was denounced by a jealous legitimate baker who did not want to suffer competition from unauthorized sources, especially late in the day when selling became more difficult and more imperative); BN, ms. fr. 21641, fol. 295 (4 May 1725); ASP, D2B⁶ 1012 (4 April 1764). Cf. the clause in Delamare forbidding bakers to sell to other bakers. Delamare, *Traité*, vol. 2 (1729), p. 763.

62. Ordinance of 20 July 1703 banning street sales, BN, ms. fr. 21638; AN, Y 15242, 26 February, 23 March 1735; Y 15241, 20 November 1731; Y 15235, 29 September 1728; Y 15242, 26 October 1735 (the time of confiscation varied from 6:00 A.M. to 6:30 P.M. La Chambre was caught twice); Y 14530, 19 June 1736; Y 9539, 12 September 1760.

63. AN, Y 15244, 7 December 1737; Y 15244, 6 March 1737; Y 15235, 29 September 1728.

64. See, for example, AN, Y 11221, 27 November 1734; "Mémoire des modernes et des jeunes maîtres boulangers," BN, Joly 400, fols. 256–68.

65. AN, Y 15063, 24 February 1752; Y 15060, 5 January, 22 September 1759, and Y 15063, 24 February 1762; Y 9377, 19 June 1725; Y 9632, 18 June 1773; Y 9370, 18 June 1771. Cf. the complaint of a privileged baker who claimed that the jurés beat her and called her "whore, bitch, and thief" when they found her delivering to clients. Y 13086, 26 October 1736.

66. AN, Y 13114, 21 September 1762; Y 9378, 12 January 1731.

67. AN, Y 11220, 2 September 1733; Y 12141, 24 June 1739.

68. Arrêt of the Paris Parlement, 21 March 1670, BN, ms. fr. 21638. The wine merchants guild also fought against this regulation. AN, Y 9434, 20 November 1733.

69. AN, Y 11221, 10 April, 27 November, 11 December 1734; Y 11223, 18 August 1736; Y 11225, 16 July 1738; Y 15224, 6 March 1737; Y 11221, 3, 9 December 1734. For examples of protestations of innocence and naïveté, see Y 11220, 2 September 1733; Y 15241, 10 April 1734; Y 11221, 27 November 1734.

70. AN, Y 13114, 21 September 1762.

71. Deliberations of the Paris municipality, AN, H* 1855, fol. 263; Gazetins, 22–27 August 1733, Arsenal, ms. Bast. 10164, fol. 108; municipal deliberations, H* 1859, fol. 326 (13–14 August 1741); ASP, D4B⁶ 103-7102, 14 March 1788; D4B⁶ 104-7356, 31 December 1788; D4B⁶ 39-2138, 30 March 1772, and D4B⁶ 88-6016, 22 October 1783; Hardy's journal, BN, ms. fr. 6680 (17 January 1767); AN, Y 15114, 26 October 1780. Mercier warned about the grave danger of baker ovens, those precious instruments of survival that "recèlent des braises innombrables." Tableau, vol. 2 (1782), p. 62. Cf. the similar experience of baker fires in Rouen, Gazetins, 7 August 1726, ms. Bast. 10156, fol. 337. The king's baker was allegedly responsible for starting the Great Fire of London in 1666. Sheppard and Newton, Story of Bread, p. 43.

72. AN, Y 13524, 19 July 1760; Y 15235, 5 November 1728.

73. Statuts et règlements, April 1783, AN, AD XI 14, art. XVI; AN, Y 14948, 2 October 1733; Y 12575, 13 June 1730; police sentence, 1 June 1736, BN, La Senne Z 2263 (10); Y 11217, 16 December 1730; AN, MC, LXXXIX-453, 9 September 1738 (receipts for "visites des fours et fourneaux").

74. AN, Y 9441, 6 May 1740.

75. See BN, Joly 1732, fol. 229; Lenoir to Dupuys, 20 July 1776, AN, Y 12830; Joly 1117, fol. 63, 20 November 1725. Cf. the Assembly of Police's remark concerning a laboureur fined for a violation of the grain trade regulations: "He was not rich enough to pay a large fine." BN, ms. fr. 11356, fol. 432, 7 July 1740; Joly 1732, fol. 229 (ca. 1786).

76. See the case of the Maubert market baker, convicted for baking "very bad" quality bread, who was ordered to pay his fine of three hundred livres within twenty-four hours. Police sentence, 12 July 1726, BN, fichier des lois, arrêts, sentences, etc.

77. AN, Y 13521, 12 July 1759; écrou, Grand Châtelet, 8–11 July 1739, APP, A B/194; A B/196, 25 November 1741 and A B/195, 2 May 1741; A B/194, 23 July, 23 August 1739; A B/195, 22 June 1740–3 February 1741.

78. The figure given in one source for total annual fines collected—48,500 livres in 1753—seems extraordinarily low to me. "Affaires de la police," BHVP, ms. 29736.

Chapter 18   Setting the Price of Bread

1. C. W. Emmet, "Famine," *Dictionary of the Bible,* ed. James Hastings, rev. ed. Frederick C. Grant and H. H. Rowley (New York, 1963), p. 293; Henri Curmond, *Le Commerce des grains et l'école physiocratique* (Paris, 1900), p. 7.

2. Narbonne, *Journal,* p. 309; Gazier, "La Police," p. 123; Poussot's journal, Arsenal, ms. Bast. 10141, fol. 309; Turgot to intendant of Champagne, 17 April 1775, in *Oeuvres de Turgot,* vol. 4, p. 491; Turgot, "Septième Lettre sur le commerce des grains, ibid., vol. 3, pp. 348–52. Letrosne shared Turgot's point of view on both the need to abolish the guilds and the imperative to sustain the tax as long as the bakers operated in an exclusivist corporate context. *De l'administration provinciale,* vol. 2, pp. 274–78. Arguing vigorously for a policed bakery, in which a pound of wheat produced a pound of bread, Mirabeau implicitly endorsed a taxing scheme. Weulersse, *Mouvement,* vol. 2, pp. 529–30. In order to safeguard and justify their *bon prix* freedom-of-grain strategy, the physiocrats were willing to put certain restraints on bakers, who, after all, were in the secondary rather than the primary sector, "producers" of an inferior order. See also Georges Weulersse, *Physiocratie (1770–74),* p. 160, and *Physiocratie (1774–1781),* p. 183.

3. Bondois, "La Disette de 1662," pp. 53–118; Samuel Bernard, October 1693, cited by Boislisle, "Grand Hiver," p. 505; Delamare, *Traité,* vol. 2 (1729), pp. 909–22.

4. Delamare notes, BN, ms. fr. 21647, fols. 64–74, and *Traité,* vol. 2 (1729), pp. 922–34; Citoyen Du Vaucelles, "Opinion sur les subsistances," AN, F^{10} 215–16; Malouin, *Description* (1761), p. 270.

5. Abbé Galiani, "Mémoire à Sartine," in *Lettres de l'abbé Galiani à Madame d'Epinay,* vol. 1, p. 416; Necker, *Législation,* pp. 282–83; Georges Weulersse, *Mouvement,* vol. 1, p. 538; Terray to jurats of Bordeaux, 16 May 1773, AD Gironde, C 1441. "Je ne crois pas qu'on puisse sans attaquer la propriété, taxer le blé," observed Monsieur Fromant. Lacombe d'Avignon, *Le Mitron de Vaugirard,* p. 41. See also Le Camus, "Mémoire sur le bled," p. 143.

6. Doumerc to procurator general, 17 February 1789, BN, Joly 1111, fols. 203–5.

7. "Mémoire," ca. 1725, Arsenal, ms. Bast. 10270; "Moyens pour empêcher l'augmentation . . . ," July 1739, BN, Joly 1120, fol. 33; Gazier, "La Police," p. 123; Narbonne, *Journal,* p. 309; Kaplan, *Provisioning Paris,* pp. 23–40, 369–70, 547–48. Cf. the contrary argument voiced in the *Encyclopédie méthodique* that one should resort to grain and (considerably later) bread taxation only in a grave penury when the people were deeply distressed. *Encyclopédie méthodique,* Jurisprudence, police et municipalités, vol. 10, p. 34. See also in the spirit of Narbonne, "Moien pour préserver tous les inconveniens de la fixation du bled," ca. 1709, BN, Joly 1111, fols. 158–67. For an illuminating discussion of taxation and attitudes toward it, see Albert Griffuel, *La Taxe du pain* (Paris, 1903).

8. Kaplan, *The Famine Plot;* Monchanin, "Deuxième Mémoire sur les subsistances," presented to the Conseil général de la Commune de Paris, February 1792, AN, F^{10} 215–16; police sentence, 6 February 1789, ASP, 2 AZ 18, pièce 50. See the call for a permanent bread tax in the debate accompanying the elaboration of the Paris cahiers de doléances in 1789. Chassin, *Les Elections et les cahiers de Paris,* vol. 3, pp. 370–71. Cf. Alfred Des Cilleuls, *Le Socialisme municipal à travers les siècles* (Paris, 1905), p. 318. On the persistence of the dilemma well into the nineteenth century, see *Enquête sur la révision de la législation des céréales,* vol. 2, p. 417. For a neat contemporaneous denunciation of the proclivity of "les imaginations faibles" to cede perpetually to fears of shortage, see *La Question du pain, ou Précis sommaire du passé et de l'avenir de la boulangerie parisienne par un correspondant de l'Indépendant Belge* (Paris, 1862), p. 29. Premising his argument on the erroneous assumption that Parisian household bread was systematically taxed for the first time after 1789, Frédéric Le Play denounced the Revolution for "setting back France three centuries" by proclaiming the legitimacy of price control and in so doing "pervertir, en matière de subsistance, l'opinion publique." Report of 1860 cited by Des Cilleuls, *Socialisme,* pp. 44, 318.

9. La Reynie to Delamare, 2 April 1694, BN, ms. fr. 21643, fol. 68; section des Enfants-Rouges, "Délibérations du comité de la section," Paris, 21 November 1790, pp. 2–3, French Revolution Collection, Kroch Library, Cornell University. This Revolutionary text asserts a sort of French subsistence exceptionalism. France is unlike many other countries because of the inordinate centrality of bread in material and psychological life. The "bonheur social" required a certain degree of regulation or police of provisioning, the Enfants-Rouges contended. Ibid.

10. Mathieu Tillet, "Projet d'un tarif," *Histoire de l'Académie royale des sciences: Avec les mémoires de mathématique et de physique, 1781* (Paris, 1784), pp. 113–14.

11. Ibid., pp. 165–66.

12. Ibid., pp. 134, 162–64, 167–68.

13. AD Calvados, C 2636; Parmentier, *Mémoire sur les avantages,* pp. 401–2; Calonne to intendant, 27 September 1984, AD Orne, C 90.

14. Baudeau, *Avis au premier besoin,* 1er traité, pp. 126–32; Bertrand's note in Malouin, *Description* (ed. Bertrand, 1771), p. 343.

15. AN, U 1440, 15 September 1781–21 December 1782 (in particular, "Nouveau tarif pour la taxe du pain").

16. Mémoire, juillet 1765, BN, ms. fr. 11347, fols. 183–211. Cf. the call for a scientific study of taxation and of the trials on which the schedules were habitually based in "Réflexions sur le commerce du pain," pp. 319–20. Like Boullemer, the author puzzled over the following imbroglio: "D'un autre côté des essais authentiques suivis avec soin par des Commissaires éclairés semblent constater des produits absolument différens. Comment accorder ces prétendus faits que l'on donne de part & d'autre pour indubitables? & dans cette incertitude, comment les Magistrats de Police pourront-ils jamais fixer d'une manière exacte le prix du pain dans des tems de calamité?"

17. Dupont cited by Des Cilleuls, *Socialisme,* p. 317; Parmentier, *Parfait Boulanger,* pp. 110–14, 606; Parmentier, *Mémoire sur les avantages,* p. 397; Béguillet, *Traité des subsistances,* pp. 175–76; *Encyclopédie méthodique,* Jurisprudence, police et municipalités, vol. 10, p. 609; "Second mémoire des boulangers," ca. 1789, Arsenal, ms. Bast. 7458; [?]

to procurator general, 29 July 1770, BN, Joly 1148, fols. 206–7, and mémoire (Troyes, 1786), Joly 1743, fol. 131; "Observations sur le projet de loi," October 1770, AD Calvados; Jean-Baptiste Biot, *Lettres écrites de la campagne sur le commerce des grains et l'approvisionnement de Paris* (Paris, 1835), p. 98. Cf. Abbé Rozier's use of Parmentier's warning in *Cours complet*, vol. 7, p. 360 ("Le Pain"). See also the description of the consequences of the tax applied in the mid-nineteenth century: "un élément aqueux, indigeste, . . . devenant insipide à mesure qu'il se déssèche." Cochut, "Le Pain à Paris," p. 424.

18. Baudeau, *Avis au premier besoin*, 1er traité, pp. 110–14, and 3e traité, 126–32; Jean-François Rewbell cited by Des Cilleuls, *Socialisme*, p. 317; *Journal de Paris*, 20 August 1791, p. 948; Galiani, *Dialogues sur le commerce*, pp. 43, 45; Morellet to Shelburne, 17 May 1775, Abbé André Morellet, *Lettres de l'abbé Morellet à lord Shelburne*, ed. Edmond Fitzmaurice (Paris, 1898), pp. 74–75.

19. Condorcet cited by Birembaut, "Problème de la panification," p. 364n; Parmentier, *Parfait Boulanger*, pp. 612–23. Cf. Victor Emion, nineteenth-century diatribe against price-fixing, *La Taxe du pain* (Paris, 1867): "Pour nombre de municipalités, ce qui détermine à taxer le pain, c'est moins . . . la croyance au préjugé populaire que l'envie de le satisfaire, et l'intérêt cherché est un intérêt d'opinion et de paix publique, bien plus qu'un intérêt réel d'abaissement de prix. Pour les gouvernements, enfin, ç'a été, souvent, avec ce calcul, un autre encore, celui de s'attacher les multitudes, en se présentant à elles comme un pouvoir protecteur, comme une intervention tutélaire, sans laquelle les prix s'éléveraient hors de portée, au grand préjudice de la vie." Cited by Des Cilleuls, *Socialisme*, p. 49.

20. On the practice of taxation in provincial France, see Gehier, subdelegate at Bar-sur-Aube, to intendant of Champagne, 10 January 1775, AD Aube, C 299; AD Seine-et-Oise, 12B 716 (concerning Pontoise in the 1780s, where bakers were required to post the official tax, as they were until relatively recently in Paris); Bernard, *La Municipalité de Brest*, pp. 318, 321; Pierre Deyon, *Amiens, capitale provinciale: Étude sur la société urbaine au XVIIe siècle* (Paris, 1967), p. 58; Fréminville, *Traité de la police*, p. 85; Jean Ricommard, *La Lieutenance générale de police de Troyes au XVIIIe siècle* (Troyes, 1934), p. 232; Coornaert, *Les Corporations en France*, p. 146; Georges Lefèbvre, *Documents relatifs à l'histoire des subsistances dans le district de Bergues* (Lille, 1914–21), vol. 1, p. xxx and graph IV. Cf. Thrupp, *Worshipful Company*, pp. 12–14; Young, *Bakers of London*, pp. 1–17; Joseph Ruwet et al., *Marché des céréales à Ruremonde, Luxembourg, Namur et Diest aux XVIIe et XVIIIe siècles* (Louvain, 1966), p. 45; Judith A. Miller, "Politics and Urban Provisioning Crises," pp. 227–62.

21. BN, Joly 1140, fols. 112–26; "Mémoire pour la communauté des boulangers de Chaalons-sur-Marne," 30 July 1785, Joly 1742, fols. 38–45; petition of the bakers of Troyes to the procurator general, ca. 1783, Joly 1743, fol. 119. Cf. Ganneron, lieutenant of police of Dammartin, to Hérault, 22 October 1725, Arsenal, ms. Bast. 10271.

22. BN, Joly 1147, fols. 63–111; Joly 1742, fols. 39, 52–53. The memorandum for the Châlons bakers detailed the operations for which the baker needed to be reimbursed and/or remunerated: shopping for grain (traveling, comparing, testing, bargaining); transporting the grain to the bakeshop; cleaning and sifting it; having it milled (transporting to and from the mill, dealing with the vagaries of weather that jeopardized rapid and successful grinding, inspecting and conditioning the flour); bolting the meal;

preparing the yeast starters; kneading, fashioning, proofing, and cooking the dough; selling it (distributing through delivery or sale in shop and/or market); maintaining fixed investment in utensils, tools, etc.; paying for wood, wages for journeymen and domestics, and their board; paying the rent for shop/house; paying taxes (in particular, the capitation and the "industrie"); sustaining the baker's family. Joly 1742, fols. 41–42.

23. ASP, D II V³ 34-2228, 1 March 1806; V. Borie, preface to Emion, *La Taxe du pain*, pp. i–iii.

24. BN, Joly 1742, fols. 14, 147ff.; Necker to intendant of Languedoc, 16 January 1778, AN, F¹¹* 1, fols. 192–93; "Observations," ca. September 1725, Joly 1117, fols. 258–59.

25. Controller general to first president of Brittany Parlement, 1 July 1725, AN, G⁷ 34; Musart, *La Réglementation du commerce des grains*, p. 158; procurator fiscal of Arpajon to procurator general, 20 April 1770, BN, Joly 1148, fol. 159; Grazet to procurator general, 15 June 1786, Joly 1743, fol. 22; *Le Figaro*, 22 April 1970. The bakers claimed that the "taxe" had not been rectified since 1968. In theory, bread was no longer "taxé" but merely "surveillé." Still, everyone agreed that there were constrictive "baromètres." See *Le Figaro*, 26 January 1968.

26. Delamare, *Traité*, vol. 2 (1729), p. 936. See also Parmentier, *Parfait Boulanger*, p. 604; Duchesne, *Code de la police* (or Code Duchesne), p. 116; Malouin, *Description* (1767), p. 271; Frédéric Le Play, in *Rapport du Conseil d'état*, no. 686 (Paris, 1857), p. 4. In addition to these three statutory sorts of bread, the police registers also enumerated a *gros pain blanc de Gonesse*, a *bis-blanc de Gonesse*, and a *bis clair*, listed just before the plain *bis*. The price of luxury breads, so-called *petits pains*, was taxed on a monthly basis. While the price remained fixed at one sou (cornu, mollet, à la Reine, and à la Ségovie or Cigovie, as it was often written in the police registers) and two sous (façon de Gonesse), the weight varied with the price of flour. Theoretically, only the masters of Paris, all of whom could easily be reached in case of the need to make more frequent changes in the tariff, could make and sell these fantasy or small loaves. In practice, the faubouriens and many forains conducted a substantial trade in small loaves. During a number of crisis points the police allowed an inordinate rise in the price of petits pains, meant to subsidize the policy of holding the line on the price of regular bread. The police fixation of the small loaves served as a model for the price that the major Parisian hospital paid contract bakers for the provision of the same type of bread. "Prix des pains," a blank form used by Commissaire Courcy, ca. 1732, AN, Y 11219; BN, ms. fr. 21638, fol. 449; Malouin, *Description* (1761), pp. 267–68; deliberations, Bureau de l'Hôtel-Dieu, 1 February 1729, AAP, no. 98. See the interesting treatment of the advantages and disadvantages of the sale and fixation by weight in the *Journal de physique*. Among the latter: it was too cumbersome for consumers to check on weight variations; the complexity of the system served as a wonderful cover for baker cheating (furnishing light-weight loaves; baking very little *bis* and thus constraining the laboring poor to take a whiter bread that weighed only 60 percent of what the dark loaf was supposed to weigh and that nourished substantially less well). "Observations sur la meilleure manière de vendre le pain," *Journal de physique*, November 1780, pp. 388–98. On the wide currency of selling breads throughout Europe at fixed prices and variable weights, see Jacob Friedrich von Bielfeld, *Institutions politiques* (The Hague, 1760–62), vol. 1, p. 127.

27. Bielfield, *Institutions politiques*, vol. 1, p. 127.

28. Gazier, "La Police," pp. 121, 126; notation of Commissaire De Facq, AN, Y 9443, 19 January 1742; *Observations sur l'approvisionnement de Paris, ou Moyens d'empecher le haut prix du pain* (Paris, 1791), pp. 5, 10, 12; ordinance of police, 3 May 1775, and sentence of police, 6 February 1789, ASP, 2 AZ 18; BN, Joly IIII, fol. 227.

29. La Reynie to Delamare, 9, 21 July 1693, BN, ms. fr. 21642, fols. 169, 181, and 15 January, 2, 6, 27 April, 24 September, ms. fr. 21643, fols. 58, 68, 70, 74, 112; de Crosne to syndic of police commissaires, 31 January 1789, AN, Y 12830; Sartine to Trudon, 24 June 1771, Y 15114; Regnier to Hérault, 18 October 1725, Arsenal, ms. Bast. 10270. On the social dimension of taxation policy, see the notion that the rich should be asked to pay more "parce que le riche est plus en état de payer que le pauvre," in "Avis économique sur la boulangerie," p. 59. Cf. the Châlons bakers' claim to credit for having offered a rye bread by the pound devised for the poor. "Mémoire pour les boulangers de Chaalons-sur-Marne," 30 July 1785, BN, Joly 1742, fol. 42.

30. Turgot to intendant of Champagne, 17 April 1775, and circular to intendants, 17 September 1775, in *Oeuvres de Turgot*, vol. 4, pp. 491, 495; Macquer, *Dictionnaire raisonné*, vol. 1, p. 302; AN, Y 10558, 4 May 1775 (Lemarchand's testimony took place in the wake of the Flour War pillaging of bakeries in Paris); "Evaluation du Prix du pain à Paris sur celuy de la farine," BHVP, série 118, cote provisoire 1482; Controller General Clugny to intendant of Caen, 11 July 1776, in Villey, "Les Boulangers de Caen," p. 184; J.-J. Edeline (syndic, paroisse de Saint-Cloud) to procurator general, ca. July 1786, BN, Joly 1743, fol. 58; arrêt du parlement, 10 February 1789, AN, AD I 23[A]. A *grand commis* working in the department of subsistence referred to the price "auquel il est taxé." "Mémoire en défense de Brochet de Saint-Priest," in *Oeuvres de Turgot*, vol. 4, p. 112. Cf. the reference of journalist-revolutionary J.-J. Rutledge to "la taxe de la police" in his discussion of the Paris bakery. "*Mémoire pour les maîtres boulangers*, p. 5.

31. "Gazetins secrets," Arsenal, ms. Bast. 10166, 12–19 October 1737; Robert Darnton, ed., "Le Lieutenant de police J.-P. Lenoir, la guerre des farines et l'approvisionnement de Paris à la veille de la Révolution," *Revue d'histoire moderne et contemporaine* 16 (October–December 1969), p. 622.

32. For a modern discussion of the validity of Old Regime trials, see Witold Kula, *Measures and Men*, trans. R. Szreker (Princeton, N.J., 1986), pp. 76–77. For a modern assessment of the *déchet*, or weight loss, that occurs during baking and that wrought havoc in life as in trials, see Alain Guerreau, "Mesures du blé et du pain à Mâcon (XIVe–XVIIIe siècles)," *Histoire et mesure* 3 (1988), pp. 178–79.

33. Parmentier, *Parfait Boulanger*, pp. 593–94, 597; report of 9 December 1740, AN, F[11] 222; Colin des Murs to Sartine, 8 January 1769, AN, Y 12618; intendant of Caen to Clugny, and the latter's reply, 26 July and 11 September 1776, quoted in Villey, "Les Boulangers de Caen," pp. 185–92. The procurator general frequently arranged for the Paris police to supervise trials held in towns in the parlement's sprawling jurisdiction. The authorities at Bar-sur-Aube carefully studied the Paris trials of 1768–69 in the preparation of their new tariff. The results of the Roissy tests conducted during Turgot's watch at the contrôle général reached many provincial towns. BN, ms. fr. 8128, fol. 36. Cf. procurator general to the mayor and échevins of Angers, 28 August 1759, BN, Joly 2417, fol. 252; police registers, Bar, 29 April 1772, AD Aube, B 299; AD Calvados, C 2628.

34. Sylvestre, *Histoire des professions alimentaires*, pp. 73–74; Arpin, *Historique*, vol. 2, p. 70; Pierre Clément, "Successeurs de Colbert," *Revue des deux mondes* 46 (15 August 1863), p. 932; BN, Joly 1116, fol. 262; Delamare, *Traité*, vol. 2 (1713), pp. 1071–78, and vol. 2 (1729), pp. 940–42. In 1316 the trial determined the price/weight characteristics of *pain de Chailly, pain coquillé,* and *pain bis.* BN, ms. fr. 8084, fols. 19–22. More than a half-century later, the Code Duchesne recommended the unvarnished Delamare tariff as a "modèle" for other cities. *Code de la police,* pp. 116–17. On the early trials in Paris and in the provinces, and on the difficulty of obtaining a (locally and/or universally) "reasonable" coefficient of remuneration for the baker in the Middle Ages, see the remarks of Françoise Desportes in her excellent *Le Pain au moyen âge*, pp. 158–59, 176–78.

35. Clément to lieutenant general, 11 December 1725 and Doubleau to lieutenant general 14 December 1725, Arsenal, ms. Bast. 10271, and Doubleau to lieutenant general, 4 January 1726, ms. Bast. 10272; Courcy to procurator general, 29 January 1739, BN, Joly 1119, fols. 41–42. Hérault was probably the first major police figure to insist on using flour as well as wheat as the index to bread price fluctuations, and on the disparity that often existed between the wheat and flour levels. Marchais to lieutenant general, 5 January 1726, and Regnier to lieutenant general, 14 March 1726, ms. Bast. 10272. Sartine betrayed a similar sensitivity; indeed, he clearly privileged flour as the critical indicator. Lieutenant general to Goupil, 1 December 1768, ms. Bast. 10277, and lieutenant general to Trudon, 21 December 1770, AN, Y 15114. Cf. the way in which the development of the flour trade upset the calculus of the London assize ("the magistrate did not trouble to ascertain the prices of meal and flour but set the assize from the price of corn, and, in London, from a cheaper grade than that which the bakers were actually using"). Thrupp, *Worshipful Company*, pp. 23–30.

36. Scipion test, March–April 1739, BN, Joly 1120, fol. 4; calculations, 15 October 1740, Joly 1122, fol. 65; procurator general to Marville, 18 September 1740, Joly 1121, fol. 118.

37. Procurator general to lieutenant general, 29 September 1740, Arsenal, ms. Bast. 10277.

38. On these questions, see Kaplan, *Bread, Politics, and Political Economy* and *Provisioning Paris.*

39. Sartine to Machurin, 28 December 1760, AN, Y 12610.

40. Arsenal, ms. Bast. 10141, fols. 430–31 (4 February 1761), 444 (17 February 1761), 447–50 (23 February 1761).

41. AN, Y 12610; *Nouvelles Ephémérides économiques* 10 (1775), pp. 156–67.

42. Laverdy to Miromesnil, 20 March 1768, in Miromesnil, *Correspondance*, vol. 5, p. 127; Scipion trial, 8 April 1767, Arsenal, ms. Bast. 7458; Sartine to Machurin, 28 November 1767, and subsequent trial, AN, F[11] 1194. Sartine's interest extended back to the stages of precultivation. On his instructions, Machurin participated in a test of a process of treating seed in order to enhance its productivity and robustness. AN, Y 12617, 14 August 1768.

43. AN, Y 12618, 23 January 1769. It is interesting to note that the bakers evinced no hostility whatsoever for economic milling, as if it were the standard method that was commonly used.

44. Sartine to syndics des commissaires, 12 March 1768, and Grimperel to lieutenant

general, ca. late March 1768, with Toupiolle's *état,* dated December 1767, AN, Y 13396. The sum that Grimperel indicated for Toupiolle's annual expenses, 3,722 livres, does not square with the figure of one hundred livres of general weekly expenses. But it is not clear if the two figures contained the same elements. Nor is it certain if the figure for Toupiolle referred to total costs or to the costs for working on the same number of setiers used in the Delamare comparison.

45. "Réflexions sur le commerce du pain," *Journal économique* (1761), pp. 319–20. Cf. an article that appeared in the same journal four years earlier proposing a quasi-national and normative standard for price control based on assumptions about grain, flour and bread type, homogeneous technology, universal consumption habits, and more or less equal costs that surely did not obtain throughout the kingdom. "Avis économique sur la boulangerie," ibid (1757), p. 65. Trials and debates over trials continued through the Revolution, though the revolutionaries were extremely impressed with Sartine's tests and with the trial-based reasoning of Tillet. See, for example, the work of the Commission des subsistances et approvisionnements de la République, and the Administration des subsistances de la Commune de Paris, AN, F^{12} 1299^B, and Birembaut, "Problème de la panification," pp. 361–67.

46. BN, Joly 1143, fol. 107; Lacombe d'Avignon, *Le Mitron de Vaugirard,* p. 35.

47. Arsenal, ms. Bast. 10275 (in a carton largely dealing with the late 1730s).

48. Delamare, *Traité,* vol. 2 (1729), p. 940; Procurator general to lieutenant general, 29 September 1740, Arsenal, ms. Bast. 10277.

49. François Dumas, *La Généralité de Tours au XVIIIe siècle* (Paris, 1894), p. 355; Turgot circular to intendants, 17 September 1775, in *Oeuvres de Turgot,* vol. 4, pp. 494–95; Para du Phanjas, notes, AN, T 466$^{1-2}$.

50. Foucaud to procurator general, 10 September 1752, BN, Joly 1113, fol. 192; Necker (officially directeur général rather than controller general) to intendant of Languedoc, 16 January 1778, AN, F^{11}* 1, fols. 197–98.

51. Cotte, *Leçons élémentaires,* p. 78; "Avis économique sur la boulangerie," p. 57; BN, Joly 1743, fols. 127, 148, 150.

Chapter 19   Policing the Price of Bread, 1725–1780

1. On this episode, see Steven L. Kaplan, "The Paris Bread Riot of 1725," *French Historical Studies* 14 (spring 1985), pp. 23–56.

2. Controller general to lieutenant general, 12, 13, 15 September and 14, 19 October 1725, AN, G^7 34; police sentence, 28 September 1725, Collection Dupré, BN, ms. fr. 8062, fol. 69; Glot, chief secretary to procurator general, to lieutenant general, 8 September 1725, Arsenal, ms. Bast. 10270; anon. police report, 25 September 1725, ms. Bast. 10270; Assembly of Police, 14, 26 September 1725, BN, Joly 1117, fols. 50, 57; AN, Y 9538, 23 October 1725; Parent to lieutenant general, 14 November 1725, ms. Bast. 10273; Delavergée to lieutenant general, 31 October 1725, ms. Bast. 10271; Gazetins de police, 16 November, 19, 26 December 1725, ms. Bast. 10155, fols. 138, 179, 186; anon. to lieutenant general, n.d. (ca. December 1725), ms. Bast. 10271.

3. Labbé and Duplessis to procurator general, BN, Joly 1118, fols. 2–200; Menyer to lieutenant general, letters and reports, n.d. (ca. January–June 1726), Arsenal, ms. Bast.

10273; Delavergée to lieutenant general, n.d. (ca. April 1726), ms. Bast. 10273; Divot to lieutenant general, n.d. (ca. May 1726), ms. Bast. 10273; lieutenant general to Divot, 23 August 1726, and Tricot to Divot, 24 August 1726, AN, Y 13634; ordinance of police, 24 August 1726, Joly 1111, fol. 176.

4. Duplessis to procurator general and also lieutenant general, 19 January, 1 February, 16, 23 March, 10 April, 18, 25 May, and 25 September 1726, BN, Joly 1118, fols. 30–31, 59–61, 98–99, 106–7, 124–25, 144–45, 147–48, 208–9.

5. Police sentence, 13 June 1727, BN, ms. fr. 21640, fols. 154–55; Divot to lieutenant general, 14 June, 30 August, 19 November 1727, Arsenal, ms. Bast. 10274; note to lieutenant general, August 1727, ms. Bast. 10152.

6. Police sentence, 20 August 1728, BN, ms. fr. 21640, fols. 158–59; Gazetins, 28 January 1728, Arsenal, ms. Bast. 10158, fol. 38; Divot to lieutenant general, with latter's marginal comment, 4 December 1728, ms. Bast. 10274.

7. Gazetins, 21 January 1729, Arsenal, ms. Bast. 10159, fol. 24; Divot to Hérault, 13 September 1730, ms. Bast. 10275; police sentence, 10 June 1735, BN, ms. fr. 21640, fols. 180–81; Gazetins, 12–13 May 1737, ms. Bast. 10166, fols. 169–70. Cf. the "monopole" that Duplessis aggressively repressed in 1729, a maneuver not on the bread-selling but on the grain-trading side of baker activity. All the objective conditions invited a decrease in grain price: "il avoit tout lieu d'esperer que cette quantité considerable de grains et de farines, joint à ce que les Bleds qui sont en terre sont très-beaux, et que la saison favorable qui s'approche nous promet une récolte abondante, du propre aveu mesme de tous les laboureurs & gens de la campagne, produiroit quelque diminution sur le prix des grains ou du moins que le prix se tiendroit pareil à celuy du precedent jour du marché." But a Paris baker named Gagnié bid up the price at the Halles by thirty sous a setier beyond the elite of the previous day. Duplessis was persuaded that this was a calculated maneuver to cover a general grain and bread augmentation, a conviction reinforced by Gagnié's reputation for having "dans les temps de disette de Bled toujours paru plus ardent à en faire hausser le prix." Brought immediately before Hérault, Gagnié was sentenced to a fine of one thousand livres, and his shop was shut down for six months. Duplessis personally supervised the walling as well as the collection of the fine. Police sentence, 13 May 1729, ms. fr. 21640, fols. 166–67.

8. BN, Joly 1117, fol. 54; ordinance, 22 September 1725, BN, ms. fr. 21651, fol. 336; sentences of 28 September 1725, ms. fr. 21640, fols. 150–51 and BHVP file; AN, Y 9538, 25 October 1725; Delavergée to lieutenant general, 31 October 1725, Arsenal, ms. Bast. 10271; Regnier to lieutenant general, 18 October 1725, ms. Bast. 10270; Assembly of Police, 20 November, 4 December 1725, Joly 1117, fols. 63, 64; ordinance, 22 December 1725, and sentences 11 January and 23 August 1726, BHVP file. In 1735 the authorities invoked the ordinance of 22 September 1725 to condemn the widow baker Parisot to a fine of three thousand livres and the loss of her market stall for having abandoned the Cimetière Saint-Jean for a full month. Police sentence, 2 September 1735, BHVP.

9. Divot to lieutenant general, 14 June 1727, Arsenal, ms. Bast. 10274; police sentences, 28 November 1727, 29 July 1729, 16 November 1731, 2 September 1735, BN, ms. fr. 21640, fols. 57, 59–62, 168–69; police sentence, 12 February 1734, cited by Des Essarts, *Dictionnaire universel,* vol. 2, p. 258.

10. Assembly of Police, 22 August and 29 September 1725, BN, Joly 1117, fols. 28, 59; Edmond-Jean-François Barbier, *Chronique de la régence et du règne de Louis XV (1718–1763), ou Journal de Barbier* (Paris, 1858), vol. 1, p. 403 (August 1725); Narbonne, *Journal,* p. 137; "état des sergents de confiance employez dans les marchez de Paris," Joly 1116, fols. 216–17; Gazetins, 5 December 1725, Arsenal, ms. Bast. 10155, fol. 166; Contody to procurator general, 31 August 1725, Joly 1117, fol. 98; police report, 24 November 1725, ms. Bast. 10271; Gazetins, 23 August 1725, cited by Funck-Brentano, "Gazette de police," p. 150; Herlaut, "Disette," pp. 14–15, 20; Louis Thuillat, *Gabriel Nicolas de la Reynie, premier lieutenant général de police de Paris* (Limoges, 1930), pp. 89–90. The miller and subsistence expert César Bucquet urged the establishment of separate flour granaries and bakeries for the watch and the guard, an institutional means of deterring these soldiers from sympathizing or even joining forces with harried consumers during crises. *Traité pratique,* p. 13. Bakery militarization reached its pinnacle in the early days of the Flour War riots (5 May 1775) when one soldier was placed in every shop in addition to the large contingents assigned to the marketplaces. Darnton, "J.-P. Lenoir," p. 617. The police proved more intrusive in the nineteenth century. At the end of 1816, an agent was sent to spend an entire night in each ovenroom in order to account for the disposition of flour and the yield into bread. Chauveau-Lagarde, *Mémoire pour les boulangers,* pp. 16–18.

11. De Guichey to lieutenant general, 8 October 1725, Arsenal, ms. Bast. 10270; BN, Joly 1116, fol. 217.

12. Gazetins, 24, 26 September 1725, Arsenal, ms. Bast. 10155, fols. 74–75, 77. Several years later Hérault again had reason to look into the cozy relationship between Labbé and certain of "his" bakers. Duplessis, a man of rigor but hardly an antagonist of the bakers, reported to the lieutenant general that Poulain, a faubourien, had purchased a muid at a price well above the current. Asked to investigate, Labbé suggested there was no need: "Je puis vous assurer Monsieur que Sr Poulain et sa femme sont de bonnes gens qui soulagent bien le peuple et qui vendent bien loyalement leur pain." Labbé to Hérault, 27 August 1727, ms. Bast. 10274. Writing in reference to the crisis of 1709, Saint-Simon related that "à Paris des commissaires y mettoient le prix à mainforte et obligeoient souvent les vendeurs à le hausser malgré eux." Here the bakers were incidental victims rather than coconspirators in a sort of famine plot. Boislisle, ed., *Mémoires,* vol. 17, p. 197.

13. Lemoyne to lieutenant general, 3 November 1725, Arsenal, ms. Bast. 10271.

14. Anon. to lieutenant general, 20 September 1725, Arsenal, ms. Bast. 10270.

15. AN, Y 12571, 27 October 1725. It is interesting to note how rapidly the "grand atroupement de populace" took place, as if the entire community had been plugged into the event as it unfolded.

16. "Mémoire de ce qui s'est passé," 9 October 1725, Arsenal, ms. Bast. 10033; Gazetins, 26 November 1725, ms. Bast. 10155, fol. 145.

17. Curé de Saint-Pierre to lieutenant general, 26 October 1725, Arsenal, ms. Bast. 10271 (in the margin of a rambling petition addressed to him by the same cleric ten days earlier, Hérault scribbled: "lettre non intelligible attendu sa qualité de pretre inutile à repondre," 10 October 1725, ms. Bast. 10270); Labbé to procurator general, 13 April 1726, BN, Joly 1118, fol. 128; baker petition, July 1727, ms. Bast. 10152; Couet de Montbayeux

to procurator general, 11 November 1725, Joly 1117, fol. 248. On the role of credit in the provisioning nexus, see Kaplan, *Provisioning Paris*, pp. 148–53, 155, 467, 473, 596, 602–4. See also the discussion of the baker market corps and baker *faillites* above.

18. Noblet to de la Magée, 18 July 1699, BN, ms. fr. 21644, fol. 194; Herlaut, "Disette," p. 87; bishop of Rennes to lieutenant general, 23 November 1725, Arsenal, ms. Bast. 10271; lieutenant general to procurator general, 3 February 1727, ms. Bast. 10003, p. 158; Des Borus to lieutenant general, 9 July 1740, ms. Bast. 10277; Forcet to lieutenant general, May 1740, and reply, 26 May 1740, ms. Bast. 10010, p. 508; Sully-Goesbriand to lieutenant general, 21 March 1742, ms. Bast. 10277.

19. Desmaze, *Métiers*, p. 154; Saint-Germain, *Vie quotidienne*, pp. 220–22; Thery, *Gonesse dans l'histoire*, pp. 89–91; police sentence, 22 June 1709, BN, ms. fr. 21646, fol. 277. Cf. an ordinance of Charles VIII in September 1439 banning the production of white bread. The restriction was commonly practiced in eighteenth-century France. See, for example, the police at Bayeux in 1725, AD Calvados, C 2643. The London assize imposed a two-sort limit in 1758. Thrupp, *Worshipful Company*, p. 30. See also the restrictions proposed by Parmentier on luxury bread making during dearth. *Recherches*, p. 416. Early in the Revolution there were calls to imitate the practice of 1709, 1725, and 1740, and to forbid the production of whites and all loaves of soft dough. To the usual arguments, citizen Du Vaucelle, vice-president of the district of Saint-Gervais, added a sansculottic, moralizing dimension: "Le pain composé de pâte ferme est plus agréable au goût, plus nourissant et beaucoup plus salubre que les pain blancs et mollets *inventés par la Sensualité*" (my emphasis). "Opinion sur les subsistances," AN, F^{10} 215–16. By Year II, the revolutionaries had shifted from a two-sort to a one-type loaf, the universal bread of equality, that survived the Terror. See the complaint in Year III that "les boulangers n'écoutant que leur intérêt et déterminés par l'appât du gain, tamisent leur farine pour en faire un pain plus délicat." The commission de police administrative to the commission de police de la section des Arcis, 3 frimaire an III, ASP, 6 AZ 701. A modern analogue to the two-sort bread restriction is the use of so-called national flour in Great Britain during the Second World War. The "grey" bread it produced provoked many complaints. Sheppard and Newton, *Story of Bread,* p. 76.

20. Ordinance, 17 August 1725, AN, AD XI 38, and BN, ms. fr. 21651, fol. 327; parlementary arrêt, Conseil secret, 21 August 1725, AN, X^{1A} 8446. Cf. an anonymous attack on bakers for putting dear *mollet* on the market even as the crisis deepened. "Mémoire et tarif du prix du pain," n.d., Arsenal, ms. Bast. 10270.

21. Assembly of Police, October 1725, BN, Joly 1117, fol. 60; police sentence, 28 September 1725, BN, ms. fr. 21640, fol. 149.

22. Parlementary arrêt, Conseil secret, 26 October 1725, AN, X^{1A} 8446; "Motifs sur lesquels Chabert . . . ," and petition to cardinal de Fleury, Arsenal, ms. Bast. 10273; BN, Joly 1117, fol. 58 (29 September 1725); Gazetins, 10 November, 1 December 1725, ms. Bast. 10155, fols. 132, 153; AN, Y 15233, 19 July 1726. Chabert, the baker in charge of the "Barbary" bakeshop, habitually sold his bread a third below the price of the Paris bakers "pour les engager à baisser le prix du leur." "Motifs sur lesquels Chabert . . . ," ms. Bast. 10273.

23. Gazetins, 10 November 1725, Arsenal, ms. Bast. 10155, fol. 125; d'Angervilliers to procurator general, 23 September 1725, BN, Joly 1116, fol. 25; police sentence, 23 October 1725, AN, Y 9538.

24. Gazetins, 14, 15 May and 27 July 1726, Arsenal, ms. Bast. 10156, fols. 210–12, 215, 313; Duplessis to procurator general, 27 July 1726, BN, Joly 1118, fol. 182; police sentence, 12 July 1726, BN, catalogue of acts. Bakers used nonwheaten grains during the crisis of 1725–26, almost always in combination with wheaten flours. See Duplessis to procurator general, 7 August 1726, Joly 1118, fol. 187, and Guery to Hérault, 5 October 1725, ms. Bast. 10270.

25. Couet de Montbayeux to procurator general, 5, 14 October 1725, BN, Joly 1116, fols. 270–71, 285–86; same to procurator general, 9 October 1725, Arsenal, ms. Bast. 10270; same to procurator general, 24 October 1725, Joly 1117, fols. 240–43; d'Angervilliers to lieutenant general, 7 November 1725; ms. Bast. 10271; Cleret to lieutenant general, 15 October 1725, ms. Bast. 10270; Bourlon to lieutenant general, 4 October 1725, ms. Bast. 10270.

26. On the genesis and structure of the crisis, see Steven L. Kaplan, "The State and the Problem of Dearth in Eighteenth-Century France: The Crisis of 1738–41 in Paris," *Food & Foodways* 4 (1990), pp. 111–41, and Marcel Lachiver et al., "La Crise de subsistances des années 1740 dans le ressort du Parlement de Paris (d'après le Fonds Joly de Fleury de la Bibliothèque nationale de Paris)," *Annales de démographie historique* (1974), pp. 281–334.

27. AN, Y 9439, 22 November 1738.

28. AN, Y 12141, 4 March 1739; Y 9440, 6 March 1739.

29. Police sentence, 29 May 1739, AN, Y 9440 and Y 9619; BN, ms. fr. 21640, fols. 192–93.

30. AN, Y 9440, 20 March 1739.

31. AN, Y 11226, 2 May 1739, and Y 9440, 2, 5 May 1739.

32. AN, Y 9441, 14, 27 May 1740; Y 9619, 20 May 1740; Gazetins, 26–27 May 1740, Arsenal, ms. Bast. 10167, fol. 118.

33. AN, Y 9441, 14 September 1740; Gazetins, 14–15 September, 28–29 December 1740, Arsenal, ms. Bast. 10167, fols. 130–31, 208; anon. to lieutenant general, n.d., ms. Bast. 10027, fol. 192.

34. Gazetins, 21–22, 26–27 October, 29–30 November, 13–14 December 1740, Arsenal, ms. Bast. 10167, fols. 166–68, 187, 195.

35. AN, Y 9441, 26 August 1740; Y 3747, 9 July 1740; Gazetins, 24 September 1740, Arsenal, ms. Bast. 10167, fol. 144; Y 12583, 18 October 1740; L.-E. Alphonse Jobez, *La France sous Louis XV, 1715–1774* (Paris, 1864–73), vol. 3, p. 171; Barbier, *Journal,* vol. 3, p. 218 (September 1740).

36. AN, Y 13747, 9 July 1740; Y 9441, 24, 26 August, 2, 9 September 1740; Y 9619, 26 August 1740; Y 9499, 2 September 1740.

37. Gazetins, 23–24, 26–27, 28–29 September 1740, Arsenal, ms. Bast. 10167, fols. 140–41, 146–48; lieutenant general to procurator general, 28 September 1740, BN, Joly 1121, fol. 110. The fickleness of (dearth-shaped) public opinion must have frustrated Marville, who worked hard to cultivate a favorable image. Cf. the enthusiasm that was said to have been felt for him at the beginning of 1740: "On croit devoir rapporter que l'on dit dans le publique que M. le lieutenant général de police tiendra la main à ce que le bon ordre soit observé dans Paris, singulièrement dans les communautés qui sont destinées par état à fournir les denrées pour la subsistance de cette ville, on adjouste

même qu'*il sera plus rigide que M. Hérault en ce qui concerne la police des boulangers*" (my emphasis). Gazetins, 9–10 February 1740, ms. Bast. 10167, fol. 24.

38. Conseil secret, 22 September 1740, AN, X^{1A} 8468; BN, Joly 1111, fols. 2–3; Joly de Fleury's marginal notes, duchesse de Ventadour to procurator general, 27 September 1740, Joly 1121, fol. 124; Commissaire Defacq, procès-verbal, 21 October 1740, Joly 1120, fols. 248–51; arrêt du Parlement, 31 December 1740, Joly 1111, fols. 4–5; lieutenant general to procurator general, 29 December 1740, Joly 1121, fol. 61; Glou to procurator general, 6 January 1741, Joly 1121, fol. 69. Administrators of Hôtel-Dieu denied patients the petits pains mollets to which they were accustomed. Deliberations, 23 September 1740, AAP, no. 109, and "Sommier," liasse 28, no. 2. Officials at Calais considered the possibility of reducing bakers to one type of bread. Draft, procurator general to fiscal procurator of Calais, Joly 1123, fol. 129. Versailles banned white and mollet loaves. Evrard, *Versailles*, p. 407n. There was no question that *all* consumers would have to endure the harsh regime of household bread. See the letter that the duchesse de Ventadour addressed to the procurator general on 27 September 1740: "Mon boulanger de Versailles dit qu'il ne peut plus me faire de pain Mollet: mais pour empescher l'indiscretion d'un jeune estomac, de se plaindre de la rigueur d'estre reduit au pain Bis, vous voudrés bien donner vos ordres pour la conservation de celle qui vous seras attachés toutte sa vie." Joly 1121, fol. 124.

39. Narbonne, *Journal*, p. 474; AN, Y 13741, 8 October 1740; Y 9441, 8, 13, 14 October 1740; procurator general to lieutenant general, 9 October 1740, and reply, 19 October 1740, BN, Joly 1121, fols. 128–29; ASP, D5Z carton 9 (Collection Lamouroux); Lenepveu, Gillet, and Noirot to procurator general, n.d., BN, Joly 1121, fol. 116; Y 14301, 7 October 1740, Y 12142, 7 October 1740; APP, écrou Grand Châtelet, 22 November 1740, A B/195. Beyond enjoining the bakers to offer large quantities of dark bread, the authorities tried to induce them to change the way in which they sold it. It was in the public interest for them to sell the dark loaves cooled and even slightly stale "parce qu'il a meilleure mine, qu'il est plus sain, plus profitable, et se coupe mieux que lorsqu'il est chaud et meme tendre." This method would spare the consumer the burden of absorbing the evaporation and loss of weight that occurred in the twenty-four hours after the bread left the oven. Joly 1122, fols. 60–61.

40. Gazetins, 11–12, 19–20, 26–27 October 1740, Arsenal, ms. Bast. 10167, fols. 156, 163–64, 167–68; AN, Y 13741, 12–13 October 1740; *Lettres de M. de Marville*, vol. 1, pp. 213, 214; Y 9441, 11, 14 October 1740. Already in late September, a police agent wrote that "le plus grand nombre dit qu'il faut que les sentences de M. le lieutenant général de police soient illusoire puisque les boulangers n'y ont point d'égard." Gazetins, 21–22 September 1740, ms. Bast. 10167, fol. 139.

41. Police sentences, 18, 25 November 1740, AN, AD I 23^A; AN, Y 9441, 18 November 1740; Y 9621, 23 December 1740 (nine convictions).

42. Gazetins, 31 January, 24–25, 28 February, 28–29 March, 7–8 April 1741, Arsenal, ms. Bast. 10168, fols. 47, 52, 54, 99, 153.

43. Gazetins, 9 March, 18–19 April 1741, Arsenal, ms. Bast. 10168, fol. 89, 168; Gazetins, 25 February 1741, BHVP, ms. 620, fol. 187; procès-verbaux, 8, 9 March, 2 April 1741, ms. Bast. 10136; AN, Y 9442, 10 March 1741; ms. Bast. 10140, fol. 252.

44. Controller general to procurator general, 10 March 1741, BN, Joly 1121, fols. 131–32; parlementary arrêt, 23 March 1741, ibid., fol. 138, and BN file; Gazetins, 17, 25 March 1741, BHVP, ms. 620, fols. 256, 268; Gazetins, 26–27 and 28–29 March 1741, Arsenal, ms. Bast. 10168, fols. 96, 99.

45. Gazetins, 11–12, 14–15, 15–16 April, 12–13, 16–17 May, 10–12 August, 27–28 September 1741, Arsenal, ms. Bast. 10168, fols. 159, 165, 167, 231, 233, 272, 332; AN, Y 9499, 14, 28 April 1741; Y 9442, 14 April 1741; Y 9619, 28 April 1741.

46. Police sentence, 19 January 1742, AN, AD I 23ᴬ; AN, Y 9443, 19 January, 13 April 1742; Marville to Maurepas, 8 February, 14 April, 12 July 1742, *Lettres de M. de Marville,* vol. 1, pp. 9, 25, 58.

47. Gazetins, 16–17 October 1740, Arsenal, ms. Bast. 10167, fols. 159, and 17–18 January 1741, ms. Bast. 10168, fol. 41; Desmaze, *Métiers,* p. 337; police sentence, 21 October 1740 (particularly defective dark bread), BN, register of acts; René-Louis de Voyer, marquis d'Argenson, *Journal et mémoires du marquis d'Argenson,* ed. E.-J.-B. Rathery (Paris, 1859–67), vol. 3, p. 169 (19 September 1740).

48. Mercier, *Tableau,* vol. 8 (1783), p. 1; deliberations of the Bureau of the Hôpital général, 26 February 1787, AAP, new series 47 (Scipion). On the popular association of Bicêtre with the devil, see the overdrawn account by Maurice Alhoy and Louis Lurine, *Les Prisons de Paris: Histoire, types, moeurs, mystères* (Paris, 1846), pp. 308–11.

49. Cadet de Vaux, cited by Arpin, *Historique,* vol. 1, p. 249; "Projet d'Ecole de Boulangerie," 1783, author's library; *Journal de Paris,* 4 July 1782, p. 756, and 12 August 1782, p. 917; Bucquet, *Observations intéressantes et amusantes,* pp. 83–84; police sentence, 13 February 1739, BN, ms. 21640, fols. 184–85. Yet even before the Ecole assumed much of the institutional baking burden, Bachaumont had announced in 1782 that prison bread was "excellent." *Mémoires secrets,* vol. 21, p. 9 (4 July 1782). Cf. Mercier's affirmation that "les prisonniers de Paris mangent un pain beaucoup meilleur que celui qu'on mange dans les cantons Helvétiques." *Tableau,* vol. 12 (1788), p. 146.

50. Gazetins, 23–24, 26–27 September 1740, Arsenal, ms. Bast. 10167, fols. 140, 145; *Mercure historique et politique* 109 (October 1740), p. 476; d'Argenson, *Mémoires,* vol. 3, p. 173 (24 September 1740); Barbier, *Journal,* vol. 3, p. 219 (September 1740); lieutenant general to procurator general, 22 September 1740, BN, Joly 1121, fol. 53; Bureau of Hôpital général to procurator general, 10 October 1740, ibid., fols. 57–58; Narbonne, *Journal,* p. 469 (20 September 1740); Luynes, *Mémoires,* vol. 3, p. 255 (September 1740); Jobez, *Louis XV,* vol. 3, p. 171. Mercier noted that "il y a de temps en temps des révoltes à Bicêtre." He referred specifically to an insurrection in 1756 provoked by a cut-back in rations and issuing in the deaths of two archers and fourteen inmates. Apparently a similar riot in 1752 cost the lives of two soldiers and fourteen inmates. Again in February 1771, a diminution of the size and a debasement of the quality of the ration incited the prisoners to riot, resulting in the deaths of around twenty persons. Two years later, the (women) prisoners at the Salpetrière rose up "sous prétexte que le pain qu'elles ont eu il y a quelques jours ne s'est pas trouvé bon." To punish the guiltiest fraction, Sartine ordered a half-dozen incarcerated "au cachot de Bicêtre," where their ration would be reduced. Mercier, *Tableau,* vol. 8 (1786), p. 11; Emile Richard, *Histoire de l'hôpital de Bicêtre, 1250–1791* (Paris, 1889), p. 93; Sartine to Guyot, 8, 9 May 1773, AN, Y 13551.

51. Poussot's journal, 27 November, 11, 28 December 1756, Arsenal, ms. Bast. 10141; Mme de *** to Mr de Mopinot, 12 June, 24 September 1757, in Lemoisie, ed., "Sous Louis le Bienaimé," pp. 771–72, and no. 13 (1 July 1905), p. 160.

52. BN, Joly 1139, fol. 129; Jacques Dupâquier et al., *Mercuriales du pays de France et du Vexin français, 1640–1792* (Paris, 1968), pp. 196–203; ASP, D5B⁶, registres 4118, 4119, 1003, 84, 2619, 4881, 3009, 3760, 3327; procurator general to controller general, 16 September 1767, and controller general to procurator general, 24 September 1767, Joly 1135, fols. 223–24. On the origins and character of the crisis of the sixties, see Kaplan, *Bread, Politics, and Political Economy,* passim.

53. Lieutenant general to syndics, 12 March 1767, and 23 January 1769, AN, Y 13396; lieutenant general to Machurin, 7 February 1770, Y 12618; Arsenal, ms. Bast. 7458, 8 April 1767; "Procès-verbal d'expérience sur 2 septiers de bled," AN, F¹¹ 1194; Sartine, mémoire, n.d., AN, F¹¹ 264; lieutenant general to Cypierre, 25 September 1768, Camille Bloch, ed., *Le Commerce des grains dans la généralité d'Orléans (1768), d'après la correspondance inédite de l'intendant Cypierre* (Orléans, 1898), p. 94; lieutenant general to Commissaire Mouricault, 28 September 1768, Y 12830; comptes and états Malisset, F¹¹ 1193 and 1194.

54. Lieutenant general to procurator general, 18 September 1767, BN, Joly 1136, fol. 12.

55. Lieutenant general to Coquelin, 28 February 1768, AN, Y 13728; lieutenant general to Thierry, 28 February 1768, Y 11255; lieutenant general to Grimperel, 28 February 1768, Y 13396. My emphasis.

56. Hardy's journal, BN, ms. fr. 6680, p. 151 (12 March 1768); lieutenant general to procurator general, 25 June 1768, BN, Joly 1137, fol. 53; ASP, D5B⁶ 3009; bulletins de sûreté, 27 July 1768, Arsenal, ms. Bast. 10123 (the jailing of a forain); lieutenant general to Coquelin, 6 September 1768, AN, Y 13728, and lieutenant general to Grimperel, 6 September 1768, Y 13396. Cf. the imprisonment of another forain and of a master. Y 12618 and Y 14095, 22 December 1768.

57. Draft, Grimperel to Sartine, 7 September 1768, AN, Y 13396.

58. Lieutenant general to procurator general, 8, 10, 14, 17, 24 September 1768, BN, Joly 1137, fols. 74, 75, 76, 77, 80.

59. BN, Joly 1137, fols. 85 (12 October 1768) and 86 (15 October 1768); lieutenant general to Goupil, 2, 22 November, 1 December 1768, Arsenal, ms. Bast. 10277; lieutenant general to procurator general, 12 November 1768, BN, Joly 1137, fol. 94; lieutenant general to procurator general, 16 November 1768, Joly 1141, fol. 95; Hardy's journal, BN, ms. fr. 6680, p. 190 (23 November 1768); Collerot Dutilleul, 23 November 1768, Y 13728; lieutenant general to Goupil, 1 December 1768, ms. Bast. 10277; lieutenant general to procurator general, 10, 21 December 1768, Joly 1137, fols. 102, 105; Hardy's journal, BN, ms. fr. 6680, fol. 197 (14 December 1768); lieutenant general to Thierry, 15 December 1768, Y 11255; lieutenant general to Coquelain, 15 December 1768, Y 13728.

60. Sartine to Grimperel, 2 November 1768, Arsenal, ms. Bast. 10277; AN, Y 13540, 13, 15, 20 December 1768; Y 13396, 19, 30 December 1768.

61. ASP, D4B⁶ 41-2260, 15 June 1771; AN, Y 13540, 16 December 1768; Y 12617, 9 September 1768. The "misère du tems" and "la cherté des grains" continued to cripple bakers in the early seventies. D4B⁶ 43-2435, 24 January 1772, and D4B⁶ 48-2926, 31 August 1773.

62. Hardy's journal, BN, ms. fr. 6680, fol. 179 (6 October 1768); lieutenant general to

Grimperel, 8, 14, 16 September, 4 October 1768, and Grimperel to lieutenant general, 15 November 1768, AN, Y 13396; lieutenant general to Thierry, 14 November 1768, Y 11225.

63. Grimperel, états, notes, and drafts to lieutenant general, September–December 1768, and lieutenant general to Grimperel, 21 November 1768, AN, Y 13396.

64. Sartine, cited in *Recueil des principales lois*, pp. 13, 132; Pierre-Joseph-André Roubaud, *Représentations aux magistrats contenant l'exposition raisonnée des faits relatifs à la liberté du commerce des grains* (Paris, 1769), pp. 410–11; Arsenal, ms. Bast. 12334. Cf. the considerable efforts in 1789, organized by Bailly under the aegis of Necker, to sustain "poor bakers" with interest-free loans. Bailly, *Mémoires*, vol. 2, p. 408; *Journal de la municipalité*, no. 10 (6 November 1789), p. 75.

65. In addition to Rutledge's famous first and second "mémoires pour les boulangers," see the dossier in AN, Y 10606, 1, 4, 5 December 1789. This story needs to be unraveled, in the context of a searching examination of the subsistence problems of the early Revolution.

66. Sartine to procurator general, 15 March 1769, BN, Joly 1143; Sartine to Goupil, 4 January 1770, Arsenal, ms. Bast. 10277; Albert to Sartine, 10 May 1770, AN, F^{12*} 153, fol. 205; Sartine to procurator general, 9, 13 June 1770, Joly 1428, fols. 287–88; Sartine to Goupil, June 1770, ms. Bast. 10277. After first instructing his commissaires to authorize officially and publicly the bakers to raise their prices, Sartine thought better of the idea: "Il ne sera pas nécessaire d'avertir les boulangers mais si les Bourgeois et le Public s'y plaignent il faudroit leur dire la raison de cetter augmentation."

67. Sartine to procurator general, 22 August 1770, BN, Joly 1428, fol. 308; Sartine to Goupil, 17 September, 14 November 1770, Arsenal, ms. Bast. 10277; Sartine to Trudon, 21, 29 December 1770, AN, Y 15114.

68. Sartine to Trudon, 22 January, 23 March, 18, 25 April, 21, 24 June, 12, 14 November 1771, AN, Y 95114; Sartine to Goupil, 22 January 1771, Arsenal, ms. Bast. 10277; Sartine to Mouricault, 23 March 1771, Y 15114; Y 9474, 5 July 1771 (Chiot and Machurin) and 18 November 1771 (Maillot); Hardy's journal, BN, ms. fr. 6680 (1 May 1771). During the summer of 1771 the bread markets were perturbed by a measure of the Cour des monnaies proscribing the use of certain coins. Sartine authorized the bakers to accept them, but only "des gens pauvres et qui achettent eux mêmes leur pain dans le marché et lorsqu'ils y seront en quelque sorte forcés pour éviter les contestations qui pourroient occasionner du désordre." Sartine to Trudon, 19 July, and Chenon to Trudon, 3 July 1771, Y 15114; Chenon to Grimperel, 29 July 1771, Y 13399. Cf. Lenoir to Gillet, 4 April 1778 (evoking the experience of 1771 during which the king "a supporté la perte" resulting from "les pièces décriées"), Y 13728.

69. Sartine to Goupil, 15 March 1772, Arsenal, ms. Bast. 10277; Sartine to Trudon, 13 March, 13 April, 11 May, 12, 14, 22, 23, 29 September 1772, and Sartine to syndics of the commissaires, 13 April, 8 May 1772, AN, Y 15114; Y 12622, 25 April 1772; Hardy's journal, BN, ms. fr. 6681, fol. 93 (5 September 1772).

70. Albert to syndics of the commissaires, 24 May, 20 October, 23 December 1775, and to Gillet, 13 July 1775, AN, Y 13728; Bachaumont, *Mémoires secrets*, vol. 30, p. 290 (19 July 1775), and p. 292 (22 July 1775). In the spirit of Turgot, Albert also suppressed protests of excessive bread prices by irate consumers on the grounds that they could "exciter la

révolte." Y 12625. Cf. the behavior of late-twentieth-century Egyptian bakers "who complained bitterly that they were trapped between rising wholesale costs and fixed retail prices." *New York Times*, 8 February 1980.

71. Jacob-Nicolas Moreau, *Mes Souvenirs*, ed. Camille Hermelin (Paris, 1898–1901), vol. 2, p. 190.

72. Lenoir to syndics, 31 January 1778, AN, Y 12830, and to Gillet, 5 July 1777 and 31 January 1778, Y 13728; Y 9479, 6, 17 May 1776. Cf. *Les Numéros parisiens:* "On mange de très-beau pain à Paris. Quoiqu'il y soit taxé, les boulangers s'autorisent presque toujours à le vendre quelque chose de plus, mais ils n'exercent gueres [*sic*] cette vexation que sur l'ouvrier et le bas peuple, qui n'a pas le tems à perdre pour courir chez le commissaire et autres préposés: ils n'iront pas sacrifier leur journée pour se plaindre de deux ou trois liards qu'on leur dérobe."

73. Lenoir papers, BN, Orléans, ms. 1421; L.-J. Bourdon-Desplanches, *Lettre à l'auteur des Observations sur le commerce des grains* (Amsterdam, 1775), p. 27; Beardé de l'Abbaye, *Recherches sur les moyens de supprimer les impôts* (Amsterdam, 1770), p. 208; Jean-Pierre Louis, marquis de Luchet, *Examen d'un livre qui a pour titre "Sur la législation et le commerce des bleds"* (Paris, 1775), pp. 61–62. On the existing emergency granary systems, see Kaplan, *Bread, Politics, and Political Economy* and "Lean Years, Fat Years."

74. This is the case I make in *Bread, Politics, and Political Economy.*

Conclusion

1. "For we being many are one bread, and one body, for we are all partakers of that one bread." 1. Corinthians 10:17, quoted by Pierre Floriot, *Traité de la messe de paroisse* (Paris, 1679), pp. 329–30, cited by Diefendorf, *Beneath the Cross*, p. 32.

2. See Kaplan, *Bread, Politics, and Political Economy*, pp. 32–33.

3. On the faubourg Saint-Antoine as a laboratory of innovation and a thorn in the corporate side, see Steven L. Kaplan, "Les Corporations, les faux-ouvriers et le faubourg Saint-Antoine," *Annales: E.S.C.* 40th year (March–April 1988), pp. 253–88.

4. R. R. Palmer, *The Age of Democratic Revolution: A Political History of Europe and America, 1760–1800* (Princeton, N.J., 1959–64), vol. 1.

5. Jacques-Louis Ménétra, *Journal of My Life*, ed. Daniel Roche and trans. Arthur Goldhammer (New York, 1986).

6. See Steven L. Kaplan, "Social Classification and Representation in the Corporate World of Eighteenth-Century France: Turgot's 'Carnival,'" in *Work in France: Representations, Meanings, Organization, and Practice*, ed. Kaplan and Cynthia J. Koepp (Ithaca, N.Y., 1986), pp. 176–228.

7. Galiani to Madame d'Epinay, 7 August 1773, in *Lettres de l'abbé Galiani à Madame d'Epinay*, vol. 2, p. 77.

8. This preoccupation with conspiracy constitutes one of the salient features of what I have called the subsistence mentality. See Kaplan, *The Famine Plot*. Given the studied vagueness of his terms, it might also correspond to what Louis Althusser calls ideology: "Le rapport imaginaire des individus à leurs conditions réelles d'existence." Cited by Michel Vovelle, *Idéologies et mentalités* (Paris, 1982), p. 6.

Bibliography

Primary Sources

Archival Materials

The documents consulted are too numerous and dispersed to enumerate individually. For sources cited, the precise references are given in the notes. Here I shall indicate the categories of materials I have used in the major institutions. I will list neither depots where I consulted a single series (e.g., Bibliothèque de l'Institut, Archives des Affaires Etrangères) nor the score of departmental archives from which I drew material.

1. Archives nationales (by series)

Y. Police archives, including minutes and registers of the audience of the chambre de police; reports and *procès-verbaux* of summonses, arrests, visits and searches, seizures, investigations, interrogations, and so on; complaints of citizens; correspondence of police officials; papers and logs of commissaires and inspectors of police; market and price data; appraisals of the state of public opinion; guild papers; sentences and ordinances; placing of seals.

X. Archives of the Parlement of Paris, including remonstrances, sentences, appeals, and especially registers and minutes of the Conseil secret.

F. Notably F^{10}, F^{11}, and F^{12}, and to a lesser degree F^{7}, which deal with police, subsistence, agriculture, commerce, industry, and elements of the correspondence between the contrôle général and the intendants in the field.

H. Deliberations of the city government as well as other materials concerning municipal administration.

V. Extraordinary commissions of the royal council. Financial records of the guilds and government audits.

O. Papers of the royal household, especially correspondence and instructions from the secretary of state, responsible for the affairs of Paris.

Q. Domanial titles, in particular the distribution of shops and residences in the city and faubourgs.

K and KK. Miscellaneous materials concerning Paris, royal legislation, public administration, and regulatory issues.

Z^{1H}. Special jurisdictions, in particular the Bureau of the municipality.

Z^{1G}. Election of Paris, fiscal documents, in particular the taille tax rolls.

AD. For the most part published legislation and regulations, including documentation concerning the guilds.

Minutier central. Notarial archives, including after-death inventories, marriage contracts, estate divisions, property transactions, apprentice contracts, contracts for loans and annuities, declarations of indebtedness, audits, and so on.

2. Bibliothèque nationale

The holdings in the Manuscript Room, especially the Joly de Fleury Collection, which deals with a rich array of administrative, social, economic, and political issues and events and contains a huge correspondence between the procurator general and officials high and low; the Delamare papers, bequeathed by the police commissaire of that name and his collaborators, touching on all aspects of the life of Paris and its regulation; and scattered documentation through the *manuscrits français* and the *nouvelles acquisitions,* including the minutes of the Assemblies of Police, diaries and journals, facta, memoranda, and correspondence. Hundreds of police sentences, parlementary arrêts, and pieces of royal legislation, all of which are printed, can be found through an incomplete index in the hemicycle of the Main Reading Room.

3. Bibliothèque de l'Arsenal

The material here, especially in the Bastille manuscripts, is an indispensable complement to the police archives in the Y series of the Archives nationales (besides the usual range of reports and minutes, there is a rich correspondence involving police officials, logs of inspections and arrests, seized papers, and evaluations of the state of opinion). In other sections there is much material concerning the guilds and the subsistence question.

4. Archives du Département de la Seine et de la Ville de Paris

In addition to a wealth of miscellaneous materials dealing with the social, administrative, and political history of Paris (especially in the AZ and DQ series), one finds here the registration archives recording vital transactions and life-cycle events, and the immense, astonishingly rich, and still underutilized consular series (from D_2B^6 through D_7B^6). The latter constitute the business history of Paris and the Paris region, containing failure petitions and audits, commercial registers and ledgers, reports of mediators, complaints of merchants, and decisions of the judges of the court.

5. Archives de la Préfecture de Police

Arrests, complaints, investigations, and a whole range of police reports for the last decades of the Old Regime and the Revolution (primarily in the AB series and the Fonds Lamoignon).

6. Archives de l'Assistance publique

A potpourri of documents, many then uncatalogued, concerning the daily life of the hospitals (in the Old Regime sense of the word) and their administration (deliberations of the board, or bureau, of the Hôtel-Dieu, correspondence of the managers of the Hôpital général), data on provisioning, and minutes of bread-making trials and other experiments.

7. Bibliothèque historique de la Ville de Paris

A panoply of documents concerning the city, including administrative records, reports on public opinion ("gazetins"), police and regulatory materials (sentences and ordinances), and public and private correspondence.

8. Museums

A host of useful iconographic and documentary materials in the Deutsches Brotmuseum at Ulm/Donau; the Musée national des arts et traditions populaires, Paris (which boasts a superb museum and archives with surprisingly pertinent ethnographic data on bread making and distribution); the Musée français du pain (S.A.M.), Charenton-le-Pont; and the Musée du bled et du pain, Verdun-sur-le-Doubs.

Anonymous, Published Works

Almanach des muses. Paris: n.p., 1770.
Almanach parisien en faveur des étrangers. Paris: Duchesne, 1772, 1786.
L'Ami du Peuple, 1791.
L'Ami du Roi, 1789–91.
Arrest notable de la Cour des Muzes . . . pour le rabais du pain. N.p., n.d.
"Avis économique sur la boulangerie." *Journal économique,* September 1757.
Le Babillard 36 (30 April, 30 June 1778).
Bulletin de la Société d'encouragement de l'industrie national, 1806.
"Chanson nouvelle sur le rabais du pain." *Magasin pittoresque* 43rd year (1875).
Courrier de France, 1790.
"De la mouture." *Journal économique,* May 1771.
Dictionnaire portatif du commerce. Paris: J. F. Bastieu, 1777.
Dictionnaire universel françois et latin, vulgairement appelé Dictionnaire de Trévoux, contenant la signification et la définition des mots de l'une et de l'autre langue. 8 vols. Paris: Compagnie des libraires associés, 1771.
Encyclopédie méthodique. 185 vols. Paris: Panckoucke, 1782–1832.
Enquête sur la boulangerie du Département de la Seine. Paris: Imprimerie Impériale, 1859.
Enquête sur la révision de la législation des céréales. 2 vols. Paris: Conseil d'Etat, 1859.
Les Ephémérides du citoyen, 1770–73.
Gazette de l'agriculture, du commerce, des arts et des finances, 1768–71.
Gazette du commerce, de l'agriculture, et des finances, 1767.
"Histoire des boulangers." *Magasin pittoresque* 25th year (1857).
Journal de l'agriculture, du commerce, des arts et des finances, 1773
Journal historique et politique des principaux événements, 1773–74.
Journal de Paris, 1782
Journal de Physique, 1775–82.
Mémoire pour les Srs. Bandy, et al. Paris: n.p., n.d.
Mémoire pour les Srs. Buvry, Buchot et al., maîtres boulangers. Paris, 1786.
Mercure de France, 1747–77.
Mercure historique et politique, 1740.

La Misère des garçons boulangers de la ville et faubourgs de Paris. Troyes: Veuve Garnier, 1786.

Nouvelles ephémérides économiques, 1775.

Les Numéros parisiens par M. D***. Paris: n.p., 1788.

Observations sur l'approvisionnement de Paris, ou Moyens d'empecher le haut prix du pain. Paris: J. Grand, 1791.

Le Patissier en colère sur les boulangers et les taverniers en vers burlesques. Paris: n.p., 1649.

La Question du pain, ou Précis sommaire du passé et de l'avenir de la boulangerie parisienne par un correspondant de l'Indépendant Belge. Paris: n.p., 1862.

Recueil des principales lois relatives au commerce des grains avec les arrêts, arrêtés, et remontrances du Parlement sur cet objet et le procès-verbal de l'assemblée générale de la police. Paris: n.p., 1769.

"Réflexions sur le commerce du pain." *Journal économique,* July 1761.

Révolutions de Paris. 10 vols. (1789–93).

Statistique de l'industrie à Paris résultante de l'enquête faite par la Chambre de commerce pour les années 1747–48. Paris: Imprimerie Impériale, 1851.

"Sur les commerces de la viande et du pain." *Précis analytique des travaux de l'Académie de Rouen.* Vol. 4. Rouen: n.p., 1777.

Tableau de Paris pour l'année 1759. Paris: n.p., 1759.

Tableau des boulangers de Paris pour l'an 1811. Paris: n.p., 1811.

Tableau universel et raisonné de la Ville de Paris. Paris: J. P. Costard, n.d. [ca. 1757].

Tablettes royales de renommée, ou Almanach général d'indication des négocians, artistes célèbres et fabricans. Paris: n.p., 1772.

Versaille et Paris (1789).

La Vie illustrée (1906).

Author Known

Alhoy, Maurice, and Louis Lurine. *Les Prisons de Paris: Histoire, types, moeurs, mystères.* Paris: G. Havard, 1846.

Antonini, Abbé Annibal. *Mémorial de Paris et de ses environs.* Ed. Abbé G. Raynal. 2 vols. Paris: Bauche fils, 1749.

Argenson, René-Louis de Voyer, marquis d'. *Journal et mémoires du marquis d'Argenson.* 9 vols. E.-J.-B. Rathery, ed. Paris: Veuve J. Renouard, 1859–67.

Aubert, F. "Observations sur la liberté du commerce des grains." *Journal du commerce,* September 1759.

Auffair, M. J. "Le Commerce abandonné à lui-même, ou Suppression des compagnies, sociétés et maîtrises." *Journal de commerce,* March 1761.

Bachaumont, Louis Petit de. *Mémoires secrets pour servir à l'histoire de la République des lettres en France depuis 1762 jusqu'à nos jours, ou Journal d'un observateur.* 30 vols. London: John Adamson, 1780–86.

Baden, Carl Friedrich von. *Carl Friedrich von Baden Brieflicher Verkehr mit Mirabeau und Dupont.* Ed. Carl Knies. Heidelberg: n.p., 1892.

Bailly, Jean-Sylvain. *Mémoires.* Ed. Saint-Albin Berville and Jean-François Barrière. 2 vols. Paris: Baudouin, 1822.

Barbier, Edmond-Jean-François. *Chronique de la régence et du règne de Louis XV (1718–1763), ou Journal de Barbier.* 8 vols. Paris: Charpentier, 1857–66.

Baudeau, Abbé Nicolas. *Avis au peuple sur son premier besoin, ou Petits Traités économiques, par l'auteur des "Ephémérides du citoyen."* Amsterdam and Paris: Hochereau Jeune, 1768.

———. *Avis aux honnêtes gens qui veulent bien faire.* Amsterdam and Paris: Desaint et al., 1768.

———. *Résultats de la liberté et de l'immunité du commerce des grains, de la farine et du pain.* Amsterdam and Paris: Desaint et al., 1768.

Beardé de l'Abbaye. *Recherches sur les moyens de supprimer les impôts.* Amsterdam: M. M. Rey, 1770.

Béguillet, Edme. *Description historique de Paris et de ses plus beaux monuments.* 3 vols. Paris: n.p., 1779–81.

———. *Discours sur la mouture économique.* Paris: Panckoucke, 1775.

———. *Mémoire sur les avantages de la mouture économique et du commerce des farines.* Dijon: L. N. Frantin, 1769.

———. *Traité de la connoissance générale des grains et de la mouture par économie.* Dijon: L. N. Frantin, 1778.

———. *Traité des subsistances et des grains qui servent à la nourriture de l'homme.* Paris: Prault Fils, 1780.

———, [with César Bucquet?]. *Manuel du meunier et du charpentier de moulins, ou Abrégé classique du traité de la mouture par économie.* Paris: Panckoucke, 1775.

Bielfeld, Jacob Friedrich von. *Institutions politiques.* 2 vols. The Hague: P. Gosse, Jr., 1760–62.

Bigot de Sainte-Croix. *Essai sur la liberté du commerce et de l'industrie.* Paris: Lacombe, 1775.

Boileau, Etienne. *Les Métiers et corporations de la ville de Paris: Le Livre des métiers d'Etienne Boileau.* Ed. René de Lespinasse and F. Bonnordot. Paris: Imprimerie Nationale, 1879.

Boislisle, Arthur-Michel de, ed. *Lettres de M. de Marville, lieutenant général de police, au ministre Maurepas (1742–1747).* 3 vols. Paris: H. Champion, 1896–1905.

———. *Mémoires des intendants sur l'état des généralités dressés pour l'instruction du duc de Bourgogne.* Paris: Imprimerie Nationale, 1881.

Boland, Antoine. *Traité pratique de la boulangerie.* Paris: E. Lacroix, 1860.

Bourdon-Desplanches, L.-J. *Lettre à l'auteur des Observations sur le commerce des grains.* Amsterdam: n.p., 1775.

———. *Projet nouveau sur la manière de faire utilement en France le commerce des grains.* Brussels and Paris: Veuve Esprit, 1785.

Briatte, Jean-Baptiste. *Offrande à l'humanité.* Amsterdam and Paris: Noyon, 1780.

Bucquet, César. *Manuel du meunier et du constructeur de moulins à eau et à grains, par M. Bucquet.* New ed. Paris: Onfroy, 1790.

——— (under pseudonym of Michel Morin). *Observations intéressantes et amusantes du Sieur César Bucquet, ancien meunier de l'Hôpital général, à MM. Parmentier et Cadet.* Revised by E. Béguillet. Paris: Chez les Marchands de Nouveautés, 1783.

————. *Traité pratique de la conservation des grains, des farines et des étuves domestiques.* Paris: Onfroy, 1783.

Bunel, Henri. "Rapport présenté par M. Bunel sur le carrefour Pirouette." *Commission municipale du Vieux Paris* (procès-verbaux, 1899), no. 10 (7 December 1899): 341–47.

Cadet de Gassicourt, Charles-Louis. "Considérations statistiques sur la santé des ouvriers." *Mémoires de la Société médicale d'émulation* 8 (1817).

Cadet de Vaux, Antoine-A. *Avis sur les blés germés par le comité de l'École gratuite de Boulangerie.* Paris: P. D. Pierres, 1782.

————. *Discours prononcés à l'ouverture de l'Ecole gratuite de Boulangerie (8 juin 1780).* Paris: P. D. Pierres, 1780.

————. *Moyens de prévenir le retour des disettes.* Paris: D. Colas, 1812.

————. *Traités divers d'économie rurale, alimentaire et domestique.* Paris: D. Colas, 1821.

Caraccioli, Louis. *Lettres récréatives et morales sur les moeurs du temps.* 2 vols. Paris: Noyon, 1767.

Chamousset, P. de. "Observations sur la liberté du commerce des grains." *Journal du commerce* 79 (September 1759).

Chauveau-Lagarde. *Mémoire ampliatif pour les boulangers.* Paris: Migneret, 1818.

————. *Mémoire pour les boulangers de Paris.* Paris: n.p., 1817.

Chevalier, Etienne. *Mémoire sur les moyens d'assurer la diminution du pain.* Paris: n.p., 1793.

Choiseul, Etienne-François, duc de. *Mémoires de M. le duc de Choiseul . . . écrits par lui-même et imprimés sous ses yeux, dans son cabinet à Chanteloup, en 1778.* Ed. Jean-Louis Soulavie. 2 vols. Chanteloup and Paris: Buisson, 1790.

Chomel, Abbé Noël. *Dictionnaire oeconomique.* 4th ed. 2 vols. Paris: Ganeau, 1740.

————. *Dictionnaire oeconomique.* Ed. M. de la Marre. 3 vols. Paris: Ganeau, 1767.

Clicquot de Blervache, Simon. *Considérations sur le commerce et en particulier sur les compagnies, sociétés et maîtrises.* Amsterdam: n.p., 1758.

———— [M. Delisle, pseud.]. *Mémoire sur les corps de métiers.* The Hague: n.p., 1758.

Colletet, François. *Le Tracas de Paris, ou la Seconde Partie de la ville de Paris, en vers burlesques.* Paris: n.p., 1650.

Condorcet, Marie Jean Antoine Nicolas Caritat, marquis de. Eulogy to Malouin. In *Histoire de l'Académie des sciences, 1778.* Paris: n.p., 1781.

————. *Le Monopole et le monopoleur* (n.d.). In vol. 14 of *Collection des principaux économistes,* ed. Eugène Daire and G. de Molinari. 1845–47. Reprint, Osnabrück: Otto Zeller, 1966.

Cotte, Louis. *Leçons élémentaires sur le choix et la conservation des grains, sur les opérations de la meunerie et de la boulangerie et sur la taxe du pain.* Paris: Frères Barbon, 1795.

Coyer, Abbé Gabriel-François. *Chinki, Histoire cochinchinoise.* London: n.p., 1768.

————. *Voyages d'Italie et de Hollande.* 2 vols. in 1. Paris: Veuve Duchesne, 1775.

Cretté de Palluel, François. "Mémoire sur les moyens d'économiser la mouture des grains et de diminuer le prix du pain." *Mémoires d'agriculture, d'économie rurale et domestique publiés par la Société royale d'agriculture,* fall 1788.

Delamare, Nicolas. *Traité de la police.* 4 vols. Paris: J.-F. Hérissant, 1705–38.

Denisart, Jean-Baptiste. *Collection de décisions nouvelles et de notions relatives à la jurisprudence actuelle.* 9th ed. 48 vols. Paris: Desaint, 1777.

Desaubiez, Vatar. *Le Bonheur public*. London: T. Hookham, 1780–82.

Des Essarts, Nicolas-Toussaint L. *Dictionnaire universel de police*. 8 vols. Paris: Moutard, 1786–90.

Diderot, Denis, et al. *Encyclopédie, ou Dictionnaire raisonné des sciences, des arts et des métiers, par une société de gens de lettres*. Ed. Diderot et Jean d'Alembert. 28 vols. Paris: Briasson, David L'Ainé, Le Breton, Durand, 1751–65.

Duchesne. *Code de la police, ou Analyse des règlements de police*. 4th ed., rev. Paris: Prault Père, 1767.

Dufrene. *La Misère des garçons boulangers de la ville et faubourgs de* Paris. Troyes: Veuve Garnier, 1715.

Duhamel du Monceau, Henri-Louis. *Traité de la conservation des grains et en particulier du froment*. Paris: H. L. Guérin and L. F. Delatour, 1753.

Du Pin, Richard. *Le Boulanger et la boulangère, ou le Pain au mois de novembre au plus tard ne vaudra que 8 sols les quatres livres*. Paris: Garnery et Volland, 1789.

Dupont de Nemours, Pierre Samuel. *De l'exportation et de l'importation des grains* (Paris, 1764). In *Collection des économistes et des réformateurs sociaux de la France*, ed. Edgard Depitre. Paris: P. Geuthner, 1911.

Dupré de Saint-Maur, Nicolas-François. *Essai sur les monnaies, ou Réflexions sur le rapport entre l'argent et les denrées*. Paris: J.-B. Coignard, 1746.

Dutillet, Jean. *Recueil des Roys de France*. Paris: I. Houze, 1602.

Essuile, Jean-François de Barandiéry-Montnayeur, comte d'. *Traité des communes, ou Observations sur leur origine et état actuel*. Paris: Colombier, 1779.

Estienne, Charles, and Jean Liébault. *L'Agriculture et maison rustique*. Lyon: C. Carteron and Cie., 1698.

Evelyn, John. *An Account of Bread: Panificium*. In the Houghton Collection for Improved Husbandry and Trade (1683).

Expilly, Abbé J.-J. *Dictionnaire géographique, historique et politique des Gaulles et de la France*. 6 vols. Paris: Desaint and Saillant, 1762–70.

———. *Tableau de la population de la France*. Paris: n.p., 1780.

Félix, C. D. *Saint-Eustache et la Chaumière, pot-pourri en deux parties à l'occasion de la Saint-Honoré en 1823*. Paris: published by author, 1823.

Ferrière, Claude-Joseph de. *Nouvelle Introduction à la pratique*. 2 vols. Brussels: n.p., 1739.

Ferrières, Charles-Elie de. *Mémoires du marquis de Ferrières*. Ed. S.-A. Berville and J.-F. Barrière. 3 vols. Paris: Baudouin Fils, 1822.

Forbonnais, Francis Veron de. *Recherches et considérations sur les finances de France, depuis l'année 1595 jusqu'à l'année 1721*. 6 vols. Liège: n.p., 1758.

Fréminville, Edmé de la Poix de. *Dictionnaire ou Traité de la police générale des villes, bourgs, paroisses et seigneuries de la campagne*. Paris: Gissey, 1758.

Funck-Brentano, F., ed. "'Gazette de police,' 24 septembre 1725." *Revue rétrospective* (December 1885–July 1886): 145–68.

Galiani, Abbé Ferdinando. *Dialogues entre M. Marquis de Roquemaure et M. le Chevalier Zanobi*. Ed. Philip Koch. Frankfurt: Klostermann, 1968.

———. *Dialogues sur le commerce des bleds, 1770*. Ed. Fausto Nicolini. Milan: Riccardi, n.d.

———. *Lettres de l'abbé Galiani à Madame d'Epinay.* Ed. Eugène Asse. 2 vols. Paris: G. Charpentier, 1881–1903.

Galippe, Victor, and G. Barré. *Le Pain: Alimentation minéralisateur—physiologie—composition—hygiène —thérapeutique.* Paris: Gauthier-Villars et fils, 1895.

———. *Le Pain: Technologie, pains divers, altérations.* Paris: Gauthier-Villars et fils, 1895.

Gazier, A., ed. "La Police de Paris en 1770, mémoire inédit, composé par ordre de Marie-Thérèse." *Mémoires de la Société de l'histoire de Paris et de l'Ile-de-France* 5 (1878): 1–131.

Geoffroy. *Histoire de l'Académie royale des sciences: Mémoires, 1732.* Paris: n.p., 1735.

Grimm, Friedrich, et al. *Correspondance littéraire, philosophique et critique par Grimm, Diderot, Raynal, Meister, etc.* Ed. Maurice Tourneux. 16 vols. Paris: Garnier Frères, 1877–82.

Grimod de la Reynière, A. B. L. *Almanach des gourmands, servant de guide dans les moyens de faire excellente chère; par un vieil amateur.* 8 vols. Paris: Chez Maradan, 1803–12.

Gosset, François. *Inventaire des arrêts du conseil du roi (January–February and March–April 1760).* 2 vols. Paris: Le Quotidien juridique, 1938.

Grosley, Pierre-Jean. *Oeuvres inédits.* Paris: C. F. Patris, 1812–13.

Guerineau de Saint-Péravi, Jean-Nicolas-Marcellin. *Principes du commerce opposé au trafic.* Paris: n.p., 1787.

Guillaute, M. *Mémoire sur la réformation de la police de France soumis au roi en 1749.* Paris: Hermann, 1974.

Henouville, Pierre Bernard d'. *Le Guide des comptables, ou l'Art de rédiger soi-même toutes sortes de comptes.* Paris: Jacques Estienne, 1709.

Herbert, Claude-Jacques. *Essai sur la police générale des grains, sur leurs prix et sur les effets de l'agriculture* (1775). In *Collection des économistes et des réformateurs sociaux de la France,* ed. Edgard Depitre. Paris: P. Geuthner, 1910.

Herbin de Halle. *Statistique générale et particulière de la France et de ses colonies.* Ed. P. Etienne. 7 vols. Paris: F. Buisson, 1803.

Houghton, John. *Husbandry and trade improv'd, being a collection of many valuable materials relating to corn, cattle, coals, hops, wool . . . with many other useful particulars, communicated by several eminent members of the Royal Society to the collector John Houghton.* London: Wooman and Lyon, 1827–28.

Hurtaut, Pierre, and P. N. Magny. *Dictionnaire historique de la ville de Paris et de ses environs.* 4 vols. Paris: Moutard, 1779.

Isambert, François-André, ed. *Recueil général des anciennes lois françaises.* 29 vols. Paris: Belin—Le Prieur, 1821–33.

Jackson, H. *An Essay on Bread.* London: L. Wilkie, 1758.

Jacquin, Abbé Armand-Pierre. *De la Santé, ouvrage utile à tout le monde.* Paris: G. Desprez, 1771.

Lacombe d'Avignon, François. *Le Mitron de Vaugirard, dialogues sur le bled, la farine et le pain, avec un petit traité de la boulangerie.* New ed. Amsterdam: n.p., 1777.

Lacroix, Sigismond, ed. *Actes de la Commune de Paris pendant la Révolution.* 7 vols. Paris: Cerf, 1894–98.

Lavoisier, Antoine-Laurent. *Résultats extraits d'un ouvrage intitulé: De la Richesse territoriale du royaume de France.* Paris: Imprimerie Nationale, 1791.

Lebois, René François. *Lettre des boulangers de Paris au peuple.* Paris: Grange, 1789.

Le Camus. "Mémoire sur le bled." *Journal économique,* November 1753.

——. "Suite du mémoire sur le pain." *Journal économique,* December 1753.

Legrand d'Aussy, Pierre J.-B. *Histoire de la vie privée des François* (1783). New ed. Ed. J.-B.-B. de Roquefort. 3 vols. Paris: Laurent-Beaupré, 1815.

Le Masson, Jean-Baptiste. *Le Calendrier des confréries de Paris.* Ed. Valentin Dufour. Paris: Willem, 1875.

Lenoir, Jean-Charles-Pierre. *Détail sur quelques établissemens de la ville de Paris demandé par sa majesté impériale la reine de Hongroi à M. Lenoir.* Paris: n.p., 1780.

Lespinasse, René de, ed. *Les Métiers et corporations de la Ville de Paris.* Vol. 1, *Ordonnances générales, métiers d'alimentation.* Paris: Imprimerie Nationale, 1886.

Lethinois, André. *Apologie du système de Colbert.* Amsterdam: Knapen et Delaguette, 1771.

Letrosne, Guillaume François. *De l'administration provinciale et de la réforme de l'impôt.* 2 vols. Basel: n.p., 1779.

Linguet, Simon-Nicolas-Henri. *Annales politiques, civiles et littéraires du XVIIIe siècle.* 19 vols. Brussels: published by author, 1777–92.

——. *Dissertation sur le bled et le pain.* Neufchâtel: n.p., 1779.

——. *Du pain et du bled.* London: n.p., 1774.

——. *Observations sur l'histoire de la Bastille.* London: T. Spilsbury, 1783.

——. *Réflexions des six corps de la ville de Paris sur la suppression des jurandes.* Paris: n.p., 1776.

——. *Réponse aux docteurs modernes, ou Apologie pour l'auteur de la "Théorie des loix" et des "Lettres sur cette théorie," avec la réfutation du système des philosophes économistes.* 2 vols. London and Paris: n.p., 1771.

——. *Théorie des loix civiles, ou Principes fondamentaux de la société.* 2 vols. London: n.p., 1767.

Lister, Martin. *A Journey to Paris in the Year 1698* (1699). Ed. Raymond P. Stearn. Urbana: University of Illinois Press, 1967.

Luchet, Jean-Pierre-Louis, marquis de. *Examen d'un livre qui a pour titre "Sur la législation et le commerce des bleds."* Paris: n.p., 1775.

Luynes, Charles Philippe d'Albert, duc de. *Mémoires du duc de Luynes sur la cour de Louis XV.* Ed. L. Dussieux and E. Soulié. 17 vols. Paris: Firmin Didot Frères, 1860–65.

Mackenzie. "Règles pour conserver la santé." *Journal économique,* November 1764.

Macquer, Philippe. *Dictionnaire portatif des arts et métiers, contenant l'histoire, la description, la police des fabriques et manufactures de France et des pays étrangers.* 2 vols. Paris: Lacombe, 1766.

——. *Dictionnaire raisonné universel des arts et métiers, contenant l'histoire, la description, la police des fabriques et manufactures de France et des pays étrangers.* New ed., rev. Ed. Pierre Jaubert. 4 vols. Paris: P.-F. Didot Jeune, 1773.

Madeleine, Philipon de la. *Vues patriotiques sur l'éducation du peuple.* Lyon: Bruyset-Ponthus, 1783.

Malouin, Paul-Jacques. *Description des arts du meunier, du vermicelier, et du boulanger, avec une histoire abrégée de la boulengerie et un dictionnaire de ces arts.* Paris: n.p., 1761.

――――. *Description des arts du meunier, du vermicellier, et du boulanger, avec une histoire abrégée de la boulangerie et un dictionnaire de ces arts.* Paris: n.p., 1767.

――――. *Description des arts du meunier, du vermicellier, et du boulanger.* New ed. Ed. Jean-Elie Bertrand. Neufchâtel: La Société Typographique, 1771.

――――. *Description et détails des arts du meunier, du vermicellier, et du boulanger.* Paris: Saillant et Noyon, 1779.

Marmontel, Jean François. *Mémoires de Marmontel.* Ed. Maurice Tourneux. 3 vols. Paris: Librairie des Bibliophiles, 1891.

Ménétra, Jacques-Louis. *Journal of My Life.* Ed. Daniel Roche. Trans. Arthur Goldhammer. New York: Columbia University Press, 1986.

Mercier, Louis-Sébastien. *Tableau de Paris.* New ed. 12 vols. Amsterdam: n.p., 1782–88.

――――. *L'An deux mille quatre-cent quarante, rêve s'il en fût jamais.* New ed. 2 vols. London: n.p., 1785.

Mildmay, William. *The Police of France, or An Account of the laws and regulations established in the kingdom for the preservation of peace and the preventing of robberies, to which is added a particular description of the police and government of the city of Paris.* London: E. Owen and T. Harrison, 1763.

Miromesnil, Armand Thomas Hue de. *Correspondance politique et administrative de Miromesnil, premier président du Parlement de Normandie.* Ed. P. LeVerdier. 5 vols. Paris and Rouen: A. Picard, 1899–1903.

M. J. D. L. *Le Pain à bon marché, ou le Monopole terrassé et le peuple vengé.* Paris: n.p., n.d. (ca. 1790).

Moheau. *Recherches et considérations sur la population de la France* (Paris, 1778). In *Collection des économistes et des réformateurs sociaux de la France,* ed. René Gonnard. Paris: P. Geuthner, 1912.

Montchrétien, Antoine de. *Traicté de l'oeconomie politique.* Ed. F. Funck-Brentano. Paris: Plon, Nourrit et Cie., 1889.

Moreau, Jacob-Nicolas. *Mes Souvenirs.* Ed. Camille Hermelin. 2 vols. Paris: E. Plon, Nourrit et Compagnie, 1898–1901.

Morellet, Abbé André. *Mémoires inédits sur le XVIIIe siècle et sur la Révolution.* Paris: Ladvocat, 1822.

――――. *Théorie du paradoxe.* Amsterdam: n.p., 1775.

――――. *Lettres de l'abbé Morellet à lord Shelburne.* Ed. Edmond Fitzmaurice. Paris: E. Plon, Nourrit et Compagnie, 1898.

Muret, Jean-Louis. *Mémoire sur la mouture des grains, et sur divers objets relatifs.* Berne: La Société Typographique, 1776.

Narbonne, Pierre. *Journal des règnes de Louis XIV et Louis XV de l'année 1701 à l'année 1744 par Pierre Narbonne, premier commissaire de police de la ville de Versailles.* Ed. J.-A. LeRoi. Versailles: Bernard, 1866.

Necker, Jacques. *Sur la législation et le commerce des grains.* 2 vols. Paris: Pissot, 1775.

Parmentier, Antoine-Augustin. *Avis aux bonnes ménagères des villes et des campagnes sur la meilleure manière de faire leur pain.* Paris: Imprimerie Royale, 1772.

――――. "Dissertation physique, chymique et économique sur la nature et la salubrité de l'eau de la Seine." *Journal de physique,* February 1775, 161–86.

———. *Dissertation sur la nature des eaux de la Seine*. Paris: Buisson, 1787.

———. *Eloge de M. Model*. Paris: Monory, 1775.

———. *Examen chymique des pommes de terre dans lequel on traite des parties constituantes du Bled*. Paris: Didot Jeune, 1773.

———. *Manière de faire le pain de pommes de terre sans mélange de farine*. Paris: Imprimerie Royale, 1779.

———. *Mémoire sur les avantages que le province de Languedoc peut retirer de ses grains, avec le mémoire sur la nouvelle manière de construire les moulins à farine par M. Dransy*. Paris: L'Imprimerie des Etats de Languedoc, 1787.

———. *Le Parfait Boulanger, ou Traité complet sur la fabrication et le commerce du pain*. Paris: Imprimerie Royale, 1778.

———. *Recherches sur les végétaux nourrissans, qui, dans les temps de disette, peuvent remplacer les aliments ordinaires, avec des nouvelles observations sur la culture des pommes de terre*. Paris: Imprimerie Royale, 1781.

Pascal, Blaise. *Pensees*. Paris: Editions Garnier Frères, 1961.

Patissier, Philibert. *Traité des maladies des artisans et de celles qui résultent des diverses professions, d'après Ramazzini*. Paris: J. B. Baillière, 1827.

Paucton, A.-J.-P. *Métrologie, ou Traité des mesures, des poids et des monnaies des anciens peuples et des modernes*. Paris: Desaint, 1780.

Perrache, Claude. *Nouvelle Méthode de tenir les livres marchands d'une manière très sûre et très facile*. Brussels: n.p., 1748.

Pidansat de Mairobert, Mathieu François. *Journal historique de la révolution opérée dans la constitution de la monarchie française, par M. de Maupeou, chancelier de France*. 7 vols. London: n.p., 1774–76.

Pigeonneau, Henri, and Alfred de Foville, eds. *L'Administration de l'agriculture au contrôle général des finances: Procès-verbaux et rapports*. Paris: Guillaumin et Cie., 1882.

Pluche, Abbé Antoine. *Le Spectacle de la nature, ou Entretiens sur les particularités de l'histoire naturelle*. 8 vols. Paris: Veuve Estienne, 1732–50.

Poncelet, Polycarpe. *Histoire naturelle du froment*. Paris: Imprimerie de G. Desprez, 1779.

———. *Mémoire sur la farine*. Paris: Pissot, 1776.

Pradel, Abraham du. *Le Livre commode des adresses de Paris pour 1692*. 2 vols. Ed. Edouard Fournier. Paris: P. Daffis, 1878.

Quesnay, François. "Fermier." In Diderot et al., *Encyclopédie*, vol. 6.

Raunié, Emile, ed. *Clairambault-Maurepas: Chansonnier historique du XVIIIe siècle*. 10 vols. Paris: A. Quentin, 1879–84.

Restif de la Bretonne, Nicolas Edmé. *Les Nuits de Paris*. In *L'Oeuvre de Restif de la Bretonne*, ed. Henri Bachelin. Paris: Editions du Trianon, 1930.

———. *La Vie de mon père*. Ed. Gilbert Rouger. Paris: Garnier, 1970.

Retz, Jean François Paul de Gondi, cardinal de. *Mémoires du Cardinal de Retz adressés à Madame du Caumartin* (1648). Ed. Aimé Louis Champollion-Figeac. 4 vols. Paris: Charpentier, 1912–13.

Richelet, Pierre. *Dictionnaire de la langue françoise ancienne et moderne*. Rev. ed. Ed. Abbé Goujet. 3 vols. Lyon: Duplain Frères, 1758.

Rivarol, Antoine de. *Mémoires de Rivarol*. Ed. S.-A. Berville. Paris: Baudouin, 1824.

Roubaud, Abbé Pierre-Joseph-André. *Récréations économiques, ou Lettres de l'auteur des "Représentations aux Magistrats" à M. le Chevalier Zanobi, principal interlocuteur des "Dialogues sur le commerce des blés."* Amsterdam and Paris: Delalain, 1770.

————. *Représentations aux magistrats contenant l'exposition raisonnée des faits relatifs à la liberté du commerce des grains*. Paris: n.p., 1769.

Roze de Chantoiseau. *Tablettes royales de rénommée, ou Almanach général d'indication*. Paris: Veuve Duchesne, 1791.

Rozier, Abbé François. *Cours complet d'agriculture théorique, pratique, économique et de médecine rurale et vétérinaire*. 9 vols. Paris: Hôtel Serpente, 1781–96.

Rubigny de Berteval. *Avis au peuple sur les premiers besoins*. Paris, ca. 1793.

Rutledge, Jean-Jacques. *Mémoire pour la communauté des maîtres boulangers de la ville et des faubourgs de Paris, présenté au roi le 19 février 1789*. Paris: Baudouin, 1789.

————. *Second Mémoire pour les maîtres boulangers, lu au bureau des subsistances de l'Assemblée nationale, par le chevalier Rutledge*. Paris: Baudouin, 1789.

Sage, Baltasar-George. *Analyse des bleds*. Paris: Imprimerie Nationale, 1776.

Saint-Priest, François-Emmanuel-Guignard, comte de. *Mémoires, la Révolution et l'émigration*. Ed. Baron de Barante. 2 vols. Paris: Calmann-Lévy, 1929.

Sauval, Henri. *Histoire et recherches des antiquités de la ville de Paris*. 3 vols. Paris: C. Moette, 1724.

Savary, Jacques. *Le Parfait Négociant*. 2 vols. 8th ed. Ed. P.-L. Savary. Paris: Claude Robustel, 1721.

Savary des Bruslons, Jacques. *Dictionnaire portatif de commerce*. 7 vols. Copenhagen: C. et A. Philibert, 1761–62.

————. *Dictionnaire portatif de commerce*. Paris: Bastien, 1777.

————. *Dictionnaire universel de commerce*. 3 vols. Geneva: Des Héritiers Cramer et Frères Philibert, 1742.

Séguier, Antoine-Louis. *Discours sur la nécessité du rétablissement des maîtrises et des corporations*. Speech delivered on 12 March 1776. Paris: Imprimerie de Fain, 1815.

Thicknesse, Phillip. *Useful Hints to Those Who Make the Tour of France*. London: R. Davis et al., 1767.

Tillet, Mathieu. "Expériences et observations sur le poids du pain au sortir du four." *Journal de physique*, February 1782, 90–103.

————. "Projet d'un tarif." *Histoire de l'Académie royale des sciences: Avec les mémoires de mathématique et de physique, 1781*. Paris: n.p., 1784. 107–68.

Tolozan, Jean-François de. *Mémoire sur le commerce de la France et de ses colonies*. Paris: Moutard, 1789.

Turgot, Anne-Robert-Jacques. *Oeuvres de Turgot et documents le concernant, avec biographie et notes*. Ed. Gustave Schelle. 5 vols. Paris: Félix Alcan, 1913–23.

Voltaire, François-Marie Arouet de. *Oeuvres complètes*. Ed. L. Moland. 69 vols. Paris: Garnier Frères, 1877–85.

————. *Voltaire's Correspondence*. Ed. Theodore Besterman. 107 vols. Geneva: Institut et Musée Voltaire, 1953–65.

Young, Arthur. *Travels in France during the Years 1787, 1788, and 1789*. Ed. Constantia Maxwell. Cambridge: University Press, 1929.

Secondary Sources

Addis, William, and Thomas Arnold. *A Catholic Dictionary*. London: Routledge and Kegan Paul, 1951.

Aldridge, Alfred O. *Franklin and His French Contemporaries*. New York: New York University Press, 1957.

Almers, Henri d.' "Le Pain à Paris." *Magasin pittoresque*, 65th year, 2d ser., 15 (1897): 315.

Amos, Percy A. *Processes of Flour Manufacture*. London: Longmans, Green, 1912.

Anne, Sylvie. *Victorine, ou le Pain d'une vie*. Paris: Presses de la Renaissance, 1985.

Apfelbaum, Marian, and Raymond Lepoutre. *Mangeurs inégaux*. Paris: Stock, 1978.

Arpin, Marcel. *Historique de la meunerie et de la boulangerie, depuis les temps préhistoriques jusqu'à l'année 1914*. 2 vols. Paris: Le Chancelier, 1948.

Ashley, William J. *The Bread of Our Forefathers: An Inquiry in Economic History*. Oxford: Clarendon Press, 1928.

Avenel, Georges d'. "Le Mécanisme de la vie moderne." *Revue des deux mondes* 147 (15 June 1895): 806–36.

———. "Paysans et ouvriers depuis sept siècles: Les Frais de nourriture aux temps modernes." *Revue des deux mondes* 168 (15 July 1898): 424–51.

Aymard, Maurice. "Pour l'histoire de l'alimentation: Quelques Remarques de méthode." *Annales: E.S.C.* 30th year (March–June 1975): 431–44.

———. "Toward the History of Nutrition: Some Methodological Remarks." In *Food and Drink in History: Selections from the Annales*. Ed. Robert Forster and Orest Ranum. Baltimore: Johns Hopkins University Press, 1979.

Babeau, Albert. *Paris en 1789*. Paris: Didot, 1892.

Bachelard, Gaston. *La Terre et les rêveries de la volonté*. Paris: J. Corti, 1948.

Bailey, Adrian. *Blessings of Bread*. New York: Paddington Press, 1975.

Barbaret, Joseph. *Le Travail en France: Monographies professionnelles*. 12 vols. Paris: Berger-Levrault et Cie., 1886–90.

Baulant, Micheline. "Niveaux de vie des paysans autour de Meaux en 1700 et en 1750." *Annales: E.S.C.* 30th year (March–June 1975): 505–18.

Bayard, Jean-Pierre. *Le Compagnonnage en France*. Paris: Payot, 1977.

Beals, Ralph L. *The Peasant Marketing System of Oaxaca, Mexico*. Berkeley: University of California Press, 1975.

Béaur, Gérard. "L'Histoire et l'économie rurale à l'époque moderne, ou les Désarrois du quantitativisme: Bilan critique." *Histoire et sociétés rurales* 1 (1994): 67–97.

Beloff, Max. *Public Order and Popular Disturbances, 1600–1714*. London: Oxford University Press, 1938.

Belshaw, Cyril. *Traditional Exchange and Modern Markets*. Englewood Cliffs, N.J.: Prentice-Hall, 1965.

Benedict, Burton. "Capital Savings and Credit among Mauritian Indians." In *Capital, Savings, and Credit in Peasant Societies*. Ed. Raymond Firth and B. S. Yang. Chicago: Aldine, 1964.

Benoist, Philippe, and Jean-Julia de Fontenelle. *Nouveau manuel complet du boulanger, du négociant en grains, du meunier et du constructeur de moulins*. Paris: Roret, 1856.

Benzacar, Joseph. *Le Pain à Bordeaux*. Bordeaux: Gonouilhou, 1905.

Bergeron, Louis. "Approvisionnement et consommation à Paris sous le premier empire." *Mémoires de la Féderation des sóciétés historiques et archéologiques de Paris et de l'Ile-de-France* 14 (1963): 197–232.

Bernard, Jean. *Le Compagnonnage. Rencontre de la jeunesse et de la tradition.* Paris: Presses universitaires de France, 1972.

Bernard, Maurice. *La Municipalité de Brest de 1750 à 1790.* Paris: Librairie Ancienne Honoré Champion, 1915.

Binet, Pierre. *La Réglementation du marché du blé en France au XVIIIe siècle et à l'époque contemporaine.* Paris: Librairie Sociale et Economique, 1939.

Biot, Jean-Baptiste. *Lettres écrites de la campagne sur le commerce des grains et l'approvisionnement de Paris.* Paris: Bachelier, 1835.

Birembaut, Arthur. "L'Ecole gratuite de boulangerie." In *Enseignement et diffusion des sciences en France au XVIIIe siècle.* Ed. René Taton. Paris: Hermann, 1964.

———. "Le Problème de la panification à Paris en 1790." *Annales historiques de la Révolution française* 37 (July–September 1965): 361–67.

Blazy, J.-P. *Gonesse, la terre et les hommes: Des origines à la Révolution.* Meaux: Imprimerie André-Pouyé, 1982.

Bloch, Camille. *L'Assistance et l'etat en France à la veille de la Révolution.* Paris: Picard et fils, 1908.

———, ed. *Le Commerce des grains dans la généralité d'Orléans (1768) d'après la correspondance inédite de l'intendant Cypierre.* Orléans: H. Herluison, 1898.

Blond, Georges, and Germaine Blond. *Histoire pittoresque de notre alimentation.* 2 vols. Paris: A. Fayard, 1960.

Boislisle, Arthur-Michel de. "Le Grand Hiver et la disette de 1709." *Revue des questions historiques* 73 (1903): 442–509; 74 (1903): 486–542.

Bondois, Paul. "La Misère sous Louis XIV: La Disette de 1662." *Revue d'histoire économique et sociale* 12 (1924): 53–118.

Bonneff, Léon, and Maurice Bonneff. *La Classe ouvrière, les boulangers . . .* Paris: "Guerre sociale," 1912.

Bortin, Meg. "Give Us Our Daily Bread." *Paris Métro,* 26 October 1977.

Bourde, André J. *Agronomie et agronomes en France au XVIIIe siècle.* 3 vols. Paris: S.E.V.P.E.N., 1967.

Bourdieu, Pierre. *La Distinction: Critique sociale du jugement.* Paris: Editions de Minuit, 1979.

———. *Outline of a Theory of Practice.* Trans. Richard Nice. Cambridge and New York: Cambridge University Press, 1977.

———. *Le Sens pratique.* Paris: Editions de Minuit, 1980.

Bouteloup, Maurice. *Le Travail de nuit dans la boulangerie.* Paris: Recueil Sirey, 1909.

Bouton, Cynthia A. *The Flour War: Gender, Class, and Community in Late Ancien Régime French Society.* University Park, Pa.: Pennsylvania State University Press, 1993.

Bouyer, Christian. *Folklore du boulanger.* Paris: Maisonneuve et Larose, 1984.

Braudel, Fernand. *Civilisation matérielle et capitalisme, XVe–XVIIIe siècles.* Vol. 1. Paris: Armand Colin, 1967.

Brian, Eric. *La Mesure de l'Etat: Administrateurs et géomètres au XVIIIe siècle.* Paris: Plon, 1994.

Bricourt, Joseph. *Dictionnaire pratique des connaissances religieuses.* 5 vols. Paris: Letou-zey et Ané, 1925–28.

Brinton, J. W. *Wheat and Politics.* Minneapolis, Minn.: Publication Office, 1931.

Bromley, R. J., and Richard Symanski. "Marketplace Trade in Latin America." *Latin American Research Review* 9 (fall 1974): 3–38.

Brunot, Ferdinand. *Histoire de la langue française des origines à 1900.* 13 vols. Paris: Armand Colin, 1905.

Bruyère, Jean. *Histoire littéraire des gens de métier de France.* Paris: Jouve, 1932.

Buré, Jean, ed. *Le Pain.* Actes du Colloque du CNERMA, Paris, November 1977. Paris: Editions du Centre National de la Recherche Scientifique, 1979.

Burnett, John. *Plenty and Want: A Social History of Diet in England.* London: Nelson, 1966.

Cacérès, Benigno. *Si le pain m'était conté.* Paris: La Découverte, 1987.

Cahen, Léon. "A propos du livre d'Afanassiev: L'Approvisionnement de Paris au début du XVIIIe siècle." *Bulletin de la Société d'histoire moderne* 22d year (5 March 1922): 162–72.

———. "La Population parisienne au milieu du XVIIIe siècle." *Revue de Paris* 5 (Sep-tember–October 1919): 146–70.

Cahier, Charles. *Caractéristiques des saints dans l'art populaire.* 2 vols. Paris: Poussielgue Frères, 1867.

Calvel, Raymond. "L'Autre Bout de la chaîne: L'evolution de la qualité du pain." *Actualités agricoles,* 30 April and 7 May 1976.

———. *Le Goût du Pain: Comment le préserver, comment le retrouver.* Paris: Jérôme Villette, 1990.

———. *Le Pain et la panification.* Paris: Presses Universitaires de France, 1964.

Camporesi, Piero. *Bread of Dreams.* Trans. David Gentilcore. Chicago: University of Chicago Press, 1989.

———. *The Magic Harvest: Food, Folklore, and Society.* Trans. J. K. Hall. Cambridge: Polity Press, 1993.

Carlier, Alfred. *Histoire du pain.* Cannes: n.p., 1938.

Carny, Lucien, ed. *Les Compagnons en France et en Europe.* 4 vols. Eyrein: R. Garry, 1973–75.

Caron, Pierre. *Commission des subsistances de l'an II: Procès-verbaux et actes.* Paris: E. Leroux, 1925.

Centre national de coordination des études et recherches sur la nutrition et l'alimenta-tion. *Le Qualité du pain, novembre 1954–août 1960.* Paris: n.p., 1962.

Chassin, Charles Louis, ed. *Les Elections et les cahiers de Paris en 1789, Vol. 2: Les Assemblées primaires et les cahiers primitifs.* Paris: Jouast et Sigaux, 1888–89.

Cherrière. *La Lutte contre l'incendie dans les halles, les marchés, et les foires de Paris sous l'ancien régime.* In *Mémoires et documents pour servir à l'histoire du commerce et de l'industrie en France.* Ed. Julien Hayem. 3d ser. Paris: Hachette, 1913.

Chevalier, Louis. *Classes laborieuses et classes dangereuses à Paris pendant la première moitié du XIXe siècle.* Paris: Plon, 1958.

Chevet, Jean-Michel. "Les Crises démographiques en France à la fin du XVIIe siècle et au XVIIIe siècle: Un Essai de mesure." *Histoire et mesure* 8 (1993): 136–44.

———. "Le Marquisat d'Ormesson: Essai d'analyse économique." Thèse de troisième cycle, Ecole des hautes études en sciences sociales, Paris, 1983.

———. "Production et productivité: Un Modèle de développement économique des campagnes de la région parisienne aux XVIIIe et XIXe siècles." *Histoire et mesure* 9 (1994): 101–45.

———, and Alain Guéry. "Consommation et approvisionnement de Paris en céréales de 1725 à 1733." In *Les Techniques de conservation des grains à long terme: Leur Rôle dans la dynamique des systèmes des cultures et des sociétés*, ed. Marceau Gast, François Sigaut, and Corinne Beutler. Paris: Editions du Centre National de la Recherche Scientifique, 1985.

Clark, Colin, and Margaret Haswell. *The Economics of Subsistence Agriculture.* 3d ed. New York: St. Martin's Press, 1967.

Clément, Pierre. "Successeurs de Colbert." *Revue des deux mondes*, 33d year, 2d period, 46 (15 August 1863): 916–45.

Cobb, Richard. *Les Armées révolutionnaires: Instrument de la terreur dans les départements.* 2 vols. Paris: Mouton, 1961–63.

Cochut, André. "Le Pain à Paris." *Revue des deux mondes*, 33d year, 2nd period, 46 (15 August 1863): 964–95; 47 (15 September 1863): 400–425.

Coffe, Jean-Pierre. *Au secours le goût.* Paris: Belfond–Le Pré aux Clercs, 1992.

Collins, E. J. T. "Why Wheat? Choice of Food Grains in Europe in the Nineteenth and Twentieth Centuries." *Journal of European Economic History* 22 (spring 1993): 7–38.

Coornaert, Emile. *Les Corporations en France avant 1789.* Paris: Gallimard, 1941.

Coquelin, Charles, and Gilbert-Urbain Guillaumin, eds. *Dictionnaire de l'économie politique.* 2 vols. Paris: Guillaumin, 1852–53.

Crossick, Geoffroy, and Heinz-Gerhardt Haupt. *Shopkeepers and Master Artisans in Nineteenth-Century Europe.* New York: Methuen, 1984.

Curmond, Henri. *Le Commerce des grains et l'école physiocratique.* Paris: A. Rousseau, 1900.

Darnton, Robert, ed. "Le Lieutenant de police J.-P. Lenoir, la guerre des farines et l'approvisionnement de Paris à la veille de la Révolution." *Revue d'histoire moderne et contemporaine* 16 (October–December 1969): 611–24.

Daumard, Adeline, and François Furet. "Méthodes de l'histoire sociale: Les Archives notariales et la mécanographie." *Annales: E.S.C.* 14th year (October–December 1959): 676–93.

———. *Structures et relations sociales à Paris au milieu du XVIIIe siècle. Cahiers des Annales*, no. 18. Paris: Armand Colin, 1961.

Davet, Hélène. *La Population de l'axe nord de Paris, 1749–74.* Thèse de troisième cycle under the direction of Pierre Goubert, Université de Paris I, n.d.

Davis, William G. *Social Relations in a Philippine Market: Self-Interest and Subjectivity.* Berkeley: University of California Press, 1973.

Deane, Phyllis, and W. A. Cole. *British Economic Growth, 1688–1959: Trends and Structure.* Cambridge: Cambridge University Press, 1967.

de Certeau, Michel. *Arts de faire.* Vol. 1 of *L'Invention du quotidien.* Ed. de Certeau. Paris: Union Générale d'Editions, 1980.

———. *La Culture au pluriel*. Paris: Union Générale d'Editions, 1974.

———. "Pratiques quotidiennes." In *Les Cultures populaires: Permanence et émergences des cultures minoritaires locales, ethniques, sociales et religieuses*. Ed. G. Poujol and R. Labourie. Toulouse: Privat, 1978.

Delumeau, Jean. *La Peur en occident (14e–18e siècles)*. Paris: Fayard, 1978.

Des Cilleuls, Alfred. *Histoire et régime de la grande industrie en France au XVIIe et XVIIIe siècles*. Paris: V. Giard and E. Brière, 1898.

———. *Le Socialisme municipal à travers les siècles*. Paris: A. Picard et fils, 1905.

Desmaze, Charles. *Les Métiers de Paris d'après les ordonnances du Châtelet*. Paris: E. Leroux, 1874.

Desnoyers, Pierre. "St. Firmin, patron des boulangers de l'Orléanais." *Mémoires de la Société archéologique et historique de l'Orléanais* 27 (1898): 155–68.

Desportes, Françoise. *Le Pain au moyen âge*. Paris: O. Orban, 1987.

Deyon, Pierre. *Amiens, capitale provinciale: Étude sur la société urbaine au XVIIe siècle*. Paris: Mouton, 1967.

Diefendorf, Barbara B. *Beneath the Cross: Catholics and Huguenots in Sixteenth-Century Paris*. New York: Oxford University Press, 1991.

Dolléans, Edouard. *Histoire du travail en France: mouvement ouvrier et législation sociale*. Paris: Domat, 1953–55.

Du Camp, Maxime. *Paris, ses organes, ses fonctions et sa vie dans la seconde moitié du XIXe siècle*. 6 vols. Paris: Hachette, 1869–75.

Dulaure, Jacques-Antoine. *Histoire physique, civile et morale de Paris*. 10 vols. Paris: Guillaume, 1823–25.

Dumas, François. *La Généralité de Tours au XVIIIe siècle*. Paris: Hachette, 1894.

Dupaigne, Bernard. *Le Pain*. Paris: La Courtille, 1979.

Dupâquier, Jacques, et al. *Mercuriales du pays de France et du Vexin français, 1640–1792*. Paris: S.E.V.P.E.N., 1968.

Dupieux, Paul. "Les Attributions de la juridiction consulaire de Paris (1563–1792)." *Bibliothèque de l'Ecole des Chartes* 95 (1934): 116–48.

Durand, Pierre. *Le Livre du pain*. Monaco: Editions du Rocher, 1973.

Emion, Victor. *La Taxe du pain*. Paris: Librairie Agricole de la Maison Rustique, 1867.

Emmet, C. W. "Famine." In *Dictionary of the Bible*. Ed. James Hastings. Rev. ed. Frederick C. Grant and H. H. Rowley. New York: Scribner's, 1963.

Evrard, Fernand. *Versailles, ville du roi (1770–1789): Etude d'économie urbaine*. Paris: E. Leroux, 1935.

Fance, Wilfred J., and B. H. Wragg. *Up to Date Breadmaking*. London: n.p., 1968.

Farb, Peter, and George Armelagos. *Consuming Passions: The Anthropology of Eating*. New York: Houghton Mifflin, 1980.

Farge, Arlette. *Délinquance et criminalité: Le Vol d'aliments à Paris au XVIIIe siècle*. Paris: Plon, 1974.

———. *La Vie fragile: Violence, pouvoirs, et solidarités à Paris au XVIIIe siècle*. Paris: Hachette, 1986.

———, and Jacques Revel. *The Vanishing Children of Paris: Rumor and Politics before the French Revolution*. Trans. C. Miéville. Cambridge: Harvard University Press, 1991.

Faure, Edgar. *La Disgrâce de Turgot*. Paris: Gallimard, 1961.

Febvre, Lucien, and Gaston Berger, eds. *Encyclopédie française.* 22 vols. Paris: Société de Gestion de l'Encyclopédie Française, 1935–66.

Feyeux, A. "Un Nouveau Livre des métiers." Parts 1 and 2. *La Science sociale,* 2d year, 4 (October 1887): 323–50 and 4 (December 1887): 518–45.

Filby, Frederick A. *A History of Food Adulteration and Analysis.* London: G. Allen and Unwin, 1934.

Foiret, F. "Deux Marchés de 1644 pour la fourniture du pain de Gonesse." *Bulletin de la Société historique du VIe arrondissement de Paris* 24 (1923): 85–87.

Foreman, Shepard, and Joyce Riegelhaupt. "Marketplace and Marketing System: Toward a Theory of Peasant Economic Integration." *Comparative Studies in Society and History* 12 (April 1970): 188–212.

Forgeais, Arthur. *Collection des plombs historiés trouvés dans la Seine.* Paris: Aubry, 1862.

Foucault, Michel. *Histoire de la folie à l'âge classique.* Paris: Gallimard, 1972.

Fournier, Edouard. *Le Roman de Molière.* Paris: E. Dentu, 1863.

Foville, Alfred de. *La France économique: Statistique raisonnée et comparative.* Paris: Armand Colin et Cie., 1890.

Franklin, Alfred. *Dictionnaire historique des arts, métiers et professions exercés dans Paris depuis le XIIIe siècle.* Paris: H. Welter, 1906.

———. *La Vie privée d'autrefois.* 23 vols. Paris: E. Plon, Nourrit et Cie., 1887–1901.

Friedmann, Karen J. "Victualling Colonial Boston." *Agricultural History* 97 (1973): 189–205.

Furet, François. "Pour une définition des classes inférieures à l'époque moderne." *Annales: E.S.C.* 18th year (May–June 1963): 459–74.

Galavaris, George. *Bread and the Liturgy: The Symbolism of Early Christian and Byzantine Bread Stamps.* Madison: University of Wisconsin Press, 1970.

Garden, Maurice. "Les Inventaires après décès: Source globale de l'histoire sociale lyonnaise ou juxtaposition de monographies familiales?" *Cahiers d'histoire* 12 (1967): 153–73.

———. *Lyon et les Lyonnais au XVIIIe siècle.* Paris: Société d'Edition les Belles Lettres, 1970.

Garrioch, David. *Neighborhood and Community in Paris, 1740–1790.* Cambridge: Cambridge University Press, 1986.

———, and Michael Sonenscher, "Compagnonnages, Confraternities and Associations of Journeyman in Eighteenth-Century Paris." *European Historical Quarterly* 16 (1986): 25–45.

Gaulle, Julien Philippe de. *Nouvelle Histoire de Paris et de ses environs.* 5 vols. Paris: P.-M. Pourrat Frères, 1839–42.

Gay, Jean-Pierre. "L'Administration de la capitale entre 1770 et 1789, la tutelle de la royauté et ses limites." *Mémoires de la Fédération des Sociétés historiques et archéologiques de Paris et de l'Ile-de-France* 8 (1956): 299–370; 9 (1957–58): 283–363; 10 (1959): 181–247; 11 (1960): 363–403; 12 (1961): 135–218.

Gazier, A., ed. "La Guerre des Farines." *Mémoires de la Société de l'histoire de Paris et de l'Ile-de-France* 6 (1879): 1–23.

Geertz, Clifford. *Peddlers and Princes: Social Change and Economic Modernization in Two Indonesian Towns.* Chicago: University of Chicago Press, 1963.

———, Hildegard Geertz, and Lawrence Rosen. *Meaning and Order in Moroccan Society.* Cambridge: Cambridge University Press, 1979.

George, Mary Dorothy. *London Life in the Eighteenth Century.* New York: Harper and Row, 1965.

Gerschenkron, Alexander. *Bread and Democracy in Germany.* Berkeley: University of California Press, 1943.

Giedion, Siegfried. *Mechanization Takes Command: A Contribution to Anonymous History.* New York: Oxford University Press, 1948.

Gille, Bertrand. *Les Origines de la grande industrie métallurgique en France.* Paris: Editions Domat Montchrestien, 1947.

Gindin, Claude. "Les Principaux Laboureurs de Gonesse à la fin du XVIIe siècle et dans les premières années du XVIIIe siècle." D.E.S. thesis, Sorbonne, May 1967.

Girod, Paul Emile. "Les Subsistances en Bourgogne et particulièrement à Dijon à la fin du XVIIIe siècle, 1774–1789." *Revue bourguignonne de l'enseignement supérieur* 16 (1906): 1–144.

Godfroid, Anne. "La Corporation des boulangers à Namur au 18e siècle." Mémoire de licencié en histoire, Université Libre de Bruxelles, 1992–93.

Goncourt, Edmond, and Jules Goncourt. *La Femme au XVIIIe siècle.* Paris: Firmin Didot Frères, 1862.

Gottschalk, Alfred. *Le Blé, la farine, et le pain.* Paris: Editions de la Tournelle, 1935.

Gould, Robert Freke. *The History of Freemasonry.* 6 vols. New York: John Beacham, 1884–87.

Graff, Harvey J. *The Literacy Myth: Literacy and Social Structure in the Nineteenth Century.* New York: Academic Press, 1979.

Graham, Sylvester. *Treatise on Bread and Bread-Making.* Boston: Light and Stearns, 1837.

Grantham, George. "Agricultural Supply during the Industrial Revolution: French Evidence and European Implications." *Journal of Economic History* 49 (1989): 43–72.

———. "The Growth of Labour Productivity in the 'Cinq Grosses Fermes' of France, 1750–1933." In *Productivity, Change, and Agricultural Development.* Ed. Bruce Campbell and Mark Overton. Manchester: n.p., 1991.

Grenier, Jean-Yves. "Croissance et déstabilisation." In *Histoire de la population française.* Ed. Jacques Dupâquier. Paris: Presses Universitaires de France, 1988. 460–65.

Griffuel, Albert. *La Taxe du pain.* Paris: L. Larose, 1903.

Grignon, Claude, and Christiane Grignon. "Styles d'alimentation et goûts populaires." *Revue française de sociologie* 21 (1980): 531–69.

Grignon, Claude, and Jean-Claude Passeron. *Le Savant et le populaire: misérabilisme et populisme en sociologie et en littérature.* Paris: Gallimard and Seuil, 1989.

Grignon, Claude, Maurice Aymard, and Françoise Sabban. "A la recherche du temps social." In *Le Temps de manger: Alimentation, emploi de temps et rythmes sociaux.* Ed. Aymard, Grignon, and Sabban. Paris: Editions de la Miason des Sciences de l'Homme, Paris [and] Institut National de la Recherche Agronomique, 1993.

Guerreau, Alain. "Mesures du blé et du pain à Mâcon (XIVe–XVIIIe siècles)." *Histoire et mesure* 3 (1988), pp. 163–219.

Gutel, Bernadette. "Les Français veulent manger du bon pain." *Hôtellerie,* no. 2197 (1991).

Gutton, Jean-Pierre. *La Société et les pauvres: L'Exemple de la Généralité de Lyon, 1534–1789.* Paris: Société d'Edition les Belles Lettres, 1970.

Harris, Seymour E. *The Assignats.* Cambridge: Harvard University Press, 1930.

Hémardinquer, Jean-Jacques. "En France aujourd'hui: Données quantitatives sur les consommations alimentaires." *Annales: E.S.C.* 16th year (May–June 1961): 553–64.

Herlaut, Auguste-Philippe. "La Disette de pain à Paris en 1709." *Mémoires de la Société de l'histoire de Paris et de l'Ile-de-France* 115 (1918): 5–100.

Hoffman, Philip T. "Land Rents and Agricultural Productivity: The Paris Basin, 1450–1789." *Journal of Economic History* 51 (December 1991): 771–806.

Horder, Thomas J., et al. *The Chemistry and Nutrition of Flour and Bread.* London: Constable, 1954.

Horn, J. E. *L'Economie politique avant les physiocrates.* Paris: Guillaumin et Cie., 1867.

Husson, Armand. *Les Consommations de Paris.* 2d ed. Paris: Librairie Hachette, 1875.

Ibels, H. G. "Le Costume et le meuble." In *La Vie parisienne au XVIIIe siècle.* Ed. Henri Bergmann. Paris: F. Alcan, 1914.

Jacob, H.-E. *Six Thousand Years of Bread: Its Holy and Unholy History.* Trans. R. Winston and C. Winston. Garden City, N.J.: Doubleday and Co., 1944.

Jacques, Jean. *Vie et mort des corporations: Grèves et luttes sociales sous l'ancien régime.* Paris: Spartacus, 1970.

Jobez, L.-E. Alphonse. *La France sous Louis XV, 1715–1774.* 6 vols. Paris: Didier, 1864–73.

Jones, A. H. "Fortune privée dans trois colonies d'Amérique." *Annales: E.S.C.* 24th year (March–April 1969): 235–49.

———. "Wealth Estimates of the American Middle Colonies." Ph.D. diss., University of Chicago, 1967.

Jorré, Georges. "Le Commerce des grains et la minoterie à Toulouse." *Revue géographique des Pyrénées* 4 (1933): 30–72.

Kaplan, Steven L. "L'Apprentissage au XVIIIe siècle: Le Cas de Paris." *Revue d'histoire moderne et contemporaine* 90 (July–September 1993): 436–79.

———. *La Bagarre: Galiani's "Lost" Parody.* The Hague: Martinus Nijhoff, 1979.

———. *Bread, Politics, and Political Economy in the Reign of Louis XV.* 2 vols. The Hague: Martinus Nijhoff, 1976.

———. "The Character and Implications of Strife among Masters inside the Guilds of Eighteenth-Century Paris." *Journal of Social History* 19 (summer 1986): 631–48.

———. "Les Corporations, les faux-ouvriers et le faubourg Saint-Antoine." *Annales: E.S.C.* 40th year, March–April, 1988: 253–88.

———. *The Famine Plot Persuasion in Eighteenth-Century France. Transactions of the American Philosophical Society* 72, part 3 (1982).

———. *Farewell, Revolution: Disputed Legacies France, 1789/1989.* Ithaca, N.Y.: Cornell University Press, 1995.

———. *Farewell, Revolution: The Historians' Feud, France, 1789/1989.* Ithaca, N.Y.: Cornell University Press, 1995.

———. "Lean Years, Fat Years: The 'Community' Granary System and the Search for Abundance in Eighteenth-Century Paris." *French Historical Studies* 10 (fall 1977): 197–230.

———. "La Lutte pour le contrôle du marché du travail à Paris au XVIIIe siècle." *Revue d'histoire moderne et contemporaine* 36 (July–September 1989): 361–412.

———. "Note sur les commissaires de police de Paris au XVIIIe siècle." *Revue d'histoire moderne et contemporaine* 28 (October–December 1981): 669–86.

———. "The Paris Bread Riot of 1725." *French Historical Studies* 14 (spring 1985): 23–56.

———. *Provisioning Paris: Merchants and Millers in the Grain and Flour Trade during the Eighteenth Century.* Ithaca, N.Y.: Cornell University Press, 1984.

———. "Réflexions sur la police du monde du travail, 1700–1815." *Revue historique* 261, no. 1 (January–March 1979): 17–77.

———. "Social Classification and Representation in the Corporate World of Eighteenth-Century France: Turgot's 'Carnival.'" In *Work in France: Representations, Meanings, Organization, and Practice.* Ed. Kaplan and Cynthia J. Koepp. Ithaca, N.Y.: Cornell University Press, 1986. 176–228.

———. "The State and the Problem of Dearth in Eighteenth-Century France: The Crisis of 1738–41 in Paris." *Food and Foodways* 4 (1990): 111–41.

Katzin, Margaret. "The Business of Higglering in Jamaica." *Social Economic Studies* 9 (September 1960): 297–331.

Kennedy, A., and B. J. Roberts. "Bread." In *Dictionary of the Bible.* Ed. James Hastings. Rev. ed. by Frederick C. Grant and H. H. Rowley. New York: Scribner's, 1963.

Kent-Jones, D. W., and John Price. *The Practice and Science of Bread-Making.* Liverpool: Northern Publishing Co., 1951.

Kerr, Clark. "Can Capitalism Dispense with Free Markets? Labor Markets: Their Characteristics and Consequences." *American Economic Review.* Papers and Proceedings 40 (May 1950): 279–83.

———. "The Balkanization of the Labor Markets." E. W. Bakke, et al. *Labor Mobility and Economic Opportunity.* New York: Technology Press of MIT and J. Wiley, 1965.

Khuri, Fuad I. "The Etiquette of Bargaining in the Middle East." *American Anthropologist* 70 (August 1968): 698–706.

Kirkland, John. *The Modern Baker, Confectioner, and Caterer.* 4 vols. London: Gresnam Publishing Co., 1927.

Kors, Alan Charles. "The Coterie Holbachique: An Enlightenment in Paris." Ph.D. diss., Harvard University, 1968.

———. *D'Holbach's Coterie: An Enlightenment in Paris.* Princeton: Princeton University Press, 1976.

Kula, Witold. *Measures and Men.* Trans. R. Szreker. Princeton, N.J.: Princeton University Press, 1986.

Labrousse, C.-Ernest. *La Crise de l'économie française à la fin de l'ancien régime et au début de la Révolution.* Paris: Presses Universitaires de France, 1944.

———. *Esquisse du mouvement des prix et des revenus en France au XVIIIe siècle.* Paris: Presses Universitaires de France, 1933.

———, and Fernand Braudel. *Histoire économique et sociale de la France.* 5 vols. Paris: Presses Universitaires de France, 1970–82.

Lachiver, Marcel. *Les Années de misère: La Famine au temps du grand roi, 1680–1720.* Paris: Fayard, 1991.

———, et al. "La Crise de subsistances des années 1740 dans le ressort du Parlement de

Paris (d'après le Fonds Joly de Fleury de la Bibliothèque nationale de Paris)." *Annales de démographie historique* (1974): 281–333.

Lafon, Jacqueline J. "L'Arbitre près la juridiction consulaire de Paris au XVIIIe siècle." *Revue historique de droit français* 51 (April–June 1973): 217–70.

Lagrange, Jean Louis, comte de. *Essai d'arithmétique politique.* In vol. 14 of *Collection des principaux économistes,* ed. Eugène Daire and G. de Molinari. 1845–47. Reprint, Osnabrück: Otto Zeller, 1966.

Lalanne, Ludovic. *Dictionnaire historique de la France.* Paris: Hachette et Cie., 1872.

Lanzac de Laborie, Léon de. *Paris sous Napoleon.* 5 vols. Paris: Plon-Nourrit et Cie., 1905–13.

Leach, Maria, ed. *Funk and Wagnall's Dictionary of Folklore.* New York: Funk and Wagnall, 1949–50.

Lebeuf, Jean. *Histoire de la ville et de tout le diocèse de Paris.* 6 vols. Paris: Fecnoz et Letouzey, 1883.

Leblond, Paul. *Le Problème de l'approvisionnement des centres urbains en denrées alimentaires en France.* Paris: Imprimerie Garnier, 1926.

Le Bras, Gabriel. *Etudes de sociologie religieuse.* Vol. 1, *Sociologie de la pratique religieuse dans les campagnes françaises.* Paris: Presses Universitaires de France, 1955.

Ledieu, Paul, ed. "Observations sur l'instruction de l'Impératrice de Russie aux députés pour la confection des loix (1774)." *Revue d'histoire économique et sociale* 8 (1920): 273–412.

Lefèbvre, Georges. *Documents relatifs à l'histoire des subsistances dans le district de Bergues.* 2 vols. Lille: C. Robbe, 1914–21.

———. "The Movement of Prices and the Origins of the French Revolution." In *New Perspectives on the French Revolution: Readings in Historical Sociology.* Ed. Jeffry Kaplow. New York: Wiley, 1965.

Lefèvre, Pierre. *Le Commerce des grains et la question du pain à Lille de 1713 à 1789.* Lille: C. Robbe, 1925.

Le Goff, Jacques. *La Civilisation de l'occident médiéval.* Paris: Arthaud, 1967.

Lelong, Maurice. *Célébration du pain.* Revest-Saint-Martin: n.p., 1963.

Lemoisie, J., ed. "Sous Louis le Bienaimé." *Revue de Paris* 12th year, no. 12 (15 June 1905): 747–94; no. 13 (1 July 1905): 153–90; no. 14 (15 July 1905): 390–430; no. 15 (1 August 1905): 559–86; no. 16 (15 August 1905): 851–68.

Le Paire, Jacques-Amédée. *Histoire de la ville de Corbeil.* 2 vols. Lagny: E. Colin, 1901–2.

Letaconnoux, Jean. *Les Subsistances et le commerce des grains en Bretagne au XVIIIe siècle.* Rennes: Oberthür, 1909.

Letrait, J.-J. "La Communauté des maîtres maçons de Paris au XVIIe et XVIIIe siècles." *Revue historique de droit français et étranger* 26 (1948): 96–136.

Levasseur, Emile. *Histoire des classes ouvrières et de l'industrie en France avant 1789.* 2d ed. 2 vols. Paris: 1901.

Levi, Giovanni. *Inheriting Power: The Story of an Exorcist.* Trans. Lydia G. Cochrane. Chicago: University of Chicago Press, 1988.

Levy, Darline G. S. *The Ideas and Careers of Simon-Nicolas-Henri Linguet: A Study in Eighteenth-Century Politics.* Urbana: University of Illinois Press, 1980.

Lick, Richard. "Les Intérieurs domestiques dans la seconde moitié du XVIIe siècle d'après les inventaires après décès à Coutance." *Annales de Normandie* 20, no. 4 (December 1970): 293–316.

Ljublinski, Vladimir S. *La Guerre des farines.* Trans. Françoise Adiba and Jacques Radiguet. Grenoble: Presses Universitaires de Grenoble, 1979.

Lockart, H. B., and R. O. Nesheim. "Nutritional Quality of Cereal Grains." In *Cereals '78: Better Nutrition for the Millions.* St. Paul, Minn.: American Association of Cereal Chemists, 1978.

Lopez, Claude Anne. *Mon Cher Papa Franklin and the Ladies of Paris.* New Haven, Conn.: Yale University Press, 1966.

Lüdtke, Alf, ed. *The History of Everyday Life: Reconstructing Historical Experiences and Ways of Life.* Trans. William Templer. Princeton: Princeton University Press, 1995.

Macherel, Claude. "Le Corps du pain et la maison du père." *L'Uomo* 3 (1990).

Mandrou, Robert. *Introduction à la France moderne, 1500–1640: Essai de psychologie historique.* Paris: A. Michel, 1974.

Mangenot, E., and A. Vacant, eds. *Dictionnaire de théologie catholique contenant l'exposé des doctrines catholiques, leurs preuves et leur histoire.* Paris: Letouzey et Ané, 1908–50.

Marion, Marcel. *Dictionnaire des institutions.* Paris: A. Picard, 1923.

———. "Les Lois de maximum et la taxation des salaires sous la Révolution." *Revue internationale de sociologie* 25 (1917): 485–501.

Markovitch, T. J. "Histoire quantitative de l'économie française." *Cahiers de l'Institut de science économique appliquée,* no. 173 (May 1966).

Marmay, Pierre. *Guide pratique de la meunerie et de la boulangerie.* Paris: E. Lecroix, 1863.

Martin, Didier. "Les Boulangers de Gonesse et leur rôle dans l'approvisionnement de Paris." Mémoire de maîtrise, Université de Paris VII, 1986.

Martin Saint-Léon, Etienne. *Histoire des corporations de métiers.* New York: Arno Press, 1975.

Marx, Karl. *Oeuvres.* Ed. Maximilien Rubel. Paris: Gallimard, 1963.

Maugarny, Albert. *La Banlieue sud de Paris: Histoire, onomastique, langage, folklore.* Le-Puy-en-Velay: "La Haute Loire," 1936.

Maurizio, A. *Histoire de l'alimentation végétale depuis la préhistoire jusqu'à nos jours.* Trans. F. Gidon. Paris: Payot, 1932.

Mayol, Pierre. "Habiter." In Luce Girard and Pierre Mayol, *Habiter.* Vol. 2 of *L'Invention du quotidien.* Ed. Michel de Certeau. Paris: Union Générale d'Editions, 1980.

McCance, Robert A., and Elsie Widdowson. *Breads, White and Brown: Their Place in Thought and Social History.* Philadelphia: Lippincott, 1956.

McCloy, Shelby. *The Humanitarian Movement in Eighteenth-Century France.* Lexington: University of Kentucky Press, 1957.

Merton, Robert King. "Intermarriage and the Social Structure: Fact and Theory." In *The Family, Its Structure and Functions.* Ed. R. L. Coser. New York: St. Martin's Press, 1964.

Meuvret, Jean. "Les Mouvements des prix de 1661 à 1715 et leurs répercussions." *Journal de la Société de statistique* 85 (May–June 1944): 109–19.

———. *Le Problème des subsistances à l'époque Louis XIV: Le Commerce des grains et la conjoncture.* 2 vols. Paris: Mouton, 1988.

Michel Montignac Magazine, no. 1 (May–June 1993).

Miller, Judith A. "Politics and Urban Provisioning Crises: Bakers, Police, and Parlements in France, 1750–93." *Journal of Modern History* 64 (June 1992): 227–62.

Mintz, Sydney W. "The Employment of Capital by Market Women in Haiti." In *Capital, Savings, and Credit in Peasant Societies.* Ed. Raymond Firth and B. S. Yang. Chicago: Aldine, 1964.

———. "Pratik: Haitian Personal Economic Relations." In *Symposium: Patterns of Land Utilization and Other Papers.* Proceedings of the 1961 Annual Spring Meeting of the American Ethnological Society. Ed. Viola E. Garfield. Seattle: University of Washington Press, 1961.

Modeste, Victor. *De la taxe de pain.* Paris: Hennuyer, 1856.

Monnier, Raymonde. *Le Faubourg Saint-Antoine, 1789–1815.* Paris: Société des Etudes Robespierristes, 1981.

Monteuuis, Albert. *Le Vrai Pain de France, ou la Question du pain sur le terrain pratique.* Paris: Malouin et Fils, 1917.

Moreau de Jonnès, M. A. "Statistique des céréales de la France: Le Blé, sa culture, sa production, sa consommation, son commerce." *Journal des économistes* 4 (1843): 129–66.

Morel, Ambroise. *Histoire abrégée de la boulangerie en France.* Paris: Imprimerie de la Bourse du Commerce, 1899.

———. *Histoire illustrée de la boulangerie en France.* Paris: Syndicat Patronal de la Boulangerie de Paris, 1924.

Moriceau, Jean-Marc. "Au rendez-vous de la 'révolution agricole' dans la France du XVIIIe siècle: A propos des régions de grande culture." *Annales:* HSS, January–February 1994: 27–63.

———. *Les Fermiers de l'Ile-de-France: L'Ascension d'un patronat agricole (XVe–XVIIIe siècles).* Paris: Fayard, 1994.

———, and Gilles Postel-Vinay. *Ferme, entreprise, famille: Grande Exploitation et changements agricoles. Les Chartier, XVIIe–XIXe siècles.* Paris: Editions de l'Ecole des Hautes Etudes en Sciences Sociales, 1992.

Murand, V. du, ed. *Inventaire sommaire des Archives communales antérieures à 1790.* Orléans: n.p., 1907.

Musart, Charles. *La Réglementation du commerce des grains en France au XVIIIe siècle: La Théorie de Delamare, étude économique.* Paris: E. Champion, 1922.

Neumann, Erich. *The Great Mother: An Analysis of the Archetype.* Trans. R. Manheim. Princeton, N.J.: Princeton University Press, 1963.

Nigeon, René. *Etat financier des corporations parisiennes d'arts et métiers au XVIIIe siècle.* Paris: Editions Rieder, 1934.

Nottin, Paul. *Le Blé, la farine, le pain, les semoules, les pâtes alimentaires.* Paris: Hachette, 1940.

Olivier, Paul. *Les Chansons des métiers.* Paris: Charpentier et Fasquelle, 1910.

Olivier-Martin, François. *L'Organisation corporative de la France de l'ancien régime.* Paris: Recueil Sirey, 1938.

Ouin-Lacroix, Charles. *Histoire des anciennes corporations d'arts et métiers et des confréries religieuses de la capitale de la Normandie.* Rouen: Lecointe Frères, 1850.

Palmer, R. R. *The Age of Democratic Revolution: A Political History of Europe and America, 1760–1800*. 2 vols. Princeton, N.J.: Princeton University Press, 1959–64.

Pardailhé-Galabrun, Annik. *La Naissance de l'intime: 3,000 foyers parisiens, XVIIe–XVIIIe siècles*. Paris: Presses Universitaires de France, 1988.

Peristiany, John G., ed. *Honour and Shame: The Values of Mediterranean Society*. London: Weidenfeld and Nicolson, 1965.

Philippe, Robert. "Une Operation pilote: l'étude du ravitaillement de Paris au temps de Lavoisier." *Annales: E.S.C.* 16th year (May–June 1961): 544–68.

Poilâne, Lionel. *Guide de l'amateur de pain*. Paris: Lionel, 1981.

Pomeranz, Y., ed. *Wheat Chemistry and Technology*. Saint Paul, Minn.: American Association of Cereal Chemists, 1978.

Poni, Carlo. "Norms and Disputes: The Shoemakers' Guild in Eighteenth-Century Bologna." Typescript.

Pontas, M. *Le Dictionnaire des cas de conscience*. Cited by Jules Gritti, "Deux Arts du vraisemblable: La Casuistique, le courrier du coeur," *Communications*, no. 11 (1968): 99–121.

Pottinger, David. *The French Book Trade in the Ancien Régime*. Cambridge: Harvard University Press, 1958.

Poulaille, Henry. *Le Pain quotidien*. Paris: B. Grasset, 1934.

Prentice, E. Parmalee. *Hunger and History: The Influence of Hunger on Human History*. New York: Harper, 1939.

Rabate, Edmond. *Le Blé, la farine, le pain*. Paris: Hachette, 1909.

Radford, Edwin, ed. *Encyclopedia of Superstitions*. New York: Philosophical Library, 1949.

Ramazzini, Bernardino. *Diseases of Workers*. Trans. W. C. Wright. New York: Hafner Publishing Co., 1964.

Renner, Hans D. *The Origin of Food Habits*. London: Faber and Faber, 1944.

Revel, Jacques. "A Capital City's Privileges: Food Supplies in Early Modern Rome." In *Food and Drink in History: Selections from the Annales*. Ed. Robert Forster and Orest Ranum. Baltimore: Johns Hopkins University Press, 1979.

Richard, Emile. *Histoire de l'hôpital de Bicêtre, 1250–1791*. Paris: G. Steinheil, 1889.

Richet, Charles, and Antonin Mans. *La Famine*. Paris: Centre de Recherches, Charles Richet, 1965.

Ricommard, Jean. *La Lieutenance générale de police de Troyes au XVIIIe siècle*. Troyes: J. L. Paton, 1934.

Roche, Daniel. *La Culture des apparences: Une Histoire du vêtement (XVIIe–XVIIIe siècles)*. Paris: Fayard, 1989.

———. *Le Peuple de Paris*. Paris: Aubier, 1981.

Rocquain, Félix. *L'Esprit révolutionnaire avant la Révolution, 1715–1789*. Paris: E. Plon, 1878.

Rudé, George F. "The Bread Riot of May 1775 in Paris and the Paris Region." In *New Perspectives on the French Revolution*. Ed. Jeffry Kaplow. New York: Wiley, 1965.

———. "Prices, Wages, and Popular Movements in Paris during the French Revolution." *Economic History Review*, 2d ser., 6 (1954): 246–67.

Ruwet, Joseph, et al. *Marché des céréales à Ruremonde, Luxembourg, Namur et Diest aux XVIIe et XVIIIe siècles*. Louvain: Publications Universitaires de Louvain, 1966.

Saineanu, Lazar. *L'Histoire naturelle et les branches connexes dans l'oeuvre de Rabelais.* Paris: E. Champion, 1921.

Saint-Germain, Jacques. *La Reynie et la police au grand siècle, d'après de nombreux documents inédits.* Paris: Hachette, 1962.

————. *La Vie quotidienne en France à la fin du grand siecle d'après les archives . . . du lieutenant générale de police, Marc-René d'Argenson.* Paris: Hachette, 1965.

Sand, George. *Questions politiques et sociales.* Paris: Calmann Levy, 1879.

Sars, Maxime de. *Lenoir, lieutenant de police, 1732–1807.* Paris: Hachette, 1948.

Schifres, Alain. "Laissez-nous notre pain quotidien." *Nouvel Observateur,* 25–31 December 1987.

————. *Les Parisiens.* Paris: J. C. Lattes, 1990.

Schuurman, Anton J. "Probate Inventory Research: Opportunities and Drawbacks." In *Inventaires après-décès et ventes de meubles: Apports à une histoire de la vie économique et quotidienne (XIVe–XIXe siècles).* Ed. Micheline Baulant, A. J. Schuurman, and Paul Servan. Louvain-la-Neuve: Academic, 1988.

Schwiedland, Eugen. "Les Industries de l'alimentation à Paris." *Revue d'économie politique* 9 (1895): 293–328.

Sébillot, Paul. *Légendes et curiosités des métiers.* Paris: E. Flammarion, 1894–95.

————. *Traditions et superstitions de la boulangerie.* Paris: LeChevalier, 1891.

Sentou, Jean. "Faillites et commerce à Toulouse en 1789." *Annales historiques de la Révolution française* 25 (July–September 1953): 217–56.

————. *Fortunes et groupes sociaux à Toulouse sous la Révolution.* Toulouse: E. Privat, 1969.

Sérand, Eugène H. L. *Le Pain: Fabrication rationnelle: Historique.* Paris: H. Dunod and E. Pinet, 1910.

Servan, Paul. "Inventaires et ventes de meubles: Apports à l'histoire économique." In *Inventaires après-décès et ventes de meubles: apports à une histoire de la vie économique et quotidienne (XIVe–XIXe siècle).* Ed. Micheline Baulant, A. J. Schuurman, and Paul Servan. Louvain-la-Neuve: Academic, 1988.

Sheppard, Ronald, and Edward Newton. *The Story of Bread.* London: Routledge and Kegan Paul, 1957.

Sherman, Harry C., and Constance S. Pearson. *Modern Bread from the Viewpoint of Nutrition.* New York: Macmillan, 1942.

Sibalis, Michael. "The Workers of Napoleonic Paris, 1800–1815." Ph.D. diss., Concordia University, 1979.

Simon, J. M., et al. "La Taille du boulanger." *Bulletin folklorique de l'Ile-de-France,* 4th ser., 31 (1969): 115–20.

Simon, Julian L. "Demographic Causes and Consequences of the Industrial Revolution." *Journal of European Economic History* 23 (spring 1994): 141–58.

Sjoberg, Gideon. *The Pre-Industrial City: Past and Present.* New York: Free Press, 1960.

Smith, S. R. "The London Apprentices as Seventeenth-Century Adolescents." *Past and Present,* no. 61 (November 1973): 148–61.

Snyder, Harry. *Bread: A Collection of Popular Papers on Wheat, Flour, and Bread.* New York: Macmillan, 1930.

Sonenscher, Michael. *Work and Wages: Natural Law, Politics, and Eighteenth-Century French Trades*. Cambridge and New York: Cambridge University Press, 1989.

Soreau, Edmond. *Ouvriers et paysans de 1789 à 1792*. Paris: Société d'Edition les Belles Lettres, 1936.

Sorokin, Pitirim. *Hunger as a Factor in Human Affairs*. Trans. Elena P. Sorokin. Ed. T. Lynn Smith. Gainesville: University Presses of Florida, 1975.

Storck, John, and Walter D. Teague. *Flour for Man's Bread—A History of Milling*. Minneapolis: University of Minnesota Press, 1952.

Stouff, Louis. *Ravitaillement et alimentation en Provence aux XIVe et XVe siècles*. Paris and The Hague: Mouton, 1970.

Subtil, A. "Réglementation municipale de la distribution des grains et de la boulangerie." In *Actes du 81e Congrès national des sociétés savantes* (congress held in Rouen-Caen, 1956). Paris: Presses Universitaires de France, 1956.

Sylvestre, A. J. *Histoire des professions alimentaires dans Paris et ses environs*. Paris: n.p., 1853.

Syndicat de la boulangerie de Paris. *Histoire*. Paris: n.p., 1900.

Szanton, Marcia C. B. *A Right to Survive: Subsistence Marketing in a Lowland Philippines Town*. University Park: Pennsylvania State University Press, 1972.

Thery, Adrien-Henri. *Gonesse dans l'histoire: Une Vieille Bourgade et son passé à travers les siècles*. Persan: Imprimerie de Persan-Beaumont, 1960.

Thomas, Keith. *Religion and the Decline of Magic*. New York: Scribner's, 1971.

Thrupp, Sylvia. *A Short History of the Worshipful Company of Bakers of London*. London: Galleon Press, 1933.

Thuillat, Louis. *Gabriel Nicolas de la Reynie, premier lieutenant général de police de Paris*. Limoges: Imprimerie Générale, 1930.

Thuillier, G. "Water Supplies in Nineteenth-Century Nivernais." *Food and Drink in History: Selections from the Annales*. Ed. Robert Forster and Orest Ranum. Baltimore: Johns Hopkins University Press, 1979.

T'ien, Ju-K'ang. *The Chinese of Sarawak: A Study of Social Structure*. London School of Economics and Political Science, Monograph on Social Anthropology no. 12. London: n.p., n.d.

Touchard-Lafosse, Georges. *Histoire de Paris et de ses environs*. Vol. 3. Paris: n.p., 1851.

Truant, Cynthia M. *The Rites of Labor: The Brotherhoods of Compagnonnage in Old and New Regime France*. Ithaca, N.Y.: Cornell University Press, 1994.

Tuetey, Alexandre. *Répertoire général des sources manuscrites de l'histoire de Paris pendant la Révolution française*. 11 vols. Paris: n.p., 1890–1914.

Vauban, Sébastien Le Prestre de. *De l'importance dont Paris est à la France et le soin que l'on doit prendre de sa conservation*. Paris: Treutel and Wurtz, 1821.

Vaury, S. *Le Guide du boulanger*. Paris: Legouix, 1834.

Vauthier, G. "Les Ouvriers de Paris sous l'empire." *Revue des études napoléoniennes* (1913).

Viard, Jean-Marie. *Le Compagnon boulanger*. Les Lilas: Editions Jérôme Villette, 1982.

Villey, Edmond. "La Taxe du pain et les boulangers de la ville de Caen en 1776: Documents." *Revue d'économie politique* 2 (1888): 178–92.

Vinçard, Pierre. *Les Ouvriers de Paris: Alimentation.* Paris: Gosselin, 1863.

Vincent, Catherine. *Les Confréries mediévales dans le royaume de France: XIIIe–XV siècle.* Paris: Albin Michel, 1994.

Vincent, François. *Histoire des famines à Paris.* Paris: Editions Politiques, Economiques et Sociales, 1946.

Vovelle, Michel. *Idéologies et mentalités.* Paris: F. Maspero, 1982.

———. Review of A. Daumard and F. Furet, *Structures et relations sociales à Paris au milieu du XVIIIe siècle* (Paris: Armand Colin, 1961). *Annales historiques de la Révolution française* 34 (January–March 1962): 99–104.

———. *Ville et campagne au 18e siècle (Chartres et la Beauce).* Paris: Editions Sociales, 1980.

Ward, Barbara. "Cash or Credit Crops? An Examination of Some Implications of Peasant Commercial Production with Specific Reference to the Multiplicity of Traders and Middlemen." *Economic Development and Cultural Change* 8 (January 1960): 148–63.

Waterbury, John. *North for the Trade: The Life and Times of a Berber Merchant.* Berkeley: University of California Press, 1972.

Weichs-Glon. "La Municipalisation de la boulangerie." *Revue d'économie politique* 10 (1897): 968–69.

Weir, David. "Les Crises économiques et les origines de la Révolution française." *Annales E.S.C.* 46th year, July–August 1991: 917–47.

Weulersse, Georges. *Le Mouvement physiocratique en France de 1756 à 1770.* 2 vols. Paris: Félix Alcan, 1910.

———. *La Physiocratie à la fin du règne de Louis XV, 1770–74.* Paris: Presses Universitaires de France, 1959.

———. *La Physiocratie sous les ministères de Turgot et de Necker (1774–1781).* Paris: Presses Universitaires de France, 1950.

Whorton, James. *Crusaders for Fitness: A History of American Health Reformers.* Princeton, N.J.: Princeton University Press, 1982.

Wihlfahrt, Julius. *A Treatise on Baking.* New York: Standard Brands, 1935.

Williams, Jane Welch. *Bread, Wine, and Money: The Windows of the Trades at Chartres Cathedral.* Chicago: University of Chicago Press, 1993.

Wood, Philip. *The Story of a Loaf of Bread.* Cambridge: Cambridge University Press, 1913.

Young, Sidney. *The Worshipful Company of Bakers of London.* London: Furnival Press, 1912.

Yourcenar, Marguerite. *With Open Eyes: Conversations with Matthieu Galey.* Trans. Arthur Goldhammer. Boston: Beacon Press, 1984.

Ziehr, W., and E. M. Buhrer. *Le Pain à travers les âges.* Paris: Editions Herme, 1985.

Zupko, Ronald E. *French Weights and Measures before the Revolution: A Dictionary of Provincial and Local Units.* Bloomington: Indiana University Press, 1978.

Index

Prepared by Mary Ann Quinn

Baker guild (*cont.*)
522, 537, 562; recruitment practices and costs of admission to, 173, 188, 191, 272, 275–83, 301; reforms of, 57, 83–84, 173, 176–77, 182, 282–83, 421; and response to edict suppressing guilds, 213–14; size and provisioning share of, 81–84, 115, 568; solidarities and conflict within, 77, 183–87, 195, 217, 572; statutes of, 155, 162–67, 170–74

Bakerooms, 284; baker absence from, 245–46; descriptions of, 61–63, 71–72, 74, 79, 92–93, 229, 353–55, 368–75, 425; technological changes in, 354–55, 371; tensions within, 240–41, 245–46, 257, 574; wells and water supply in, 489–90; workers and working conditions in, 79, 229–34, 249, 355–56, 371, 373

Bakers: accounting and bookkeeping practices of, 93, 140–42, 243, 360, 388–90; adages on, 227, 231; assets of, 346–52, 358–62, 576; associations and lending of names by, 110–11, 418; balance between profits and public duty of, 464, 501, 515–18, 559, 565–66, 577, 579; as bankers, 139–40, 366; bread accounts receivable of, 145–46, 359–60, 420; business assets of, 353–58; case studies of, 368–75; classes and types of, 16, 81–84, 453; collection difficulties of, 146–50, 408–9, 420, 535, 571; concern for reputation of, 17, 102–3, 139, 147, 219, 423–36, 470, 476, 576; costumes and clothing of, 339, 352, 368–69, 372–75; credit practices and needs of, 137–51, 321–22, 365, 377, 411, 533–34; and customer loyalty, 114–15, 137–40, 144, 146, 323, 415, 418, 426–27, 429, 569; daily production and supply by, 77, 103–5, 125–27, 356, 559, 569; debt recovery against, 380–87, 391–94, 397–98, 571; debts and liabilities of, 17, 363–67, 380–81, 392, 413, 419, 576; decline in numbers of, 121–22, 569; establishment of, 271–301; family networks and dynasties of,

128, 163, 301, 305–7, 325, 335, 575; fortunes of, 337, 341–76; grain and flour stocks of, 62–63, 128, 190, 356–58, 365, 371, 405, 577; households and residences of, 96, 234, 346–52, 368–75; insults and innuendo against, 425–28, 434–35, 576–77; literacy of, 140, 315; local reputations of, 425–28, 431–34; marriage endogamy of, 302–5, 335, 575; marriage fortunes and later net worth of, 307–15, 317–21, 335–36, 362–63, 575–76; marriage strategies of, 17, 293–94, 301–2, 575; mortality of, 231, 323; negative representations of, 423–25, 436, 463, 467, 470–71, 534, 576; and obligations to poor, 138–40, 145, 541; other professions of, 112–14, 127–28, 375, 415, 464–65; overhead and production costs of, 189, 442, 508–10, 514–15; pension and retirement arrangements of, 292–94, 299, 301, 312; personal lives and social status of, 264, 330–36, 350–52, 376, 428; police assistance to, 465, 533–35, 556–58; police views of, 525–27, 577; public service and accountability of, 34–35, 96–99, 231, 425, 463–67, 492, 529, 535, 565–66, 577; public surveillance of, 402, 552, 576; and relations with creditors, 390–91, 399, 410; and relations with suppliers, 110, 191, 377–81, 410–11; rents of, 347–48, 392; reputed dishonesty and greed of, 51, 68, 424–25, 463, 527, 536, 538, 556, 559; rivalries and professional jealousies of, 102–3, 112, 115, 429–35, 498, 569; sanctions against, 490–91, 546, 548, 566, 577; solidarity and mutual assistance among, 115, 416, 418, 429; subsistence scientists and, 51–53, 57; succession strategies and inheritance disputes of, 291–93, 301, 324, 435–36, 575; supply obligations of, 117–18, 464, 528–30, 535, 566, 578; as victims of dearth, 533, 538–39, 548, 555–58; women, 123–24, 126. *See also* Illicit bakers

Baker widows, 99, 110, 123–24, 126, 164,

283; continuation of business operations by, 297–98, 323–28; numbers of, 83; remarriages and fortunes of, 251, 254, 315–17, 324, 326–27; and renunciation of estate debts, 397–98

Baker wives: bookkeeping and credit management by, 321–23, 383, 396; business roles and profession of, 93, 255, 302, 321–23, 336, 575–76; clothing of, 350, 352, 368, 372–75; literacy of, 315, 369; and relations with customers, 146–47, 321–23; role of in quarrels and disputes, 431–34

Bakery workers. *See* Journeymen

Bakeshops, 16, 114–15, 575; and client fidelity, 89–91; credit offered by, 89–90, 93, 139; data on, 451, 453; descriptions of, 86, 92–93, 141, 369–75; equipment and furnishings of, 285–86, 353–54; inspections of, 213, 217, 544–45; prices at, 88–90; products at, 41, 88, 91–92; provisioning share of, 87; as social nexus, 87, 91, 139; as temptation, 93

Balladur, Edouard, 26

Balzac, Honoré de, 568

Banalités, 117

Bananas, 27

Bankruptcy: fraudulent, 383–84, 400–401, 426

Barbary: wheat from, 537

Barbier, Edmond-Jean-François, 543, 550

Bar-le-Duc, 498

Barley, 33, 45, 48, 49, 452, 544

Barm. *See* Brewer's yeast

Barter, 143–44

Baskets (*bannetons*), 63, 285

Bassinage, 70

Baudeau, Abbé Nicolas: on bread and bread making, 46, 49, 70, 444, 479; on Parisian tastes and consumption, 34, 37, 446; on police of provisioning, 51–52, 188, 498, 500, 518

Beans, 48

Beaumont, 509

Beauvais, 55

Beccaria, Cesare Bonesana, marchese di, 395

Beer, 68, 118, 351

Béguillet, Edme, 34, 37, 43, 444, 446–47, 449–50

Bercy, 454

Bèrnard, Samuel, 390, 494, 537

Bertier de Sauvigny, Louis-Jean, 49

Bertrand, Jean-Elie, 45, 610 n.53

Bethmont (lawyer), 35

Bicêtre, 419–20, 549–50

Bigot de Sainte-Croix, Louis-Claude, 196

Birch, 76

Blanvilain family, 331–34

Bonneuil, 118–19

Bookkeeping, 183, 400–401

Books and book ownership, 259, 351–52, 368, 373, 375

Bordeaux, 31

Boullanger, commissaire, 220–21

Boullemer (lieutenant general of Alençon), 499

Bourbon, Louis-Henri, duc de, 532–33

Bourdieu, Pierre, 19

Bourgeois de Paris, 116, 146, 350, 428

Bran, 43–44, 57, 91–92, 129, 443, 469, 500, 514–15

Bread, 16, 78–79; changes in, 33, 569; classes and color of, 2–3, 33–36, 40–45, 71, 503–4; dependency on, 9, 26–30; and everyday order, 9–12, 15, 564–65, 567, 570, 577; as France, 4, 29; home baking of, 50–51, 81, 91, 454; as luxury, 27–28, 36; markings on, 464, 471, 475, 488; as nexus of community, 91, 115; nostalgia for, 8, 59, 130; nutritional value of, 23, 29–30, 469; sacrality and symbolic charge of, 3, 23–25, 59, 563, 567, 580; and socioeconomic stratification, 25, 36, 40, 42, 46, 90, 134, 483, 567. *See also* Breadways; Parisian breadways

Bread carriers. *See* Porteuses

Bread distribution: and home deliveries, 134–35, 139, 172. *See also* Bakeshops; Bread markets; Police of bread and bakers: and off-market sales

Bread making, 1, 6, 80, 249; carelessness or sabotage in, 245–48; ingredients of, 63–70; odors and, 72, 470; physical labor and tempo of, 67, 72–74, 77, 80, 227, 233, 249; sanitary conditions of, 71–72; and shapes and weights of loaves, 72–74, 479; skill and judgment in, 61, 66, 72, 77–80; stages and techniques of, 16, 65–79, 227, 233, 569; and subsistence scientists, 51–59; tools of, 63; trials of, 49–50, 442–45, 508–16; variability of weight and elements in, 477–78; water and, 63–65; yield of, 47, 50, 52, 442–45. *See also* Ecole gratuite de la boulangerie; Fermentation; Kneading

Bread markets, 114–15; atmosphere, pace, and rhythms of, 89–90, 100–103, 139, 519, 528, 570; bakers and supply at, 82–83, 85–87, 119–31, 441, 523; bread confiscations at, 103–4, 481; bread varieties at, 41–42; credit at, 89, 139; data collection and reports from, 104, 451, 523–25; de facto, 94–95, 115, 121, 441; forain role in, 84–87, 121–35; militarization of, 524, 530–31, 542–43; number and growth of, 84, 93–96; police and regulation of, 89, 94–100, 506, 541; popular clientele of, 89–90, 134, 506, 523; prices and discounting at, 41–42, 88–89, 135, 480–81, 530, 542; and removal of unsold bread, 481–84, 530, 542–43; reorganization of, 95–96, 124–27; social and psychological patterns of sales at, 481; supply obligations at, 97–99; women bakers at, 123–24. *See also* Halles; Market stalls

Bread proverbs and metaphors, 25–26, 33, 78, 469

Bread quality, 165, 415, 500, 504; and color, 33–37; consumer sensitivity to in crisis periods, 538, 548; and quality of ingredients, 63, 79, 247, 468–70; as right and obligation, 467–70

Bread riots, 34, 89, 97, 132, 465, 521, 560–61

Bread *taxis*, 10

Breadways, 1, 16, 24, 60, 580. *See also* Parisian breadways

Breast-feeding, 81

Brewer's yeast, 37–39, 67–69, 71, 129, 569

Briatte, Abbé Jean-Baptiste, 30, 447

Bricoteau, 49–50, 54, 443, 511–13

Brion, 449

Brissac family, 177

Brocq, Jean-Baptiste, 54, 58, 248–49

Brodin de la Jutais, Pierre, 50

Brokers and brokerage system, 377, 379, 382, 386, 557

Buckwheat, 27, 33

Bucquet, César, 34, 42, 47, 49, 549; consumption and yield estimates of, 444–45, 447; and subsistence scientists, 52–55

Budé, Guillaume, 445

Burgundy, parlement of, 460

Business failures, 556; baker assets, liabilities, and net worth in, 403–9; case studies of, 410–16; causes and consequences of, 144–45, 381–83, 392, 403, 411, 417–19, 571; and creditors, 392–93, 400–402, 410, 571; and debt consolidation, 399; and fraudulent bankruptcy, 400–401; incidence of, 400–405, 421; and reputation, 17, 392, 400–402, 421–22; value and reliability of records of, 144, 400–402

Butchers, 377, 420

Butter, 88, 91, 135

Cabals and strikes, 206, 239–40, 267–68, 278, 580

Cadet de Vaux, Antoine-Alexis, 53, 55, 549

Cadot, commissaire, 539, 545

Caen, 500, 508

Calonne, Charles-Alexandre de, 498

Calvel, Raymond, 4–5

Capitalism, 580

Capitation tax, 171, 200, 361–62

Carcassonne, 502

Carters, 113, 375

Cash, 350, 377

Cassava, 27

Garçons. *See* Journeymen

Gâteaux du roi, 544

Geindres, 71, 78, 205, 227, 236, 570; and derivation of name, 231–33

Généralités, 459

Geneva, 31, 356

Gens sans aveu. See Floating population

Germany, 27, 43, 45

Gluten, 32, 72

Godparents, 379

Goncourt, Edmond and Jules, 329

Gonesse, 117–18, 541; bakers of, 53, 88, 98, 101, 126–31, 133–35, 524, 534–37, 543; bread of, 40, 64, 128–30, 488; and Paris supply, 126–31

Goussainville, 117

Government, 174–77, 185, 188, 281; apprehensive paternalism of, 564; and political rationalization, 157–58; as public administration, 439, 458; and social contract of subsistence, 492; and subsistence science research, 58

Graham, Sylvester, 79

Grain, 24, 28; and yield estimates, 499

Grain and flour merchants: and bakers, 354–55, 377, 379, 381, 383, 385, 387–88, 408, 412; public accountability of, 463, 577

Grain and flour trade, 451, 494; baker involvement in, 113, 166, 377, 419, 455, 465; corruption in, 379; and credit, 377, 398; *droiture* traffic in, 441, 451–55; liberalization of, 30, 85; variables in, 494

Granaries and flour reserves, 13, 58, 421, 450, 452, 465, 512, 521, 552, 558; and baker stocks, 356–58, 562, 577

Grand Châtelet, 394–97

Grand Pannetier, 155–60, 175, 177

Grande Panneterie, 121, 179

Grands-Augustins, quai des: guild headquarters on, 167

Great Chain of Being, 164, 575

Greece, 231

Greenland, 30

Grignon, Claude, 19

Grimod de la Reynière, A. B. L., 59, 325, 480

Grimperel, commissaire, 442–43, 514–15, 554, 556–57, 566

Gros pain, 40, 46, 503. *See also* Household breads

Gruaux. *See* Middlings

Guerineau de Saint-Péravi, Jean-Nicolas-Marcellin, 400

Guillaute (constabulary officer), 216

Guingamp, 503

Guyot, commissaire, 476

Haggling, 88–90, 480–81, 504, 530, 570

Haiti, 377

Halles bread market, 2, 94, 96, 110, 433, 523, 530; baker turnover at, 122; flour weighing station at, 390; importance of, 124; journeyman demonstrations at, 206–7; market renaissance of, 455; market stalls at, 99–101; regrating at, 485; riots at, 183; rumor at, 426

Hardy, Siméon-Prosper, 139, 213, 424

Hawking, 139

Hay and straw, 118, 415

Henri IV, 128

Hérault, René, 25, 210, 463, 465, 509, 535, 538–39, 543; and bakers, 521–30; and boycott by forain bakers, 506–7; and guild abuses, 189; nostalgia for, 544; and police corruption, 531–34; sporadic liberalism of, 528

Herbert, Claude-Jacques, 445

Hinterland, 31, 118, 461–62, 498

Hippocrates, 38

Hoarding, 371, 528–30, 534, 547, 558, 562

Home deliveries, 134–35, 139, 172

Homelessness, 418

Honor, 224, 242, 287, 393, 423, 433; stakes of, 436

Hôpital général, Scipion bakery of, 49, 54, 299, 419, 443, 549; bread and flour trials at, 509–13

Hospitals. *See* Prisons and hospitals

Hôtel-Dieu, 145, 397

Houdouart, Nicolas, 185
Household breads (People's bread), 37, 39–40, 42, 45–50, 57–58, 134, 444, 508, 538
Huissiers, 397
Hunger, 424, 481
Husbands: authority of, 329–30
Husson, Armand, 19

Idleness, 208–9, 225. *See also* Floating population; Loitering
Illegitimate births, 256
Illicit bakers, 115, 121, 174, 271, 300–301, 326–28, 441; production of, 453, 455; women, 126
Illiteracy, 140
Indian corn. *See* Maize
Inoculation, 39
Insects, 229
Institutional provisioning, 451–52, 515; bread quality of, 54, 58; and economic bread, 49. *See also* Granaries and flour reserves; Prisons and hospitals
Insults and innuendo, 423, 425–28, 431, 576
Intendants, 459
Invalides, Hôtel Royal des, 54, 248
Italy, 51

Jackson, Henry, 43, 468
Jansenism, 132
Jewelry, 351, 371–72, 396, 412
Joly de Fleury, Guillaume-François, 516, 550; price schedules of, 509–10
Journeyman inns, 260–62
Journeymen, 220; ages of, 203; and abolition of guilds, 213–14, 240, 283, 574–75; blacklisting of, 219, 246; career opportunities for, 202–3, 225, 271; constraints on personal and family life of, 16, 250, 255–59, 270; drinking and drunkenness of, 245, 247, 265–67; fatigue and demanding labor of, 16, 227, 230, 249–50, 255, 264–65, 574; health and illnesses of, 229–31, 265; indiscipline and insubordination of, 205–7, 213–14, 239–40,

245–49, 572, 574–75; job placement and illicit employment of, 166, 204–7, 216–18, 296–301; literacy of, 203, 210; lodging and "homes" of, 215, 219, 255–56, 259–61; marriages of, 250–57; mistreatment of, 218, 241, 248; nicknames and origins of, 203–4, 215; organization and collective action of, 217, 239, 249, 262, 267–68, 574–75; possessions of, 258–59; and reestablishment of guilds, 283–84; registration and documentation of, 207, 210–13, 215–16, 218–22, 248–49; and relations with masters, 201, 204–5, 219, 225–26, 239–42, 245, 249, 259; and right to control of their labor, 204, 208–13, 215, 574; scapegoating of, 242, 245–49, 430, 476–77; secret associations of, 250, 268–70 (*see also* Compagnonnages and confréries); sociability and solidarity among, 201–3, 213, 261–63; social segregation of, 247, 255, 259, 264–65; status and self-perceptions of, 201–3, 249, 262, 279, 573–74; training and supervision of, 208, 248–49; turnover among, 208–9, 235, 249; unemployment among, 222–25; violence and theft by, 212–13, 242–45, 265; wages of, 235–39, 365–67; working conditions of, 231, 265, 574. See also *Geindres*
Journeymen's wire, 261
Journeywomen, 109
Jurés, 39; and baker-debt disputes, 388–89; conflicts among, 183–84; financial abuses by, 178–180, 185; forains and, 133, 165, 482, 488; infractions of bread restrictions by, 544–45; inspection zeal of, 475; police role of, 134–35, 164–65, 505; responsibilities and functions of, 163–67, 170–71
Justice, 436, 520
Just price, 493, 503, 506, 561

King, 164, 575; as baker of last resort, 492, 578; subsistence responsibilities of, 12–14, 458, 580

Mills, 166, 361
Mirabeau, Victor de Riquetti, marquis de, 471
Mitterand, François, 26
Mniszech (Polish traveler), 450
Mobile bakeries, 172
Moheau (writer-administrator), 446
Molière, Jean-Baptiste Poquelin de, 38
Mollesse, 45–46
Mollet bread. See *Pain mollet*
Monetary instability, 533
Monopoly, 28, 188, 278, 380, 499, 527, 551, 568
Montdidier, 55
Montignac, Michel, 5
Moral economy, 30
Morellet, Abbé André, 30, 501
Mortellerie merchants. *See* Ports and port merchants
Most-favored client relations. See *Pratik* relations
Mouricault, commissaire, 539

Naples, 56
Napoléon Bonaparte, 36, 59, 135, 215, 564
Narbonne, Pierre de, commissaire, 94, 493, 544
Natural law, 500, 551, 580
Nature, 47
Necker, Jacques, 3, 54, 56, 446–47, 562; on price schedules, 494, 502; on subsistence imperative, 458
Neighborhood, 91, 139, 314
Nervous disorders, 230
Nicknames and aliases, 203, 215
Nicolai (lord of Goussainville), 117
Night work, 227, 265
Notaries and notarial conventions, 143, 307–8, 337–38

Oats, 45, 48, 118, 415, 452
Occult, 398
Odors, 72, 470
Ombreval, Nicolas-Jean-Baptiste Ravot d', 532–33

Orléans, 55
Orry, Philibert, 546–48
Ovens, 63, 286, 353; construction and maintenance of, 57, 74, 290–91; loading and unloading of, 77–78, 233; public use of, 51, 81, 465; temperature of, 64, 78; and water supply, 489–90

Pain à la reine, 37
Pain brié, 71
Pain broyé, 273
Pain de cabaret, 41
Pain de Chailly, 40
Pain de Chapitre, 130
Pain de ménage. *See* Household bread
Pain mollet, 41, 71, 88, 91, 103, 134–35, 475, 483, 513, 544–45, 569; debate over, 67–68; introduction and trial of, 37–40, 60
Paintings and artwork, 351, 373, 374
Panic buying, 521
Para du Phanjas, Abbé, 517
Parent, commissaire, 522, 539
Paris, 14; bread markets in, 82, 93; bread types and qualities in, 40–45, 51; exemption from liberalization of, 552, 564; flour stocks in, 356–57; government concern for tranquility of, 458–59, 518–19, 547; growth of, 157, 579–80; population and consumption estimates for, 440–55; privileged areas of, 211
Paris Faculty of Medicine, 38, 51, 64
Paris guard and watch, 543
Parisian breadways, 19; extravagance of, 36–37; and Gonesse bread, 128–30; provincial view of, 35–36, 567–68; quality demands of, 2, 30, 35, 59, 452; and reliance on bakers for daily supply, 79, 81; and "right" to white, 32, 33, 34, 41, 45, 50, 567; tenacity of, 580; unifying aspects of, 60
Paris provisioning: bakers involved in, 81–84; changes in patterns of, 455; government attitudes and assumptions in, 439–40, 494; guild reforms as threat to, 214; supply and consumption data for,

447–48, 451, 456; supply zones of, 118, 461–62, 495, 503, 519, 550

Parlements, 460

Parlement of Paris, 36, 38, 83, 502; and forains, 135; power and influence of, 460–61

Parmentier, Antoine-Augustin, 32, 57, 549; on bakers and baking, 51–53, 61, 64–69, 75, 77–78, 189, 425; on bread varieties and merits, 30, 43, 48, 67–69; consumption and yield estimates of, 444–47; on French breadways, 23–24; on Gonesse bread, 130; on journeyman irresponsibility, 247–49; on police of bread and prices, 467, 478, 500–501; and school for baking, 54, 56–58

Pastry-makers guild, 41

Patin, Gui, 38–39

Paucton, A.-J.-P., 446, 448–49, 451

Pawning, 417

Pellerin, commissaire, 490

People's bread. See Household bread

Petit, Dr. Léon, 71

Petitions, 133

Petit pains, 40, 156. *See also* Luxury breads

Philippines, 377

Physiocrats, 27–30, 85, 115, 518, 570, 580; on prices and price-fixing, 494, 499–500, 563

Picardy, 39, 69; canal of, 451

Pillory, 244

Piochart (Fbg Saint-Antoine baker), 45

Plantains, 27

Pliny, 46

Pluche, Abbé Antoine, 46–47, 52

Poilâne, Lionel, 8

Police archives: negative bias of, 223

Police of bread and bakers: aims and approaches of, 17–18, 456–58, 463–64, 491–92, 519, 556–57; and assistance to bakers, 383, 535, 556–57, 577; and baker guild, 164–65, 189–90; and baker supply obligations, 97–99, 464, 466, 492, 528, 530, 542; as balancing of baker and consumer interests, 147, 420–21, 492,

504, 519, 521, 523, 528, 564–65, 577–79; and bread quality, 467–70; consumer role in, 476, 530–33, 538, 540, 542, 545–47, 578, 580; and crisis restrictions on bread types, 536–37, 544; data collection and surveillance as basis of, 216, 439–40, 451–57, 504, 520, 530, 579; and distinctions between periods of calm and crisis in, 464; as everyday local police, 91, 458, 577; flexibility and evolution of, 456–58, 518–20; and forains, 295, 474–75, 480, 484, 506, 522, 541–42, 553; inspections and, 475, 507; liberal critique of, 466–67; and obligation to mark loaves, 464; and off-market sales, 139, 171–72, 174, 485–88; organization of, 458–63; and police view of bakers, 463, 525–27; role of commissaires in, 17, 462–63, 532–33; and sales by weight and weight violations, 247, 466, 470–80, 537, 578; socially differential policies of, 506; and solicitude and assistance for bakers, 117–18, 383, 465, 533–35, 539, 556–58, 577; tactics and sanctions of, 465–66, 472–73, 490–92, 524–25; visibility as principle of, 216, 485

Police of prices: consumer role in, 540, 547; and distinctions between periods of calm and of crisis, 525–27, 539, 565; flexibility and improvisation of, 518–20, 579; and limits of intervention, 521, 528, 577; as negotiation and balance, 523, 565, 579; and price monitoring, 550; tactics and sanctions of, 524–25, 527, 541

Political consciousness and socialization, 187, 191, 572

Political economy, 519

Poncelet, Polycarpe, 32, 43

Pontoise, 44, 49, 390, 509

Poor consumers, 562; bread distribution to, 165, 473, 475, 539, 546; and credit, 137–38; hunger and impatience of, 93, 481, 519; needs of, 42, 44–48, 471; purchasing patterns and psychology of, 480–81

Population estimates, 356, 448–50

Porteuses, 105–9, 112, 132, 482–83
Ports and port merchants, 183, 382–83, 412, 462; decline in, 455
Potatoes, 24, 27, 48, 468
Pots de vin, 195, 285–86

Poussot, inspector, 103, 124, 206–7, 334–35; and baker credit needs, 417; on flour yields, 442, 510–11; inspections by, 550–51; on markets, 95–96, 120
Poverty, 28. *See also* Laboring poor; Poor consumers
Pratik relations, 91, 377–81, 385, 430; concessions and reciprocities in, 378; as market regulator, 378
Préciput, 312–13, 324, 328
Premontval, commissaire, 489
Prévôt des marchands, 97, 156; provisioning role of, 461–62
Prévôté de Paris. See Châtelet
Price ceilings, 525, 535; bakers and, 560–61; police attitude toward, 520; as "tacit," 504–7, 550; under liberalization, 553–55
Price-fixing, 12, 17, 57; arguments against, 445, 498–503; baker opposition to, 501–3, 511, 522, 525, 531–33; and bread trials as basis for, 508–16, 520, 579; consumer people and, 465–66; and current or normative price, 504–7, 516, 531; and distinctions between periods of calm and dearth, 494–96, 501–2, 512, 518–20; as equilibrium between consumers and bakers, 497, 499, 502, 506; impossibility of, 503; as imputation of government responsibility, 497, 501, 579; and just price, 493, 506, 579; necessity for, 467, 493, 495–98; as protection for bakers, 465, 496–97, 578; quality, weight, and supply compensations for, 474, 500, 539, 542–43, 547–48, 551; ratio of bread to wheat prices as rule of, 494, 498, 507, 512, 514, 516–19, 527; socially differential use of, 501, 506; as threat to supply, 493, 501, 529. *See also* Current price

Prices: baker-consumer negotiation of, 550, 553; baker records of, 141–42; communication of, 505; and credit sales, 139; and current or normative price, 504–7; and distinction between raw materials and manufactured goods, 563; fluctuations in, 485, 518–21, 523, 545–48, 553–55; immutability of, 580; issues and symbolic charge of, 563; psychological thresholds of, 524–25, 548; social differentiation in, 90, 519, 541; variables and determinants of, 88, 466, 495, 497–98, 504, 515, 518–19, 523, 526, 566, 579. *See also Mercuriales*
Prie, Madame de, 532
Prisons and hospitals, 391, 468; bread riots at, 34–35, 54, 549–50
Privileged bakers, 298–99
Procurator general, 460–61
Property separation, 384
Provence, parlement of, 447
Provins, 508
Prudhomme, Louis-Marie, 35
Prussia, 56, 76
Public Administration, 439, 458
Public health, 71
Public opinion, 48, 496, 531–32, 539
Public order, 138, 458–59, 482, 518–19, 547
Pumpernickel bread, 43

Quesnay, François, 177, 446
Quinze-Vingt, cour des, 47

Regnard, commissaire, 220
Regnaudet, commissaire, 247, 417
Regnault Senior, commissaire, 539
Regrating, 91, 112, 485, 530
Religious and devotional objects, 351–52, 372
Religious communities, 446
Rennes, 503, 535
Restif de la Bretonne, Nicolas Edme, 42, 139
Revolution, 18, 47, 396, 501, 558, 575; bread

quality during, 3; and breadways, 35–36, 42, 59–60; and workers, 207, 214
Rewbell, Jean-François, 500
Rice, 9, 27–28
Richelieu, Armand-Jean du Plessis, cardinal, duc de, 450–51
Rivarol, Antoine de, 469
Rochefort, 498–99
Roissy, 516
Roland, commissaire, 476
Rome, 46, 98, 231
Roubaud, Abbé Pierre-Joseph-André, 558
Rousseau, Guillaume César, 41
Rousseau, Jean-Jacques, 24
Rural history, 19
Rutledge, Jean-Jacques, 13, 425, 449, 558
Rye, 31–33, 44–45, 452

Saint-Antoine, faubourg, 568–69; establishment in, 119, 271, 294–95; merchant bakers of, 576; nonguild shop production in, 453, 455; privileges of, 81, 83–87, 116, 155, 157, 159; riot in, 521–22, 531
Saint-Eustache church, 455
Saint Honoré, confrérie, 167–69
Saint-Honoré, faubourg, market, 94
Saint Lazare, confrérie, 167–69, 275
Saint-Marcel, faubourg, 72
Saint-Paul bread market, 523
Saint-Pierre, Charles-Irénée Castel, abbé de, 394
Saint Pierre des Lions, 168
Salt, 66, 69–70, 135, 545
Sand, George, 227, 231, 249
Sanitation, 71, 93, 229
Sartine, Antoine Raymond de, 489, 564, 566, 579; bakers and, 465, 552–60; bread trials of, 49–50, 442, 456–57, 508, 516; price policies of, 494, 506, 553–55, 559
Saulet, commissaire, 476
Saumur, 502
Savary des Bruslons, Jacques, 209
Scams, 398
Schifres, Alain, 6
Science, 59

Scipion bakery. *See* Hôpital général, Scipion bakery of
Secretary of state for the royal household, 459, 462
Séguier, Antoine-Louis, 36
Senlis, 502
Sexual aspersions and assaults, 224, 428, 434–35
Silver, 371, 373
Sociability, 114, 151, 250, 265, 270, 570
Social contract of subsistence, 12–14, 20, 492, 497, 519, 551, 567, 580
Social question, 574–75
Société d'agriculture, 57
Société des philosophes économistes, 49
Soft-dough breads. See *Pain mollet*
Sorcery, 429
Sorokin, Pitirim, 10
Soudure, 528
Spain, 27
Speculation, 375, 421
Starch makers, 91–92, 118, 164, 165, 500, 514; as illicit bakers, 111; restrictions on grain use by, 544
Strikes. *See* Journeymen: organization and collective action of
Subsistence anxieties, 8, 10, 523, 543–46; and consumer fastidiousness, 79; and weather, 525
Subsistence crises. *See* Dearths
Subsistence imperative, 8–9, 458; as premise of public policy, 439
Subsistence mentality, 14, 20; and price-fixing, 496, 501
Subsistence question, 8–9; as government focus, 439, 458, 510, 575, 577; periodization of, 19; politicization of, 460, 564
Subsistence science, 32, 52–56, 60, 444, 552, 568
Subsistence unrest, 117, 183, 468, 544, 580
Suicide, 231, 397, 420, 559
Suki, 377
Sweden, 45, 56
Switzerland, 43
Syndics, 173

Tailles (tally sticks), 93, 137, 140–43, 321, 338, 360; destruction of, 489; falsifying of, 243

Taste, 19, 38

Taverns and inns, 149, 198, 200; as base for collective organization and sedition, 266–68; demand for luxury breads at, 41, 129, 488, 546; forain sales to, 134–36, 488; journeymen's frequentation of, 210, 214, 219, 250, 261, 266–68; police view of, 220, 267–68; provisioning of, 37, 134–36, 165, 174, 488; as social forums, 266–67, 379, 428, 432–33

Taxation. See Price-fixing

Taxes, 171

Tax farms, 451

Taylorism, 53

Technological innovation, 15, 279, 371; and mechanization of bakery, 71, 80; and provisioning trade, 510

Te Deums, 180

Terray, Joseph-Marie, abbé, 468, 495, 516

Theft, 223–24, 243–45, 262–64, 414–15

Thellusson, Isaac, 32, 36, 44; population estimates of, 449

Tillet, Mathieu, 444, 496, 499, 501, 518; baking tests by, 479–80

Toulouse, 31

Toupiolle, Pierre, commissaire, 566; bread trials by, 50, 442–43, 514–15

Tourton, commissaire, 532

Transport, 101, 113, 355, 368, 375

Trennin (police official at Versailles), 450

Trousseau, 317–20, 576

Troyes, 75, 77, 502

Trudaine de Montigny, Jean-Charles-Philibert, 48

Trudon, commissaire, 287, 560

Turgot, Anne-Robert-Jacques: abolition of guilds by, 83, 213–14, 574; and bakers, 85, 188–89, 560; and controller general's loaf, 45; critique of corporate system by, 279–80; fall of, 282, 561; and journeyman response to guild abolition, 214, 240; and liberalization of grain and flour trade, 2, 30, 52; and price-fixing, 493, 495, 507, 516–17

Tuscany, 56

Unemployment, 208–9, 223–25

Urban consumption, 40, 492; and rural breads, 46–47; and socioeconomic stratification, 60. *See also* Parisian breadways

Vauban, Sébastien Le Prestre, seigneur de, 446

Vaugirard, 49

Vaugirard, the mitron de. *See* Lacombe d'Avignon, François

Venal offices, 361, 372

Venereal disease, 230, 243

Verbal confrontations, 426, 436

Vérité, Sieur, 44

Versailles, 529

Viennese artisans: bakery design by, 92–93

Vingtième de l'industrie, 171

Voltaire, François-Marie Arouet de, 13, 24–25, 27, 32, 42, 208

Wages, 235–39, 365–67; of apprentices, 194–95; and "going" rate, 195, 217, 235; of journeymen, 208–9; lag in real, 405

Water, 63–65, 129, 429, 470, 489–90

Wealthy consumers, 547; and credit, 137–38; and obligations to poor, 46, 501–2

Weather, 233, 525, 542, 551; and prices, 499, 513–15, 523

Weights and measures, 104, 166; and baker requirement to provide scales, 464; and bread sales by weight, 474, 477–80; discrepancies in, 440–42. *See also* Police of bread and bakers: and sales by weight and weight violations

Wet nurses, 81

Wheat, 39; flour and bread yields of, 103, 441–42, 515; nutritional qualities of, 32, 43; preparation of, 79; sprouted, 56

Wheaten bread, 16, 30–32, 39

White bread, 2, 16, 19, 33–36, 50, 504; consumed at expense of the poor, 36–37; as metaphor, 25, 33; nutritional deficiencies of, 47; prohibitions against, 536, 544

Widows, 336

Wigs, 24, 258, 372–73, 431; and correlations with culture and fortune, 350–51; and flour wastage, 36

Wine, 371–72

Wine merchants: bakers as, 113–14

Women, 55, 106–9, 324, 448; assaults against, 109, 146–47, 434–35; as bakers, 110–12, 123–24, 126, 213, 218; march on Versailles by, 12, 469, 578; and subsistence anxiety, 465, 523, 547

Wood, 76–77, 91, 186–87, 351

Wood merchants: bakers as, 118, 415

Work, 59, 279, 572; and right to quit, 207–13; as social connection and control, 204–5, 207–9, 223, 265, 574

Workers: as servants, 240; documentation of, 215–16; police of, 165, 206–7, 215–16, 221, 239–40

Yourcenar, Marguerite, 1

Steven Laurence Kaplan is Goldwin Smith Professor of History at Cornell University. He is the author of *Farewell, Revolution: Disputed Legacies, France, 1789/1989*, *Farewell, Revolution: The Historians' Feud, France, 1789/1989*, and *Provisioning Paris: Merchants and Millers in the Grain and Flour Trade During the Eighteenth Century*. He is also the coeditor of *Work in France: Representations, Meaning, Organization, and Practice*.

Library of Congress Cataloging-in-Publication Data

Kaplan, Steven Laurence
The bakers of Paris and the bread question, 1700–1775 / by Steven
Laurence Kaplan.
p.   cm.
Includes bibliographical references and index.
ISBN 0-8223-1706-0 (cloth : alk. paper). —
ISBN 0-8223-1715-X (pbk. : alk. paper)
1. Bread industry—France—Paris Region—History—18th century.
2. Bakers and bakeries—France—Paris Region—History—18th century.
3. Food supply—France—Paris Region—History—18th century.
4. Bread—Prices—France—Paris Region—History—18th century.
I. Title.
HD9058.B743F85   1996
664'.7523'0944361—dc20   95-23182   CIP